Heterocyclic Chemistry

Heterocyclic Chemistry

Fifth Edition

John A. Joule
School of Chemistry, The University of Manchester, UK

Keith Mills
Chemistry Consultant, Ware, UK

A John Wiley & Sons, Ltd., Publication

This edition first published 2010
© 2010 Blackwell Publishing Ltd

Registered office
John Wiley & Sons Ltd, The Atrium, Southern Gate, Chichester, West Sussex, PO19 8SQ, United Kingdom

For details of our global editorial offices, for customer services and for information about how to apply for permission to reuse the copyright material in this book please see our website at www.wiley.com.

The right of the author to be identified as the author of this work has been asserted in accordance with the Copyright, Designs and Patents Act 1988.

5   2014

Library of Congress Cataloging-in-Publication Data
Joule, J. A. (John Arthur)
    Heterocyclic chemistry / John A. Joule, Keith Mills. – 5th ed.
        p.   cm.
    Includes bibliographical references and index.
    ISBN 978-1-4051-9365-8 (pbk.) – ISBN 978-1-4051-3300-5 (pbk.)   1. Heterocyclic chemistry.   I. Mills, K.
(Keith)   II. Title.
    QD400.J59 2009
    547′.59–dc22   2009028759

ISBN Cloth: 978-1-405-19365-8
ISBN Paper: 978-1-405-13300-5
A catalogue record for this book is available from the British Library.
Set in 10 on 12 pt Times by Toppan Best-set Premedia Limited

Contents

Preface to the Fifth Edition

Heterocyclic compounds have a wide range of applications but are of particular interest in medicinal chemistry, and this has catalysed the discovery and development of much heterocyclic chemistry and methods. The preparation of a fifth edition has allowed us to review thoroughly the material included in the earlier editions, to make amendments in the light of new knowledge, and to include recent work. Within the restrictions that space dictates, we believe that all of the most significant heterocyclic chemistry of the 20th century and important more recent developments, has been covered or referenced.

We have maintained the principal aim of the earlier editions – to teach the fundamentals of heterocyclic reactivity and synthesis in a way that is understandable by undergraduate students. However, in recognition of the level at which much heterocyclic chemistry is now normally taught, we include more advanced and current material, which makes the book appropriate both for post-graduate level courses, and as a reference text for those involved in heterocyclic chemistry in the work place.

New in this edition is the use of colour in the schemes. We have highlighted in red those parts of products (or intermediates) where a change in structure or bonding has taken place. We hope that this both facilitates comprehension and understanding of the chemical changes that are occurring and, especially for the undergraduate student, quickly focuses attention on just those parts of the molecules where structural change has occurred. For example, in the first reaction below, only changes at the pyridine nitrogen are involved; in the second example, the introduced bromine resulting from the substitution and its new bond to the heterocycle, are highlighted. We also show all positive and negative charges in red.

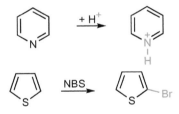

In recognition of the enormous importance of organometallic chemistry in heterocyclic synthesis, we have introduced a new chapter dealing exclusively with this aspect. Chapter 4, 'Organometallic Heterocyclic Chemistry', has: (i) a general overview of heterocyclic organometallic chemistry, but most examples are to be found in the individual ring chapters, (ii) the use of transition metal-catalysed reactions that, as a consequence of a regularity and consistency that is to a substantial degree independent of the heterocyclic ring, is best treated as a whole, and therefore most examples are brought together here, with relatively few in the ring chapters.

Other innovations in this fifth edition are discussions in Chapter 5 of the modern techniques of: (i) solid-phase chemistry, (ii) microwave heating and (iii) flow reactors in the heterocyclic context. Reflecting the large part that heterocyclic chemistry plays in the pharmaceutical industry, there are entirely new chapters that deal with 'Heterocycles in Medicine' (Chapter 33) and 'Heterocycles in Biochemistry; Heterocyclic Natural Products' (Chapter 32).

We devote a new chapter (31) to some important topics: fluorinated heterocycles, isotopically labelled heterocycles, the use of bioprocesses in heterocyclic transformations, 'green chemistry' and the somewhat related topic of ionic liquids, and some the applications of heterocyclic compounds in every-day life.

1. The main body of factual material is to be found in chapters entitled 'Reactions and synthesis of...' a particular heterocyclic system. Didactic material is to be found partly in advanced general discussions of heterocyclic reactivity and synthesis (Chapters 3, 4 and 6), and partly in six short summary chapters (such as 'Typical Reactivity of Pyridines, Quinolines and Isoquinolines'; Chapter 7), which aim to capture the essence of that typical reactivity in very concise resumés. These last are therefore suitable as an introduction to the chemistry of that heterocyclic system, but they are insufficient in themselves and should lead the reader to the fuller discussions in the 'Reactions and Synthesis of ...' chapters. They will also serve the undergraduate student as a revision summary of the typical chemistry of that system.

2. More than 4000 references have been given throughout the text: the references to original work have been chosen as good leading references and are, therefore, not necessarily the first or last mention of that particular topic or method or compound; some others are included as benchmark papers and others for their historical interest. The extensive list of references is most relevant to post-graduate teaching and to research workers, however we believe that the inclusion of references does not interfere with the readability of the text for the undergraduate student. Many review references are also included: for these we give the title of the article; titles are also given for the books to which we refer. The majority of journals are available only on a subscription (personal or institutional) basis, but most of their web sites give free access to abstracts and a few, such as *Arkivoc* and *Beilstein Journal of Organic Chemistry* give free access to full papers. Free access to the full text of patents, with a search facility, is available via government web sites. *Organic Syntheses*, the 'gold standard' for practical organic chemistry, has totally free online access to full procedures.

3. Exercises are given at the ends of most of the substantive chapters. These are divided into straightforward, revision exercises, such as will be relevant to an undergraduate course in heterocyclic chemistry. More advanced exercises, with solutions given on line at www.wiley.com/go/joule, are designed to help the reader to develop understanding and apply the principles of heterocyclic reactivity. References have not been given for the exercises, though all are real examples culled from the literature.

4. We largely avoid the use of 'R' and 'Ar' for substituents in the structures in schemes, and instead give actual examples. We believe this makes the chemistry easier to assimilate, especially for the undergraduate reader. It also avoids implying a generality that may not be justified.

5. Structures and numbering for heterocyclic systems are given at the beginnings of chapters. Where the commonly used name differs from that used in *Chemical Abstracts*, the name given in square brackets is the official *Chemical Abstracts* name, thus: indole [1*H*-indole]. We believe that the systematic naming of heterocyclic substances is of importance, not least for use in computerised databases, but it serves little purpose in teaching or for the understanding of the subject and, accordingly, we have devoted only a little space to nomenclature. The reader is referred to an exposition on this topic[1] and also to the Ring Index of *Chemical Abstracts* in combination with the Chemical Substances Index, from whence both standardised name and numbering can be obtained for all known systems. Readers with access to electronic search facilities such as *SciFinder* and *Crossfire* can easily find the various names for substances via a search on a drawn structure.

6. There are several general reference works concerned with heterocyclic chemistry, which have been gathered together as a set at the end of this chapter, and to which the reader's attention is drawn. In order to save space, these vital sources are not repeated in particular chapters, however all the topics covered in this book are covered in them, and recourse to these sources should form the early basis of any literature search.

7. The literature of heterocyclic chemistry is so vast that the series of nine listings – 'The Literature of Heterocyclic Chemistry', Parts I–IX[2] – is of considerable value at the start of a literature search. These listings appear in *Advances in Heterocyclic Chemistry*,[3] itself a prime source for key reviews on heterocyclic topics; the journal, *Heterocycles*, also carries many useful reviews specifically in the heterocyclic area. *Progress in Heterocyclic Chemistry*[4] published by the International Society of Heterocyclic Chemistry[5] also carries reviews, and monitors developments in heterocyclic chemistry over a calendar year. Essential at the beginning of a literature search is a consultation with the appropriate chapter(s) of *Comprehensive Heterocyclic Chemistry*, the original[6a] and its two updates,[6b,6c] or, for a useful introduction and overview, the handbook[7] to the series. It is important to realize that particular topics in the three parts of *Comprehensive Heterocyclic Chemistry* must be read together – the later parts update, but do not repeat, the earlier material. Finally, the *Science of Synthesis* series, published over the period 2000–2008, contains authoritative discussions of information organized in a hierarchical system.[8] Volumes 9–17 discuss aromatic heterocycles.

8. There are three comprehensive compilations of heterocyclic facts: the early series[9] edited by Elderfield, discusses pioneering work. The still-continuing and still-growing series of monographs[10] dealing with particular heterocyclic systems, edited originally by Arnold Weissberger, and latterly by Edward C. Taylor and Peter Wipf, is a vital source of information and reviews for all those working with heterocyclic compounds. Finally, the heterocyclic volumes of *Rodd's Chemistry of Carbon Compounds*[11] contain a wealth of well-sifted information and data.

P.1 Hazards

This book is designed, in large part, for the working chemist. All chemistry is hazardous to some degree and the reactions described in this book should only be carried out by persons with an appropriate degree of skill, and after consulting the original papers and carrying out a proper risk assessment. Some major hazards are highlighted (*Explosive*: general discussion (5.4), sodium azide (29.1.1.5.3), tetrazoles: diazonium salts and others (29.1.1.3), perchlorates (5.4; 11 (introductory paragraph)), tosyl azide (5.4). *Toxicity*: general (31.6.1), fluoroacetate (31.1.1.4), chloromethylation (e.g. 14.9.2.1)),[12] but this should not be taken to mean that every possible hazard is specifically pointed out. Certain topics are included only as information and are not suitable for general chemistry laboratories – this applies particularly to explosive compounds.

P.2 How to Use This Textbook

As indicated above, by comparison with earlier editions, this fifth edition of *Heterocyclic Chemistry* contains more material, including more that is appropriate to study at a higher level, than that generally taught in a first degree course. Nevertheless we believe that undergraduates will find the book of value and offer the following modus operandi as a means for undergraduate use of this text.

The undergraduate student should first read Chapter 2, which will provide a structural basis for the chemistry that follows. We suggest that the material dealt with in Chapters 3 and 4 be left for study at later stages, and that the undergraduate student proceed next to those chapters (7, 10, 13, 15, 19 and 23) that explain heterocyclic principles in the simplest terms and which should be easily understandable by students who have a good grounding in elementary reaction chemistry, especially aromatic chemistry.

The student could then proceed to the main chapters, dealing with 'Reactions and Synthesis of...' in which will be found full discussions of the chemistry of particular systems – pyridines, quinolines, etc. These utilise many cross references that seek to capitalise on that important didactical strategy – comparison and analogy with reactivity already learnt and understood.

Chapters 3, 4 and 6 are advanced essays on heterocyclic chemistry. Sections can be sampled as required – 'Electrophilic Substitution' could be read at the point at which the student was studying electrophilic substitutions of, say, thiophene – or Chapter 3 can be read as a whole. We have devoted considerable space in Chapter 3 to discussions of radical substitution, and Chapter 4, because of their great significance, is devoted entirely to metallation and the use of organometallic reagents, and to transition metal-catalysed reactions. These topics have grown enormously in importance since the earlier editions, and are of great relevance to heterocyclic chemistry.

Acknowledgements

We thank Richard Davies, Sarah Hall and Gemma Valler and their colleagues at Wiley, and earlier Paul Sayer at Blackwell, for their patience and support during the preparation of this fifth edition. We acknowledge many significant comments and corrections by Rob Young and Paul Beswick, and thank Mercedes Álvarez, Peter Quayle, Andrew Regan and Ian Watt for their views on the use of colour in schemes. We are greatly indebted to Jo Tyszka for her meticulous and constructive copy-editing. JAJ thanks his wife Stacy for her encouragement and patience during the writing of Heterocyclic Chemistry, Fifth Edition.

References

[1] 'The nomenclature of heterocycles', McNaught, A. D., *Adv. Heterocycl. Chem.*, **1976**, *20*, 175.

[2] Katritzky, A. R. and Weeds, S. M., *Adv. Heterocycl. Chem.*, **1966**, *7*, 225; Katritzky, A. R. and Jones, P. M., *ibid.*, **1979**, *25*, 303; Belen'kii, L. I., *ibid.*, **1988**, *44*, 269; Belen'kii, L. I. and Kruchkovskaya, N. D., *ibid.*, **1992**, *55*, 31; *idem, ibid.*, **1998**, *71*, 291; Belen'kii, L. I., Kruchkovskaya, N. D., and Gramenitskaya, V. N., *ibid.*, **1999**, *73*, 295; *idem, ibid.*, **2001**, *79*, 201; Belen'kii, L. I. and Gramenitskaya, V. N., *ibid.*, **2005**, *88*, 231; Belen'kii, L. I., Gramenitskaya, V. N., and Evdokimenkova, Yu. B., *ibid.*, **2004**, *92*, 146.

[3] *Adv. Heterocycl. Chem.*, **1963–2007**, 1–94.

[4] *Progr. Heterocycl. Chem.*, **1989–2009**, 1–21.

[5] http://euch6f.chem.emory.edu/ishc.html and the related Royal Society of Chemistry site: http://www.rsc.org/lap/rsccom/dab/perk003.htm

[6] (a) 'Comprehensive heterocyclic chemistry. The structure, reactions, synthesis, and uses of heterocyclic compounds', Eds. Katritzky, A. R. and Rees, C. W., Vols 1–8, Pergamon Press, Oxford, **1984**; (b) 'Comprehensive heterocyclic chemistry II. A review of the literature 1982–1995', Ed. Katritzky, A. R., Rees, C. W., and Scriven, E. F. V., Vols 1–11, Pergamon Press, **1996**; (c) 'Comprehensive heterocyclic chemistry III. A review of the literature 1995–2007', Eds. Katritzky, A. R., Ramsden, C. A., and Scriven, E. F. V., and Taylor, R. J. K., Vols 1–15, Elsevier, **2008**.

[7] 'Handbook of heterocyclic chemistry, 2nd edition 2000', Katritzky, A. R. and Pozharskii, A. F., Pergamon Press, Oxford, **2000**; 'Handbook of heterocyclic chemistry. Third edition 2010', Katritzky, A. R., Ramsden, C. A., Joule, J. A., and Zhdankin, V. V., Elsevier, **2010**.

[8] 'Science of Synthesis', Vols. 9–17, 'Hetarenes', Thieme, **2000–2008**.

[9] 'Heterocyclic compounds', Ed. Elderfield, R. C., Vols. 1–9, Wiley, **1950–1967**.

[10] 'The chemistry of heterocyclic compounds', Series Eds. Weissberger, A., Wipf, P., and Taylor, E. C., Vols. 1–64, Wiley-Interscience, **1950–2005**.

[11] 'Rodd's chemistry of carbon compounds', Eds., Coffey, S. then Ansell, M. F., Vols IVᴀ–IVʟ, and Supplements, **1973–1994**, Elsevier, Amsterdam.

[12] United States Department of Labor, Occupational Safety & Health Administration Reports: Chloromethyl Methyl Ether (CMME) and Bis-Chloromethyl Ether (BCME); see also: Berliner, M. and Belecki, K., *Org. Synth.*, **2007**, *84*, 102 (discussion).

Web Site

Power Point slides of all figures from this book, along with the solution to the exercises, can be found at http://www.wiley.com/go/joul.

Biography

John Arthur Joule was born in Harrogate, Yorkshire, England, but grew up and attended school in Llandudno, North Wales, going on to study for BSc, MSc, and PhD (1961; with George F. Smith) degrees at The University of Manchester. Following post-doctoral periods in Princeton (Richard K. Hill) and Stanford (Carl Djerassi) he joined the academic staff of The University of Manchester where he served for 41 years, retiring and being appointed Professor Emeritus in 2004. Sabbatical periods were spent at the University of Ibadan, Nigeria, Johns Hopkins Medical School, Department of Pharmacology and Experimental Therapeutics, and the University of Maryland, Baltimore County. He was William Evans Visiting Fellow at Otago University, New Zealand.

Dr. Joule has taught many courses on heterocyclic chemistry to industry and academe in the UK and elsewhere. He is currently Associate Editor for *Tetrahedron Letters*, Scientific Editor for *Arkivoc*, and Co-Editor of the annual *Progress in Heterocyclic Chemistry*.

Keith Mills was born in Barnsley, Yorkshire, England and attended Barnsley Grammar School, going on to study for BSc, MSc and PhD (1971; with John Joule) degrees at The University of Manchester.

Following post-doctoral periods at Columbia (Gilbert Stork) and Imperial College (Derek Barton/ Philip Magnus), he joined Allen and Hanburys (part of the Glaxo Group) at Ware and later Stevenage (finally as part of GSK), working in Medicinal Chemistry and Development Chemistry departments for a total of 25 years. During this time he spent a secondment at Glaxo, Verona. Since leaving GSK he has been an independent consultant to small pharmaceutical companies.

Dr. Mills has worked in several areas of medicine and many areas of organic chemistry, but with particular emphasis on heterocyclic chemistry and the applications of transition metal-catalysed reactions.

Heterocyclic Chemistry was first published in 1972, written by George Smith and John Joule, followed by a second edition in 1978. The third edition (Joule, Mills and Smith) was written in 1995 and, after the death of George Smith, a fourth edition (Joule and Mills) appeared in 2000; these authors also published *Heterocyclic Chemistry at a Glance* in 2007.

Definitions of Abbreviations

acac = acetylacetonato [MeCOCHCOMe$^-$]

adoc = adamantanyloxycarbonyl

Aliquat® = tricaprylmethylammonium chloride [MeN(C$_8$H$_{17}$)$_3$Cl]

p-An = $para$-anisyl [4-MeOC$_6$H$_4$]

aq. = aqueous

atm = atmosphere

9-BBN = 9-borabicyclo[3.3.1]nonane [C$_8$H$_{15}$B]

BINAP = 2,2′-bis(diphenylphosphino)-1,1′-binaphthalene [C$_{44}$H$_{32}$P$_2$]

BINOL = 1,1′-bi(2-naphthol) [C$_{20}$H$_{14}$O$_2$]

Bn = benzyl [PhCH$_2$]

Boc = $tertiary$-butoxycarbonyl [Me$_3$COC=O]

BOM = benzyloxymethyl [PhCH$_2$OCH$_2$]

BOP = (benzotriazol-1-yloxy)tris(dimethylamino)phosphonium hexafluorophosphate

BSA = N,O-bis(trimethylsilyl)acetamide [MeC(OSiMe$_3$)=NSiMe$_3$]

Bt = benzotriazol-1-yl [C$_6$H$_4$N$_3$]

i-Bu = iso-butyl [Me$_2$CHCH$_2$]

n-Bu = $normal$-butyl [Me(CH$_2$)$_3$]

s-Bu = $secondary$-butyl [MeCH$_2$C(Me)H]

t-Bu = $tertiary$-butyl [Me$_3$C]

Bus = $tertiary$-butylsulfonyl [Me$_3$CSO$_2$]

c. = concentrated

c = $cyclo$ as in c-C$_5$H$_9$ = cyclopentyl [C$_5$H$_9$]

CAN = cerium(IV) ammonium nitrate [Ce(NH$_4$)$_2$(NO$_3$)$_6$]

Cbz = benzyloxycarbonyl (PhCH$_2$OC=O)

CDI = 1,1′-carbonyldiimidazole [((C$_3$H$_3$N$_2$)$_2$C=O]

Chloramine T = N-chloro-4-methylbenzenesulfonamide sodium salt [TsN(Cl)Na]

cod = cycloocta-1,5-diene [C$_8$H$_{12}$]

coe = cyclooctene [C$_8$H$_{14}$]

cp = cyclopentadienyl anion [c-C$_5$H$_5^-$]

cp* = pentamethylcyclopentadienyl anion [Me$_5$-c-C$_5$]

m-CPBA = $meta$-chloroperbenzoic acid [3-ClC$_6$H$_4$CO$_3$H]

CSA = camphorsulfonic acid

CuTC = thiophene-2-carboxylic acid copper(I) salt [C$_5$H$_3$CuO$_2$S]

Cy = cyclohexyl [C$_6$H$_{11}$]

DABCO = 1,4-diazabicyclo[2.2.2]octane [C$_6$H$_{12}$N$_2$]

dba = $trans,trans$-dibenzylideneacetone [PhCH=CHCOCH=CHPh]

DBU = 1,8-diazabicyclo[5.4.0]undec-7-ene [C$_9$H$_{16}$N$_2$]

DCC = N,N'-dicyclohexylcarbodiimide [c-C$_6$H$_{11}$N=C=N-c-C$_6$H$_{11}$]

DCE = 1,2-dichloroethane [Cl(CH$_2$)$_2$Cl]

DDQ = 2,3-dichloro-5,6-dicyano-1,4-benzoquinone [C$_8$Cl$_2$N$_2$O$_2$]

de = diastereomeric excess

DEAD = diethyl azodicarboxylate [$EtO_2CN=NCO_2Et$]

DIAD = diisopropyl azodicarboxylate [$i-PrO_2CN=NCO_2i-Pr$]

DIBALH = diisobutylaluminium hydride [$(Me_2CHCH_2)_2AlH$]

DMA = *N,N*-dimethylacetamide [$MeCONMe_2$]

DMAP = 4-dimethylaminopyridine [$C_7H_{10}N_2$]

DME = 1,2-dimethoxyethane [$MeO(CH_2)_2OMe$]

DMF = *N,N*-dimethylformamide [$Me_2NCH=O$]

DMFDMA = dimethylformamide dimethyl acetal [$Me_2NCH(OMe)_2$]

DMSO = dimethylsulfoxide [$Me_2S=O$]

DoM = directed *ortho*-metallation

DPPA = diphenylphosphoryl azide [$(PhO)_2P(O)N_3$]

dppb = 1,4-bis(diphenylphosphino)butane [$Ph_2P(CH_2)_4PPh_2$]

dppf = 1,1′-bis(diphenylphosphino)ferrocene [$C_{34}H_{28}FeP_2$]

dppp = 1,3-bis(diphenylphosphino)propane [$Ph_2P(CH_2)_3PPh_2$]

EDTA = ethylenediaminetetracetic acid [$(HO_2CCH_2)_2N(CH_2)_2N(CH_2CO_2H)_2$]

ee = enantiomeric excess

El$^+$ = general electrophile

eq = equivalent(s)

ESR = electron spin resonance

Et = ethyl [CH_3CH_2]

f. = fuming

Fur = furyl as in 2-Fur = 2-furyl (furan-2-yl) [C_4H_3O]

FVP = flash vacuum pyrolysis

Het = general designation for an aromatic heterocyclic nucleus

HMDS = 1,1,1,3,3,3-hexamethyldisilazane [$Me_3SiNHSiMe_3$]

hplc = high pressure liquid chromatography

HOMO = highest occupied molecular orbital

hν = ultraviolet or visible irradiation

hy = high yield

Kryptofix 2.2.2 = 4,7,13,16,21,24-hexaoxa-1,10-diazabicyclo[8.8.8]hexacosane [$C_{18}H_{36}N_2O_6$]

LDA = lithium diisopropylamide [$LiNi-Pr_2$]

LiTMP = lithium 2,2,6,6-tetramethylpiperidide [$LiN(CMe_2(CH_2)_3CMe_2)$]

liq. = liquid

LR = Lawesson's reagent [$C_{14}H_{14}O_2P_2S_4$]

LUMO = lowest unoccupied molecular orbital

Me = methyl [CH_3]

MOM = methoxymethyl [CH_3OCH_2O]

mp = melting point

MS = molecular sieves

MTBD = 1,3,4,6,7,8-hexahydro-1-methyl-2*H*-pyrimido[1,2-*a*]pyridine [$C_8H_{15}N_3$]

Ms = mesyl (methanesulfonyl) [$MeSO_2$]

MSH = *O*-(mesitylenesulfonyl)hydroxylamine [$H_2NOSO_2C_6H_2-2,4,6-Me_3$]

MW = reaction heated by microwave irradation

NBS = *N*-bromosuccinimide [$C_4H_4BrNO_2$]

NDA = sodium diisopropylamide [$NaNi-Pr_2$]

NIS = *N*-iodosuccinimide [$C_4H_4INO_2$]

NMP = *N*-methylpyrrolidone [C_4H_9NO]

NPE = 2-(4-nitrophenyl)ethyl [$4-O_2NC_6H_4CH_2CH_2$]

Nu^- = general nucleophile

n-Oct = *normal*-octyl[$Me(CH_2)_7$]

OXONE® = potassium peroxymonosulfate [$2KHSO_5.KHSO_4.K_2SO_4$]

Ph = phenyl [C_6H_5]

PhH = benzene [C_6H_6]

Phosphorus oxychloride (phosphoryl chloride) = $POCl_3$

Phth = phthaloyl [$1,2\text{-}COC_6H_4CO$]

PIFA = phenyliodine(III) bis(trifluoroacetate) [$PhI(OCOCF_3)_3$]

PMB = *para*-methoxybenzyl [$4\text{-}MeOC_6H_4CH_2$]

PMP = 1,2,2,6,6-pentamethylpiperidine [$C_{10}H_{21}N$]

ⓅⓅ = pyrophosphate [$OP(=O)(OH)OP(=O)OH$]

PPA = polyphosphoric acid

i-Pr = *iso*-propyl [Me_2CH]

n-Pr = *normal*-propyl [$CH_3CH_2CH_2$]

proton sponge = 1,8-bis(dimethylamino)naphthalene [$C_{14}H_{18}N_2$]

PSSA = polystyrenesulfonic acid

py = pyridine, usually as a solvent

Py = pyridyl, as in 2-Py = 2-pyridinyl (pyridin-2-yl), 3-Py, 4-Py [C_5H_4N]

Pybox = 2,6-bis[(4*S*,5*S*)-4,5-diphenyl-2-oxazolin-2-yl]pyridine [$C_{35}H_{27}N_3O_2$]

R_f = general designation of perfluoroalkyl [C_nF_{2n+1}]

R_F = $R_f(CH_2)_n$

rp = room (atmospheric) pressure

rt = room temperature

salcomine = *N,N'*-bis(salicylidene)ethylenediaminocobalt(II) [$C_{16}H_{14}N_2O_2Co$]

SDS = sodium dodecylsulfate [$C_{12}H_{25}SO_3Na$]

SEM = trimethylsilylethoxymethyl [$Me_3Si(CH_2)_2OCH_2$]

SES = 2-(trimethylsilyl)ethanesulfonyl [$Me_3Si(CH_2)_2SO_2$]

SET = single electron transfer

SOMO = singly occupied molecular orbital

TASF = tris(dimethylamino)sulfur (trimethylsilyl)difluoride [$(Me_2N)_3S(Me_3SiF_2)$]

TBAF = tetra-*normal*-butylammonium fluoride [$n\text{-}Bu_4N^+\ F^-$]

TBAS = tetra-*normal*-butylammonium hydrogen sulfate [$n\text{-}Bu_4N^+\ HSO_4^-$]

TBDMS = *tertiary*-butyldimethylsilyl [$Me_3C(Me)_2Si$]

TBTA = tris[(1-benzyl-1*H*-1,2,3-triazol-4-yl)methyl]amine

TfO^- = triflate [$CF_3SO_3^-$]

tfp = trifuran-2-ylphosphine [$P(C_4H_3O)_3$]

THF = tetrahydrofuran (2,3,4,5-tetrahydrofuran) [C_4H_8O]

THP = tetrahydropyran-2-yl [C_5H_9O]

TIPS = tri-*iso*-propylsilyl [$i\text{-}Pr_3Si$]

TMEDA = *N,N,N',N'*-tetramethylethylenediamine [$Me_2N(CH_2)_2NMe_2$]

TMP = 2,2,6,6-tetramethylpiperidine [$C_9H_{19}N$]

TMS = trimethylsilyl [Me_3Si]

TMSOTf = trimethylsilyl triflate [$Me_3SiOSO_2CF_3$]

TolH = toluene [$C_6H_5CH_3$]

p-Tol = *para*-tolyl [$4\text{-}MeC_6H_4$]

o-Tol = *ortho*-tolyl [$2\text{-}MeC_6H_4$]

TosMIC = tosylmethyl isocyanide [$4\text{-}MeC_6H_4SO_2CH_2NC$]

triflate = trifluoromethanesulfonate [$CF_3SO_3^-$]

Ts = tosyl [4-MeC$_6$H$_4$SO$_2$]

Ⓓ® = β-D-2-deoxyribofuranosyl

Ⓡ = β-D-ribofuranosyl

Ⓢ = a sugar, usually a derivative of ribose or deoxyribose, attached to heterocyclic nitrogen, in which the substituents have not altered during the reaction shown.

))) = sonication

1

Heterocyclic Nomenclature

A selection of the structures, names and standard numbering of the more common heteroaromatic systems and some common non-aromatic heterocycles are given here as a necessary prelude to the discussions which follow in subsequent chapters. The aromatic heterocycles have been grouped into those with six-membered rings and those with five-membered rings. The names of six-membered aromatic heterocycles that contain nitrogen generally end in 'ine', though note that 'purine' is the name for a very important bicyclic system which has both a six- and a five-membered nitrogen-containing heterocycle. Five-membered heterocycles containing nitrogen general end with 'ole'. Note the use of italic '*H*' in a name such as '9*H*-purine' to designate the location of an *N*-hydrogen in a system in which, by tautomerism, the hydrogen could reside on another nitrogen (e.g. N-7 in the case of purine). Names such 'pyridine', 'pyrrole', 'thiophene', originally trivial, are now the standard, systematic names for these heterocycles; names such as '1,2,4-triazine' for a six-membered ring with three nitrogens located as indicated by the numbers, are more logically systematic.

A device that is useful, especially in discussions of reactivity, is the designation of positions as 'α', 'β', or 'γ'. For example, the 2- *and* the 6-positions in pyridine are equivalent in reactivity terms, so to make discussion of such reactivity clearer, each of these positions is referred to as an 'α-position'. Comparable use of α and β is made in describing reactivity in five-membered systems. These useful designations are shown on some of the structures. Note that carbons at angular positions do not have a separate number, but are designated using the number of the preceding atom followed by 'a' – as illustrated (only) for quinoline. For historical reasons purine does not follow this rule.

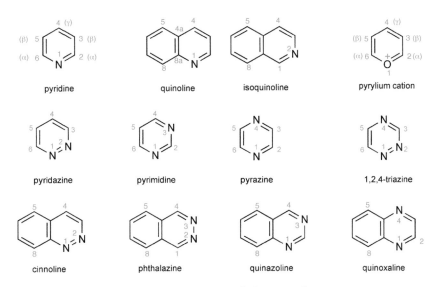

Six-membered aromatic heterocycles

Heterocyclic Chemistry 5th Edition   John Joule and Keith Mills
© 2010 Blackwell Publishing Ltd

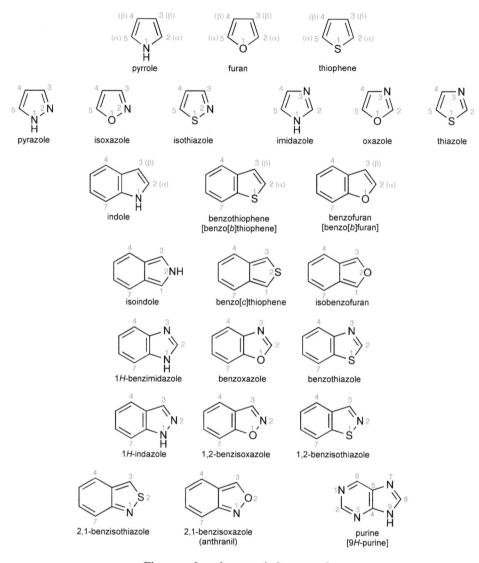

Five-membered aromatic heterocycles

A detailed discussion of the systematic rules for naming polycyclic systems in which several aromatic or heteroaromatic rings are fused together is beyond the scope of this book, however, a simple example will serve to illustrate the principle. In the name 'pyrrolo[2,3-*b*]pyridine', the numbers signify the positions of the first-named heterocycle, numbered as if it were a separate entity, which are the points of ring fusion; the italic letter, '*b*' in this case, designates the *side* of the second-named heterocycle to which the other ring is fused, the lettering deriving from the numbering of that heterocycle as a separate entity, i.e. side *a* is between atoms 1 and 2, side *b* is between atoms 2 and 3, etc. Actually, this particular heterocycle is more often referred to as '7-azaindole' – note the use of the prefix 'aza' to denote the replacement of a ring carbon by nitrogen, i.e. of C-7–H of indole by N.

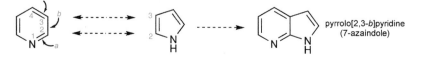

The main thrust of this book concerns the aromatic heterocycles, exemplified above, however Chapter 30 explores briefly the chemistry of saturated or partially unsaturated systems, including three- and four-membered heterocycles.

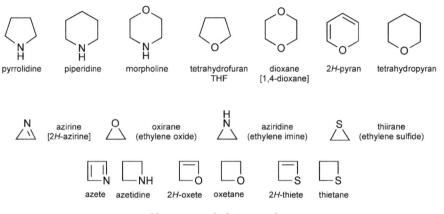

Non-aromatic heterocycles

2

Structures and Spectroscopic Properties of Aromatic Heterocycles

This chapter describes the structures of aromatic heterocycles and gives a brief summary of some physical properties.[1] The treatment we use is the valence-bond description, which we believe is appropriate for the understanding of all heterocyclic reactivity, perhaps save some very subtle effects, and is certainly sufficient for a general textbook on the subject. The more fundamental, molecular-orbital description of aromatic systems is less relevant to the day-to-day interpretation of heterocyclic reactivity, though it is necessary in some cases to utilise frontier orbital considerations,[2] however such situations do not fall within the scope of this book.

2.1 Carbocyclic Aromatic Systems

2.1.1 Structures of Benzene and Naphthalene

The concept of aromaticity as represented by benzene is a familiar and relatively simple one. The difference between benzene on the one hand and alkenes on the other is well known: the latter react with electrophiles, such as bromine, easily by addition, whereas benzene reacts only under much more forcing conditions and then typically by substitution. The difference is due to the cyclic arrangement of six π-electrons in benzene: this forms a conjugated molecular-orbital system which is thermodynamically much more stable than a corresponding non-cyclically conjugated system. The additional stabilisation results in a diminished tendency to react by addition and a greater tendency to react by substitution for, in the latter manner, survival of the original cyclic conjugated system of electrons is ensured in the product. A general rule proposed by Hückel in 1931 states that aromaticity is observed in cyclically conjugated systems of $4n + 2$ electrons, that is with 2, 6, 10, 14, etc., π-electrons; by far the majority of monocyclic aromatic and heteroaromatic systems are those with six π-electrons.

In this book we use the pictorial valence-bond resonance description of structure and reactivity. Even though this treatment is not rigorous, it is still the standard means for the understanding and learning of organic chemistry, which can at a more advanced level give way to the more complex, and mathematical, quantum-mechanical approach. We begin by recalling the structure of benzene in these terms.

In benzene, the geometry of the ring, with angles of 120 °, precisely fits the geometry of a planar trigonally hybridised carbon atom, and allows the assembly of a σ-skeleton of six sp^2 hybridised carbon atoms in a strainless planar ring. Each carbon then has one extra electron which occupies an atomic p orbital orthogonal to the plane of the ring. The p orbitals interact to generate π-molecular orbitals associated with the aromatic system.

Benzene is described as a 'resonance hybrid' of the two extreme forms which correspond, in terms of orbital interactions, to the two possible spin-coupled pairings of adjacent p electrons: structures **1** and **2**. These are known as 'resonance contributors', or 'mesomeric structures', have no existence in their own right, but serve to illustrate two extremes which contribute to the 'real' structure of benzene. Note the standard use of a double-headed arrow to inter-relate resonance contributors. Such arrows must never be confused with the use of opposing straight 'fish-hook' arrows that are used to designate an equilibrium

Heterocyclic Chemistry 5th Edition   John Joule and Keith Mills
© 2010 Blackwell Publishing Ltd

between two species. Resonance contributors have no separate existence; they are not in equilibrium one with the other.

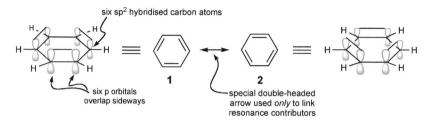

Structure of benzene; resonance contributors (mesomeric structures)

Sometimes, benzenoid compounds (and also, occasionally six- and five-membered heterocyclic systems) are represented using a circle inside a hexagon (pentagon); although this emphasises their delocalised nature and the close similarity of the ring bond lengths (all exactly identical only in benzene itself), it is not helpful in interpreting reactions, or in writing 'mechanisms', and we do not use this method in this book.

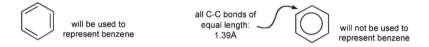

Treating naphthalene comparably reveals three resonance contributors, **3**, **4** and **5**. The valence-bond treatment predicts quite well the non-equivalence of the bond lengths in naphthalene: in two of the three contributing structures, C-1–C-2 is double and in one it is single, whereas C-2–C-3 is single in two and double in one. Statistically, then, the former may be looked on as 0.67 of a double bond and the latter as 0.33 of a double bond: the measured bond lengths confirm that there indeed is this degree of bond fixation, with values closely consistent with statistical prediction.

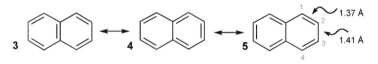

Structure of naphthalene; resonance contributors (mesomeric structures)

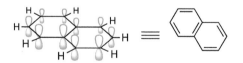

2.1.2 Aromatic Resonance Energy[3]

The difference between the ground-state energy of benzene and that of hypothetical, non-aromatic, 1,3,5-cyclohexatriene corresponds to the degree of stabilisation conferred to benzene by the special cyclical interaction of the six π-electrons. This difference is known as aromatic resonance energy. Quantification depends on the assumptions made in estimating the energy of the 'non-aromatic' structure, and for this reason and others, a variety of values have been calculated for the various heteroaromatic systems; their absolute values are less important than their relative values. What one can say with certainty is that the resonance energy of bicyclic aromatic compounds, like naphthalene, is considerably less than twice that of the corresponding monocyclic system, implying a smaller loss of stabilisation energy on conversion to a reaction intermediate which still retains a complete benzene ring, for example during electrophilic substitution (see 3.2). The resonance energy of pyridine is of the same order as that of benzene; that of thiophene is lower, with pyrrole and lastly furan of lower stabilisation energy still. Actual values for the stabilisations of these systems vary according to assumptions made, but are in the same relative order (kJ mol^{-1}): benzene (150), pyridine (117), thiophene (122), pyrrole, (90), and furan (68).

2.2 Structure of Six-Membered Heteroaromatic Systems

2.2.1 Structure of Pyridine

The structure of pyridine is completely analogous to that of benzene, being related by replacement of CH by N. The key differences are: (i) the departure from perfectly regular hexagonal geometry caused by the presence of the heteroatom, in particular the shorter carbon–nitrogen bonds, (ii) the replacement of a hydrogen in the plane of the ring with an unshared electron pair, likewise in the plane of the ring, located in an sp^2 hybrid orbital and not at all involved in the aromatic π-electron sextet; it is this nitrogen lone pair which is responsible for the basic properties of pyridines, and (iii) a strong permanent dipole, traceable to the greater electronegativity of nitrogen compared with carbon.

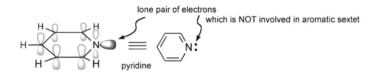

It is important to realise that the electronegative nitrogen causes inductive polarisation, mainly in the σ-bond framework, and additionally stabilises those polarised mesomeric contributors in which nitrogen is negatively charged – **8**, **9**, and **10** – which, together with contributors **6** and **7**, which are strictly analogous to the Kekulé contributors to benzene, represent pyridine. The polarised contributors also imply a permanent polarisation of the π-electron system.

Structure of pyridine; resonance contributors (mesomeric structures)

The polarisations resulting from inductive and mesomeric effects are in the same direction in pyridine, resulting in a permanent dipole towards the nitrogen atom. This also means that there are fractional positive charges on the carbons of the ring, located mainly on the α- and γ-positions. It is because of this general electron-deficiency at carbon that pyridine and similar heterocycles are referred to as 'electron-poor', or sometimes 'π-deficient'. A comparison with the dipole moment of piperidine, which is due wholly to the induced polarisation of the σ-skeleton, gives an idea of the additional polarisation associated with distortion of the π-electron system.

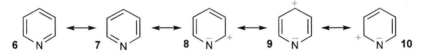

2.2.2 Structure of Diazines

The structures of the diazines (six-membered systems with two nitrogen atoms in the ring) are analogous, but now there are two nitrogen atoms and a corresponding two lone pairs; as an illustration, the main contributors (**11–18**) to pyrimidine are shown below.

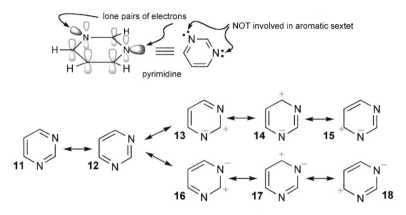

Structure of pyrimidine; resonance contributors (mesomeric structures)

2.2.3 Structure of Pyridinium and Related Cations

Electrophilic addition to the pyridine nitrogen generates pyridinium ions, the simplest being 1*H*-pyridinium formed by addition of a proton. 1*H*-Pyridinium is actually isoelectronic with benzene, the only difference being the nuclear charge of nitrogen, which makes the system, as a whole, positively charged. Thus pyridinium cations are still aromatic, the diagram making clear that the system of six p orbitals required to generate the aromatic molecular orbitals is still present, though the formal positive charge on the nitrogen atom severely distorts the π-system, making the α- and γ-carbons in these cations carry fractional positive charges which are higher than in pyridine, the consquence being increased reactivity towards nucleophiles. Electron density at the pyridinium β-carbons is also reduced relative to these carbons in pyridines.

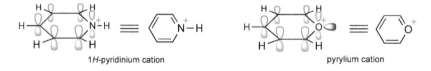

In the pyrylium cation, the positively charged oxygen also has an unshared electron pair, in an sp² orbital in the plane of the ring, exactly as in pyridine. Once again, a set of resonance contributors, **19–23**, makes clear that this ion is strongly positively charged at the 2-, 4- and 6-positions; in fact, because the more electronegative oxygen tolerates positive charge much less well than nitrogen, the pyrylium cation is certainly a less stabilised system than a pyridinium cation.

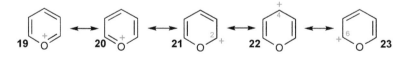

Structure of pyrylium cation; resonance contributors (mesomeric structures)

2.2.4 Structures of Pyridones and Pyrones

Pyridines with an oxygen at either the 2- or 4-position exist predominantly as carbonyl tautomers, which are therefore known as 'pyridones'[4] (see also 2.5). In the analogous oxygen heterocycles, no alternative tautomer is possible; the systems are known as 'pyrones'. The extent to which such molecules are aromatic has been a subject for considerable speculation and experimentation, and estimates have varied considerably. The degree of aromaticity depends on the contribution that dipolar structures, **25** and **27**, with a 'complete' pyridinium (pyrylium) ring make to the overall structure. Pyrones are less aromatic than pyridones, as can be seen from their tendency to undergo addition reactions (11.2.2.4), and as would be expected from a consideration of the 'aromatic' contributors, **25** and **27**, which have a positively charged ring heteroatom, oxygen being less easily able to accommodate this requirement.

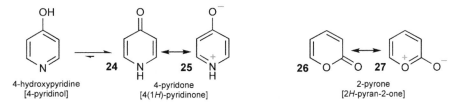

4-hydroxypyridine
[4-pyridinol]   4-pyridone   2-pyrone
                       [4(1*H*)-pyridinone]   [2*H*-pyran-2-one]

2.3 Structure of Five-Membered Heteroaromatic Systems[5]
2.3.1 Structure of Pyrrole
Before discussing pyrrole it is necessary to recall the structure of the cyclopentadienyl anion, which is a six π-electron aromatic system produced by the removal of a proton from cyclopentadiene. This system serves to illustrate nicely the difference between aromatic stabilisation and reactivity, for it is a very reactive, fully negatively charged entity, and yet is 'resonance stabilised' – everything is relative. Cyclopentadiene, with a pK_a of about 14, is much more acidic than a simple diene, just because the resulting anion is resonance stabilised. Five equivalent contributing structures, **28–32**, show each carbon atom to be equivalent and hence to carry one fifth of the negative charge.

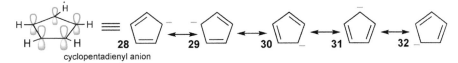

cyclopentadienyl anion

Structure of cyclopentadienyl anion; resonance contributors (mesomeric structures)

Pyrrole is isoelectronic with the cyclopentadienyl anion, but is electrically neutral because of the higher nuclear charge on nitrogen. The other consequence of the presence of nitrogen in the ring is the loss of radial symmetry, so that pyrrole does not have five equivalent mesomeric forms: it has one with no charge separation, **33**, and two pairs of equivalent forms in which there is charge separation, indicating electron density drift *away* from the nitrogen. These forms do not contribute equally; the order of importance is: **33** > **35,37** > **34,36**.

lone pair of electrons
IS part of the aromatic sextet

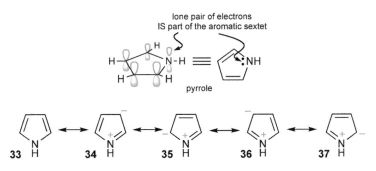

pyrrole

Structure of pyrrole; resonance contributors (mesomeric structures)

Resonance leads, then, to the establishment of partial negative charges on the carbons and a partial positive charge on the nitrogen. Of course the inductive effect of the nitrogen is, as usual, towards the heteroatom and away from carbon, so that the electronic distribution in pyrrole is a balance of two opposing effects, of which the mesomeric effect is probably the more significant, and this results in a dipole moment directed *away* from the nitrogen. The lengths of the bonds in pyrrole are in accord with this exposition, thus the 3,4-bond is very much longer than the 2,3-/4,5-bonds, but appreciably shorter than a normal single bond between sp² hybridised carbons, in accord with contributions from the polarised structures **34–37**. It is because of this electronic drift away from nitrogen and towards the ring carbons that five-membered heterocycles of the pyrrole type are referred to as 'electron-rich', or sometimes 'π-excessive'.

It is most important to recognise that the nitrogen lone pair in pyrrole forms part of the aromatic six-electron system.

2.3.2   Structures of Thiophene and Furan

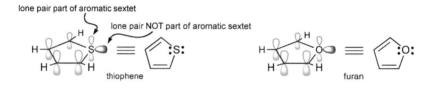

The structures of thiophene and furan are closely analogous to that discussed in detail for pyrrole above, except that the NH is replaced by S and O, respectively. A consequence is that the heteroatom in each has one lone pair as part of the aromatic sextet, as in pyrrole, but also has a second lone pair that is not involved, and is located in an sp² hybrid orbital in the plane of the ring. Mesomeric forms exactly analogous to those (above) for pyrrole can be written for each, but the higher electronegativity of both sulfur and oxygen means that the polarised forms, with positive charges on the heteroatoms, make a smaller contribution. The decreased mesomeric electron drift away from the heteroatoms is insufficient, in these two cases, to overcome the inductive polarisation towards the heteroatom (the dipole moments of tetrahydrothiophene and tetrahydrofuran, 1.87 D and 1.68 D, respectively, both towards the heteroatom, are in any case larger than that of pyrrolidine) and the net effect is that the dipoles are directed towards the heteroatoms in thiophene and furan.

The larger bonding radius of sulfur is one of the influences making thiophene more stable (more aromatic) than pyrrole or furan – the bonding angles are larger and angle strain is somewhat relieved, but in addition, a contribution to the stabilisation involving sulfur d-orbital participation may be significant.

2.3.3   Structures of Azoles
The 1,3- and 1,2-azoles, five-membered rings with two heteroatoms, present a fascinating combination of heteroatom types – in all cases, one heteroatom must be of the five-membered heterocycle (pyrrole, thiophene, furan) type and one of the imine type, as in pyridine; imidazole with two nitrogen atoms illustrates this best. Contributor **39** is a particularly favourable one.

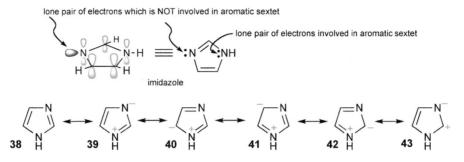

Structure of imidazole; resonance contributors (mesomeric structures)

2.3.4 Structures of Pyrryl and Related Anions

Removal of the proton from an azole *N*–hydrogen generates an N-anion, for example the pyrryl anion. Such species are still aromatic, but now have a lone pair of electrons at the nitrogen, in an sp^2 hybrid orbital, in the plane of the ring and *not* part of the aromatic sextet.

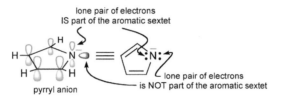

Even in the simplest example, pyrrole itself, the acidity (pK_a 17.5) is very considerably greater than that of its saturated counterpart, pyrrolidine (pK_a ~ 44); similarly the acidity of indole (pK_a 16.2) is much greater than that of aniline (pK_a 30.7). One may rationalise this relatively increased acidity on the grounds that the charge is not localised, and this is illustrated by resonance forms which show the delocalisation of charge around the heterocycle. With the addition of electron-withdrawing substituents, or with the inclusion of extra heteroatoms, especially imine groups, the acidity is enhanced. A nice, though extreme, example is tetrazole, for which the pK_a is 4.8, i.e. of the same order as a carboxylic acid!

pyrryl anion
pK_a 17.5

tetrazolyl anion
pK_a 4.8

2.4 Structures of Bicyclic Heteroaromatic Compounds

Once the concepts of the structures of benzene, naphthalene, pyridine and pyrrole, as prototypes, have been assimilated, it is straightforward to extrapolate to those systems which combine two (or more) of these types, thus quinoline is like naphthalene, only with one of the rings a pyridine, and indole is like pyrrole, but with a benzene ring attached.

Resonance representations must take account of the pattern established for benzene and the relevant heterocycle. Contributors in which both aromatic rings are disrupted make a very much smaller contribution and are shown in parentheses.

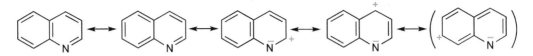

Structure of quinoline; resonance contributors (mesomeric structures)

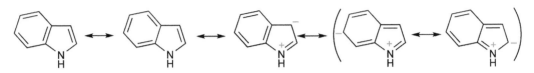

Structure of indole; resonance contributors (mesomeric structures)

2.5 Tautomerism in Heterocyclic Systems[6,7]

A topic which has attracted a large research effort over the years is the determination of the precise structure of heterocyclic molecules which are potentially tautomeric – the pyridinol/pyridone relationship (2.2.4) is one such situation. In principle, when an oxygen is located on a carbon α or γ to nitrogen, two tautomeric forms can exist; the same is true of amino groups.

Early attempts to use the results of chemical reactions to assess the form of a particular compound were misguided, since these can give entirely the wrong answer: the minor partner in such a tautomeric equilibrium may be the one that is the more reactive, so a major product may be actually derived from the minor component in the tautomeric equilibrium. Most secure evidence on these questions has come from comparisons of spectroscopic data for the compound in question with unambiguous models – often N- and O-methyl derivatives.

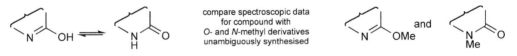

Determination of tautomeric equilibrium positions

In summary, α and γ oxy-heterocycles generally prefer the carbonyl form; amino-heterocycles nearly always exist as amino tautomers. Sulfur analogues – potentially thiol or thione – tend to exist as thione in six-membered situations, but as thiol in five-membered rings.

The establishment of tautomeric form is perhaps of most importance in connection with the purine and pyrimidine bases which form part of DNA and RNA, and, through H-bonding involving carbonyl oxygen, provide the mechanism for base pairing (cf. 32.4).

2.6 Mesoionic Systems[8]

There are a substantial number of heterocyclic substances for which no plausible, unpolarised mesomeric structure can be written: such systems are termed 'mesoionic'. Despite the presence of a nominal positive and negative charge in all resonance contributors to such compounds, they are not salt-like, are of course overall neutral, and behave like 'organic' substances, dissolving in the usual solvents. Examples of mesoionic structures occur throughout the text. Amongst the earliest mesoionic substances to be studied were the sydnones, for which several contributing structures can be drawn.

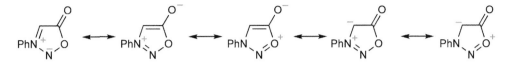

Structure of a sydnone; resonance contributors (mesomeric structures)

Mesoionic structures occur amongst six-membered systems too – one example is illustrated below.

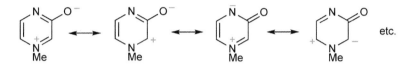

Structure of a pyrazinium-3-olate; resonance contributors (mesomeric structures)

If there is any one feature that characterises mesoionic compounds it is that their dipolar structures lead to reactions in which they serve as 1,3-dipoles in cycloadditions.

2.7 Some Spectroscopic Properties of Some Heteroaromatic Systems

The use of spectroscopy is at the heart of chemical research and analysis, but a knowledge of the particular chemical shift of, say, a proton on a pyridine, or the particular UV absorption maximum of, say, an indole, is only of direct relevance to those actually pursuing such research and analysis, and adds nothing to the understanding of heteroaromatic reactivity. Accordingly, we give here only a brief discussion, with relatively little data, of the spectroscopic properties of heterocyclic systems, anticipating that those who may be involved in particular research projects will turn to reviews[1] or the original literature for particular data.

The ultraviolet and infrared spectra of heteroaromatic systems are in accord with their aromatic character. Spectroscopic investigation, particularly ultraviolet/visible (UV/VIS) and nuclear magnetic resonance (NMR) spectroscopies, is particularly useful in the context of assessing the extent of such properties, in determining the position of tautomeric equilibria, and in testing for the existence of non-isolable intermediates.

2.7.1 Ultraviolet/Visible (Electronic) Spectroscopy

The simple unsubstituted heterocyclic systems show a wide range of electronic absorption, from the simple 200 nm band of furan, for example, to the 340 nm maximum shown by pyridazine. As is true for benzenoid compounds, the presence of substituents that can conjugate causes profound changes in electronic absorption, but the many variations possible are outside the scope of this section.

The UV spectra of the monocyclic azines show two bands, each with fine structure: one occurs in the relatively narrow range of 240–260 nm and corresponds to the $\pi \rightarrow \pi^*$ transitions, analogous with the $\pi \rightarrow \pi^*$ transitions in the same region in benzene (see Table 2.1). The other band occurs at longer wavelengths, from 270 nm in pyridine to 340 nm in pyridazine and corresponds to the interaction of the heteroatom lone pair with aromatic π electrons, the $n \rightarrow \pi^*$ transitions, which of course cannot occur in benzene. The absorptions due to $n \rightarrow \pi^*$ transitions are very solvent dependent, as is exemplified in Table 2.1 by the case of pyrimidine. With pyridine, this band is only observed in hexane solution, for in alcoholic solution the shift to shorter wavelengths results in masking by the main $\pi \rightarrow \pi^*$ band. Protonation of the ring nitrogen naturally quenches the $n \rightarrow \pi^*$ band by removing the heteroatom lone pair; protonation also has the effect of considerably increasing the intensity of the $\pi \rightarrow \pi^*$ band, without changing its position significantly, the experimental observation of which has diagnostic utility.

Table 2.1 *Ultraviolet spectra of monocyclic azines (fine structure not given)*

Heterocycle (solvent)	N → π* λ_{max} (nm)	ε	π → π* λ_{max} (nm)	π → π* λ_{max} (nm)	ε	ε
Pyridine (hexane)	270	450	195	251	7500	2000
Pyridine (ethanol)	–	–	–	257	–	2750
Pyridinium (ethanol)	–	–	–	256	–	5300
Pyridazine (hexane)	340	315	–	246	–	1400
Pyrimidine (hexane)	298	326	–	243	–	2030
Pyrazine (hexane)	328	1040	–	260	–	5600
Pyrimidine (water)	271	410	–	243	–	3210
Pyrimidinium (water)	–	–	–	242	–	5500
Pyrylium (90% aq. HClO4)	–	–	220	269	1400	8500
Benzene (hexane)	–	–	204	254	7400	200

Table 2.2 *Ultraviolet spectra of bicyclic azines (fine structure not given)*

Heterocycle	λ_{max} (nm)	λ_{max} (nm)	λ_{max} (nm)	ε	ε	ε
Quinoline	313	270	226	2360	3880	35500
Quinolinium	313	–	233	6350		34700
Isoquinoline	317	266	217	3100	4030	37000
Isoquinolinium	331	274	228	4170	1960	37500
Quinolizinium	324	284	225	14500	2700	17000
Naphthalene	312	275	220	250	5600	100000

Table 2.3 *Ultraviolet spectra of monocyclic five-membered heterocycles*

Heterocycle	λ_{max} (nm)	λ_{max} (nm)	ε	ε
Pyrrole	210	–	5100	–
Furan	200	–	10000	–
Thiophene	235	–	4300	–
Imidazole	206	–	3500	–
Oxazole	205	–	3900	–
Thiazole	235	–	3000	–
Cyclopentadiene	200	239	10000	3400

The bicyclic azines have much more complex electronic absorption, and the n → π* and π → π* bands overlap; being much more intense, the latter mask the former. Broadly, however, the absorptions of the bicyclic azines resemble that of naphthalene (Table 2.2).

The UV spectra of the simple five-membered heteroaromatic systems all show just one medium-to-strong low-wavelength band with no fine structure. Their absorptions have no obvious similarity to that of benzene, and no detectable n → π* absorption, not even in the azoles, which contain a pyridine-like nitrogen (Tables 2.3 and 2.4).

2.7.2   Nuclear Magnetic Resonance (NMR) Spectroscopy[9]

The chemical shifts[10] of protons attached to, and in particular of the carbons in, heterocyclic systems, can be taken as relating to the electron density at that position, with lower fields corresponding to electron-deficient carbons. For example, in the ¹H spectrum of pyridine, the lowest-field signals are for the α-protons (Table 2.5), the next lowest is that for the γ-proton and the highest-field signal corresponds to the β-protons, and this is echoed in the corresponding ¹³C shifts (Table 2.6). A second generality relates to the inductive

Table 2.4 *Ultraviolet spectra of bicyclic compounds with five-membered heterocyclic rings*

Heterocycle	λ_{max} (nm)	λ_{max} (nm)	λ_{max} (nm)	ε	ε	ε
Indole	288	261	219	4900	6300	25000
Benzo[b]thiophene	288	257	227	2000	5500	28000
Benzo[b]furan	281		244	2600		11000
2-t-Bu-isoindole	223, 266	270, 277	289, 329	48000, 1800	1650, 1850	1250, 3900
Isobenzofuran	215, 244, 249	254, 261, 313	319, 327, 334, 343	14800, 2500, 2350	2250, 1325, 5000	5000, 7400, 4575, 6150
Indolizine	347	295	238	1950	3600	32000
Benzimidazole	259	275		5620	5010	
Benzothiazole	217, 251	285	295	18620, 5500	1700	1350
Benzoxazole	231, 263	270	276	7940, 2400	3390	3240
2-Methyl-2H-indazole	275	292	295	6310	6170	6030
2,1-Benzisothiazole	203, 221	288sh, 298	315sh	14450, 16220	7590, 2880	3980
Purine	263	–	–	7950	–	–

Table 2.5 ^1H *chemical shifts (ppm) for heteroaromatic ring protons*

Heterocycle	δ_1	δ_2	δ_3	δ_4	δ_5	δ_6	δ_7	δ_8	Others
Pyridine	–	8.5	7.1	7.5	–	–	–	–	–
2-Pyridone	–	–	6.6	7.3	6.2	7.3	–	–	–
Quinoline	–	8.8	7.3	8.0	7.7	7.4	7.6	8.1	–
Quinoline N-oxide	–	8.6	7.3	7.7	–	–	–	8.8	–
Isoquinoline	9.1	–	8.5	7.5	7.7	7.6	7.5	7.9	–
Isoquinoline N-oxide	8.8	–	8.1	–	–	–	–	–	–
Pyridazine	–	–	9.2	7.7	–	–	–	–	–
Pyrimidine	–	9.2	–	8.6	7.1	–	–	–	–
Pyrimidine N-oxide	–	9.0	–	8.2	7.3	8.4	–	–	–
Pyrazine	–	8.5	–	–	–	–	–	–	–
1,2,4-Triazine	–	–	9.6	–	8.5	9.2	–	–	–
1,3,5-Triazine	–	9.2	–	–	–	–	–	–	–
Cinnoline	–	–	9.15	7.75	–	–	–	–	–
Quinazoline	–	9.2	–	9.3	–	–	–	–	–
Quinoxaline	–	9.7	–	–	–	–	–	–	–
Phthalazine	9.4	–	–	–	–	–	–	–	–
Pyrylium	–	9.6	8.5	9.3	–	–	–	–	in SO$_2$ (liq.)
Pyrrole	–	6.6	6.2	–	–	–	–	–	–
Thiophene	–	7.2	7.1	–	–	–	–	–	–
Furan	–	7.4	6.3	–	–	–	–	–	–
Indole	–	6.5	6.3	7.5	7.0	7.1	7.4	–	–
Benzo[b]furan	–	7.5	6.7	7.5	7.1	7.2	7.4	–	–
Benzo[b]thiophene	–	7.3	7.3	7.7	7.3	7.3	7.8	–	–
Indolizine	6.3	6.6	7.1	–	7.8	6.3	6.5	7.2	–
Imidazole	–	7.9	–	7.25	–	–	–	–	–
1-Methylimidazole	–	7.5	–	7.1	6.9	–	–	–	–
Pyrazole	–	–	7.6	6.3	–	–	–	–	–
1-Methylpyrazole	–	–	7.5	6.2	7.4	–	–	–	3.8 (CH$_3$)
Thiazole	–	8.9	–	8.0	7.4	–	–	–	–
Oxazole	–	7.95	–	7.1	7.7	–	–	–	–
Benzimidazole	–	7.4	–	7.0	6.9	–	–	–	–
Benzoxazole	–	7.5	–	7.7	7.8	7.8	7.7	–	–
Pyrazole	–	–	7.6	7.3	–	–	–	–	–
Isothiazole	–	–	8.5	7.3	8.7	–	–	–	–
Isoxazole	–	–	8.1	6.3	8.4	–	–	–	–
Indazole	–	–	8.1	7.8	7.1	7.35	7.55	–	–
1,2,3-Triazole	–	–	–	7.75	–	–	–	–	–
1,2,4-Triazole	–	–	7.9	–	8.85	–	–	–	–
Tetrazole	–	–	–	–	9.5	–	–	–	–
Purine	–	9.0	–	–	–	9.2	–	8.6	–
Benzene	7.27	–	–	–	–	–	–	–	–
Anisole	–	6.9	7.2	6.9	–	–	–	–	–
Aniline	–	6.5	7.0	6.6	–	–	–	–	–
Nitrobenzene	–	8.2	7.4	7.6	–	–	–	–	–
Naphthalene	7.8	7.5	–	–	–	–	–	–	–

Table 2.6   *^{13}C chemical shifts (ppm) for heteroaromatic ring carbons*

Heterocycle	δ_1	δ_2	δ_3	δ_4	δ_5	δ_6	δ_7	δ_8	δ ring junction	δ ring junction	Other
Pyridine	–	150	124	136	–	–	–	–	–	–	–
1-*H*-pyridinium	–	143	129	148	–	–	–	–	–	–	–
Pyridine *N*-oxide	–	139	126	126	–	–	–	–	–	–	–
1-Me-pyridinium	–	146	129	146	–	–	–	–	–	–	50 (CH$_3$)
2-Pyridone	–	165	121	142	107	136	–	–	–	–	–
4-Pyridone	–	140	116	176	–	–	–	–	–	–	–
Quinoline	–	151	122	136	1289	127	130	131	129 (4a)	149 (8a)	–
Isoquinoline	153	–	143	120	126	130	127	128	136 (4a)	129 (8a)	–
Pyridazine	–	–	153	128	–	–	–	–	–	–	–
1*H*-pyridazinium	–	–	152	138	–	–	–	–	–	–	–
pyrimidine	–	158	–	156	121	–	–	–	–	–	–
1-*H*-pyrimidinium	–	152	–	159	125	–	–	–	–	–	–
pyrazine	–	146	–	–	–	–	–	–	–	–	–
1*H*-pyrazinium	–	143	–	–	–	–	–	–	–	–	–
Cinnoline	–	–	146	125	128	132	132	130	127 (4a)	151 (8a)	–
Quinazoline	–	161	–	156	127	128	134	129	135 (4a)	150 (8a)	–
Quinoxaline	–	146	–	–	130	130	–	–	143 (4a)	–	–
Phthalazine	152	–	–	–	127	133	–	–	126 (4a)	–	–
1,2,3-Triazine	–	–	–	150	118	–	–	–	–	–	–
1,2,4-Triazine	–	–	158	–	150	151	–	–	–	–	–
1,3,5-Triazine	–	166	–	–	–	–	–	–	–	–	–
Pyrylium (BF$_4^-$)	–	169	128	161		–	–	–	–	–	–
2-Pyrone	–	162	117	143	106	152	–	–	–	–	–
2,6-Me$_2$-4-pyrone	–	166	114	180		–	–	–	–	–	20 (CH$_3$)
Coumarin	–	161	117	144	129	124	132	117	119 (4a)	154 (8a)	–
Chromone	–	156	113	177	125	126	134	118	125 (4a)	156 (8a)	–
Pyrrole	–	117	108	–	–	–	–	–	–	–	–
Thiophene	–	126	127	–	–	–	–	–	–	–	–
Furan	–	144	110	–	–	–	–	–	–	–	–
Indole	–	124	102	121	122	120	111	–	128 (3a)	136 (7a)	–
Oxindole	–	179	36	124	122	128	110	–	125 (3a)	143 (7a)	–
Benzo[*b*]furan	–	145	107	122	123	125	112	–	128 (3a)	155 (7a)	–
Benzo[*b*]thiophene	–	126	124	124	124	124	123	–	140 (3a)	140 (7a)	–
Indolizine	100	114	113	–	126	111	117	120	133 (8a)	–	–
Imidazole	–	135	–	122		–	–	–	–	–	–
1-Methylimidazole	–	138	–	130	120	–	–	–	–	–	33 (CH$_3$)
Thiazole	–	154	–	143	120	–	–	–	–	–	–
Oxazole	–	151	–	125	138	–	–	–	–	–	–
Benzimidazole	–	144	–	110	123	122	119	–	–	–	–
Benzothiazole	–	155	–	123	126	125	122	–	153 (3a)	134 (7a)	–
Benzoxazole	–	153	–	121	125	124	111	–	140 (3a)	150 (7a)	–
Pyrazole	–	–	135	106	135	–	–	–	–	–	–
Isothiazole	–	–	157	123	148	–	–	–	–	–	–
Isoxazole	–	–	150	105	159	–	–	–	–	–	–
Indazole	–	–	133	120	120	126	110	–	123 (3a)	140 (7a)	–
3-Methyl-1,2-benzisothiazole	–	163	–	–	–	–	–	–	152 (7a)	–	–
Purine	–	152	–	155	131	146	–	146	–	–	–
Uracil	–	151	–	142	100	164	–	–	–	–	–
Benzene	129	–	–	–	–	–	–	–	–	–	–
Anisole	160	114	130	121	–	–	–	–	–	–	–
Aniline	149	114	129	116	–	–	–	–	–	–	–
Nitrobenzene	149	124	130	135	–	–	–	–	–	–	–
Naphthalene	128	126	–	–	–	–	–	–	133 (4a)	–	–

electron withdrawal by the heteroatom – for example it is the hydrogens on the α-carbons of pyridine that are at lower field than that at the γ-carbon, and it is the signals for protons at the α-positions of furan that are at lower field than those at the β-positions. Protons at the α-positions of pyrylium cations present the lowest-field ^{1}H signals. In direct contrast, the chemical shifts for *C*-protons on electron-rich heterocycles, such as pyrrole, occur at much higher fields.

Coupling constants between 1,2-related (*ortho*) protons on heterocyclic systems vary considerably. Typical values round six-membered systems show smaller values closer to the heteroatom(s). In five-membered heterocycles, altogether smaller values are typically found, but again those involving a hydrogen closer to the heteroatom are smaller, except in thiophenes, where the larger size of the sulfur atom influences the coupling constant. The magnitude of such coupling constants reflects the degree of double-bond character (bond fixation) in a particular C–C bond.

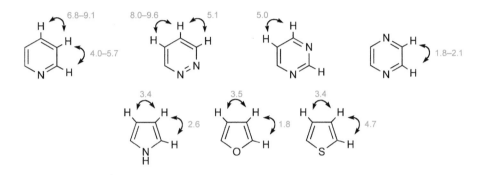

The use of ^{15}N NMR spectroscopy is of obvious relevance to the study of nitrogen-containing heterocycles – it can, for example, be used to estimate the hybridisation of nitrogen atoms.[11]

References

[1] 'Physical Methods in Heterocyclic Chemistry', Vols *1–5*, Ed. Katritzky, A. R., Academic Press, New York, **1960–1972**; 'Comprehensive Heterocyclic Chemistry. The Structure, Reactions, Synthesis, and Uses of Heterocyclic Compounds', Ed. Katritzky, A. R. and Rees, C. W., Vols *1–8*, Pergamon Press, Oxford, **1984**; 'Comprehensive Heterocyclic Chemistry II. A Review of the Literature 1982–1995', Ed. Katritzky, A. R., Rees, C. W. and Scriven, E. F. V., Vols *1–11*, Pergamon Press, **1996**; 'Comprehensive Heterocyclic Chemistry III. A Review of the Literature 1995–2007', Eds. Katritzky, A. R., Ramsden, C. A., Scriven, E. F. V. and Taylor, R. J. K., Vols *1–15*, Elsevier, **2008**.

[2] 'Frontier Orbitals and Organic Chemical Reactions', Fleming, I., Wiley-Interscience, **1976**.

[3] 'Aromaticity of heterocycles', Cook, M. J., Katritzky, A. R., and Linda, P., *Adv. Heterocycl. Chem.*, **1974**, *17*, 257; 'Aromaticity of heterocycles: experimental realisation of Dewar–Breslow definition of aromaticity', Hosmane, R. A. and Liebman, J. F., *Tetrahedron Lett.*, **1991**, *32*, 3949; 'The relationship between bond type, bond order and bond lengths. A re-evaluation of the aromaticity of some heterocyclic molecules', Box, V. G. S., *Heterocycles*, **1991**, *32*, 2023; 'Heterocyclic aromaticity', Katritzky, A. R., Karelson, M. and Malhotra, N., *Heterocycles*, **1991**, *32*, 127; 'The concept of aromaticity in heterocyclic chemistry', Simkin, B. Ya., Minkin, V. I. and Glukhovtsev, M. N., *Adv. Heterocycl. Chem.*, **1993**, *56*, 303.

[4] 'In solution at high dilution, or in the gas phase, hydroxypyridine tautomers are more important or even dominant', Beak, P., Covington, J. B., Smith, S. G., White, J. M. and Zeigler, J. M., *J. Org. Chem.*, **1980**, *45*, 1354.

[5] Fringuelli, F., Marino, G., Taticchi, A. and Grandolini, G., *J. Chem. Soc., Perkin Trans. 2*, **1974**, 332.

[6] 'The tautomerism of heterocycles', Elguero, J., Marzin, C., Katritzky, A. R. and Linda, P., *Adv. Heterocycl. Chem.*, Supplement 1, **1976**; 'Energies and alkylations of tautomeric heterocyclic compounds: old problems – new answers', Beak, P., *Acc. Chem. Res.*, **1977**, *10*, 186; 'Prototropic tautomerism of heteroaromatic compounds', Katritzky, A. R., Karelson, M. and Harris, P. A., *Heterocycles*, **1991**, *32*, 329.

[7] 'Recent developments in ring-chain tautomerism. I. Intramolecular reversible addition reactions to the C=O group', Valters, R. E., Fülöp, F. and Korbonits, D., *Adv. Heterocycl. Chem.*, **1995**, *64*, 251; 'Recent developments in ring-chain tautomerism. II. Intramolecular reversible addition reactions to the C=N, C=C=C and C=C groups', *idem, ibid.*, **1997**, *66*, 1; 'Tautomerism of heterocycles: five-membered rings with two or more heteroatoms', Minkin, V. I., Garnovskii, A. D., Elguero, J., Katritzky, A. R. and Denisko, O. V., *Adv. Heterocycl. Chem.*, **2000**, *76*, 159; 'Tautomerism involving other than five- and six-membered rings', Claramunt, R. M., Elguero, J. and Katritzky, A. R., *ibid.*, **2000**, *77*, 1; 'Tautomerism of heterocycles: condensed five-six, five-five and six-six ring systems with heteroatoms in both rings', Shcherbakova, I., Elguero, J. and Katritzky, A. R., *ibid.*, **2000**, *77*, 52; 'The tautomerism of heterocycles. Six-membered heterocycles: Annular tautomerism', Stanovnik, B., Tisler, M., Katritzky, A. R. and Denisko, O. V., *ibid.*, **2001**, *81*, 254; 'The tautomerism of heterocycles: substituent tautomerism of six-membered heterocycles', *ibid.*, **2006**, *91*, 1.

[8] 'Mesoionic compounds', Ollis, W. D. and Ramsden, C. A., *Adv. Heterocycl. Chem.*, **1976**, *19*, 1; 'Heterocyclic betaine derivatives of alternant hydrocarbons', Ramsden, C. A., *ibid.*, **1980**, *26*, 1; 'Mesoionic heterocycles (1976–1980)', Newton, C. G. and Ramsden, C. A., *Tetrahedron*, **1982**,

38, 2965; 'Six-membered mesoionic heterocycles of the *m*-quinodimethane dianion type', Friedrichsen, W., Kappe, T. and Böttcher, A., *Heterocycles*, **1982**, *19*, 1083.

[9] 'Applications of Nuclear Magnetic Resonance Spectroscopy in Organic Chemistry', Jackman, L. M. and Sternhell, S., Pergamon Press, **1969**; 'Carbon-13 NMR Spectroscopy', Breitmaier, E. and Voelter, W., VCH, **1990**.

[10] Both proton and carbon chemical shifts are solvent dependent – the figures given in the tables are a guide to the relative shift positions of proton and carbon signals in these heterocycles.

[11] von Philipsborn, W. and Müller, R., *Angew. Chem., Int. Ed. Engl.*, **1986**, *25*, 383.

3

Substitutions of Aromatic Heterocycles

This chapter describes in general terms the types of reactivity found in the typical six- and five-membered aromatic heterocycles. We discuss electrophilic addition (to nitrogen) and electrophilic, nucleophilic and radical substitution chemistry. This chapter also has discussion of *ortho*-quinodimethanes, in the heterocyclic context. Organometallic derivatives of heterocycles, and transition metal (especially palladium)-catalysed chemistry of heterocycles, are so important that we deal with these aspects separately, in Chapter 4. Emphasis on the typical chemistry of individual heterocyclic systems is to be found in the summary chapters (7, 10, 13, 15, 19 and 23), and a more detailed examination of typical heterocyclic reactivity and many more examples for particular heterocyclic systems are to be found in the chapters – 'Pyridines: Reactions and Synthesis', etc.

3.1 Electrophilic Addition at Nitrogen

Many heterocyclic compounds contain a ring nitrogen. In some, especially five-membered heterocycles, the nitrogen may carry a hydrogen. It is vital to the understanding of the chemistry of such nitrogen-containing heterocycles to know whether, and to what extent, they are basic – will form salts with protic acids or complexes with Lewis acids – and for heterocycles with N-hydrogen, to what extent they are acidic – will lose the N-hydrogen as a proton to an appropriately strong base (see 3.5). As a measure of these properties, we use pK_a values to express the acidity of heterocycles with N-hydrogen and pK_{aH} values to express base strength. *The lower the pK_a value the more acidic; the higher the pK_{aH} value the more basic.* It may be enough to simply remember this trend, but a little more detail is given below.

For an acid AH dissociating in water:

$$AH \; + \; H_2O \; \rightleftharpoons \; A^- \; + \; H_3O^+$$

$$K_a \; = \; \frac{[A^-][H_3O^+]}{[AH]} \qquad pK_a \; = \; -\log[K_a]$$

The corresponding equation for a base involves the *dissociation of the conjugate acid* of the base, so we use pK_{aH}:

$$BH^+ \; + \; H_2O \; \rightleftharpoons \; B \; + \; H_3O^+$$

Heterocycles which contain an imine unit (C=N) as part of their ring structure, pyridines, quinolines, isoquinolines, 1,2- and 1,3-azoles, etc., do not utilise the nitrogen lone pair in their aromatic π-system (cf. 2.2) and therefore it is available for donation to electrophiles, just as in any simpler amine. In other words, such heterocycles are basic and will react with protons, or other electrophilic species, by addition at nitrogen. In many instances the products from such additions – salts – are isolable.

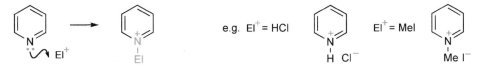

Heterocyclic Chemistry 5th Edition John Joule and Keith Mills
© 2010 Blackwell Publishing Ltd

For the reversible addition of a proton, the position of equilibrium depends on the pK_{aH} of the heterocycle,[1] and this in turn is influenced by the substituents present on the ring: electron-releasing groups enhance the basicity and electron-withdrawing substituents reduce the basic strength. The pK_{aH} of simple pyridines is of the order of 5, while those for 1,2- and 1,3-azoles depend on the character of the other heteroatom: pyrazole and imidazole, with two nitrogen atoms, have values of 2.5 and 7.1, respectively.

Related to basicity, but certainly not always mirroring it, is the N-nucleophilicity of imine-containing heterocycles. Here, the presence of substituents adjacent to the nitrogen can have a considerable effect on how easily reaction with, for example, alkyl halides takes place, and indeed whether nitrogen attacks at carbon, forming N^+-alkyl salts,[2] or by deprotonation, bringing about a 1,2-dehydrohalogenation of the halide, the heterocycle then being converted into an N^+-hydrogen salt. The classical study of the slowing of N-alkylation by the introduction of steric interference at α-positions of pyridines showed one methyl to slow the rate by about threefold, whereas 2,6-dimethyl substitution slowed the rate between 12 and 40 times.[3] Taking this to an extreme, 2,6-di-*t*-butylpyridine will not react at all with iodomethane; the very reactive methyl fluorosulfonate will N-methylate it, but only under high pressure.[4] The quantitative assessment of reactivity at nitrogen must always take into account both steric (especially at the α-positions) and electronic effects: 3-methylpyridine reacts faster (×1.6), but 3-chloropyridine reacts slower (×0.14) than pyridine. In bicyclic molecules, *peri* substituents have a significant effect on the relative rates of reaction with iodomethane: for pyridine, isoquinoline (no *peri* hydrogen), quinoline and 8-methylquinoline, rates are 50, 69, 8 and 0.008, respectively.

Other factors can influence the rate of quaternisation: all the diazines react with iodomethane more slowly than does pyridine. Pyridazine, much more weakly basic (pK_{aH} 2.3) than pyridine, reacts with iodomethane faster than the other diazines, a result which is ascribed to the 'α effect', i.e. the increased nucleophilicity is deemed to be due to electron repulsion between the two immediately adjacent nitrogen lone pairs.[5] Reaction rates for iodomethane with pyridazine, pyrimidine and pyrazine are respectively 0.25, 0.044 and 0.036, relative to the rate with pyridine.

3.2 Electrophilic Substitution at Carbon[6]
The study of aromatic heterocyclic reactivity can be said to have begun with the results of electrophilic substitution processes – these were traditionally the means for the introduction of substitutents onto heterocylic rings. To a considerable extent, that methodology has been superseded, especially for the introduction of carbon substituents, by methods relying on the formation of organometallic nucleophiles (4.1) and on palladium-catalysed processes (4.2). Nonetheless, the reaction of heterocycles with electrophilic reagents is still extremely useful in many cases, particularly for electron-rich, five-membered heterocycles.

3.2.1 Aromatic Electrophilic Substitution: Mechanism
Electrophilic substitution of aromatic (and heteroaromatic) molecules proceeds *via* a two-step sequence, initial *addition* (of El^+) giving a positively charged intermediate (a σ-complex, or Wheland intermediate), then *elimination* (normally of H^+), of which the former is usually the slower (rate-determining) step. Under most circumstances such substitutions are irreversible and the product ratio is determined by kinetic control.

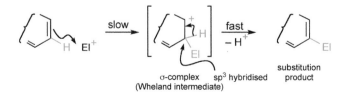

Electrophilic substitution of aromatic compounds

3.2.2 Six-Membered Heterocycles

An initial broad division must be made in considering heteroaromatic electrophilic substitution, into those heterocycles that are basic and those that are not, for, in the case of the former, the interaction of the nitrogen lone pair with the electrophile (cf. 3.1), or indeed with any other electrophilic species in the proposed reaction mixture (protons in a nitrating mixture, or aluminium chloride in a Friedel–Crafts combination), will take place far faster than any *C*-substitution, thus converting the substrate into a positively charged salt and therefore enormously reducing its susceptibility to attack by El^+ at carbon. It is worth recalling the rate reduction attendant upon the change from benzene to the *N,N,N*-trimethylanilinium cation (PhN^+Me_3), where the electrophilic substitution rate goes down by a factor of 10^8, even though in this instance the charged atom is only attached to, and not a component of, the aromatic ring. Thus all heterocycles with a pyridine-type nitrogen (i.e. those containing C=N) do not easily undergo *C*-electrophilic substitution, unless: (i) there are other substituents on the ring which 'activate' it for attack or (ii) the molecule has another, fused benzene ring in which substitution can take place. For example, simple pyridines do not undergo many useful electrophilic substitutions, but quinolines and isoquinolines undergo substitution in the benzene ring. It has been estimated that the intrinsic reactivity of pyridine (i.e. not protonated) to electrophilic substitution is around 10^7 times less than that of benzene, that is to say, about the same as that of nitrobenzene.

When quinoline or isoquinoline undergo nitration in the benzene ring, the actual species attacked is the *N*-protonated heterocycle and even though substitution is taking place in the benzene ring, it must necessarily proceed through a doubly charged intermediate; this results in a much slower rate of substitution than for naphthalene, the obvious comparison – the 5- and 8-positions of quinolinium are attacked at a rate about 10^{10} times slower than the 1-position of naphthalene, and it is estimated that the nitration of pyridinium cation is at least 10^5 slower still.[7] A study of the bromination of methylpyridines in acidic solution allowed an estimate of 10^{-13} for the partial rate factor for bromination of a pyridinium cation.[8]

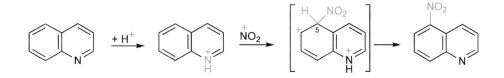

'Activating' substitutents,[9] i.e. groups that can release electrons either inductively or especially mesomerically, make the electrophilic substitution of pyridine rings to which they are attached faster; for example 4-pyridone nitrates at the 3-position *via* the *O*-protonated salt.[10] In order to understand the activation, it is helpful to view the species attacked as a (protonated) phenol-like substrate. Electrophilic attack on neutral pyridones is best visualised as attack on a carbonyl-conjugated enamine (N–C=C–C=O). Dimethoxypyridines also undergo nitration *via* their cations, but the balance is often delicate, for example 2-aminopyridine brominates at C-5, in acidic solution, *via* the free base.[11]

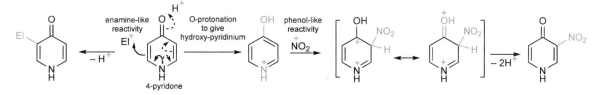

Electrophilic attack on 4-pyridones at C-3/5

Pyridines carrying activating substituents at C-2 are attacked at C-3/C-5, those with such groups at C-3 are attacked at C-2/C-6, and not at C-4, whilst those with substituents at C-4 undergo attack at C-3.

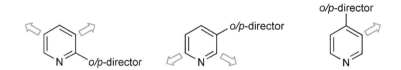

Positions of electrophilic attack on pyridines carrying activating substituents

Substituents that reduce the basicity of a pyridine nitrogen can also influence the susceptibility of the heterocycle to electrophilic substitution, in these cases by increasing the proportion of neutral (more reactive) pyridine present at equilibrium: 2,6-dichloropyridine nitrates at C-3, as the free base, and only 10^3 times more slowly than does 1,3-dichlorobenzene. As a rule-of-thumb: (i) pyridines with a $pK_{aH} > 1$ will nitrate as cations, slowly unless strongly activated, and at a position dictated by the substituent, (ii) weakly basic pyridines, $pK_{aH} < -2.5$, nitrate as free bases, the position of attack again depending on the influence of the substituent.[11] Pyridines carrying strongly electron-withdrawing substituents, or heterocycles with additional heteroatoms, diazines for example, are so deactivated that electrophilic substitutions do not take place, but again with the caveat that activating substituents do allow such substitutions in oxy- and amino-diazines.

3.2.3   Five-Membered Heterocycles

For five-membered, electron-rich heterocycles, the utility of electrophilic substitutions is much greater.[12] Heterocycles such as pyrrole, thiophene and furan undergo a range of electrophilic substitutions with great ease, at either type of ring position, but with a preference for attack adjacent to the heteroatom – at their α-positions.

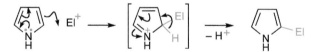

Electrophilic subsitution of pyrrole at an α-position

These substitutions are facilitated by electron release from the heteroatom: pyrroles are more reactive than furans, which are in turn more reactive than thiophenes. Quantitative comparisons[13] of the relative reactivities of the three heterocycles vary from electrophile to electrophile, but for trifluoroacetylation, for example, the pyrrole:furan:thiophene ratio is: $5 \times 10^7 : 1.5 \times 10^2 : 1$;[14] in formylation, furan is 12 times more reactive than thiophene,[15] and for acetylation, the value is 9.3.[16] In hydrogen exchange (deuteriodeprotonation), the partial rate factors for the α and β positions of N-methylpyrrole[17] are 3.9×10^{10} and 2.0×10^{10} respectively; for this same process, the values for furan are 1.6×10^8 and 3.2×10^4 and for thiophene, 3.9×10^8 and 1.0×10^5 respectively,[18] and in a study of thiophene, α:β ratios ranging from 100:1 to 1000:1 were found for different electrophiles.[19] Relative substrate reactivity parallels positional selectivity i.e. the α:β ratio decreases in the order furan > thiophene > pyrrole.[20] Nice illustrations of these relative reactivities are found in acylations of compounds containing two different systems linked together.[21]

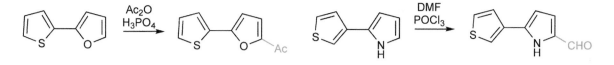

The positional selectivity of attack on pyrroles can be completely altered by the presence of bulky groups on nitrogen: 1-(*t*-butyldimethylsilyl)pyrrole and 1-(tri-*i*-propylsilyl)pyrrole are attacked exclusively at their β-positions.[22]

Indoles are only slightly less reactive than pyrroles, electrophilic substitution taking place in the hetero-cyclic ring, at a β-position; in acetylation using a Vilsmeier combination (*N,N*-dimethylacetamide/phosgene), the rate ratio compared with pyrrole is 1:3.[23] In contrast to pyrrole, there is a very large difference in reactivity between the two hetero-ring positions in indoles: 2600:1, β:α in Vilsmeier acylation. With reference to benzene, indole reacts at its β-position around 5×10^{13} times as fast.[24] Again, these differences can be illustrated conveniently using an example[25] that contains two types of system linked together.

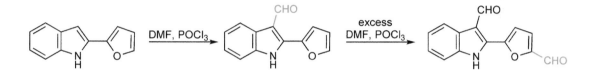

The reactivity of an indole is very comparable to that of a phenol: typical of phenols is their ability to be substituted even by weak electrophiles, like benzenediazonium cations, and indeed indoles (and pyrroles) also undergo such couplings; depending on p*H*, indoles can undergo such processes *via* a small equilibrium concentration of anion formed by loss of the *N*-proton (cf. 3.5); of course this is an even more rapid process, shown to be 10^8 faster than for the neutral heterocycle.[26] The Mannich substitution (electrophile: $CH_2=N^+Me_2$) of 5- and 6-hydroxy-indoles, takes place *ortho* to the phenolic activating group on the benzene ring, and not at the indole β-position.[27] Comparisons of the rates of substitution of the pairs furan/benzo[*b*]furan and thiophene/benzo[*b*]thiophene showed the bicyclic systems to be less reactive than the monocyclic heterocycles, the exact degree of difference varying from electrophile to electrophile.[28]

Finally, in the 1,2- and 1,3-azoles there is a fascinating interplay of the propensities of an electron-rich five-membered heterocycle with an imine basic nitrogen. This latter reduces the reactivity of the heterocycle towards electrophilic attack at carbon, both by inductive and mesomeric withdrawal, and importantly by addition of electrophilic species to the imine nitrogen (e.g. salt formation in acidic media). As an example, depending on acidity, the nitration of pyrazole can proceed by attack on the pyrazolium cation[29] or on the free base.[30] A study of acid-catalysed exchange showed the order: pyrazole > isoxazole > isothiazole, paralleling pyrrole > furan > thiophene, but each diazole is much less reactive than the corresponding heterocycle without the azomethine nitrogen, but, equally, each is still more reactive than benzene, the partial rate factors for exchange at their 4-positions being 6.3×10^9, 2.0×10^4 and 4.0×10^3 respectively. Thiophene is 3×10^5 times more rapidly nitrated than 4-methylthiazole.[31] The mono- and dinitration of a 2-(thien-2-yl) thiazole illustrates the relative reactivities.[32]

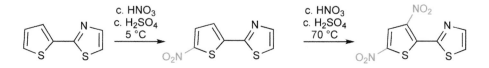

3.3 Nucleophilic Substitution at Carbon[33]
3.3.1 Aromatic Nucleophilic Substitution: Mechanism

Nucleophilic substitution of aromatic compounds proceeds *via* an *addition* (of Nu⁻) then *elimination* (of a negatively charged entity, most often Hal⁻) two-step sequence, of which the former is usually rate-determining (the $S_N(AE)$ mechanism: **S**ubstitution **N**ucleophilic **A**ddition **E**limination). It is the stabilisation (delocalisation of charge) of the negatively charged intermediates (Meisenheimer complexes) that is the key to such processes, for example in reactions of *ortho-* and *para*-chloronitro-benzenes, the nitro group is involved in the charge dispersal.

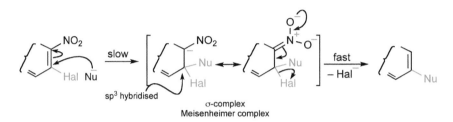

Aromatic nucleophilic substitution *via* an addition/elimination sequence

3.3.2 Six-Membered Heterocycles

In the heterocyclic field, the displacement of a good leaving group, often halide, by a nucleophile is a very important general process, especially for six-membered systems. In the chemistry of five-membered aromatic heterocycles, such processes only come into play in situations such as where, as in benzene chemistry, the leaving group is activated by an *ortho-* or *para*-nitro group, or in the azoles, where the leaving group is attached to the carbon of the imine unit in analogy with the six-membered imines.

The α- and γ-positions of a six-membered halo-azine, a 2-, 4- or 6-halo-pyridine being the prototype, are activated for the initial nucleophilic addition step by two factors: (i) inductive and mesomeric withdrawal of electrons by the nitrogen and (ii) inductive withdrawal of electrons by the halogen. Additionally, in the intermediates formed, the negative charge resides largely on the nitrogen: α- and γ-halides are *much* more reactive to nucleophilic displacement than β-halides.

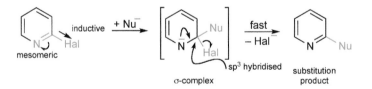

A quantitative comparison for displacements of chloride with sodium methoxide in methanol showed the 2- and 4-chloropyridines to react at roughly the same rate as 4-chloronitrobenzene, with the γ-isomer somewhat more reactive than the α-halide.[34] It is notable that even 3-chloropyridine, where only inductive activation can operate, is appreciably more reactive than chlorobenzene.

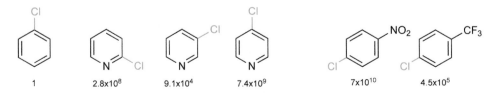

Rates of displacement of chloride by MeO⁻ relative to chlorobenzene, at 50 °C

The presence of a formal positive charge on the nitrogen, as in *N*-oxides and pyridinium salts, has a further very considerable enhancing effect on the rate of nucleophilic substitutions, *N*-oxidation having a smaller effect than quaternisation: in the latter there is a full formal positive charge on the molecule but *N*-oxides are overall electrically neutral. In reactions with methoxide, the 2-, 3- and 4-chloropyridine *N*-oxides are 1.9×10^4, 1.1×10^5, and 1.1×10^3 times more reactive than the corresponding chloropyridines, and displacements of halide in the 2-, 3- and 4-chloro-1-methylpyridinium salts are 4.6×10^{12}, 2.9×10^8, and 5.7×10^9 times more rapid. Another significant point to emerge from these rate studies concerns the relative rate enhancements, at the three ring positions: the effect of the charge is much greater at an α- than at a γ-position, such that in the salts the order is 2 > 4 > 3, as opposed to both neutral pyridines, where the order of reactivity is 4 > 2 > 3, and *N*-oxides, where the α-positions have about the same reactivity as the γ-positions.[35] The utility of a nitro group as a leaving group (nitrite) in heterocyclic chemistry is emphasised by a comparison of its relative reactivity to nucleophilic displacement: 4-nitropyridine is about 1100 times more reactive than 4-bromopyridine. Sulfones are also highly reactive and widely used leaving groups. A comparison of the rates of displacement of 4-methylsulfonylpyridine with its *N*-methyl quaternary salt showed a rise in rate by a factor of 7×10^8.[36] Although methoxide is not generally a good leaving group, when attached to a pyridinium salt it is only about four times less easily displaced than iodide, bromide and chloride; fluoride in the same situation is displaced about 250 times faster than the other halides.[37]

A substantial study of the activating effects of other substituents on the displacement of 2-halo-pyridines is very instructive and some examples are shown below. The activating effect of trifluoromethyl is particularly notable.[38]

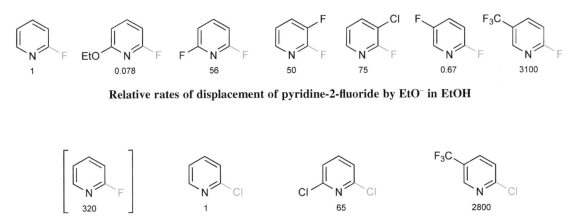

Relative rates of displacement of pyridine-2-fluoride by EtO⁻ in EtOH

Relative rates of displacement of pyridine-2-chloride by EtO⁻ in EtOH

In certain situations, particularly with relatively poor nucleophiles such as anilines, reaction rates can be enhanced considerably by the addition of acids, such as HCl, CF_3CO_2H or BF_3, to the reaction mixture, so that the much more reactive protonated haloazine is the substrate. Due to the relatively low basicity of anilines, sufficient free base is present to act as the nucleophile.

Turning to bicyclic systems, and a study of reaction with ethoxide, a small increase in the rate of reaction relative to pyridines is found for chloroquinolines at comparable positions.[39] In the bicyclic compounds, quaternisation again greatly increases the rate of nucleophilic substitution, having a larger effect (~10^7) at C-2 than at C-4 (~10^5).[40]

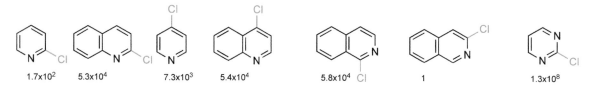

1.7x10²   5.3x10⁴   7.3x10³   5.4x10⁴   5.8x10⁴   1   1.3x10⁸

Relative rates of displacement of chloride by EtO⁻ at 20 °C

Diazines with halogen α and γ to nitrogen are much more reactive than similar pyridines, for example 2-chloropyrimidine is ~10^6 times more reactive than 2-chloropyridine.

3.3.3   Vicarious Nucleophilic Substitution (VNS Substitution)[41]

A process known as 'Vicarious Nucleophilic Substitution' (VNS) of hydrogen has been widely applied to carboaromatic and to heteroaromatic compounds. In general form, the process requires the presence of a nitro group on the substrate, which permits the addition of a carbon nucleophile, of the form $(X)(Y)(R)C^-$, where X is a potential leaving group and Y is an anion-stabilising group that permits the formation of the carbanion. Most often X is a halogen and Y can be arylsulfonyl, ester or benzotriazole (which can serve both as the anion stabilizing substituent and also as leaving group). A typical sequence is shown below: following addition, *ortho* or *para* to the nitro group, elimination of HX takes place to form a conjugated, non-aromatic nitronate, which on reprotonation returns the molecule to aromaticity and produces the substituted product. Excess of the base used to generate the initial carbanion must be employed in order to drive the process forward by subsequently bringing about the irreversible elimination of HX from the nitronate salt.

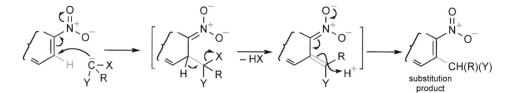

Vicarious nucleophilic substitution (VNS) of aromatic compounds

Three VNS sequences are shown below, each illustrating a different aspect. In the first example, the anion-stabilising group (Y) (trifluoromethanesulfonyl) also serves as the leaving group (X).[42] The second example shows the operation of a VNS substitution in a five-membered heterocycle with the nucleophile (X=Cl; Y=SO₂Ph) attacking at C-5, vinylogously conjugated to the nitro group.[43] The third example is somewhat unusual in that the attacking nucleophile (X=Cl; Y=SO₂*p*-Tol) does not even attack the nitro-substituted ring: addition occurs at C-2 in 6-nitroquinoxaline, for this produces an anion stabilised by delocalisation involving both N-1 and the nitro group.[44]

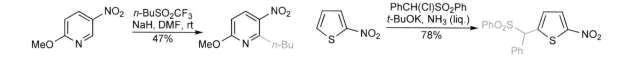

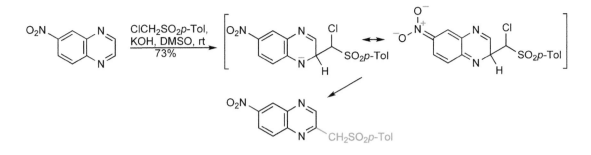

3.4 Radical Substitution at Carbon[45]

Both electron-rich and electron-poor heterocyclic rings are susceptible to substitution of hydrogen by free radicals. Although electrically neutral, radicals exhibit varying degrees of nucleophilic or electrophilic character and this has a very significant effect on their reactivity towards different heterocyclic types. These electronic properties are a consequence of the interaction between the SOMO (**S**ingly **O**ccupied **M**olecular **O**rbital) of the radical and either the HOMO, or the LUMO, of the substrate, depending on their relative energies; these interactions are usefully compared with charge-transfer interactions.

Nucleophilic radicals carry cation-stabilising groups on the radical carbon, allowing electron density to be transferred from the radical to an electron-deficient heterocycle; they react, therefore, only with electron-poor heterocycles and will not attack electron-rich systems: examples of such radicals are ˙CH$_2$OH, alkyl˙, and acyl˙. Substitution by such a radical can be represented in the following general way:

$$H \!-\!(\text{Het}) + R^{\bullet} \quad \longleftrightarrow \quad \left[H \!-\!(\stackrel{\bullet\,-}{\text{Het}}) \; R^{+} \right] \quad \longrightarrow \quad \left[H \!-\!(\text{Het}) \!-\! R \right] \xrightarrow[\text{[O]}]{-H^{\bullet}} \quad (\text{Het}) \!-\! R$$

electron-poor heterocycle   nucleophilic radical

Electrophilic radicals, conversely, are those which would form stabilised anions on gaining an electron, and therefore react readily with electron-rich systems; examples are ˙CF$_3$ and ˙CH(CO$_2$Et)$_2$. Substitution by such a radical can be represented in the following general way:

$$H \!-\!(\text{Het}) + R^{\bullet} \quad \longleftrightarrow \quad \left[H \!-\!(\stackrel{\bullet\,+}{\text{Het}}) \; R^{-} \right] \longrightarrow \quad \left[H \!-\!(\text{Het}) \!-\! R \right] \xrightarrow[\text{[O]}]{-H^{\bullet}} \quad (\text{Het}) \!-\! R$$

electron-rich heterocycle   electrophilic radical

Aryl radicals can show both types of reactivity. A considerable effort (mainly older work) was devoted to substitutions by aryl radicals; they react with electron-rich and electron-poor systems at about the same rate, but often with poor regioselectivity.[46]

3.4.1 Reactions of Heterocycles with Nucleophilic Radicals
The Minisci Reaction[47]

The reaction of nucleophilic radicals, under acidic conditions, with heterocycles containing an imine unit is by far the most important and synthetically useful radical substitution of heterocyclic compounds. Pyridines, quinolines, diazines, imidazoles, benzothiazoles and purines are amongst the systems that have been shown to react with a wide range of nucleophilic radicals, selectively at positions α and γ to the nitrogen, with replacement of hydrogen. Acidic conditions are essential because *N*-protonation of the heterocycle both greatly increases its reactivity and promotes regioselectivity towards a nucleophilic radical, most of which hardly react at all with the neutral base. A particularly useful feature of the process is that it can be used to introduce acyl groups, directly, i.e. to effect the equivalent of a Friedel–Crafts substitution – impossible under normal conditions for such systems (cf. 3.2.2). Tertiary radicals are more stable, but also more nucleophilic and therefore more reactive than methyl radicals in Minisci reactions. The majority of Minisci substitutions have been carried out in aqueous, or at least partially aqueous, media, making isolation of organic products particularly convenient.

Several methods have been employed to generate the required carbon-centred radical, many depending on the initial formation of oxy or methyl radicals, which then abstract hydrogen or iodine from suitable substrates, as illustrated below.[48] The re-aromatisation of the intermediate radical-cation is usually brought about by its reaction with excess of the oxidant used to form the initial radical.

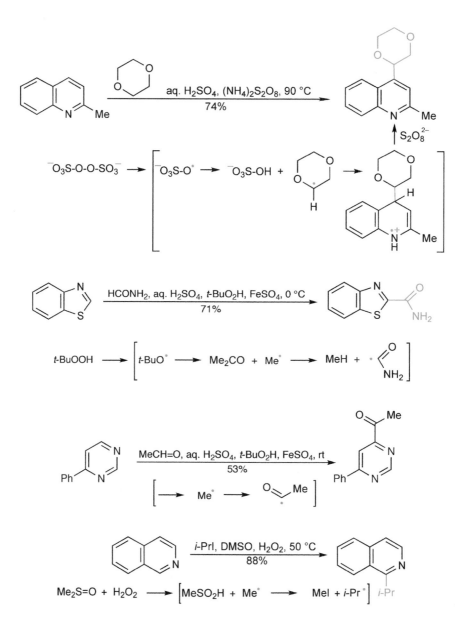

In contrast to the oxidative generation of radicals described above, reductions of alkyl iodides using tris(trimethylsilyl)silane also produces alkyl radicals under conditions suitable for Minisci-type substitution.[49] Carboxylic acids (α-keto acids) are also useful precursors for alkyl[50] and/or acyl[51] radicals *via* silver-catalysed peroxide oxidation, or from their 1-hydroxypyridine-2-thione derivatives,[52] the latter in non-aqueous conditions.

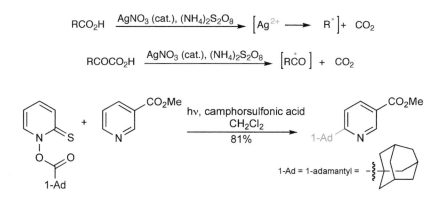

N,N-Dialkyl-formamides can be converted into either alkyl or acyl radicals, depending on the conditions.[53]

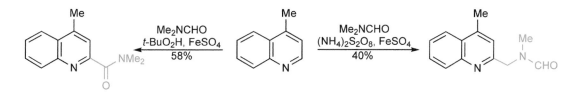

An instructive and useful process is the two-component coupling of an alkene with an electrophilic radical: the latter will of course not react with the protonated heterocycle, but after addition to the alkene, a nucleophilic radical is generated, which will react.[54]

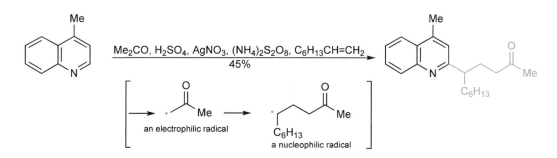

When more than one reactive position is available in a heterocyclic substrate, as is often the case for pyridines for example, there are potential problems with regioselectivity or/and disubstitution (since the product of the first substitution is often as reactive as the starting material). Regioselectivity is dependent to a certain extent on the nature of the attacking radical and the solvent, but may be difficult to control satisfactorily.[55]

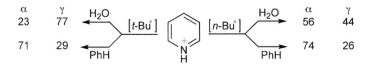

A point to note is that for optimum yields, radical substitutions are often not taken to full conversion (of starting heterocycle), but as product and starting material are often easily separated this is usually not a problem. Ways of avoiding disubstitution include control of pH (when the product is less basic than the starting material) or the use of a two-phase medium to allow removal of a more lipophilic product from the aqueous acidic reaction phase.

Very selective monosubstitution can also be achieved by the ingenious use of an N^+-methoxy quaternary salt, in place of the usual protonic salt. Here, re-aromatisation is the result of loss of methanol, leaving as a product a much less reactive, neutral pyridine.[56]

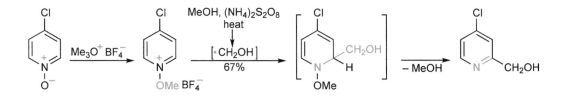

In addition to substitution of hydrogen, *ipso* replacement of nitro, sulfonyl and acyl substituents can occur, and may compete with normal substitution.[57]

3.4.2 Reactions with Electrophilic Radicals

Although much less well developed than the Minisci reaction, substitution with electrophilic radicals can be used in some cases to achieve selective reaction in electron-rich heterocycles.[58]

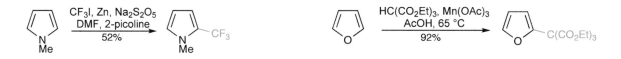

3.5 Deprotonation of *N*-Hydrogen[59]

Pyrroles, imidazoles, pyrazoles and benzo-fused derivatives that have a free *N*-hydrogen have pK_a values for the loss of the *N*-hydrogen as a proton in the region of 14–18. This is to say that they can be completely converted into *N*-anions by reaction with strong bases like sodium hydride or *n*-butyllithium. In reactivity terms, these *N*-anions are nucleophilic at the nitrogen, in direct contrast to the neutral heterocycle, and thus provide the means by which the nitrogen of azoles can be substituted, for example by reaction with alkyl halides, or with other electrophiles that can provide protection/masking of the nitrogen, the *N*-substituent to be subsequently removed (see 4.2.10 for palladium-catalysed azole *N*-arylations). Similar *N*-substitutions can also be achieved with bases that generate only an equilibrium (low) concentration of the *N*-anion.

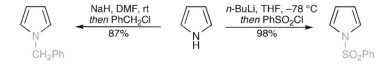

3.6 Oxidation and Reduction[60] of Heterocyclic Rings

Generally speaking, the electron-poor heterocycles are more resistant to oxidative degradation than are electron-rich systems – it is usually possible to oxidise alkyl side-chains attached to electron-poor heterocycles whilst leaving the ring intact; this is not generally true of electron-rich, five-membered systems.

   The conversion of monocyclic heteroaromatic systems into reduced, or partially reduced derivatives is generally possible, especially in acidic solutions, where it is a cation that is the actual species reduced. It follows that the six-membered types, which usually have a basic nitrogen, are more easily reduced than the electron-rich, five-membered counterparts. Heteroaromatic quaternary salts are likewise easily reduced.

3.7 *ortho*-Quinodimethanes in Heterocyclic Compound Synthesis[61]

The generation then trapping of *ortho*-quinodimethanes, in both intermolecular and intramolecular reactions, is a significant method for the construction of polycyclic heterocyclic compounds. This section describes the most important methods for the generation of such species, and gives some examples of their trapping. From the point of view of ring construction, the most important trapping reactions are those in which the *ortho*-quinodimethane acts as a diene in Diels–Alder cycloadditions, thereby regaining a fully aromatic heterocyclic ring, as illustrated below.[62] The unstable and reactive *ortho*-quinodimethanes are not isolated, but are generated in the presence of the trapping reactant. Their adducts with sulfur dioxide can be a convenient way in which to store *ortho*-quinodimethanes generated by other means.[61]

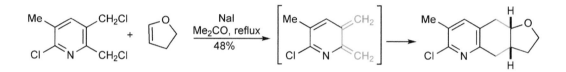

The ease with which an *ortho*-quinodimethane can be formed is related to the stability of the aromatic heterocycle from which it is derived and to the degree of double-bond character between the *ortho* ring carbons. The first of these aspects can be nicely illustrated by comparing the thiophene 2,3-quinodimethane[63] with its furan counterpart[64] – the latter is more stable than the former – the thiophene-derived species has much more to lose in its formation from an aromatic thiophene (and much more to gain by reacting to regain that aromaticity) than does the latter.

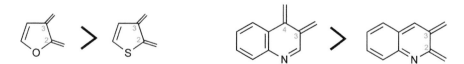

Relative stabilities of heterocyclic *ortho*-quinodimethanes

   ortho-Quinodimethanes are much easier to produce if the bond between the *ortho* ring carbons in the precursor has appreciable double-bond character. Thus, in five-membered heterocycles, it is much easier to produce a 2,3-quinodimethane, than a 3,4-quinodimethane. Similarly, in bicyclic six-membered systems, for example quinolines,[65] it is much easier to produce 3,4-quinodimethanes than 2,3-quinodimethanes, structures for which imply a loss of resonance stabilisation in the second ring.

   Three main strategies have been employed for the production of heterocyclic *ortho*-quinodimethanes: a 1,4-elimination, the chelotropic loss of sulfur dioxide from a 2,5-dihydrothiophene *S,S*-dioxide and the electrocyclic ring opening of a cyclobuteno-heterocycle; each of these is illustrated diagramatically below.

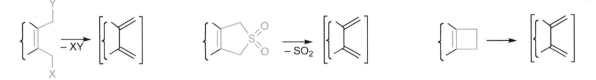

Generation of heterocyclic *ortho*-quinodimethanes

The use of cyclobuteno-heterocycles is of course dependent on a convenient synthesis (for an example, see 14.13.2.5), but when available, they are excellent precursors, only rather moderate heating being required for ring opening, as shown by the example below, in which the initial Diels–Alder adduct is aromatised by reaction with excess quinone.[66]

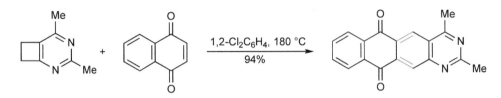

1,4-Eliminations have involved 1,2-bis(bromomethyl)-heterocycles with iodide,[67] *ortho*-(trimethylsilylmethyl) heterenemethyl ammonium salts,[68] *ortho*-(trimethylsilylmethyl) heterenecarbinol mesylates, each with a source of fluoride, and *ortho*-(tri-*n*-butylstannylmethyl) heterenecarbinol acetates with a Lewis acid.[69]

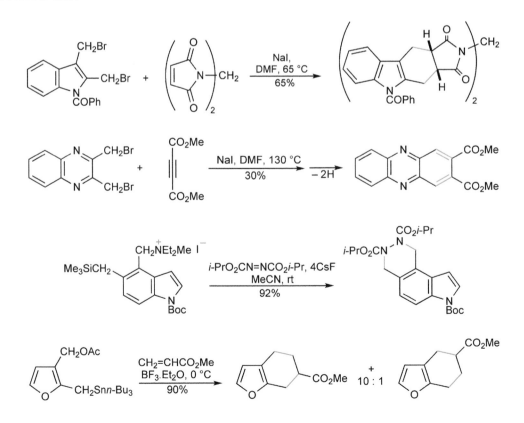

An extensively developed route involves loss of a proton from indol-3-ylcarboxaldehyde imines (or their pyrrolic counterparts[70]), following reaction with an acylating agent, as illustrated below.[71]

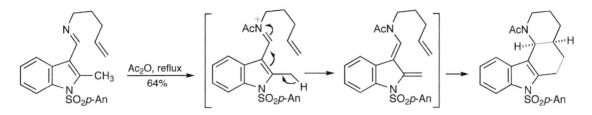

The extrusion of sulfur dioxide from heterocyclic sulfones is probably the most generally used method for the generation of *ortho*-quinodimethanes, and many examples have been reported. Such sulfones are generally stable and easy to synthesise by various routes. In addition, the acidity of the protons adjacent to the sulfone unit allows for base-promoted introduction of substituents, before thermolytic extrusion and the Diels–Alder step. Two examples of sulfur dioxide extrusion are shown below.[72]

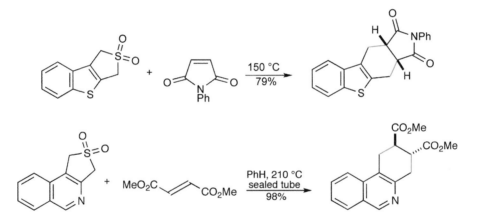

References

[1] Gas-phase proton affinities (PAs) (cf. 'The reactivity of heteroaromatic compounds in the gas phase', Speranza, M., *Adv. Heterocycl. Chem.*, **1986**, *40*, 25) are rather similar for all bases; such measurements, though of considerable theoretical interest, are of limited value in considerations of solution chemistry.

[2] 'The quaternisation of heterocyclic compounds', Duffin, G. F., *Adv. Heterocycl. Chem.*, **1964**, *3*, 1; 'Quaternisation of heteroaromatic compounds: quantitative aspects', Zoltewicz, J. A. and Deady, L. W., *Adv. Heterocycl. Chem.*, **1978**, *22*, 71; 'The quantitative analysis of steric effects in heteroaromatics', Gallo, R., Roussel, C. and Berg, U., *Adv. Heterocycl. Chem.*, **1988**, *43*, 173.

[3] Brown, H. C. and Cahn, A., *J. Am. Chem. Soc.*, **1955**, *77*, 1715.

[4] Okamoto, Y. and Lee, K. I., *J. Am. Chem. Soc.*, **1975**, *97*, 4015.

[5] Zoltewicz, J. A. and Deady, L. W., *J. Am. Chem. Soc.*, **1972**, *94*, 2765.

[6] 'Electrophilic substitution of heterocycles: quantitative aspects'; 'Part I, Electrophilic substitution reactions; Part II, Five-membered heterocyclic rings; Part III, Six-membered heterocyclic rings', Katritzky, A. R. and Taylor, R., *Adv. Heterocycl. Chem.*, **1990**, *47*, 1; 'Halogenation of heterocyclic compounds', Eisch, J. J., *Adv. Heterocycl. Chem.*, **1966**, *7*, 1; 'Halogenation of heterocycles: I. Five-membered rings', Grimmett, M. R., *ibid.*, **1993**, *57*, 291; 'II. Six- and seven-membered rings', *ibid.*, *58*, 271.

[7] Austin, M. W. and Ridd, J. H., *J. Chem. Soc.*, **1963**, 4204.

[8] Gilow, H. M. and Ridd, J. H., *J. Org. Chem.*, **1974**, *39*, 3481.

[9] 'Substitution in the pyridine series: effect of substituents', Abramovitch, R. A. and Saha, J. G., *Adv. Heterocycl. Chem.*, **1966**, *6*, 229.

[10] 'Mechanisms and rates of the electrophilic substitution reactions of heterocycles', Katritzky, A. R. and Fan, W.-Q., *Heterocycles*, **1992**, *34*, 2179.

[11] 'Electrophilic substitution of heteroaromatic six-membered rings', Katritzky, A. R. and Johnson, C. D., *Angew. Chem., Int. Ed. Engl.*, **1967**, *6*, 608.

[12] 'Electrophilic substitutions of five-membered rings', Marino, G., *Adv. Heterocycl. Chem.*, **1971**, *13*, 235.

[13] Marino, G., *J. Heterocycl. Chem.*, **1972**, *9*, 817.

[14] Clementi, S. and Marino, G., *Tetrahedron*, **1969**, *25*, 4599.

[15] Clementi, S., Fringuelli, F., Linda, P., Marino, G., Savelli, G. and Taticchi, A., *J. Chem. Soc., Perkin Trans. 2*, 1973, **2097**.

[16] Linda, P. and Marino, S., *Tetrahedron*, **1967**, *23*, 1739.

17 Quantitative comparisons must not ignore the considerable activating effect of a methyl group on an aromatic ring, whether attached to carbon or to nitrogen.

18 Bean, G. P., *J. Chem. Soc., Chem. Commun.*, **1971**, 421; Clementi, S., Forsythe, P. P., Johnson, C. D. and Katritzky, A. R., *J. Chem. Soc., Perkin Trans. 2*, **1973**, 1675; Clementi, S., Forsythe, P. P., Johnson, C. D., Katritzky, A. R. and Terem, B., *ibid.*, **1974**, 399.

19 Clementi, S., Linda, P. and Marino, G., *J. Chem. Soc. (B)*, **1970**, 1153.

20 Clementi, S. and Marino, G., *J. Chem. Soc., Perkin Trans. 2*, **1972**, 71.

21 Gol'dfarb, Y. L. and Danyushevskii, Y. L., *J. Gen. Chem. USSR (Engl. Transl.)*, **1961**, *31*, 3410; Boukou-Poba, J.-P., Farnier, M. and Guilard, R., *Can. J. Chem.*, **1981**, *59*, 2962.

22 Muchowski, J. M. and Naef, R., *Helv. Chim. Acta*, **1984**, *67*, 1168; Simchen, G. and Majchrzak, M. W., *Tetrahedron*, **1985**, *26*, 5035.

23 Cipiciani, A., Clementi, S., Linda, P., Marino, G. and Savelli, G., *J. Chem. Soc., Perkin Trans. 2*, **1977**, 1284.

24 Laws, A. P. and Taylor, R., *J. Chem. Soc., Perkin Trans. 2*, **1987**, 591.

25 Holla, B. S. and Ambekar, S. Y., *Indian J. Chem., Sect. B*, **1976**, *14B*, 579.

26 Challis, B. C. and Rzepa, H. S., *J. Chem. Soc., Perkin Trans. 2*, **1975**, 1209; Butler, A. R., Pogorzelec, P. and Shepherd, P. R., *idem.*, **1977**, 1452.

27 Monti, S. A. and Johnson, W. O., *Tetrahedron*, **1970**, *26*, 3685.

28 Clementi, S., Linda, P. and Marino, G., *J. Chem. Soc., (B)*, **1971**, 79.

29 Austin, M. W., Blackborrow, J. R., Ridd, J. H. and Smith, B. V., *J. Chem. Soc.*, **1965**, 1051.

30 Austin, M. W., *Chem. Ind.*, **1982**, 57.

31 Poite, C., Roggero, J., Dou, H. J. M., Vernin, G. and Metzsger, J., *Bull. Soc. Chim. Fr.*, **1972**, 162.

32 Chauvin, P., Morel, J., Pastour, P. and Martinez, J., *Bull. Soc. Chim. Fr.*, **1974**, 2099.

33 'Nucleophilic heteroaromatic substitution', Illuminati, G., *Adv. Heterocycl. Chem.*, **1964**, *3*, 285; 'Reactivity of azine, benzoazine, and azinoazine derivatives with simple nucleophiles', Shepherd, R. G. and Fedrick, J. L., *ibid.*, **1965**, *4*, 145; 'Formation of anionic σ-adducts from heteroaromatic compounds: structures, rates and equilibria', Illuminati, G. and Stegel, F., *ibid.*, **1983**, *34*, 306.

34 Liveris, M. and Miller, J., *J. Chem. Soc.*, **1963**, 3486; Miller, J. and Kai-Yan, W., *ibid.*, 3492.

35 Johnson, R. M., *J. Chem. Soc. (B)*, **1966**, 1058.

36 Barlin, G. B. and Benbow, J. A., *J. Chem. Soc., Perkin Trans. 2*, **1974**, 790.

37 O'Leary, M. H. and Stach, R. W., *J. Org. Chem.*, **1972**, *37*, 1491.

38 Schlosser, M. and Rausis, T., *Helv. Chim. Acta*, **2005**, *88*, 1240.

39 Chapman, N. B. and Russell-Hill, D. Q., *J. Chem. Soc.*, **1956**, 1563.

40 Barlin, G. B. and Benbow, J. A., *J. Chem. Soc., Perkin Trans. 2*, **1975**, 298.

41 'Vicarious nucleophilic substitution of hydrogen', Makosza, M. and Winiarski, J., *Acc. Chem. Res.*, **1987**, *20*, 282; 'Applications of vicarious nucleophilic substitution in organic synthesis', Makosza, M. and Wojciechowski, K., *Liebigs Ann./Receuil*, **1997**, 1805.

42 Wróbel, Z. and Makosza, M., *Org. Prep. Proc. Int.*, **1990**, 575.

43 Wojciechowski, K., *Synth. Commun.*, **1997**, *27*, 135.

44 Ostrowski, S. and Makosza, M., *Tetrahedron*, **1988**, *44*, 1721.

45 'Radicals in organic synthesis: formation of carbon-carbon bonds', Giese, B., Pergamon Press, **1986**; 'Free radical substitution of heteroaromatic compounds', Norman, R. O. C. and Radda, G. K., *Adv. Heterocycl. Chem.*, **1963**, *2*, 131.

46 Klemm, L. H. and Dorsey, J., *J. Heterocycl. Chem.*, **1991**, *28*, 1153.

47 Minisci, F., Galli, R., Cecere, M., Malatesta, V. and Caronna, T., *Tetrahedron Lett.*, **1968**, 5609; 'Substitutions by nucleophilic free radicals: a new general reaction of heteroaromatic bases', Minisci, F., Fontana, F. and Vismara, E., *J. Heterocycl. Chem.*, **1990**, *27*, 79; Minisci, F., Citterio, A., Vismara, E. and Giordano, C., *Tetrahedron*, **1985**, *41*, 4157; 'Advances in the synthesis of substituted pyridazines via introduction of carbon functional groups into the parent heterocycle', Heinisch, G., *Heterocycles*, **1987**, *26*, 481; 'Recent developments of free radical substitutions of heteroaromatic bases', Minisci, F., Vismara, E. and Fonatana, F., *Heterocycles*, **1989**, *28*, 489.

48 Buratti, W., Gardini, G. P., Minisci, F., Bertini, F., Galli, R. and Perchinunno, M., *Tetrahedron*, **1971**, 3655; Minisci, F., Gardini, G. P., Galli, R. and Bertini, F., *Tetrahedron Lett.*, **1970**, 15; Sakamoto, T., Sakasai, T. and Yamanaka, H., *Chem. Pharm. Bull.*, **1980**, *28*, 571; Minisci, F., Vismara, E. and Fonatana, F., *J. Org. Chem.*, **1989**, *54*, 5224.

49 Togo, H., Hayashi, K. and Yokoyama, M., *Chem. Lett.*, **1993**, 641.

50 Fontana, F., Minisci, F., Nogueira-Barbosa, M. C. and Vismara, E., *Tetrahedron*, **1990**, *46*, 2525.

51 Fontana, F., Minisci, F., Nogueira-Barbosa, M. C. and Vismara, E., *J. Org. Chem.*, **1991**, *56*, 2866.

52 Barton, D. H. R., Garcia, B., Togo, H. and Zard, S. Z., *Tetrahedron Lett.*, **1986**, *27*, 1327; Barton D. H. R., Chern, C.-Y. and Jaszberenyi, J. Cs., *ibid.*, **1992**, *33*, 5013.

53 Gardini, G. P., Minisci, F., Palla, G., Arnone, A. and Galli, R., *Tetrahedron Lett.*, **1971**, 59; Citterio, A., Gentile, A., Minisci, F., Serravalle, M. and Ventura, S., *J. Org. Chem.*, **1984**, *49*, 3364.

54 Citterio, A., Gentile, A. and Minisci, F., *Tetrahedron Lett.*, **1982**, *23*, 5587.

55 Minisci, F., Vismara, E., Fontana, F., Morini, G., Serravalle, M. and Giordano, G., *J. Org. Chem.*, **1987**, *52*, 730.

56 Katz, R. B., Mistry, J. and Mitchell, M. B., *Synth. Commun.*, **1989**, *19*, 317.

57 'Radical ipso attack and ipso substitution in aromatic compounds', Tiecco, M., *Acc. Chem. Res.*, **1980**, *13*, 51.

58 Tordeaux, M., Langlois, B. and Wakselman, C., *J. Chem. Soc., Perkin Trans. 1*, **1990**, 2293; Cho, I.-S. and Muchowski, J. M., *Synthesis*, **1991**, 567.

59 'Basicity and acidity of azoles', Catalan, J., Abboud, J. L. M. and Elguero, J., *Adv. Heterocycl. Chem.*, **1987**, *41*, 187.

60 'The reduction of nitrogen heterocycles with complex metal hydrides', Lyle, R. E. and Anderson, P. S., *Adv. Heterocycl. Chem.*, **1966**, *6*, 46; 'The reduction of nitrogen heterocycles with complex metal hydrides', Keay, J. G., *ibid.*, **1986**, *39*, 1.

61 'Heterocyclic ortho-quinodimethanes', Collier, S. J. and Storr, R. C., *Prog. Heterocycl. Chem.*, **1998**, *10*, 25.

62 Carly, P. R., Cappelle, S. L., Compernolle, F. and Hoornaert, G. J., *Tetrahedron*, **1996**, *52*, 11889.

63 Munzel, N. and Schweig, A., *Chem. Ber.*, **1988**, *121*, 791.

64 Trahanovsky, W. S., Cassady, T. J. and Woods, T. L., *J. Am. Chem. Soc.*, **1981**, *103*, 6691.

65 White, L. A., O'Neill, P. M., Park, B. K. and Storr, R. C., *Tetrahedron Lett.*, **1995**, *37*, 5983.

[66] Herrera, A., Martinez, R., González, B., Illescas, B., Martin, N. and Seoane, C., *Tetrahedron Lett.*, **1997**, *38*, 4873.

[67] Mertzanos, G. E., Stephanidou-Stephanatou, J., Tsoleridis, C. A. and Alexandrou, N. E., *Tetrahedron Lett.*, **1992**, *33*, 4499; Alexandrou, N. E., Mertzanos, G. E., Stephanidou-Stephanatou, J., Tsoleridis, C. A. and Zachariou, P., *ibid.*, **1995**, *36*, 6777; Pindur, U., Gonzalez, E. and Mehrabani, F., *J. Chem. Soc., Perkin Trans. 1*, **1997**, 1861.

[68] Kinsman, A. C. and Snieckus, V., *Tetrahedron Lett.*, **1999**, *40*, 2453.

[69] Liu, G.-B., Mori, H. and Katsumura, S., *Chem. Commun.*, **1996**, 2251.

[70] Leusink, F. R., ten Have, R., van der Berg, K. J. and van Leusen, A. M., *J. Chem. Soc., Chem. Commun.*, **1992**, 1401.

[71] Magnus, P., Gallagher, T., Brown, P. and Pappalardo, P., *Acc. Chem. Res.*, **1984**, *17*, 25.

[72] Ko, C.-W. and Chou, T., *Tetrahedron Lett.*, **1997**, *38*, 5315; Tomé, A. C., Cavaleiro, J. A. S. and Storr, R. C., *Tetrahedron*, **1996**, *52*, 1723; Chen, H.-C and Chou, T.-s, *Tetrahedron*, **1998**, *54*, 12609.

4

Organometallic Heterocyclic Chemistry

Heterocyclic 'organometallics' cover a wide range of types and reactivities, and can be prepared for any metal, although relatively few are of practical importance for the synthetic chemist. The most significant are: (i) nucleophilic (and often basic) compounds, mainly lithium and magnesium compounds, (ii) nucleophilic, generally non-basic compounds, such as those of zinc, aluminium and titanium, (iii) compounds of tin, and the 'metalloids' boron and silicon, which have relatively low classical nucleophilicity, but are particularly important as notionally 'nucleophilic' partners in transition-metal-catalysed coupling reactions, but also have interesting chemistry in their own right, (iv) transition metals, of which the most important are intermediates in catalytic cycles, particularly palladium, copper, nickel and rhodium.

In this chapter, for general organometallics we give an overview, with most of the examples for particular heterocyclic rings in the main chapters (with further discussions); however, for the transition metals, because of the regularity of reactivity across the whole range of heterocycles, most of the examples are given in this chapter.

4.1 Preparation and Reactions of Organometallic Compounds

The general methods of preparation of these compounds[1] are:

1. *Direct metallation.* Direct C–H metallations are of several types, of which the most important is reaction ('deprotonation') with a strongly basic reagent, usually a lithium compound, but is also possible for magnesium and zinc. Electrophilic metallation can be carried out with palladium(II) and mercury(II) salts, and neutral C–H insertion by other transition metals is becoming increasingly important, usually for catalytic reactions.
2. *Halogen–metal exchange.* The simplest type of halogen metal exchange is by reaction of metal with a halide, such as in the preparation of a Grignard reagent, but the most common way is by reaction of the halide with an organometallic reagent, particularly an alkyllithium. Exchanges using organomagnesium and organozinc compounds are now very well developed and offer advantages in selectivity and functional group compatibility.
3. *Metal–metal exchange.* Metal–metal exchange usually involves the reaction of an organometallic reagent with an electrophilic metal source, such as a salt, halo or alkoxy derivative. This is most widely used for the preparation of organo-boron, -tin, -zinc and -silicon compounds by reaction with organolithium or magnesium reagents.

4.1.1 Lithium[2]

Lithio-heterocycles have proved to be the most useful organometallic derivatives: they react with the whole range of electrophiles in a manner exactly comparable to that of aryllithiums and can often be prepared by direct metallation (*C*-hydrogen deprotonation), as well as by halogen exchange between a halo-heterocycle and an alkyllithium. As well as reaction with carbon electrophiles, lithiated species are often the most convenient source of heterocyclic derivatives of less electropositive metals, such as zinc, boron, silicon and tin, as will be seen in the following sections.

Heterocyclic Chemistry 5th Edition   John Joule and Keith Mills
© 2010 Blackwell Publishing Ltd

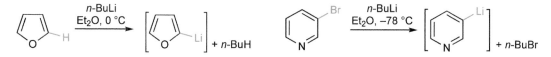

The two main routes to hetero-organolithiums exemplified

4.1.1.1 Direct Lithiation (C-Hydrogen Deprotonation)

Many heterocyclic systems react directly with alkyllithiums or with lithium amides to give the lithio-heterocycle *via* abstraction of a proton. Although a 'free' anion is never formed, the ease of lithiation correlates well with *C*-hydrogen acidity and, of course, with the stability of the corresponding conjugate base (carbanion).[3] Lithiations by deprotonation are therefore directly related to base-catalysed proton exchange[4] using reagents such as sodium methoxide, at much higher temperatures, which historically provided the first indication that preparative deprotonations might be regioselective and thus of synthetic value. It must be remembered that kinetic and equilibrium acidities may be different; thermodynamic products are favoured by higher temperatures and by more polar solvents.

The details of the mechanism of this type of metallation are still under discussion, but can be represented as involving a four-centre transition state, although higher-level complexes with more than one metal atom and complexation with the ring heteroatom are probably involved.

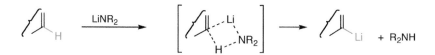

The main factor giving increased acidity of heterocyclic *C*-hydrogen relative to benzenoid *C*-hydrogen is the inductive effect of the heteroatom(s), thus metallation occurs at the carbon α to the heteroatom, where the inductive effect is felt most strongly, unless other factors, with varying degrees of importance, intervene. These include the following:

1. *Mesomerism.* Except in the case of side-chain anions, the 'anion' orbital is orthogonal to the π-system and so it is not mesomerically delocalised. However, electron density, and therefore *C*-hydrogen acidity at ring carbons, is affected by resonance effects.
2. *Coordination of the metal to the heteroatom.* Stronger coordination between the metal of the base and a heteroatom leads to enhanced acidity of the adjacent *C*-hydrogen due to increased inductive withdrawal of electron density – it is proportionately stronger, for example, for oxygen than for sulfur.
3. *Lone-pair interactions.* Repulsion between the electrons in the orbital of the 'anion' or incipient anion. This interaction is thought to be important in pyridines and other azines, and may be a kinetic rather than equilibrium effect, at least in the case of lithiation.[5]
4. *Polarisability of the heteroatom.* More polarisable atoms, such as sulfur, are able to disperse charge more effectively.
5. *Substituent effects.* Directed *ortho*-metallation (DoM)[6] is extremely useful in heterocyclic chemistry, just as in carbocyclic chemistry. Metallation *ortho* to the directing group is promoted by either inductive effects (e.g. Cl, F), or chelation (e.g. $CH_2OH \rightarrow CH_2OLi$), or a combination of these, and may overcome the intrinsic regioselectivity of metallation of a particular heterocycle. When available, this is by far the most important additional factor influencing the regioselectivity of lithiation.

Lithiating Agents

Lithiations are normally carried out with alkyllithiums or lithium amides. *n*-Butyllithium is the most widely used alkyllithium, but *t*-butyllithium and occasionally *s*-butyllithium are used when more powerful reagents are required. Phenyllithium was used in older work, but is uncommon now, although it can be of value when a less reactive, more selective base is required.[7] A very powerful metallating reagent is formed from a mixture of *n*-butyllithium and potassium *t*-butoxide: this produces the potassium derivative of the heterocycle.

Lithium diisopropylamide (LiN(*i*-Pr)$_2$; LDA) is the most widely used lithium amide, but lithium 2,2,6,6-tetramethylpiperidide (LiTMP) is rather more basic and less nucleophilic – it has found particular use in the metallation of diazines. Alkyllithiums are stronger bases than the lithium amides, but usually react at slower rates. Metallations with the lithium amides are reversible, so for efficient conversion, the heterocyclic substrate must be more acidic (>4 pK_a units) than the corresponding amine.

Solvents
Ether solvents – Et$_2$O and THF – are normally used. The more strongly coordinating THF increases the reactivity of the lithiating agent by increasing its dissociation. A mixture of ether, THF and pentane (Trapp's solvent) can be employed for very low temperature reactions (<100 °C) (THF freezes at this temperature). To increase the reactivity of the reagents even further, ligands such as TMEDA (*N,N,N',N'*-tetramethylethylenediamine; Me$_2$N(CH$_2$)$_2$NMe$_2$) or HMPA ((Me$_2$N)$_3$PO) (**CAUTION**: *carcinogen*) are sometimes added – these strongly and specifically coordinate the metal cation.

4.1.1.2 Halogen Exchange
Bromo- and iodo-heterocycles react rapidly with alkyllithiums, even at temperatures as low as −100 °C, to give the lithio-heterocycle. Exchange of fluorine is unknown and of chlorine, rare, but is known under special circumstances, such as in polychloro compounds. The apparent preparation of a number of lithio-heterocycles by direct reaction of a chloro heterocycle using lithium metal with naphthalene has been reported. Since such lithiations have to be carried out in the presence of the electrophile, it may be that the process is more complex than it seems.[8]

Positional selectivity in halogen exchange is governed by factors similar to those in direct deprotonation, such as acidity of the corresponding C–H and the presence of directing groups.

$$(\text{Het} \rightarrow \text{Br}) \quad + \quad \text{RLi} \quad \longrightarrow \quad (\text{Het} \rightarrow \text{Li}) \quad + \quad \text{RBr}$$

Mechanistically, the exchange process may involve a four-membered or other cyclic concerted transition state, or may possibly proceed *via* an electron-transfer sequence, however direct nucleophilic attack, at least on iodine, has been demonstrated in the case of iodobenzene,[9] and cannot therefore be dismissed as a more general mechanism.

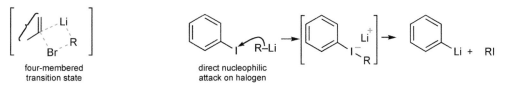

four-membered transition state

direct nucleophilic attack on halogen

Possible mechanisms for metal–halogen exchange

Halogen Exchange Reagents
n-Butyllithium is the usual exchange reagent; the *n*-butyl bromide byproduct does not usually interfere with subsequent steps. When the presence of an alkyl bromide is undesirable, two equivalents of *t*-butyllithium

can be employed – the initially formed *t*-butyl bromide is consumed by reaction with the second equivalent of alkyl-lithium, producing isobutene.

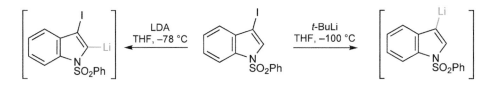

It is very important to differentiate between pure bases, such as lithium diisopropylamide, which act only by deprotonation, and alkyllithiums, which can act as bases *or* take part in halogen exchange. When using alkyllithiums, exchange is favoured over deprotonation by the use of lower temperatures. The reaction of 3-iodo-1-phenylsulfonylindole with the two types of reagent is illustrative.[10]

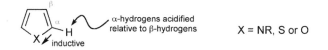

4.1.1.3 Regiostability
When an organolithium is prepared by halogen exchange, or a kinetic C–H metallation, there may be the possibility of equilibration involving a more 'acidic' hydrogen in the molecule. This occurs more readily in strongly coordinating solvents and at higher temperatures, and may be promoted by the presence of electron-withdrawing *N*-protecting groups. Derivatives of magnesium and zinc are much more resistant to such migrations and are totally regio-stable for all practical purposes.

Similar equilibria, *via* halogen migration, are also possible and, indeed, can be used to advantage (see for example, 17.4.2).

4.1.1.4 Ring Lithiation of Five-Membered Heterocycles

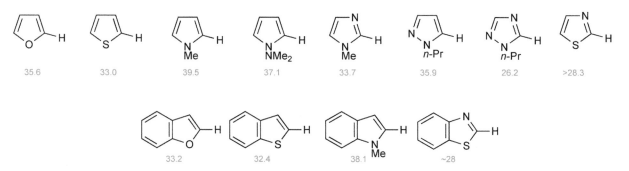

Equilibrium pK_a values[#] for deprotonation of some five-membered heterocycles in THF[11]
Measured pK_a values vary according to solvent etc.

Despite the lower electronegativity of sulfur, and hence a weaker inductive effect, thiophene metallates about as readily as furan, probably in part because the higher polarisability of sulfur allows more efficient charge distribution.[12] The lithiation of 2-(2-furyl)thiophene, in either ring depending on conditions, is instructive;[13] preferential lithiation of the furan ring in the non-polar solvent is probably due to stronger coordination of lithium to the oxygen, thus increasing the inductive effect on the α-hydrogen in the furan ring.

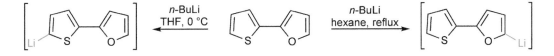

The use of stronger bases can result in dimetallation.[14]

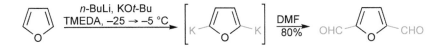

Directing groups can overcome the normal tendency for α-lithiation in five-membered heterocycles, as shown in the thiophene example below, however the use of lithium diisopropylamide does allow 'normal' α-lithiation in this case.[15]

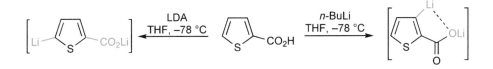

Lithiation of pyrroles is complicated by the presence of a much more acidic hydrogen on nitrogen, however 1-methylpyrrole lithiates, at C-2, albeit under slightly more vigorous conditions than for furan.[16] Removable protecting groups on the pyrrole nitrogen allow α-lithiation, *t*-butoxycarbonyl (Boc) is an example; it has additional advantages: not only is it easily hydrolytically removed, but it also withdraws electrons, thus acidifying the α-hydrogen further, and finally it provides chelation assistance.[17]

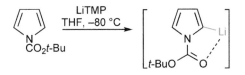

Benzo[*b*]thiophenes, benzo[*b*]furans and *N*-blocked indoles lithiate on the heterocyclic ring, α to the heteroatom.[18] Lithiation at the other hetero-ring position can be achieved *via* halogen exchange, but low temperatures must be maintained to prevent equilibration to the more stable 2-lithiated heterocycle.[10]

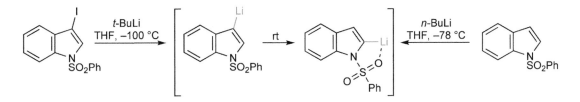

Benzene-ring-lithiated intermediates can be prepared by metal–halogen exchange, even, in the case of indoles, without protection of the NH, i.e. it is possible to produce an *N,C*-dimetallated species (20.5.2).[19]

The 1,3-azoles lithiate very readily, at C-2. One may understand this in terms of a combination of the acidifying effects seen at an α-position of pyridine (both inductive and mesomeric electron withdrawal) with that at the α-positions of thiophene, furan and pyrrole (inductive only). 2-Substituted-1,3-azoles generally lithiate at C-5.[20]

For imidazoles, it is usual for the *N*-hydrogen first to be masked,[21] and a variety of protecting groups have been used for that purpose, many of which provide additional stabilisation and an additional reason for regioselective α-lithiation by coordinating the lithium: trimethylsilylethoxymethyl (Me₃Si(CH₂)₂OCH₂; SEM) is one such group.[22]

It is a significant comment on the relative ease of α-lithiation in six- and five-membered systems that (*N*-protected) pyrazoles lithiate at C-5, i.e. at the pyrrole-like α-position, though, again chelation assistance from the *N*-protecting group also directs to C-5.[23]

One must be aware that hetero-ring cleavage[24] can occur in β-lithiated five-membered systems, because the heteroatom can act as a leaving group, if the temperature is allowed to rise.[25]

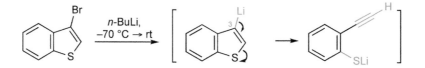

Higher azoles – triazoles, tetrazoles, heterodiazoles – lithiate readily, but some of the lithio-compounds are relatively unstable.

4.1.1.5 Ring Lithiation of Six-Membered Heterocycles

The preparation of lithiated derivatives of six-membered heterocycles like pyridines, quinolines and diazines must overcome the problem that they are intrinsically susceptible to nucleophilic addition/substitution (3.3.2) by the lithium reagents. In contrast to the selective lithiation of five-membered rings, the direct metallation of pyridine is quite difficult and complex, but it can be achieved using the very strong base combination *n*-butyllithium/potassium *t*-butoxide. In relatively non-polar solvents (ether/hexane) kinetic 2-metallation predominates, but in a polar solvent (THF/HMPA/hexane), or under equilibrating conditions, the 4-isomer is the major product. The pyridine α- and γ-positions, being more electron-deficient than a β-position, have the kinetically most acidic protons, and of the two former anions, location of negative charge at the γ-position is the more stable situation, perhaps due to unfavourable repulsion between the coplanar nitrogen lone pair and the α-'anion'. In non-polar solvents, stronger coordination of the metal cation with the nitrogen lone pair will reduce this repulsive interaction and thus increase the relative stability of the α-'anion'.[26] As a corollary of this, pyridine can be selectively lithiated at C-2 when the lone pair is tied up as a complex with boron trifluoride[27] or using special reagents which act *via* complexation to the nitrogen and then intramolecular delivery of base to the α-proton (8.4.1). This is consistent with much earlier studies of base-catalysed exchange, when it was demonstrated that *N*-oxides and *N*[+]-alkyl quaternary salts exchange more rapidly at an α-position.[28]

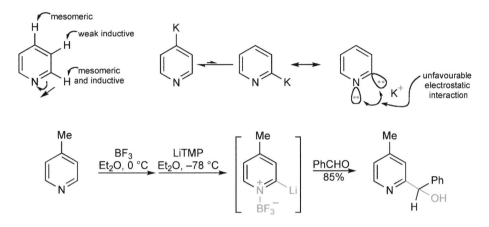

All the isomerically pure lithio-pyridines can be prepared by halogen exchange, though 3-bromopyridine requires a lower temperature, or a change to a less-dissociating solvent, to discourage nucleophilic addition; bromopicolines can be similarly converted, without deprotonation at the methyl groups.

Pyridines carrying groups which direct metallation *ortho*, using chelation and/or inductive influences, can be directly lithiated without risk of nucleophilic addition. When the group is at a 2-[29] or 4-position,[30] lithiation must occur at a β-carbon; pyridines with *ortho*-directing groups located at a β-position usually lithiate at C-4: this is true, for example, of chloro- and fluoro-pyridines;[31] 3-methoxymethoxy-,[32] 3-pivaloylamino-,[33] 3-trimethylsilylethoxymethoxy-[34], 3-*t*-butylaminosulfonyl-,[35] 3-diethylaminocarbonyloxy- or 3-diethylaminothiocarbonyloxy-pyridines,[36] and the adduct from 3-formylpyridine and Me₂N(CH₂)₂NMeLi;[37] 3-ethoxypyridine, however, metallates at C-2.[38]

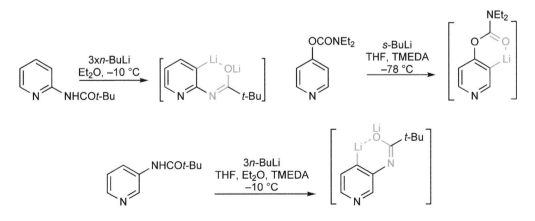

Quinolines react like pyridines, but are more susceptible to nucleophilic addition;[39] this is also an increased problem with pyrimidines, relative to pyridines, but nevertheless they can be lithiated by deprotonation or by halogen exchange at low temperatures, around −100 °C. The presence of 2- and/or 4-substituents adds some stability to lithiated pyrimidines.[40]

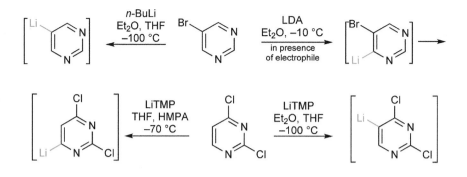

Pyrazines and pyridazines react in accord with the principles discussed above.[41]

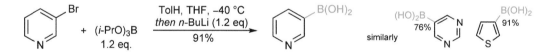

4.1.1.6 Generation of the Organolithium in the Presence of the Electrophile

Rather than a sequential lithiation, followed by addition of an electrophile, it may be better to generate the organolithium compound, either by halogen exchange or direct deprotonation, *in the presence of* the electrophile. The reason for doing this may be that the organolithium intermediate is unstable, or that it avoids side reactions, such as lithium migration. It is also more efficient, particularly on larger scale, as only one temperature-controlled addition is required and it often allows the use of higher reaction temperatures. There is a balance, which may be temperature dependent, between the rates of formation of the organometallic, its reaction with the electrophile and the reaction of the lithiating agent with the electrophile. Examples of this approach include the preparation of heterocyclic boronic acids *via* halogen exchange[42] and indole-2-boronates and 2-silanes *via* direct deprotonation.[43] In the latter case, the low reactivity of the base (LDA) to trialkyl borates and silyl chlorides allows temperatures as high as 0 °C to be used.

4.1.1.7 Lithiation in the Presence of Functional Groups

N-Hydrogen

When a ring nitrogen bears a hydrogen, it is normal to use a protecting group during lithiation (by halogen exchange), but in a number of cases the 'protection' involves simply deprotonating the nitrogen. In some cases, for example, 2-iodoindole[44] or 4-bromopyrazole,[45] excess of the alkyl lithium can be used, but the process is more complex than might be thought and internal protonation may occur, leading to overall dehalogenation. The initial use of a separate strong base to deprotonate the NH before carrying out the lithium exchange may be better, as in the reaction of six-ring bromo-indoles, where potassium hydride is used for the deprotonation (20.5.2).

Iodine-magnesium exchange has been carried out for 5- and 6-iodouracils in the presence of lithium chloride, using methylmagnesium chloride to deprotonate both N-hydrogens, then addition of isopropylmagnesium chloride to exchange the halogen.[46]

Protecting Groups

The standard solution for ketones and aldehydes is to use protecting groups such as acetals, but this can be inconvenient or maybe incompatible with the product due to the conditions required for deprotection. However, *in situ* protection as carbinolamine anions is an ideal solution, being very efficient time-wise and process-wise.[47] (Note that intermediates of the same type are formed during quenching of lithio-compounds with DMF and can be used in the same way.)

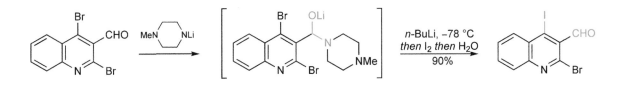

Another, very useful, *in situ* protection of indoles and pyrroles, involves deprotonation of the nitrogen then addition of carbon dioxide to give the lithium *N*-carboxylate salt.[48]

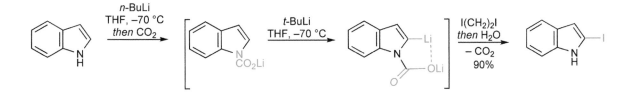

4.1.2 Magnesium[49]

Heterocyclic magnesium compounds (Grignard reagents), which are often difficult to prepare by the standard method (halides with magnesium metal), can be readily prepared by metal–halogen exchange with an alkyl Grignard. These exchanges show the same selectivity as lithium exchanges: 2 > 3 in five-membered rings and 3 > 2 in six-membered rings (for the same halogen) and I > Br, as illustrated by the dihalothiophenes shown below.[50]

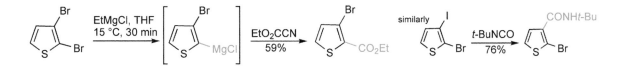

Isopropylmagnesium halides are the most widely used reagents and are effective over a very wide range of heterocyclic systems; they can even be used in the presence of esters if the temperature is kept low and are also suitable for solid-state synthesis.[51]

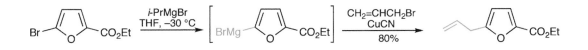

The relatively modest directing effect of a carboxylate overrides the normally higher reactivity in thiazoles of C-2 > C-5.[51]

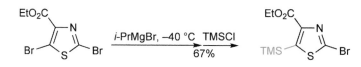

The exchange of the bromomethyl oxazole shown below had to be carried out in the presence of the electrophile, due to the high instability of the intermediate magnesium compound.[52]

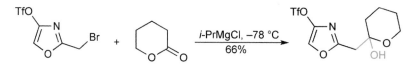

Electron-withdrawing substituents increase the rate of exchange, as shown above, but, generally, exchange of bromine is quite slow. However, the addition of lithium salts brings about dramatic increases in reactivity of Grignard reagents for halogen exchange and also increases the reactivity of the resulting (hetero)aryl organometallic. This is best conducted by using the complex *i*-PrMgCl.LiCl, which probably exists as a magnesiate *i*-PrMgCl$_2^-$ Li$^+$, which is more reactive than the (oligomeric) Grignard. The pyridine example below demonstrates the difference in reactivity.[53]

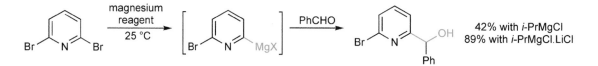

LiCl seems to be a 'magic ingredient' for enhancing reactivity during magnesium and zinc halogen–metal exchanges, and direct metallation, and also of the subsequent reactions of the products. This 2004 discovery may supersede previous methods. The addition of (THF-soluble) CeCl$_3$.LiCl to Grignard reactions (including some pyridyl Grignard reagents) with hindered or readily enolisable ketones greatly increases the yields.[54]

The addition of lithium chloride to magnesium amide bases, for example TMPMgCl, greatly increases solubility (in THF) and reactivity for direct metallation, even for sensitive substrates such as pyrimidines.[55] Amongst a range of heterocyclic substrates, it is notable that 2,6-dichloropyridine gives clean 4-substitution, whereas 'standard' amide bases give mixtures of C-3 and C-4 products. Magnesium bis-amides, such as (TMP)$_2$Mg.2LiCl, are more successful in some situations, such as where *t*-butyl esters are used as directing groups.[56]

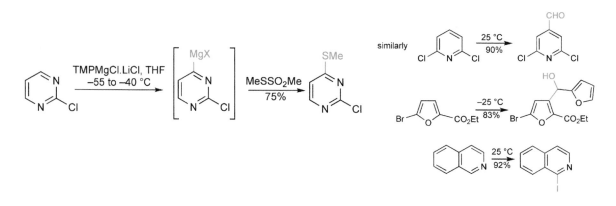

Direct C–H magnesiation can be carried out with lithium tri-*n*-butylmagnesiate on oxazoles,[57] thiophenes and fluoro- and chloro-pyridines, the intermediates being used for trapping electrophiles and in coupling reactions. The use of the highly hindered neopentylmagnesium bromide allows iodine–magnesium exchange, even in the presence of ketones.[58]

Bromine exchange of bromo-2-furoic acids can be carried out after formation of the magnesium salt of the acid by reaction with methyl Grignard in the presence of lithium chloride.[59] This procedure can also be used for iodine exchange in imidazoles, without protection of the NH.[60]

4.1.3 Zinc[61]

Heteroaryl zinc compounds are particularly useful in palladium-catalysed coupling, being compatible with many functional groups. They are often prepared *in situ via* lithiation, followed by reaction with zinc halides, but direct zincation of halides can be carried out, using either Riecke zinc, as in the example below,[62] or, more conveniently, ordinary zinc dust,[63] with various means of activation.

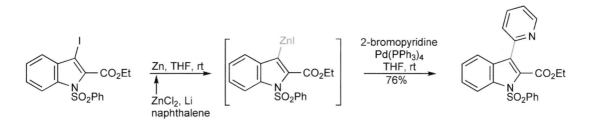

Activation with cobalt chloride and allyl chloride has been used for chlorothiophenes (2-Cl is more reactive than 3-Cl) and activated aryl chlorides.[64] Commercial zinc dust and the heteroaryl halide can be used to make the heteroarylzinc in both electron-rich and electron-poor systems.[65]

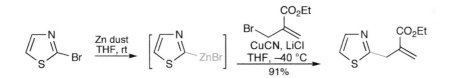

The addition of LiCl to zinc dust (activated with 1,2-dibromoethane plus TMSCl) has a dramatic effect on the reaction rates.[66] The reaction is successful with iodo-heterocycles and some activated bromo-compounds, such as the furan ester shown below.

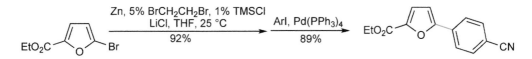

Organoindium(III) compounds can be prepared under very similar conditions to these organozincs and have an even greater tolerance of functional groups: they are compatible with alcohols and phenols.[67]

Di-(heteroaryl) zincs, for use in coupling reactions, can be prepared by direct exchange of iodides, for example 5-iodofurfural, with (*i*-Pr)$_2$Zn in the presence of a Li(acac) catalyst.[68]

The direct C–H zincation of a number of heterocyclic systems can be carried out using (TMP)$_2$Zn.2MgCl$_2$.2LiCl (prepared by reacting TMPMgCl.LiCl with ZnCl$_2$), including 1,3,4-oxadiazoles, 1,2,4-triazoles and compounds bearing sensitive functional groups, such a nitro or aldehyde.[69] In less reactive systems, such as benzofuran and benzothiophene, the use of microwave irradiation allows efficient conversion.[70]

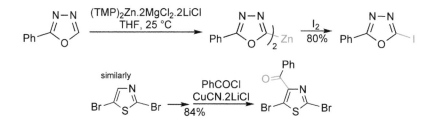

4.1.4 Copper

Copper derivatives, in the form of cuprates (RCu(X)Li), are usually prepared *in situ* from lithium, magnesium or zinc compounds, by reaction with a Cu(I) source such as CuCN. They are often used to improve conversions using, for example, an acid chloride or an allylic halide as electrophile – a number of examples appear in other sections of this book. They are also useful as the organometallic partners in some cross-coupling reactions (using cobalt catalysts).[71]

A more convenient direct cupration of iodides can be carried out using highly hindered cuprates, such as (Nphyl)$_2$CuLi.[72] The reaction is faster in electron-deficient rings, and chelating groups allow the use of bromides as substrates.

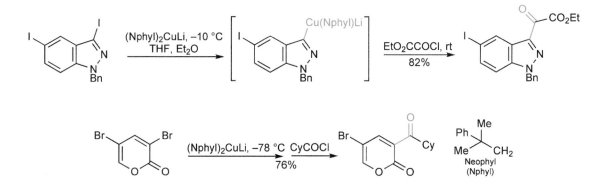

4.1.5 Boron

Practically all the organoboron compounds of interest in the current context are boronic acids or closely related compounds.[73] Boronic acids are relatively weak acids that ionise by association, not dissociation – stable salts of tetrahedral aryl trihydroxyboronates can be isolated as stable solids and used in coupling reactions.[74] Some typical pK_as are shown below.[75] Pyridine boronic acids exist as zwitterions in water at pH 7.[76] Electron-withdrawing groups and the inductive effects of heteroatoms increase acidity, and formation of esters with 1,2- or 1,3-diols can reduce the pK_a by up to 2.5 units.[77]

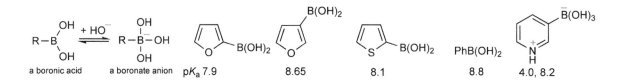

Boronic acids readily dehydrate, eventually giving a boroxin, but the conversion is easily reversible and the interconversion can be brought about simply by dissolution in wet or dry solvents. The solid 'acids' very commonly occur as mixtures of acid and anhydrides, which makes precise measurement of molar quantities difficult. Simple esters, such as with methanol, readily form on dissolution in the alcohol, but also hydrolyse very rapidly in air. Cyclic esters are more stable and, particularly, pinacol esters are widely used in coupling reactions, as they are reasonably stable on storage, have a known stoichiometry and, of course, react well; they are the most common form of boronate available commercially. More stable esters are formed with substituted diethanolamines, due to the extra coordination afforded by the basic nitrogen. *N*-Methyldiethanolamine has long been used for isolation and characterisation of boronic acids and its esters can be used in coupling reactions, although somewhat variably. The free acid is readily liberated by treatment with aqueous acid or ammonium chloride. The *N*-phenyl analogue has found particular use as a stable source of 2-pyridylboronic acid, the weaker donation by the aniline nitrogen giving a good balance of stability and reactivity.

An important application is the use of *N*-methyliminodiacetic acid (MIDA) esters as protecting groups (4.2.8). These MIDA esters are readily cleaved in mild basic aqueous conditions, but are stable to many standard functional-group transformations, even chromic acid oxidations.

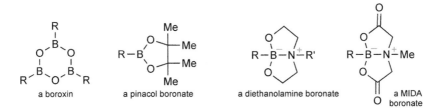

a boroxin   a pinacol boronate   a diethanolamine boronate   a MIDA boronate

4.1.5.1 Trifluoroborates

Boron has a high affinity for fluoride, and boronic acids can be converted, *via* reaction with KHF$_2$, into trifluoroborates (RBF$_3$K), the fluorine analogues of the boronate anion. These compounds are very stable, but can be reactive under the appropriate conditions and are very useful in palladium-catalysed couplings.

4.1.5.2 Protodeboronation

Boronic acids are potentially susceptible to acid- or base-catalysed protodeboronation, but the conditions necessary vary widely. The ease of cleavage of the C–B bond under basic or acidic conditions correlate with the corresponding carbanion stability or ease of protonation of the ring, respectively. When a relatively stable carbanion can be formed, such as in furan boronic acids containing electron-withdrawing groups, base-catalysed deboronation can become an important unwanted side reaction during palladium-catalysed boronic acid couplings.[78] Indeed, imidazole and oxazole 2-boronic acids have not yet been isolated, possibly due to their very ready deboronation.

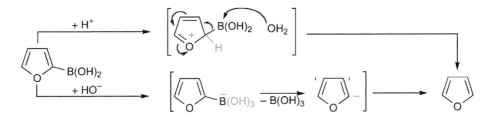

Pyridine 2-boronic acid is rather unstable (unlike the 3- and 4-isomers) and can only be isolated as esters, *N*-substituted diethanolamine esters being the most stable. A possible rationale for this instability may be the parallel with the mechanism for the ready decarboxylation of pyridine 2-carboxylic acid *via* a transient ylide intermediate (8.11).

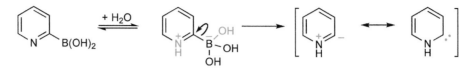

4.1.5.3 Preparation

Boronates (i.e. boronic acids and esters) are usually prepared by one of two methods: reaction of organolithiums or Grignard reagents with a trialkyl borate, usually tri-*iso*-propyl borate, or palladium-catalysed boronation.

Palladium-catalysed boronation can be carried out using either 4,4,4′,4′,5,5,5′,5′-octamethyl-2,2′-bi-1,3,2-dioxaborolane (pinacol diboron or bis(pinacolato)diboron) or pinacol borane,[79] the latter being preferred because of the lower cost of the reagent. The mechanisms of the conversions are very similar to cross-coupling reactions, the difference being the transfer of boron, instead of carbon, to palladium in the transmetallation step. The mechanism (see below) of transfer of boron from the diboron compound seems straightforward, but exactly how it is transferred from the borane is less clear.

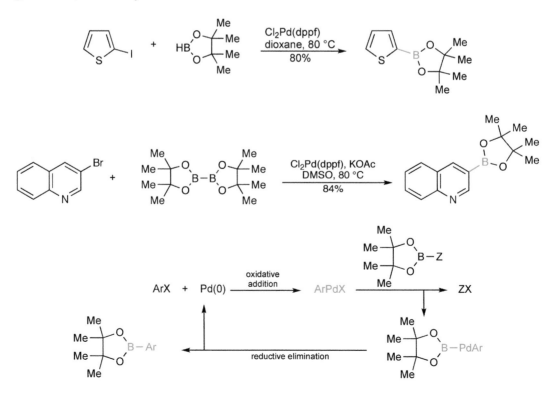

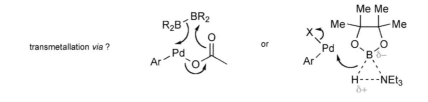

transmetallation *via* ?

Iridium-catalysed C–H-boronation can be carried out using either pinacol borane or pinacol diborane, both methods giving comparable results.[80] Reaction occurs at α-positions of five-membered rings and is compatible with halogen substituents, as exemplified below.[81]

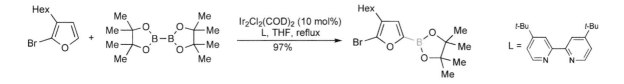

Minor methods for the synthesis of boronic acids involve transmetallation with silicon or mercury.[82]

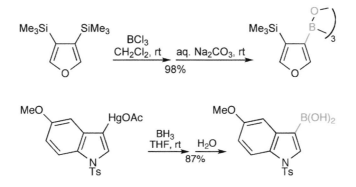

As is also true for silicon and tin compounds, the high stability of boronates, particularly cyclic esters, allows them to be incorporated into and carried through as substituents in a range of reaction types, such as the synthesis of pyrazole boronates for cross couplings (see 4.2.7.4).

In addition to the very important palladium-catalysed reactions, boronic acids undergo a number of useful reactions that do not require transition-metal catalysis, particularly those involving electrophilic *ipso*-substitutions by carbon electrophiles. The Petasis reaction involves *ipso*-replacement of boron under Mannich-like conditions and is successful with electron-rich heterocyclic boronic acids. A variety of quinolines and isoquinolines, activated by ethyl pyrocarbonate, have been used as the 'Mannich reagent'.[83] A Petasis reaction on indole 3-boronic acids under standard conditions was an efficient route to very high de α-indolylglycines.[84]

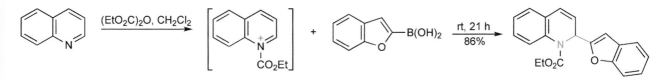

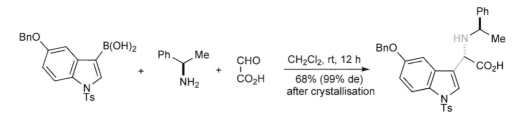

Furan and indole trifluoroborates undergo HF-catalysed *ipso*-substitution reactions with enones, which can also be made highly enantioselective.[85]

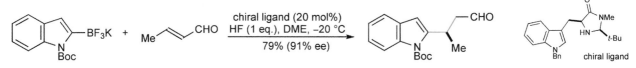

A long-standing reaction is the oxidation of aryl boronic acids to 'phenols' by alkaline peroxide, usually in the work-up of a borate-organolithium reaction, without isolation of the boronic acid, i.e. an efficient ArBr → ArOH conversion. A variant under milder conditions uses sodium perborate (for the conversion of 5-bromopyrimidines),[86] and, using oxone, oxindoles can be prepared from 1-Boc indoles *via* direct 2-lithiation.[87]

4.1.6 Silicon[88] and Tin[89]

(**CAUTION**: *see the discussion of organotin toxicity on page 67*)

Silicon and tin compounds have many similarities to organoborons, both in preparation, stability and reactivity. Reaction of organolithiums with silyl and stannyl halides is straightforward, and the preparations *via* palladium-catalysed reactions of distannanes[90] and disilanes with aryl halides exactly follow the boron analogues, though coupling with hexaalkyldisilanes requires rather more vigorous conditions.[91] The disilane method can be used for the preparation of trimethylsilyl compounds from aryl chlorides[92] and dimethylsilanols for cross couplings.[93]

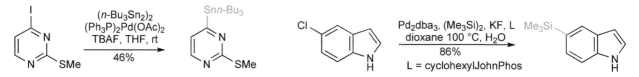

Aryl silicon compounds can be prepared by metal-catalysed reaction of halides with silanes, as in the rhodium-catalysed reaction below.[94] The mechanistic details of this reaction (probably) differ from the palladium-catalysed borane reaction. (**NOTE**: *triethoxysilane is extremely toxic!*)

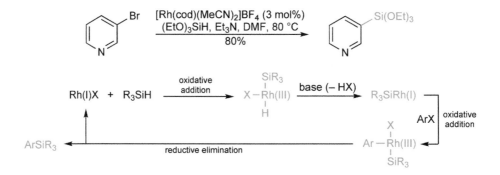

Useful alternative preparations of stannanes include palladium-catalysed decarboxylation of stannyl esters.[90] Trialkylstannyl and trialkylsilyl anions are highly reactive and will displace halogen without the use of a catalyst.[95] It is possible to directly silylate indoles and pyrroles *via* electrophilic substitution.[96]

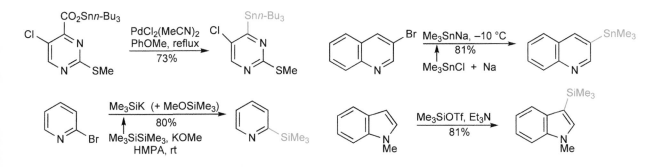

The relatively high stability of carbon-silicon/-boron/-tin bonds allows the 'metal' to be carried through many heterocyclic syntheses as an inert substitutent: some examples are shown below.[97]

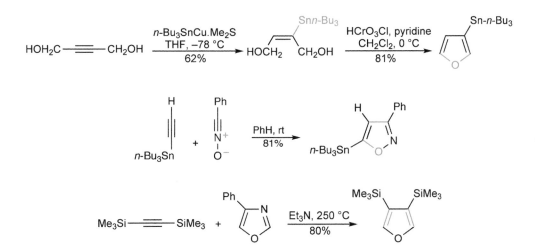

Silicon and tin are both subject to *ipso*-replacement by electrophiles, *via* an electrophilic addition/metal elimination mechanism analogous to other aromatic substitutions, but at a much faster rate than the corresponding replacement of hydrogen.[98] *Ipso*-substitutions also take place on heterocycles and, in the case of electron-rich systems, probably *via* the same type of mechanism.

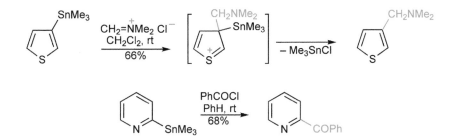

4-Trimethylsilyl-pyridines will also react with aldehydes under fluoride catalysis; an intramolecular example is shown below.[99]

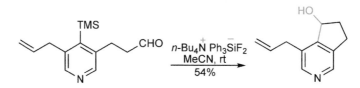

4.1.7 Mercury[100]

Only Hg(II) compounds are of interest in the current context and they can be prepared by exchange reactions of other organometallics with mercuric salts or, more usefully from the heterocyclic viewpoint, by direct electrophilic mercuration with mercuric salts, particularly mercuric acetate.

There is a lot of information on the mercuration of heterocycles in the old literature,[101] but it is seldom used nowadays due to the major disadvantages of toxicity and associated waste management; it can be, however, a very useful reaction. The advantages of mercuration are that it can be carried out in hydroxylic and acidic solvents and in the presence of air, and that mercury in the product is easily replaced by *ipso*-substitution with other electrophiles, such as halogens, and gives boronic acids by reaction with borane.

There is a differential reactivity to nitrogen heterocycles between Hg(II) and other types of electrophile, such as bromine, possibly due to a weaker coordination of the mercuric ion to a ring nitrogen; for example mercuration is more successful in electrophilic substitutions for oxazoles. In both oxazoles and thiazoles, the preferred position for mercuration is C-5,[102] but in the latter, trimercuration occurs quite readily.

4.1.8 Palladium

Organopalladium compounds can be prepared by electrophilic palladation, oxidative addition to aryl halides or reaction of Pd(II) with organometallic reagents. These transformations are all vital for the palladium-catalysed reactions discussed later in this chapter.

4.1.9 Side-Chain Metallation ('Lateral Metallation')[103]

4.1.9.1 Side-Chain Metallation of Six-Membered Heterocycles

Anions that are immediately adjacent to the ring on alkyl side-chains are subject to varying degrees of stabilisation by interaction with the ring. The most favourable situation is where the side-chain is linked directly to a C=N, as in the 2- and 6-positions of a pyridine, or at a 4-position of a pyridine. Such anions are stabilised in much the same way as an enolate (conjugated enolate). We use the word 'enaminate' to describe this nitrogen-containing, enolate-like anion.

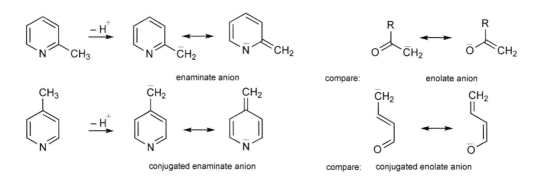

Quantitative measures for some methyl deprotonations are: 2-methylpyridine (pK_a 34), 3-methylpyridine (pK_a 37.7), 4-methylpyridine (pK_a 32.2), 4-methylquinoline (pK_a 27.5).[104] These values can be usefully compared with those typical for ketone α-deprotonation (19–20) and toluene side-chain deprotonation (~41). Thus, strong bases can be used to convert methyl-pyridines quantitatively into side-chain anions, however the enolate-like stabilisation of the anion is sufficient that reactions can often be carried out using weaker bases under equilibrating conditions, i.e. under conditions where there is only a small percentage of anion present at any one time. It may be that under such conditions, side-chain deprotonation involves *N*-hydrogen-bonded or *N*-coordinated pyridines.

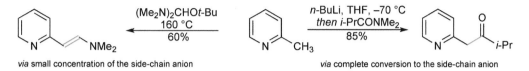

via small concentration of the side-chain anion   *via* complete conversion to the side-chain anion

An alternative means for effecting reaction at a side-chain depends on a prior electrophilic addition to the nitrogen: this acidifies further the side-chain hydrogens, then deprotonation generates an enamine or an enamide, each being nucleophilic at the side-chain carbon; the condensation of 4-picoline with benzaldehyde using acetic anhydride illustrates this.

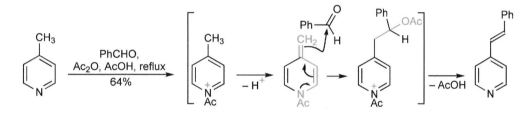

4.1.9.2 Side-Chain Metallation of Five-Membered Heterocycles

The metallation of a side-chain on a simple five-membered heterocycle is much more difficult than in the six-membered series, because no enaminate stabilising resonance is available. Nonetheless, it also is selective for an alkyl adjacent to the heteroatom, because the heteroatom acidifies by induction. Relatively more forcing conditions need to be applied, especially if an *N*-hydrogen is present,[105] but an elegant method has been developed for indoles, in which the first-formed *N*-anion is blocked with carbon dioxide, the lithium carboxylate thus formed then neatly also facilitating 2-methyl lithiation by intramolecular chelation; this device has the further advantage that, following reaction of the side-chain anion with an electrophile, the *N*-protecting group is removed simply, during aqueous processing.[106]

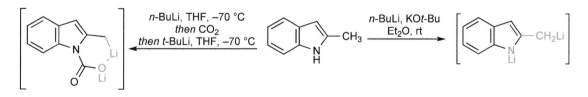

Side-chains at C-2 on 1,3-azoles are activated in a manner analogous to pyridine α-alkyl groups, and can be metallated, but more care is needed to avoid ring metallation.[107]

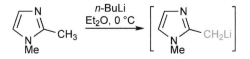

4.2 Transition Metal-Catalysed Reactions[108]

Transition-metal-catalysed reactions are probably the most important area in synthetic organic chemistry and they have been used extensively in both the ring synthesis and the functionalisation of heterocycles. As well as completely new modes of reactivity, variants of older synthetic methods have been developed using the milder and more selective processes that attach to the use of transition-metal catalysts. Although this section is devoted to reactions catalysed by a range of transition metals, palladium-catalysed processes vastly outnumber the others (Ni, Rh, Cu, Fe). Therefore, the following discussions will be concerned with palladium-catalysed processes, with occasional diversions, where appropriate, into other metals. In fact, many of the processes and mechanistic details of the 'minor' metals are very similar to those of palladium.

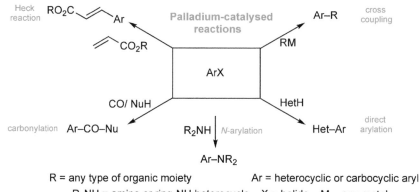

R = any type of organic moiety   Ar = heterocyclic or carbocyclic aryl
R$_2$NH = amine or ring-NH heterocycle   X = halide   M = any metal

In general, heterocyclic compounds undergo palladium-catalysed reactions in ways exactly analogous to carbocycles; heterocyclic sulfur and nitrogen atoms seldom interfere with these (homogeneous) palladium catalysts, which must be contrasted with the well-known poisoning of hydrogenation catalysts, such as palladium metal on carbon, by sulfur- and nitrogen-containing molecules.

Palladium-catalysed processes typically utilise only 1–5 mol% of the catalyst and proceed through small concentrations of transient palladium species: there is a sequence of steps, each with an organopalladium intermediate, and it is important to become familiar with these basic organopalladium processes in order to rationalise the overall conversion. Concerted, rather than ionic, mechanisms are the rule, so it is misleading to compare them too closely with apparently similar classical organic mechanisms, however curly arrows can be used as a memory aid (in the same way as one may use them for cycloaddition reactions), and this is the way in which palladium-catalysed reactions are explained in the following discussion. (For convenience, an organometallic component can be referred to as the *nucleophilic partner* and the halide as the *electrophilic partner*, but this should not necessarily be taken to imply reactivity as defined in classical chemistry. Also, references to 'the halide' should be understood to include all related substrates, such as triflates.)

4.2.1 Basic Palladium Processes

NOTE: For clarity, ligands that are not involved in the transformation under consideration are omitted from the following schemes, however it is important to understand that most organopalladium compounds normally exist as 4-coordinate, square-planar complexes, although the more reactive key intermediates may have lower degrees of coordination. The equilibration of these various degrees of ligand binding plays an important role in the overall reaction sequences, both in the individual reactions and in *cis–trans* isomerisation of the square planar complexes. Ligands are major determinants of the rates of all the individual steps and can be 'tailored' for specific purposes.

Despite an apparent similarity between RPdX and RMgX, their chemical properties are very different. The former are usually stable to air and water, and unreactive to the usual electrophilic centres, such as carbonyl, whereas RMgX do react with oxygen, water and carbonyl compounds.

4.2.1.1 Concerted Reactions
Oxidative Addition
Aromatic and vinylic halides react with Pd(0) to give an organopalladium halide: aryl(or alkenyl)PdHal. This is formally similar to the formation of a Grignard reagent from magnesium metal, Mg(0), and a halide, but mechanistically, a concerted, direct 'insertion' of palladium into the carbon–halogen bond is believed to be involved. The ease of reaction: X = I > Br ~ OTf >> Cl >> F, explains why chloro and fluoro substituents can normally be tolerated, not interfering in palladium-catalysed processes, however the use of highly reactive catalysts does allow the use of chloro compounds as substrates for these reactions. As a simple illustration, Pd(PPh$_3$)$_4$ reacts with iodobenzene at room temperature, but requires heating to 80 °C for a comparable insertion into bromobenzene. Although alkyl halides will undergo oxidative addition to Pd(0), the products are generally much less stable.

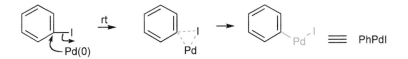

Oxidative addition involves a concerted nucleophilic-like attack by Pd(0), but differs from a standard two-step aromatic nucleophilic displacement in that direct attack at the carbon–halogen bond occurs and mesomeric stabilisation of an intermediate is not involved. That being said, those same mesomeric relationships do contribute, together with inductive effects, to the total electron density at the carbons involved and, all other things being equal, *the tendency is for oxidative addition to select the carbon with the lowest electron density.* In simple systems, there seems to be a good correlation with total electron density at carbon: pyridine, furan and thiophene show highest reactivity at C-2. In pyridines, for oxidative addition, the order is C-2 > C-4 > C-3, whereas for nucleophilic displacement it is C-4 > C-2 > C-3, showing the greater effect of induction at C-2 in the former.

Reductive Elimination
Organopalladium species with two organic units attached to the metal, R^{1}PdR^2, are generally unstable: extrusion of the metal, in a zero oxidation state, takes place, with the consequent linking of the two organic units. Because this is again a concerted process, stereochemistry in the organic moieties is conserved.

1,2-Insertion
Organopalladium halides add readily to double and triple bonds in a concerted, and therefore *syn,* manner (*via* a π-complex, not shown for clarity).

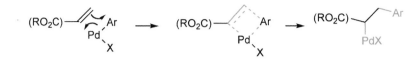

This process works best with electron-deficient alkenes, such as ethyl acrylate, but will also take place with isolated or even with electron-rich alkenes. In reactions with acrylates, the palladium becomes attached to the carbon adjacent to the ester, i.e. the aromatic moiety becomes attached to the carbon β to the ester.

1,1-Insertion

Carbon monoxide, and isonitriles, will insert into a carbon–palladium bond, subsequent reaction with a nucleophile generates the product.

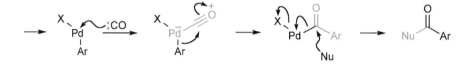

β-Hydride Elimination

When a *syn* β-hydrogen is present in an alkylpalladium species, a rapid elimination of a palladium hydride occurs, generating an alkene. This reaction is much faster in RPdX than in R_2Pd and is the reason that attempted palladium-catalysed reactions of alkyl halides often fail.

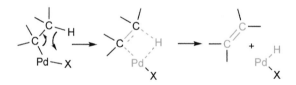

Transmetallation

Palladium(II) compounds, such as ArPdX and PdX_2, generally react readily with a wide variety of organometallic reagents, of varying nucleophilicity, such as R_4Sn, $RB(OH)_2$, RMgX and RZnX, transferring the R group to palladium with overall displacement of X. The details of the reactions are not fully understood and will vary from metal to metal, but a concerted transfer is probably the best means for their interpretation. It should be noted that the reactivity of these organometallic compounds towards palladium (or at least in the overall reaction) does not parallel their reactivity in nucleophilic additions, for example to carbonyl groups; indeed the less reactive metals (B, Sn) are generally the most effective.

The process probably involves coordination between the metal and the palladium *via* a bridging oxygen (boronic acids, silanols) or halogen (tin, zinc, magnesium), followed by internal transfer of the organic residue. The diagram shows a simple four-centre transition state, but more complex arrangements are possible.

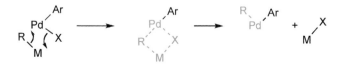

For boronic acids, coordination of the boron with a nucleophile, such as hydroxide, fluoride or an amine, giving a tetrahedral boronate anion, is necessary to drive transmetallation.

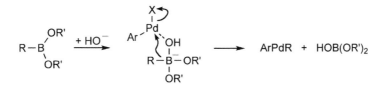

4.2.1.2 Ionic Reactions
Addition to Palladium–Alkene π-Complexes
Like those of Hg^{2+} and Br$^+$, Pd^{2+}–alkene complexes are very susceptible to attack by nucleophiles. In contrast to 1,2-insertion, this process exhibits *anti* stereospecificity.

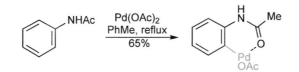

Aromatic Palladation
In reactions like aromatic mercuration, palladium(II) compounds will metallate aromatic rings *via* an electrophilic substitution, hence electron-rich systems are the most reactive.[109] *ortho*-Palladation assisted by electron-releasing chelating groups has been used frequently.[110]

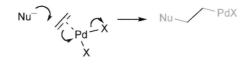

4.2.2 Catalysts[111]
The catalyst (or catalytic system) is generally composed of a metal and a ligand – most commonly a phosphine, but sometimes an amine or imidazole carbene. For most reactions, the active catalyst is the zerovalent metal i.e. Pd(0), and can be added as such, as a stable complex such as Pd(PPh$_3$)$_4$ – tetrakis(triphenylphosphine) palladium(0) – referred to colloquially as 'tetrakis'.

On the other hand, a Pd(II) pre-catalyst, such as palladium acetate, together with a ligand (or as a pre-formed complex) can be used and has the benefit of better stability for storage.

It is sometimes a cause of confusion that the added 'catalyst' is a Pd(II) compound, but it must be remembered that an initiation step – reduction of Pd(II) to Pd(0) – is required before the catalytic cycle can start. This reduction is usually brought about by a component of the reaction, as shown below, but sometimes a separate reducing agent, such as DIBALH, can be used.

$$2RM + PdX_2 \rightarrow R_2Pd \rightarrow Pd(0) + R–R$$

$$PdX_2 + Ph_3P + H_2O \rightarrow Pd(0) + Ph_3PO + 2HX$$

There are a very large number of catalyst systems in the literature and every new issue of a journal seems to contain yet more! Many of these catalysts have merit in specific situations, but a relatively small number will suffice for the large majority of reactions. For library synthesis, a single catalyst has to be generally active over a range of reactants, but not necessarily optimum for all. In more critical situations, such as scale-up, a variety of catalysts can be screened to optimise yields or other features of the reaction.

The ligand is the main variable in the catalyst system. Phosphines can be varied in steric bulk or in their donor strength, or finely tuned as chelating diphosphines. Alkyl groups on phosphorus increase the donor strength, increasing the electron density on the metal and thus the reactivity of the catalyst to less reactive substrates, such as chlorides. Steric bulk decreases the number of ligands that can coordinate to the metal atom, therefore increasing its reactivity. Tri-*o*-tolylphosphine is moderately bulky and moderate in its donor effects. Very bulky ligands that also contain alkyl groups on phosphorus, such as the (Buchwald–Hartwig) biphenyl compounds, form very powerful catalysts that are effective for poorly reactive substrates and can often be used at very low concentrations.

Carbene ligands are usually derived from very hindered imidazoles and are strong donors, so also form powerful catalysts. The carbenes themselves are very unstable to air, so are often generated *in situ* by reaction of precursor imidazolium salts with base. Alternatively, additionally stabilized preformed complexes, such as the 'PEPPSI' group (stabilized by the additional pyridine ligand) can be used.

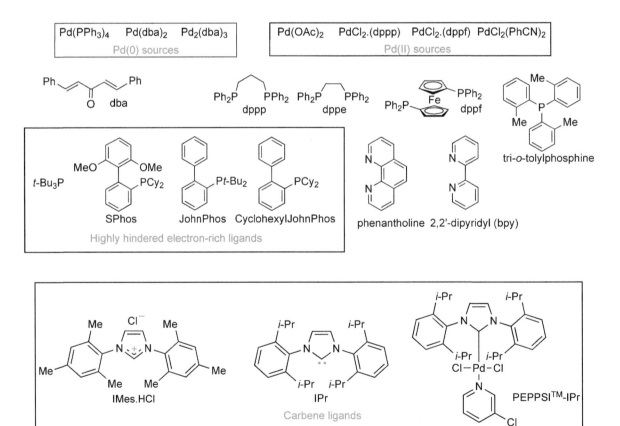

Palladium on charcoal, in the presence of a phosphine, can be used as the catalyst in Sonogashira and Suzuki reactions,[112] but a phosphine-free method, shown below, is effective with a wide range of heterocyclic partners.[113]

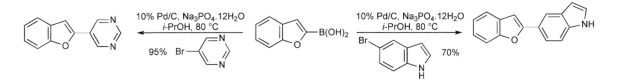

NOTE: There are often descriptions in the literature of 'ligand-free' or 'ligandless' catalysts. What is usually meant is that a standard ligand, such as a phosphine or carbene, has not been added. There are always ligands present – a ligand can be halide, hydroxide, amines in the reacting molecule or solvents, such as water or THF, counter ions (if a Pd(II) compound is used), and so on. Moreover, halide anions have been shown to be very influential ligands.[114]

4.2.2.1 Additives

In addition to the catalyst and base (if required), the use of metal-salt additives is very common for enhancing cross coupling and other reactions. CuI is the most common additive and in many cases it may operate as an intermediate transmetallating agent: RM → RCu → RPdR′. In other cases, it has been said to remove excess phosphine, thus increasing the reactivity of the palladium. Ag_2O is sometimes used and may act by transmetallation or as a halide trap.

4.2.2.2 Less Common Catalysts

Iron, cobalt and manganese are effective catalysts for cross coupling and other reactions. They were studied in the very early days of transition-metal catalysis and are now being resurrected.[115]

Fe(III) salts catalyse the coupling of Grignard reagents with alkenyl and aryl halides. The mechanism is not fully understood, but probably resembles the standard palladium sequence, through either an Fe(0)–Fe(II) or an Fe(I)–Fe(III) cycle. A particular feature is that chlorides are superior, in terms of yield, to bromides and iodides as substrates. Triflates also give very high yields and couple selectively in the presence of chlorides. Heterocyclic examples include 6-chloropurines, 4-chloropyrimidines, 6-chloro-1,3-dimethyluracil and 2-chloropyridines,[116] and the dichloropyridine example below.[117]

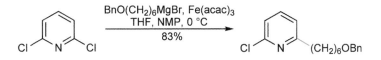

Iron ($FeCl_2.(py)_4$) also catalyses Suzuki coupling,[118] but this has not yet been applied to heterocycles.

Manganese(II) chloride (2–5 mol%) as the catalyst gave generally high yields in the coupling of Grignard reagents with 2- and 4-chloroquinolines, 1-chloroisoquinoline and 4-chloro-2-phenylquinazoline. Other substrates, which gave somewhat lower yields, included 2-chloropyrimidine, 6-chloropurine and 2-chlorobenzthiazole.[119]

Cobalt(II) catalyses the coupling of Grignard reagents with chloropyridines[120] and of aryl cuprates with aryl halides.[121] This reaction shows unusual substrate reactivity patterns and the mechanism is thought to involve a radical intermediate at the oxidative addition step.

4.2.3 The Electrophilic Partner; The Halides/Leaving Groups

Finding a suitable electrophilic substrate is not usually very challenging as a wide range is generally available or readily prepared at any (carbon) position of all heterocyclic rings. The main additional consideration is occasional instability of the halide.

The most common leaving groups for these reactions are halides and triflates, but some useful alternatives for specific situations are described later. Bromide is usually the first choice, having sufficient reactivity for use with the common catalyst–ligand combinations, and being readily available in many cases. Chlorides have advantages of availability and cost, but are more catalyst-dependent, although this is not too much of a problem following the advent of the hindered, electron-rich ligands, such as the Buchwald–Hartwig group and carbenes. Iodides may be required for relatively unreactive substrates or catalysts, but in the general case, this does not necessarily imply a better or faster overall reaction or outcome for all reactions, as the steps following oxidative addition may be less efficient in the presence of iodide ions. A very useful and general conversion of bromine to iodine can be carried out.[122]

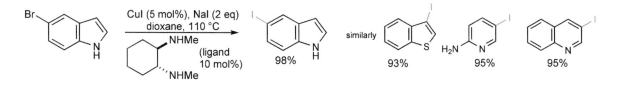

Oxy compounds are readily available for a number of ring systems and can be used as precursors for halides, but conversion into triflates or other oxygen-linked leaving groups is generally preferred. For example, oxindoles are very accessible sources of C-2 triflates as electrophilic indole components,[123] preparation of the 2-halides from the indole usually involving lithiation. In the example shown below, the *N*-protecting group is readily removed by potassium carbonate in methanol. An alternative 1-phenylsulfonyl protecting group can be used, but is more difficult to remove.[124] The isomeric C-3 counterpart can be a halide, prepared by direct electrophilic substitution, or the indoxyl triflate.[125]

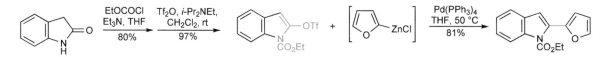

There are very few examples of fluoride acting as a leaving group in oxidative addition, although it is a very good leaving group in two-step aromatic nucleophilic substitution. The examples that are known are in electron-deficient systems and generally use nickel catalysts, exemplified by the reactions of fluoroazines with Grignard reagents.[126]

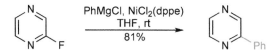

4.2.3.1 *Leaving Group Selectivity*[127]
Achieving a selective coupling reaction involving just one particular halogen in a polyhalo-compound can be very useful synthetically. This can take two main forms – competition between identical halogens and between different halogens.

Selectivity in the coupling reaction is determined by selectivity in oxidative addition and normally the differences between halogens are dominant: I > Br > Cl. Triflate is usually more reactive than bromide, but it may not always override other effects (see the examples below). When the halogens are the same, differences in positional reactivity come into play. The tendency is for selective oxidative addition to occur at the carbon of lowest electron density and this can be determined by ^{13}C NMR spectroscopy.[128] Other effects may be involved, such as chelation or steric hindrance, particularly when there is competition between two otherwise identical halogens.

Although it is possible to predict the result from physical measurements and calculations, the patterns are well established experimentally for most heterocyclic systems. These general selectivity patterns for the reactivity of leaving groups in palladium-catalysed reactions, and some interesting specific examples, are indicated below, the point of first reaction indicated in red. The differences between the two pairs of pyridine 2/3-bromo-3/2-triflates are intriguing, one set seeming to be position selective and the other leaving-group selective. Further specific examples will be found later, in the sections dealing with the particular ring systems.

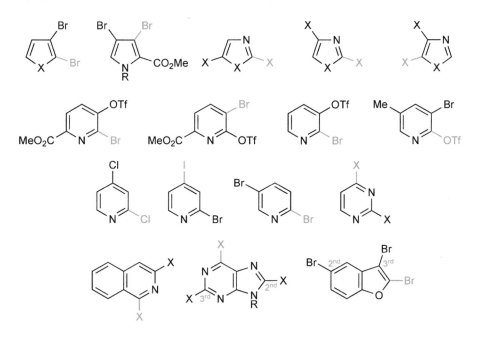

4.2.3.2 Less Common Leaving Groups

Methylthio is a good leaving group in Pd-catalysed cross couplings of azines (pyridine, pyrimidine, pyrazine) with benzylzinc reagents. The reaction is also successful for 2-methylthiobenzimidazole and -benzthiazole. In pyrimidines, the reaction shows high selectivity for C-2 over C-4; the corresponding sulfones are much less reactive – the reverse of nucleophilic substitutions.[129] Methylthio-1,2,4,5-tetrazines are good electrophilic partners for boronic acid and stannane couplings, if a thiophilic copper additive (CuTC) is used to activate-capture the thiolate leaving group.[130] An unusual feature in this method is that exactly the same conditions are used for the reactions of both boronic acids and stannanes, with no base being required for the former.

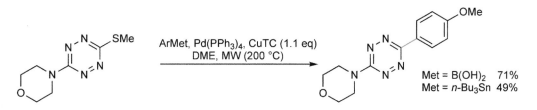

A similar method has been used for boronic acid couplings of 3-methylthio-1,2,4-triazine, using Cu(I) 3-methylsalicylate as the additive[131] and for Stille couplings of a wider range of heterocyclic substrates.[132]

Even thiones react similarly with boronic acids[133] and also in the Sonogashira reaction, where less than one equivalent of copper was required, as the sulfur is eventually converted into $Et_3N.H_2S$.[134]

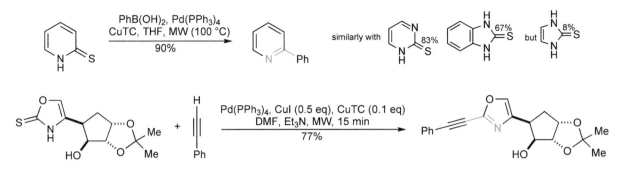

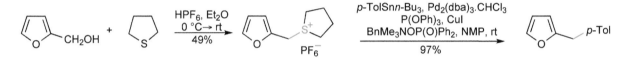

Benzylic-type sulfonium salts derived from thiophene-2- and 3- and furan-2-methanols, where halides are not stable, have been used as substrates for Suzuki, Stille and Negishi couplings. The 1-Boc-pyrrole-2-methanol analogue was only successful in Stille couplings. An important feature of this method is that triphenyl phosphite was required as the ligand to overcome the problem of reaction of the usual phosphine ligands with the sulfonium salts.[135]

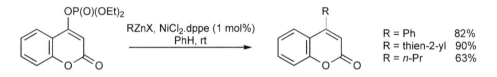

Phosphates are useful milder alternative leaving groups, for example in coumarins,[136] and furans and indolizines.[137]

OP(O)(OEt)₂

$$\text{OP(O)(OEt)}_2 \xrightarrow[\text{PhH, rt}]{\text{RZnX, NiCl}_2.\text{dppe (1 mol\%)}}$$

R = Ph	82%
R = thien-2-yl	90%
R = *n*-Pr	63%

Other fluorinated sulfonates, such as nonaflates (ArOSO₂C₄F₉), can be used as substitutes for triflates and are said to be less susceptible to S–O bond cleavage by nucleophiles.[138]

Malonate anion, with a palladium catalyst, displaces the acetoxy of 3- and 4-(acetoxymethyl)quinolines, probably *via* a three-centered benzylic equivalent of a π-allyl complex. The reaction fails with the 2-isomer, but also works with the 3- and 4-(1-acetoxyethyl) compounds and 4-(1-acetoxyethyl)isoquinoline.[139]

4.2.4 Cross-Coupling Reactions

$$RM + ArX \rightarrow R{-}Ar$$

Cross-coupling reactions[140] – the reaction of a (hetero)aryl halide (ArX), or its equivalent, with an organometallic reagent (RM), resulting in the formation of a carbon–carbon bond – are undoubtedly the most widely used transition-metal-catalysed reactions in general organic and heterocyclic chemistry.

4.2.4.1 Mechanism

The catalytic cycle for all these reactions is: (i) oxidative addition, (ii) transmetallation and (iii) reductive elimination to regenerate Pd(0). However, the reaction conditions required can vary dramatically for different metals, as can compatibility with functional groups. Boron and silicon reagents require the presence of a base and the transmetallations take place *via* anionic intermediates, while tin, zinc and magnesium transmetallations probably go *via* neutral halogen-bridged species.

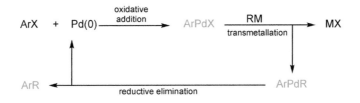

4.2.4.2 Side Reactions

A number of side reactions may occur in cross-coupling experiments and can sometimes consume a considerable proportion of the reactants:

(a) Reduction of the halide: ArX $\rightarrow$ ArH
(b) Protonolysis of the organometallic: RM $\rightarrow$ RH
(c) Homo-coupling of the halide: ArX $\rightarrow$ Ar–Ar
(d) Homo-coupling of the organometallic: RM $\rightarrow$ R–R
(e) Transfer of groups from the ligand (particularly Ph from Ph_3P): ArX $\rightarrow$ ArPh
(f) Oxidation of the organometallic (particularly boronates), by air or peroxides: RM $\rightarrow$ ROH.

The mechanisms of some of these side reactions are not always clear, particularly the source of the reducing agent in (a) and (c). It is possible that radical reactions may sometimes be involved. Homo-coupling of the organometallic (d) will always occur to some extent if a pre-catalyst-Pd(II) is used. Homo-coupling of the halide (c) is sometimes desired and can be achieved efficiently by using a Pd catalyst in the presence of a reducing agent, such as indium.[141]

An example of transfer of the phenyl group from triphenylphosphine (e) is seen in the coupling of the phenyldiethanolamine ester of 2-pyridylboronic acid, where this side reaction represents about 20% of the product, however using tri-*o*-tolylphosphine circumvents the problem (4.2.7.6).

4.2.5 The Nucleophilic (Organometallic) Partner

The preparations of the various types of organometallic compounds that can be used as cross-coupling partners are described earlier in this chapter. Most magnesium and zinc compounds have to be prepared, then used, as needed. However, boronates (i.e. boronic acids and esters), stannanes and silanes are much more stable to air and water, and many of them can be stored for long periods.

A large number of heterocyclic boronates and, to a lesser degree, stannanes are available commercially. Included in the suppliers' lists are many compounds, the preparations and properties of which do not appear in the literature. They are often noted in papers just as reagents for syntheses of libraries for biological testing and, although this may be a practical approach, it is scientifically unsatisfactory because there is no 'trail' of characterization, particularly as some of these compounds, for example azine α-boronates, are significant from a theoretical viewpoint.

Other than ability to perform the required reaction, factors to be taken into consideration when choosing the organometallic include functional group compatibility, selectivity and reaction conditions. When considering scale-up, disposal of metal residues, particularly tin and zinc, and by-products of other additives, such as fluoride, can be significant problems.

Grignard and zinc reagents require dry, generally non-polar, solvents, the main difference between the two being that zinc organometallics are much more tolerant of functional groups than are Grignard reagents. However, some zinc reagents may have a relatively low solubility in solvents such as THF, which can result in slow reactions. Stille reactions are usually carried out in non-aqueous solvents, but the reagents are quite stable to water and some mono-organotin reactions can be carried out in aqueous solution. The major problems with tin reagents are toxicity and removal of organotin impurities from the product. Boronic acid (and their trifluoroborate equivalent) couplings are very tolerant to a variety of conditions and functional groups, and can be conducted in aqueous and non-aqueous conditions, and with a variety of bases: aqueous bases of various strengths, anhydrous bases in non-polar solvents, triethylamine in DMF, and so on. Silanes can be a useful substitute for boronic acids, but do not have the large range of conditions or availability of reagents.

Carbanionic reagents, such as enolates and cyanide can also be used in place of the organometallic component.

4.2.5.1 The Cross-Coupling Reactions

There are a number of 'named reactions', which are specific to the organometallic used and, in chronological order of introduction, these are:

Kumada–Corriu (Grignard reagents) (1972)
Sonogashira (*in situ* copper acetylides) (1975)
Negishi (zinc reagents) (1977)
Stille (tin reagents) (1977)
Suzuki–Miyaura (boron reagents, particularly boronic acids) (1979)
Hiyama–Denmark (silicon reagents) (1988).

The Suzuki–Miyaura reaction is certainly the most widely used, but each of the others has its own particular advantages (and in some cases disadvantages!). Other metals, such as aluminium,[142] zirconium[143] and indium[144] are occasionally used in variants of Kumada/Negishi-type reactions.

The Suzuki–Miyaura Reaction[73]

(This process is sometimes referred to simply as the 'Suzuki reaction'.) Boronic acids are by far the most versatile coupling partners, and most suited to combinatorial chemistry and library synthesis due their stability, ease of handling and ease of removal of by-products on work-up. They are generally considered to have low toxicity, although they may have some enzyme-inhibiting activity. The cyclic esters may be preferred for enhanced stability and consistent stoichiometry. The corresponding trifluoroborates[145] show even greater stability.

The Suzuki reaction usually involves heating the boronic acid (or ester), halide, catalyst and a base in a suitable solvent. The presence of base is crucial, but it can vary from very weak to strong. The original, and still popular, conditions use $Pd(PPh_3)_4$ as catalyst, with aqueous base and an immiscible solvent, such as toluene with ethanol,[146] but DME is an advantageous solvent in some cases.[147] Dioxane is also a popular solvent and anhydrous bases, such as potassium phosphate, can be beneficial, particularly when deboronation is a problem. Triethylamine in DMF is also a useful base–solvent combination.[148]

Two biphenyl-derived phosphines (shown below) are proposed[149] as effective ligands for general heterocyclic Suzuki couplings, covering a wide range of ring systems, both as boronic acids and/or the halide component. Examples include thiophene-2- and -3-boronates, pyrrole-2- and -3-boronates, furan-2-boronates, indole-5-boronates and pyridine-3- and -4-boronates.

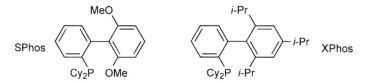

The use of trifluoroborate salts in couplings,[150] which are very easily prepared from boronic acids by reaction with KHF_2, is a useful variant of the Suzuki reaction. These salts have the advantage of enhanced (often considerably) stability compared to boronic acids and this is particularly notable for alkenyl compounds, which can be stored for a considerable time. The coupling conditions are very similar to those for boronates and are applicable to a wide range of heterocyclic substrates,[149,151,152]

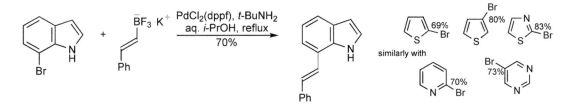

Very few functional groups interfere with boronate couplings, but free NH, either in the ring or attached to it, is sometimes said to block the reaction. In other cases, for no obvious reason, there is no problem, for example unprotected 2-chlorobenzimidazole couples nicely, under microwave conditions, with a variety of aryl boronic acids and aryl trifluoroborates, the latter being the favoured reagents.[153] 3-Amino-2-chloropyridine does not react with phenyboronic acid, but gives high yields after protection of the NH$_2$ as the acetyl derivative (86%) or as the benzylidene compound (90%).[154] On the other hand, a number of unprotected amino-chloro-pyridines and -pyrimidines undergo Suzuki couplings in high yield, using a highly hindered ferrocenylphosphine ligand.[155]

The Stille Reaction[89,156]

The Stille coupling involves heating a halide, a stannane and a catalyst in a suitable solvent, such as toluene or DMF. No base is required. The conditions are relatively straightforward, with little overall variation, apart from the catalyst. However, if a triflate is used instead of a halide, the reaction may not succeed, as transmetallation of aryltin compounds with arylpalladium triflates is often difficult. Addition of a halide source, such as lithium chloride, usually solves this problem as it allows the formation of the arylpalladium chloride, which can undergo transmetallation.

The combination of CsF with CuI is said to work synergistically to enhance reactivity over a range of coupling partners in general Stille couplings.[157] In this method, Pd(PPh$_3$)$_4$ is the preferred catalyst for iodides and triflates, and PdCl$_2$ plus *t*-Bu$_3$P the preferred catalyst for chlorides and bromides.

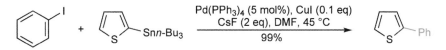

Organotin reagents have the advantage of greater stability compared to boronates in certain situations, for example they resist protodemetallation better at the 2-position of 1,3-azoles. In these cases the boronates are unknown, but the tin derivatives are easily prepared and couple well.

The Stille reaction is a fine synthetic method, but substantial problems are associated with the use of organotin compounds. Trialkyltin reagents and their by-products show a range of toxic effects.[158] In particular, *trimethyltin derivatives are potent neurotoxins that are readily absorbed through the skin and require extreme care in handling.* Tri-*n*-butyltin derivatives are considerably less toxic than trimethyltin, but may show enzyme-inhibiting and immunological effects. Organotin compounds generally are very damaging to the environment, particularly in watercourses, even at very low levels. Because of this, there are severe controls on their release in aqueous effluents.

In addition, it is notoriously difficult to remove traces of organotin reagents and by-products from the product of a reaction. Although purification to a level that is very satisfactory from a chemical viewpoint (<0.1% i.e. 1000 ppm Sn) may be possible, this amount is potentially very important biologically and if used in synthesis for biological screening, there is a possibility that these levels could significantly affect the results. Modified techniques, such as the use of solid-supported[159] or fluorous tin reagents can help to minimise residues.

The use of mono-organotin compounds, which have much lower toxicity, for couplings offers major potential advantages, but further development of this area is required. Phenyltrichlorotin couples with some halo-heterocyclic substrates in aqueous solution, using palladium and a water-soluble ligand, the reacting species actually being the trihydroxytin compound produced by hydrolysis.[160] A promising method uses bis-(silylamino)tin reagents, prepared by the reaction of a commercially available Sn(II) compound with halides.[161] A notable application of this method is the efficient transfer of [11]CH$_3$ in couplings with several bromo-quinoline isomers.[162]

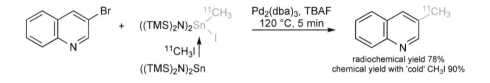

radiochemical yield 78%
chemical yield with 'cold' CH$_3$I 90%

The Hiyama–Denmark Reaction

Several types of organosilicon compounds can be used in cross coupling, fluoride usually being required to activate the silicon for the transmetallation in the earlier examples. Silanols are newer and more versatile reagents, in particular dimethylsilanols, which are used with bases – either reversible (alkoxides and cesium carbonate) or irreversible (sodium hydride and sodium hexamethyldisilazide). In the latter case, the silanoate salts can be isolated and stored.[163]

A range of 2-dimethylsilanol derivatives of *N*-protected and substituted indoles, *N*-Boc-pyrrole, furan and thiophene couple in high yields with iodides, bromides and chlorides, the latter two requiring the use of highly activating ligands, such as SPhos.[164] Isoxazolyl silanols are readily available *via* cycloaddition ring synthesis and couple well.[165] Other conditions for silanol couplings include Pd(PPh$_3$)$_4$ with silver oxide and catalytic TBAF.[166]

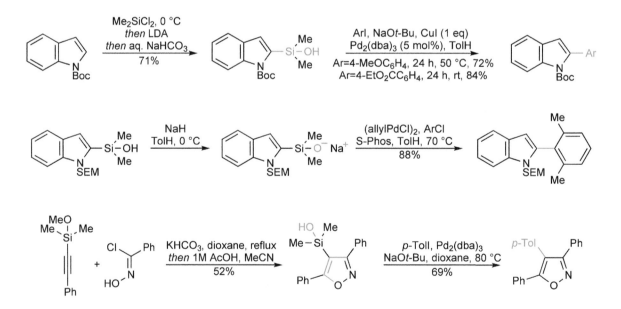

Aryltrimethoxysilanes, for use in cross couplings,[167] can be prepared by a palladium-catalysed reaction of aryl iodides with trimethoxysilane,[168] *but this silane is extremely toxic!*

Trimethylsilyl compounds generally do not react in attempted cross couplings, but 2-trimethylsilylpyridines will react in the presence of TBAF and silver oxide.[169] Also, 2-trimethylsilylpyridines containing chloro, fluoro or methoxy substituents will couple in the presence of a CuI additive.[170]

The Kumada–Corriu Reaction[171]

Grignard reagents are easy to prepare on a large scale, but they are incompatible, at least under standard conditions, with many functional groups, and even add to the rings of the more reactive azines. However, efficient couplings of a wide range of halo-heterocycles can be carried out, for example α-chloro- and -bromo-, and β-bromo-azines with phenylmagnesium chloride, as shown below. Moreover, the preparation and coupling of some functionalised Grignards can also be carried out by maintaining low temperatures.[172]

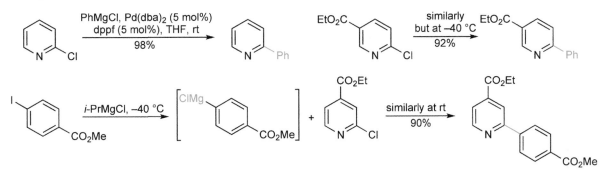

2-, 3- and 4-Tri(quinolyl)magnesates (Q₃MgLi), prepared by bromine exchange with *n*-Bu₃MgLi, give moderate yields in cross couplings with heteroaryl bromides.[173]

The Negishi Reaction[174]

Zinc reagents are more tolerant of functional groups than Grignard reagents and they can be used to introduce quite complex alkyl chains (which is difficult to do with the other organometallics), as demonstrated by the example below.[175] When used as a heterocyclic organometallic component, zinc reagents can be prepared with relative ease at practically any position of any heterocyclic ring.

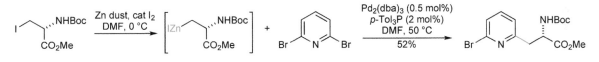

The Sonogashira Reaction

A typical Sonogashira reaction involves heating an aryl halide, catalyst and a terminal alkyne with cuprous iodide and triethylamine. This reaction does not require the preparation of an organometallic reagent, because transmetallation occurs *via* an alkynylcopper, formed *in situ*. The main use of the Sonogashira reaction in heterocyclic chemistry is probably for the synthesis of intermediates for ring synthesis.[176]

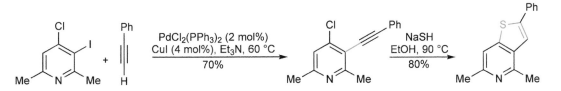

Alternatives to the Sonogashira reaction involve the use of preformed alkynyl organometallics,[177] such as trifluoroborates,[178] or the reverse coupling of haloalkynes with organometallics.[179]

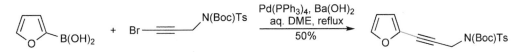

4.2.6 Other Nucleophiles

Boronic acids[180] and 9-BBN derivatives are often used as alkyl partners,[181] but trifluoroborates may be more convenient and stable. Introduction of a protected aminoethyl chain, using a trifluoroborate, has substantial potential for bioactive molecules – the reaction shown below using 3-bromopyridine works equally well for pyrimidine (C-5), thiophene (C-2), and indole (C-5).[182]

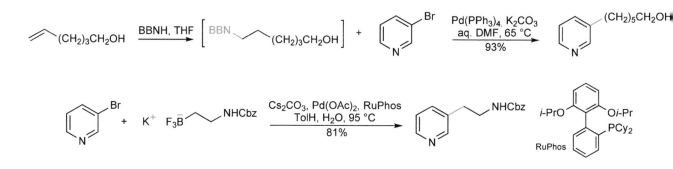

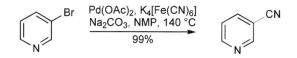

A cyano group can be introduced using palladium and zinc cyanide,[183] potassium cyanide or potassium ferrocyanide, the last having the significant advantage of low toxicity, and can be carried out with either a palladium[184] or copper catalyst.[185] Tri-*n*-butyltin cyanide can be used similarly, but a modification using catalytic tri-*n*-butyltin chloride with potassium cyanide is much to be preferred.[186]

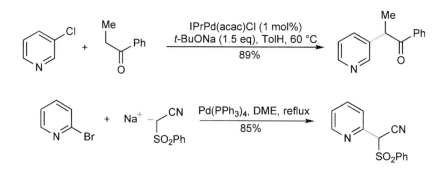

Pd(0) catalyses the coupling of a variety of stabilised enolates, for example the reaction of bromopyridines with the sodium salts of sulfonyl and phosphonyl acetonitriles.[187] Simple enolates can be arylated using Pd-carbene catalysts[188] or using highly hindered phosphine ligands.[189]

Phosphorus can be introduced *via* Pd-, Ni- or Cu-catalysed reactions.[190]

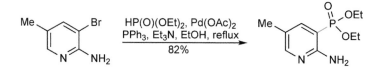

Hydroxymethyl can be introduced *via* coupling with benzoyloxymethylzinc iodide. This reaction shows an unusual pattern, with chloroazines being superior to the corresponding bromides and iodides; 2-bromonaphthalene and 3-bromoquinoline do not react.

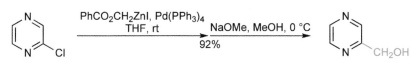

4.2.7 The Ring Systems in Cross-Coupling Reactions

As reactions in benzo-fused rings are generally similar to simple carbocyclic compounds, our discussion of these is limited.

4.2.7.1 Pyrrole,[191] Furan and Thiophene

2-Boronates and -stannanes are usually made *via* direct lithiation and the 3-analogues *via* lithium exchange on 3-bromo or -iodo compounds. Boronic acids at α- and β-positions of all three systems generally couple well, but some α-derivatives are quite susceptible to base-catalysed cleavage, which is enhanced by the presence of β-electron-withdrawing groups.[192] In the reverse couplings, using the heterocyclic halide, the only potential problem is the relative instability of some halo-pyrroles, but this can be alleviated to some degree by the presence of electron-withdrawing protecting groups on nitrogen. Examples below include the double coupling of 2,5-dibromothiophene,[193] the reaction of a thiophene-2-stannane with a vinyl iodide,[194] and a nickel(0)-catalysed alkylation of 3-bromothiophene.[195]

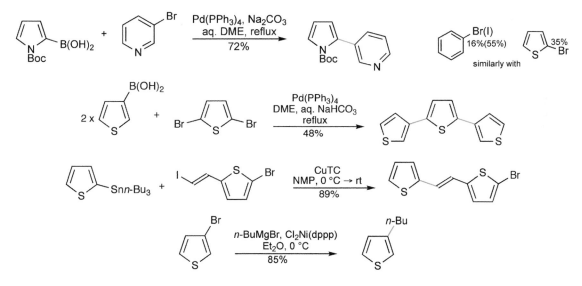

4.2.7.2 Indole, Benzofuran, Benzothiophene

Lithiation of *N*-protected indoles is the main route to 2-boronic acids, which have found wide use, although the usual protecting groups (Boc and arylsulfonyl) increase the ease of deboronation. With Boc-protected derivatives, the indole is less consistent than the pyrrole in coupling reactions.[192] *In situ* protection of the indole NH as the *N*-carboxylate can be used to prepare both the 2-boronic acid[196] and 2-tri-*n*-butylstannane[197] for one-pot coupling procedures.

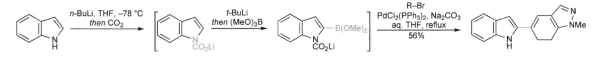

1-TBS and 1-TIPS indole-3-boronic acids, prepared *in situ via* halogen exchange, couple well with a range of heterocyclic substrates.[198]

A synthesis of the natural product meridianin D demonstrates selectivity in halogen–lithium exchange and substrate selectivity, using a halogenated boronic acid. The presence of the *N*-tosyl probably controls the halogen exchange selectivity.[199]

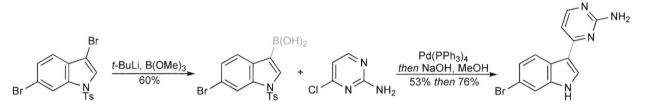

1-Tosyl-3-tri-*n*-butylstannylindole, available *via* palladium-catalysed stannylation, undergoes coupling with a range of halides and triflates.[200]

1-Methylindol-2-ylzinc chloride can be coupled with alkenyl halides[201] and the couplings of *N*-protected indol-2-ylsilanols are discussed above.

Unprotected 5- and 6-tri-*n*-butylstannylindoles (and comparable boronic acids) are prepared *via* lithium–bromine exchange, using the KH then *t*-butyllithium procedure, and couple cleanly with acid chlorides.[202]

2,3-Diiodo-1-phenysulfonylindole can be converted by sequential, *in situ*, magnesium–iodine exchanges into various 3-substituted indol-2-ylboronates, which participate in Suzuki couplings.[203]

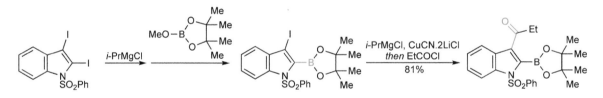

Benzothiophene-3-boronates can be prepared *via* both lithiation[204] and palladium-catalysed boronation.[205]

4.2.7.3   *1,3-Azoles*[206]

Imidazole boronate esters at C-4 and C-5 are known, but attempts to prepare imidazole 2-boronic acid or boronate esters, either *via* lithio compounds or palladium-catalysed boronation, fail. The products of these attempts indicate that the boronate may have formed, but rapidly deboronates.[207] Oxazole-[208] and thiazole-2-stananes[209] are much more stable and work well in coupling reactions.

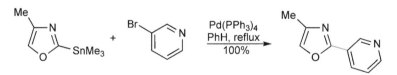

2-Stannyl imidazoles[210] can be used, but the corresponding zinc compounds give high yields and are preferable.[211]

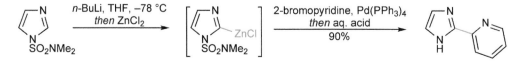

N-Protected 2-substituted imidazole 5-boronates are readily prepared *via* direct lithiation, but the use of 2-silyl compounds as precursors for the corresponding 2*H*-5-boronates fail at the desilylation stage. However, catalytic reduction of a 2-chloro analogue gives a high yield of the 1-THP protected boronate, but yields of its coupling products are quite modest.[212]

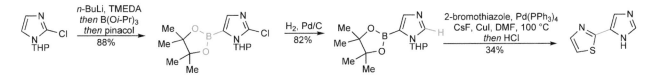

2-Phenyl-5- and -4-stannyl-thiazoles, prepared *via* lithium-halogen exchange, are equally reactive in Stille couplings.[213] 2-Phenyloxazole 4-pinacolboronates are prepared by palladium-catalysed boronation of the 5-*H*-triflate or *via* lithium-bromine exchange on the 5-methyl-4-bromide.[214]

Unprotected 4(5)-bromoimidazole couples well with arylboronic acids under phase-transfer conditions.[215] Boronic acid couplings of 2-chloro- and 4-trifloxy-oxazoles are equally successful.[216]

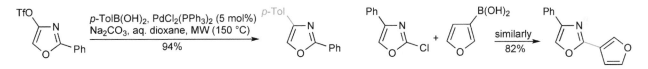

In a 5-bromo-4-trifloxy-oxazole, the triflate reacted selectively in a Stille coupling with PhSnMe₃.[217] 2,4-Dibromothiazole reacts selectively at C-2 in Negishi couplings, the products being convertible into zinc or tin derivatives for further couplings.[218] The coupling of 4,5-diiodoimidazole protected with SEM on N-1, is completely selective for the 5-halogen.[219]

4.2.7.4   1,2-Azoles
1-SEM and -THP-protected pyrazole-5-boronates are easily prepared *via* direct lithiation and undergo cross couplings under standard conditions.[220] Pyrazole-5/3-stannanes (with no *N*-protection) can be prepared by ring synthesis from alkynyl stannanes. The example shown below, bearing a 4-fluorine, requires iodo-arenes as substrates for cross coupling.[221]

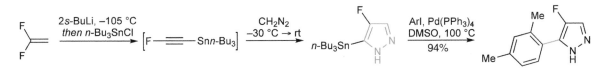

A wide range of isoxazole-4-boronates and some 5-boronates can be prepared by dipolar cycloaddition of nitrile oxides with alkynylboronates.[222]

3-Iodoindazoles require Boc-protection of the NH for successful Sonogashira,[223] Suzuki and Heck reactions.[224] 3,5-Dichloro- and dibromo-isothiazole-4-carbonitriles undergo selective cross coupling at C-5 for a range of reaction types; a further coupling of the remaining 3-halogen is possible for Stille and Negishi reactions, but Suzuki coupling requires iodine.[225]

4.2.7.5   Higher azoles
There seem to be few examples of organometallics derived from higher azoles (triazoles, tetrazole, oxadiazoles etc.) being used in coupling reactions, possibly due to the relative instability of boronates and stannanes, although there is an example of a coupling of a 5-stannyltetrazole.[226]

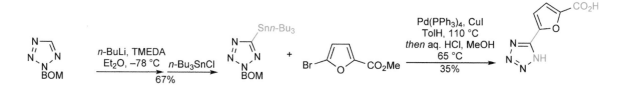

Zinc derivatives seem to be more amenable, for example 1-benzyloxy-1,2,3-triazol-5-ylzinc iodide (prepared *via* direct lithiation) couples well, but the corresponding stannane is less stable and can only be used for palladium-catalysed acylations.[227]

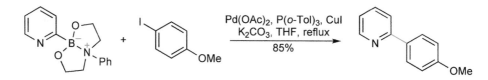

Generally the reverse-mode coupling, using the halo-heterocycle is better, for example 1-benzyl-5-bromotetrazole gives high yields with a range of boronic acids, the yield of a single example of a Stille coupling being significantly lower.[228]

4.2.7.6 *Pyridines and Quinolines*

Boronic acids, stannanes, zinc derivatives and Grignard reagents at all positions of the pyridine ring generally couple well. Silane couplings also work well in some cases. The 3- and 4-pyridine-boronic acids have been widely used, but the 2-isomer is rather unstable, undergoing ready deboronation, although some couplings have been carried out using crude material. However, the *N*-phenyldiethanolamine ester of pyridin-2-ylboronic acid[229] has the right balance between stability and reactivity and is commercially available. Tri-*o*-tolylphosphine is used as the ligand in this case because with triphenylphosphine, substantial transfer of the phenyl from phosphorus occurs.

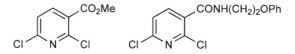

Both 2- and 3-pyridyl stannanes can be prepared by palladium-catalysed stannylation.[230] The *N*-oxides and quaternary salts of pyridyl stannanes can be used in Stille couplings.[231]

The coupling of pyridin-3-ylzinc bromide requires a solvent exchange from the ether used for lithiation and metal exchange, to THF, for the coupling.[232]

The Sonogashira coupling of 2,4-dichloroquinoline in water with a palladium–carbon catalyst (triphenylphosphine, triethylamine, CuI, 80 °C) shows complete selectivity for the 2-position[233] and Suzuki couplings of 1,3-dichloroisoquinoline are selective for C-1.[234]

The 2,6-dichloronicotinic ester shown below gives catalyst-variable selectivity in couplings with phenylboronic acid, but the chelating amide is more selective: 9:1 compared to 2.5:1 for the ester.[235]

The selectivity (C-2:C-4:2,4-disubstitution) of Suzuki couplings of 2,4-dibromopyridine varies greatly with catalyst and conditions. The most selective C-2 mono-coupling is obtained using Pd(PPh₃)₄ with thallium hydroxide as base.[236] (*Caution! Thallium hydroxide is very toxic.*)

4.2.7.7 Diazines[237]

By analogy with pyridin-2-ylboronic acid, the azine boronic acids with the boron α to nitrogen would be expected to be relatively unstable. Pyrimidine-5-boronic acids have been quite widely used and pyridazine-4-boronic acids are also well characterised. However, all the other isomeric diazine boronic acids are uncharacterised in the literature, but have been mentioned in patents and in suppliers' catalogues. The only references to their use in couplings seem to be as listed examples in library syntheses, without individual experimental details, particularly pyrazin-2-ylboronate, 2-methylthiopyrimidin-4-ylboronate, and pyrimidine-2- and -4-boronates.

Pyridazin-4-ylboronates are available *via* Diels–Alder reactions of alkynyl boronates and 1,2,4,5-tetrazines,[238] and the 4-stannanes similarly from 1,2,4-triazines[239] and 1,2,4,5-tetrazines.[240] The stannylpyridazine, without other substituents, can also be obtained by this latter method, as shown below.[241]

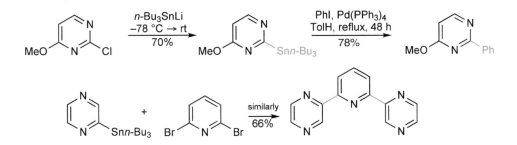

A range of stannylpyrimidines and pyrazines can be prepared by reaction of the halodiazines with a lithiostannane, however attempts to prepare 3-stannylpyridazines fail. Their Stille reactions are consistently successful.[242]

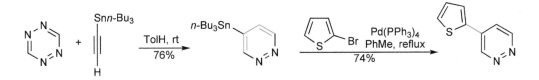

2,4-Di-*t*-butoxy-pyrimidines are used as protected forms of uracil 5-boronic acids and 5-halides for Suzuki and Stille couplings, respectively, but suffer from the problem that an intermediate in their preparation (2,4-dichloro-5-bromopyrimidine) is highly allergenic. Unprotected 5-iodouracil will couple in some cases, but the Stille coupling of the 5-bromo-2,4-bis(trimethylsilyloxy) derivative is more consistently successful.[243]

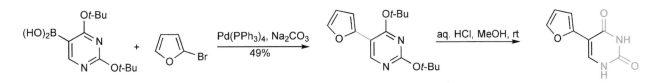

Couplings of 5-iododeoxyuridine (with protection of the sugar oxygens) are successful with a range of heteroaryl trimethylstannanes.[244]

2,4-Dichloropyrimidine gives a selective Sonogashira coupling at C-4[245] and 2,4,6-trichloropyrimidine shows clean sequential Suzuki couplings: C-4, then C-6, then C-2.[246] In Sonogashira reactions on dihalopyrimidines, a 4-chloro reacts in preference to a 5-bromo, but not a 5-iodo.[247]

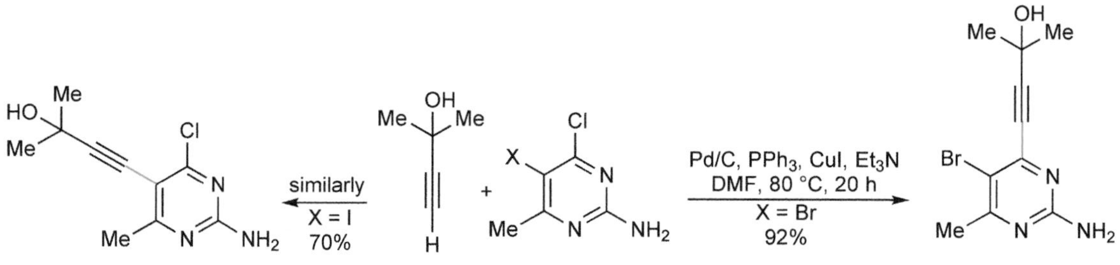

The use of ultrasonic activation is notable in the Negishi couplings of diazinyl zinc halides.[248]

4.2.7.8 *Higher Azines*

3,5-Di-(benzylamino)-6-chloro-1,2,4-triazine undergoes Sonogashira coupling under mild conditions (CuI, Et_3N, MeOH, reflux, 3 h; 50–77%).[249]

4.2.7.9 *Oxygen Heterocycles*

4-Tosyloxycoumarins are good substrates for Sonogashira and Negishi couplings.[250] 3-Bromo-4-sulfonyloxycoumarins showed complementary selectivity in Suzuki couplings, depending on the sulfonyl group, with relative reactivity of 4-TfO > 3-Br > 4-TsO.[251]

2-Methyl-4-pyrone-3-triflate gives good yields in Stille couplings, under microwave heating, with a range of heterocyclic stannanes.[252]

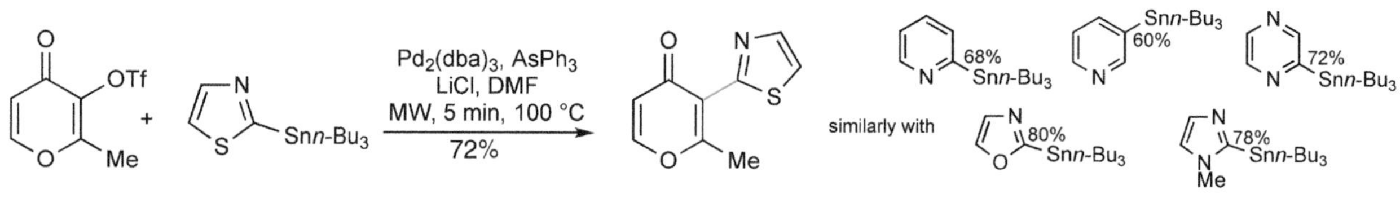

3,5-Dibromo-2-pyrone reacts selectively at C-3 under standard Suzuki conditions, but this is switched to C-5 by a change of solvent and addition of cuprous iodide.[253] Stille couplings show the same C-3 selectivity, but this is enhanced, rather than reversed by the addition of cuprous iodide.[254] 4,6-Dichloro-2-pyrone shows reasonable selectivity for C-6 in Sonogashira reactions.[255]

4.2.7.10 *Purines*[256]

Most commonly, palladium-catalysed substitutions on purines are carried out on the halo-purine, but some metallated purines are useful. 2-Stannyl-6-chloropurines can be prepared *via* direct (C–H) lithiation, without protection of C-8 (27.7.1). 6-Purinyl zinc compounds can be prepared by reaction of the iodide with activated zinc metal.[256]

O-Tosyl, -mesitylenesulfonyl and -2,4,6-tri-*iso*-propylphenylsulfonyl derivatives of deoxyguanosine are good substrates for Suzuki couplings, when used with dicyclohexyl-JohnPhos as ligand.[257] Stille and Negishi couplings on 6,8-dichloropurines are highly selective for C-6.[258]

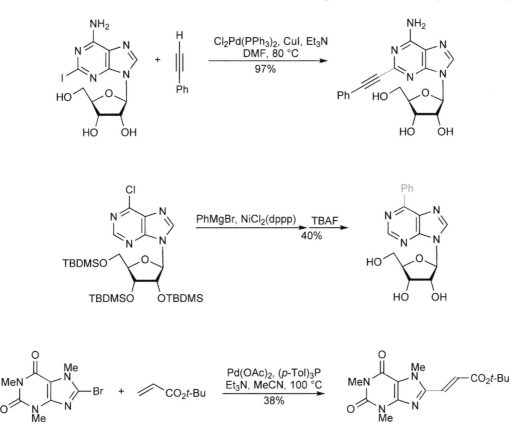

5-Bromo- and -iodo-indolizines give consistently good results in Suzuki couplings using very simple conditions (0.5 mol% PdCl$_2$, K$_2$CO$_3$, aq. dioxane, 80 °C; no ligand added).[259]

4.2.8 Organometallic Selectivity

The selective reaction of one organometallic in the presence of another is very useful in building up complex organometallic reagents. This is best achieved by using metals that have very different requirements for their coupling conditions. The greatest differences are between boronates and related reagents, which require added base and often aqueous conditions, and stannanes or organozincs, for which non-polar solvents with no added base tend to be used. In compounds containing both a metal and a halide, the challenge is to minimize self-coupling, although in the absence of a competitive substrate this is a useful way of synthesizing oligomers and polymers. Illustrative examples are shown below in the coupling of 2-bromo-3-tri-*n*-butylstannyl pyridine with 3-diethylborylpyridine[260] and the synthesis of a nucleoside boronic acid analogue.[261]

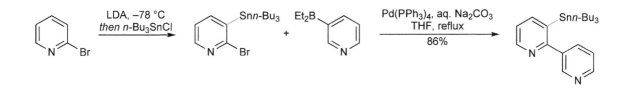

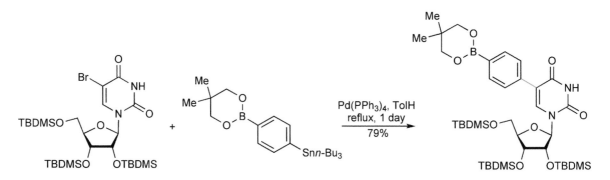

Negishi couplings on 2-bromo-5- and -6-tri-*n*-butylstannyl pyridines are possible due to the relatively high reactivity of 2-bromopyridines (the isomeric 5-bromo-2-stannylpyridine gave only low yields under the same conditions) and the low temperature of the reaction.[262]

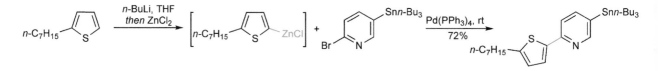

For Suzuki reactions using haloboronic acids and another halide, the obviously best course is to have a more reactive halide in the substrate than in the boronate, so there are quite a few examples of chloro-heteroaryl-boronic acids coupling efficiently with aryl bromides. This difference is clearly shown by the coupling reactions of 2-chloro- and 2-bromopyridine-5-boronic acids with bromo-heterocycles.[263]

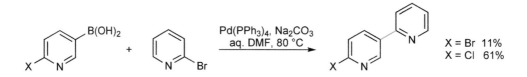

High-yielding Suzuki substitutions of halogen in halo-boronates are also possible using *N*-methylimino-diacetic acid (MIDA) to form the protected boronate, which resists coupling under anhydrous conditions. This approach has been applied to aryl and heteroaryl (only thiophene and benzofuran, shown below) systems,[264] and also for polyene synthesis.[265]

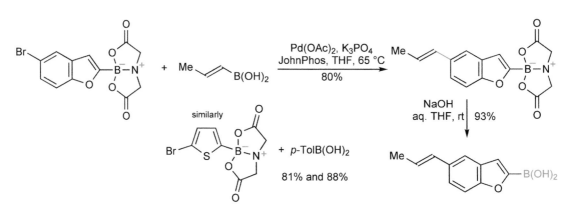

In couplings of 5-metallo-2-chlorothiazoles, stannanes are preferred to boronates due to the relative instability of the latter. The corresponding zinc derivatives are unsatisfactory.[266]

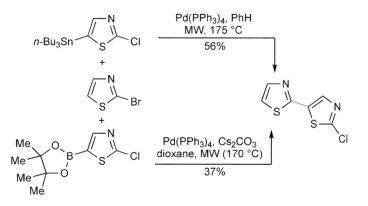

4.2.9 Direct C–H Arylation

$$HetH + ArX \rightarrow HetAr$$

The direct arylation, with substitution of hydrogen, mainly of electron-rich heterocycles, such as mono- and di-hetero 5-membered rings, by reaction with aryl[267] or alkenyl halides, is a very useful supplement to cross coupling, eliminating the need for the preparation of an organometallic partner.

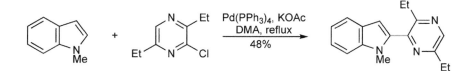

Both palladium and rhodium are effective catalysts, but the rhodium reactions seem to be more subject to steric effects. The reaction proceeds *via* a catalytic cycle similar to cross coupling, except that the formation of the diorgano(palladium) is brought about by electrophilic attack by the arylpalladium halide, rather than a transmetallation reaction, hence the need for electron-rich systems.[268,269] The equivalent alkynylation can also be carried out using haloalkynes.[270]

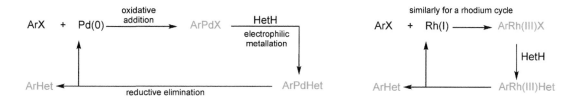

It is significant that these types of reaction may be easier than standard electrophilic substitutions in 1,3-azoles, reflecting the different nature of electrophilic metal cations compared to simple electrophiles, such as bromine.

Oxazoles can be arylated at either C-2 or C-5, the method shown being notable for the use of water as solvent.[271] 2-Substitution of 5-aryl oxazoles, using palladium acetate, can be carried out without an added ligand.[272]

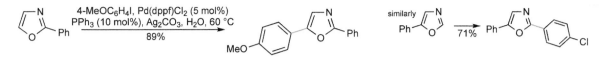

The direct coupling of *t*-butyl thiazole-4-carboxylate at C-2 is successful with a wide range of (hetero) aryl halides, the hindered ester (rather than the methyl ester), together with the use of tri-*o*-tolylphosphine as ligand, giving optimum selectivity for C-2 *vs.* C-5 for iodoarenes and heteroarenes. Changing the ligand to JohnPhos allows extension to chloro- and bromo-heterocycles.[273]

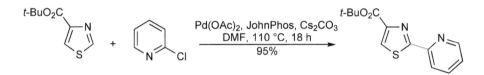

The reaction of 3-methoxythiophene is highly selective for C-2 and has been put to use for the synthesis of thiophene oligomers.[274] The substitution of thiophenes, using aryl iodides, can also be carried out without interference from bromo substituents in the substrate.[275]

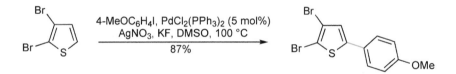

The reaction is selective, as would be expected, for C-8 in purines[276] and their nucleosides.[277] Even totally unprotected deoxyadenosine reacts well with a range of aryl iodides, without arylation of the 6-NH$_2$ except, curiously, for 2-iodonitrobenzene, which was completely selective for the amine group![278]

Alkenylations can also be carried out: the optimum conditions for 2-bromopropene with palladium acetate include the use of triphenylarsine with silver carbonate and triethylamine, but the advantages over more amenable conditions, using triphenylphosphine with potassium or cesium carbonate, are marginal.[279]

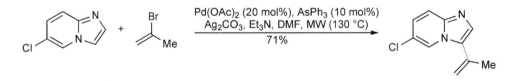

A rhodium catalyst, under microwave heating, is similarly successful with NH imidazoles, benzimidazole, benzoxazole, and a 1,2,4-triazole.[280]

Indoles show a strong tendency to give 2-substituted products, using either rhodium[281] or palladium catalysts, in contrast to normal C-3 electrophilic substitution.

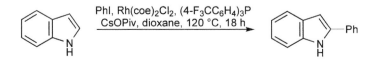

N-Substituted indoles can be similarly arylated at C-2, using modified conditions, the reactions being successful even when the indole contains strong electron-withdrawing groups.[282]

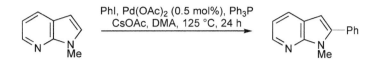

This 'abnormal' regioselectivity can be explained by equilibration of the intermediate palladated indole cation, followed by a relatively slow deprotonation.

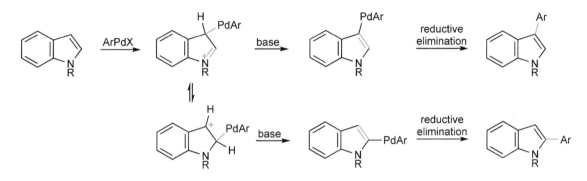

The reaction using NH indoles and a magnesium base can be controlled to give either C-2 or C-3 arylation.[283] With magnesium oxide, a mild reversible base, only 2-substitution occurs, but with pre-formed strongly coordinating magnesium derivatives, as formed by reaction with Grignard reagents or best, magnesium hexamethyldisilazide, high C-3:C-2 ratios result.

Alternatively, very clean 3-substitution of NH indoles by aryl bromides can be achieved by use of phosphine-free conditions, but the reaction is inhibited by electron-withdrawing groups on the indole. Similar reactions using phosphines give 1- or 3-arylation depending on the phosphine.[284]

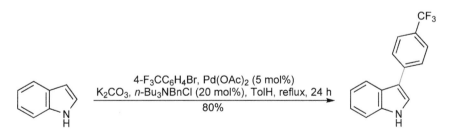

N-oxides of pyridines[285] and diazines[286] react well and with complete selectivity for the position α to the *N*-oxide. Mechanistic studies indicate that the reaction is not based on normal (two step: addition then proton loss) electrophilic palladation of the *N*-oxide, but possibly by a simultaneous palladation–deprotonation that was used to explain the success of the reaction in very electron-deficient (non-heterocyclic) systems, where there is also considerable acidification of the hydrogen.[268]

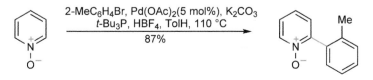

Although azines do not usually react under these conditions, the arylation of pyrazine and pyridine is possible using a gold(I) catalyst with *t*-BuOK.[287]

Consideration could also be given to possible involvement of an (ylide ↔ carbene)/Pd complex, perhaps assisted by coordination with the oxygen. (Carbene intermediates are well known in azoles, cf. the rhodium reactions shown later.)

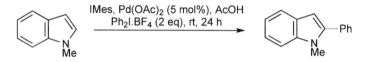

(Stable metal (Ir, Ru, Os, Au)-carbene complexes of such pyridinium and related ylides have been isolated, including an iridium derivative by direct preparation from pyridine.[288])

A reaction using diaryl iodonium salts is thought to proceed *via* a Pd(II)–Pd(IV) cycle (simple halides are not sufficiently reactive to carry out oxidative addition on Pd(II)).[289]

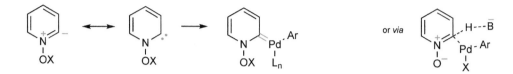

4.2.9.1 C–H insertion

A different type of metallation, directed by an acyl group at either the pyridine 3- or 4-position, uses a catalytic ruthenium complex and results in a reductive Heck-type substitution, as illustrated below. The mechanism involves insertion of the metal into a C–H bond. The process is non-polar and works equally well with electron-rich heterocycles, for example indole.[290]

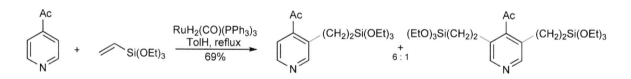

The rhodium-catalysed reaction between 1,3-azoles and terminal alkenes is thought to proceed *via* a carbene complex.[291]

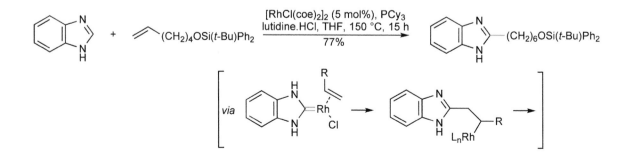

4.2.9.2 Oxidative Coupling of Arenes

The catalytic oxidative coupling of two dissimilar arenes is also possible, for example, *N*-acetyl indoles, or benzofurans, with benzene and other simple arenes. The mechanism involves sequential metallations in the two rings.[292] The regioselectivity can be controlled by the choice of oxidant, Cu(OAc)$_2$ favouring C-3 and AgOAc, C-2 substitution in 1-acetylindole.[293]

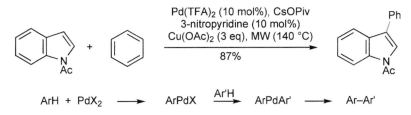

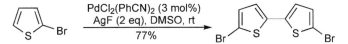

The similar oxidative dimerisation of 2-bromothiophene illustrates selectivity in the presence of halogen. The silver fluoride seems to be the oxidant as it is reduced to silver metal during the reaction.[294]

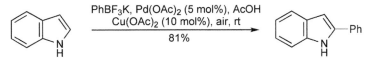

2-Arylation of indoles can also be carried out *via* arylpalladium acetates generated from boronic acids[295] or trifluoroborate salts[296] and palladium acetate. The reactions are catalytic in palladium, cycling of the Pd(II) being effected by the use of a re-oxidant (Cu(II)/air). The reaction works well on NH and *N*-methyl indoles but fails with the *N*-acetyl derivative.

4.2.10 *N*-Arylation

$$ArX + R_2NH \rightarrow ArNR_2$$

Transition-metal-catalysed reactions can be used to introduce aryl or heteroaryl groups onto the ring NH, or attached amino groups, of heterocycles. They can also be used for the displacement of leaving groups by amines in all types of heterocyclic systems, including the use of milder conditions for substitutions at relatively activated positions, such as α- and γ-positions in pyridines, where nucleophilic substitutions can be carried out.

There are two general ways in which to carry out this process: (i) reaction with an aryl halide using a Pd, Cu or Ni catalyst, (ii) reaction with an aryl boronic acid catalysed by Cu(II). Minor methods include reactions with diaryl iodonium salts or high oxidation state aryl metals.

4.2.10.1 Buchwald–Hartwig Reaction (Palladium-Catalysed Amination)

Although occasional examples had been described earlier, the design and development of new highly active ligands for palladium gave new impetus to transition-metal-catalysed aminations.[297] A number of relatively complex ligands were used in earlier work, but simpler versions, such as JohnPhos, have now become prominent.[298] These methods work well with heterocycles, for example *N*-arylation of indoles, using triflates, bromides and chlorides.[299]

The mechanism of palladium/aryl halide amination is very closely related to that of cross coupling, with displacement of the halide on palladium (or copper or nickel) by an amine or *N*-anion instead of the trans-metallation step. In the case of Cu and Ni catalysis, it may proceed through M(0)–M(II) or M(I)–M(III) cycles.

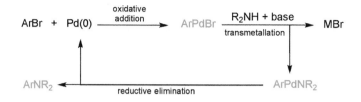

Arylation of a pendant amino group in a wide range of amino heterocycles, including pyrazoles, thiazoles, thiadiazoles,[300] oxazoles, isoxazoles, pyridines and diazines can be carried out using Xantphos as the ligand.[301] In another method, with a number of halo-azines as the arylating agents, the use of sodium phenoxide as base is a key feature. This reaction works equally well under classical thermal conditions (80 °C, 2 h) or with microwave heating (170 °C, 2 h).[301]

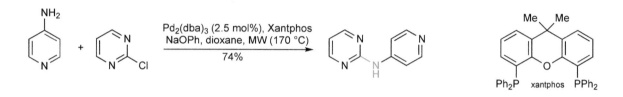

The conversion of purines into arylamino derivatives,[302] particularly with polycyclic arylamines, is of significance for investigations of mutagenesis. Displacement of bromine on purine nucleosides can be carried out at C-6[303] and C-8 of deoxyguanosine, with protection of the 2-amino, the sugar hydroxyl groups, and the 6-oxo as an ether.[304] 8-Bromoadenosine, with protection of the sugar, but not the amino group, couples well with anilines.[305] The reverse method is also possible, for example reaction on the 2-amino of a 6-benzyloxy-deoxyriboside (i.e. a protected deoxyguanosine) with bromopyrenes.[306]

Intramolecular reactions can also be carried out, such as cyclisation of an N-1-COCH₂NHCbz indole displacing a 2-iodo group.[307]

Aryl hydrazines, can be prepared *via* arylation of benzophenone hydrazone, Boc-hydrazide or bis-Boc-hydrazide.[308] Such transformations can also be carried out using copper catalysts.[309]

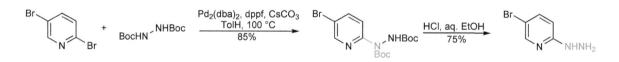

N-Alkenylation can also be carried out. For indoles, pyrroles and carbazoles, the *N*-lithio-compound is the preferred reactant, the magnesium derivative or a mixture of the indole with potassium phosphate giving significantly lower yields.[310]

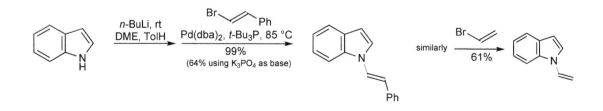

4.2.10.2 Nickel-Catalysed Amination

Nickel is particularly useful for reactions of aryl chlorides, for example, 2-, 3- and 4-chloropyridines are aminated in the presence of a carbene ligand.[311]

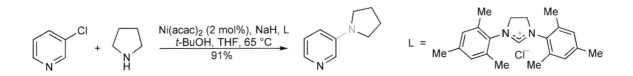

4.2.10.3 Copper-Catalysed Amination

A copper-catalysed amination – the Ullmann reaction – was the forerunner (1904) of all transition-metal-catalysed couplings, but the vigorous conditions that are required limited its use. In recent years there has been a resurgence in copper catalysis, due in part to the development of better ligands and understanding of mechanisms.[312]

Copper catalysts have the advantages of lower cost, low toxicity and, often, less need for complex ligands. An early example, using phenanthroline as ligand, is shown below.[313]

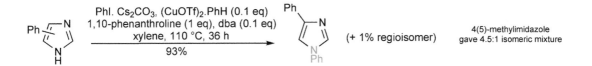

A very simple system, using cuprous iodide with no added ligand, gives good results with a wide range of aryl and heteroaryl bromides (and a few chlorides and iodides), reacting for example with imidazole, 1,2,4-triazole, pyrazole and pyrrole.[314]

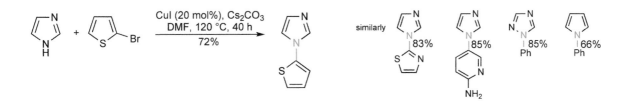

4,7-Dimethoxy-1,10-phenanthroline is a superior ligand for the arylation[315] and heteroarylation[316] of imidazoles using a cuprous oxide catalyst.

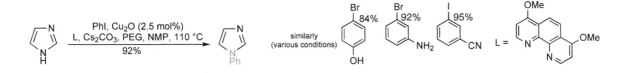

Lithium chloride is an effective promoter (no organic ligand added) for the CuI-catalysed N-1 arylation of 5- and 7-azaindoles.[317] Very clean and high-yielding *N*-arylation results from the use of the tetra-*n*-butylammonium salt of 2-pyridone as substrate for CuI-catalysed reactions with iodides.[318]

Proline is a highly effective ligand for the CuI-catalysed displacement of bromine in 3-bromopyridines, 2-bromothiazoles and 5-bromo-1-phenylsulfonylindole by primary amines, morpholine and pyrazole.[319]

4.2.10.4 Chan–Lam Reaction

This conversion employs a boronic acid with a copper(II) catalyst. The method was developed for reactions of heterocycles, including the *N*-arylation of isatin[320] and of pyrazole, imidazole and their benzo derivatives (1,2,3-triazole, 1,2,4-triazole and 5-phenyltetrazole gave only very modest yields).[321] These conditions also apply to the *N*-arylation of 2-pyridone (and various fused derivatives), 3-pyridazinones, indole-2-carboxylates and pyrrole-2-carboxylates.[322]

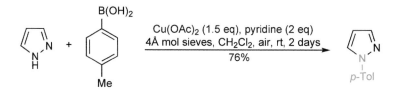

The reaction mechanism proceeds *via* a transmetallation giving an aryl-copper-nitrogen species, followed by a reductive elimination, but the difference from the aryl halide reaction is that this generates a species (Cu(0)) that cannot enter into a catalytic cycle. However, methods that are catalytic in copper have been developed, using an oxidant to regenerate Cu(II), although here there is the possibility for a variation in mechanism, involving Cu(III).[323]

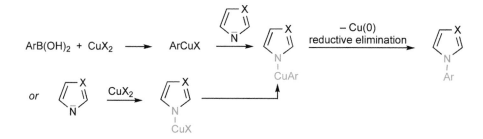

The use of TMEDA as ligand gives the highest yield in the catalytic reaction, using air as re-oxidant for copper. It also gives the highest regioselectivity in the reaction with 4(5)-phenyl imidazole.[323] (Note that even in the stoichiometric reaction above, the presence of air is considered to be beneficial.)

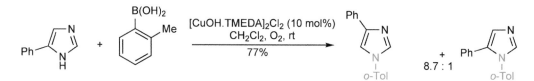

Arylation of the amidic nitrogen in oxy-purine and oxy-pyrimidine nucleosides[302] is consistently successful. Arylation on exocyclic amino groups, such as in deoxyadenosine, is also possible, but less reliable.[324]

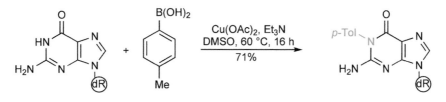

Similar conditions are used for the arylation, with suitable protection of other positions, at N-1 of uracil and cytosine derivatives and at N-9 of purines.[325]

An analogous reaction using alkenyl boronic acids is one of the best processes for the *N*-alkenylation of pyridones and amides.[326] Indoles can be *N*-cyclopropylated using cyclopropylboronic acid/cupric acetate.[327]

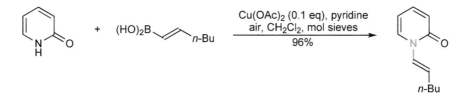

4.2.10.5   Other Variations[327]

As is the case for cross-coupling reactions, arylstannanes[328] and aryltrialkoxysilanes[329] can be substituted for boronic acids in this method, but would appear to offer few advantages. A number of other, usually Cu(II)-catalysed, reagents can be used to arylate azoles and indoles, of which diaryliodonium salts are the most useful.[330] Aryllead triacetates[331] and triarylbismuth diacetates [332] may find very occasional use, but *N*-cyclopropylation using tricyclopropylbismuth with cupric acetate is possibly more interesting.[333]

4.2.10.6   O-Arylation

O-Arylations can be carried out under conditions very similar to those for *N*-arylations, using Pd or Cu catalysts, examples being displacement of 5-bromo in *N*-protected indoles by phenoxides,[334] and reactions of 2-chloro- and 2-bromo-pyridines with 2-aryl-ethanols.[335]

Copper powder can also be used to catalyse the reaction of activated halides (in pyridine, quinoline, pyrimidine, benzothiazole) with phenols. Microwave heating is far superior to conventional heating for chlorides and bromides, but there is little difference for iodide.[336]

Highly selective CuI-catalysed *O*- or *N*-arylation of aminoalcohols can be carried out by choice of ligand: N using 3,4,7,8-tetramethylphenantholine; O using 2-isobutyrylcyclohexanone.[337]

4.2.11   Heck Reactions[338]

$$ArBr + CH{=}CHR \rightarrow ArCH{=}CHR$$

A standard Heck reaction, as shown in the example below,[339] involves the palladium-catalysed reaction of a halide with an alkene, most commonly an electron-deficient alkene such as an acrylate, but other types can also be used. Heck-type cyclisation onto olefins is a useful reaction for ring synthesis.

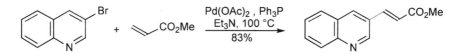

The sequence involves an initial oxidative addition of palladium(0) to the halide, followed by in insertion of the arylpalladium halide into the double bond and finally a β-hydride elimination. Both of the latter two reactions are concerted-*syn*, therefore a rotation around the carbon–carbon single bond must precede the hydride elimination.

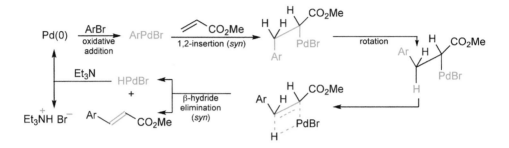

Electron-deficient alkenes, such as acrylates, show a strong regiochemical bias, with the aryl group becoming attached to the β-position. Terminal, unactivated alkenes tend to substitute on the terminal carbon and the regiochemistry of reactions with enol ethers is controllable. The α-substitution of enol ethers is a useful means, following hydrolysis, of introducing acyl groups.[340]

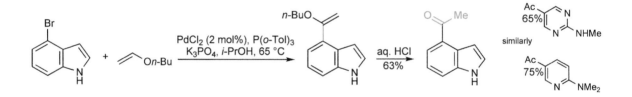

The electron-rich nature of heterocycles such as indoles, furans and thiophenes allows a different type of Heck reaction to be carried out.[341] In this 'oxidative' modification, the aryl palladium derivative is generated by electrophilic palladation with a palladium(II) reagent.

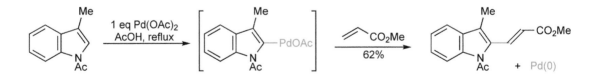

This process is not catalytic in the standard way, as the Pd(0), generated in the final hydride elimination, cannot effect the first (electrophilic) ring palladation. However, the addition of an oxidant selective for Pd(0), such as a peroxide or Cu(II), allows the cycle to continue by re-formation of the reactive Pd(II) species.[342]

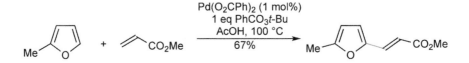

In indoles, the reaction generally occurs at C-3, unless that is blocked, as in the example above. However, the position of substitution can be influenced by choice of solvent and possibly, to a lesser extent, the re-oxidant. Equilibration of the intermediate palladated indole is probably responsible for the variations in regioselectivity.[343]

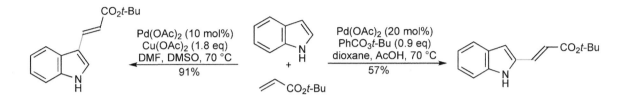

The palladation, and therefore substitution, can also be directed to C-2 in indoles by the use of a chelating 2-pyridylmethyl group on the nitrogen.[344] Direction by a carboxy group (*via* the Pd carboxylate) is generally useful in indole, furan, thiophene and pyrrole, although it is not always completely selective. The carboxyl group is lost during the reaction.[345]

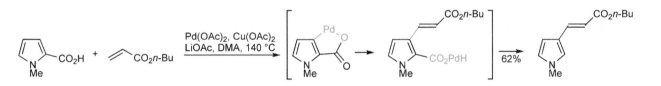

4.2.12 Carbonylation Reactions

$$ArX + CO + Nu(H) \rightarrow ArCONu$$

Palladium-catalysed carbonylation of halides, with carbon monoxide, can be used to prepare esters, amides and ketones by trapping the intermediate acylpalladium halide with alcohols,[346] amines[347] and organometallics, respectively. Boronic acids are probably the best organometallics for the preparation of ketones, but conditions must be adjusted to give the best selectivity between the acylation reaction and simple Suzuki coupling of the boronic acid with the starting halide.[348]

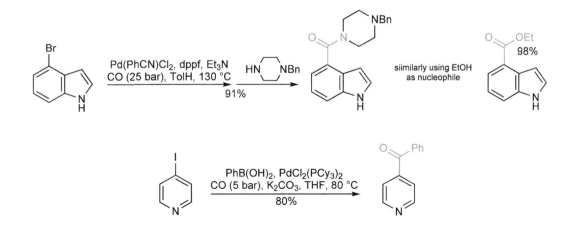

References

[1] 'Handbook of Functionalized Organometallics', Knochel, P. (Ed.), Wiley-VCH, **2005**.

[2] 'Preparative Polar Organometallic Chemistry', Brandsma, L. and Verkruijsse, H. D., Vol. 1, Springer-Verlag, **1987**; 'The Chemistry of Organolithium Compounds', Wakefield, B. J., Pergamon Press, **1974**; 'Generation and reactions of sp²-carbanionic centers in the vicinity of heterocyclic nitrogen atoms', Rewcastle, G. W. and Katritzky, A. R., *Adv. Heterocycl. Chem.*, **1993**, *56*, 155; 'Ring and Lateral Metallation of Heteroaromatic Substrates using Strong Base Systems', FMC Lithium Link, Spring **1993**, FMC Lithium Division, 449, North Cox Road, Gastonia, NC 28054, USA.

[3] von Ragué Schleyer, P., Chandrasekhar, J., Kos, A. J., Clark, T. and Spitznagel, G. W., *J. Chem. Soc., Chem. Commun.*, **1981**, 882.

[4] 'Base catalysed hydrogen exchange', Elvidge, J. A., Jones, J. R., O'Brien, C., Evans, E. A. and Sheppard, H. C., *Adv. Heterocycl. Chem.*, **1974**, *16*, 1.

[5] Zoltewicz, J. A., Grahe, G. and Smith, C. L., *J. Am. Chem. Soc.*, **1969**, *91*, 5501; Meot-Ner (Mautner), M and Kafafi, S. A., *ibid.*, **1988**, *110*, 6297.

[6] 'Heteroatom-facilitated lithiations', Gschwend, H. W. and Rodriguez, H. R., *Org. React.*, **1979**, *26*, 1; 'Directed metallation of π-deficient azaaromatics: strategies of functionalisation of pyridines, quinolines and diazines', Quéguiner, G., Marsais, F., Snieckus, V. and Epsztajn, J., *Adv. Heterocycl. Chem.*, **1991**, *52*, 187; 'Metallation and metal-assisted bond formation in π-electron deficient heterocycles', Undheim, K. and Benneche, T., *Heterocycles*, **1990**, *30*, 1155; 'Directed *ortho* metallation. Tertiary amide and *O*-carbamate directors in synthetic strategies for polysubstituted aromatics', Snieckus, V., *Chem. Rev.*, **1990**, *90*, 879.

[7] Mallet, M., *J. Organomet. Chem.*, **1991**, *406*, 49.

[8] Gómez, I., Alonso, E., Ramón, D. J. and Yus, M., *Tetrahedron*, **2000**, *56*, 4043.

[9] Reich, H. J., Phillips, N. H. and Reich, I. L., *J. Am. Chem. Soc.*, **1985**, *107*, 4101.

[10] Saulnier, M. G. and Gribble, G. W., *J. Org. Chem.*, **1982**, *47*, 757.

[11] Fraser, R. R., Mansour, T. S. and Savard, S., *Can. J. Chem.*, **1985**, *63*, 3505.

[12] von Ragué Schleyer, P., Clark, T., Kos, A. J., Spitznagel, G. W., Rohde, C., Arad, D., Houk, K. N. and Rondan, N. G., *J. Am. Chem. Soc.*, **1984**, *106*, 6467.

[13] Carpita, A., Rossi, R. and Veracini, C. A., *Tetrahedron*, **1985**, *41*, 1919.

[14] Feringa, B.L., Hulst, R., Rikers, R. and Brandsma, L., *Synthesis*, **1988**, 316.

[15] Carpenter, A. J. and Chadwick, D. J., *Tetrahedron Lett.*, **1985**, *26*, 1777.

[16] Gronowitz, S. and Kada, R., *J. Heterocycl. Chem.*, **1984**, *21*, 1041.

[17] Hasan, I., Marinelli, E. R., Lin, L.-C. C., Fowler, F. W. and Levy, A. B., *J. Org, Chem.*, **1981**, *46*, 157.

[18] e.g. for benzo[*b*]thiophen and benzo[*b*]furan: Kerdesky, F. A. J. and Basha, A., *Tetrahedron Lett.*, **1991**, *32*, 2003.

[19] Moyer, M. P., Shiurba, J. F. and Rapoport, H., *J. Org, Chem.*, **1986**, *51*, 5106.

[20] 'Metallation and metal–halogen exchange reactions of imidazoles', Iddon, B., *Heterocycles*, **1985**, *23*, 417.

[21] See, however, Katritzky, A. R., Slawinski, J. J., Brunner, F. and Gorun, S., *J. Chem. Soc., Perkin Trans. 1*, **1989**, 1139.

[22] Lipshutz, B. H., Huff, B. and Hagen, W., *Tetrahedron Lett.*, **1988**, *29*, 3411.

[23] Katritzky, A. R., Rewcastle, G. W. and Fan, W.-Q., *J. Org. Chem.*, **1988**, *53*, 5685; Heinisch, G., Holzer, W. and Pock S., *J. Chem. Soc., Perkin Trans. 1*, **1990**, 1829.

[24] 'Ring-opening of five-membered heteroaromatic anions', Gilchrist, T. L., *Adv. Heterocycl. Chem.*, **1987**, *41*, 41.

[25] Dickinson, R. P. and Iddon, B., *J. Chem. Soc. (C)*, **1971**, 3447.

[26] Verbeek, J., George, A. V. E., de Jong, R. L. P. and Brandsma, L., *J. Chem. Soc., Chem. Commun.*, **1984**, 257; Verbeek, J. and Brandsma, L., *J. Org. Chem.*, **1984**, *49*, 3857.

[27] Kessar, S. V., Singh, P., Singh, K. N. and Dutt, M., *J. Chem. Soc., Chem. Commun.*, **1991**, 570.

[28] Zoltewicz, J. A. and Helmick, L. S., *J. Am. Chem. Soc.*, **1970**, *92*, 7547; Zoltewicz, J. A. and Cantwell, V. W., *J. Org. Chem.*, **1973**, *38*, 829; Zoltewicz, J. A. and Sale, A. A., *J. Am. Chem. Soc.*, **1973**, *95*, 3928.

[29] Mallet, M., *J. Organomet. Chem.*, **1991**, *406*, 49.

[30] Miah, M. A. J. and Snieckus, V., *J. Org. Chem.*, **1985**, *50*, 5436.

[31] Gribble, G. W. and Saulnier, M. G., *Tetrahedron Lett.*, **1980**, *21*, 4137.

[32] Ronald, R. C. and Winkle, M. R., *Tetrahedron*, **1983**, *39*, 2031.

[33] Estel, L., Linard, F., Marsais, F., Godard, A. and Quéguiner, G., *J. Heterocycl. Chem.*, **1989**, *26*, 105.

[34] Sengupta, S. and Snieckus, V., *Tetrahedron Lett.*, **1990**, *31*, 4267.

[35] Alo, B. I., Familoni, O. B., Marsais, F. and Quéguiner, G., *J. Heterocycl. Chem.*, **1992**, *29*, 61.

[36] Beaulieu, F. and Snieckus, V., *Synthesis*, **1992**, 112; Tsukazaki, M. and Snieckus, V., *Heterocycles*, **1992**, *33*, 533.

[37] Kelly, T. R., Xu, W. and Sundaresan, J., *Tetrahedron Lett.*, **1993**, *34*, 6173.

[38] Marsais, F., Le Nard, G. and Quéguiner, G., *Synthesis*, **1982**, 235.

[39] Godard, A., Jaquelin, J.-M. and Quéguiner, G., *J. Organomet. Chem.*, **1988**, *354*, 273.

[40] Kress, T. J., *J. Org. Chem.*, **1979**, *44*, 2081; Frissen, A. E., Marcelis, A. T. M., Buurman, D. G., Pollmann, C. A. M. and van der Plas, H. C., *Tetrahedron*, **1989**, *45*, 5611; Turck, A., Plé, N., Majovic, L. and Quéguiner, G., *J. Heterocycl. Chem.*, **1990**, *27*, 1377.

[41] Turk, A., Mojovic, L. and Quéguiner, G., *Synthesis*, **1988**, 881; Ward, J. S. and Merritt, L., *J. Heterocycl. Chem.*, **1991**, *28*, 765.

[42] Li, W., Nelson, D. P., Jensen, M. S., Hoerrner, R. S., Cai, D., Larsen, R. D. and Reider, P. J., *J. Org. Chem.*, **2002**, *67*, 5394.

[43] Vazquez, E., Davies, I. W. and Payack, J. F., *J. Org. Chem.*, **2002**, *67*, 7551.

[44] Herbert, J. M and Maggiani, M., *Synth. Commun.*, **2001**, 947.

[45] Hahn, M., Heinisch, G., Holzer, W. and Schwarz, H., *J. Heterocycl. Chem.*, **1991**, *28*, 1189.

[46] Kopp, F. and Knochel, P., *Org. Lett.*, **2007**, *9*, 1639.

[47] Comins, D. L., Nolan, J. M. and Bori, I. D., *Tetrahedron Lett.*, **2005**, *36*, 6697; see also 'The synthetic utility of α-amino alkoxides'; Comins, D. L., *Synlett*, **1992**, 615.

[48] Bergman, J. and Venemalm, L., *J. Org. Chem.*, **1992**, *57*, 2495.

[49] 'Functionalised magnesium organometallics as versatile intermediates for the synthesis of polyfunctionalised heterocycles', Ila, H., Baron, A., Wagner, A. J. and Knochel, P., *Chem. Commun.*, **2006**, 583.

[50] Christophersen, C., Begtrup, M., Ebdrup, S., Petersen, H. and Vedsø, P., *J. Org. Chem.*, **2003**, *68*, 9513.

[51] Abarbi, M., Thibonnet, J., Bérillon, L., Dehmel, F., Rottländer, M. and Knochel, P., *J. Org. Chem.*, **2000**, *65*, 4618.

[52] Smith, A. B., Minbiole, K. P., Verhoest, P. R. and Schelhaas, M., *J. Am. Chem. Soc.*, **2001**, *123*, 10942.

[53] Krasovskiy, A. and Knochel, P., *Angew. Chem. Int. Ed.*, **2004**, *43*, 3333.

[54] Krasovskiy, A, Kopp, F. and Knochel, P., *Angew. Chem. Int. Ed.*, **2006**, *45*, 497.

[55] Krasovskiy, A., Krasovszkaya, V. and Knochel, P.. *Angew. Chem. Int. Ed.*. **2006**, *45*, 2958.

[56] Clososki, G. C., Rohbogner, C. J. and Knochel, P., *Angew. Chem. Int. Ed.*, **2007**, *46*, 7681.

[57] Bayh, O., Awad, H., Mongin, F., Hoarau, C., Bischoff, L., Trécourt, F., Quéguiner, G., Marsais, F., Blanco, F., Abarca, B. and Ballesteros, R., *J. Org. Chem.*, **2005**, *70*, 5190.

[58] Kneisel, F. F. and Knochel, P., *Synlett*, **2002**, 1799.

[59] Kopp, F., Wunderlich, S. and Knochel, P., *Chem. Commun.*, **2007**, 2075.

[60] Kopp, F. and Knochel, P., *Synlett*, **2007**, 980.

[61] 'Preparations and applications of functionalized organozinc compounds', Knochel, P., Millot, N., Rodriguez, A. and Tucker, C. E., *Org. React.*, **2001**, *58*, 417; 'The Chemistry of Organozinc Compounds', Ed. Rappoport, Z. and Marek, I. John Wiley & Sons, **2006**.

[62] Sakamoto, T., Kondo, Y., Takazawa, N. and Yamanaka, H., *Tetrahedron Lett.*, **1993**, *34*, 5955; Sakamoto, T., Kondo, Y., Murata, N. and Yamanaka, H., *Tetrahedron*, **1993**, *49*, 9713.

[63] Prasad, A. S. B., Stevenson, T. M., Citineni, J. R., Nyzam, V. and Knochel, P., *Tetrahedron*, **1997**, *53*, 7237.

[64] Gosmini, C., Amatore, M., Claudel, S. and Périchon, J., *Synlett*, **2005**, 2171.

[65] Prasad, A. S. B., Stevenson, T. M., Citineni, J. R., Nyzam, V. and Knochel, P., *Tetrahedron*, **1997**, *53*, 7237.

[66] Krasovskiy, A., Malakhov, V., Gavryushin, A. and Knochel, P., *Angew. Chem. Int. Ed.*, **2006**, *45*, 6040.

[67] Chen, Y.-H. and Knochel, P., *Angew. Chem. Int. Ed.*, **2008**, *47*, 7648.

[68] Gong, L.-Z. and Knochel, P., *Synlett*, **2005**, 267.

[69] Wunderlich, S. H. and Knochel, P., *Angew. Chem. Int. Ed.*, **2007**, *46*, 7685.

[70] Wunderlich, S. and Knochel, P., *Org. Lett.*, **2008**, *10*, 4705.

[71] Korn, T. J., Schade, M. A., Cheemala, M. N., Wirth, S., Guevara, S. A., Cahiez, G. and Knochel, P., *Synthesis*, **2006**, 3547.

[72] Yang, X. and Knochel, P., *Synthesis*, **2006**, 2618.

[73] 'Boronic Acids', Ed. Hall, D. G., Wiley-VCH, **2005**; 'The Synthesis and applications of heterocyclic boronic acids', Tyrrell, E. and Brookes, P., *Synthesis*, **2003**, 469.

[74] Cammidge, A. N., Goddard, V. H. M., Gopee, H., Harrison, N. L., Hughes, D. L., Schubert, C. J., Sutton, B. M., Watts, G. L. and Whitehead, A. J., *Org. Lett.*, **2006**, *8*, 4071.

[75] Florentin, D., Fournié-Zaluski, M. C., Callanquin, M. and Roques, B. P., *J. Heterocycl. Chem.*, **1976**, *13*, 1265.

[76] Fischer, F.C. and Havinga, E., *Recl. Trav. Chim. Pays-Bas*, **1974**, *93*, 21.

[77] Suenaga, H., Nakashima, K. and Sinkai, S., *J. Chem. Soc., Chem. Commun.*, **1995**, 29.

[78] Florentin, D., Fournié-Zaluski, M. C., Callanquin, M. and Roques, B. P., *J. Heterocycl. Chem.*, **1976**, *13*, 1265; Brandão, M. A., de Oliveira, A. B. and Snieckus, V., *Tetrahedron Lett.*, **1993**, *34*, 2437.

[79] Murata, M., Oyama, T., Watanabe, S. and Masuda, Y., *J. Org. Chem.*, **2000**, *65*, 164.

[80] Ishiyama, T. and Miyaura, N., *Pure Appl. Chem.*, **2006**, *78*, 1369; Ishiyama, T., Takagi, J., Nobuta, Y. and Miyaura, N., *Org. Synth.*, **2005**, *82*, 126.

[81] Liversedge, I. A., Higgins, S. J., Giles, M., Heeney, M. and McCulloch, I., *Tetrahedron Lett.*, **2006**, *47*, 5143.

[82] Song, Z. Z., Zhan, Z. Y., Mak, T. C. W. and Wong, H. N. C., *Angew. Chem., Int. Ed. Engl.*, **1993**, *32*, 432; Zheng, Q., Yang, Y. and Martin, A. R., *Tetrahedron Lett.*, **1993**, *34*, 2235.

[83] Chang, Y. M., Lee, S. H., Nam, M. H., Cho, M. Y., Park, Y. S. and Yoon, C. M., *Tetrahedron Lett.*, **2005**, *46*, 3053.

[84] Jiang, B., Yang, C.-G. and Gu, X.-H., *Tetrahedron Lett.*, **2001**, *42*, 2545.

[85] Lee, S. and MacMillan, D. W. C., *J. Am. Chem. Soc.*, **2007**, *129*, 15438.

[86] Medina, J. R., Henry, T. A. and Axten, J. M., *Tetrahedron Lett.*, **2006**, *47*, 7363.

[87] Vazquez, E. and Payack, J. F., *Tetrahedron Lett.*, **2004**, *45*, 6549.

[88] 'Silicon Reagents in Organic Synthesis', Colvin, E., Academic Press, **1988**.

[89] 'Tin in Organic Synthesis', Pereyre, M., Quintard, J.-P. and Rahm, A., Butterworths, **1987**; 'Organotin Chemistry', 2nd Edn., Davies, A. G., Wiley InterScience, **2004**; 'Tin Chemistry: Fundamentals, Frontiers and Applications', Gielen, M., Davies, A. G., Pannell, K. and Tiekink, E. (Eds), Wiley InterScience, **2008**.

[90] Majeed, A. J., Antonsen, Ø., Benneche, T. and Undheim, K., *Tetrahedron*, **1989**, *45*, 993.

[91] Shippey, M. A. and Dervan, P. B., *J. Org. Chem.*, **1977**, *42*, 2654; Babin, P., Bennetau, B., Theurig, M. and Dunogues, J., *J. Organometal. Chem.*, **1993**, *446*, 135.

[92] McNeill, E., Barder, T. E. and Buchwald, S. L., *Org. Lett.*, **2007**, *9*, 3785.

[93] Denmark, S. E. and Kallemeyn, J. M., *Org. Lett.*, **2003**, *5*, 3483.

[94] Murata, M., Ishikura, M., Nagata, M., Watanabe, S. and Masuda, Y., *Org. Lett.*, **2002**, *4*, 1843.

[95] Yamamoto, Y. and Yanagi, A., *Chem. Pharm. Bull.*, **1982**, *30*, 1731.

[96] Frick, U. and Simchen, G., *Synthesis*, **1984**, 929.

[97] Fleming, I. and Taddei, M., *Synthesis*, **1985**, 898; Kondo, Y., Uchiyama, D., Sakamaoto, T. and Yamanaka, H., *Tetrahedron Lett.*, **1989**, *30*, 4249.

[98] 'Non-conventional electrophilic aromatic substitutions and related reactions', Hartshorn, S. R., *Chem. Soc. Rev.*, **1974**, *3*, 167; 'Unusual electrophilic substitution in the aromatic series *via* organosilicon intermediates', Bennetau, B. and Dunogues, J., *Synlett*, **1993**, 171; 'Tin for organic synthesis. VI. The new role for organotin reagents in organic synthesis', Neumann, W. P., *J. Organometal. Chem.*, **1992**, *437*, 23; Cooper, M. S., Fairhurst, R. A., Heaney, H., Papageorgiou, G. and Wilkins, R. F., *Tetrahedron*, **1989**, *45*, 1155.

[99] Beierle, J. M., Osimboni, E. B., Metallinos, C., Zhao, Y. and Kelly, T. R., *J. Org. Chem.*, **2003**, *68*, 4970.
[100] 'Organomercury Compounds in Organic Synthesis', Larock, R. C., Springer: Berlin, **1985**.
[101] 'The Organic Compounds of Mercury', Makarova, L. G. and Nesmeyanov, A. N., North-Holland, Amsterdam, **1967**.
[102] Oxazoles: Shvaika, O. P. and Klimisha, G. P., *Chem. Heterocycl. Compd.*, **1966**, *2*, 14; Thiazoles: Gusinskaya, S. L., Telly, V. Y. and Makagonova, T. P., *Chem. Heterocycl. Compd.*, **1970**, *6*, 322.
[103] 'Lateral lithiation reactions promoted by heteroatomic substitutents', Clark, R. D. and Jahinger, A., *Org. React.*, **1995**, *47*, 1.
[104] Fraser, R. R., Mansour, T. S. and Savard, S., *J. Org. Chem.*, **1985**, *50*, 3232.
[105] Naruse, Y., Ito, Y. and Inagaki, S., *J. Org. Chem.*, **1991**, *56*, 2256.
[106] Katritzky, A. R. and Akutagawa, K., *J. Am. Chem. Soc.*, **1986**, *108*, 6808.
[107] Noyce, D. S., Stowe, G. T. and Wong, W., *J. Org. Chem.*, **1974**, *39*, 2301.
[108] 'Recent developments in microwave-assisted transition metal-catalysed C–C and C–N bond-forming reactions', Appukkuttan, P. and Van der Eycken, E., *Eur. J. Org. Chem.*, **2008**, 1133; 'Palladium in Heterocyclic Chemistry: A Guide for the Synthetic Chemist', 2nd Edn., Li, J. J. and Gribble, G. W., Elsevier Science, **2006**; 'Pd-Assisted multicomponent synthesis of heterocycles', Balme, G., Bossharth, E. and Monteiro, N., *Eur. J. Org. Chem.*, **2003**, 4101; 'Handbook of Organopalladium Chemistry', Ed. Negishi, E., Wiley Interscience, **2002**; 'Connection between metallation of azines and diazines and cross-coupling strategies for the synthesis of natural and biologically active molecules', Godard, A., Marsais, F., Plé, N., Trécourt, F., Turck, A. and Quéguiner, G., *Heterocycles*, **1995**, *40*, 1055; 'Carbon–carbon bond formation in heterocycles using Ni- and Pd-catalysed reactions', Kalinin, V. N., *Synthesis*, **1992**, 413; 'Transition metals in the synthesis and functionalisation of indoles', Hegedus, L. S., *Angew. Chem., Int. Ed. Engl.*, **1988**, *27*, 1113.
[109] Stock, L. M., Tse, K., Vorvick, L. J. and Walstrum, S. A., *J. Org. Chem.*, **1981**, *46*, 1757.
[110] 'Cyclopalladated complexes in organic synthesis', Ryabov, A. D., *Synthesis*, **1985**, 233; 'Mechanisms of intramolecular activation of C–H bonds in transition metal complexes', idem, *Chem. Rev.*, **1990**, *90*, 403.
[111] 'Pd-N-Heterocyclic carbene (NHC) catalysts for cross-coupling reactions', Kantchev, E. A. B., O'Brian, V. J. and Organ, M. G., *Aldrichimica Acta*, **2006**, *39*, 97; 'Development and application of highly active and selective palladium catalysts', *Tetrahedron*, **2005**, *61*, Issue 41 (Ed: Fairlamb, I. J. S.); 'Palladium-catalysed coupling reactions of aryl chlorides', Littke, A. F. and Fu, G. C., *Angew. Chem. Int. Ed.*, **2002**, *41*, 4176; 'Cross Coupling'. Aldrich, *ChemFiles*, Vol. 7, No. 10.
[112] Tagata, T. and Nishida, M., *J. Org. Chem.*, **2003**, *68*, 9412; see also Felpin, F-X, Ayad, T. and Mitra, S., *Eur. J. Org. Chem.*, **2006**, 2679.
[113] Kitamura, Y., Sako, S., Udzu, T., Tsutsui, A., Maegawa, T., Monguchi, Y. and Sajiki, H., *Chem. Commun.*, **2007**, 5069.
[114] For a discussion see Amatore, C. and Jutand, A., *Acc. Chem. Res.*, **2000**, *33*, 314.
[115] 'The promise and challenge of iron-catalysed cross coupling', Sherry, B. D. and Fürstner, A., *Acc. Chem. Res.*, **2008**, *41*, 1500.
[116] Fürstner, A., Leitner, A., Méndez, M. and Krause, H., *J. Am. Chem. Soc.*, **2002**, *124*, 13856.
[117] Scheiper, B., Glorius, F., Leitner, A. and Fürstner, A., *Proc. Natl. Acad. Sci. USA*, **2004**, *101*, 11960.
[118] Kylmälä, T., Valkonen, A., Rissanen, K., Xu, Y. and Franzén, R., *Tetrahedron Lett.*, **2008**, *49*, 6679.
[119] Rueping, M. and Ieawsuwan, W., *Synlett*, **2007**, 247.
[120] Ohmiya, H., Yorimitsu, H. and Oshima, K., *Chem. Lett.*, **2004**, *33*, 1240.
[121] Kom, T., Schade, M., Schade, S. and Knochel, P., *Org. Lett.*, **2006**, *8*, 725.
[122] Klapars, A. and Buchwald, S. L., *J. Am. Chem. Soc.*, **2002**, *124*, 14844.
[123] Rossi, E., Abbiati, G., Canevari, V., Celentano, G. and Magri, E., *Synthesis*, **2006**, 299.
[124] Benoît, J., Malapel, B. and Mérour, J-Y., *Synth. Commun.*, **1996**, *26*, 3289.
[125] Gribble, G. and Conway S. C., *Synth. Commun.*, **1992**, *22*, 2129.
[126] Mongin, F., Mojovic, L., Guillamet, B., Trécourt, F. and Quéguiner, G., *J. Org. Chem.*, **2002**, *67*, 8991.
[127] 'Regioselective cross-coupling reactions of multiply halogenated nitrogen-, oxygen- and sulfur-containing heterocycles', Schröter, S., Stock, C. and Bach, T., *Tetrahedron*, **2005**, *61*, 224.
[128] Handy, S. T. and Zhang, Y., *Chem. Commun.*, **2006**, 299.
[129] Angioletti, M. E., Casalnuovo, A. L. and Selby, T,.P., *Synlett* **2000**, 905.
[130] Leconte, N., Keromnes-Wuillaume, A., Suzenet, F. and Guillaumet, G., *Synlett*, **2007**, 204.
[131] Alphonse, F.-A., Suzenet, F., Keromnes, A., Lebret, B. and Guillaumet, G., *Synlett*, **2002**, 447.
[132] Egi, M. and Liebeskind, L. S., *Org. Lett.*, **2003**, *5*, 801.
[133] Prokopcová, H. and Kappe, C. O., *J. Org. Chem.*, **2007**, *72*, 4440.
[134] Silva, S., Sylla, B., Suzenet, F., Tatibouët, A., Rauter, A. P. and Rollin, P., *Org. Lett.*, **2008**, *10*, 853.
[135] Zhang, S., Marshall, D. and Liebeskind, L. S., *J. Org. Chem.*, **1999**, *64*, 2796.
[136] Wu, J. and Yang, Z., *J. Org. Chem.*, **2001**, *66*, 7875.
[137] Schwier, T., Sromek, A. W., Yap, D. M. L., Chernyak, D. and Gevorgyan, V., *J. Am. Chem. Soc.*, **2007**, *129*, 9868.
[138] Denmark, S. E. and Sweis, R. F., *Org. Lett.*, **2002**, *4*, 3771.
[139] Legros, J.-Y., Primault, G., Toffano, M., Rivière, M.-A. and Fiaud, J.-C., *Org. Lett.*, **2000**, *2*, 433.
[140] 'Cross coupling', *Acc. Chem. Res.* **2008**, *41*, Special issue (11), 1439–1564; 'Metal-Catalyzed Cross-Coupling Reactions', 2nd Edn, de Meijere, A. and Diederich, F., Eds., Wiley-VCH, Weinheim, **2004**; 'Catalytic cross-coupling reactions in biaryl synthesis', Stanforth, S. P., *Tetrahedron*, **1998**, *54*, 263; 'Organometallics in coupling reactions in π-deficient azaheterocycles', Undheim, K. and Benneche, T., *Adv. Heterocycl. Chem.*, **1995**, *62*, 305.
[141] Lee, K. and Lee, P. H., *Tetrahedron Lett.*, **2008**, *49*, 4302.
[142] Baba, S. and Negishi, E., *J. Am. Chem. Soc.*, **1976**, *98*, 6729.
[143] Negishi, E. and Van Horn, D. E., *J. Am. Chem. Soc.*, **1977**, *99*, 3168.
[144] Jaber, N., Schumann, H. and Blum, J., *J. Heterocycl. Chem.*, **2003**, *40*, 565.
[145] 'Recent advances in organotrifluoroborates chemistry', Stefani, H. A., Cella, R. and Vieira, A. S., *Tetrahedron*, **2007**, *63*, 3623.
[146] Miyaura, N., Yanagi, T. and Suzuki, A., *Synth. Commun.*, **1981**, *11*, 513.
[147] Gronowitz, S. and Lawitz, K., *Chem. Scripta*, **1984**, *23*, 120; Gronowitz, S., Bobosik, K. and Lawitz, K. *Chem. Scripta*, **1984**, *24*, 5.

[148] Thompson, W. J. and Gaudino, J., *J. Org. Chem.*, **1984**, *49*, 5237.

[149] Billingsley, K. and Buchwald, S. L., *J. Am. Chem. Soc.*, **2007**, *129*, 3385.

[150] Molander, G. A. and Figuero, R., *Aldrichim. Acta*, **2005**, *38*, 49.

[151] Molander, G. A. and Bernardi, C. R., *J. Org. Chem.* **2002**, *67*, 8424.

[152] Cañeque, T., Cuadro, A.. M., Alvarez-Builla, J. and Vaquero, J. J., *Tetrahedron Lett.*, **2009**, *50*, 1419.

[153] Savall, B. M. and Fontimayor, J. R., *Tetrahedron Lett.*, **2008**, *49*, 6667.

[154] Caron, S., Massett, S. S., Bogle, D. E., Castaldi, M. J. and Braish, T. F., *Org. Proc. Res. Dev.*, **2001**, *5*, 254.

[155] Itoh, T. and Mase, T., *Tetrahedron Lett.*, **2005**, *46*, 3573.

[156] 'The mechanism of the Stille reaction', Espinet, P. and Echavarren, A. M., *Angew. Chem. Int. Ed.*, **2004**, *43*, 4704.

[157] Mee, S. P., Lee, V. and Baldwin, J. E., *Angew. Chem. Int. Ed.*, **2004**, *43*, 1132.

[158] A search of the internet will find many references to the toxicity of tin compounds.

[159] Kuhn, H. and Neumann, W. P., *Synlett*, **1994**, 123.

[160] Roschin, A. I., Bumagin, N. A. and Beletskaya, I. P., *Tetrahedron Lett.*, **1995**, *36*, 125; Rai, R., Aubrecht, K. B. and Collum, D. B., *Tetrahedron Lett.*, **1995**, *36*, 3111.

[161] Fouquet, E. and Rodriguez, A. L., *Synlett*, **1998**, 1323; Fouquet, E., Pereyre, M. and Rodriguez, A. L., *J. Org. Chem.*, **1997**, *62*, 5242.

[162] Huiban, M., Huet, A., Barré, L., Sobrio, F., Fouquet, E. and Perrio, C., *Chem. Commun.*, **2006**, 97.

[163] 'Palladium-catalysed cross-coupling reactions of organosilanols and their salts: practical alternatives to boron- and tin-based methods', Denmark, S. E. and Regens, C. S., *Acc. Chem. Res.*, **2008**, *41*, 1486.

[164] Denmark, S. E., Baird, J. D. and Regens, C. S., *J. Org. Chem.*, **2008**, *73*, 1440.

[165] Denmark, S. E. and Kallemeyn, J. M., *J. Org. Chem.*, **2005**, *70*, 2839.

[166] Napier, S., Marcuccio, S. M., Tye, H. and Whittaker, M., *Tetrahedron Lett.*, **2008**, *49*, 3939.

[167] Mowery, M. E. and De Shong, P., *J. Org. Chem.*, **1999**, *64*, 1684.

[168] Lam, P. Y. S., Deudon, S., Averill, K. M., Li, R., He, M. Y., DeShong, P. and Clark, C. G., *J. Am. Chem. Soc.*, **2000**, *122*, 7600.

[169] Napier, S., Marcuccio, S. M., Tye, H. and Whittaker, M., *Tetrahedron Lett.*, **2008**, *49*, 6314.

[170] Pierrat, P., Gros, P. and Fort, Y., *Org. Lett.*, **2005**, *7*, 697.

[171] Kumada, M., Tamao, K. and Sumitani, K., *Org. Synth.*, **1988**, *Coll. Vol. 6*, 407.

[172] Bonnet, V., Mongin, F., Trécourt, F., Quéguiner, G. and Knochel, P., *Tetrahedron*, **2002**, *58*, 4429.

[173] Dumouchel, S., Mongin, F. Trécourt, F. and Quéguiner, G., *Tetrahedron Lett.*, **2003**, *44*, 3877.

[174] 'Palladium-catalyzed alkenylation by the Negishi coupling', Negishi, E., Hu, Q., Huang, Z., Qian, M. and Wang, G., *Aldrichimica Acta*, **2005**, *38*, 71.

[175] Jackson, R. F. W. and Perez-Gonzalez, M., *Org. Synth.*, **2005**, *81*, 77.

[176] Sakamoto, T., Kondo, Y., Watanabe, R. and Yamanaka, H., *Chem. Pharm. Bull.*, **1986**, *34*, 2719.

[177] 'Palladium-catalysed alkynylation', Negishi, E. and Anastasia, L., *Chem. Rev.*, **2003**, *103*, 1979.

[178] Molander, G. A., Katona, B. W. and Machrouhi, F., *J. Org. Chem.*, **2002**, *67*, 8416.

[179] Yang, X., Zhu, L., Zhou, Y., Li, Z. and Zhai, H., *Synthesis*, **2008**, 1729.

[180] 'Suzuki-Miyaura cross-coupling reactions of alkylboronic acid derivatives or alkyltrifluoroborates with aryl, alkenyl or alkyl halides and triflates', Doucet, H., *Eur. J. Org. Chem.*, **2008**, 2013.

[181] Iglesias, B., Alvarez, R. and de Lera, A. R., *Tetrahedron*, **2001**, *57*, 3125.

[182] Molander, G. A. and Ludivine, J.-G., *J. Org. Chem.*, **2007**, *72*, 8422.

[183] Martin, M. T., Liu, B., Cooley, B. E. and Eaddy, J. F., *Tetrahedron Lett.*, **2007**, *48*, 2555.

[184] Schareina, T., Zapf, A. and Beller, M., *J. Organomet. Chem.*, **2004**, *689*, 4576.

[185] Schareina, T., Zapf, A. and Beller, M. J., *Tetrahedron Lett.*, **2005**, *46*, 2585.

[186] Yang, C. and Williams, J. M., *Org. Lett.*, **2004**, *6*, 2837.

[187] Sakamoto, T., Katoh, E., Kondo, Y. and Yamanaka, H., *Heterocycles*, **1988**, *27*, 1353.

[188] Navarro, O., Marlon, N., Scott, N. M., González, J., Amoroso, D., Bell, A. and Nolan, S. P., *Tetrahedron*, **2005**, *61*, 9716.

[189] Kawatsura, M. and Hartwig, J. F., *J. Am. Chem. Soc.*, **1999**, *121*, 1473.

[190] Bessmertnykh, A., Douaihy, C. M., Muniappan, S. and Guilard, R., *Synthesis*, **2008**, 1575 and references therein.

[191] 'Palladium-catalysed cross-coupling reactions involving pyrroles' Banwell, M. G., Goodwin, T. E., Ng, S., Smith, J. A. and Wong, D. J., *Eur. J. Org. Chem.*, **2006**, 3043.

[192] Johnson, C. N., Stemp, G., Anand, N., Stephen, S. C. and Gallagher, T., *Synlett*, **1998**, 1025.

[193] Gronowitz, S. and Peters, D., *Heterocycles*, **1990**, *30*, 645.

[194] Allred, G. D. and Liebskind, L. S., *J. Am. Chem. Soc.*, **1996**, *118*, 2748.

[195] Li, G., Wang, X., Li, J., Zhao, X. and Wang, F., *Tetrahedron*, **2006**, *62*, 2576.

[196] Dandu, R., Tao, M., Josef, K. A., Bacon, E. R. and Hudkins, R. L., *J. Heterocycl. Chem.*, **2007**, *44*, 437.

[197] Hudkins, R. L., Diebold, J. L. and Marsh, F. D., *J. Org. Chem.*, **1995**, *60*, 6218.

[198] Nishida, A., Miyashita, N., Fuwa, M. and Nakagawa, M., *Heterocycles*, **2003**, *59*, 473.

[199] Jiang, B. and Yang, C., *Heterocycles*, **2000**, *53*, 1489.

[200] Ciattini, P. G., Morera, E. and Ortar, G., *Tetrahedron Lett.*, **1994**, *35*, 2405.

[201] Herz, H. G., *Synthesis*, **1999**, 1013.

[202] Cherry, K., Lebegue, N., Leclerc, V., Carato, P., Yous, S. and Berthelot, P., *Tetrahedron Lett.*, **2007**, *48*, 5751.

[203] Baron, O. and Knochel, P., *Angew. Chem. Int. Ed.*, **2005**, *44*, 3133.

[204] Thompson, W.J., Jones, J.H., Lyle, P.A. and Thies, J.E. *J. Org. Chem.* **1988**, *53*, 2052.

[205] Ishiyama, T., Murata, M. and Miyaura, N., *J. Org. Chem.*, **1995**, *60*, 7508.

[206] 'Cross-coupling reactions on azoles with two or more heteroatoms', Schnürch, M. Flasik, R., Khan, A. F., Spina, M., Mihovilovic, M. D. and Stanetty, P., *Eur. J. Org. Chem.* **2006**, 3283.

[207] Jones, N. A., Antoon, J. W., Bowie, A. L., Borak, J. B. and Stevens, E. P,. *J. Heterocycl. Chem.*, **2007**, *44*, 363.
[208] Dondoni, A., Fantin, G., Fogagnolo, M., Medici, A. and Pedrini, P., *Synthesis*, **1987**, 693.
[209] Bailey, T. R., *Tetrahedron Lett.*, **1986**, *27*, 4407.
[210] Cliff, M. D. and Pyne, S. G., *Tetrahedron*, **1996**, *52*, 13703.
[211] Cliff, M. D. and Pyne, S. G., *Synthesis*, **1994**, 681
[212] Primas, N., Mahatsekake, C., Bouillon, A., Lancelot, J-C., Sopkovà-de Oliveira Santos, J., Lohier, J.-F. and Rault, S., *Tetrahedron*, **2008**, *64*, 4596.
[213] Hämmerle, J., Schnürch, M. and Stanetty, P., *Synlett*, **2007**, 2975.
[214] Araki, H., Katoh, T. and Inoue, M., *Synlett*, **2006**, 555.
[215] Bellina, F., Cauteruccio, S. and Rossi, R., *J. Org. Chem.*, **2007**, *72*, 8543.
[216] Flegeau, E. F., Popkin, M. E. and Greaney, M. F., *Org. Lett.*, **2006**, *8*, 2495.
[217] Kelly, T. R. and Lang, F., *J. Org. Chem.*, **1996**, *61*, 4623.
[218] Bach, T. and Heuser, S., *J. Org. Chem.*, **2002**, *67*, 5789.
[219] Kawasaki, I., Katsuma, H., Nakayama, Y., Yamashita, M. and Ohto, S., *Heterocycles*, **1998**, *48*, 1887.
[220] Gérard, A-L., Bouillon, A., Mahatsekake, C., Collot, V. and Rault, S., *Tetrahedron Lett.*, **2006**, *47*, 4665.
[221] Hanamoto, T., Koga, Y., Kido, E., Kawanami, T., Furuno, H. and Inanaga, J., *Chem. Commun.*, **2005**, 2041.
[222] Moore, J. E., Davies, M. W., Goodenough, K. M., Wybrow, R. A. J., York, M., Johnson, C. N. and Harrity, J. P. A., *Tetrahedron*, **2005**, *61*, 6707.
[223] Arnautu, A., Collot, V., Ros, J. C., Alayrac, C., Witulski, B. and Rault, S., *Tetrahedron Lett.*, **2002**, *43*, 2695.
[224] Crestey, F., Collot, V., Stiebing, S. and Rault, S., *Synthesis*, **2006**, 3506.
[225] Christoforou, I. C. and Koutentis, P. A., *Org. Biomol. Chem.*, **2006**, *4*, 3681.
[226] Bookser, B. C., *Tetrahedron Lett.*, **2000**, *41*, 2805.
[227] Felding, J., Uhlmann, P., Kristensen, J., Vedso, P. and Begtrup, M., *Synthesis*, **1998**, 1181.
[228] Yi, K.Y. and Yoo, S., *Tetrahedron Lett.*, **1995**, *36*, 1679.
[229] Hodgson, P. B. and Salingue, F. H., *Tetrahedron Lett.*, **2004**, *45*, 685; Jones, N. A., Antoon, J. W., Bowie, A. L., Borak, J. B.,and Stevens, E. P., *J. Heterocycl. Chem.*, **2007**, *44*, 363.
[230] Benaglia, M., Toyota, S., Woods, C. R. and Siegel, J. S., *Tetrahedron Lett.*, **1997**, *38*, 4737.
[231] Zoltewicz, J. A. and Cruskie, M. P., *J. Org. Chem.*, **1995**, *60*, 3487.
[232] Simkovsky, N. M., Ermann, M., Roberts, S. M., Parry, D. M. and Baxter, A. D., *J. Chem. Soc., Perkin Trans. 1*, **2002**, 1847.
[233] Reddy, E. A., Barange, D. K., Islam, A., Mukkanti, K. and Pal, M., *Tetrahedron*, **2008**, *64*, 7143.
[234] Ford, A., Sinn, E. and Woodward, S., *J. Chem. Soc., Perkin Trans. 1*, **1997**, 927.
[235] Yang, W., Wang, Y. and Corte, J. R., *Org. Lett.*, **2003**, *5*, 3131.
[236] Sicre, C., Alonso-Gómez, J.-L. and Cid, M. M., *Tetrahedron*, **2006**, *62*, 11063.
[237] 'Palladium-catalysed cross-coupling reactions on pyridazine moieties', Nara, S., Martinez, J., Wermuth, C.-G. and Parrot, I. *Synlett*, **2006**, 3185.
[238] Helm. M. D., Moore, J. E., Plant, A. and Harrity, J. P. A., *Angew. Chem. Int. Ed.*, **2005**, *44*, 3889.
[239] Sauer, J., Heldmann, D. K. and Pabst, G,.R., *Eur. J. Org. Chem.*, **1999**, 313.
[240] Sauer, J. and Heldmann, D. K., *Tetrahedron*, **1998**, *54*, 4297.
[241] Heldmann, D. K. and Sauer, J., *Tetrahedron Lett.*, **1997**, *38*, 5791.
[242] Darabantu, M., Boully, L., Turck, A. and Plé, N., *Tetrahedron*, **2005**, *61*, 2897.
[243] Peters, D., Hornfeldt, A.-B. and Gronowitz, S., *J. Heterocycl. Chem.*, **1990**, *27*, 2165.
[244] Peters, D., Hornfeldt, A.-B., Gronowitz, S. and Johansson, N. G., *J. Heterocycl. Chem.* **1991**, *28*, 529.
[245] Deng, X. and Mani, N. S., *Org. Lett.*, **2006**, *8*, 269.
[246] Delia, T. J., Schomaker, J. M. and Kalinda, A. S., *J. Heterocycl. Chem.*, **2006**, *43*, 127.
[247] Pal, M., Batchu, V. R., Swamy, N. K. and Padakanti, S., *Tetrahedron Lett.*, **2006**, *47*, 3923.
[248] Turck, A., Plé, N., Leprêtre-Gaquère, A. and Quéguiner, G., *Heterocycles*, **1998**, *49*, 205.
[249] Nyffenegger, C., Fournet, G. and Joseph, B., *Tetrahedron Lett.*, **2007**, *48*, 5069.
[250] Wu, J., Liao, Y. and Yang, Z., *J. Org. Chem.*, **2001**, *66*, 3642.
[251] Zhang, L., Meng, T., Fan, R. and Wu, J., *J. Org. Chem.*, **2007**, *72*, 7279.
[252] Lopez, O. D., Goodrich, J. T., Yang, F. and Snyder, L. B., *Tetrahedron Lett.*, **2007**, *48*, 2063.
[253] Ryu, K.-M., Gupta, A. K., Han, J. W., Oh, C. H. and Cho, C.-G., *Synlett*, **2004**, 2197.
[254] Kim, W.-S., Kim, H.-J. and Cho, C.-G., *Tetrahedron Lett.*,**2002**, *43*, 9015.
[255] Fairlamb, I. J. S., O'Brien, C. T., Lin, Z. and Lam, K. C., *Org. Biomol. Chem.*, **2006**, *4*, 1213.
[256] 'Purines bearing carbon substituents in positions 2, 6, or 8 by metal or organometal-mediated C-C bond-forming reactions', Hocek, M., *Eur. J. Org. Chem.*, **2003**, 245.
[257] Lakshman, M. K., Gunda, P. and Pradhan, P., *J. Org. Chem.*, **2005**, *70*, 10329.
[258] Nolsøe, J., Gundersen, L.-L. and Rise, F., *Acta. Chem. Scand.*, **1999**, *53*, 366.
[259] Kuznetsov, A. G., Bush, A. A. and Babaev, E. V., *Tetrahedron*, **2008**, *64*, 749.
[260] Bouillon, A., Voisin, A. S., Robic, A., Lancelot, J.-C., Collot, V. and Rault, S., *J. Org. Chem.*, **2003**, *68*, 10178.
[261] Yamamoto, Y., Seko, T. and Nemoto, H., *J. Org. Chem.*, **1989**, *54*, 4734.
[262] Getmanenko, Y. A. and Twieg, R. J., *J. Org Chem.*, **2008**, *73*, 830.
[263] Parry, P. R., Wang, C., Batsanov, A. S., Bryce, M. R. and Tarbit, B., *J. Org. Chem.*, **2002**, *67*, 7541.
[264] Gillis, E. P. and Burke, M. D., *J. Am. Chem. Soc.*, **2007**, *129*, 6716.
[265] Lee, S. J., Gray, K. C., Paek, J. S. and Burke, M. D., *J. Am. Chem. Soc.*, **2008**, *130*, 466; Gillis, E. P. and Burke, M. D., *J. Am. Chem. Soc.*, **2008**, *130*, 14084.
[266] Stanetty, P., Schnürch, M. and Mihovilovic, M. D., *J. Org. Chem.*, **2006**, *71*, 3754.
[267] Akita, Y., Itagaki, Y., Takizawa, S. and Ohta, A., *Chem. Pharm. Bull.*, **1989**, *37*, 1477.
[268] 'Recent advances in intermolecular direct arylation reactions', Campeau, L.-C., Stuart, D. R. and Fagnou, K., *Aldrichimica Acta*, **2007**, *40*, 35.

[269] 'Palladium-catalysed direct arylation of simple arenes in the synthesis of biaryl molecules', Campeau, L.-C. and Fagnou, K., *Chem. Commun.*, **2006**, 1253.

[270] Seregin, I. V., Ryabova, V. and Gevorgyan, V., *J. Am. Chem. Soc.*, **2007**, *129*, 7742.

[271] Ohnmacht, S. A., Mamone, P., Culshaw, A. J. and Greaney, M. F., *Chem. Commun.*, **2008**, 1241.

[272] Besselièvre, F., Mahuteau-Betzer, F., Grierson, D. S. and Piguel, S., *J. Org. Chem.*, **2008**, *73*, 3278.

[273] Martin, T., Verrier, C., Hoarau, C. and Marsais, F., *Org. Lett.*, **2008**, *10*, 2909.

[274] Borghese, A., Geldhof, G. and Antoine, L., *Tetrahedron Lett.*, **2006**, *47*, 9249.

[275] Kobayashi, K., Sugie, A., Takahashi, M., Masui, K. and Mori, A., *Org. Lett.*, **2005**, *7*, 5083.

[276] Čerňa, I., Pohl, R., Klepetářová, B. and Hocek, M., *Org. Lett.*, **2006**, *8*, 5389.

[277] Cerna, I., Pohl, R. and Hocek, M., *Chem. Commun.*, **2007**, 4729.

[278] Storr, T. E., Firth, A. G., Wilson, K., Darley, K., Baumann, C. G. and Fairlamb, I. J. S., *Tetrahedron*, **2008**, *64*, 6125.

[279] Koubachi, J., El Kazzouli, S., Berteina-Raboin, S., Mouaddib, A. and Guillaumet, G., *Synthesis*, **2008**, 2537.

[280] Lewis, J. C., Wu, J. Y., Bergman, R. G. and Ellman, J. A., *Angew. Chem. Int. Ed.*, **2006**, *45*, 1589.

[281] Wang, X., Lane, B. S. and Sames, D., *J. Am. Chem. Soc.*, **2005**, *127*, 4996.

[282] Lane, B. S and Sames, D., *Org. Lett.*, **2004**, *6*, 2897.

[283] Lane, B. S., Brown, M. A. and Sames, D., *J. Am. Chem. Soc.*, **2005**, *127*, 8050.

[284] Bellina, F., Benelli, F. and Rossi, R., *J. Org. Chem.*, **2008**, *73*, 5529.

[285] Campeau, L.-C., Rousseaux, S. and Fagnou, K., *J. Am. Chem. Soc.*, **2005**, *127*, 18020.

[286] Leclerc, J.-P. and Fagnou, K., *Angew. Chem. Int. Ed.*, **2006**, *45*, 7781.

[287] Li, M. and Hua, R, *Tetrahedron Lett.*, **2009**, *50*, 1478.

[288] Alvarez, E., Conejero, S., Lara, P., López, J. A., Paneque, M., Petronilho, A., Poveda, M. L., del Rio, D., Serrano, O. and Carmona, E., *J. Am. Chem. Soc.*, **2007**, *129*, 14130.

[289] Deprez, N. R., Kalyani, D., Krause, A. and Sanford, M. S., *J. Am. Chem. Soc.*, **2006**, *128*, 4972.

[290] Grigg, R. and Savic, V., *Tetrahedron Lett.*, **1997**, *38*, 5737.

[291] Tan, K. L., Bergman, R. G. and Ellman, J. A., *J. Am. Chem. Soc.*, **2002**, *124*, 13964.

[292] Stuart, D. R. and Fagnou, K., *Science*, **2007**, *316*, 1172.

[293] Dwight, T. A., Rue, N. R., Charyk, D., Josselyn, R. and DeBoef, B., *Org. Lett.*, **2007**, *9*, 3137; Potavathri, S., Dumas, A. S., Dwight, T. A., Naumiec, G. R., Hammann, J. M. and DeBoef, B., *Tetrahedron Lett.*, **2008**, *49*, 4050.

[294] Masui, K., Ikegami, H. and Mori, A., *J. Am. Chem. Soc.*, **2004**, *126*, 5074.

[295] Yang, S.-D., Sun, C.-L., Fang, Z., Li, B.-J., Li, Y.-Z. and Shi, Z.-J., *Angew. Chem. Int. Ed.*, **2008**, *47*, 1473.

[296] Zhao, J., Zhang, Y. and Cheng, K., *J. Org. Chem.*, **2008**, *73*, 7428.

[297] Harmann, B. C. and Hartwig, J. F., *J. Am. Chem. Soc.*, **1998**, *120*, 7369; Old, D. W., Wolfe, J. P. and Buchwald, S. L., *J. Am. Chem. Soc.*, **1998**, *120*, 9722.

[298] Wolfe, J. P., Tomori, H., Sadighi, J. P., Yin, J. and Buchwald, S. L., *J. Org. Chem.*, **2000**, *65*, 1158.

[299] Old, D. W., Harris, M. C. and Buchwald, S. L., *Org. Lett.*, **2000**, *2*, 1403.

[300] Yin, J., Zhao, M. M., Huffman, M. A. and McNamara, J. M., *Org. Lett.*, **2002**, *4*, 3481.

[301] Schulte, J. P. and Tweedie, S. R., *Synlett*, **2007**, 2331.

[302] 'Applications of boronic acids in selective C–C and C–N arylation of purines', Strouse, J. J., Jeselnik, M. and Arterburn, J. B., *Acta. Chim. Slov.*, **2005**, *52*, 187.

[303] Lakshman, M. K., Keeler, J. C., Hilmer, J. H. and Martin, J. Q., *J. Am. Chem. Soc.*, **1999**, *121*, 6090.

[304] Gillet, L. C. J. and Schärer, O. D., *Org. Lett.*, **2002**, *4*, 4205.

[305] Schoffers, E., Olsen, P. D. and Means, J. C., *Org. Lett.*, **2001**, *3*, 4221.

[306] Chakraborti, D., Colis, L., Schneider, R. and Basu, A. K., *Org. Lett.*, **2003**, *5*, 2861.

[307] Snider, B. B. and Zeng, H., *Org. Lett.*, **2000**, *2*, 4103.

[308] Arterburn, J. B., Rao, K. V., Ramdas, R. and Dible, B. R., *Org. Lett.*, **2001**, *3*, 1351.

[309] Lam, M. S., Lee, H. W., Chan, A. S. C. and Kwong, F. Y., *Tetrahedron Lett.*, **2008**, *49*, 6192; Wolter, M., Klapars, A. and Buchwald, S. L., *Org. Lett.*, **2001**, *3*, 3803.

[310] Lebedev, A. Y., Izmer, V. V., Kazyul'kin, D. N., Beletskaya, I. P. and Voskoboynikov, A. Z., *Org. Lett.*, **2002**, *4*, 623.

[311] Gradel, B., Schneider, R. and Fort, Y., *Tetrahedron Lett.*, **2001**, *42*, 5689.

[312] 'Renaissance of Ullmann and Goldberg reactions – progress in copper catalyzed C–N-, C–O- and C–S-coupling', Kunz, K., Scholz, U. and Ganzer, D., *Synlett*, **2003**, 2429.

[313] Kiyomori, A., Marcoux, J.-F. and Buchwald, S. L., *Tetrahedron Lett.*, **1999**, *40*, 2657.

[314] Zhu, L., Guo, P., Li, G., Lan, J., Xie, R. and You, J., *J. Org. Chem.*, **2007**, *72*, 8535.

[315] Altman, R. A. and Buchwald, S. L., *Org.Lett.*, **2006**, *8*, 2779.

[316] Altman, R. A., Koval, E. D. and Buchwald, S.,L., *J.Org. Chem.*, **2007**, *72*, 6190.

[317] Hong, C. S., Seo, J. Y. and Yum, E. K., *Tetrahedron Lett.*, **2007**, *48*, 4831.

[318] Zhang, H., Chen, B.-C., Wang, B., Chao, S. T., Zhao, R., Lim, N. and Balasubramanian, B., *Synthesis*, **2008**, 1523.

[319] Yeh, V. S. C. and Wiedeman, P. E., *Tetrahedron Lett.*, **2006**, *47*, 6011.

[320] Chan, D. M. T., Monaco, K. L., Wang, R.-P. and Winters, M. P., *Tetrahedron Lett.*, **1998**, *39*, 2933.

[321] Lam, P. Y. S., Clark, C. G., Saubern, S., Adams, J., Winters, M. P., Cham, D. M. T. and Combs, A., *Tetrahedron Lett.*, **1998**, *39*, 2941.

[322] Mederski, W. W. K. R., Lefort, M., Germann, M. and Kux, D., *Tetrahedron*, **1999**, *55*, 12757.

[323] Collman, J. P., Zhong, M., Zhang,C. and Costanzo, S., *J. Org. Chem.*, **2001**, *66*, 7892.

[324] Dai, Q., Ran, C. and Harvey, R. G., *Tetrahedron*, **2006**, *62*, 1764.

[325] Jacobsen, M. F., Knudsen, M. M. and Gothelf, K. V., *J. Org. Chem.*, **2006**, *71*, 9183.

[326] Lam, P. Y. S., Vincent, G., Bonne, D. and Clark, C. G., *Tetrahedron Lett.*, **2003**, *44*, 4927.

327 Tsuritani, T., Strotman, N. A., Yamamoto, Y., Kawasaki, M., Yasuda, N. and Mase, T., *Org. Lett.*, **2008**, *10*, 1653.

328 Lam, P. Y. S., Vincent, G., Bonne, D. and Clark, C. G., *Tetrahedron Lett.*, **2002**, *43*, 3091; Lam, P. Y. S., Clark, C. G., Saubern, S., Adams, J., Averill, K. M., Chan, D. M. T. and Combs, A., *Synlett*, **2000**, 674.

329 Song, R.-J., Deng, C.-L., Xie, Y.-X. and Li, J.-H., *Tetrahedron Lett.*, **2007**, *48*, 7845.

330 Kang, S.-K., Lee, S.-H. and Lee, D., *Synlett*, **2000**, 1022.

331 Elliott, G. I. and Konopelski, J. P., *Org.Lett.*, **2000**, 3055.

332 Fedorov, A. Y. and Finet, J.-P., *Tetrahedron Lett.*, **1999**, *40*, 2747.

333 Gagnon, A., St-Onge, M., Little, K., Duplessis, M. and Barabé, F., *J. Am. Chem. Soc.*, **2007**, *129*, 44.

334 Schwarz, N., Pews-Davtyan, A., Alex, K., Tillack, A. and Beller, M., *Synthesis*, **2007**, 3722.

335 Humphries, P. S., Bailey, S., Do, Q.-Q. T., Kellum, J. H., McClellan, G. A. and Wilhite, D. M., *Tetrahedron Lett.*, **2006**, *47*, 5333.

336 D'Angelo, N. D., Peterson, J. L., Booker, S. K., Fellows, I., Dominguez, C., Hungate, R., Reider, P. J. and Kim, T.-S., *Tetrahedron Lett.*, **2006**, *47*, 5045.

337 Shafir, A., Lichtor, P. A. and Buchwald, S. L., *J. Am. Chem. Soc.*, **2007**, *129*, 3490.

338 'Palladium catalysed vinylation of organic halides', Heck, R. F., *Org. React.*, **1982**, *27*, 345; 'Advances in the Heck chemistry of aryl bromides and chlorides', Whitcombe, N. J., Hii, K. K. and Gibson, S. E., *Tetrahedron*, **2001**, *57*, 7449.

339 Frank, W. C., Kim, Y. C. and Heck, R. F., *J. Org. Chem.*, **1978**, *43*, 2947.

340 He, T., Tao, X., Wu, X., Cai, L. and Pike, V.W., *Synthesis*, **2008**, 887.

341 Itahara, T., Ikeda, M. and Sakakibara, T., *J. Chem. Soc., Perkin Trans. 1*, **1983**, 1361.

342 Tsuji, J. and Nagashima, H., *Tetrahedron*, **1984**, *46*, 2699.

343 Grimster, N. P., Gauntlett, C., Godfrey, C. R. A. and Gaunt, M. J., *Angew. Chem. Int. Ed.*, **2005**, *44*, 3125.

344 Capito, E., Brown, J. M. and Ricci, A., *Chem. Commun.*, **2005**, 1854.

345 Maehara, A., Tsurugi, H., Satoh, T. and Miura, M., *Org. Lett.*, **2008**, *10*, 1159.

346 Bessard, Y. and Crettaz, R., *Heterocycles*, **1999**, *51*, 2589; Albaneze-Walker, J., Bazaral, C., Leavey, T., Dormer, P. G. and Murray, J. A., *Org. Lett.*, **2004**, *6*, 2097.

347 Kumar, K., Zapf, A., Michalik, D., Tillack, A., Heinrich, T., Böttcher, H., Arlt, M. and Beller, M., *Org. Lett.*, **2004**, *6*, 7.

348 Couve-Bonnaire, S., Carpentier, J-F., Mortreux, A. and Castanet, Y., *Tetrahedron*, **2003**, *59*, 2793.

5

Methods in Heterocyclic Chemistry

The traditional methods of organic synthesis have been supplemented, and often supplanted, by several newer techniques, all of which are relevant to heterocyclic chemistry, and are discussed in this chapter.

One important approach is to carry out reactions with tethered substrates, to simplify manipulation, purification and isolation, avoiding the often tedious and wasteful standard techniques such as liquid–liquid extractions and chromatography. The original way of doing this was by carrying out the reactions on solid phase – that is, where the substrate is attached (tethered) to an insoluble polymeric solid support – and this remains the most popular method. Soluble polymeric supports can also be used, but are not so convenient. A more recent approach is the use of (non-polymeric) 'phase tags', which are auxiliary groups that have high affinities for particular solvents or adsorbents, allowing selective capture of the tagged components. These methods have also been extended to the use of tethered reagents, and scavengers for removal of by-products, rather than immobilisation of the substrate.

This chapter also includes a discussion of the use of microwave heating which can greatly accelerate reactions, cutting down on reaction times and often allowing transformations that would otherwise not be practicable. The use of flow reactors is a burgeoning area, with advantages at all scales and is discussed in the heterocyclic context.

5.1 Solid-Phase Reactions[1] and Related Methods
5.1.1 Solid-Phase Reactions
This method was originally developed by Merrifield for peptide synthesis, with the link (tether) being achieved by alkylation of the carboxyl oxygen of protected amino acids with chloromethylated polystyrene resins. It is now very widely used in general and heterocyclic chemistry, but modified resins are much more common, although still normally based on the polystyrene backbone, but with various intermediate chains and functional linking groups, which allow easier control over conditions for cleavage of the product. Popular variants are based on Wang resins, containing an intermediate alkoxybenzyl group and the related Rink resins. Solid-supported reactions are particularly amenable to combinatorial, high throughput and automated synthesis.[2] (**NOTE**: In schemes, the resin backbone, including any spacer group, may be indicated generically as '**PS**' in reagent listings, but in structural diagrams it is shown as shaded circles.)

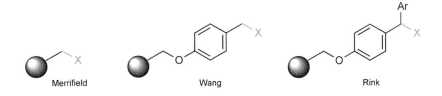

Merrifield   Wang   Rink

Heterocyclic Chemistry 5th Edition   John Joule and Keith Mills
© 2010 Blackwell Publishing Ltd

The process of solid-phase synthesis (SPS) involves linking the substrate to the resin, carrying out various reactions, washing away by-products, and finally cleaving the product from the support with minimal impurities. The polymer-bound intermediates may simply be filtered off and washed after each stage or, more conveniently, by using a special flask containing a sinter, controlled by a stopcock, at the base.

Individual reactions may be slower on solid phase, due to congestion at the reacting site, and it is common to use substantial excesses of reagents to ensure complete reaction. However, overall processes are generally faster because only one isolation step is needed. Modified polymers, for example with polyethylene glycol spacers between the polymeric benzyl and the reacting group, are said to give reactions that are more similar to solution reactions.

Practically all types of reaction can be carried out, although some may limit the choice of linking group, including those that may include aggressive or sensitive reagents, such as lithiation, halogenation or metal-catalysed couplings. Solid-phase synthesis can be adapted to most standard heterocyclic reactions and syntheses.

A major question is how to attach the substrate to the polymer in such a way that it can be selectively and easily removed at the end of the sequence. In carboaromatic and aliphatic chemistry this is often done through a functional group such as a carboxylic acid or an amine. However, this can restrict choice of substrate; an alternative method is through a 'traceless link' such as a silane, which can be removed, for example by protonolysis, to leave a hydrogen at the point of attachment, but this may not be particularly convenient. Here, heterocycles have the advantage! Attachment to the support can be by methods similar to those described above, but also *via* the ring nitrogen, or other potential ring heteroatoms.

Some illustrative and self-explanatory examples are shown below.[3,4,5] For a discussion of the heterocyclic reactivity involved in the examples shown, the reader should consult the relevant ring-system chapter.

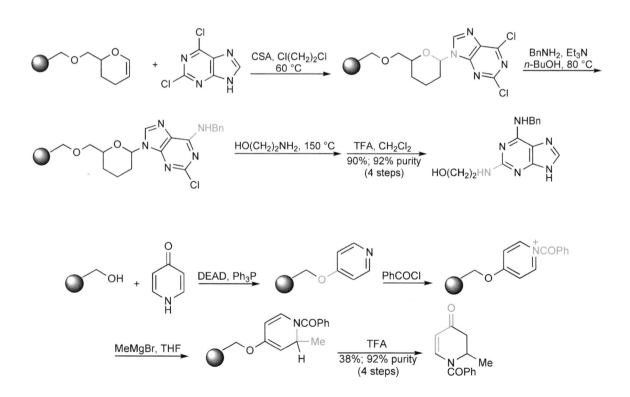

An alternative attachment is through a heteroatom when the heterocyclic ring formation is the final step – it is often easy to incorporate a final cyclisation (heterocycle formation) step in such a way that it results in cleavage of the product from the support at the same time.[6]

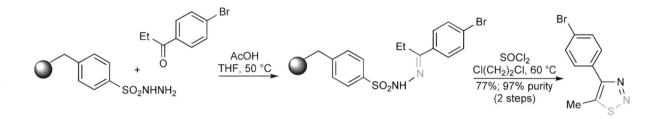

Sulfur is a useful link for heterocycles because its use as a leaving group (or better, after conversion to sulfoxide[7] or sulfone) can bring about cleavage from the support combined with addition of a nucleophile. This method has been used for both azoles[8] and azines,[9] as shown below.

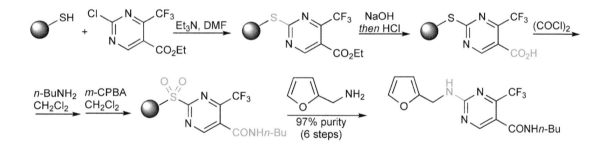

An interesting (traceless) example combines the linking with a first-stage reaction by use of a resin acid chloride in a Reissert-reaction–alkylation sequence from isoquinolines, the normal Reissert final-stage hydrolysis being the means of cleavage from the resin.[10]

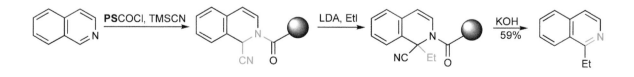

5.1.2 Solid-Supported Reagents and Scavengers

Another application of solid-phase chemistry is the use of polymer-bound reagents for reactions with substrates in solution, which offers similar advantages in requiring minimal purification: the thiazole synthesis shown below, which involved the use of a polymer-bound brominating agent and secondly a polymer-bound base, gave the intermediate and product in greater than 95% purity, without the need for any chromatography.[11]

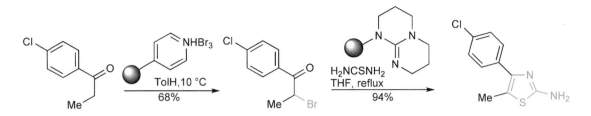

Other examples include reactions, such as the Mitsunobu condensation, where polystyrene-diphenylphosphine can be used instead of triphenylphosphine, avoiding production of the difficult-to-remove by-product, triphenylphosphine oxide.

Solid-phase 'scavengers' can also be used both for removal of excess reagents and trace impurities, such as metal residues, for example palladium from cross-coupling reactions. A range of resins is available to remove a variety of metal residues, usually involving binding to a thiol or amino group.

Instead of simply filtering off the resin between stages, an alternative approach is to carry out reactions in flow systems with sequential chambers containing catalysts and/or reagents and/or scavengers. An illustrative example is the triazole synthesis shown below, *via* a copper-catalysed azide–acetylene cycloaddition, where excess azide was used to ensure complete reaction of the acetylene substrate. The mixture of azide and acetylene was first passed through a chamber containing the reaction catalyst – CuI.Me$_2$NCH$_2$**PS** – then though a polystyrene-thiourea resin to remove a small amount of copper that had leached from the catalyst. Finally, passage though a polystyrene-phosphine resin to eliminate the excess azide, and removal of the solvent, gave essentially pure product.[12]

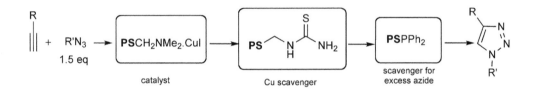

While all the above methods have been used very successfully on standard laboratory scale, they also have significant potential for rapid synthesis on medium scale (up to several kg) and possibly even larger. High-loading resins can bind an equal weight of substrate and so are efficient in terms of volume.[13]

At the other end of the scale, products can be isolated on single beads of resin produced by combinatorial chemistry, involving sequential splitting and mixing batches of beads between reaction steps such that single beads contain only one product. Spectroscopic methods have been developed for the non-destructive characterisation of products while still on the bead.[14]

5.1.3 Solid-Phase Extraction (SPE)
Rather than carry out liquid–liquid extractions, which are time-consuming and lead to large quantities of waste, solid supports can be constructed that have high affinity for selected types of molecules. Simple filtration of a solution through a column of the solid-phase medium results in efficient extraction of the product. This method is quite widely used and is of particular interest for fluorous chemistry (see below).

5.1.4 Soluble Polymer-Supported Reactions[15]
Soluble polymers, such as polyethylene glycol (PEG), can be used in similar ways to the insoluble supports, but they generally require a more elaborate work-up, often involving precipitation of the polymeric

complex. They have not been widely used for heterocyclic synthesis, however in one example, thiophenes were made from PEG cyanoacetic ester, ketones and sulfur.[16]

5.1.5 Phase Tags

A phase tag can be as simple as incorporation of a protecting group containing a carboxyl or amino group, allowing selective extraction into base or acid respectively. However, general methods using non-reactive tags are more versatile, of which the most important and best-developed group are fluorous compounds, with a significant number of protecting groups, reagents and scavengers being commercially available.

5.1.5.1 Fluorous Tags[17]

A fluorous compound is one that contains one or more perfluoroalkyl groups that confer special physical properties on the molecule, particularly selective affinity for solvents and adsorbants, without changing the chemical properties to a great degree. They can be differentiated into 'light fluorous', which generally have a single perfluoroalkyl group (usually C_6 to C_{10}) and 'heavy fluorous', which may contain multiple perfluoroalkyl groups. Light fluorous compounds are generally much more useful in the current context. The perfluoroalkyl group is usually 'insulated' from a functional group by a number of methylenes to avoid altering its reactivity too much, for example where it is a nucleophile or where formation of a cation is involved in cleavage, as for the Boc analogue shown below. However, in certain cases, for example sulfonate leaving groups, it may be beneficial to have the stronger electron withdrawal of a fully fluorinated group.

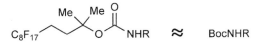

Fluorous chemistry can be applied in many ways that are similar to solid-phase methodology, including the use of traceless and reactive attachment points for substrates, or as reagents or scavengers. Notable examples of reagents where the by-products are difficult to separate in the 'normal' form, but easy in the fluorous form, are azodicarboxylates and phosphines, for Mitsunobu reactions, and 'heavy' fluorous tin reagents, for various purposes, including Stille couplings.[18]

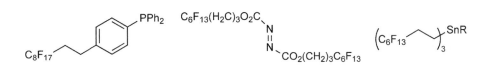

Fluorous tagging can facilitate separation by standard purification methods, but more specific methods are available. A particularly useful technique is SPE onto fluorous silica gel, which contains a fluorinated chain bonded to the silica. The non-fluorous components are washed off with a fluorophobic solvent then the fluorous component is eluted with a fluorophilic solvent. Note that ordinary organic solvents are used; fluorinated solvents are seldom required. Many organic solvents are fluorophilic, but can be switched dramatically, in the case of water-miscible solvents, by the addition of relatively small amounts of water, into fluorophobes. The selectivity is such that the fluorophobic solvent can be 20% water in methanol, followed, as the fluorophilic solvent, by pure methanol.

The exact size of the perfluoroalkyl group can be critical. In the example shown below, fluorous versions of the Mukaiyama 2-chloropyridinium reagent were used to couple acids and amines in DMF, the fluorous pyridone by-product being removed by precipitation on addition of 20% water. Some of the pyridone was retained in the solution when R_f was C_8F_{17}, but none when it was $C_{10}F_{21}$.[19]

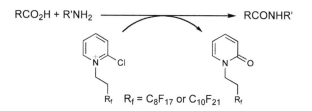

The scheme shown below illustrates a typical fluorous sequence, including the use of a fluorous tagged intermediate where the tag can be converted into a leaving group, in analogy with solid-phase sulfur links.[20]

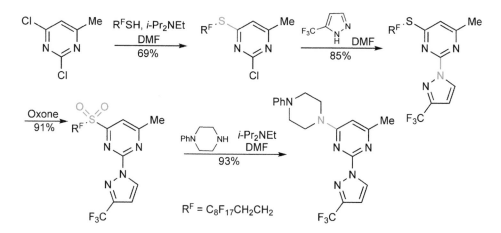

5.1.5.2 Ionic-Liquid Tags

Some syntheses using intermediates with ionic-liquid tags have been reported, but they are not as versatile as solid or fluorous supports. The tagged intermediates and products are often isolated by precipitation and an advantage of ionic liquids is that the affinity and solubility of the tagged compounds can be altered by exchange of the associated anion.

An example of an efficient ionic liquid-phase tag is for Biginelli reactions, where a potentially very versatile tagged acetoacetate was used. This ester was prepared straightforwardly from a pre-formed ionic-liquid-tagged alcohol.[21]

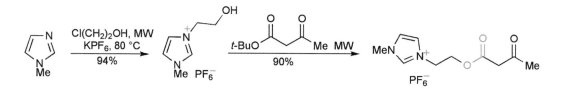

In other cases, a less convenient approach to the preparation of the tagged intermediate has been used, involving quaternisation, followed by anion exchange, of complex haloalkyl substrates, for example a C-6–linked glycoside.[22]

5.2 Microwave Heating[23]

Microwave heating (commonly designated 'MW' or 'μW' in reaction schemes) is very widely used in heterocyclic chemistry, and organic chemistry generally, as it is much quicker and more convenient than conventional heating. It is also much easier to control – heat input can be cut off at the flick of a switch – and is very suitable for automated systems, for example in combinatorial chemistry and high-throughput synthesis. It can also improve the outcome of many reactions, when compared to conventional heating.[24]

In contrast to conventional heating, which involves a slow and inefficient heat transfer through the vessel walls from an external heat source, heating by microwaves occurs directly within the reaction mixture by interaction of the radiation with dipoles. Therefore, one of the reaction components, or the solvent, must have polar bonds[25] (there may be subtle variations in the reaction depending on differential heating of the solvent and reactants). Alternatively, passive heating elements, such as graphite or silicon carbide can be added, which absorb the radiation, generating heat, and thus allowing reactions of weakly absorbing substrates in low-polarity solvents.[26] Addition of these passive heating elements to standard microwave reactions can also be used to generate higher temperatures.

There are two basic types of microwave equipment – multimode and monomode. Multimode is the type found in domestic microwave ovens, where reflections inside a metal casing may produce a non-homogenous field, leading to uneven heating with hot spots. However, with proper design, a completely random field can be generated, which will give even heating. Monomode uses a waveguide to generate a standing wave with a homogenous field, at least when not perturbed by a reaction, and the waveguide can also be designed to focus the radiation, giving a more intense local field. Monomode reaction chambers are limited to a relatively small volume (about 100 ml or so) due to the precise positional requirements for the reaction vessel. Consequently, it is necessary to use multimode for large equipment, such as multi-reaction plates and arrays, and for scale-up.

The use of domestic microwave ovens is to be greatly discouraged – although cheap, they allow for very little control and give no information about precise conditions and therefore fall short of proper scientific requirements. They are also dangerous, with the possibility of fires and explosions! Purpose-made commercial equipment is much more versatile and, with feedback control from sensors for temperature and pressure, is much safer, and provides proper control and output data.

Reaction methodology usually needs to be modified for microwave reactions. Solvent-free reactions are popular and eliminate the risk from flammable solvents. Reactions can also be carried out on pastes or with the reactants adsorbed onto solid supports, such as alumina or bentonite.

The apparently perverse method of microwave heating with concomitant cooling (either by blowing air onto the vessel or using more sophisticated jacket systems) has been shown to be beneficial, reducing side reactions, and such cooling systems are now common on commercial microwave equipment designed for synthesis. A demonstration of this principle is a study of Suzuki couplings using an encapsulated palladium catalyst. Here, the catalyst absorbs most of the microwave radiation (an ideal situation, as it is the only place where the reaction occurs!), but much of the heat produced is transferred to the solvent causing unwanted and unproductive heating of the reactants and products, leading to decomposition of sensitive components, for example some thiophene and benzofuran boronic acids. When the reaction is carried out with cooling, reducing the bulk temperature during irradiation, the yields from problematic substrates are considerably improved.[27]

The existence, or not, of non-heating effects of microwave radiation on reactions is somewhat controversial, as the energy of microwaves is too low to break chemical bonds directly. However, microwave irradiation certainly can produce different results to conventional heating, for example isomer ratios in *N*-alkylation of azoles. It can be argued that this is just due to the different rates of heating, but it can be a useful feature.[28]

Batch scale-up of microwave reactions can be carried out to a certain extent (1 kg or so[29]), but for larger reactions it is difficult, due to the limited depth of penetration of the radiation. However, combination of microwave heating with flow reactors is very promising, due to the relatively small volume of the reaction zone.

5.3   Flow Reactors

Flow reactors are currently under intense investigation and development. They can be used in several different ways, but the most important application here is for carrying out 'normal' chemistry in a continuous (rather than batch) mode, and can be applied at all scales from microsynthesis up to process development and production.

Classically, a reaction is carried out as a batch process, i.e. a batch of starting material is placed in a vessel, the reaction is carried out, worked up and the product isolated. If necessary, the vessel is then reloaded and the sequence repeated, and so on. There are advantages and disadvantages in this approach. The main advantage is that it is fairly straightforward to do and may use relatively simple equipment. The disadvantages include inefficient use of time, limited throughput and, particularly as the scale increases, increasing difficulty of heating and cooling efficiently, with longer reagent addition times, the worst outcome being a runaway reaction. Batch scale-up synthesis is notorious for not following on from the conditions used for the small-scale work. At the other extreme, library and analogue synthesis on a small-scale is also very time consuming and tedious. Flow reactors can provide solutions for both these disadvantages.

In a flow reactor, the reaction occurs in relatively small diameter glass, plastic or metal tubes or channels up to a few mm in diameter, with the reaction zone confined to a small heating/cooling chamber. Input of starting materials and output of product is continuous, being fed simply by combining streams of reactants in a T-junction or in a more sophisticated mixing chamber. Control valves can be added for operations under pressure. The reactors can also be miniaturised, with channels down to a few tens of microns in diameter, formed by etching onto small glass, ceramic or metal plates, similar to silicon chips – hence the phrase 'lab-on-a-chip'. These chip reactors are sometimes referred to as microreactors[30] and the technique as microfluidics (differentiation into microfluidics for sub-millimetre and mesofluidics for mm–cm diameters of flow reactors is useful).

Due to the small volumes of the reaction zone, high rates of heat transfer, both in and out, are possible. Microwave heating is particularly suitable, provided metal tubing is not used, and low-temperature reactions are easy to carry out – the exceptional efficiency of cooling may allow reactions to be carried out at a higher temperature than in a batch reactor. It is also relatively easy to devise safe high-pressure systems. Overall, flow reactors allow a much higher degree of control than classical methods. Highly sophisticated integrated bench-top flow systems are available commercially that have the capacities to work up to 350 °C and 2900 psi.[31]

There are many ingenious sophisticated variations on the reactor cells,[32] but a simple, though not very versatile, flow cell has been described for use in a microwave heater, where the reaction solution was percolated through sand in a test tube as the equivalent of multiple microchannels. This set-up was used to carry out a Bohlmann–Rahtz pyridine synthesis and a Fischer indolisation.[33]

Flow reactors have another major asset – safety. Although capable of producing substantial aggregate quantities of product, the amount in the reaction zone at any one time is very small, therefore any *reaction* hazard, that is where the hazard arises during the reaction rather than hazardous starting materials or products, is minimised.

In theory and usually in practice, the scale-up of flow reactions is relatively straightforward – just carry on for longer and/or add more reactors of the same size in parallel. As the individual reaction scale doesn't change, neither do reaction conditions.

A scale-up preparation of ionic liquids is illustrative. Here the reaction of 1-methylimidazole with *n*-butyl bromide under solvent-free conditions is highly exothermic and prone to runaway in a batch reaction of any significant scale. However, using a flow reactor with reaction tubes between 2 mm and 6 mm diameter, fed by a micromixer with 0.45 mm channels, a continuous flow reaction at 85 °C was able to produce over

9 kg of product per day.[34] A flow reactor has also been used to tame the very dangerous diazotisation of aminotetrazole (see 29.1.1.3).[35]

Small-scale combinatorial chemistry for library and analogue generation is easily automated with micro-reactors, using similar technology to that used for automated hplc, where sequential syntheses can be carried out on the same 'chip', for example in the preparation of analogues of ciprofloxacin.[36]

A significant limitation of flow reactors is that solids, either as suspensions of starting materials or precipitated products, easily block the channels, although solid-supported reagents and catalysts can be used by immobilisation onto the walls or as packing.

5.4 Hazards: Explosions

Many compounds used or prepared by the chemist are hazardous, but most of the hazards can be controlled by proper working practice. However, explosive hazards are the exception, as explosions are very difficult to contain in the normal laboratory.

Physical explosions, such as those due to excessive pressure build-up in a closed vessel or ignition of flammable gas mixtures, should not occur if procedures are followed. The main hazard is from intrinsically explosive compounds detonation of which can be initiated by shock, heat or friction; the main substances of concern in heterocyclic chemistry are azides, perchlorate salts and some high-nitrogen compounds. These last are discussed in the appropriate ring chapters and simple inorganic and organic azides are discussed in Section 29.1.1.3. A widely used reagent – tosyl azide – presents a significant explosive hazard and although we describe reactions using such reagents, as they appear in the literature, safer substitutes, for example 4-acetamidobenzenesulfonyl azide,[37] which is commercially available, should be used whenever possible. Dodecyl- and naphthylsulfonyl azides are other possible alternatives.[38]

The iminium perchlorate shown below is a useful reagent for the preparation of heterocycles, but is shock sensitive and has the explosive power of TNT! Fortunately, the much safer tetrafluoroborate salt is a suitable substitute.[39]

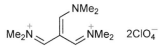

Where an explosive *reaction* hazard (as opposed to an explosive product) is present, the use of flow reactors can be particularly useful for mitigating risk.

References

[1] 'Recent advances in the preparation of heterocycles on solid support', Franzén, R. G., *J. Comb. Chem.*, **2000**, *2*, 195; 'Recent progress in solid phase heterocycle synthesis', Corbett, J. W., *Org. Prep. Proc. Int.*, **1998**, *30*, 489; 'Solid phase organic reactions, III (for Nov 1996–Dec 1997)' [and previous articles in the series], Booth, S., Hermkens, P. H. H., Ottenheijm, H. C. J. and Rees, D. C., *Tetrahedron*, **1998**, *54*, 15385.

[2] An example of a high throughput robotic synthesis: Brooking, P., Crawshaw, M., Hird, M. W., Jones, C., MacLachlan, W. S., Readshaw, S. A. and Wilding, S., *Synthesis*, **1999**, 1986.

[3] Nugiel, D. A., Cornelius, A. M. and Corbett, J. W., *J. Org. Chem.*, **1997**, *62*, 201.

[4] Chen, C. and Munoz, B., *Tetrahedron Lett.*, **1998**, *39*, 6781.

[5] Huang, W. and Scarborough, R. M., *Tetrahedron Lett.*, **1999**, *40*, 2665.

[6] Hu, Y., Baudart, S., and Porco, J. A., *J. Org. Chem.*, **1999**, *64*, 1049.

[7] Masquelin, T., Meunier, N., Gerber, F. and Rosse, G., *Heterocycles*, **1998**, *48*, 2489.

[8] Lee, I. Y., Lee, J. Y., Lee, H. J. and Gong, Y-D., *Synlett*, **2006**, 2483.

[9] Gayo, L. M. and Suto, M. J., *Tetrahedron Lett.*, **1997**, *38*, 211.

[10] Lorsbach, B. A., Bagdanoff, J. T., Miller, R. B. and Kurth, M. J., *J. Org. Chem.*, **1998**, *63*, 2244.

[11] Habermann, J., Ley, S. V., Scicinski, J. J., Scott, J. S., Smits, R. and Thomas, A. W., *J. Chem. Soc., Perkin Trans. 1*, **1999**, 2425.

[12] Smith, C. D., Baxendale, I. R., Lanners, S., Hayward, J. J., Smith, S. C. and Ley, S. V., *Org. Biomol. Chem.*, **2007**, *5*, 1559.

[13] Raillard, S. P., Ji, G., Mann, A. D. and Baer, T. A., *Org. Proc. Res. Dev.*, **1999**, *3*, 177.

[14] Freeman, C. E. and Howard, A. G., *Analyst*, **2001**, *126*, 538; Swali, V. and Bradley, M., *Anal. Commun.*, **1997**, *34*, 15H.

[15] 'Soluble polymer-supported organic synthesis', Toy, P. H. and Janda, K. D., *Acc. Chem. Res.*, **2000**, *33*, 546.

[16] Zhang, H., Yang, G., Chen, J. and Chen, Z., *Synthesis*, **2004**, 3055.

[17] 'Handbook of Fluorous Chemistry', Gladysz, J. A., Curran, D. P., and Horvath, I. T. Eds., Wiley-VCH, **2004**; 'Fluorous synthesis of heterocyclic systems', Zhang, W. *Chem. Rev.*, **2004**, *104*, 2531; 'Organic synthesis with light-fluorous reagents, reactants, catalysts and scavengers', Curran, D. P., *Aldrichimica Acta*, **2006**, *36*, 3. Fluorous Technologies Inc. are the main suppliers of fluorous compounds and have a very informative web site: www.fluorous.com.

[18] Hoshino, M., Degenkolb, P. and Curran, D. P., *J. Org. Chem.*, **1997**, *62*, 8341.

[19] Matsugi, M., Suganuma, M., Yoshida, S., Hasebe, S., Kunda, Y., Hagihara, K. and Oka, S., *Tetrahedron Lett.*, **2008**, *49*, 6573.

[20] Zhang, W., *Org. Lett.*, **2003**, *5*, 1011.

[21] Legeay, J. C., Vanden Ende, J. J., Toupet, L. and Bazureau, J. P., *Arkivoc*, **2007**, *iii*, 13.

[22] Pathak, A. K., Yerneni, C. K., Young, Z. and Pathak, V., *Org. Lett.*, **2008**, *10*, 145.

[23] Microwave manufacturers web sites often provide libraries of references to applications, and informative videos or slide presentations: (September 2008) www.cem.com; www.biotage.com; www.milestonesci.com.

[24] General reviews: Tierney, J. P. and Lidstrom, P. (Eds), 'Microwave-Assisted Synthesis', Blackwell, **2005**; 'Microwave irradiation for accelerating organic reactions. Part I: Three-, four- and five-membered heterocycles', El Ashry, E. S. H., Ramadan, E., Kassem, A. A. and Hagar, M., *Adv. Heterocycl. Chem.*, **2005**, *88*, 1; 'Part II: Six-, seven-membered, spiro, and fused heterocycles', El Ashry, E. S. H., Ramadan, E. and Kassem, A. A., *Adv. Heterocycl. Chem.*, **2006**, *90*, 1.

[25] For an analysis of the theoretical and technical aspects of microwaves in chemistry see Nüchter, M., Ondruschka, B., Bonrath, W. and Gum, A., *Green Chem.*, **2004**, *6*, 128.

[26] Kremsner, J. M. and Kappe, C. O., *J. Org. Chem.*, **2006**, *71*, 4651.

[27] Baxendale, I. R., Griffiths-Jones, C. M., Ley, S. V. and Tranmer, G. K., *Chem. Eur. J.*, **2006**, *12*, 4407.

[28] J. Cléophax, J., Liagre, M., Loupy, A. and Petit, A., *Org. Proc. Res. Dev.* **2000**, *4*, 498; Perreux, L. and Loupy, A. (Tetrahedron Report 588) *Tetrahedron* **2001**, *57*, 9199.

[29] For example, the Biotage Advancer Kilobatch.

[30] Microreactor suppliers: www.corning.com; www.syrris.com; www.micronit.com. 'Application of microreactor technology in process development', Zhang, X., Stefanick, S. and Villani, F. J., *Org. Proc. Res. Dev.* **2004**, *8*, 455; 'Microreactors in Organic Synthesis and Catalysis', Ed. Wirth, T., Wiley-VCH: Weinheim, **2008**; 'Recent advances in synthetic micro reaction technology', Watts, P. and Wiles, C. *Chem. Commun.* **2007**, 443.

[31] www.uniqsis.com; www.thalesnano.com

[32] Leading ref: Hornung, C. H., Mackley, M. R., Baxendale, I. R. and Ley. S. V., *Org. Proc. Res. Dev.*, **2007**, *11*, 399.

[33] Bagley, M. C., Jenkins, R. L., Lubinu, M. C., Mason, C. and Wood, R., *J. Org. Chem.* **2005**, *79*, 703.

[34] Waterkemp, D. A., Heiland, M., Schlüter, M., Sauvageau, J., Beyersdorf, T. and Thöming, J., *Green Chem.*, **2007**, *9*, 1084.

[35] Kralj, J. G., Murphy, E. R., Jensen, K. F., Williams, M. D. and Renz, R., 41st AIAA/ASME/SAE/ASEE Joint Propulsion Conference & Exhibit, 10–13 July 2005, Tucson, Arizona. Paper AIAA 2005-3516. See also Renz, R. N., Williams, M. D., and Fronabarger, J. W., US patent 7253288 (publ. 08/07/2007) (Note: The introduction to the patent has a good discussion of the concept).

[36] Schwalbe, T., Kadzimirsz, D. and Jas, G., *QSAR Comb. Sci.* **2005**, *24*, 758.

[37] Baum, J. S., Shook, D. A., Davies, H. M. L. and Smith, D., *Synth. Commun.*, **1987**, *17*, 1709.

[38] Hazen, G. G., Weinstock, L. M., Connell, R. and Bollinger, F. W., *Synth. Commun.*, **1981**, *11*, 947.

[39] Ragan, J. A., McDermott, R. E., Jones, B. P., am Ende, D. J., Clifford, P. J., McHardy, S. J., Heck, S. D., Liras, S. and Segelstein, B. E., *Synlett*, **2000**, 1172.

6

Ring Synthesis of Aromatic Heterocycles

The preparation of benzenoid compounds nearly always begins with an appropriately substituted, and often readily available, benzene derivative. The preparation of heteroaromatic compounds presents a very different picture, for it often involves ring synthesis.[1] Of course, when first considering a suitable route to a desired target, it is always important to give thought to the possibility of utilising a commercially available compound that contains the heterocyclic nucleus and which could be modified by manipulation, introduction and/or elimination of substituents[2] – a synthesis of tryptophan, for example, would start from indole – however if there is no obvious starting material, a ring synthesis has to be designed that leads to a heterocyclic intermediate appropriately substituted for further elaboration into the desired target.

This chapter shows how just a few general principles allow one to understand the methods which are used in the construction of the heterocyclic ring of an aromatic heterocyclic compound from precursors that do not have that ring. It discusses the principles, and analyses the types of reaction frequently used in constructing an aromatic heterocycle, and also the way in which appropriate functional groups are placed in the reactants, in order to achieve the desired ring synthesis.

6.1 Reaction Types Most Frequently Used in Heterocyclic Ring Synthesis

By far the most frequently used process is the addition of a nucleophile to a carbonyl carbon (or the more reactive carbon of an O-protonated carbonyl). When the reaction leads to C–C bond formation, then the nucleophile is the β-carbon of an enol or an enolate anion, or of an enamine.

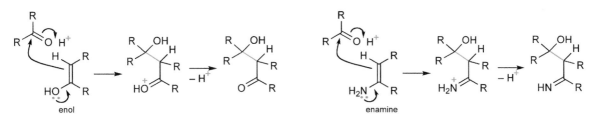

Typical C–C bonding processes in heteroaromatic ring synthesis

When the process leads to C–heteroatom bond formation, then the nucleophile is an appropriate heteroatom, either anionic (-X⁻) or neutral (-XH):

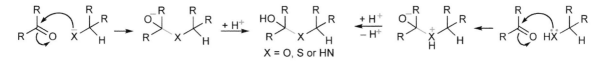

Typical C–Heteroatom bonding processes in heteroaromatic ring synthesis

Heterocyclic Chemistry 5th Edition   John Joule and Keith Mills
© 2010 Blackwell Publishing Ltd

In all cases, subsequent loss of water produces a double bond, either a C–C or a C–heteroatom double bond, i.e. formation of an aldol condensation product, or the formation of an imine or enamine.

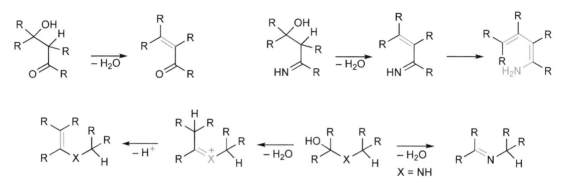

Dehydrations produce alkenes, imines, or enamines or enol/thioenol ethers

These two basic processes, with minor variants, cover the majority of the steps involved in classical heteroaromatic ring synthesis. We shall show below how a sequence of such simple steps leads, *via* a set of equilibria, to the final product, driven to completion by the formation of an aromatic stabilised system. In a few instances, displacements of halide, or other leaving groups, from saturated carbon are also involved.

In completely separate categories are heterocyclic ring syntheses that involve electrocyclic processes (see 6.4) and in some transition metal-catalysed ring-forming steps.

6.2 Typical Reactant Combinations
Although there are some examples of nearly all possible retrosynthetic dissections and synthetic recombinations of five- and six-membered aromatic heterocycles, yet by far the majority of ring syntheses fall into two categories; in the first, for each ring size, only C–heteroatom bonding is needed, i.e. the rest of the skeleton is present, intact, in one starting component; in the second, for each ring size, one C–C bond and one C–heteroatom linkage are required.

Typical disconnections for the synthesis of five- and six-membered aromatic heterocycles

6.2.1 Typical Ring Synthesis of a Pyrrole Involving Only C–Heteroatom Bond Formation
We can now look at specific examples, and see how the principles above can lead to the aromatic heterocycles. In the first of the two broad categories, where only C–heteroatom bonds need to be formed, and for the synthesis of five-membered heterocycles, precursors with two carbonyl groups related 1,4 are required, thus 1,4-diketones react with ammonia or primary amines to give 2,5-disubstituted pyrroles; two successive heteroatom-to-carbonyl carbon additions and loss of two molecules of water produce the aromatic ring, though the exact order of these several steps is never certain.

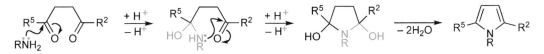

Typical sequence for the synthesis of a pyrrole from a 1,4-dicarbonyl compound

6.2.2 Typical Ring Synthesis of a Pyridine Involving Only C–Heteroatom Bond Formation

For six-membered rings, the 1,5-dicarbonyl precursor has to contain a C–C double bond in order to lead directly to the aromatic system.

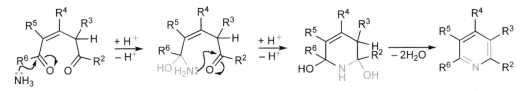

Typical sequence for the synthesis of a pyridine from an unsaturated 1,5-dicarbonyl compound

The use of an otherwise saturated 1,5-dicarbonyl compound does not lead directly to an aromatic pyridine, though it is easy to dehydrogenate the dihydro-heterocycle.

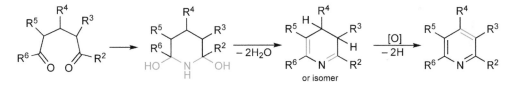

Typical sequence for the synthesis of a pyridine from a saturated 1,5-dicarbonyl compound

6.2.3 Typical Ring Syntheses Involving C–Heteroatom C–C Bond Formations

In the second broad category, needing both C–C and C–heteroatom links to be made, one component must contain an enol/enolate/enamine, or the equivalent thereof, while the second obviously must have electrophilic centres to match. The following general schemes show how this works out for the two ring sizes.

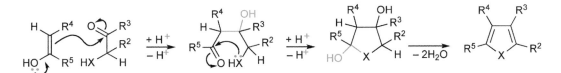

Typical sequence for the synthesis of a five-membered heterocycle from an enol and a C_2XH compound
(note: R^4 must be an acidifying group: ketone, ester, nitrile, or nitro)

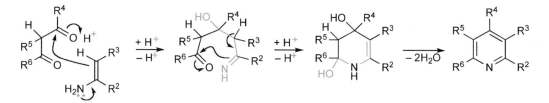

Typical sequence for the synthesis of a pyridine from a 1,3-dicarbonyl compound and an enamine

Where a carbonyl component at the oxidation level of an acid is used then the resultant product carries an oxygen substituent at that carbon. Similarly, if a nitrile group is used instead of a carbonyl group, as an electrophilic centre, then the resulting heterocycle carries an amino group at that carbon, thus:

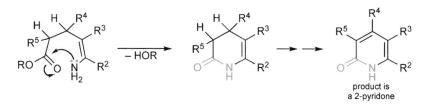

Cyclisation onto a carbonyl group at the carboxylic acid oxidation level gives 2-pyridones

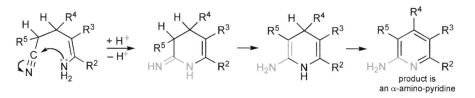

Cyclisation onto a cyano group gives α-amino-pyridines

- The exact sequence of nucleophilic additions, deprotonations/protonations, and dehydrations is never known with certainty, but the sequences shown here, and indeed in the rest of the book, are reasonable ones; the exact order of steps almost certainly varies with conditions,[3] particularly pH.
- Some of the components shown in the examples above have two electrophilic centres and some have a nucleophilic and an electrophilic centre; in other situations components with two nucleophilic centres are required. In general, components in which the two reacting centres are either 1,2- or 1,3-related are utilised most often in heterocyclic synthesis, but 1,4- (e.g. HX–C–C–YH) (X and Y are heteroatoms) and 1,5-related (e.g. O=C–(C)$_3$–C=O) bifunctional components, and reactants that provide one-carbon units (formate, or a synthon for carbonic acid – phosgene, Cl$_2$C=O, or a safer equivalent) are also important.
- Amongst many examples of 1,2-difunctionalised compounds are 1,2-dicarbonyl compounds, enols (which first react in a nucleophilic sense at carbon and then provide an electrophilic centre (the carbonyl carbon), Hal–C–C=O, and systems with HX–YH units.
- Amongst often used 1,3-difunctionalised compounds are the doubly electrophilic 1,3-dicarbonyl compounds and α,β-unsaturated carbonyl compounds (C=C–C=O), doubly nucleophilic HX–C–YH (amidines and ureas are examples), and α-amino- or α-hydroxycarbonyl compounds (HX–C–C=O), which have an electrophilic and a nucleophilic centre.

The two nucleophilic centres can both be heteroatoms, as in syntheses of pyrimidines and pyrazoles.

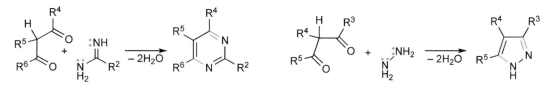

Typical reactant combinations for the synthesis of pyrimidines and pyrazoles

In syntheses of benzanellated systems, phenols can take the part of enols, and anilines react in the same way as enamines.[4]

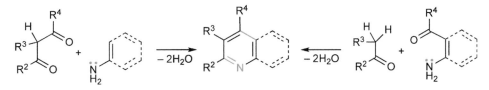

In quinoline syntheses, anilines are like the enamines in pyridine syntheses

6.3 Summary

The chemical steps involved in heteroaromatic synthesis are mostly simple and straightforward, even though a first look at the structures of starting materials and product might make the overall effect seem very mysterious. In devising a sequence of sensible steps it is important to avoid obvious pitfalls, like suggesting that an electrophile react with electrophilic centre, or a nucleophile with a source of electrons.

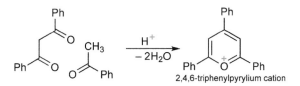

As an illustration, a complete step-by-step analysis of the reaction of 1,3-diphenylpropane-1,3-dione with acetophenone giving 2,4,6-triphenylpyrylium is presented below. Note that although many separate steps are involved, each of them is very simple when considered individually. Note also, that the dicarbonyl tautomer of the diketone is shown, which is in equilibrium with an appreciable percentage of mono-enol tautomer.

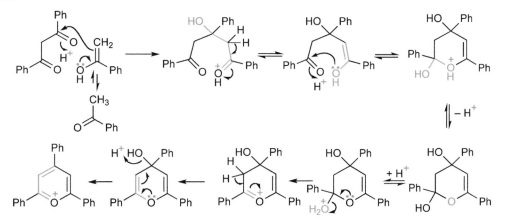

The ring synthesis of 2,4,6-triphenylpyrylium cation, step by step

The sequence shows an initiating step as nucleophilic attack by acetophenone enolate on the protonated diketone, however an equally plausible sequence, shown below, starts with the nucleophilic addition of the enolic hydroxyl of the diketone to protonated acetophenone. We show this alternative to emphasise the uncertainty of the detailed order of events in such multi-step syntheses.

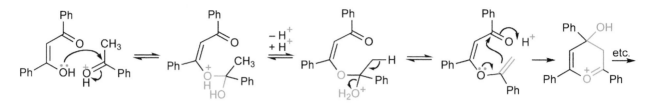

Alternative interpretation for the ring synthesis of 2,4,6-triphenylpyrylium cation, step by step

A final important point to be made is that most of the steps in such sequences are reversible; the overall sequence proceeds to product nearly always because the product is the thermodynamically most stable molecule in the sequence, or because the product is removed from the equilibria by a step which is irreversible under the conditions used. A nice example is the inter-relationship between 1,4-diketones and furans; the latter can be synthesised by heating the former, in acid, under conditions which lead to the distillation of the furan (18.13.1.1), but in the reverse sense, furans are hydrolysed to 1,4-diketones by aqueous acid (18.1.1.1).

6.4 Electrocyclic Processes in Heterocyclic Ring Synthesis

There is a type of electrocyclic process that is of considerable value for heterocyclic ring synthesis: 1,3-dipolar cycloadditions producing five-membered heterocycles.

1,3-Dipoles always contain a heteroatom as the central atom of the trio, either sp or sp^2 hybridised. Amongst other examples, cycloadditions have been demonstrated with azides ($N\equiv N^+-N^--R$), nitrile oxides ($R-C\equiv N^+-O^-$) and nitrile ylides ($R-C\equiv N^+-C^-R_2$), where the central atom is sp-hybridised nitrogen, and with nitrones ($R_2C=N^+(R)-O^-$), carbonyl ylides ($R_2C=O^+-C^-R_2$) and azomethine ylides ($R_2C=N^+(R)-C^-R_2$), where the central atom is sp^2 hybridised.

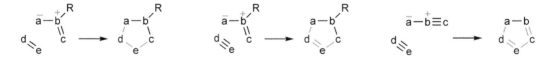

General combinations to produce five-membered heterocycles *via* 1,3-dipolar cycloadditions

Dipolar cycloadditions[5] can, of course, only produce five-membered rings. Addition of dipolarophiles can generate tetrahydro, dihydro or aromatic oxidation level heterocycles, as illustrated above. Alkene dipolarophiles, with a group that can be eliminated following cycloaddition, give the same result as equivalent alkyne dipolarophiles, for example enamines as the dipolarophile, interact with azides, as the 1,3-dipole, with subsequent elimination of the amine, affording 1,2,3-triazoles.[6]

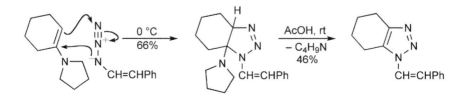

Many mesoionic substances (2.6) can act as 1,3-dipoles, and, after elimination of a small molecule – carbon dioxide in the example shown – produce aromatic heterocycles.[7]

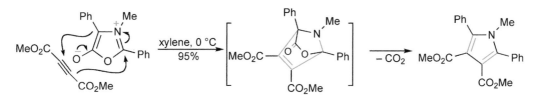

6.5 Nitrenes in Heterocyclic Ring Synthesis[8]

The insertion of a nitrene into a C–H bond has been made the key step in several synthetic routes to both five- and six-membered aromatic systems. A nitrene is a monovalent, six-electron, neutral nitrogen, most often generated by thermolysis or photolysis of an azide ($RN_3 \rightarrow RN + N_2$), or by deoxygenation of a nitro group. The insertion process can be written in a general way:

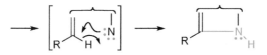

Nitrene insertion can form a ring

The preparation of an indole[9] (nitrene generated from an azide – the Hemetsberger–Knittel synthesis) and of carbazole[10] (nitrene generated by deoxygenation of a nitro group) illustrate the power of the method.

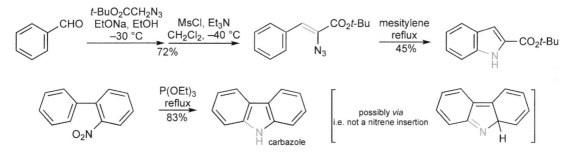

6.6 Palladium Catalysis in the Synthesis of Benzo-Fused Heterocycles

Nucleophilic cyclisations onto palladium-complexed alkenes have been used to prepare indoles, benzofurans and other fused systems. The process can be made catalytic in some cases by the use of reoxidants such as *p*-benzoquinone or copper(II) salts.

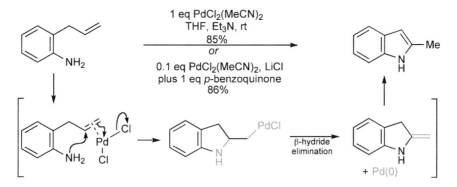

References

[1] 'Synthesis of aromatic heterocycles', Gilchrist, T. L., *J. Chem. Soc., Perkin Trans. 1*, **1999**, 2849, and previous reviews in the series.

[2] '*C*-substitution of nitrogen heterocycles', Vorbrüggen, H. and Maas, M., *Heterocycles*, **1988**, *27*, 2659 (discusses electrophilic and radical substitutions, lithiations and the use of *N*-oxides); 'Regioselective substitution in aromatic six-membered nitrogen heterocycles', Comins, D. L. and O'Connor, S., *Adv. Heterocycl. Chem.*, **1988**, *44*, 199 (discusses electrophilic, nucleophilic and radical substitution, and metallation).

[3] 'The mechanisms of heterocyclic ring closures', Katritzky, A. R., Ostercamp, D. L., and Yousaf, T. I., *Tetrahedron*, **1987**, *43*, 5171.

[4] 'Heteroannelations with *o*-aminoaldehydes', Caluwe, P., *Tetrahedron*, **1980**, *36*, 2359.

[5] '1,3-Dipolar cycloaddition chemistry', Vols. 1 and 2, Ed. Padwa, A., Wiley-Interscience, **1984**.

[6] Nomura, Y., Takeuchi, Y., Tomoda, S. and Ito, M. M., *Bull. Chem. Soc. Jpn.*, **1981**, *54*, 261.

[7] Huisgen, R., Gotthardt, H., Bayer, H. O. and Schaefer, F. C., *Chem. Ber.*, **1970**, *103*, 2611; Potts, K. T. and McKeough, D., *J. Am. Chem. Soc.*, **1974**, *96*, 4268.

[8] 'Synthesis of heterocycles through nitrenes', Kametani, T., Ebetino, F. F., Yamanaka, T. and Nyu, Y., *Heterocycles*, **1974**, *2*, 209.

[9] Kondo, K., Morohoshi, S., Mitsuhashi, M. and Murakami, Y., *Chem. Pharm. Bull.*, **1999**, 1227.

[10] 'Recent advances in the chemistry of carbazoles', Joule, J. A., *Adv. Heterocycl. Chem.*, **1984**, *35*, 84; 'Phosphite-reduction of aromatic nitro-compounds as a route to heterocycles', Cadogan, J. I. G., *Synthesis*, **1969**, 11.

7

Typical Reactivity of Pyridines, Quinolines and Isoquinolines

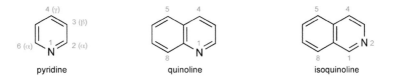

pyridine   quinoline   isoquinoline

The detailed descriptions of the chemistry of the heterocyclic systems covered in this book are preceded at intervals, by six highly condensed and simplified discussions (Chapters 07, 10, 13, 15, 19 and 23) of the types of reaction, ease of such reactions and regiochemistry of such reactions for groups of related heterocycles. In this chapter the group comprises pyridine, as the prototype electron-poor six-membered heterocycle, and its benzo-fused analogues, quinoline and isoquinoline. As in each of these summary chapters, reactions are shown in brief and either as the simplest possible example, or in general terms.

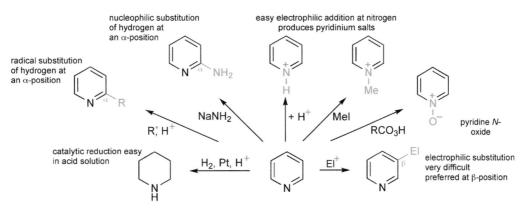

Typical reactions of pyridine

The formal replacement of a CH in benzene by N leads to far-reaching changes in typical reactivity: pyridines are much less susceptible to electrophilic substitution than benzene and much more susceptible to nucleophilic attack. However, pyridine undergoes a range of simple electrophilic additions, some reversible, some forming isolable products, each involving donation of the nitrogen lone pair to an electrophile, and thence the formation of 'pyridinium' salts which, of course, do not have a counterpart in benzene chemistry at all. The ready donation of the pyridine lone pair in this way does not destroy the aromatic

Heterocyclic Chemistry 5th Edition   John Joule and Keith Mills
© 2010 Blackwell Publishing Ltd

sextet (compare with pyrrole, Chapters 15 and 16) – pyridinium salts are still aromatic, though much more polarised than neutral pyridines.

Pyridines react with electrophiles by donation of the nitrogen lone pair

Electrophilic substitution of aromatic compounds proceeds *via* a two-step sequence – addition (of X⁺) then elimination (of H⁺), of which the former is usually the slower (rate-determining) step. Qualitative predictions of relative rates of substitution at different ring positions can be made by inspecting the structures of the σ-complexes (Wheland intermediates) formed in the first step, on the assumption that their relative stabilities reflect the relative energies of the transition states that lead to them.

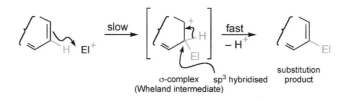

σ-complex   sp³ hybridised   substitution product
(Wheland intermediate)

Aromatic electrophilic substitution *via* an addition/elimination sequence

Electrophilic substitution at carbon, in simple pyridines at least, is very difficult, in contrast to the reactions of benzene – Friedel–Crafts acylations, for example, do not occur at all with pyridines. This unreactivity can be traced to two factors:

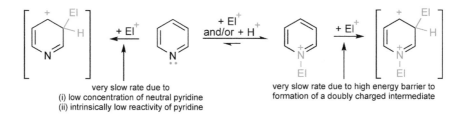

very slow rate due to
(i) low concentration of neutral pyridine
(ii) intrinsically low reactivity of pyridine

very slow rate due to high energy barrier to
formation of a doubly charged intermediate

- Exposure of a pyridine to a medium containing electrophilic species immediately converts the heterocycle into a pyridinium cation, with the electrophile (or a proton from the medium) attached to the nitrogen. The extent of conversion depends on the nature and concentration of the electrophile (or protons) and the basicity of the particular pyridine, and is usually nearly complete. Obviously, the positively charged pyridinium cation is many orders of magnitude less easily attacked by the would-be electrophile, at carbon, than the original neutral heterocycle. The electrophile, therefore, has Hobson's choice – it must either attack an already positively charged species, or seek out a neutral pyridine from the very low concentration of uncharged pyridine molecules.
- The carbons of a pyridine are, in any case, electron-poor, particularly at the α- and γ-positions: formation of a σ-complex between a pyridine and an electrophile is intrinsically disfavoured. The least disfavoured, i.e. best option, is attack at a β-position – resonance contributors to the cation thus produced do not include one with the particularly unfavourable sextet, positively-charged nitrogen situation (shown in

parentheses for the α- and γ-intermediates). The situation has a direct counterpart in benzene chemistry, where a consideration of possible intermediates for electrophilic substitution of nitrobenzene provides a rationalisation of the observed *meta*-selectivity.

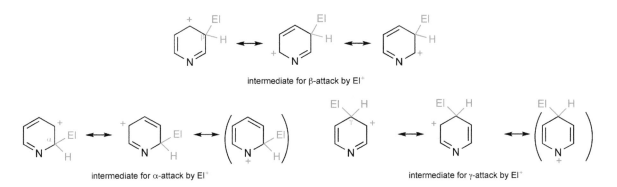

intermediate for β-attack by El⁺

intermediate for α-attack by El⁺   intermediate for γ-attack by El⁺

Substituents can exert a significant influence on the ease of electrophilic attack, just as in benzene chemistry. Strongly electron-withdrawing substituents simply render the pyridine even more inert, however activating groups – amino and oxy, and even alkyl – allow substitution to take place, even though by way of the protonated heterocycle i.e. *via* a dicationic intermediate. The presence of halogen substituents, which have a base-weakening effect and are only weakly deactivating, can allow substitution to take place in a different way – by allowing an appreciably larger concentration of the free, neutral pyridine to be present.

Pyridine rings are resistant to oxidative destruction, as are benzene rings. In terms of reduction, however, the heterocyclic system is much more easily catalytically reduced, especially in acidic solution. Similarly, pyridinium salts can be easily reduced both with hydrogen over a catalyst, and by nucleophilic chemical reducing agents.

Nucleophilic substitution of aromatic compounds proceeds *via* an addition (of Nu⁻) then elimination (of a negatively charged entity, most often Hal⁻) two-step sequence, of which the former is usually rate-determining. Rates of substitution at different ring positions can be assessed by inspecting the structures of the negatively charged intermediates (Meisenheimer complexes) thus formed, on the assumption that their relative stabilities (degree of delocalisation of negative charge) reflect the relative energies of the transition states that lead to them. For example, 2- and 4-halonitro-benzenes are substituted by nucleophiles because the anionic adduct derives stabilisation by delocalisation of the charge onto the nitro group(s).

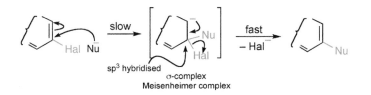

sp³ hybridised
σ-complex
Meisenheimer complex

Aromatic nucleophilic substitution *via* an addition/elimination sequence

The electron-deficiency of the carbons in pyridines, particularly α- and γ-carbons, makes nucleophilic addition and, especially nucleophilic displacement of halide (and other good leaving groups), a very important feature of pyridine chemistry.

Nucleophilic displacement of pyridine α- and γ-leaving groups, e.g. halide, is easy

Such substitutions follow the same mechanistic route as the displacement of halide from 2- and 4-halo-nitrobenzenes, i.e. the nucleophile first adds and then the halide departs. By analogy with the benzenoid situation, the addition is facilitated by: (i) the electron-deficiency at α- and γ-carbons, further increased by the halogen substituent, and (ii) the ability of the heteroatom to accommodate negative charge in the intermediate thus produced. A comparison of the three possible intermediates makes it immediately plain that this latter is not available for attack at a β-position, and thus β nucleophilic displacements are very much slower – for practical purposes they do not occur (see, however, reactions with palladium catalysis, 4.2)

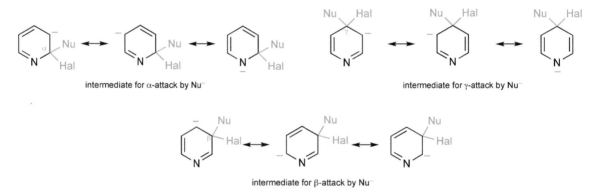

Intermediates explain selectivity of nucleophilic attack on halopyridines

It is useful to compare the reactivity of α- and γ-halopyridines with the reaction of acid halides and β-halo-α,β-unsaturated ketones, respectively, both of which also interact easily with nucleophiles and also by an addition/elimination sequence resulting in overall displacement of the halide by the nucleophile.

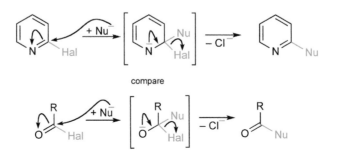

Comparison of the reactivity of an α-halopyridine with an acid chloride

In the absence of an α- or γ-halogen, pyridines are less reactive and, of course, do not have a substituent suitable for leaving as an anion to complete a nucleophilic substitution. Nucleophilic additions do however take place, but the resultant dihydropyridine adduct requires removal of 'hydride' in some way, to complete an overall substitution. Such reactions, for example with sodium amide or with organometallic reagents, are selective for an α-position, possibly because the nucleophile is delivered *via* a complex involving interaction of the ring nitrogen with the metal cation associated with the nucleophile. The addition of organometallic or hydride reagents to N^+-acylpyridinium salts is an extremely useful process: the products, 1,2- or 1,4-dihydropyridines, are stable because the nitrogen electron pair is involved with resonance in the carbamate unit.

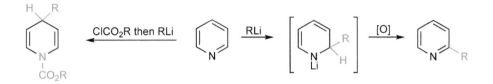

The generation and use of metallated aromatics has become extremely important for the introduction of substituents, especially carbon substituents, by subsequent reaction with an electrophile. Despite the ease of nucleophilic addition and substitution discussed above, iodine and bromine at all positions of a pyridine undergo metal/halogen exchange, at low temperature, *without* nucleophilic displacement or addition, thus forming the corresponding pyridyllithiums. Low-temperature direct lithiation of pyridines at an α-position, or elsewhere *via* directed *ortho*-metallation, is also possible. Similarly, useful pyridyl Grignard reagents are available by reaction of bromopyridines with *iso*-propylmagnesium chloride *at room temperature*.

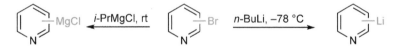

Formation of pyridyllithiums and pyridyl Grignard reagents

Radical substitution of pyridines, in acid solution, is a preparatively useful process. For efficient reaction, the radicals must be 'nucleophilic', like ˙CH_2OH, alkyl˙, and acyl˙. A hydroxymethylation provides the example shown.

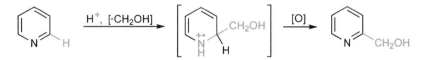

Pyridines carrying oxygen at an α- or γ-position exist as tautomers having carbonyl groups – pyridones. Nonetheless, there is considerable parallelism between their reactions and those of phenols: pyridones are activated towards electrophilic substitution, attack taking place *ortho* and *para* to the oxygen. They readily form anions, by loss of the *N*-hydrogen, which are analogous in structure and reactivity to phenolates, though in the heterocyclic system, the anion can react with an electrophile at either oxygen or nitrogen, depending on conditions.

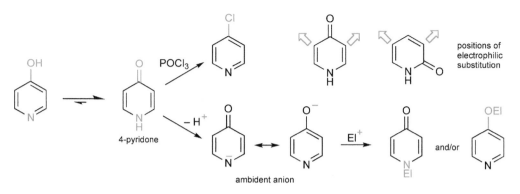

Typical reactions of pyridones, illustrated for 4-pyridone

Where pyridones differ from phenols is in their interaction with reagents such as POCl₃, where transformation of the oxygen substituent into halide occurs. Here, the pyridones react in an amide-like fashion, the inorganic reagent reacting first at the oxygen.

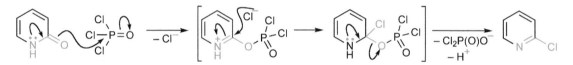

Mechanism of reaction of pyridones with phosphoryl chloride, illustrated for 2-pyridone

The special properties associated with pyridine α- and γ-positions are evident again in the reactions of alkyl-pyridines: protons on alkyl groups at those positions are particularly acidified because the 'enaminate' anions formed by side-chain deprotonation are delocalised. The ability to form side-chain anions provides a useful means for the manipulation of α- and γ-side-chains.

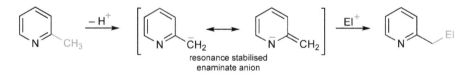

Deprotonation of α- and γ-alkyl groups is relatively easy

Pyridinium salts show the properties that have been discussed above, but in extreme, thus they are highly resistant to electrophilic substitution but, conversely, nucleophiles add very easily. Especially useful are the adducts formed from N^+-CO₂R salts with alkyl- or aryllithiums (see above). The hydrogens of pyridinum α- and γ-alkyl side-chains are further acidified compared with an uncharged alkyl pyridine.

N-oxide chemistry, which self-evidently has no parallel in benzenoid chemistry, is an extremely important and useful aspect of the chemistry of azines. The structure of *N*-oxides means that they are both more susceptible to electrophilic substitution *and* react more easily with nucleophiles – an extraordinary concept when first encountered. On the one hand, the formally negatively charged oxygen can release electrons to stabilise an intermediate for electrophilic attack and, on the other, the positively charged ring nitrogen can act as an electron sink to encourage nucleophilic addition.

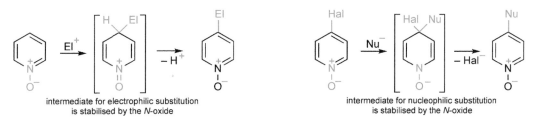

| intermediate for electrophilic substitution is stabilised by the *N*-oxide | intermediate for nucleophilic substitution is stabilised by the *N*-oxide |

The *N*-oxide group facilitates both electrophilic *and* nucleophilic substitutions

There are a number of very useful processes in which the *N*-oxide function allows the introduction of substituents, usually at an α position, and in the process the oxide function is removed; reaction with phosphoryl chloride is an example.

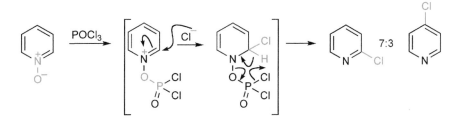

Conversion of pyridine *N*-oxide into halopyridines (mechanism shown for α-substitution)

Quinoline and isoquinoline, the two possible structures in which a benzene ring is annelated to a pyridine ring, represent an opportunity to examine the effect of fusing one aromatic ring to another. Clearly, both the effect that the benzene ring has on the reactivity of the pyridine ring, and *vice versa*, as well as comparison with the chemistry of naphthalene must be considered. Firstly, it will be clear from the discussion of pyridine with electrophiles that, of the two rings, electrophilic substitution favours the benzenoid ring, rather than the pyridine ring. Regioselectivity, which in naphthalene favours an α-position, is mirrored in quinoline/isoquinoline chemistry by preferred substitution at the 5- and 8-positions. It should be noted that such substitutions usually involve attack on the species formed by electrophilic addition (often protonation) at the nitrogen, which has the effect of further discouraging (preventing) attack on the heterocyclic ring.

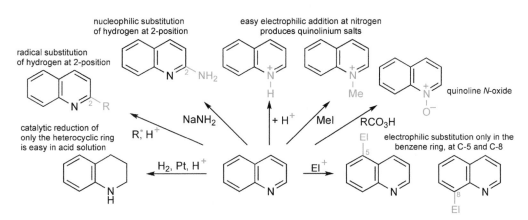

Typical reactions of quinoline (isoquinoline is very similar)

Just as for naphthalene, the regiochemistry of attack is readily interpreted by looking at the possible intermediates: those for attack at C-5/8 allow delocalisation of charge, while an intermediate for attack at C-6/7 would have a localised charge.

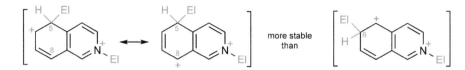

Electrophilic substitution of quinoline is selective for the benzene ring at 5- (shown) and 8-positions

Just as quinoline and isoquinoline are reactive towards electrophiles in their benzene rings, so they are reactive to nucleophiles in the pyridine ring, especially (see above) at the positions α and γ to the nitrogen and, further, are more reactive in this sense than pyridines. This is consistent with the structures of the intermediates for, in these, a full and complete aromatic benzene ring is retained. Since the resonance stabilisation of the bicyclic aromatic is considerably less than twice that of either benzene or pyridine, the loss in resonance stabilisation in proceeding from the bicyclic system to the intermediate is considerably less than in going from pyridine to an intermediate adduct. There is an obvious analogy: the rate of electrophilic substitution of naphthalene is greater than that of benzene for, in forming a σ-complex from the former, less resonance energy is sacrificed.

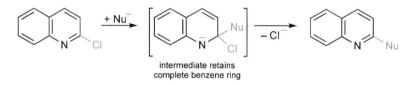

Intermediate for nucleophilic substitution of 2- and 4-halo-quinolines retains a complete benzene ring

A significant difference in this typical behaviour applies to the isoquinoline 3-position – the special reactivity that the discussion above has developed for positions α to pyridine nitrogen, and that also applies to the isoquinoline 1-position, does not apply at C-3. In the context of nucleophilic displacements, for example, an intermediate for reaction of a 3-halo-isoquinoline cannot achieve delocalisation of negative charge onto the nitrogen unless the aromaticity of the benzene ring is disrupted. Therefore, such intermediates are considerably less stabilised and reactivity considerably tempered.

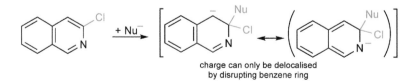

3-Haloisoquinolines do not undergo easy nucleophilic substitution

The displacement of halogen at all positions of the pyridine and quinoline/isoquinoline nucleus is achievable using palladium(0) catalysis (see 4.2 for a detailed discussion). Couplings with alkenes (Heck reactions), with alkynes, with alkenyl- or aryltin or -boron species are complemented by couplings in the opposite sense using pyridinyl/quinolinyl/isoquinolinyl metal reagents, with alkenyl or aryl halides or triflates. The application of this extremely useful methodology allows transformations in one step that would otherwise require extensive sequences of steps. Two examples, chosen at random, are shown below.

Transition-metal-catalysed processes are very important in pyridine/quinoline/isoquinoline chemistry

A great variety of methods is available for the ring synthesis of pyridines: the most obvious approach is to construct a 1,5-dicarbonyl compound, preferably also having further unsaturation, and allow it to react with ammonia, when loss of two mole equivalents of water produces the pyridine. 1,4-Dihydropyridines, which can easily be dehydrogenated to the fully aromatic system, result from the interaction of saturated 1,5-dicarbonyl compounds and ammonia.

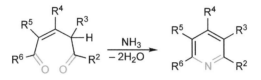

Typical pyridine ring synthesis

Nearly all quinoline syntheses begin from an arylamine; that shown generally below – the acid-catalysed interaction with a 1,3-diketone – involves addition of the amine nitrogen to one of the carbonyl groups and a ring closure onto the aromatic ring having the character of an aromatic electrophilic substitution. Another much-used route utilises the interaction of an *ortho*-aminoaraldehyde (or -ketone) with a ketone having an α methylene.

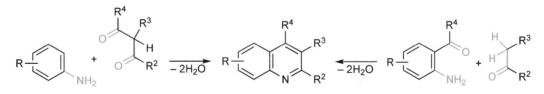

Typical quinoline ring syntheses

Amides of 2-(aryl)ethanamines can be made to ring close producing 3,4-dihydroisoquinolines (which can be easily dehydrogenated to the aromatic systems) using reagents such as phosphoryl chloride; again, the ring-closure step is an intramolecular electrophilic substitution of the aromatic ring.

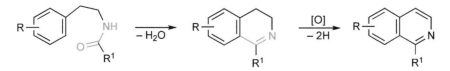

A typical isoquinoline ring synthesis

8

Pyridines: Reactions and Synthesis

pyridine

Pyridine and its simple derivatives are stable and relatively unreactive liquids, with strong penetrating odours that are unpleasant to some people. They are much used as solvents and bases, especially pyridine itself, in reactions such as *N*- and *O*-acylation and -tosylation. Pyridine and the three monomethyl pyridines (picolines) are completely miscible with water.

Pyridine was first isolated, like pyrrole, from bone pyrolysates: the name is constructed from the Greek for fire, 'pyr', and the suffix 'idine', which was at the time being used for all aromatic bases – phenetidine, toluidine, etc. Pyridine and its simple alkyl derivatives were for a long time produced by isolation from coal tar, in which they occur in quantity. In recent years this source has been displaced by synthetic processes: pyridine itself, for example, can be produced on a commercial scale in 60–70% yields by the gas-phase high-temperature interaction of crotonaldehyde, formaldehyde, steam, air and ammonia over a silica–alumina catalyst. Processes for the manufacture of alkyl-pyridines involve reaction of acetylenes and nitriles over a cobalt catalyst.

8.1 Reactions with Electrophilic Reagents
8.1.1 Addition to Nitrogen
In reactions that involve bond formation using the lone pair of electrons on the ring nitrogen, such as protonation and quaternisation, pyridines behave just like tertiary aliphatic or aromatic amines. When a pyridine reacts as a base or a nucleophile it forms a 'pyridinium', cation in which the aromatic sextet is retained and the nitrogen acquires a formal positive charge.

8.1.1.1 Protonation of Nitrogen
Pyridines form crystalline, frequently hygroscopic, salts with most protic acids. Pyridine itself, with pK_{aH} 5.2 in water, is a much weaker base than saturated aliphatic amines which have pK_{aH} values mostly between 9 and 11. Since the gas-phase proton affinity of pyridine is actually very similar to those of aliphatic amines, the observed solution values reflect relatively strong solvation of aliphatic ammonium cations;[1] this difference may in turn be related to the mesomerically delocalised charge in pyridinium ions and the consequent reduced requirement for external stabilisation *via* solvation.

Electron-releasing substituents generally increase the basic strength; 2-methyl- (pK_{aH} 5.97), 3-methyl (5.68) and 4-methylpyridine (6.02) illustrate this. The basicities of pyridines carrying groups that can interact mesomerically as well as inductively vary in more complex ways, for example 2-methoxypyridine (3.3) is a weaker, but 4-methoxypyridine (6.6) a stronger base than pyridine; the effect of inductive with-

Heterocyclic Chemistry 5th Edition   John Joule and Keith Mills
© 2010 Blackwell Publishing Ltd

drawal of electrons by the electronegative oxygen is felt more strongly when it is closer to the nitrogen, i.e. at C-2.

Large α-substituents impede solvation of the protonated form: 2,6-di-*t*-butylpyridine is less basic than pyridine by one pK_{aH} unit and 2,6-bis(tri-*iso*-propylsilyl)pyridine will not dissolve even in 6M hydrochloric acid.[2]

8.1.1.2 Nitration at Nitrogen (see also 8.1.2.2)

This occurs readily by reaction of pyridines with nitronium salts, such as nitronium tetrafluoroborate.[3] Protic nitrating agents such as nitric acid of course lead exclusively to *N*-protonation.

1-Nitro-2,6-dimethylpyridinium tetrafluoroborate is one of several *N*-nitro-pyridinium salts that can be used as non-acidic nitrating agents with good substrate and positional selectivity. The 2,6-disubstitution serves to sterically inhibit resonance overlap between nitro group and ring and consequently increase reactivity as a nitronium ion donor, however the balance between this advantageous effect and hindering approach of the aromatic substrate is illustrated by the lack of transfer nitration reactivity in 2,6-dihalo-analogues.[4]

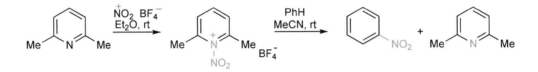

8.1.1.3 Amination of Nitrogen

The introduction of nitrogen at a different oxidation level can be achieved with hydroxylamine *O*-sulfonic acid[5] or using [*N-para*-tolylsulfonylimino]phenyliodinane with copper(II) triflate;[6] the attacking species is a nitrene.

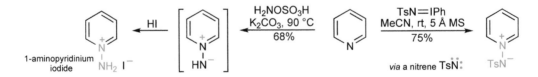

8.1.1.4 Oxidation of Nitrogen

In common with other tertiary amines, pyridines react smoothly with percarboxylic acids to give *N*-oxides, which have their own rich chemistry (8.13). There are many other ways to *N*-oxidise pyridines: oxygen with ruthenium trichloride as catalyst is one example;[7] hydrogen-peroxide–urea with trifluoroacetic anhydride *N*-oxidises pyridines carrying electron-withdrawing groups.[8] Similarly, there are many ways to deoxygenate pyridine *N*-oxides: samarium iodide, chromous chloride, stannous chloride with low-valent titanium, ammonium formate with palladium and catalytic hydrogenation all do the job at room temperature,[9] molybdenum hexacarbonyl in hot ethanol is another alternative.[10] The most frequently used methods have involved oxygen transfer to trivalent phosphorus[11] or divalent sulfur.[12] Ammonium formate with a palladium-on-carbon catalyst removes the oxygen *and* reduces the ring, smoothly giving piperidines.[13]

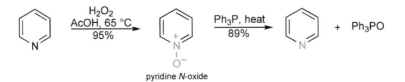

pyridine *N*-oxide

8.1.1.5 Sulfonation at Nitrogen

Pyridine reacts[14] with sulfur trioxide to give the crystalline, zwitterionic pyridinium-1-sulfonate, usually known as the pyridine sulfur trioxide complex. This compound is hydrolysed in hot water to sulfuric acid and pyridine (for its reaction with hydroxide see 8.12.3), but more usefully it can serve as a mild sulfonating agent (for examples see 16.1.1.3 and 18.1.1.3) and as an activating agent for dimethylsulfoxide in Moffat oxidations.

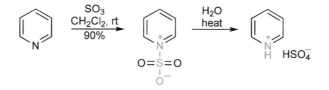

When pyridine is treated with thionyl chloride, a synthetically useful dichloride salt is formed, which can, for example, be transformed into pyridine-4-sulfonic acid. The reaction is believed to involve initial attack by sulfur at nitrogen, followed by nucleophilic addition of a second pyridine at C-4 (cf. 8.12.2).[15]

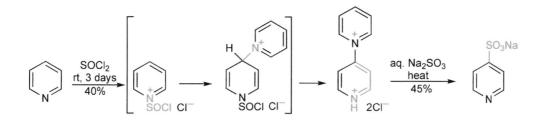

8.1.1.6 Halogenation at Nitrogen

Pyridines react easily with halogens and inter-halogens[16] to give crystalline compounds, largely undissociated when dissolved in solvents such as carbon tetrachloride. Structurally they are best formulated as resonance hybrids related to trihalide anions. 1-Fluoropyridinium triflate is also crystalline and serves as an electrophilic fluorinating agent (31.1).[17]

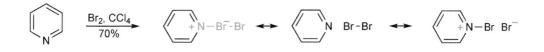

These salts must be distinguished from pyridinium tribromide, obtained by treating pyridine hydrobromide with bromine, which does not contain an *N*-halogen bond, but does have a trihalide anion. The stable, crystalline, commercially available salt can be used as a source of molecular bromine, especially where small accurately known quantities are required.

pyridinium bromide
(pyridine hydrobromide)   pyridinium tribromide

8.1.1.7 Acylation at Nitrogen

Carboxylic, and arylsulfonic acid halides react rapidly with pyridines generating 1-acyl- and 1-arylsulfonyl-pyridinium salts in solution, and in suitable cases some of these can even be isolated as crystalline, non-hygroscopic solids.[18] Solutions of these salts, generally in excess pyridine, are commonly used for the preparation of esters and sulfonates from alcohols, and of amides and sulfonamides from amines. 4-Dimethylaminopyridine[19] (DMAP) is widely used (in catalytic quantities) to activate anhydrides in a similar manner. The salt derived from DMAP and *t*-butyl chloroformate is stable even in aqueous solution at room temperature.[20] The more stable these salts are, the higher their catalytic activity in acylation reactions.[21]

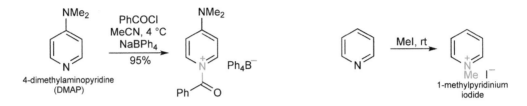

8.1.1.8 Alkylation at Nitrogen

Alkyl halides and sulfates react readily with pyridines at room temperature, giving quaternary *N*-substituted pyridinium salts. As with aliphatic tertiary amines, increasing substitution around the nitrogen, or around the halogen-bearing carbon, causes an increase in the alternative, competing, elimination process, which gives alkene and *N*-proto-pyridinium salt, thus 2,4,6-trimethylpyridine (collidine) is used as a base in dehydrohalogenation reactions. A process which is useful when an alcohol, but not the halide, is available is to use the protonic borofluoride of the pyridine (this also works with imidazoles) and react this with the alcohol in a Mitsunobu reaction.[22]

8.1.2 Substitution at Carbon

In most cases, electrophilic substitution of pyridines occurs very much less readily than for the correspondingly substituted benzene. The main reason is that the electrophilic reagent, or a proton in the reaction medium, adds first to the pyridine nitrogen, generating a pyridinium cation, which is naturally very resistant to attack by an electrophile. When it does occur, electrophilic substitution at carbon must involve *either* highly unfavoured attack on a pyridinium cation *or* a relatively easier attack, but on a very low equilibrium concentration of uncharged free pyridine base.

Some of the typical benzene electrophilic substitution reactions do not occur at all: Friedel–Crafts alkylation and acylation fail because pyridines form complexes with the Lewis-acid catalyst required, involving donation of the nitrogen lone pair to the metal centre. Milder electrophilic species, such as Mannich cations, diazonium ions or nitrous acid, which in any case require activated benzenes for success, naturally fail with pyridines.

Electrophilic *C*-substitution in pyridines carrying strongly activating substituents (nitrogen and oxygen) is discussed in Sections 8.9.3.1 and 8.9.2.1.

8.1.2.1 Proton Exchange

H–D exchange *via* an electrophilic substitution process, such as will operate for benzene, does not take place with pyridine. A special mechanism allows selective exchange at the two α-positions in DCl–D$_2$O, or even in water at 200 °C, the key species being an ylide formed by 2/6-deprotonation of the 1*H*-pyridinium cation (see also 8.11).[23] Efficient exchange at all positions can be achieved at 110 °C in D$_2$O in the presence of hydrogen and palladium-on-carbon (a method which also works for other heterocycles, including indoles).[24]

8.1.2.2   Nitration

Pyridine itself can be converted into 3-nitropyridine only inefficiently by direct nitration, even with extremely vigorous conditions,[25] however a couple of ring methyl groups facilitate electrophilic substitution sufficiently to allow nitration;[26] both collidine (2,4,6-trimethylpyridine) and its *N*-methyl quaternary salt are nitrated at similar rates under the same conditions, showing that the former reacts *via* its *N*-protonic salt.[27] Steric or/and inductive inhibition of *N*-nitration allows *C*-3-substitution using nitronium tetrafluoroborate; an example is the nitration of 2,6-dichloropyridine[4] or of 2,6-difluoropyridine using tetramethylammonium nitrate with trifluoromethansulfonic anhydride.[28]

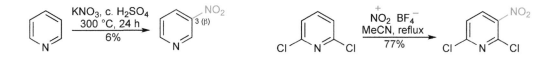

3-Nitropyridine itself, and substituted derivatives, *can*, however, be prepared efficiently from *N*-nitropyridinium salts. Initial reaction with dinitrogen pentoxide at nitrogen is followed by sulfur dioxide, when this is used as solvent or co-solvent, or hydrogensulfite, addition at C-2 forming a 1,2-dihydropyridine. Transfer of the nitro group to a β-position, *via* a [1,5]-sigmatropic migration, is then followed by elimination of the nucleophile, regenerating the aromatic system.[29]

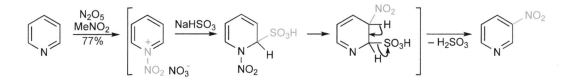

The same effect can be achieved more conveniently using a mixture of nitric acid and trifluoroacetic anhydride (NO_2OCOCF_3 adds NO_2^+ to the nitrogen) then the addition of sodium metabisulfite.[30] Incidentally, if cyanide is added instead of metabisulfite, elimination of nitrous acid produces 2-cyanopyridines.[31]

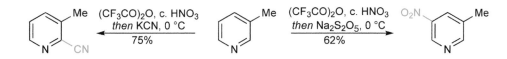

8.1.2.3   Sulfonation

Pyridine is very resistant to sulfonation using concentrated sulfuric acid or oleum, only very low yields of the 3-sulfonic acid being produced after prolonged reaction periods at 320 °C. However, addition of mercuric sulfate in catalytic quantities allows smooth sulfonation at a somewhat lower temperature. The role of the catalyst is not established; one possibility is that *C*-mercuration is the first step (cf. 8.1.2.5).[32]

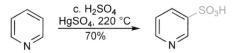

The *C*-sulfonation of 2,6-di-*t*-butylpyridine with sulfur trioxide[33] is a good guide to the intrinsic reactivity of a pyridine ring, for in this situation the bulky alkyl groups effectively prevent addition of sulfur trioxide to the ring nitrogen, allowing progress to a 'normal' electrophilic *C*-substitution intermediate, at about the same rate as for sulfonation of nitrobenzene. A maximum conversion of 50% is all that is achieved, because for every *C*-substitution a proton is produced, which deactivates a molecule of starting material by *N*-protonation.

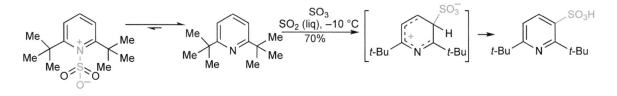

8.1.2.4   Halogenation

3-Bromopyridine is produced in good yield by the action of bromine in oleum.[34] The process is thought to involve pyridinium-1-sulfonate as the reactive species, since no bromination occurs in 95% sulfuric acid. 3-Chloropyridine can be produced by chlorination at 200 °C, or at 100 °C in the presence of aluminium chloride.[35]

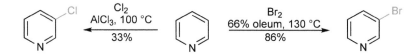

2-Bromo- and 2-chloro-pyridines can be made efficiently by reaction of pyridine with the halogen, at 0–5 °C in the presence of palladium(II) chloride.[36]

8.1.2.5   Acetoxymercuration

The salt formed by the interaction of pyridine with mercuric acetate at room temperature can be rearranged to 3-acetoxymercuripyridine by heating to only 180 °C.[37] This process, where again there is *C*-attack by a relatively weakly electrophilic reagent, like that described for mercuric-sulfate-catalysed sulfonation, may involve attack on an equilibrium concentration of free pyridine.

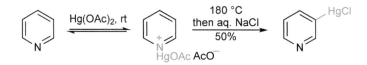

8.2   Reactions with Oxidising Agents

The pyridine ring is generally resistant to oxidising agents, vigorous conditions being required for its breakdown, thus pyridine itself is oxidised by neutral aqueous potassium permanganate at about the same rate as benzene (sealed tube, 100 °C), to give carbon dioxide. In acidic solution, pyridine is more resistant, but in alkaline media more rapidly oxidised, than benzene.

In most situations, carbon substituents can be oxidised with survival of the ring, thus alkyl-pyridines can be converted into pyridine carboxylic acids with a variety of reagents.[38] Some selectivity can be achieved: only α- and γ-groups are attacked by selenium dioxide; the oxidation can be halted at the aldehyde oxidation level.[39]

8.3 Reactions with Nucleophilic Reagents

Just as electrophilic substitution is the characteristic reaction of benzene and electron-rich heteroaromatic compounds (pyrrole, furan etc.), so substitution reactions with nucleophiles can be looked on as characteristic of pyridines.

Nucleophilic substitution of hydrogen differs in an important way from electrophilic substitution: whereas the last step in electrophilic substitution is loss of proton, an easy process, the last step in nucleophilic substitution of hydrogen has to be a hydride transfer, which is less straightforward and generally needs the presence of an oxidising agent as hydride acceptor. Nucleophilic substitution of an atom or group at an α- or γ-position, that is a good leaving group, is, however, usually an easy and straightforward process.

8.3.1 Nucleophilic Substitution with 'Hydride' Transfer[40]

8.3.1.1 Alkylation and Arylation

Reaction with alkyl- or aryl-lithiums proceeds in two discrete steps: addition to give a dihydro-pyridine *N*-lithio-salt which can then be converted into the substituted aromatic pyridine by oxidation (e.g. by air), disproportionation or elimination of lithium hydride.[41] The *N*-lithio-salts can be observed spectroscopically and in some cases isolated as solids.[42] Attack is nearly always at an α-position; reaction with 3-substituted-pyridines usually takes place at both available α-positions, but predominantly at C-2;[43] this regioselectivity may be associated with relief of strain when C-2 rehybridises to sp³ during addition.

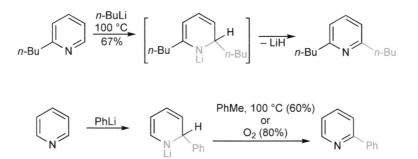

It is possible to trap the 2-substituted lithium derivatives produced by organometallic addition with electrophiles: for example the use of di-*t*-butyl azodicarboxylate leads, after a simple aerial oxidative re-aromatisation, to 2,5-disubstituted pyridines.[44]

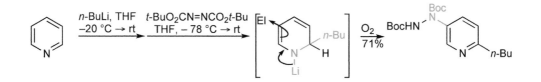

From the preparative viewpoint, nucleophilic alkylations can be *greatly* facilitated by the device of prior quaternisation of the pyridine in such a way that the *N*-substituent can be subsequently removed – these processes are dealt with in 8.12.2.

8.3.1.2 Amination
Amination of pyridines and related heterocycles, generally at a position α to the nitrogen, is called the Chichibabin reaction,[45] the pyridine reacting with sodamide with the evolution of hydrogen. The 'hydride' transfer and production of hydrogen probably involve interaction of amino-pyridine product, acting as an acid, with the anionic intermediate. The preference for α-substitution may be associated with an intramolecular delivery of the nucleophile, perhaps guided by complexation of ring nitrogen with metal cation.

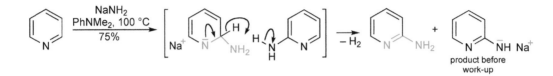

More vigorous conditions are required for the amination of 2- or 4-alkyl-pyridines, since proton abstraction from the side-chain (cf. 8.10) by the amide occurs first, and ring attack must therefore involve a dianionic intermediate.[46] Amination of 3-alkyl-pyridines is regioselective for the 2-position.[47]

Vicarious nucleophilic substitution (3.3.3) permits the introduction of amino groups *para* (or *ortho* if *para* blocked) to nitro groups by reaction with methoxyamine[48] or 1-amino-1,2,4-triazole.[49] In contrast, VNS substitution of 3-nitropyridine with benzyl chloroacetate proceeds at C-4.[50]

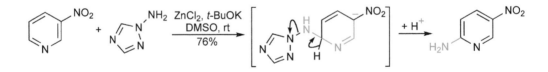

8.3.1.3 Silylation
In an exceptionally efficient process, pyridine is converted into 4-trimethylsilylpyridine on reaction with trimethylsiliconide anion. This process probably proceeds *via* a 1,4-dihydro-adduct (which can be trapped as its *N*-CO$_2$Et derivative by addition of ethyl chloroformate), the fully aromatic product arising *via* hydride shift to silicon.[51]

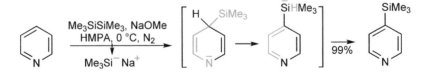

8.3.1.4 Hydroxylation
Hydroxide ion, being a much weaker nucleophile than amide, attacks pyridine only at very high temperatures to produce a low yield of 2-pyridone,[52] which can be usefully contrasted with the much more efficient reaction of hydroxide with quinoline and isoquinoline (9.3.1.3) and with pyridinium salts (8.12.3).

8.3.2 Nucleophilic Substitution with Displacement of Good Leaving Groups

Halogen, and also, though with fewer examples, nitro,[53] alkoxysulfonyloxy[54] and methoxy[55] substituents at α- or γ-positions, but not at β-positions, are relatively easily displaced by a wide range of nucleophiles *via* an addition–elimination mechanism facilitated by: (i) electron withdrawal by the substituent and (ii) the good leaving ability of the substituent. γ-Halo-pyridines are more reactive than the α-isomers; β-halo-pyridines are *very* much less reactive, being much closer to, but still somewhat more reactive than halo-benzenes, but even 3-halogen can be displaced with heteroatom nucleophiles under microwave irradiation.[56]

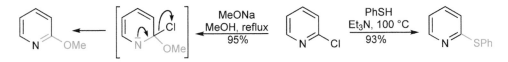

Fluorides are more reactive than the other halides,[57] (cf. 3.3.2) for example 2-fluoropyridine can be converted into 2-dialkylamino-pyridines using lithium amides at room temperature.[58] This could be compared with the 130 °C required to displace α-bromine using the potassium salt of pyrazole.[59] Displacement of nitro can be made the means for the synthesis of α- and γ-fluoro-pyridines.[53] Of the five fluorines in pentafluoropyridine, the γ-fluorine is displaced most rapidly.[60]

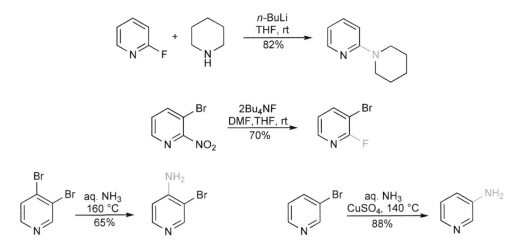

Replacements of halide by reaction with ammonia can be achieved at considerably lower temperatures than those illustrated, under 6–8 kbar pressure.[61]

The sulfonic acid substituent in 5-nitropyridine-2-sulfonic acid[62] can be displaced by alcohols, amines or chloride.[63]

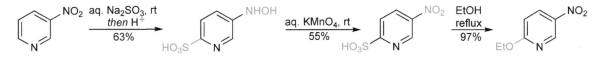

In some displacements, an alternative mechanism operates. For example the reaction of either 3- or 4-bromopyridine with secondary amines in the presence of sodamide/sodium *t*-butoxide, produces the same mixture of 3- and 4-dialkylamino-pyridines; this proceeds *via* an elimination process (S$_N$(EA) – **S**ubstitution **N**ucleophilic **E**limination **A**ddition) and the intermediacy of 3,4-didehydropyridine (3,4-pyridyne).[64] That no 2-aminated pyridine is produced shows a greater difficulty in generating 2,3-pyridyne; it can, however, be formed by reaction of 3-bromo-2-chloro-pyridines with *n*-butyllithium[65] or *via* the reaction of 3-trimethylsilyl-2-trifluoromethanesulfonyloxypyridine with fluoride.[66] Significantly, a 4-aryloxy or 4-phenylthio-substituent stabilises a 2,3-pyridyne.[67]

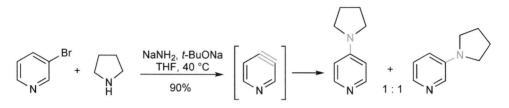

It is possible to replace α-chlorine with bromine or iodine by reaction with the halotrimethylsilane; no doubt this involves an intermediate pyridinium salt, as shown (see also 8.12.2).[68]

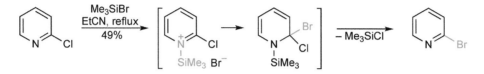

Carbon nucleophiles can also be used: deprotonated nitriles will displace a halogen;[69,70] electron-rich aromatic compounds will displace α-halogen *ortho* or *para* to nitro or cyano, using aluminium chloride catalysis.[71]

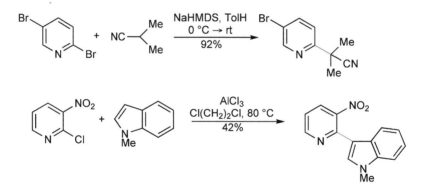

8.4 Metallation and Reactions of C-Metallated-Pyridines

8.4.1 Direct Ring C–H Metallation[72]

When pyridine is heated to 165 °C in MeONa–MeOD, H–D exchange occurs at all positions *via* small concentrations of deprotonated species, at the relative rates α:β:γ, 1:9.3:12.[73] Some pyridines have been selectively lithiated at C-2 *via* complexes with hexafluoroacetone[74] or boron trifluoride;[75] complexation removes the lone pair and additionally provides inductive (and in the former case also chelation) effects to assist the regioselective α-metallation. The selective 2-lithiation of pyridine *N*-oxides can also be achieved in favourable circumstances: one instructive example is the regioselective 6-lithiation of 2-pivaloylaminopyridine *N*-oxide, i.e. adjacent to the *N*-oxide group, and not at C-3, *ortho* to the directing 2-substitutent. The regioselective C-2-lithiation of 3,4-dimethoxypyridine *N*-oxide also shows the influence of the *N*-oxide functionality.[76]

Regioselective metallation at an α-position of a pyridine can be achieved with the mixed base produced from two mole equivalents of *n*-butyllithium with one of dimethylaminoethanol i.e. it is a 1:1 mixture of *n*-BuLi and Me₂N(CH₂)₂OLi (BuLi-LiDMAE).[77] The regioselectivity is ascribed to intramolecular delivery of the butyllithium, as shown. With this complex base, the regioselectivity is maintained despite the presence of other groups which, with bases such as LDA, would direct an *ortho*-lithiation process: 2-chloro-, 2-methoxy-, 2-methylthio- and 2-dimethylamino-pyridines all lithiate efficiently at C-6; 3-chloro- and 3-methoxypyridines are lithiated at C-2; 4-chloro-, 4-dimethylamino- and 4-methoxypyridines, again lithiate at an α-carbon;[78] 2-, 3- or 4-phenylpyridines are metallated only at an α position;[79] it is even possible to lithiate 4-picoline at C-2,[80] despite potential loss of a side-chain proton (cf. 8.10). An extrapolation of this idea is the use of Me₃SiCH₂Li-LiDMAE, which is less nuclophilic and so is a lesser problem with respect to a subsequently added electrophile.[81]

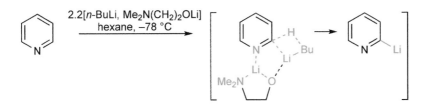

A nice example of the use of α-lithiated pyridines is their nucleophilic addition to azines, oxidation during work-up then producing bihetaryls.[82]

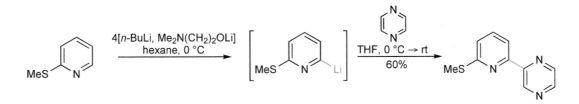

There are many example of direct lithiation with the assistance of *ortho*-directing groups. Halo-, particularly chloro-, or better, fluoro-pyridines, but even bromo-pyridines undergo lithiation *ortho* to the halogen, using lithium di-*iso*-propylamide. 3-Halopyridines react mainly at C-4, and 2- and 4-halopyridines necessarily lithiate at a β-position. In the lithiation of methoxy-pyridines, using mesityllithium the 3-isomer metallates at C-2,[83] whereas 3-methoxymethoxypyridine,[84] 3-di-*iso*-propylaminocarbonyl-[85] 3-tetrahydropyranyloxy-[86] and 3-*t*-butylcarbonylamino-[87] -pyridines all lithiate at C-4. Each of the three pyridine carboxylic acids undergoes *ortho*-lithiation, without protection, using lithium tetramethylpiperidide,[88] which can also be used for the three isomeric cyanopyridines, trapped to give the corresponding boronic acids.[89] Metallation using lithium magnesates (Bu₃MgLi) can be conducted at a somewhat higher temperature, −10 °C.[90]

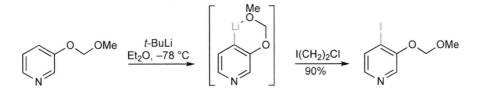

Lithiation of 2- and 4-*t*-butoxycarbonylaminopyridines can only take place at C-3; a neat sequence involving, first, ring lithiation to allow introduction of a methyl group and, second, side-chain methyl lithiation (8.10), provides one route to azaindoles (20.16), as illustrated below for the synthesis of 5-azaindole (pyrrolo[3,2-*c*]pyridine).[91]

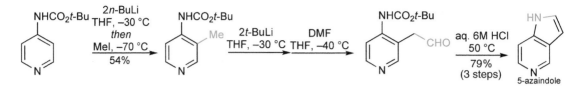

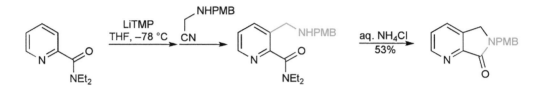

An electrophile that can be used to introduce an aminomethyl unit is a formimine generated *in situ* from *N*-(cyanomethyl)-*para*-methoxybenzylamine; subsequent ring-closure can produce pyrrolopyridinones.[92]

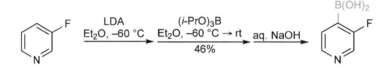

Lithiated pyridines can be converted into boronic acids, or esters, one example being shown below.[93]

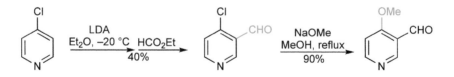

The use of halogen to direct lithiation[94] can be combined with the ability to subsequently displace the halogen with a nucleophile.[95]

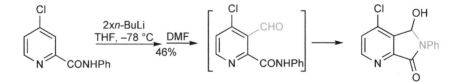

Two directing groups, 1,3-related, cause lithiation to occur between the two groups.[96] A nice example in which two directing groups do not direct to the same carbon is that of 2,5-dimethoxypyridine: lithiation takes place 11:3 at C-4 *versus* C-3, by changing the protecting group on the 5-oxygen to methoxymethyl, exclusive 4-lithiation is achieved and on changing to tri-*iso*-propylsilyl, exclusive 3-lithation results.[97]

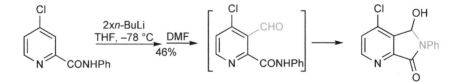

Bromine and iodine also direct lithiations, but isomerisation – 'halogen dance' (see discussion in 17.4.2) – can be a problem, however advantage can be taken of the isomerisation (to the more stable lithio-derivative) in suitable cases. In the example below, the more stable lithio compound is that in which the formally negatively charged ring carbon is located between two halogen-bearing carbon atoms.[98]

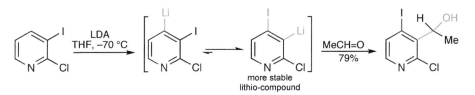

8.4.2 Metal–Halogen Exchange

Lithium derivatives are easily prepared by standard procedures and behave as typical organometallic nucleophiles, thus, for example, 3-bromopyridine undergoes efficient exchange with *n*-butyllithium in ether at −78 °C. With the more basic tetrahydrofuran as solvent, and at this temperature, the alkyllithium becomes more nucleophilic and only addition to the ring occurs, although the exchange *can* be carried out in tetrahydrofuran at lower temperatures.[99]

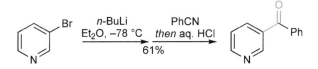

Lithio-pyridines can also be prepared from bromo-pyridines at 0 °C *via* exchange using trimethylsilylmethyllithium (TMSCH₂Li) and lithium dimethylaminoethoxide (LiDMAE) in toluene (2,5-dibromopyridine[100] and 2,3-dibromopyridine[101] react selectively at C-2) and from *chloro*-pyridines, *via* naphthalene-catalysed reductive metallation.[102] Metal–halogen exchange with 2,5-dibromopyridine can also lead efficiently to 2-bromo-5-lithiopyridine;[103] the example below illustrates its trapping with the 'Weinreb amide' of formic acid as a formyl-transfer reagent,[104] however the regioselectivity can be reversed by reaction in toluene.[105] Monolithiation of 2,6-dibromopyridine is best achieved by 'inverse addition' – dibromide to *n*-butyllithium, or by using dichloromethane as solvent.[106] Reaction in toluene is also favourable for exchange at a β-position and 3-bromopyridine can be converted into the 3-boronic acid in toluene;[107] 3- and 4-stannanes can be made from the bromides using ether as solvent.[108]

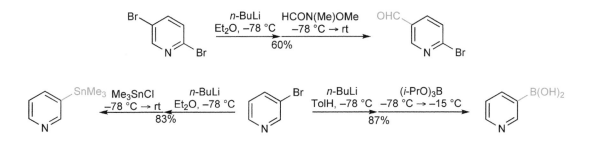

The combination of metal–halogen exchange with the presence of a directing substituent can lead to regioselective metallation.[109]

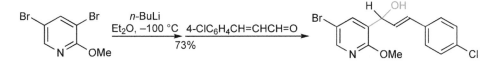

Pyridyl Grignard reagents are readily prepared by exchange of bromine or iodine using *iso*-propyl Grignard reagents.[110] In 2,5-dibromopyridine, the exchange is selective at C-5; other dibromo-pyridines, including 2,6-dibromopyridine, also give clean mono-exchange.[111] Formation of pyridyl Grignard species in this way will even tolerate functional groups such as esters and nitriles, provided the temperature is kept low.

The regioselectivity noted above for 2,5-dibromopyridine in both lithiations and Grignard formation, can be contrasted with the regioselectivity observed with 5-bromo-2-iodopyridine with *iso*-propylmagnesium chloride, where the greater reactivity of iodine overcomes the tendency for β- over α-exchange.[112] The use of *iso*-PrMgCl.LiCl gives very fast exchange with bromopyridines,[113] and pyridine Grignard reagents can also be obtained using 'active magnesium'.[114]

8.5 Reactions with Radicals; Reactions of Pyridyl Radicals

8.5.1 Halogenation

At temperatures where bromine (500 °C) and chlorine (270 °C) are appreciably dissociated into atoms, 2- and 2,6-dihalo-pyridines are obtained *via* radical substitution.[115]

8.5.2 Carbon Radicals[116]

This same preference for α-attack is demonstrated by phenyl-radical attack, but the exact proportions of products depend on the method of generation of the radicals.[117] Greater selectivity for phenylation at the 2- and 4-positions is found in pyridinium salts.[118] Aryl radicals will add intramolecularly, to a neutral pyridine, at any of the pyridine ring positions.[119]

Of more preparative value are the reactions of nucleophilic radicals, such as HOCH$_2\bullet$ and R$_2$NCO$\bullet$, which can be easily generated under mild conditions, for example HOCH$_2\bullet$ from ethylene glycol by persulfate oxidation with silver nitrate as catalyst.[120] These substitutions are carried out on the pyridine protonic salt, which provides both increased reactivity and selectivity for an α-position; the process is known as the Minisci reaction (cf. 3.4.1).[121] It is accelerated by electron-withdrawing substituents on the ring.

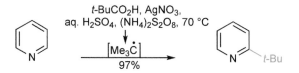

8.5.3 Dimerisation

Both sodium and nickel bring about 'oxidative' dimerisations,[122] despite the apparently reducing conditions, the former giving 4,4'-bipyridine and the latter 2,2'-bipyridine.[123] Each reaction is considered to involve the same anion-radical resulting from transfer of an electron from metal to heterocycle, and the species has been observed by ESR spectroscopy, when generated by single electron transfer (SET) from lithium diisopropylamide.[124] In the case of nickel, the 2,2'-mode of dimerisation may be favoured by chelation to the metal surface. Bipyridyls are important for the preparation of Paraquat-type weedkillers.

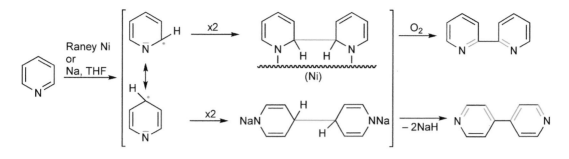

Intermediate, reduced dimers can be trapped under milder conditions,[125] and reduced monomers when the pyridine carries a 4-substituent.[126]

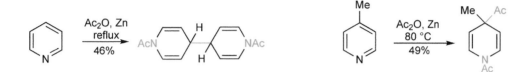

8.5.4 Pyridinyl Radicals

Irradiation of iodopyridines generates pyridinyl radicals, which will effect radical substitution of aromatic compounds.[127] Pyridinyl radicals can be generated from halo-pyridines, using tin hydrides, and participate in typical radical cyclisation reactions.[128,129] Each of the three bromo-pyridines is converted, by tris(trimethylsilyl)silane and azobis(isobutyronitrile), into a radical which substitutes benzene.[130]

8.6 Reactions with Reducing Agents

Pyridines are much more easily reduced than benzenes, for example catalytic reduction proceeds easily at atmospheric temperature and pressure, usually in weakly acidic solution, but also in dilute alkali over nickel.[131] Reduction in neutral solution is accelerated by microwave heating.[132]

Of the hydride reagents, sodium borohydride is without effect on pyridines, though it does reduce pyridinium salts (8.12.1), lithium aluminium hydride effects the addition of one hydride equivalent to pyridine,[133] but lithium triethylborohydride reduces it to piperidine efficiently.[134]

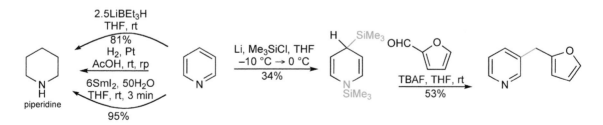

The combination lithium/chlorotrimethylsilane produces a 1,4-dihydro doubly silylated product, the enamine character in which can be utilised for the introduction of 3-alkyl groups *via* reaction with aldehydes.[135]

Metal/acid combinations, which in other contexts do bring about reduction of iminium groups, are without effect on pyridines. Samarium(II) iodide in the presence of water smoothly reduces pyridine to piperidine.[136] Sodium in liquid ammonia, in the presence of ethanol, affords the 1,4-dihydropyridine[137] and 4-pyridones are reduced to 2,3-dihydro derivatives.[138] Birch reduction of pyridines carrying esters, followed by trapping with alkyl halides, produces dihydro-pyridines, and methyl 4-methoxypicolinate quaternary salts are comparably reduced and trapped giving 1,2-dihydro-derivatives, which can be smoothly hydrolysed to enones.[139]

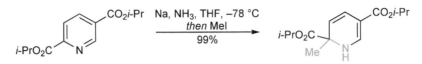

8.7 Electrocyclic Reactions (Ground State)

There are no reports of thermal electrocyclic reactions involving simple pyridines. 2-Pyridones, however, participate as 4π components in Diels–Alder additions, especially under high pressure.[140]

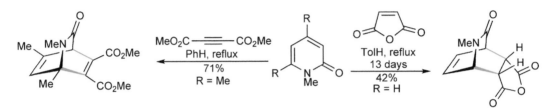

N-Tosyl-2-pyridones with a 3-alkoxy or 3-arylthio substituent, undergo cycloaddition with electron-deficient alkenes under milder conditions, as illustrated below,[141] and the cycloaddition of the corresponding 3-hydroxypyridone is promoted by *O*-deprotonation.[142]

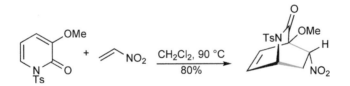

The quaternary salts of 3-hydroxy-pyridines are converted by mild base into zwitterionic, organic-solvent-soluble species, for which no neutral resonance form can be drawn. These pyridinium-3-olates undergo a number of dipolar cycloaddition reactions, especially across the 2,6-positions.[143]

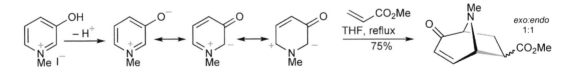

8.8 Photochemical Reactions

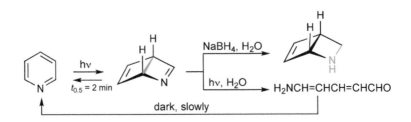

Ultraviolet irradiation of pyridines can produce highly strained species that can lead to isomerised pyridines or can be trapped. The three picolines and the three cyano-substituted pyridines constitute 'photochemical triads': irradiation of any isomer, in the vapour phase at 254 nm, results in the formation of all three isomers.[144] From pyridines[145] and from 2-pyridones[146] 2-azabicyclo[2.2.0]-hexadienes and -hexenones can be obtained; in the case of pyridines these are usually unstable and revert thermally to the aromatic heterocycle. Pyridone-derived bicycles are relatively stable, 4-alkoxy- and -acyloxy-pyridones are converted in particularly good yields. Irradiation of *N*-methyl-2-pyridone in aqueous solution produces a mixture of regio- and stereoisomeric 4π plus 4π photo-dimers.[147]

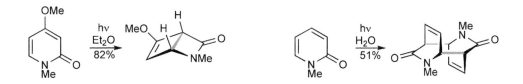

Photocatalysed 2π plus 2π cycloadditions between a pair of tethered 4-pyridones[148] can generate spectacularly complex rings systems easily, as shown.

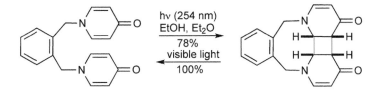

The photoreactions of pyridinium salts in water give 6-azabicyclo[3.1.0]hex-3-en-2-ols or the corresponding ethers, which can undergo regio- and stereoselective ring-openings of the aziridine by attack of nucleophiles under acidic conditions. These products are useful starting materials for synthesis.[149]

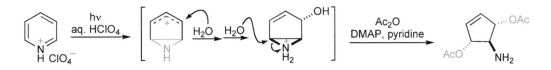

Photolysis of pyridine *N*-oxides in alkaline solution induces ring opening to cyano-dienolates.[150]

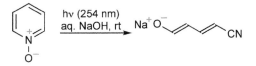

8.9   Oxy- and Amino-Pyridines
8.9.1   Structure

The three oxy-pyridines are subject to tautomerism involving hydrogen interchange between oxygen and nitrogen, but with a significant difference between α- and γ- on the one hand and β-isomers on the other.

Under all normal conditions, α- and γ-isomers exist almost entirely in the carbonyl tautomeric form, and are accordingly known as 'pyridones'; the hydroxy-tautomers are detected in significant amounts only in very dilute solutions in non-polar solvents like petrol, or in the gas phase, where, for the α-isomer, 2-hydroxypyridine is actually the dominant tautomer by 2.5:1.[151] The polarised pyridone form is favoured by solvation.[152] 3-Hydroxypyridine exists in equilibrium with a corresponding zwitterionic tautomer, the exact ratio depending on solvent.

In this chapter we utilise the generally accepted terms '2-pyridone', '4-pyridone' rather than the strictly correct '2(1*H*)-pyridinone', etc.

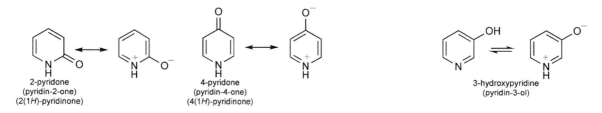

very dilute solutions in non-polar solvents like petrol, or in the gas phase, where, for the α-isomer,

2-pyridone
(pyridin-2-one)
(2(1*H*)-pyridinone)

4-pyridone
(pyridin-4-one)
(4(1*H*)-pyridinone)

3-hydroxypyridine
(pyridin-3-ol)

All three amino-pyridines exist in the amino form; the α- and γ-isomers are polarised in a sense opposite to that in the pyridones.

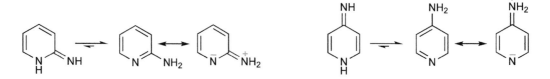

8.9.2   Reactions of Pyridones
8.9.2.1   *Electrophilic Addition and Substitution*
3-Hydroxypyridine protonates on nitrogen, with a typical pyridine pK_{aH} of 5.2; the pyridones are much less basic and, like amides, protonate on oxygen.[153] However, the reaction of 4-pyridone with acid chlorides produces *N*-acyl derivatives. 1-Acetyl-4-pyridone subsequently equilibrates in solution affording a mixture with 4-acetoxypyridine.[154]

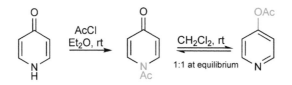

Electrophilic substitution at carbon can be effected much more readily with the three oxy-pyridines than with pyridine itself, and it occurs *ortho* and *para* to the oxygen function, as indicated below. Acid catalysed exchange of 4-pyridone in deuterium oxide, for example, gives 3,5-dideuterio-4-pyridone, *via* C-protonation of the neutral pyridone.[155]

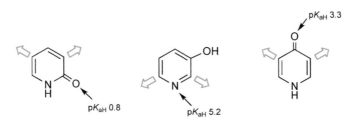

Positions of electrophilic substitution of oxy-pyridines

Substitutions usually proceed *via* attack on the neutral pyridone,[156] but in very strong acid, where there is almost complete *O*-protonation, 4-pyridone undergoes a slower nitration, *via* attack on the salt, but with the same regioselectivity.[157]

Electrophilic substitutions of 3-hydroxypyridine take place at C-2, for example nitration,[158] Mannich substitution[159] and iodination.[160] Its phenol-like character is nicely illustrated by efficient 2,4,6-tribromination with *N*-bromosuccinimide.[161] 2-Methoxypyridine brominates at C-5[162] and 4-methoxypyridine at C-3.[161]

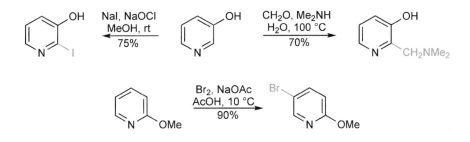

8.9.2.2 Deprotonation and Reaction of Salts

N-Unsubstituted pyridones are acidic, with pK_a values of about 11 for *N*-deprotonation giving mesomeric anions. These ambident anions can be alkylated on either oxygen or nitrogen, producing alkoxy-pyridines or *N*-alkyl-pyridones, respectively, the relative proportions depending on the reaction conditions;[163] *N*-alkylation is usually predominant for primary halides; *O*-alkylation for secondary halides.[164] The reagent combination sodium hydride with lithium bromide in dimethylformamide and dimethoxyethane gives mainly *N*-alkylation.[165] A clean method for the synthesis of *N*-alkylated 4-pyridones is to convert the pyridone first into the *O*-trimethylsilyl ether[166] which can then be reacted selectively at nitrogen, subsequent removal of the silicon giving the *N*-alkylpyridone.[138] Alternatively, 2-alkoxy-pyridines, generated by alkoxide displacement on a 2-halo-pyridine, can be isomerised: for example 2-benzyloxypyridine is converted into 1-benzyl-2-pyridone on heating with lithium iodide at 100 °C.[167] 2-Pyridone is sufficiently acidic to take part in Mitsunobu reactions with alcohols, though, again, mixtures of *O*- and *N*-alkylation products result.[168]

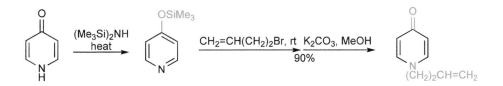

8.9.2.3 Replacement of Oxygen

The conversion of the carbonyl group in pyridones into a leaving group has a very important place in the chemistry of pyridones, the most frequently encountered examples involving reaction with phosphoryl chloride and/or phosphorus pentachloride leading to the chloro-pyridine, *via* an assumed dichlorophosphate intermediate as indicated below. Conversion into bromo derivatives is possible with phosphorus oxybromide but can be more conveniently achieved with *N*-bromosuccinimide and triphenylphosphine in refluxing dioxane[169] or with phosphorus pentoxide with tetra-*n*-butylammonium bromide in hot toluene.[170] Similarly, treatment with phosphorus pentoxide and a secondary amine, or of 2- or 4-trimethylsilyloxypyridines (prepared *in situ*) with secondary amines,[166] produces dialkylamino-pyridines. Pyridones are converted into triflates by reaction with trifluoromethanesulfonic anhydride and a base;[171] these derivatives are of particular interest in the context of palladium(0)-catalysed cross-couplings (4.2).

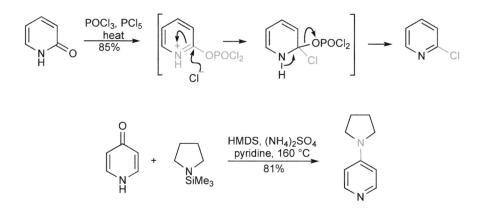

The usual way to remove oxygen completely from a pyridone is by conversion, as described, into halogen followed by catalytic hydrogenolysis.[172] Alternatively, reaction of the pyridone salt with 5-chloro-1-phenyltetrazole then hydrogenolysis of the resulting ether can be used.[173]

8.9.3 Reactions of Amino-Pyridines

8.9.3.1 Electrophilic Addition and Substitution

The three amino-pyridines are all more basic than pyridine itself and form crystalline salts by protonation at the ring nitrogen. The α- and γ-isomers are monobasic only, because charge delocalisation over both nitrogen atoms, in the manner of an amidinium cation, prevents the addition of a second proton. The effect of the delocalisation is strongest in 4-aminopyridine (pK_{aH} 9.1) and much weaker in 2-aminopyridine (pK_{aH} 7.2). Delocalisation is not possible for the β-isomer, which thus *can* form a di-cation in strong acid (pK_{aH}s 6.6 and −1.5).[174]

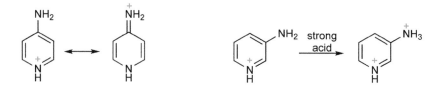

Whereas alkylation of amino-pyridines, irreversible at room temperature, gives the product of kinetically controlled attack at the most nucleophilic nitrogen, the ring nitrogen,[175] acetylation gives the product of reaction at a side-chain amino group. The acetylamino-pyridine which is isolated probably results from *N*-deprotonation of an *N*-acyl-pyridinium salt followed by side-chain *N*-acylation, with loss of the ring *N*-acetyl during aqueous work-up, as suggested below.

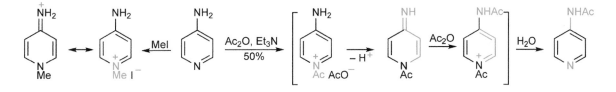

As in benzene chemistry, electron-releasing amino groups facilitate electrophilic substitution, so that, for example, 2-aminopyridine undergoes 5-bromination in acetic acid even at room temperature; this product can then be nitrated, at room temperature, forming 2-amino-5-bromo-3-nitropyridine.[176] Bromination of all three amino-pyridines is best achieved with *N*-bromosuccinimide at room temperature, products being 2-amino-5-bromo-, 3-amino-2-bromo- and 4-amino-3-bromopyridines.[161] Similarly, chlorination of 3-amino-pyridines affords 3-amino-2-chloro-pyridines.[177] Nitration of amino-pyridines in acid solution is also relatively easy, with selective attack of 2- and 4-isomers at β-positions. A mechanistic study of dialkylamino-pyridines showed nitration to involve attack on the salts.[178]

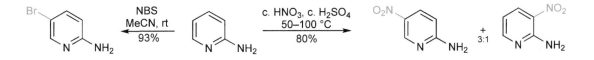

A limited number of examples of *C*-alkylations of aminopyridines have been reported. With 1-hydroxymethylbenzotriazole (29.3) in the presence of acid, 2-aminopyridine reacts at C-5.[179] 4-Dimethylaminopyridine is trifluoroacetylated at C-3 with trifluoroacetic anhydride; the example shows the subsequent intramolecular nucleophilic displacement of the dimethylamino group and thence formation of a pyrazolo[4,3-*c*]pyridine.[180]

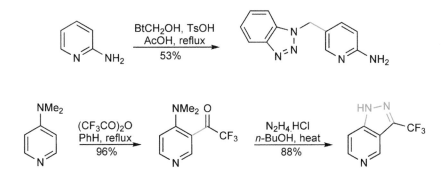

8.9.3.2 Reactions of the Amino Group

β-Amino-pyridines give normal diazonium salts on reaction with nitrous acid, but with α- and γ-isomers, unless precautions are taken, the corresponding pyridones are then produced *via* easy hydrolysis,[171,181] water addition at the diazonium-bearing carbon being rapid.[182] With care, however, this same susceptibility to nucleophilic displacement can be harnessed in effecting Sandmeyer-type reactions, without the use of copper, of diazonium salts from either 2- or 4-aminopyridines.[181,183,184]

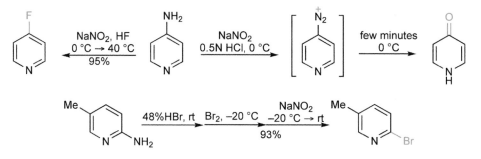

8.10 Alkyl-Pyridines

The main feature of the reactivity of alkyl-pyridines is deprotonation of the alkyl group at the carbon adjacent to the ring.[185] Measurements of side-chain exchange in methanolic sodium methoxide, 4:2:3, 1800:130:1,[186] and of pK_a values in tetrahydrofuran[187] each have the γ-isomer more acidic than the α-isomer, both being much more acidic than the β-isomer, though the actual carbanion produced in competitive situations can depend on both the counter ion and the solvent. Alkyllithiums selectively deprotonate an α-methyl whereas amide bases produce the more stable γ-anion.[188] The much greater ease of deprotonation[189] of the α- and γ-isomers is related to mesomeric stabilisation of the anion involving the ring nitrogen, not available to the β-isomer, for which there is only inductive facilitation, but deprotonation can be effected at a β-methyl under suitable conditions;[190] the difference in acidity between 2- and 3-methyl groups allows selective reaction at the former.[191]

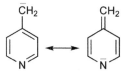

Resonance stabilisation of 'enaminate' anions formed by deprotonating the methyl groups of 4- and 2-picolines

The 'enaminate' anions produced by deprotonating α- and γ-alkyl-pyridines can participate in a wide range of reactions,[192] being closely analogous to enolate anions; some examples are given below. Note that dimethyl-pyridines are often referred to as 'lutidines' – thus the example below is 2,4-lutidine (2,4-dimethylpyridine).

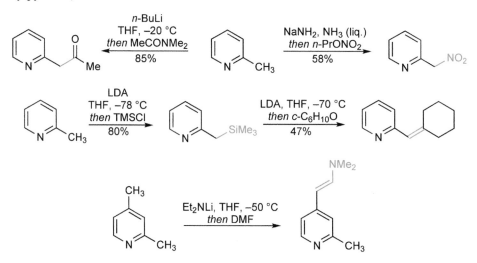
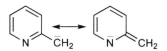

An important aspect is the oxidation of side-chain methyl to hydroxymethyl or aldehyde oxidation levels. The former can be achieved by reaction of the lithiated species with molecular oxygen, then quenching with dimethyl sulfide in acetic acid.[193] Conversion of methyl to the aldehyde oxidation level (see also 8.2) can be achieved by dibromination, then hydrolysis.[194]

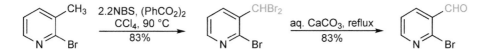

It is even possible to lithiate a methyl in the presence of a free amino group and this can be made the means to synthesise 6-azaindoles (pyrrolo[2,3-*c*]pyridines).[195]

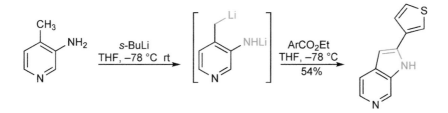

In the quaternary salts of alkyl-pyridines, the side-chain hydrogens are considerably more acidic and condensations can be brought about under quite mild conditions, the reactive species being a dienamine.[196] Dienamides are the reacting nucleophiles in aldol-type condensations brought about with acetic anhydride or in side-chain trifluoroacetylation of 2-picoline.[197]

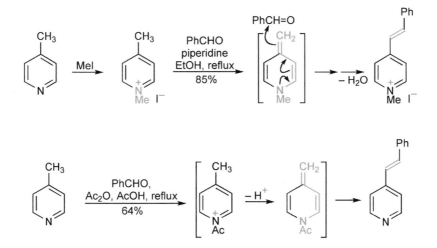

A further consequence of the stabilisation of carbanionic centres at pyridine α- and γ-positions is the facility with which vinyl-pyridines,[198] and alkynyl-pyridines, add nucleophiles, in Michael-like processes (mercury-catalysed hydration of alkynyl-pyridines goes in the opposite sense[199]).

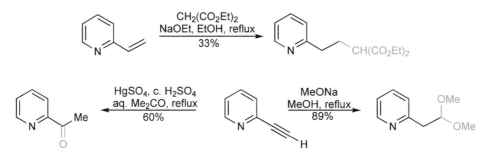

8.11 Pyridine Aldehydes, Ketones, Carboxylic Acids and Esters

These compounds all closely resemble the corresponding benzene compounds in their reactivity because the carbonyl group cannot interact mesomerically with the ring nitrogen. The pyridine 2- (picolinic), 3- (nicotinic), and 4- (isonicotinic) acids exist almost entirely in their zwitterionic forms in aqueous solution; they are slightly stronger acids than benzoic acid. Decarboxylation of picolinic acids is relatively easy and results in the transient formation of the same type of ylide that is responsible for specific proton α-exchange of pyridine in acid solution (see 8.1.2.1).[200] This transient ylide can be trapped by aromatic or aliphatic aldehydes in a reaction known as the Hammick reaction.[201] As implied by this mechanism, quaternary salts of picolinic acids also undergo easy decarboxylation.[202] The Hammick reaction can also be carried out by heating a silyl ester of picolinic acid in the presence of a carbonyl electrophile.[203]

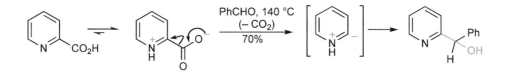

8.12 Quaternary Pyridinium Salts

The main features of the reactivity of pyridinium salts are: (i) the greatly enhanced susceptibility to nucleophilic addition and displacement at the α- and γ-positions, sometimes followed by ring opening[236] and (ii) the easy deprotonation of α- and γ-alkyl groups (see also 8.10).

8.12.1 Reduction and Oxidation

The oxidation of pyridinium salts[204] to pyridones by alkaline ferricyanide is presumed to involve a very small concentration of hydroxide adduct. 3-Substituted pyridinium ions are transformed into mixtures of 2- and 6-pyridones; for example oxidation of 1,3-dimethylpyridinium iodide gives a 9:1 ratio of 1,3-dimethyl-2- and -6-pyridones.

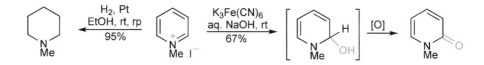

Catalytic reduction of pyridinium salts to piperidines is particularly easy in ethanol at room temperature and pressure; they are also susceptible to hydride addition by complex metal hydrides[205] or formate,[206] and lithium/ammonia reduction.[207] In the reduction with sodium borohydride in protic media, the main product is a tetrahydro derivative with the double bond at the allylic, 3,4-position, formed by initial hydride addition at C-2, followed by enamine β-protonation and a second hydride addition. Some fully reduced material is always produced and its relative percentage increases with increasing *N*-substituent bulk, consistent with a competing sequence having initial attack at C-4, generating a dienamine, which can then undergo two successive proton-then-hydride addition steps. When 3-substituted pyridinium salts are reduced with sodium borohydride, 3-substituted-1,2,5,6-tetrahydropyridines result.

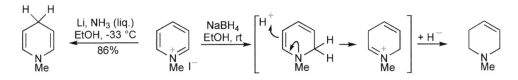

N-Alkoxy- or *N*-aryloxycarbonyl-pyridiniums can be reductively trapped as dihydro derivatives by borohydride;[208] no further reduction occurs because the immediate product is an enamide and not an enamine and therefore does not protonate under the conditions of the reduction.[209] The 1,2-dihydro-isomers, which can be produced essentially exclusively by reduction at −70 °C in methanol, can serve as dienes in Diels–Alder reactions. Irradiation causes conversion into 2-azabicyclo[2.2.0]hexenes; removal of the carbamate and *N*-alkylation gives derivatives that are synthons for unstable *N*-alkyl-dihydropyridines, and convertible into the latter thermally.[210]

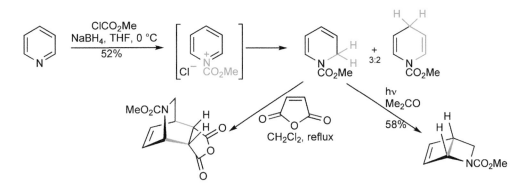

The easy specific reduction of 3-acyl-pyridinium salts giving stable 3-acyl-1,4-dihydropyridines using sodium dithionite ($Na_2S_2O_4$) is often quoted, because of its perceived relevance to nicotinamide coenzyme activity (32.2.1). The mechanism involves addition of oxygen at C-4 as its first step; the first intermediate protonates on sulfur and the subsequent *C*-4-protonation may involve intramolecular hydrogen transfer from the sulfur, with sulfur dioxide loss.[211] 1,4-Dihydropyridines are normally air-sensitive, easily re-aromatised molecules; the stability of 3-acyl-1,4-dihydropyridines is related to the conjugation between ring nitrogen and side-chain carbonyl group (see also Hantzsch synthesis, 8.14.1.2). However, even simple pyridinium salts, provided the *N*-substituent is larger than propyl, or for example benzyl, can be reduced to 1,4-dihydropyridines with sodium dithionite.[212]

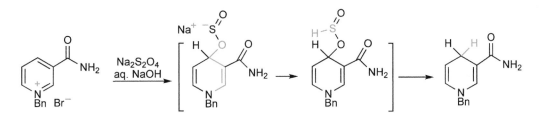

8.12.2   Organometallic and Other Nucleophilic Additions

Organometallic reagents add very readily to *N*-alkyl-, *N*-aryl- and, with important synthetic significance, *N*-alkoxy- or *N*-aryloxycarbonyl-pyridinium salts. In *N*-alkyl- or *N*-aryl-pyridinium cations, addition is to an α-carbon; the resulting 2-substituted-1,2-dihydropyridines are unstable, but can be handled and spectroscopically identified, with care, and more importantly can be easily oxidised to a 2-substituted pyridinium salt.[213]

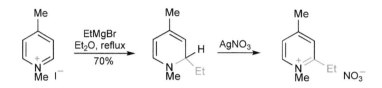

The great significance of the later discovery, that exactly comparable additions to *N*-alkoxycarbonyl- or *N*-aryloxycarbonyl-pyridinium cations, generated and reacted *in situ*, is that the dihydro-pyridines that result are stable, and can be further manipulated. If re-aromatisation[214] is required, the *N*-substituent can be easily removed to give a substituted pyridine. It is worth noting the contrast to the use of *N*-acyl-pyridinium salts for reaction with alcohol, amine nucleophiles (8.1.1.7), when attack is at the carbonyl carbon; the use of an *N*-alkoxy/aryloxycarbonyl-pyridinium salt in the present context diverts attack to a ring carbon.

Generally, organometallic addition to *N*-alkoxycarbonyl- or *N*-aryloxycarbonyl-pyridinium salts[192] takes place at either the 2- or 4-positions,[215] however higher selectivity for the 4-position can be achieved using copper reagents.[216] High selectivity for the 2-position is found in the addition of phenyl,[217] alkenyl and alkynyl organometallics,[218] including ethoxycarbonylmethyl[219] and alkynyl[220] tin reagents.

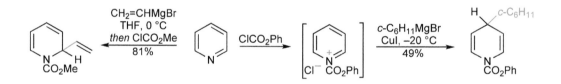

The dihydropyridines produced by the methods described above are multifunctional and can be manipulated, for example the enamide character in these products can be utilised by interaction with an electrophile (iodine in the example) which brings about intramolecular attack and formation of a lactone.[221]

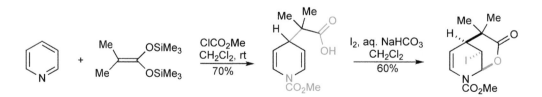

Silylation at nitrogen with *t*-butyldimethylsilyl triflate, generates pyridinium salts which, because of the size of the *N*-substitutent, react with Grignard reagents exclusively at C-4.[222] Similarly, 4-substituted *N*-tri-*iso*-propylsilyl-pyridinium salts react with hindered dialkylmagnesiums at C-4, providing a route through to 4,4-dialkyl-piperidines, as shown.[223]

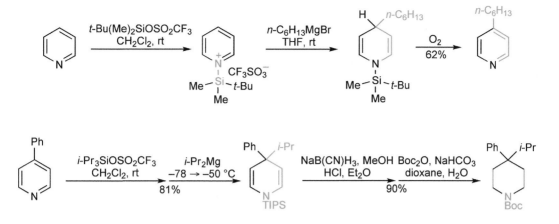

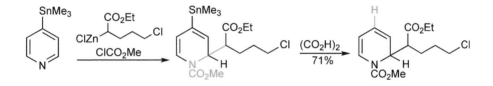

4-Substituents tend to direct attack to an α-carbon;[224,225] the use of a removable 4-blocking group – trimethyltin in the example below – can be made the means for the production of 2-substituted isomers.[226]

The use of chiral chloroformates, such as that derived from *trans*-2-(α-cumyl)cyclohexanol, allows diastereoselective additions to 4-methoxypyridine. The introduction of a tri-*iso*-propylsilyl group at C-3 greatly enhances the diastereoselectivity. The products of these reactions are multifunctional chiral piperidines which have found use in the asymmetric synthesis of natural products.[227]

Some nucleophiles add to *N*-fluoro-pyridinium salts to give dihydropyridines in which elimination of fluoride occurs *in situ* to give the 2-substituted pyridine.[228] However, the preparation of the pyridinium salts requires the use of elemental fluorine (31.1) and also, some carbanions are subject to competitive reactions such as C-fluorination. However, silyl enol ethers do react efficiently; stabilised heteronucleophiles (phenolate, azide) can also be used, and isonitriles produce picolinamides.[229]

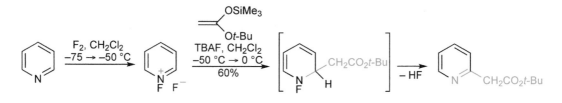

In a similar way, pyridine phosphonium salts and phosphonates can be prepared by reaction of trivalent phosphorus compounds with the more accessible *N*-trifluoromethanesulfonyl-pyridinium salts, when trifluoromethanesulfone is the leaving group from nitrogen (as sulfinate anion); attack is normally at C-4, as illustrated below.[230] The *N*-trifluoromethanesulfonyl-pyridinium salts also react with ketones[231] or with electron-rich aromatic compounds[232] to give 1,4-dihydropyridine adducts. Subsequent treatment with potassium *t*-butoxide brings about elimination of trifluoromethanesulfinic acid, and thus aromatisation. It is also possible to utilise phosphonates in reaction with aldehydes, leading finally to 4-substituted pyridines.[233]

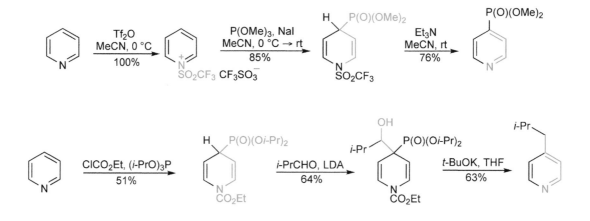

8.12.3   Nucleophilic Addition Followed by Ring Opening[234]

There are many examples of pyridinium salts, particularly, but not exclusively, those with powerful electron-withdrawing *N*-substituents, adding a nucleophile at C-2 and then undergoing a ring opening. The classic example is addition of hydroxide to the pyridine sulfur trioxide complex, which produces the sodium salt of glutaconaldehyde, as shown below.[235]

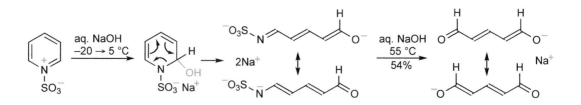

Another intriguing example is a synthesis of azulene that utilises the bis(dimethylamine) derivative of glutaconaldehyde produced with loss of 2,4-dinitroaniline from 1-(2,4-dinitrophenyl)pyridinium chloride (Zincke's salt).[236,237]

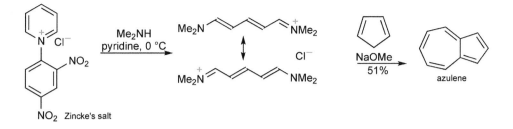

The reaction of Zincke's salt with primary amines, including α-amino acid esters, is a useful synthesis of variously *N*-substituted pyridinium salts; the nitrogen of the final product is the nitrogen of the primary amine reactant.[238,239]

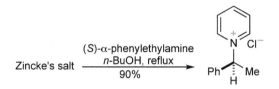

8.12.4   Cyclisations Involving an α-Position or an α-Substituent

It is often possible to convert pyridinium salts into bicyclic, neutral products, with nitrogen at a ring junction, in which the ring closure involves an α-substituent or the electrophilic nature of the α-position – Sections 28.1.2 and 28.2.3 give examples.

8.12.5   *N*-Dealkylation

The conversion of *N*-alkyl- or -aryl-pyridinium salts into the corresponding pyridine, i.e. the removal of the *N*-substitutent, is generally not an easy process; however triphenylphosphine[240] or simply heating the iodide salt[241] can work for metho-salts. 1-Triphenylmethyl-4-dimethylaminopyridinium chloride[242] and 1-trialkylsilyl-pyridinium triflates[243] are isolable and relatively stable salts; *O*-tritylations and *O*-silylations involving transfer of trityl or trialkylsilyl from the positively charged nitrogen in such salts are usually carried out without isolation, using mixtures of 4-dimethylaminopyridine (DMAP) with chlorotriphenyl-methane or, for example, chloro-*t*-butyldimethylsilane.[244]

8.13   Pyridine *N*-oxides[245]

The reactions of pyridine *N*-oxides are of great interest,[246] differing significantly from those of both neutral pyridines and pyridinium salts.

A striking difference between pyridines and their *N*-oxides is the susceptibility of the latter to electro-philic nitration. This can be understood in terms of mesomeric release from the oxide oxygen, and is parallel to electron release by oxygen and hence increased reactivity towards electrophilic substitution in phenols and phenoxides. One can find support for this rationalisation by a comparison of the dipole moments of trimethylamine and its *N*-oxide, on the one hand, and pyridine and its *N*-oxide, on the other: the difference of 2.03 D for the latter pair is much smaller than the 4.37 D found for the former. The smaller difference signals significant contributions from those canonical forms in which the oxygen is neutral and the ring negatively charged. Clearly, however, the situation is subtle, as those contributors carrying formal positive charges on α- and γ-carbons suggest a polarisation in the opposite sense and thus an increased susceptibility to nucleophilic attack too, compared with the neutral pyridine, and this is indeed found to be the case. Summarising: the *N*-oxide function in pyridine *N*-oxides serves to facilitate, on demand, both electrophilic and nucleophilic addition to the α- and γ-positions.

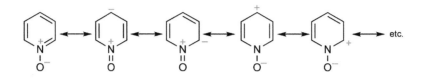

8.13.1   Electrophilic Addition and Substitution

Pyridine *N*-oxides protonate and are alkylated at oxygen; stable salts can be isolated in some cases.[247] *O*-Alkylation with benzylic and allylic halides in the presence of silver oxide produces the corresponding aldehydes, the oxygen being derived from the *N*-oxide.[248]

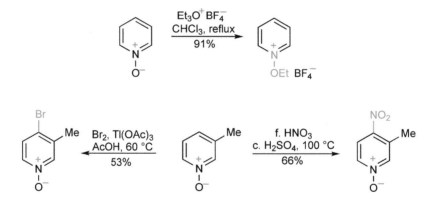

Electrophilic nitration and bromination of pyridine *N*-oxides can be controlled to give 4-substituted products[249] by way of attack on the free *N*-oxide.[250] Under conditions where the *N*-oxide is *O*-protonated, substitution follows the typical pyridine/pyridinium reactivity pattern thus, in fuming sulfuric acid, bromination shows β-regioselectivity.[251] Mercuration takes place at the α-position,[252] however mercuric-catalysed sulfonation produces the 3-sulfonic acid.[253]

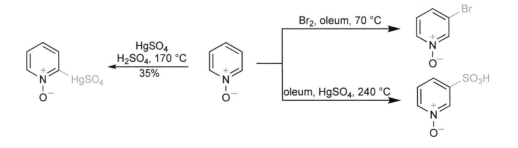

8.13.2 Nucleophilic Addition and Substitution

The *N*-oxide function enhances the rate of nucleophilic displacement of halogen from α- and γ-positions. The relative rates 4 > 2 > 3 found for pyridines are echoed for the *N*-oxides (but significantly are 2 > 4 > 3 in pyridinium salts).[254]

Grignard reagents add to pyridine *N*-oxide, forming adducts, which can be characterised from a low-temperature reaction, but which at room temperature undergo disrotatory ring opening, the isolated product being an acyclic, unsaturated oxime. Heating with acetic anhydride brings about re-aromatisation, *via* electrocyclic ring closure rendered irreversible by the loss of acetic acid.[255] Comparable additions/ring openings are observed with 1-alkoxy-pyridiniums.[256,257]

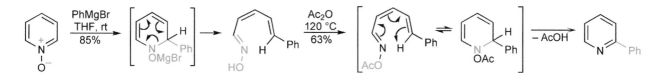

The direct introduction of an acetylide moiety, using pyridine *N*-oxide (or quinoline, diazine and triazine *N*-oxides) can be achieved in a comparable way, by reaction with potassium phenylacetylide; reaction with the lithium salt requires addition of acetyl chloride at the end of the reaction to aromatise.[258] At low temperature, and using *i*-PrMgCl, 2-metallation of pyridine *N*-oxides can be achieved, and thus, the introduction of electrophiles at the 2-position.[259]

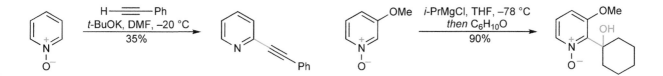

8.13.3 Addition of Nucleophiles then Loss of Oxide

A range of synthetically useful rearrangements convert pyridine *N*-oxides into variously substituted pyridines in which an α-(γ-)position, or an α-substituent has been modified.

Reaction with phosphorus oxychloride[260] or with acetic anhydride leads to the formation of 2-chloro- or 2-acetoxy-pyridines, respectively. Mechanistically, electrophilic addition to oxide is followed by nucleophilic addition to an α- or γ-position, the process being completed by an elimination. Similarly, conversions of pyridine *N*-oxides into 2-cyanopyridines depend on prior conversion of oxide into silyloxy or carbamate.[261]

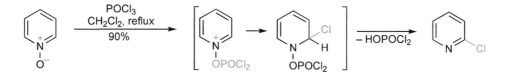

It is also possible to introduce a nitrogen function using these types of process: *O*-tosylation then *t*-butylamine as a synthon for ammonia leads to 2-aminopyridines[262] and oxalyl chloride, together with a secondary amide, produces 2-amino-pyridine amides.[263]

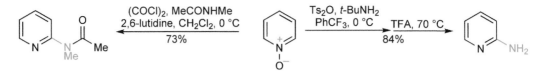

2-Methyl-pyridine *N*-oxides react with hot acetic anhydride and produce 2-acetoxymethyl-pyridines; trifluoroacetic anhydride reacts at room temperature, with fewer by-products.[264] Repetition of the sequence affords 2-aldehydes after hydrolysis.[265] The course[266] of the rearrangement would seem to be most simply explained by invoking an electrocyclic sequence, as shown below.

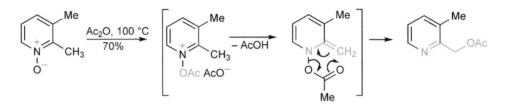

The following sequence illustrates several aspects of *N*-oxide chemistry, including easy nucleophilic substitution (of nitro) at a γ-position.[267]

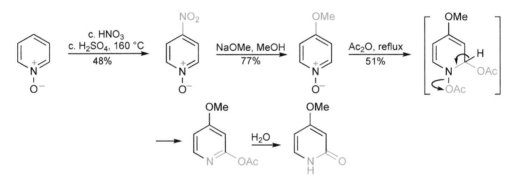

8.14   Synthesis of Pyridines
8.14.1   Ring Synthesis[268]
There are very many ways of achieving the synthesis of a pyridine ring; in this section, the main general methods and some less general sequences are described and exemplified.

8.14.1.1   From 1,5-Dicarbonyl Compounds and Ammonia
Ammonia reacts with 1,5-dicarbonyl compounds to give 1,4-dihydropyridines, which are easily dehydrogenated to pyridines. With unsaturated 1,5-dicarbonyl compounds, or their equivalents (e.g. pyrylium ions), ammonia reacts to give pyridines directly.

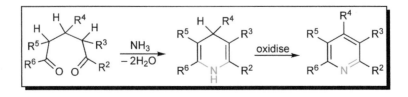

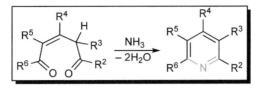

1,5-Diketones are accessible *via* a number routes, for example by Michael addition of enolate to enone (or precursor Mannich base[269]) or by ozonolysis of a cyclopentene precursor. They react with ammonia, with loss of two mole equivalents of water to produce cyclic bis-enamines, i.e. 1,4-dihydro-pyridines, which are generally unstable, but can be easily and efficiently dehydrogenated to the aromatic heterocycle.

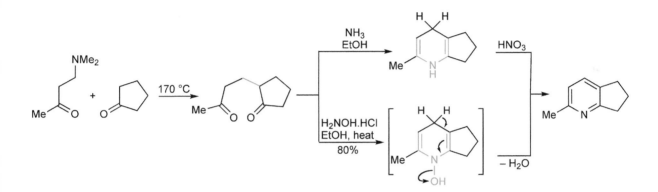

The oxidative final step can be neatly avoided by the use of hydroxylamine[270] instead of ammonia, when a final 1,4-loss of water produces the aromatic heterocycle. In an extension of this concept, the construction of a 1,5-diketone equivalent by tandem Michael addition of an *N,N*-dimethylhydrazone anion to an enone, then acylation, has loss of dimethylamine from nitrogen as the final aromatisation step.[271]

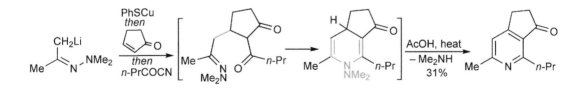

The use of an unsaturated 1,5-dicarbonyl compound will afford an aromatic pyridine directly; a number of methods are available for the assembly of the unsaturated diketone, including the use of pyrylium ions or pyrones[272] (see Chapter 11) as synthons, or the alkylation of an enolate with a 3,3-bis(methylthio)-enone.[273] 2,2':6',2''-Terpyridine can be synthesised in one pot from 2-acetylpyridine, dimethylformamide dimethylacetal (DMFDMA) and ammonia; the first step is presumed to be dimethyl-aminomethylenation of the ketone methyl group, followed then by addition/elimination by the enolate of the starting ketone.[274]

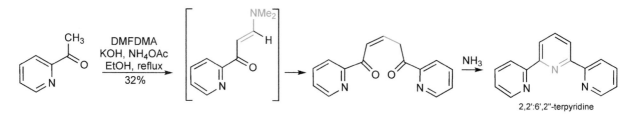

2,2':6',2''-terpyridine

When one of the carbonyl carbons is at the oxidation level of acid (as in a 2-pyrone), then the product, reflecting this oxidation level, is a 2-pyridone.[275] Similarly, 4-pyrones react with ammonia or primary amines to give 4-pyridones[276] and the bis-enamines which can be obtained directly from ketones by condensation on both sides of the carbonyl group with DMFDMA, produce 4-pyridones on reaction with primary amines.[277] When one of the 'carbonyl' units is actually a nitrile, then an amino-pyridine results.[278]

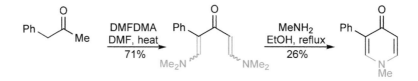

8.14.1.2   *From an Aldehyde, Two Equivalents of a 1,3-Dicarbonyl Compound and Ammonia*

Symmetrical 1,4-dihydropyridines, which can be easily dehydrogenated, are produced from the interaction of ammonia, an aldehyde and two equivalents of a 1,3-dicarbonyl compound, which must have a central methylene.

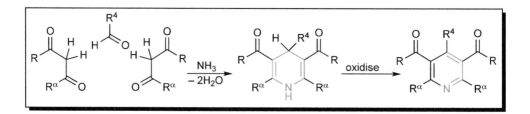

The Hantzsch Synthesis[279]

The product from the classical Hantzsch synthesis is necessarily a symmetrically substituted 1,4-dihydro-pyridine, since two mole equivalents of one dicarbonyl component are utilised, the aldehyde carbonyl carbon becoming the pyridine C-4. The precise sequence of intermediate steps is not known for certain, and may indeed vary from case to case, for example the ammonia may become involved early or late, but a reasonable sequence would be: aldol condensation followed by Michael addition generating, *in situ*, a 1,5-dicarbonyl compound.

The 1,4-dihydropyridines produced in this approach, carrying conjugating substituents at each β-position, are stable, and can be easily isolated before dehydrogenation; classically the oxidation has been achieved with nitric acid, or nitrous acid, but other oxidants such as ceric ammonium nitrate, cupric nitrate or manganese dioxide on Montmorillonite, amongst many, also achieve this objective smoothly.[280] Iodine with potassium hydroxide at 0 °C is amongst the mildest.[281]

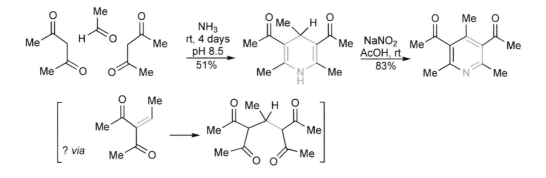

Hantzsch reactions to produce 5,6,7,8-tetrahydroquinolines, i.e. unsymmetrically substituted pyridines, work well using ceric ammonium nitrate (CAN) as a catalyst.[282]

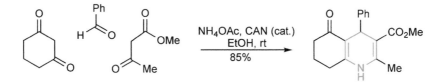

More often, unsymmetrical 1,4-dihydropyridines are produced by conducting the Hantzsch synthesis in two stages, i.e. by making the (presumed) aldol condensation product separately, then reacting with ammonia and a different 1,3-dicarbonyl component, or an enamino-ketone, in a second step.[283]

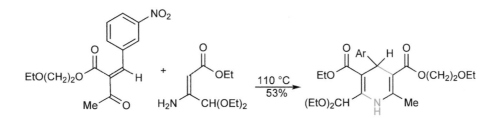

This strategy can also be applied for the synthesis of 2,2′:6′,2″-terpyridines with *in situ* aromatisation, there being no β-carbonyl groups to stabilise the dihydro-pyridine.[284]

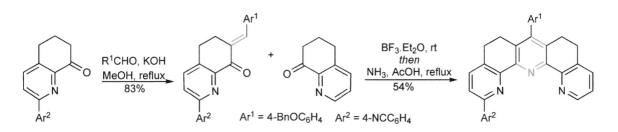

Ar^1 = 4-BnOC$_6$H$_4$ Ar^2 = 4-NCC$_6$H$_4$

8.14.1.3 From 1,3-Dicarbonyl Compounds (or Synthons) and 3-Amino-Enones or -Nitriles
Pyridines are formed from the interaction between a 1,3-dicarbonyl compound and a 3-amino-enone or 3-amino-acrylate; 3-cyano-2-pyridones result if cyanoacetamide is used instead of an amino-enone.

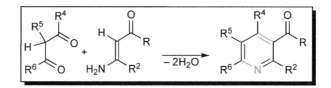

This approach, in its various forms, is one of the most versatile and useful, since it allows the construction of unsymmetrically substituted pyridines from relatively simple precursors. Again, in this pyridine-ring construction, intermediates are not isolated and it is usually difficult to be sure of the exact sequence of events.[285]

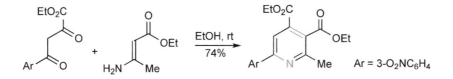

3-Amino-enones or 3-amino-acrylates can be prepared by the straightforward reaction of ammonia with a 1,3-diketone or a 1,3-keto-ester. The simplest 1,3-dicarbonyl compound, malondialdehyde, is too unstable to be useful, but its acetal enol ether can be used instead, as shown below.[286]

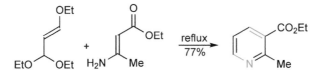

Vinamidinium ($R_2NCH=CR^5CH=N^+R_2$) salts (best as non-hygroscopic hexafluorophosphates) will serve as synthons for substituted malondialdehydes in these syntheses; this is one case in which the intermediate is known.[287] Using hydroxylamine instead of ammonia leads to *N*-oxides.[288] The vinamidinium salts are available by reaction of the relevant substituted acetic acid ($R^5CH_2CO_2H$) with phosphorus oxychloride and dimethylformamide.[289]

The Guareschi Synthesis
This variation makes use of cyanoacetamide as the nitrogen-containing component and thus leads to 3-cyano-2-pyridones.

Providing the two carbonyl groups are sufficiently different in reactivity, only one of the two possible isomeric pyridine/pyridone products is formed *via* reaction of the more electrophilic carbonyl group with the central carbon of the 3-amino-enone, 3-amino-acrylate, or cyanoacetamide.[290,291]

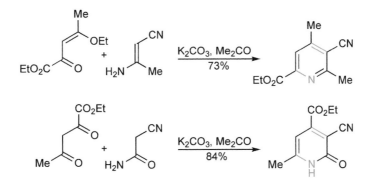

Variations include the use of 3-alkoxy-enones (i.e. the enol ethers of 1,3-diketones) when the initial Michael-type interaction dictates the regiochemistry.[292] Using nitroacetamide instead of cyanoacetamide produces 3-nitro-2-pyridones[293] and using $H_2NCOCH_2C(NH_2)=N^+H_2$ Cl^- gives 2-aminopyridine-3-carboxamides.[294]

Ring closures to produce pyridines and pyridones can also be carried out with starting materials at a lower oxidation level, with *in situ* dehydrogenation by air or added oxygen, i.e. instead of using a 1,3-dicarbonyl component, an α,β-unsaturated ketone/aldehyde is employed, as illustrated below.[295] If such condensations are carried out in the *absence* of oxygen, loss of hydrogen cyanide brings about aromatisation giving 2-pyridones with no substituent at C-3, particularly when the pyridone-4-substituent is aryl.[296]

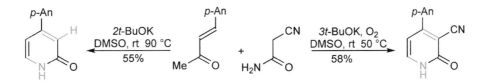

The reaction of yne-ones (also synthons for 1,3-dicarbonyl compounds) with 3-amino-enones or 3-amino-acrylates (the *Bohlmann–Rahtz reaction*) is regioselective, since conjugate addition of the ketone enamine is the first step; the intermediates thus produced can be isolated from reactions in ethanol and converted on to the aromatic pyridine.[297,298] Acetic acid or ytterbium triflate[299] give good results.

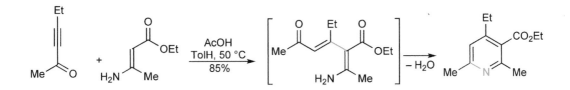

As a final example, malonate anions will add to yne-imines to produce pyridones directly.[300]

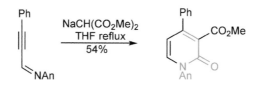

8.14.1.4 Via Cycloadditions

A number of 6π cycloadditions, some with inverse electron-demand, some with subsequent extrusion of a small molecule to achieve aromaticity, have been used to construct pyridines.

From Oxazoles

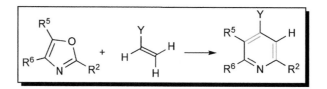

Historically, the first of these was the addition of a dienophile to an oxazole; using acrylonitrile, hydrogen cyanide is lost to aromatise and the oxazole oxygen is retained (giving 3-hydroxypyridines) and using acrylic acid, the oxygen is lost as water, as illustrated below.[301]

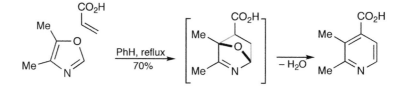

From Triazines

1,2,3-[302] and 1,2,4-Triazines, acting as inverse electron-demand azadienes, add to enamines (sometimes prepared *in situ*[303]) and thus, following extrusion of nitrogen and loss of amine, a pyridine is produced (see 29.2.1).[304] 1,2,4-Triazines will also react with other dienophiles: reaction with ethynyltributyltin, for example, gives 4-stannyl-pyridines;[305] norbornadiene is useful as an acetylene equivalent, cyclopentadiene being lost finally.[306] Oxazinones can also be used as the 'diene' component, with carbon dioxide as a final loss.[307]

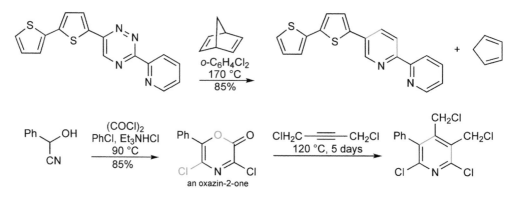

an oxazin-2-one

From Acyclic Azadienes

The *O,O'*-bis-*t*-butyldimethylsilyl derivative of an imide serves as an azadiene in reaction with dienophiles; 2-pyridones are the result, following desilylation.[308]

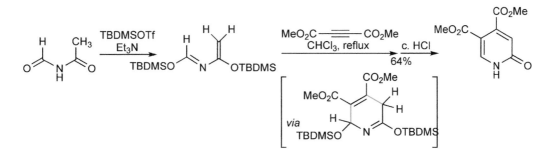

Unsaturated *O*-silylated oximes or unsaturated hydrazones, in particular those with a silylated oxygen at C-3 as well, take part in cycloadditions, with loss of the *N*-substituent (see also 8.14.1.1) giving 3-hydroxypyridines.[309,310] The starting 1-azadienes are easily available *via* α-nitrosation of a ketone, then oxime and enol *O*-silylations.

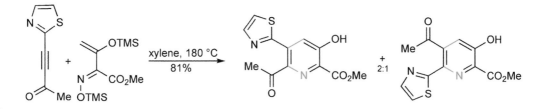

A three-component, one-pot process involving formation of a vinyl imine from a palladium(0)-catalysed coupling followed by the cycloaddition and then aromatisation *via* toluenesulfinate elimination gives bicyclic 2-amino-pyridines.[311]

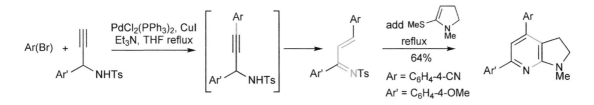

By Thermal Electrocyclisation of Aza-1,3,5-Trienes

Electrocyclisation of 1-aza-1,3,5-trienes generates dihydropyridines, which can be oxidised to pyridines; however, if an oxime or hydrazine derivative is used, elimination of water or an amine *in situ* gives the pyridine directly. This method is particularly useful for fusion of pyridines to other ring systems and is illustrated by the example below.[312]

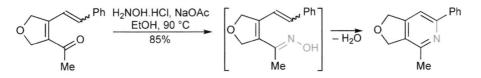

By Metal-Mediated [2 + 2 + 2] Cycloadditions[313]

The cobalt-catalysed interaction of a nitrile and two equivalents of an acetylene (or one equivalent of each of two different acetylenes) brings three components together to form an aromatic pyridine ring.[314]

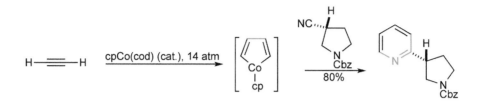

Cobalt has been used most often, in some cases on solid support,[315] but the ring construction can be brought about using a titanium(II) alkoxide,[316] or with a nickel(0) catalyst.[317] Isocyanates with a ruthenium catalyst generate 2-pyridones.[318]

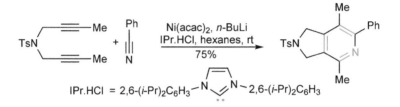

8.14.1.5 Miscellaneous Methods

This section includes a selection of examples that are of interest both mechanistically and preparatively.

From Furans

Ring-opening and reclosure processes using furans include several significant methods for the construction of pyridines. 2,5-Dihydro-2,5-dimethoxy-furans (see 18.1.1.4) carrying as a C-2 side-chain an aminoalkyl group, give rise to 3-hydroxy-pyridines.[319]

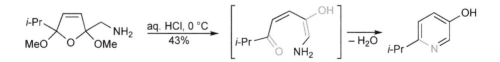

From Propargylamine

Following the interaction of propargylamine amino group with a ketone, producing an enamine, ring closure can be effected with a gold catalyst, the whole process being conducted in one pot.[320]

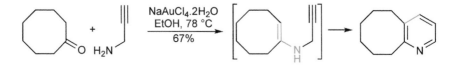

From Enamides

Enamides, easily available by *N*-acylation of imines, can be converted into 2-chloronicotinaldehydes by exposure to the Vilsmeier reagent: the example shows the putative intermediate.[321]

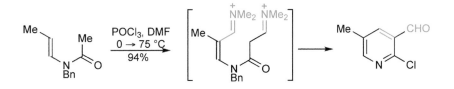

8.14.1.6 *Industrial Syntheses*

Many alkyl-pyridines are manufactured commercially by chemically complex processes that often produce them as mixtures. A good example is the extraordinary *Chichibabin synthesis*, in which paraldehyde and ammonium hydroxide react together at 230 °C under pressure to afford 52% of 5-ethyl-2-methylpyridine; so here, four mole equivalents of acetaldehyde and one of ammonia combine.[322]

8.14.2 Examples of Notable Syntheses of Pyridine Compounds

8.14.2.1 *Fusarinic Acid*

Fusarinic acid is a mould metabolite with antibiotic and antihypertensive activity. Two syntheses of this substance employ cycloadditions, one[323] to produce a 1,5-diketone and the other[324] to generate a 1-dimethylamino-1,4-dihydropyridine.

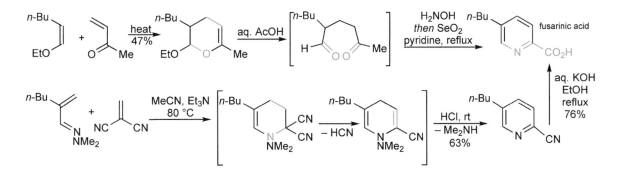

8.14.2.2 *Pyridoxine*

Pyridoxine, vitamin B$_6$, has been synthesised by several routes, including one that utilises a Guareschi ring synthesis, as shown below.[325] Another utilises a cycloaddition to an oxazole (8.14.1.4).[326]

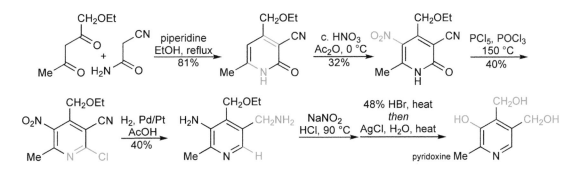

8.14.2.3   *Nemertelline*

The total synthesis of nemertelline, a hoploemertin worm toxin, illustrates the use of metallation and palladium-catalysed couplings.[327]

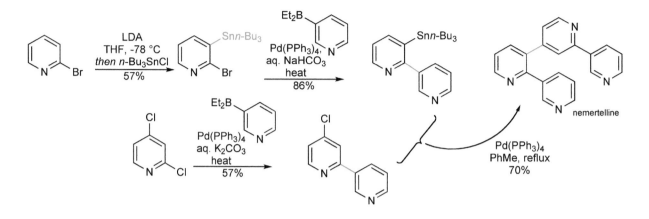

8.14.2.4   *Louisanin A*

Louisianin A is one of a family of *Streptomyces*-derived inhibitors of the growth of testosterone-responsive Shionogi carcinoma cells.[328]

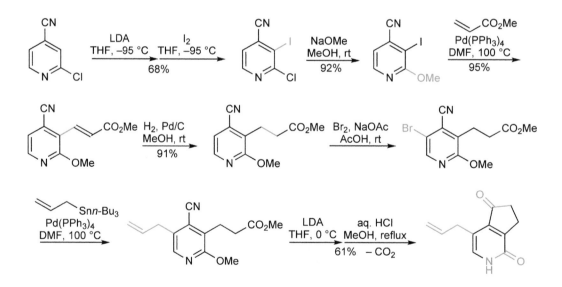

Exercises

Straightforward revision exercises (consult Chapters 7 and 8):

(a) In what way does pyridine react with electrophilic reagents such as acids and alkyl halides?

(b) What factors make it much more difficult to bring about electrophilic substitution of pyridine than benzene?

(c) How do pyridines compare with benzenes with regard to: (i) oxidative destruction of the ring and (ii) reduction of the ring?

(d) Give two examples of pyridines reacting with nucleophilic reagents with substitution of a hydrogen.

(e) What are the relative reactivities of bromobenzene, 2-bromopyridine, 3-bromopyridine towards replacement of the halide with ethoxide on treatment with NaOEt?

(f) How could one generate 2-lithiopyridine?

(g) What would result from treatment of 3-chloropyridine with LDA at low temperature?

(h) Draw the main tautomeric forms of 2-hydroxypyridine (2-pyridone), 3-hydroxypyridine and 2-aminopyridine.

(i) How could one convert 4-pyridone cleanly into 1-ethyl-4-pyridone?

(j) What would be the result of treating a 1:1 mixture of 2- and 3-methylpyridines with 0.5 equivalents of LDA and then 0.5 equivalents of MeI?

(k) Draw the structure of the product(s) you would expect to be formed if pyridine were reacted successively with methyl chloroformate and then phenyllithium.

(l) In pyridine *N*-oxides, both electrophilic substitution and nucleophilic displacement of halide from C-4 go more rapidly than in pyridine – explain.

(m) Describe two important methods for the synthesis of pyridines from precursors that do not contain the ring.

(n) What compounds would result from the following reagent combinations: (i) H_2NCOCH_2CN (cyanoacetamide) with $MeCOCH_2COMe$; (ii) $MeC(NH_2)=CHCO_2Et$ (ethyl 3-aminocrotonate) with $MeCOCH_2COMe$; (iii) $PhCH=O$, $MeCOCH_2COMe$ and NH_3?

More advanced exercises:

1. Suggest a structure for the products: (i) $C_7H_8N_2O_3$ produced by treating 3-ethoxypyridine with f. HNO_3/c. H_2SO_4 at 100 °C, (ii) $C_6H_4BrNO_2$ produced by reaction of 4-methylpyridine first with Br_2/H_2SO_4/oleum then with hot $KMnO_4$.

2. Deduce a structure for the product $C_9H_{15}N_3$ produced by reacting pyridine with the potassium salt of $Me_2N(CH_2)_2NH_2$.

3. Deduce structures for the product formed by: (i) reacting 2-chloropyridine with (a) hydrazine $\rightarrow C_5H_7N_3$, (b) water $\rightarrow C_5H_5NO$; (ii) 4-nitropyridine heated with water at 60 °C $\rightarrow C_5H_5NO$.

4. Deduce structures for the products formed in turn by reacting 4-chloropyridine with: (i) sodium methoxide $\rightarrow C_6H_7NO$, A, this with iodomethane $\rightarrow C_7H_{10}INO$, then this heated at 185 °C $\rightarrow C_6H_7NO$, isomeric with A.

5. Treatment of 4-bromopyridine with $NaNH_2$ in NH_3 (liq.) gives two products (isomers, $C_5H_6N_2$) but reaction with sodium methoxide gives a single product, C_6H_7NO. What are the products and why is there a difference?

6. Write structures for the products to be expected in the following sequences: (i) 4-diisopropylaminocarbonyl pyridine with LDA then with benzophenone, then with hot acid $\rightarrow C_{19}H_{13}NO_2$; (ii) 2-chloropyridine with LDA then iodine $\rightarrow C_5H_3ClNI$; (iii) 3-fluoropyridine with LDA, then with acetone $\rightarrow C_8H_{10}FNO$; (iv) 2-bromopyridine with butyllithium at −78 °C, then chlorotrimethylstannane $\rightarrow C_8H_{13}NSn$.

7. A crystalline solid $C_9H_{11}BrN_2O_3$ is formed when 2-methyl-5-nitropyridine is reacted with bromoacetone. Subsequent treatment with $NaHCO_3$ affords $C_9H_8N_2O_2$ – deduce the structures and write out a mechanism.

8. When the salt, $C_9H_{13}IN^+$ I^- produced by reacting pyridine with 1,4-diiodobutane is then treated with Bu_3SnH in the presence of AIBN, a new salt, $C_9H_{12}N^+$ I^- is formed, which has ^1H NMR signals for four aromatic protons. Suggest structures for the two salts and a mechanism of formation of the latter.

9. Deduce a structure for the product, $C_6H_{11}NO_3$, produced by exposing 4-methyl-2-pyridone to the following sequence: (i) irradiation at 310 nm, (ii) O_3/MeOH/-78 °C then $NaBH_4$.

10. Write structures for the compounds produced at each stage in the following sequence: 4-methylpyridine reacted with $NaNH_2 \rightarrow C_6H_8N_2$, this then with $NaNO_2/H_2SO_4$ at 0 °C $\rightarrow$ rt $\rightarrow C_6H_7NO$, then this with sodium methoxide and iodomethane $\rightarrow C_7H_9NO$ and finally this with $KOEt/(CO_2Et)_2 \rightarrow C_{11}H_{13}NO_4$.

11. Nitration of aniline is not generally possible, yet nitration of 2- and 4-aminopyridines can be achieved easily – why?

12. When 3-hydroxypyridine is reacted with 5-bromopent-1-ene, a crystalline salt $C_{10}H_{14}NBrO$ is formed. Treatment of the salt with mild base gives a dipolar substance $C_{10}H_{13}NO$, which on heating provides a neutral, non-aromatic isomer. Deduce the structures of these compounds.

13. Give an explanation for the relatively easy decarboxylation of pyridine-2-acetic acid; what is the organic product?

14. Suggest a structure for the product, $C_{16}H_{22}N_2O_5$ resulting from the interaction of 4-vinylpyridine with diethyl acetamidomalonate ($AcNHCH(CO_2Et)_2$) and base.

15. Write structures for the products of reacting: (i) 2,3-dimethylpyridine with butyllithium then diphenyldisulfide $\rightarrow C_{13}H_{13}NS$; (ii) 2,3-dimethylpyridine with NBS then with PhSH $\rightarrow C_{13}H_{13}NS$ isomeric with the product in (i).

16. Write structures for the isomeric compounds $C_7H_6N_2O$ (formed in a ratio of 4:3) when 3-cyanopyridine methiodide is reacted with alkaline potassium ferricyanide.

17. Predict the sites at which deuterium would be found when 1-butylpyridinium iodide is reduced with $NaBD_4$ in EtOH forming (mainly) 1-butyl-1,2,5,6-tetrahydropyridine.

18. Deduce structures for the final product, and intermediate, in the following sequence: pyridine with methyl chloroformate and sodium borohydride gave $C_7H_9NO_2$, then this irradiated gave an isomer which had NMR signals for only two alkene protons – what are the compounds?

19. When pyridine *N*-oxide is heated with c. H_2SO_4 and c. HNO_3, a product $C_5H_4N_2O_3$ is formed; separate reactions of this with PCl_3 then H_2/Pd-C produces $C_5H_4N_2O_2$ and $C_5H_6N_2$ sequentially. What are the three products?

20. Write a structure for the cyclic product, $C_{18}H_{21}NO_4$, from the reaction of ammonia, phenylacetaldehyde ($PhCH_2CH=O$), and two mole equivalents of methyl acetoacetate. How might it be converted into a pyridine?

21. 2,3-Dihydrofuran reacts with acrolein to give $C_7H_{10}O_2$; reaction of this with aq. H_2NOH/HCl gives a pyridine, C_7H_9NO: deduce structures.

22. What pyridines or pyridones would be produced from the following combinations of reactants: (a) H_2NCOCH_2CN (cyanoacetamide) with: (i) $EtCOCH_2CO_2Et$; (ii) 2-acetylcyclohexanone; (iii) ethyl propiolate; (b) $MeC(NH_2)=CHCO_2Et$ (ethyl 3-aminocrotonate) with (i) but-3-yne-2-one; (ii) $MeCOC(CO_2Et)=CHOEt$.

23. When the sodium salt of formyl acetone ($MeCOCH=CHO^- Na^+$) is treated with ammonia, a pyridine C_8H_9NO is formed. Deduce a structure and explain the regiochemistry of reaction.

References

[1] Arnett, E. M., Chawla, B., Bell, L., Taagepera, M., Hehre, W. J. and Taft, R. W., *J. Am. Chem. Soc.*, **1977**, *99*, 5729.

[2] Corey, E. J. and Zheng, G. Z., *Tetrahedron Lett.*, **1998**, *39*, 6151.

[3] Olah, G. A., Narang, S. C., Olah, J. A., Pearson, R. L. and Cupas, C. A., *J. Am. Chem. Soc.*, **1980**, *102*, 3507.

[4] Duffy, J. L. and Laali, K. K., *J. Org. Chem.*, **1991**, *56*, 3006.

[5] Gösl, R. and Meuwsen, A., *Org. Synth., Coll. Vol. V*, **1973**, 43.

[6] Jain, S. L., Sharma, V. B. and Sain, B., *Tetrahedron Lett.*, **2003**, *44*, 4385.

[7] Jain, S. L. and Sain, B., *Chem. Commun.*, **2002**, 1040.

[8] Caron, S., Do, N. M. and Sieser, J. E., *Tetrahedron Lett*, **2000**, *41*, 2299.

[9] Zhang, Y. and Lin, R., *Synth Commun.*, **1987**, *17*, 329; Akita, Y., Misu, K., Watanabe, T. and Ohto, A., *Chem. Pharm. Bull.*, **1976**, *24*, 1839; Malinowski, M. and Kaczmarek, L., *Synthesis*, **1987**, 1013; Balicki, R., Kaczmarek, L. and Malinowski, M., *Synth. Commun.*, **1989**, 897; Balicki, R., *Synthesis*, **1989**, 645.

10 Yoo, B. W., Choi, J. W. and Yoon, C. M., *Tetrahedron Lett.*, **2006**, *47*, 125.

11 'Heterocyclic *N*-Oxides', Katritzky, A.R. and Lagowski, J. M., Methuen, London, **1967**; 'Aromatic Amine Oxides', Ochiai, E., Elsevier, New York, **1967**; 'Heterocyclic *N*-Oxides', Albini, A. and Pietra, S., CRC Press Wolfe Publishing, London, **1991**; 'Heterocyclic *N*-oxides and *N*-imides', Katritzky, A. R. and Lam, J. N., *Heterocycles*, **1992**, *33*, 1011.

12 Olah, G. A., Arvanaghi, M. and Vankar, Y. D., *Synthesis*, **1980**, 660.

13 Zacharie, B., Moreau, N. and Dockendorff, C., *J. Org. Chem.*, **2001**, *66*, 5264.

14 Baumgarten, P., *Chem. Ber.*, **1926**, *59*, 1166.

15 Evans, R. F., Brown, H. C. and van der Plas, H. C., *Org. Synth., Coll. Vol. V*, **1973**, 977.

16 Popov, A. I. and Rygg, R. H., *J. Am. Chem. Soc.*, **1957**, *79*, 4622.

17 Umemoto, T., Tomita, K. and Kawada, K., *Org. Synth.*, **1990**, *69*, 129.

18 King, J. A. and Bryant, G. L., *J. Org. Chem.*, **1992**, *57*, 5136.

19 '4-Dialkylaminopyridines as highly active acylation catalysts', Höfle, G., Steglich, W. and Vorbrüggen, H., *Angew. Chem., Int. Ed. Engl.*, **1979**, *17*, 569; '4-Dialkylaminopyridines: super acylation and alkylation catalysts', Scriven, E. F. V., *Chem. Soc. Rev.*, **1983**, *12*, 129.

20 Guibé-Jampel, E., and Wakselman, M., *J. Chem. Soc., Chem. Commun.*, **1971**, 267.

21 Held, I., Villinger, A. and Zipse, H., *Synthesis*, **2005**, 1425.

22 Petit, S., Azzouz, R., Fruit, C., Bischoff, L. and Marsais, F., *Tetrahedron Lett.*, **2008**, *49*, 3663.

23 Zoltewicz, J. A. and Smith, C. L., *J. Am. Chem. Soc.*, **1967**, *89*, 3358; Zoltewicz, J. A. and Cross, R. E., *J. Chem. Soc., Perkin Trans. 2*, **1974**, 1363 and 1368; Werstuik, N. H. and Ju, C., *Can. J. Chem.*, **1989**, *67*, 5; Zoltewicz, J. A. and Meyer, J. D., *Tetrahedron Lett.*, **1968**, 421.

24 Esaki, H., Ito, N., Sakai, S., Maegawa, T., Monguchi, Y. and Sajika, H., *Tetrahedron*, **2006**, *62*, 10954.

25 Den Hertog, H. J. and Overhoff, J., *Recl. Trav. Chim. Pays-Bas*, **1930**, *49*, 552.

26 Brown, E. V. and Neil, R. H., *J. Org. Chem.*, **1961**, *26*, 3546.

27 Johnson, C. D., Katritzky, A. R., Ridgewell, B. J. and Viney, M., *J. Chem. Soc., B*, **1967**, 1204.

28 Shackelford, S. A., Anderson, M. B., Christie, L. C., Goetzen, T., Guzman, M. C., Hananel, M. A., Kornreich, W. D., Li, H., Pathak, V. P., Rabinovich, A. K., Rajapakse, R. J., Truesdale, L. K., Tsank, S. M. and Vazir, H. N., *J. Org. Chem.*, **2003**, *68*, 267.

29 'Nitropyridines, their synthesis and reactions', Bakke, J. M., *J. Heterocycl. Chem.*, **2005**, *42*, 463.

30 Katritzky, A. R., Scriven, E. F. V., Majumder, S., Akhmedova, R. G., Vakulenko, A. V., Akhmedov, N. G., Murugan, R. and Abboud, K. A., *Org. Biomol. Chem*, **2005**, 538.

31 Katritzky, A. R., Scriven, E. F. V., Majumder, S., Tu, H., Vakulenko, A. V., Akhmedov, N. G. and Murugan, R., *Synthesis*, **2005**, 993.

32 McElvain, S. M. and Goese, M. A., *J. Am. Chem. Soc.*, **1943**, *65*, 2233.

33 '2,6-Di-*t*-butylpyridine – an unusual base', Kanner, B., *Heterocycles*, **1982**, *18*, 411.

34 den Hertog, H. J., den Does, L. V. and Laandheer, C. A., *Recl. Trav. Chim. Pays-Bas*, **1962**, *91*, 864.

35 Pearson, D. E., Hargreave, W. W., Chow, J. K. T. and Suthers, B. R., *J. Org. Chem.*, **1961**, *26*, 789.

36 Paraskewas, S., *Synthesis*, **1980**, 378.

37 McCleland, N. P. and Wilson, R. H., *J. Chem. Soc.*, **1932**, 1263.

38 Bartok, W., Rosenfeld, D. D. and Schriesheim, A., *J. Org. Chem.*, **1963**, *28*, 410; Black, G., Depp, E. and Corson, B. B., *J. Org. Chem.*, **1949**, *14*, 14.

39 Jerchel, D., Heider, J. and Wagner, H., *Justus Liebigs Ann. Chem.*, **1958**, *613*, 153.

40 'Nucleophilic Aromatic Substitution of Hydrogen', Chupakhin, O. N., Charushin, V. N. and van der Plas, H. C., Academic Press, San Diego, **1994**.

41 Evans, J. C. W. and Allen, C. F. H., *Org. Synth., Coll. Vol. II*, **1943**, 517.

42 'Formation of anionic σ-adducts from heterocyclic compounds: structures, rates and equilibria', Illuminati, G., and Stegel, F., *Adv. Heterocycl. Chem.*, **1983**, *34*, 305.

43 Abramovitch, R. A. and Giam, C.-S., *Canad. J. Chem.*, **1964**, *42*, 1627; Abramovitch, R. A. and Poulton, G. A., *J. Chem. Soc., B*, **1969**, 901.

44 Zhang, L.-H and Tan, Z., *Tetrahedron Lett.*, **2000**, *41*, 3025 and references to trapping with other electrophiles, therein.

45 'Amination of heterocyclic bases by alkali amides', Leffler, M. T., *Org. Reactions*, **1942**, *1*, 91; 'Advances in the Chichibabin reaction', McGill, C. K. and Rappa, A., *Adv. Heterocycl. Chem.*, **1988**, *44*, 2; 'Advances in the amination of nitrogen heterocycles', Vorbrüggen, H., *Adv. Heterocycl. Chem.*, **1990**, *49*, 117.

46 Viscardi, G., Savarino, P., Quagliotto, P., Barni, E. and Bottam M., *J. Heterocycl. Chem.*, **1996**, *33*, 1195.

47 Abramovitch, R. A., Helmer, F. and Saha, J. G., *Chem. Ind.*, **1964**, 659; Ban, Y. and Wakamatsu, T., *ibid.*, 710.

48 Seko, S. and Miyake, K., *Chem. Commun.*, **1998**, 1519.

49 Bakke, J. M., Svensen, H. and Trevisan, R., *J. Chem. Soc., Perkin Trans. 1*, **2001**, 376.

50 Andreassen, E. J. and Bakke, J. M., *J. Heterocycl. Chem.*, **2006**, *43*, 49.

51 Postigo, A. and Rossi, R. A., *Org. Lett.*, **2001**, *3*, 1197; Postigo, A., Vaillard, S. E. and Rossi, R. A., *J. Phys. Org. Chem.*, **2002**, *15*, 889.

52 Chichibabin, A. E., *Chem. Ber.*, **1923**, *56*, 1879.

53 Kuduk, S. D., DiPardo, R. M. and Bock, M. G., *Org. Lett.*, **2005**, *7*, 577.

54 Hanessian, S. and Kagotani, M., *Synthesis*, **1987**, 409.

55 Yamanaka, H. and Ohba, S., *Heterocycles*, **1990**, *31*, 895.

56 Cherng, Y.-J., *Tetrahedron*, **2002**, *58*, 4931.

57 DuPriest, M. T., Schmidt, C. L., Kuzmich, D. and Williams, S. B., *J. Org. Chem.*, **1986**, *51*, 2021.

58 Pasumansky, L., Hernández, A. R., Gamsey, S., Goralski, C. T. and Singaram, B., *Tetrahedron Lett.*, **2004**, *45*, 6417.

59 Elhaïk, J., Pask, C. M., Kilner, C. A. and Halcrow, M. A., *Tetrahedron*, **2007**, *63*, 291.

60 Chambers, R. D., Sandford, G. and Trmcic, J., *J. Fluorine Chem.*, **2007**, *128*, 1439.

61 Hashimoto, S., Otani, S., Okamoto, T. and Matsumoto, K., *Heterocycles*, **1988**, *27*, 319.

62 Bakke, J. M., Ranes, E., Romming, C. and Sletvold, I., *J Chem. Soc., Perkin Trans. 1*, **2000**, 1241.

63 Bakke, J. M. and Sletvold, I., *Org. Biomol. Chem.*, **2003**, *1*, 2710.

[64] Jamart-Gregoire, B., Leger, C. and Caubere, P., *Tetrahedron Lett.*, **1990**, *31*, 7599; 'Hetarynes', den Hertog, H. J. and van der Plas, H. C., *Adv. Heterocycl. Chem.*, **1965**, *4*, 121; 'Hetarynes', Reinecke, M. G., *Tetrahedron*, **1982**, *38*, 427.

[65] Mallet, M. and Queguiner, G., *C. R. Acad. Sci., Ser. C*, **1972**, *274*, 719; Walters, M. A., Carter, P. H. and Banerjee, S., *Synth. Commun.*, **1992**, *22*, 2829.

[66] Walters, M. A. and Shay, J. J., *Synth. Commun.*, **1997**, *27*, 3573.

[67] Connon, S. J. and Hegarty, A. F., *Tetrahedron Lett.*, **2001**, *42*, 735.

[68] Schlosser, M. and Cottet, F., *Eur. J. Org. Chem.*, **2002**, 4181.

[69] Skerlj, R. T., Bogucki, D. and Bridger, G. J., *Synlett*, **2000**, 1488.

[70] Klapars, A., Waldman, J. H., Campos, K. R., Jensen, M. S., McLaughlin, M., Chung, J. Y. L., Cvetovich, R. J. and Chen, C.-y., *J. Org. Chem.*, **2005**, *70*, 10186.

[71] Pal, M., Batchu, V. R., Dager, I., Swamy, N. K. and Padakanti, S., *J. Org. Chem.*, **2005**, *70*, 2376.

[72] 'Advances in the directed metallation of azines and diazines. Part I: Metallation of pyridines, quinolines and carbolines', Mongin, F. and Quéguiner, G., *Tetrahedron*, **2001**, *57*, 4059.

[73] Zoltewicz, J. A., Grahe, G. and Smith, C. L., *J. Am. Chem. Soc.*, **1969**, *91*, 5501.

[74] Taylor, S. L., Lee, D. Y. and Martin, J. C., *J. Org. Chem.*, **1983**, *48*, 4157.

[75] Tagawa, Y., Nomura, M., Yamagata, K., Teshima, D., Shibata, K. and Goto, Y., *Heterocycles*, **2004**, *63*, 2863.

[76] Mongin, O., Rocca, P., Thomas-dit-Dumont, L., Trécourt, F., Marsais, F., Godard, A. and Quéguiner, G., *J. Chem. Soc., Perkin Trans. 1*, **1995**, 2503.

[77] Gros, P., Fort, Y., and Caubère, P., *J. Chem. Soc., Perkin Trans. 1*, **1997**, 3597.

[78] '*n*-BuLi/lithium aminoalkoxide aggregates: new and promising lithiating agents for pyridine derivatives', Gros, P. and Fort, Y., *Eur. J. Org. Chem.*, **2002**, 3375.

[79] Gros, P. and Fort, Y., *J. Org. Chem.*, **2003**, *68*, 2028.

[80] Kaminski, T., Gros, P. and Fort, Y., *Eur. J. Org. Chem.*, **2003**, 3855.

[81] Doudouh, A., Gros, P. C., Fort, Y. and Woltermann, C., *Tetrahedron*, **2006**, *62*, 6166.

[82] Gros, P. and Fort, Y., *J. Chem. Soc., Perkin Trans. 1*, **1998**, 3515.

[83] Commins, D. L. and LaMunyon, D. H., *Tetrahedron Lett.*, **1988**, *29*, 773.

[84] Ronald, R. C. and Winkle, M. R., *Tetrahedron*, **1983**, *39*, 2031.

[85] Epsztajn, J, Berski, Z., Brzezinski, J. Z. and Józwick, A., *Tetrahedron Lett.*, **1980**, *21*, 4739.

[86] Azzouz, R., Bischoff, L., Fruit, A. and Marsais, F., *Synlett*, **2006**, 1908.

[87] Güngör, T., Marsais, F. and Quéguiner, G., *Synthesis*, **1982**, 499.

[88] Mongin, F., Trécourt, F. and Quéguiner, G., *Tetrahedron Lett.*, **1999**, *40*, 5483.

[89] Cailly, T., Fabis, F., Bouillon, A., Lemaître, S., Sopkova, J. and de Santos, O., *Synlett*, **2006**, 53.

[90] Awad, H., Mongin, F., Trécourt, F., Quéguiner, G. and Marsais, F., *Tetrahedron Lett.*, **2004**, *45*, 7873.

[91] Hands, D., Bishop, B., Cameron, M., Edwards, J. S., Cottrell, I. F. and Wright, S. H. R., *Synthesis*, **1996**, 877.

[92] Deguest, G., Devineau, A., Bischoff, L., Fruit, C. and Marsais, F., *Org. Lett.*, **2006**, *8*, 5889.

[93] Bouillon, A., Lancelot, J.-C., Collot, V., Bovy, P. R. and Rault, S., *Tetrahedron*, **2002**, *58*, 4369.

[94] Marzi, E., Bigi, A. and Schlosser, M., *Eur. J. Org. Chem.*, **2001**, 1371; 'The organometallic approach to molecular diversity – halogens as helpers', Schlosser, M., *Eur. J. Org. Chem.*, **2001**, 3975.

[95] Marsais, F., Trécourt, F., Bréant, P. and Quéguiner, G., *J. Heterocycl. Chem.*, **1988**, *25*, 81.

[96] Epsztajn, J., Bieniek, A. and Kowalska, J. A., *Tetrahedron*, **1991**, *47*, 1697.

[97] Lomberget, T., Radix, S. and Barret, R., *Synlett*, **2005**, 2080.

[98] Guillier, F., Nivoliers, F., Cochennee, A., Godard, A., Marsais, F. and Queguiner, G., *Synth. Commun.*, **1996**, *26*, 4421.

[99] Bell, A. S., Roberts, D. A. and Ruddock, K. S., *Synthesis*, **1987**, 843

[100] Doudouh, A., Woltermann, C. and Gros, P. C., *J. Org. Chem.*, **2007**, *72*, 4978.

[101] Grios, P. C. and Elaachbouni, F., *Chem. Commun.*, **2008**, 4813.

[102] Kondo, Y., Murata, N. and Sakamoto, T., *Heterocycles*, **1994**, *37*, 1467; Gómez, I., Alonso, E., Ramón, D. J., and Yus, M., *Tetrahedron*, **2000**, *56*, 4043.

[103] Parham, W. E. and Piccirilli, R. M., *J. Org. Chem.*, **1977**, *42*, 257.

[104] Lipshutz, B. H., Pfeiffer, S. S. and Chrisman, W., *Tetrahedron Lett.*, **1999**, *40*, 7889.

[105] Wang, X., Rabbat, P., O'Shea, P., Tillyer, R., Grabowski, E. J. J. and Reider, P. J., *Tetrahedron Lett.*, **2000**, *41*, 4335.

[106] Cai, D., Hughes, D. L. and Verhoeven, T. R., *Tetrahedron Lett.*, **1996**, *37*, 2537; Paterson, M. A. and Mitchell, J. R., *J. Org. Chem.*, **1997**, *62*, 8237.

[107] Cai, D., Larsen, R. D., and Reider, P. J., *Tetrahedron Lett.*, **2002**, *43*, 4285.

[108] Janka, M., Anderson, G. K. and Rath, N. P., *Inorg. Chim. Acta*, **2004**, *357*, 2339.

[109] Bargar, T. M., Wilson, T. and Daniel, J. K., *J. Heterocycl. Chem.*, **1985**, *22*, 1583.

[110] Trecourt, F., Breton, G., Bonnet, V., Mongin, F., Marsais, F. and Quéguiner, G., *Tetrahedron*, **2000**, *56*, 1349; 'Functionalised magnesium organometallics as versatile intermediates for the synthesis of polyfunctional heterocycles', Ila, H., Baron, O., Wagner, A. J., and Knochel, P., *Chem. Commun.*, **2006**, 583.

[111] Getmanenko, Y. A. and Twieg, R. J., *J. Org. Chem.*, **2008**, *73*, 830.

[112] Song, J. J., Yee, N. K., Tan, Z., Xu, J., Kapadia, S. R. and Senenayake, C. H., *Org. Lett.*, **2004**, *6*, 4905.

[113] Ren, H. and Knochel, P., *Chem. Commun.*, **2006**, 726.

[114] Sugimoto, O., Yamada, S. and Tanji, K.-i., *J. Org. Chem.*, **2003**, *68*, 2054.

[115] Wibaut, J. P. and Nicolaï, J. R., *Recl. Trav. Chim. Pays-Bas*, **1939**, *58*, 709; McElvain, S. M., and Goese, M. A., *J. Am. Chem. Soc.*, **1943**, *65*, 2227.

[116] 'Radical addition to pyridines, quinolines and isoquinolines', Harrowven, D. C. and Sutton, B. J., *Prog. Heterocycl. Chem.*, **2004**, *16*, 27.

[117] Elofson, R. M., Gadallah, F. F. and Schutz, K. F., *J. Org. Chem.*, **1971**, *36*, 1526; Gurczynski, M. and Tomasik, P., *Org. Prep. Proc. Int.*, **1991**, *23*, 438.

[118] Bonnier, J. M. and Court, J., *Bull. Soc. Chim. Fr.*, **1972**, 1834; Minisci, F., Vismara, E., Fontana, F., Morini, G., Serravelle, M. and Giordano, C., *J. Org. Chem.*, **1986**, *51*, 4411.

[119] Harrowven, D. C., Sutton, B. J. and Coulton, S., *Org. Biomol. Chem.*, **2003**, *1*, 4047.

[120] Minisci, F., Porta, O., Recurpero, F., Punta, C., Gambarotti, C., Pruna, B., Pierini, M. and Fontana, F., *Synlett*, **2004**, 874.

[121] 'Recent developments of free radical substitutions of heteroaromatic bases', Minisci, F., Vismara, E., and Fontana, F., *Heterocycles*, **1989**, *28*, 489.

[122] 'The action of metal catalysts on pyridines', Badger, G. M. and Sasse, W. H. F., *Adv. Heterocycl. Chem.*, **1963**, *2*, 179; 'The bipyridines', Summers, L. A., *Adv. Heterocycl. Chem.*, **1984**, *35*, 281.

[123] Sasse, W. H. F., *Org. Synth., Coll. Vol. V*, **1973**, 102.

[124] Newkome, G. R. and Hager, D. C., *J. Org. Chem.*, **1982**, *47*, 599.

[125] Nielsen, A. T., Moore, D. W., Muha, G. M. and Berry, K. H., *J. Org. Chem.*, **1964**, *29*, 2175; Frank, R. L. and Smith, P. V., *Org. Synth., Coll. Vol. III*, **1955**, 410.

[126] Atlanti, P. M., Biellmann, J. F. and Moron, J., *Tetrahedron*, **1973**, *29*, 391; Beddoes, R. L., Arshad, N. and Joule, J. A., *Acta Crystallogr., Sect. C*, **1996**, *C52*, 654.

[127] Ryang, H.-S. and Sakurai, H., *J. Chem. Soc., Chem. Commun.*, **1972**, 594; Ohkura, K., Terashima, M., Kanaoka, Y. and Seki, K. *Chem. Pharm. Bull.*, **1993**, *41*, 1920.

[128] Jones, K. and Fiumana, A., *Tetrahedron Lett.*, **1996**, *38*, 8049.

[129] Zhao, J., Yang, X., Jia, X., Shengjun Luo, S. and Zha, H., *Tetrahedron*, **2003**, *59*, 9379.

[130] Martínez-Barrasa, V., de Viedma, A. G., Burgos, C. and Alvarez-Builla, J., *Org. Lett.*, **2000**, *2*, 3933.

[131] Lunn, G. and Sansome, E. B., *J. Org. Chem.*, **1986**, *51*, 513.

[132] Piras, L., Genesio, E., Ghiron, C. and Taddei, M., *Synlett*, **2008**, 1125.

[133] Giam, C. S., and Abbott, S. D., *J. Am. Chem. Soc.*, **1971**, *93*, 1294.

[134] Blough, B. E. and Carroll, F. I., *Tetrahedron Lett.*, **1993**, *34*, 7239.

[135] Tsuge, O., Kanemasa, S., Naritomi, T. and Tanaka, J., *Bull. Chem. Soc. Jpn.*, **1987**, *60*, 1497.

[136] Kamochi, Y. and Kudo, T., *Heterocycles*, **1993**, *36*, 2383.

[137] Birch, A. J. and Karakhanov, E, A., *J. Chem. Soc., Chem. Commun.*, **1975**, 480.

[138] Guerry, P. and Neier, R., *Synthesis*, **1984**, 485.

[139] Donohoe, T. J., Johnson, D. J., Mace, L. H., Thomas, R. E., Chiu, J. Y. K., Rodrigues, J. S., Compton, R. G., Banks, C. E., Tomcik, P., Bamford, M. J. and Ichihara, O., *Org. Biomol. Chem.*, **2006**, *4*, 1071.

[140] Heap, U., *Tetrahedron*, **1975**, *31*, 77; Tomisawa, H. and Hongo, H., *Chem. Pharm. Bull.*, **1970**, *18*, 925; Matsumoto, K., Ikemi-Kono, Y., Uchida, T. and Acheson, R. M., *J. Chem. Soc., Chem. Commun.*, **1979**, 1091; 'Diels-Alder cycloadditions of 2-pyrones and 2-pyridones', Afarinkia, K., Viader, V., Nelson, T. D. and Posner, G. H., *Tetrahedron*, **1992**, *48*, 9111.

[141] Posner, G. H., Vinader, V. and Afarinkia, K., *J. Org. Chem.*, **1992**, *57*, 4088.

[142] Okamura, H., Nagaike, H., Iwagawa, T. and Nakatani, M., *Tetrahedron Lett.*, **2000**, *41*, 8317; Kipassa, N. T., Okamura, H., Kina, K., Hamada, T. and Iwagawa, T., *Org. Lett.*, **2008**, *10*, 815.

[143] 'Synthetic applications of heteroaromatic betaines with six-membered rings', Dennis, N., Katritzky, A. R. and Takeuchi, Y., *Angew. Chem., Int. Ed. Engl.*, **1976**, *15*, 1; 'Cycloaddition reactions of heteroaromatic six-membered rings', Katritzky, A. R. and Dennis, N., *Chem. Rev.*, **1989**, *89*, 827; Lomenzo, S. A., Enmon, J. L., Troyer, M. C. and Trudell, M. L., *Synth. Commun.*, **1995**, *25*, 3681.

[144] Pavlik, J. W., Laohhasurayotin, S. and Vongnakorm, T., *J. Org. Chem.*, **2007**, *72*, 7116.

[145] Joussot-Dubien, J. and Houdard, J., *Tetrahedron Lett.*, **1967**, 4389; Wilzbach, K. E. and Rausch, D. J., *J. Am. Chem. Soc.*, **1970**, *92*, 2178.

[146] De Selms, R. C. and Schleigh, W. R., *Tetrahedron Lett.*, **1972**, 3563; Kaneko, C., Shiba, K., Fujii, H. and Momose, Y., *J. Chem. Soc., Chem. Commun.*, **1980**, 1177.

[147] Nakamura, Y., Kato, T. and Morita, Y., *J. Chem. Soc., Perkin Trans. 1*, **1982**, 1187.

[148] Johnson, B. L., Kitahara, Y., Weakley, T. J. R. and Keana, J. F. W., *Tetrahedron Lett.*, **1993**, *34*, 5555.

[149] Kaplan, L., Pavlik, J. W. and Wilzbach, K. E., *J. Am. Chem. Soc.*, **1972**, *94*, 3283; Ling, R., Yoshida, M. and Mariano, P. S., *J. Org. Chem.*, **1996**, *61*, 4439; Ling, R. and Mariano, P. S., *ibid.*, **1998**, *63*, 6072; Feng, X., Duesler, E. N. and Mariano, P. S., *J. Org. Chem.*, **2005**, *70*, 5618.

[150] Buchardt, O., Christensen, J. J., Nielsen, P. E., Koganty, R. R., Finsen, L., Lohse, C. and Becher, J., *Acta Chem. Scand., Ser. B*, **1980**, *34*, 31; Lohse, C., Hagedorn, L., Albini, A. and Fasani, E., *Tetrahedron*, **1988**, *44*, 2591.

[151] Beak, P., Fry, F. S., Lee, J. and Steele, F., *J. Am. Chem. Soc.*, **1976**, *98*, 171; Stefaniak, L., *Tetrahedron*, **1976**, *32*, 1065; Beak, P., Covington, J. B., Smith, S. G., White, J. M. and Zeigler, J. M., *J. Org. Chem.*, **1980**, *45*, 1354.

[152] Bensaude, O., Chevrier, M. and Dubois, J.-E., *Tetrahedron Lett.*, **1978**, 2221.

[153] Schoffner, J. P., Bauer, L. and Bell, C. L., *J. Heterocycl. Chem.*, **1970**, *7*, 479 and 487.

[154] Flemming, I. and Philippides, D., *J. Chem. Soc. C*, **1970**, 2426; Effenberger, F., Mück, A. O. and Bessey, E., *Chem. Ber.*, **1980**, *113*, 2086.

[155] Bellingham, P., Johnson, C. D. and Katritzky, A. R., *Chem. Ind.*, **1965**, 1384.

[156] Burton, A. G., Halls, P. J. and Katritzky, A. R., *J. Chem. Soc., Perkin Trans. 2*, **1972**, 1953.

[157] Brignell, P. J., Katritzky, A. R. and Tarhan, H. O., *J. Chem. Soc., B*, **1968**, 1477.

[158] De Selms, R. C., *J. Org. Chem.*, **1968**, *33*, 478.

[159] Stempel, A. and Buzzi, E. C., *J. Am. Chem. Soc.*, **1949**, *71*, 2969.

[160] Sheldrake, P. W., Powling, L. C. and Slaich, P. K., *J. Org. Chem.*, **1997**, *62*, 3008.

[161] Cañibano, V., Rodríguez, J. F., Santos, M., Sanz-Tejedor, M. A., Carreño, M. C., González, G. and García-Ruano, J. L., *Synthesis*, **2001**, 2175.

[162] Windscheif, P.-M. and Vögtle, F., *Synthesis*, **1994**, 87.

[163] Kornblum, N. and Coffey, G. P., *J. Org. Chem.*, **1966**, *31*, 3449; Hopkins, G. C., Jonak, J. P., Minnemeyer, H. J. and Tieckelmann, H., *J. Org. Chem.*, **1967**, *32*, 4040; Dou, H. J.-M, Hassanaly, P. and Metzger, J., *J. Heterocycl. Chem.*, **1977**, *14*, 321.

172 *Heterocyclic Chemistry*

164 Liu, H., Ko, S.-B., Josien, H. and Curran, D. P., *Tetrahedron Lett.*, **1995**, *36*, 845; Sato, T., Yoshimatsu, K. and Otera, J., *Synlett*, **1995**, 845.
165 Liu, H., Ko, S.-B. and Curran, D. P., *Tetrqhedron Lett.*, **1995**, *36*, 8917.
166 Vorbrüggen, H. and Krolikiewicz, K., *Chem. Ber.*, **1984**, *117*, 1523.
167 Lanni, E. L., Bosscher, M. A., Ooms, B. D., Shandro, C. A., Ellsworth, B. A. and Anderson, C. E., *J. Org. Chem.*, **2008**, *7*, 6425.
168 Comins, D. L. and Jianhua, G., *Tetrahedron Lett.*, **1994**, *35*, 2819.
169 Sugimoto, O., Mori, M. and Tanji, K., *Tetrahedron Lett.*, **1999**, *40*, 7477.
170 Kato, Y., Okada, S., Tomimoto, K. and Mase, T., *Tetrahedron Lett.*, **2001**, *42*, 4849.
171 Smith, A. P., Savage, S. A., Love, J. C. and Fraser, C. L., *Org. Synth.*, **2002**, *78*, 51.
172 Isler, O., Gutmann, H., Straub, U., Fust, B., Böhni, E. and Studer, A., *Helv. Chim. Acta*, **1955**, *38*, 1033.
173 Lowe, J. A., Ewing, F. E. and Drozda, S. E., *Synth. Commun.*, **1989**, *19*, 3027.
174 Girault, G., Coustal, S. and Rumpf, P., *Bull Soc. Chim. France*, **1972**, 2787; Forsythe, P., Frampton, R., Johnson, C. D. and Katritzky, A. R., *J. Chem. Soc., Perkin Trans. 2*, **1972**, 671.
175 Frampton, R., Johnson, C. D. and Katritzky, A. R., *Justus Liebigs Ann. Chem.*, **1971**, *749*, 12.
176 Fox, B. A. and Threlfall, T. L., *Org. Synth., Coll. Vol. V*, **1973**, 346.
177 v. Schickh, O., Binz, A. and Schule, A., *Chem. Ber.*, **1936**, *69*, 2593; Bakke, J. M. and Riha, J., *J. Heterocycl. Chem.*, **2001**, *38*, 99.
178 Burton, A. G., Frampton, R. D., Johnson, C. D. and Katritzky, A. R., *J. Chem. Soc., Perkin Trans. 2*, **1972**, 1940.
179 Katritzky, A. R., El-Zemity, S. and Lang, H., *J. Chem. Soc., Perkin Trans. 1*, **1995**, 3129.
180 Kawase, M., Koyanagi, J. and Saito, S., *Chem. Pharm. Bull.*, **1999**, *47*, 718.
181 Kalatzis, E., *J. Chem. Soc., B*, **1967**, 273 and 277.
182 Bunnett, J. F. and Singh, P., *J. Org. Chem.*, **1981**, *46*, 4567.
183 Allen, C. F. H. and Thirtle, J. R., *Org. Synth., Coll. Vol. III*, **1955**, 136; Windscheif, P.-M. and Vögtle, *Synthesis*, **1994**, 87.
184 Yoneda, N. and Fukuhara, T., *Tetrahedron*, **1996**, *52*, 23; Coudret, C., *Synth. Commun.*, **1996**, *26*, 3543.
185 'Methylpyridines and other methylazines as precursors of bicycles and polycycles', Abu-Shanab, F. A., Wakefield, B. J. and Elnagdi, M. H., *Adv. Heterocycl. Chem.*, **1997**, *68*, 181.
186 White, W. N. and Lazdins, D., *J. Org. Chem.*, **1969**, *34*, 2756.
187 Fraser, R. R., Mansour, T. S. and Savard, S., *J. Org. Chem.*, **1985**, *50*, 3232.
188 Levine, R., Dimmig, D. A. and Kadunce, W. M., *J. Org. Chem.*, **1974**, *39*, 3834; Kaiser, E. M., Bartling, G. J., Thomas, W. R., Nichols, S. B. and Nash, D. R., *J. Org. Chem.*, **1973**, *38*, 71.
189 Beumel, O. F., Smith, W. N. and Rybalka, B., *Synthesis*, **1974**, 43.
190 Kaiser, E. M. and Petty, J. D., *Synthesis*, **1975**, 705; Davis, M. L., Wakefield, B. J. and Wardell, J. A., *Tetrahedron*, **1992**, *48*, 939.
191 Lochte, H. L. and Cheavers, T. H., *J. Am. Chem. Soc.*, **1957**, *79*, 1667; Ghera, E., David, Y. B. and Rapoport, H., *J. Org. Chem.*, **1981**, *46*, 2059.
192 Williams, J. L. R., Adel, R. E., Carlson, J. M., Reynolds, G. A., Borden, D. G. and Ford, J. A., *J. Org. Chem.*, **1963**, *28*, 387; Cassity, R. P., Taylor, L. T. and Wolfe, J. F., *J. Org. Chem.*, **1978**, *43*, 2286; Konakahara, T. and Takagi, Y., *Heterocycles*, **1980**, *14*, 393; Konakahara, T. and Takagi, Y., *Synthesis*, **1979**, 192; Woodward, R. B. and Kornfield, E. C., *Org. Synth., Coll. Vol. III*, **1973**, 413; Feuer, H. and Lawrence, J. P., *J. Am. Chem. Soc.*, **1969**, *91*, 1856; Markovac, A., Stevens, C. L., Ash, A. B. and Hackley, B. E., *J. Org. Chem.*, **1970**, *35*, 841; Bredereck, H., Simchen, G. and Wahl, R., *Chem. Ber.*, **1968**, *101*, 4048; Pasquinet, E., Rocca, P., Godard, A., Marsais, F. and Quéguiner, G., *J. Chem. Soc., Perkin Trans. 1*, **1998**, 3807; Ragan J. A., Jones, B. P., Meltz, C. N. and Teixeira, J. J., *Synthesis*, **2002**, 483.
193 Berlin, M., Aslanian, R., de Lera Ruiz, M. and McCormick, K. D., *Synthesis*, **2007**, 2529.
194 Mandal, A. B., Augustine, J. K., Quattropani, A. and Bombrun, A., *Tetrahedron Lett.*, **2005**, *46*, 6033.
195 Song, J. J., Tan, Z., Gallou, F., Xu, J., Yee, N. K. and Senanayake, C. H., *J. Org. Chem.*, **2005**, *70*, 6512.
196 Jerchel, D. and Heck, H. E., *Justus Liebigs Ann. Chem.*, **1958**, *613*, 171.
197 Kawase, M., Teshima, M., Saito, S. and Tani, S., *Heterocycles*, **1998**, *48*, 2103.
198 Doering, W. E. and Weil, R. A. N., *J. Am. Chem. Soc.*, **1947**, *69*, 2461; Leonard, F. and Pschannen, W., *J. Med. Chem.*, **1966**, *9*, 140.
199 Sakamoto, T., Kondo, Y., Shiraiwa, M. and Yamanaka, H., *Synthesis*, **1984**, 245.
200 Moser, R. J. and Brown, E. V., *J. Org. Chem.*, **1972**, *37*, 3938.
201 Brown, E. V. and Shambhu, M. B., *J. Org. Chem.*, **1971**, *36*, 2002; Rapoport, H. and Volcheck, E. J., *J. Am. Chem. Soc.*, **1956**, *78*, 2451
202 Quast, H. and Schmitt, E., *Justus Liebigs Ann. Chem.*, **1970**, *732*, 43; Katritzky, A. R. and Faid-Allah, H. M., *Synthesis*, **1983**, 149.
203 Effenberger, F. and König, J., *Tetrahedron*, **1988**, *44*, 3281; Bohn, B., Heinrich, N. and Vorbrüggen, H., *Heterocycles*, **1994**, *37*, 1731.
204 Prill, E. A. and McElvain, S. M., *Org. Synth., Coll. Vol II*, **1943**, 419; 'Oxidative transformation of heterocyclic iminium salts', Weber, H., *Adv. Heterocycl. Chem.*, **1987**, *41*, 275.
205 Anderson, P. S., Kruger, W. E. and Lyle, R. E., *Tetrahedron Lett.*, **1965**, 4011; Holik, M. and Ferles, M., *Coll. Czech. Chem. Commun.*, **1967**, *32*, 3067.
206 Cervinka, O. and Kriz, O., *Coll. Czech. Chem. Commun.*, **1965**, *30*, 1700.
207 de Koning, A. J., Budzelaar, P. H. M., Brandsma, L., de Bie, M. J. A. and Boersma, J., *Tetrahedron Lett.*, **1980**, *21*, 2105.
208 'N-Dienyl amides and lactams. Preparation and Diels–Alder reactivity', Smith, M. B., *Org. Prep. Proc. Int.*, **1990**, *22*, 315.
209 Fowler, F. W., *J. Org. Chem.*, **1972**, *37*, 1321.
210 Beeken, P., Bonfiglio, J. N., Hassan, I., Piwinski,. J. J., Weinstein, B., Zollo, K. A. and Fowler, F. W., *J. Am. Chem. Soc.*, **1979**, *101*, 6677; Comins, D. L. and Mantlo, N. B., *J. Org. Chem.*, **1986**, *51*, 5456.
211 Carelli, V., Liberatore, F., Scipione, L., Di Rienzo, B. and Tortorella, S., *Tetrahedron*, **2005**, *61*, 10331.
212 Wong, Y. S., Marazano, C., Grecco, D. and Das, B. C., *Tetrahedron Lett.*, **1994**, *35*, 707.
213 Thiessen, L. M., Lepoivre, J. A. and Alderweireldt, F. C., *Tetrahedron Lett.*, **1974**, 59.
214 e.g. with chloranil: Chia, W.-L. and Cheng, Y. W., *Heterocycles*, **2008**, *75*, 375.
215 Comins, D. L. and Abdullah, A. H., *J. Org. Chem.*, **1982**, *47*, 4315.

[216] Piers, E. and Soucy, M., *Canad. J. Chem.*, **1974**, *52*, 3563; Akiba, K., Iseki, Y. and Wada, M., *Tetrahedron Lett.*, **1982**, *23*, 429; Comins, D. L. and Mantlo, N. B., *Tetrahedron Lett.*, **1983**, *24*, 3683; Comins, D. L., Smith, R. K. and Stroud, E. D., *Heterocycles*, **1984**, *22*, 339; Shiao, M.-J., Shih, L.-H., Chia, W.-L. and Chau, T.-Y., *Heterocycles*, **1991**, *32*, 2111.

[217] Lyle, R. E., Marshall, J. L. and Comins, D. L., *Tetrahedron Lett.*, **1977**, 1015.

[218] Yamaguchi, R., Nakazono, Y. and Kawanisi, M., *Tetrahedron Lett.*, **1983**, *24*, 1801

[219] Dhar, T. G. and Gluchowski, C., *Tetrahedron Lett.*, **1994**, *35*, 989.

[220] Itoh, T., Hasegawa, H., Nagata, K. and Ohsawa, A., *Synlett*, **1994**, 557.

[221] Xu, Y., Rudler, H., Denise, B., Parlier, A., Chaquin, P. and Herson, P., *Tetrahedron Lett.*, **2006**, *47*, 4541; Rudler, H., Denise, B., Xu, Y., Parlier, A. and Vaissermann, J., *Eur. J. Org. Chem.*, **2005**, 3724.

[222] Akiba, K., Iseki, Y. and Wada, M., *Tetrahedron Lett.*, **1982**, *23*, 3935.

[223] Bräckow, J. and Wanner, K. T., *Tetrahedron*, **2006**, *62*, 2395.

[224] Fraenkel, G., Cooper, J. W. and Fink, C. M., *Angew. Chem., Int. Ed. Engl.*, **1970**, *7*, 523.

[225] Comins, D. L., Weglarz, M. A. and O'Connor, S., *Tetrahedron Lett.*, **1988**, *29*, 1751.

[226] Commins, D. L. and Brown, J. D., *Tetrahedron Lett.*, **1986**, *27*, 2219.

[227] Comins, D. L., Joseph, S. P., Hong, H., Al-awar, R. S., Foti, C. J., Zhang, Y., Chen, X., LaMunyon, D. H. and Weltzien, M., *Pure Appl. Chem.*, **1997**, *69*, 477; Comins, D. L., *J. Heterocycl. Chem.*, **1999**, *36*, 1491; Kuethe, J. T. and Comins, D. L., *J. Org. Chem.*, **2004**, *69*, 2863; Comins, D. L. and Sahn, J. J., *Org. Lett.*, **2005**, *7*, 5227.

[228] Kiselyov, A. S. and Strekowski, L., *J. Org. Chem.*, **1993**, *58*, 4476; *idem*, *J. Heterocycl. Chem.*, **1993**, *30*, 1361.

[229] Kiselyov, A. S., *Tetrahedron Lett.*, **2005**, *46*, 2279.

[230] Haase, M., Goerls, H. and Anders, E., *Synthesis*, **1998**, 195.

[231] Katritzky, A. R., Zhang, S., Kurz, T., Wang, M. and Steel, P. J., *Org. Lett.*, **2001**, *3*, 2807.

[232] Corey, E. J. and Tian, Y., *Org. Lett.*, **2005**, *7*, 5535.

[233] Lee, P. H., Lee, K., Shim, J. H., Lee, S. G. and Kim, S., *Heterocycles*, **2006**, *67*, 777.

[234] 'Pyridine ring nucleophilic recyclisations', Kost, A. N., Gromov, S. P. and Sagitullin, R. S., *Tetrahedron*, **1981**, *37*, 3423; 'Synthesis and reactions of glutaconaldehyde and 5-amino-2,4-pentadienals', Becher, J., *Synthesis*, **1980**, 589.

[235] Becher, J., *Org. Synth.*, **1979**, *59*, 79.

[236] 'Ring transformation of pyridines and benzo derivatives under the action of C-nucleophiles', Gromov, S. P., *Heterocycles*, **2000**, *53*, 1607.

[237] Hafner, K. and Meinhardt, K.-P., *Org. Synth.*, **1984**, *62*, 134.

[238] Genisson, Y., Marazano, C., Mehmandoust, M., Grecco, D. and Das, B. C., *Synlett*, **1992**, 431.

[239] Yamaguchi, I., Higashi, H., Shigesue, S., Shingai, S. and Sato, M., *Tetrahedron Lett.*, **2007**, *48*, 7778.

[240] Kutney, J. P. and Greenhouse, R., *Synth. Commun.*, **1975**, 119; Berg, U., Gallo, R. and Metzger, J., *J. Org. Chem.*, **1976**, *41*, 2621.

[241] Aumann, D. and Deady, L. W., *J. Chem. Soc., Chem. Commun.*, **1973**, 32.

[242] Bhatia, A. V., Chaudhury, S. K. and Hernandez, O., *Org. Synth.*, **1997**, *75*, 184.

[243] Olah, G. A. and Klumpp, D. A., *Synthesis*, **1997**, 744.

[244] Chaudhury, S. K. and Hernandez, O., *Tetrahedron Lett.*, **1979**, 95; *ibid.*, 99.

[245] 'Heterocyclic N-Oxides', Katritzky, A.R. and Lagowski, J. M., Methuen, London, **1967**; 'Aromatic Amine Oxides', Ochiai, E., Am. Elsevier, New York, **1967**; 'Heterocyclic N-Oxides', Albini, A. and Pietra, S., CRC Press Wolfe Publishing, London, **1991**; 'Heterocyclic N-oxides and N-imides', Katritzky, A. R. and Lam, J. N., *Heterocycles*, **1992**, *33*, 1011.

[246] 'Rearrangements of *t*-amine oxides', Oae, S. and Ogino, K., *Heterocycles*, **1977**, *6*, 583.

[247] Reichardt, C., *Chem. Ber.*, **1966**, *99*, 1769.

[248] Chen, D. X., Ho, C. M., Wu, Q. Y. R., Wu, P. R., Wong, F. M. and Wu, W., *Tetrahedron Lett.*, **2008**, *49*, 4147.

[249] Taylor, E. C. and Crovetti, A. J., *Org. Synth., Coll. Vol. IV*, **1963**, 654; Saito, H. and Hamana, M., *Heterocycles*, **1979**, *12*, 475.

[250] Johnson, C. D., Katritzky, A. R., Shakir, N. and Viney, M., *J. Chem. Soc. (B)*, **1967**, 1213.

[251] van Ammers, M., den Hertog, H. J. and Haase, B., *Tetrahedron*, **1962**, *18*, 227.

[252] Van Ammers, M. and Den Hertog, H. J., *Recl. Trav. Chim. Pays Bas*, **1962**, *81*, 124.

[253] Mosher, H. S. and Welch, F. J., *J. Am. Chem. Soc.*, **1955**, *77*, 2902.

[254] Liveris, M. and Miller, J., *J. Chem. Soc*, **1963**, 3486; Johnson, R. M., *J. Chem. Soc. C*, **1966**, 1058.

[255] Van Bergen, T. J. and Kellogg, R. M., *J. Org. Chem.*, **1971**, *36*, 1705; Andersson, H., Wang, X., Björklund, M., Olsson, R. and Almqvist, F., *Tetrahedron Lett.*, **2007**, *48*, 6941; Andersson, H., Almqvist, F. and Olsson, R., *Org. Lett.*, **2007**, *9*, 1335.

[256] Schnekenburger, J. and Heber, D., *Chem. Ber.*, **1974**, *107*, 3408.

[257] Nishiwaki, N., Minakata, S., Komatsu, M. and Ohshiro, Y., *Chem. Lett.*, **1989**, 773.

[258] Prokhorov, A. M., Makosza, M. and Chupakhin, O. N., *Tetrahedron Lett.*, **2009**, *50*, 1444.

[259] Andersson, H., Gustafsson, M., Olsson, R. and Almqvist, F., *Tetrahedron Lett.*, **2008**, *49*, 6901.

[260] Jung, J.-C., Jung, Y.-J. and Park, O.-S., *Synth. Commun.*, **2001**, *31*, 2507.

[261] Fife, W. K., *J. Org. Chem.*, **1983**, *48*, 1375; Vorbrüggen, H. and Krolikiewicz, K., *Synthesis*, **1983**, 316.

[262] Yin, J., Xiang, B., Huffman, M. A., Raab, C. E. and Davies, I. W., *J. Org. Chem.*, **2007**, *72*, 4554.

[263] Manley, P. J. and Bilodeau, M. T., *Org. Lett.*, **2002**, *4*, 3127.

[264] Fontenas, C., Bejan, E., Ait Haddou, H. and Balavoine, G. G. A., *Synth. Commun.*, **1995**, *25*, 629.

[265] Ginsberg, S. and Wilson, I. B., *J. Am. Chem. Soc.*, **1957**, *79*, 481.

[266] Bodalski, R. and Katritzky, A. R., *J. Chem. Soc. (B)*, **1968**, 831; Koenig, T., *J. Am. Chem. Soc.*, **1966**, *88*, 4045.

[267] Walters, M. A. and Shay, J. J., *Tetrahedron Lett.*, **1995**, *36*, 7575.

[268] 'De novo synthesis of substituted pyridines', Henry, G. D., *Tetrahedron*, **2004**, *60*, 6043.

[269] Gill, N. S., James, K. B., Lions, F. and Potts, K. T., *J. Am. Chem. Soc.*, **1952**, *74*, 4923; Keuper, R., Risch, N., Flörke, U. and Haupt, H.-J., *Liebigs Ann.*, **1996**, 705; Keuper, R., Risch, N., *ibid*, 717.

[270] Knoevenagel, E., *Justus Liebigs Ann. Chem.*, **1894**, *281*, 25; Stobbe, H. and Vollard, H., *Chem. Ber.*, **1902**, *35*, 3973; Stobbe, H., *ibid.*, 3978.

[271] Kelly, T. R. and Liu, H., *J. Am. Chem. Soc.*, **1985**, *107*, 4998.

[272] Katritzky, A. R., Murugan, R. and Sakizadeh, K., *J. Heterocycl. Chem.*, **1984**, *21*, 1465.

[273] Potts, K. T., Cipullo, M. J., Ralli, P. and Theodoridis, G., *J. Am. Chem. Soc.*, **1981**, *103*, 3584 and 3585; Potts, K. T., Ralli, P., Theodoridis, G. and Winslow, P., *Org. Synth.*, **1986**, *64*, 189.

[274] Cooke, M. W., Wang, J., Theobald, I. and Hanan, G. S., *Synth. Commun.*, **2006**, *36*, 1721.

[275] Kvita, V., *Synthesis*, **1991**, 883.

[276] Bickel, A. F., *J. Am. Chem. Soc.*, **1947**, *69*, 1805; Campbell, K. N., Ackermann, J. F. and Campbell, B. K., *J. Org. Chem.*, **1950**, *15*, 221.

[277] Abdulla, R. F., Fuhr, K. H. and Williams, J. C., *J. Org. Chem.*, **1979**, *44*, 1349.

[278] Kurihara, H. and Mishima, H., *J. Heterocycl. Chem.*, **1977**, *14*, 1077; Johnson, F., Panella, J. P., Carlson, A. A. and Hunneman, D. H., *J. Org, Chem.*, **1962**, *27*, 2473.

[279] '4-Aryldihydropyridines, a new class of highly active calcium antagonists', Bossart, F., Meyer, H. and Wehinger, E., *Angew. Chem., Int. Ed. Engl.*, **1981**, *20*, 762.

[280] Pfister, J. R., *Synthesis*, **1990**, 689; Maquestian, A., Mayence, A. and Eynde, J.-J. V., *Tetrahedron Lett.*, **1991**, *32*, 3839; Alvarez, C., Delgado, F., García, O., Medina, S. and Márquez, C., *Synth. Commun.*, **1991**, *21*, 619.

[281] Yadav, J. S., Reddy, B. V. S., Sabitha, G. and Reddy, G. S. K. K., *Synthesis*, **2000**, 1532.

[282] Ko, S. and Tao, C.-F., *Tetrahedron*, **2006**, *62*, 7293.

[283] Satoh, Y., Ichihashi, M. and Okumura, K., *Chem. Pharm. Bull.*, **1992**, *40*, 912.

[284] Kelly, T. R. and Lebedev, R. L., *J. Org. Chem.*, **2002**, *67*, 2197; Cave, G. W. V. and Raston, C. L., *Tetrahedron Lett.*, **2005**, *46*, 2361; Tu, S., Li, T., Shi, F., Wang, Q., Zhang, J., Xu, J., Zhu, X., Zhang, X., Zhu, S. and Shi, D., *Synthesis*, **2005**, 3045.

[285] Oka, Y., Omura, K., Miyake, A., Itoh, K., Tomimoto, M., Tada, N. and Yurugi, S., *Chem. Pharm. Bull.*, **1975**, *23*, 2239.

[286] Brenner, D. G., Halczenko, W. and Shepard, K. L., *J. Heterocycl. Chem.*, **1982**, *19*, 897.

[287] Petrich, S. A., Hicks, F. A., Wilkinson, D. R., Tarrant, J. G., Bruno, S. M., Vargas, M., Hosein, K. N., Gupton, J. T. and Sikorski, J. A., *Tetrahedron*, **1995**, *51*, 1575; Marcoux, J.-F., Marcotte, F.-A., Wu, J., Dormer, P. G., Davies, I. W., Hughes, D. and Reider, P. J., *J. Org. Chem.*, **2001**, *66*, 4194.

[288] Davies, I. W., Marcoux, J.-F. and Reider, P. J., *Org. Lett.*, **2001**, *3*, 209.

[289] Davies, I. W., Taylor, M., Marcoux, J.-F., Wu, J., Dormer, P. G., Hughes, D. and Reider, P. J., *J. Org. Chem.*, **2001**, *66*, 251.

[290] Henecke, H., *Chem. Ber.*, **1949**, *82*, 36.

[291] Mariella, R. P., *Org. Synth., Coll. Vol. IV*, **1963**, 210.

[292] Bottorff, E. M., Jones, R. G., Kornfield, E. C. and Mann, M. J., *J. Am. Chem. Soc.*, **1951**, *73*, 4380.

[293] Wai, J. S., Williams, T. M., Bamberger, D. L., Fisher, T. E., Hoffman, J. M., Hudcosky, R. J., MacTough, S. C., Rooney, C. S. and Saari, W. S., *J. Med. Chem.*, **1993**, *36*, 249.

[294] Dornow, A. and Neuse, E., *Chem. Ber.*, **1951**, *84*, 296.

[295] Matsui, M., Oji, A., Kiramatsu, K., Shibata, K. and Muramatsu, H., *J. Chem. Soc., Perkin Trans 2*, **1992**, 201; Jain, R., Roschangar, F. and Ciufolini, M. A., *Tetrahedron Lett.*, **1995**, *36*, 3307.

[296] Carles, L., Narkunan, K., Penlou, S., Rousset, L., Bouchu, D. and Ciufolini, M. A., *J. Org. Chem.*, **2002**, *67*, 4304.

[297] Bohlmann, F. and Rahtz, D., *Chem. Ber.*, **1957**, *90*, 2265.

[298] 'The Bohlmann-Rahtz pyridine synthesis: from discovery to applications', Bagley, M. C., Glover, C. and Merritt, E.A., *Synlett*, **2007**, 2459.

[299] Davis, J. M., Truong, A. and Hamilon, A. D., *Org. Lett.*, **2005**, *7*, 5405.

[300] Hachiya, I., Ogura, K. and Shimizu, M., *Org. Lett.*, **2002**, *4*, 2755.

[301] Naito, T., Yoshikawa, T., Ishikawa, F., Isoda, S., Omura, Y. and Takamura, I., *Chem. Pharm. Bull.*, **1965**, *13*, 869; Kondrat'eva, G. Ya. and Huan, C.-H., *Dokl. Akad. Nauk, SSSR*, **1965**, *164*, 816 (*Chem. Abstr.*, **1966**, *64*, 2079).

[302] Okatani, T., Koyama, J., Suzata, Y. and Tagahara, K., *Heterocycles*, **1988**, *27*, 2213.

[303] Sainz, Y. F., Raw, S. A. and Taylor, R. J. K., *J. Org. Chem.*, **2005**, *70*, 10086.

[304] Boger, D. L. and Panek, J. S., *J. Org. Chem.*, **1981**, *46*, 2179.

[305] Sauer, J. and Heldmann, D. K., *Tetrahedron Lett.*, **1998**, *39*, 2549.

[306] Pfüller, O. C. and Sauer, J., *Tetrahedron Lett.*, **1998**, *39*, 8821.

[307] Carly, P. R.,, Cappelle, S. L., Compernolle, F. and Hoornaert, G. J., *Tetrahedron*, **1996**, *52*, 11889; Meerpoel, L. and Hoornaert, G., *Tetrahedron Lett.*, **1989**, *30*, 3183.

[308] Sainte, F., Serckx-Poncin, B., Hesbain-Frisque, A.-M. and Ghosez, L., *J. Am. Chem. Soc.*, **1982**, *104*, 1428.

[309] Fletcher, M. D., Hurst, T. E., Miles, T. J. and Moody, C. J., *Tetrahedron*, **2006**, *62*, 5454.

[310] Lu. J.-Y. and Arndt, H.-D., *J. Org. Chem.*, **2007**, *72*, 4205.

[311] Schramm, O. G., Oeser, T. and Müller, T. J. J., *J. Org. Chem.*, **2006**, *71*, 3494.

[312] Trost, B. M. and Gutierrez, A. C., *Org. Lett.*, **2007**, *9*, 1473.

[313] 'Construction of pyridine rings by metal-mediated [2 + 2 + 2] cycloaddition', Varela, J. A. and Saá, C., *Chem. Rev.*, **2003**, *103*, 3787.

[314] Chelucci, G., Faloni, M. and Giacomelli, G., *Synthesis*, **1990**, 1121; 'Organocobalt-catalysed synthesis of pyridines', Bönnemann, H. and Brijoux, W., *Adv. Heterocycl. Chem.*, **1990**, *48*, 177.

[315] Senaiar, R. S., Young, D. D. and Deiters, A., *Chem. Commun.*, **2006**, 1313.

[316] Suzuki, D., Tanaka, R., Urabe, H. and Sato, F., *J. Am. Chem. Soc.*, **2002**, *124*, 3518.

[317] Tekavec, T. N., Zuo, G., Simon, K. and Louie, J., *J. Org. Chem.*, **2006**, *71*, 5834.

[318] Yamamoto, Y., Takagishi, H. and Itoh, K., *Org. Lett.*, **2001**, *3*, 2117.

[319] Clauson-Kaas, N., Elming, N. and Zdeněk, *Acta Chem. Scand.*, **1955**, *9*, 1, 9, 14, 23, and 30; Clauson-Kaas, N., Petersen, J. B., Sorensen, G. O., Olsen, G., and Jansen, G., *Acta Chem. Scand.*, **1965**, *19*, 1146.

[320] Abbiati, G., Arcadi, A., Bianchi, G., Giuseppe, S. D., Marinelli, F. and Rossi, E., *J. Org. Chem.*, **2003**, *68*, 6959.

[321] Gangadasu, B., Narender, P., Kumar, S. B., Ravinder, M., Rao, B. A., Ramesh, Ch., Raju, B. C. and Rao, V. J., *Tetrahedron*, **2006**, *62*, 8398.

[322] Frank, R. C., Pilgrim, F. J. and Riener, E. F., *Org. Synth., Coll. Vol. IV*, **1963**, 451.

[323] Chumakov, Yu I. and Sherstyuk, V. P., *Tetrahedron Lett.*, **1965**, 129.

[324] Waldner, A., *Synth. Commun.*, **1989**, *19*, 2371.

[325] Harris, S. A. and Folkers, K., *J. Am. Chem. Soc.*, **1939**, *61*, 1245.

[326] Harris, E. E., Firestone, R. A., Pfister, K., Boettcher, R. R., Cross, F. J., Currie, R. B., Monaco, M., Peterson, E. R. and Reuter, W., *J. Org. Chem.*, **1962**, *27*, 2705; Doktorova, N. D., Ionova, L. V., Karpeisky, M. Ya., Padyukova, N. Sh., Turchin, K. F. and Florentiev, V. L., *Tetrahedron*, **1969**, *25*, 3527.

[327] Cruskie, M. P., Zoltewicz, J. A. and Abboud, K. A., *J. Org. Chem.*, **1995**, *60*, 7491.

[328] Chang, C.-Y., Liu, H.-M. and Chow, T. J., *J. Org. Chem.*, **2006**, *71*, 6302.

9

Quinolines and Isoquinolines:
Reactions and Synthesis

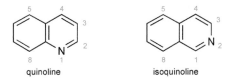

quinoline   isoquinoline

Quinoline is a high-boiling liquid; isoquinoline is a low-melting solid; each has a sweetish odour. Both bases have been known for a long time: quinoline was first isolated from coal tar in 1834, isoquinoline from the same source in 1885. Shortly after the isolation of quinoline from coal tar, it was also recognised as a pyrolytic degradation product of cinchonamine, an alkaloid closely related to quinine, from which the name quinoline is derived; the word quinine, in turn, derives from *quina*, a Spanish version of a local South American name for the bark of quinine-containing Cinchona species.

9.1 Reactions with Electrophilic Reagents

9.1.1 Addition to Nitrogen

All the reactions noted in this category for pyridine (see 8.1.1), which involve donation of the nitrogen lone pair to electrophiles, also occur with quinoline and isoquinoline and little further comment is necessary, for example the respective pK_{aH} values, 4.94 and 5.4, show them to be of similar basicity to pyridine. Each, like pyridine, readily forms an *N*-oxide and quaternary salts.

9.1.2 Substitution at Carbon

9.1.2.1 Proton Exchange

Benzene ring *C*-protonation, and thence exchange, *via N*-protonated quinoline, requires strong sulfuric acid and occurs fastest at C-8, then at C-5 and C-6; comparable exchange in isoquinoline takes place somewhat faster at C-5 than at C-8.[1] At lower acid strengths each system undergoes exchange α to nitrogen, at C-2 for quinoline and C-1 for isoquinoline. These processes involve a zwitterion produced by deprotonation of the *N*-protonated heterocycle.

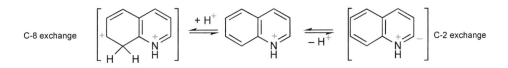

9.1.2.2 Nitration (see also 9.3.1.2)

The positional selectivity for proton exchange is partly mirrored in nitrations, quinoline gives approximately equal amounts of 5- and 8-nitro-quinolines, whereas isoquinoline produces almost exclusively the

5-nitro-isomer;[2] mechanistically the substitutions involve nitronium ion attack on the *N*-protonated heterocycles. Nitration in the pyridine ring, at a position β to the heteroatom, can be achieved *via* the Baake–Katritzky protocol (8.1.2.2).[3] 7-Nitroisoquinoline can be obtained by nitration of 1,2,3,4,-tetrahydroisoquinoline and then dehydrogenation of the hetero ring with potassium nitrosodisulfonate.[4]

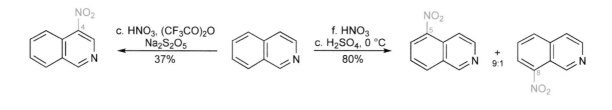

9.1.2.3 Sulfonation

Sulfonation of quinoline gives largely the 8-sulfonic acid, whereas isoquinoline affords the 5-acid.[5] Reactions at higher temperatures produce other isomers, under thermodynamic control, for example both quinoline 8-sulfonic acid and quinoline 5-sulfonic acid are isomerised to the 6-acid.[6]

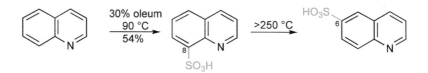

9.1.2.4 Halogenation

Ring substitution of quinoline and isoquinoline by halogens is rather complex, products depending on the conditions used.[7] In concentrated sulfuric acid, quinoline gives a mixture of 5- and 8-bromo derivatives; comparably, isoquinoline is efficiently converted into the 5-bromo-derivative in the presence of aluminium chloride,[8] or with *N*-bromosuccinimide in concentrated sulfuric acid.[9]

Introduction of halogen to the hetero-rings occurs under remarkably mild conditions in which halide addition to a salt initiates the sequence. Thus treatment of quinoline or isoquinoline hydrochlorides with bromine produces 3-bromoquinoline and 4-bromoisoquinoline, respectively, as illustrated below for the latter.[10]

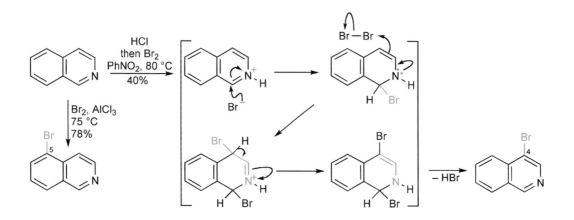

9.1.2.5 Acylation and Alkylation

There are no generally useful processes for the introduction of carbon substituents by electrophilic substitution of quinolines or isoquinolines, except for a few examples in which a ring has a strong electron-releasing substituent, for example 4-dimethylaminoquinoline undergoes smooth trifluoroacetylation at C-3.[11]

9.2 Reactions with Oxidising Agents

It requires vigorous conditions to degrade a ring in quinoline and isoquinoline: examples of attack at both rings are known, though degradation of the benzene ring, generating pyridine diacids, should be considered usual;[12] ozonolysis can be employed to produce pyridine dialdehydes,[13] or after subsequent hydrogen peroxide treatment, diacids.[14] Electrolytic oxidation of quinoline is the optimal way to convert quinoline into pyridine-2,3-dicarboxylic acid ('quinolinic acid')[15]; alkaline potassium permanganate converts isoquinoline into a mixture of pyridine-3,4-dicarboxylic acid ('cinchomeronic acid') and phthalic acid.[16]

9.3 Reactions with Nucleophilic Reagents
9.3.1 Nucleophilic Substitution with 'Hydride' Transfer
Reactions of this type occur fastest at C-2 in quinoline and at C-1 in isoquinolines.

9.3.1.1 Alkylation and Arylation

The immediate products of addition of alkyl and aryl Grignard reagents and alkyl- and aryllithiums are dihydro-quinolines and -isoquinolines and can be characterised as such, but can be oxidised to afford the *C*-substituted, re-aromatised heterocycles; illustrated below is a 2-arylation of quinoline.[17]

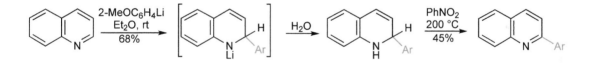

Vicarious nucleophilic substitution (3.3.3) allows the introduction of substituents into nitroquinolines: cyanomethyl and phenylsulfonylmethyl groups, for example, can be introduced *ortho* to the nitro group, in 5-nitroquinolines at C-6 and in 6-nitroquinolines at C-5.[18,19]

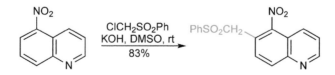

9.3.1.2 Amination[20] and Nitration

Sodium amide reacts rapidly and completely with quinoline and isoquinoline, even at −45 °C, to give dihydro-adducts with initial amide attack at C-2 (main) and C-4 (minor) in quinoline, and C-1 in isoquinoline. The quinoline 2-adduct rearranges to the more stable 4-aminated adduct at higher temperatures.[21] Oxidative trapping of the quinoline adducts provides 2- or 4-aminoquinoline;[22] isoquinoline reacts with potassium amide in liquid ammonia at room temperature to give 1-aminoisoquinoline.[23]

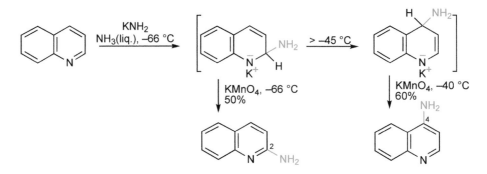

Oxidative aminations are possible at other quinoline and isoquinoline positions, even on the benzene ring, providing a nitro group is present to promote the nucleophilic addition.[24]

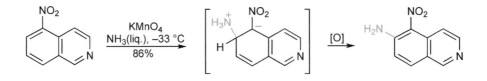

The introduction of a nitro group at C-1 in isoquinolines can be achieved using a mixture of potassium nitrite, dimethylsulfoxide and acetic anhydride.[25] The key step is the nucleophilic addition of nitrite to the heterocycle previously quaternised by reaction at nitrogen with a complex of dimethylsulfoxide and the anhydride.

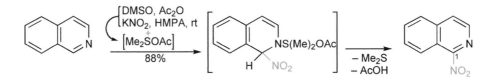

9.3.1.3 Hydroxylation

Both quinoline and isoquinoline can be directly hydroxylated with potassium hydroxide at high temperature with the evolution of hydrogen.[26] 2-Quinolone ('carbostyril') and 1-isoquinolone ('isocarbostyril') are the isolated products.

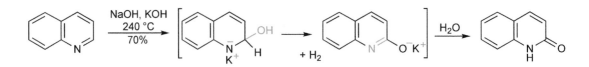

9.3.2 Nucleophilic Substitution with Displacement of Good Leaving Groups

The main principle here is that halogen on the homocyclic rings of quinoline and isoquinoline, and at the quinoline-3- and the isoquinoline-4 positions, behaves as would a halo-benzene. In contrast, 2- and 4-halo-quinolines and 1-halo-isoquinolines have the same susceptibility as α- and γ-halopyridines (see 8.3.2). 3-Halo-isoquinolines are intermediate in their reactivity to nucleophiles.[27]

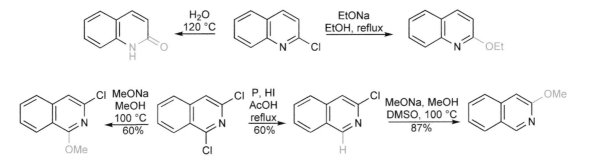

An apparent exception to the relative unreactivity of 3-halo-isoquinolines is provided by the reaction of 3-bromoisoquinoline with sodium amide. Here, a different mechanism, known by the acronym ANRORC (**A**ddition of **N**ucleophile, **R**ing **O**pening and **R**ing **C**losure), leads to the product, apparently of direct displacement, but in which a switching of the ring nitrogen to become the substituent nitrogen, has occurred.[28]

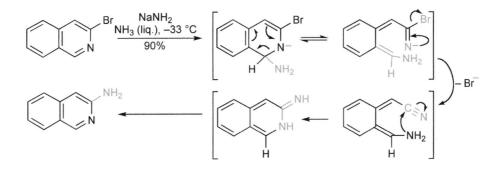

A useful method for the conversion of quinoline-2- and -4- and isoquinoline-1-chlorides into iodides utilizes the hydrochloride salt of the heterocycle in reaction with sodium iodide in hot acetonitrile; presumably it is the *N*-protonated species that is attacked by the iodide.[29]

The reaction of 2,4-dichloro-quinolines with an equivalent of sodium azide results in selective displacement at the 4-position, but, if an acid is added, the 2-position is preferred; the 2-azides exist as a ring/chain mixture, the tricyclic tetrazolo[1,5-*a*]quinoline predominating.[30]

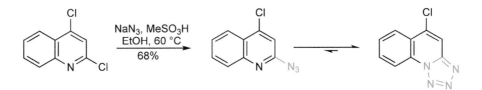

9.4 Metallation and Reactions of *C*-Metallated Quinolines and Isoquinolines
9.4.1 Direct Ring C–H Metallation
Direct lithiation, i.e. *C*-deprotonation of quinolines[31] requires an adjacent substituent, such as chlorine, fluorine or alkoxy. Historically, what is probably the first ever strong base *C*-lithiation of a six-membered heterocycle was the 3-lithiation of 2-ethoxyquinoline.[32] Both 4- and 2-dimethylaminocarbonyloxyquinolines lithiate at *C*-3; 4-pivaloylaminoquinoline lithiates at the *peri* position, *C*-5. Quinolines with an *ortho*-directing

group at C-3 lithiate at C-4, not at C-2.[33] Quinoline 2-, 3- and 4-carboxylic acids *C*-lithiate (*C*-3, C-4 and C-3 respectively) using two mole equivalents of lithium tetramethylpiperidide.[34]

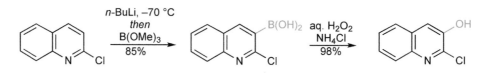

2-Lithiation of 1-substituted 4-quinolones[35] and 3-lithiation of 2-quinolone[36] provides derivatives with the usual nucleophilic propensity, as illustrated below.

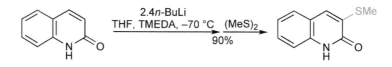

9.4.2 Metal–Halogen Exchange

The preparation of lithio-quinolines and -isoquinolines *via* metal–halogen exchange is complicated by competing nucleophilic addition, however the use of low temperatures does allow metal–halogen exchange at both pyridine[37] and benzene ring positions[38] in quinolines, and the isoquinoline-1-[36] and 4-positions,[39] subsequent reaction with electrophiles generating *C*-substituted products. It seems that for benzene ring lithiation, two mole equivalents of butyllithium are necessary so that one equivalent can associate with the ring nitrogen, as suggested below.

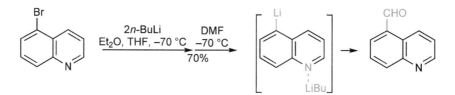

Quinolinylzinc reagents can be produced by reaction of a halide with activated zinc,[40] and 2-, 3- and 4-bromoquinolines can be converted into the corresponding lithium tri(quinolyl)magnesates at −10 °C by treatment with *n*-Bu₃MgLi in THF or toluene, the 3-isomer giving the best yields of final products.[41]

9.5 Reactions with Radicals

Regioselective substitutions can be achieved α to the nitrogen, with nucleophilic radicals, in acid solution – the Minisci reaction (3.4.1).

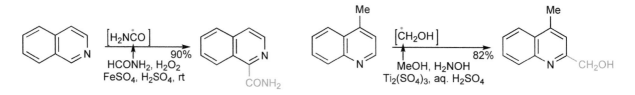

9.6   Reactions with Reducing Agents

Selective reduction of either the pyridine or the benzene rings in quinoline and isoquinoline can be achieved: the heterocyclic ring is reduced to the tetrahydro level by sodium cyanoborohydride in acid solution,[42] by sodium borohydride in the presence of nickel(II) chloride,[43] by zinc borohydride[44] or, traditionally, by room temperature and room pressure catalytic hydrogenation in methanol. In strong acid solution it is the benzene ring which is selectively saturated;[45] longer reaction times can then lead to decahydro derivatives. Treatment of quinoline and isoquinoline with sodium borohydride in a mixture of acetic acid and acetic anhydride gives good yields of *N*-acetyl-1,2-dihydro derivatives.[46]

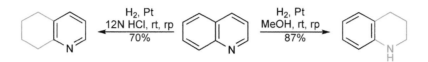

Lithium in liquid ammonia conditions can produce 1,4-dihydroquinoline[47] and 3,4-dihydroisoquinoline.[48] Conversely, lithium aluminium hydride reduces generating 1,2-dihydroquinoline[49] and 1,2-dihydroisoquinoline.[50] These dihydro-heterocycles[51] can be easily oxidised back to the fully aromatic systems, or disproportionate,[52] especially in acid solution, to give a mixture of tetrahydro and re-aromatised compounds. Stable dihydro-derivatives (see also 9.13) can be obtained by trapping following reduction, as a urethane, by reaction with a chloroformate.[53] Quaternary salts of quinoline and isoquinoline are particularly easily reduced, either catalytically or with a borohydride in protic solution, giving 1,2,3,4-tetrahydro-derivatives.

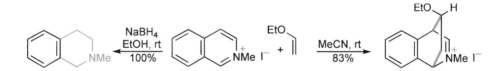

9.7   Electrocyclic Reactions (Ground State)

The tendency for relatively easy nucleophilic addition to the pyridinium ring in isoquinolinium salts is echoed in the cycloaddition (shown above) of electron-rich dienophiles such as ethoxyethene, which is reversed on refluxing in acetonitrile.[54]

9.8   Photochemical Reactions

Of a comparatively small range of photochemical reactions described for quinolines and isoquinolines, perhaps the most intriguing are some hetero-ring rearrangements of quaternary derivatives, which can be illustrated by the ring expansions of their *N*-oxides to 3,1-benzoxazepines.[55] As with 2-pyridones, 2-quinolones undergo 2+2 photo-dimerisation involving the C-3–C-4 double bond.[56]

9.9   Oxy-Quinolines and Oxy-Isoquinolines

Quinolinols and isoquinolinols in which the oxygen is at any position other than C-2 or C-4 for quinolines and C-1 or C-3 for isoquinolines are true phenols i.e. have an hydroxyl group, though they exist in equilibrium with variable concentrations of zwitterionic structures, with the nitrogen protonated and the oxygen deprotonated. They show the typical reactivity of naphthols.[57] 8-Quinolinol has long been used in analysis as a chelating agent, especially for Zn(II), Mg(II) and Al(III) cations; the Cu(II) chelate is used as a fungicide.

2-Quinolone (strictly 2(1*H*)-quinolinone), 4-quinolone[58] and 1-isoquinolone are completely in the carbonyl tautomeric form[59] for all practical purposes – the hydroxyl tautomers lack a favourable polarised resonance contribution, as illustrated below for 1-isoquinolone.

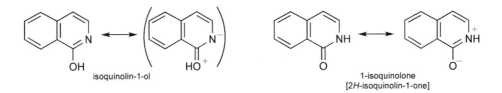

isoquinolin-1-ol

1-isoquinolone
[2*H*-isoquinolin-1-one]

In 3-oxy-isoquinoline there is an interesting and instructive situation: here the two tautomers are of comparable stability. 3-Isoquinolinol is dominant in dry ether solution, 3-isoquinolone is dominant in aqueous solution. A colourless ether solution of 3-isoquinolinol turns yellow on addition of a little methanol because of the production of some of the carbonyl tautomer. The similar stabilities are the consequence of the balancing of two opposing tendencies: the presence of an amide unit in 3-isoquinolone forces the benzene ring into a less favoured quinoid structure; conversely, the complete benzene ring in isoquinolinol necessarily means loss of the amide unit and its contribution to stability. One may contrast this with 1-isoquinolone which has an amide, as well as a complete benzene unit.[60]

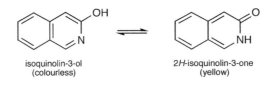

isoquinolin-3-ol
(colourless)

2*H*-isoquinolin-3-one
(yellow)

The position of electrophilic substitution of quinolones and isoquinolones depends upon the pH of the reaction medium. Each type protonates on carbonyl oxygen, so reactions in strongly acidic media involve attack on this cation and therefore in the benzene ring: the contrast is illustrated below by the nitration of 4-quinolone at different acid strengths.[61] The balance between benzene ring and unprotonated heterocyclic ring selectivity is small, for example 2-quinolone chlorinates preferentially, as a neutral molecule, at C-6, and only secondly at C-3.

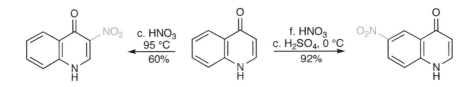

Strong acid-catalysed H-exchange of 2-quinolone proceeds fastest at C-6 and C-8; of 1-isoquinolone at C-4, then at C-5~C-7.[62] This is echoed in various electrophilic substitutions, for example formylation.[63]

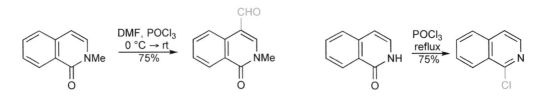

The carbonyl tautomers deprotonate at N–H, generating ambient anions that can react at either oxygen or nitrogen, depending on the exact conditions; for example *O*-alkylation can be achieved with silver carbonate.[64] They are converted, as with the pyridones, into halo-quinolines and halo-isoquinolines[65] by reaction with phosphorus halides.

9.10 Amino-Quinolines and Amino-Isoquinolines

Amino-quinolines and -isoquinolines exist as amino tautomers and all protonate on ring nitrogen. Only 4-aminoquinoline shows appreciably enhanced basicity (pK_{aH} 9.2); the most basic amino-isoquinoline is the 6-isomer (pK_{aH} 7.2), indeed this is the most basic of all the benzene-ring-substituted amino-quinolines and -isoquinolines.

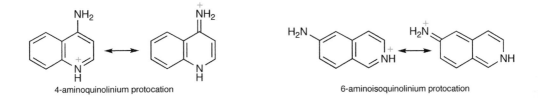

4-aminoquinolinium protocation   6-aminoisoquinolinium protocation

9.11 Alkyl-Quinolines and Alkyl-Isoquinolines

The particular acidity of the protons of pyridine α- and γ-alkyl groups is echoed by quinoline-2-[66] and 4-alkyl groups and by alkyl at the isoquinoline 1-position, but to a much lesser extent by alkyl at isoquinoline C-3. Condensation reactions with alkyl groups at these activated positions can be achieved in either basic or acidic media; the key nucleophilic species in the latter cases is probably an enamine,[67] or enamide,[68] and in the former, a side-chain carbanion.[69]

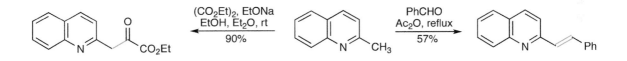

The selectivity is nicely illustrated by the oxidation of only the 2-methyl of 2,3,8-trimethylquinoline with selenium dioxide, giving the 2-aldehyde.[70]

9.12 Quinoline and Isoquinoline Carboxylic Acids and Esters

There is little to differentiate these derivatives from benzenoid acids and esters, save for the easy decarboxylation of quinoline-2- and isoquinoline-1-acids, *via* an ylide that can be trapped with aldehydes as electrophiles – the Hammick reaction.[71] Loss of carbon dioxide from *N*-methylquinolinium-2- and -isoquinolinium-1-acids, and trapping of resulting ylides, can be achieved with stronger heating.[72]

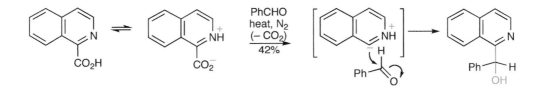

9.13 Quaternary Quinolinium and Isoquinolinium Salts

The predominant property of these salts is the ease with which nucleophiles add to the quinolinium-2- and the isoquinolinium-1-positions. Such additions are favoured in these bicyclic compounds since the products retain a complete aromatic benzene ring. Hydroxide, hydride[73] and organometallic nucleophiles all add with facility, though the resulting dihydroaromatic products require careful handling if they are not to disproportionate or be oxidised.[74] This approach can give 2-trifluoromethyl-quinolines: CF_3 carbanion (from trifluoro(trimethylsilyl)methane and fluoride) is added to an N-(*para*-methoxybenzyl)-quinolinium salt, then the N-substituent is removed and oxidation to the aromatic level achieved with ceric ammonium nitrate.[75]

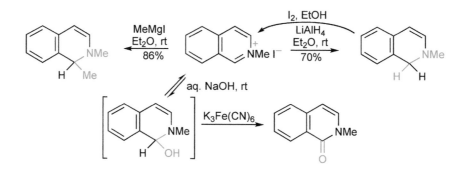

The position of fastest addition to quinolinium salts is C-2, but with reversible reactions, a thermodynamic adduct with the addend at C-4 and the residual double bond in conjugation with the nitrogen, can be obtained.[76]

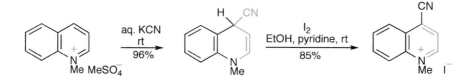

The Zincke salt (N-(2,4-dinitrophenyl) salt) of isoquinoline[77] is easily transformed into chiral isoquinolinium salts on reaction with chiral amines – an ANRORC sequence in which the nitrogen of the chiral amine ends up as the nitrogen of the isoquinolinium product[78] – nucleophilic addition of Grignard reagents to these salts shows good stereoselectivity.[79]

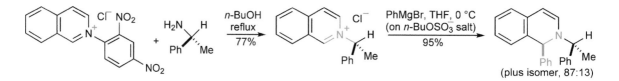

More practically significant, are the many examples of nucleophilic addition to salts in which an N-substituent conjugates with the nitrogen, and thus stabilizes the product – Reissert compounds were the first examples. These are produced by cyanide addition to an N^+-acyl-quinolinium or -isoquinolinium salt; in the classical process[80] the acylating agent is benzoyl chloride. Reissert compounds[81] are traditionally prepared using a dichloromethane–water two-phase medium; improvements include utilising phase-transfer catalysts with ultrasound[82] or crown ether catalysis.[83] Stereoselectivity can be achieved by the use of chiral acylating agents.[84]

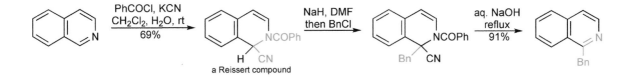

Reissert compounds have utility in a number of ways: deprotonation, alkylation and removal of acyl and cyanide groups leads to the corresponding substituted heterocycles. *N*-Sulfonyl analogues of Reissert adducts easily eliminate arylsulfinate, thus providing a method for the introduction of a cyano group.[85]

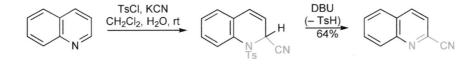

Allylsilanes will also trap *N*-acyl-quinolinium[86] and *N*-acyl-isoquinolinium[87] salts, silyl alkynes will add with silver ion catalysis,[88] and terminal alkynes using copper(I) iodide.[89]

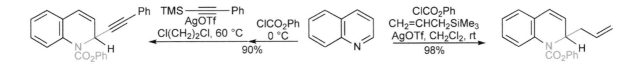

There are many examples of the preparation of stable 1,2-dihydro-isoquinolines *via* N^+-quaternisation with dimethyl acetylenedicarboxylate and then nucleophile addition to the generated iminium unit or its involvement in a subsequent cycloaddition[90] – two examples are shown below.[91,92]

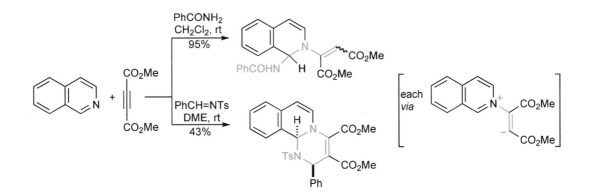

9.14 Quinoline and Isoquinoline *N*-Oxides

N-Oxide chemistry in these bicyclic systems largely parallels the processes described for pyridine *N*-oxide, with the additional possibility of benzene ring electrophilic substitution, for example mixed acid nitration of quinoline *N*-oxide takes place at C-5 and C-8 *via* the *O*-protonated species, but at C-4 at lower acid strength;[93] nitration of isoquinoline *N*-oxide takes place at C-5.[94]

Diethyl cyanophosphonate converts quinoline and isoquinoline *N*-oxides into the 1- and 2-cyano-heterocycles in high yields in a process which must have *O*-phosphorylation as a first step, and in which the elimination of diethylphosphate may proceed *via* a cyclic transition state;[95] trimethylsilyl cyanide and diazabicycloundecene effect the same transformation.[96] A chloroformate and an alcohol convert the *N*-oxides into ethers, as illustrated below for isoquinoline *N*-oxide,[97] a chloroformate and a Grignard reagent produce 2-substituted quinolines,[98] and a chloroformate then an isonitrile produce 2-carbamoyl-1,2-dihydro-isoquinolines.[99]

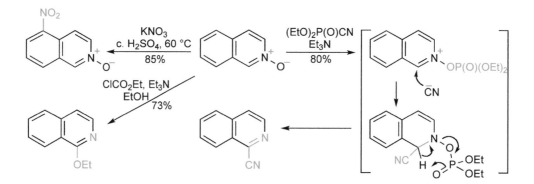

9.15 Synthesis of Quinolines and Isoquinolines

9.15.1 Ring Syntheses

The more generally important approaches to quinoline and isoquinoline compounds from non-heterocyclic precursors are summarised in this section.

9.15.1.1 Quinolines from Aryl-Amines and 1,3-Dicarbonyl Compounds

Anilines react with 1,3-dicarbonyl compounds to give intermediates which can be cyclised with acid.

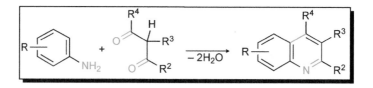

The Combes Synthesis

Condensation of a 1,3-dicarbonyl compound with an arylamine gives a high yield of a β-amino-enone, which can then be cyclised with concentrated acid.[100] Mechanistically, the cyclisation step is an electrophilic substitution by the *O*-protonated amino-enone, followed by loss of water to give the aromatic quinoline.

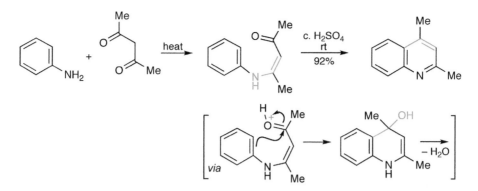

In order to access 4-unsubstituted quinolines, a 1,3-ketoaldehyde, in protected form, guarantees the required regioselectivity; the example below produces a 1,8-naphthyridine[101] (pyrido[2,3-*b*]pyridine).[102]

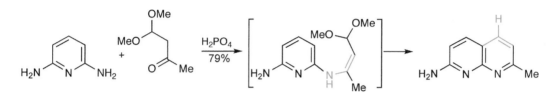

Conrad–Limpach–Knorr Reaction

If the 1,3-dicarbonyl component is at the 1,3-keto acid oxidation level, then the product is a quinolone.[103] Anilines and β-keto esters react at lower temperatures to give the kinetic product, a β-aminoacrylate, cyclisation of which gives a 4-quinolone. At higher temperatures, β-keto acid anilides are formed and cyclisation of these affords 2-quinolones.

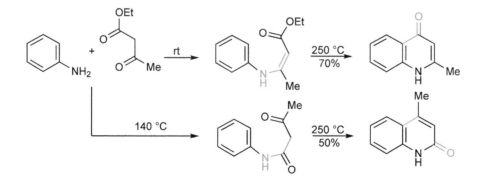

β-Aminoacrylates, for cyclisation to 4-quinolones, are also available *via* the addition of anilines to acetylenic esters[104,105] or by displacement of ethoxy from ethoxymethylenemalonate (EtOCH=C(CO$_2$Et)$_2$).[106]

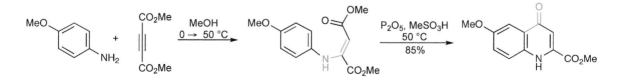

Usefully functionalised quinolines are easily accessible from anilines: the *N*-acetyl derivative is simply reacted with the Vilsmeier reagent and a 2-chloro-3-formyl-quinoline results. One may speculate that a 3-formyl-anilide, or an equivalent (shown), is involved, placing this useful reaction into the Combes category.[107,108]

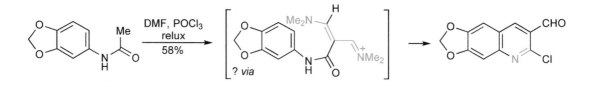

Cyclisations where the benzene ring carries an electron-withdrawing group, which would disfavour the electrophilic cyclising step, can be effected using the variant shown below – the substrate is simply heated strongly – the mechanism of ring closure is probably best viewed as the electrocyclisation of a 1,3,5-3-azatriene.[109]

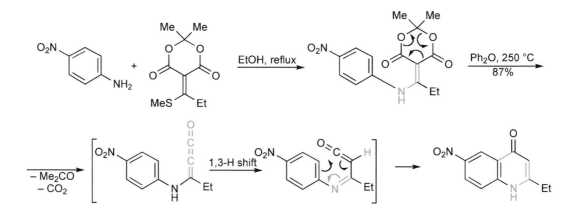

9.15.1.2 *Quinolines from Aryl-Amines and α,β-Unsaturated Carbonyl Compounds*

Arylamines react with an α,β-unsaturated carbonyl compound in the presence of an oxidising agent to give quinolines. When glycerol is used as an *in situ* source of acrolein, quinolines carrying no substituents on the heterocyclic ring are produced.

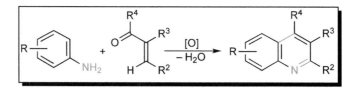

The Skraup Synthesis[110]

In this extraordinary reaction, quinoline is produced when aniline, concentrated sulfuric acid, glycerol and a mild oxidising agent are heated together.[111] The reaction has been shown to proceed *via* dehydration of the glycerol to acrolein, to which aniline then adds in a conjugate fashion. Acid-catalysed cyclisation produces a 1,2-dihydro-quinoline, finally dehydrogenated by the oxidising agent – the corresponding nitro-benzene or arsenic acid have been used classically. The Skraup synthesis is best for the ring synthesis of quinolines unsubstituted on the hetero-ring.[112]

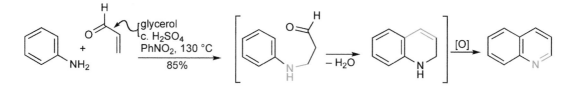

In principle, *meta*-substituted arylamines could give rise to both 5- and 7-substituted quinolines. In practice, electron-donating substituents direct ring closure *para*, thus producing 7-substituted-quinolines; *meta*-halo-aryl-amines produce mainly the 7-halo-isomer. Arylamines with a strong electron-withdrawing *meta*-substituent give rise mainly to the 5-substituted quinoline.

Skraup syntheses sometimes become very vigorous and care must be taken to control their potential violence; pre-forming the Michael adduct and using an alternative oxidant (*p*-chloranil is the best) has been shown to be advantageous in terms of yield and as a better means for controlling the reaction.[113]

A more controlled, stepwise variant utilises an *N*-tosyl-aniline, carrying out the conjugate addition first, the ring closure second, and the aromatisation *via* the elimination of toluenesulfinate, each of the intermediates being isolated.[114]

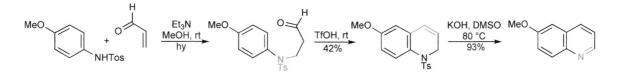

Doebner–Miller Synthesis

The use of an enone confirms the mechanism, showing that interaction of the aniline amino group with the carbonyl group is *not* the first step, and this variation is known as the Doebner–Miller synthesis.

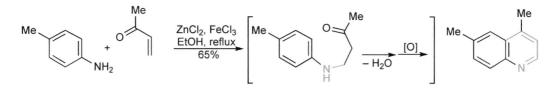

Improvements to the regime for Doebner–Miller ring closures include the use of a two-phase organic/aqueous acid system[115] to minimize alkene polymerization and the use of indium(III) chloride on silica with microwave irradiation.[116] It is significant that the accepted and proved regiochemistry for these cyclisations is *reversed* when the reaction is carried out in trifluoroacetic acid, imine formation being the first step, at least for unsaturated 2-keto esters.[117]

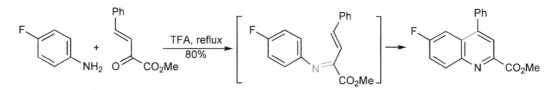

1-(Arylamino)prop-2-ynes are at the same oxidation level as the intermediates in the Skraup and Doebner–Miller strategies. The cyclising step for such substrates requires electrophilic activation of the alkyne, the electrophile ending at the quinoline 3-position.[118]

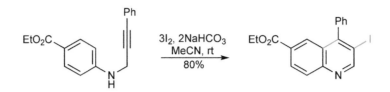

9.15.1.3 Quinolines from ortho-Acyl-Arylamines and Carbonyl Compounds
ortho-Acyl-arylamines react with ketones having an α-methylene to give quinolines.

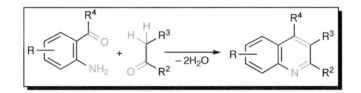

The Friedländer Synthesis[119]
This route has been used extensively for the synthesis of substituted quinolines. In the original sequence, an *ortho*-acyl-arylamine[120] is condensed with a ketone or aldehyde (which must contain an α-methylene group) by base or acid catalysis to yield the quinoline. The orientation of condensation depends on the regioselectivity of enolate or enol formation.[121] Control of regiochemistry can be obtained by using a removable phosphonate, to direct enolisation, as in $RCOCH_2P(O)(OMe)_2$.[122] 2-Substituted quinolines can be obtained regioselectively from methyl ketones using pyrrolidine as catalyst.[123]

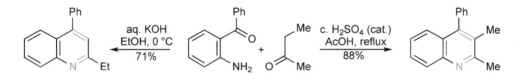

Several improved conditions are available: the use of toluenesulfonic acid,[124] molecular iodine,[125] chlorotrimethylsilane,[126] dodecylphophonic acid[127] and sodium tetrachloroaurate(III) ($NaAuCl_4$) all produce excellent yields of structurally varied quinolines. *ortho*-Aminobenzyl-alcohol can serve as starting material, being oxidized to the amino-aldehyde *in situ*, which then takes part in the condensation, using catalytic copper(II) chloride and potassium hydroxide under oxygen.[128] One can also utilize *ortho*-nitrobenzaldehyde, carrying out reduction to amine *and* condensation in one pot utilising a mixture of tin(II) chloride and zinc chloride.[129]

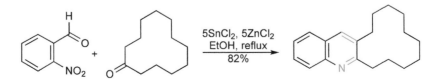

Naphthyridines can also be obtained utilising the Friedländer strategy. [122,123,130]

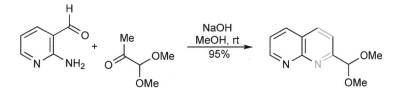

The Pfitzinger Synthesis

Hydrolysis of isatins, which are easy to synthesise (20.16.4), gives *ortho*-aminoaryl-glyoxylates, which react with ketones affording quinoline-4-carboxylic acids.[131] The carboxylic acid group can be removed, if required, by pyrolysis with calcium oxide.

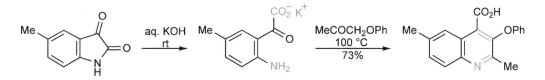

9.15.1.4 *Quinolines by Forming the N–C-2 Bond*

Conjugate addition of a nucleophile to an alkynyl ketone unit *ortho* to amino allows interaction of the amine and carbonyl groups and thus the formation of a quinoline.[132]

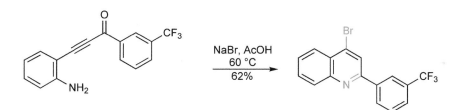

The ring closure of *ortho*-nitroaryl-dimethylaminomethylene ketones is related and produces 4-quinolones *via* selective reduction of the nitro group and then cyclisation.[133]

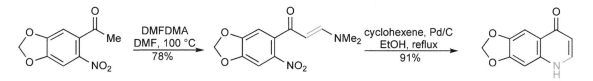

Also in this category is the generation of the cyclisation intermediate by reacting an *ortho*-lithiated *N*-acyl-aniline with a vinamidinium salt.[134] A dimethylaminomethylene-ketone was used for the related reaction of 3-lithiated 2- or 4-acylamino-pyridines, probably *via* the intermediate shown.[135]

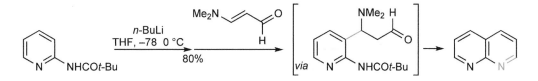

9.15.1.5 Other Methods for Quinolines

Ring closure of *ortho*-aminoaryl-alkynyl-carbinols, readily available by acetylide addition to an aryl-ketone or -aldehyde, can be achieved with copper or palladium catalysis.[136] Comparable *ortho*-nitroaryl-carbinols undergo nitro group reduction *and* ring closure simply by treatment with a metal/acid combination.[137]

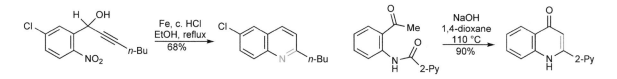

Palladium-catalysed amidation of halo-arenes allows simple assembly of precursors (above) to 4-quino-lones, in which the 2,3-bond is formed by base-catalysed condensation.[138]

9.15.1.6 Isoquinolines from Aryl-aldehydes and Aminoacetal

Aromatic aldehydes react with aminoacetal (2,2-diethoxyethanamine) to generate imines that can be cyclised with acid to isoquinolines carrying no substituents on the heterocyclic ring.

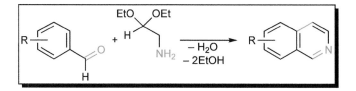

The Pomeranz–Fritsch Synthesis

This synthesis[139] is normally carried out in two stages. Firstly, an aryl aldehyde is condensed with aminoacetal to form an aryl-aldimine. This stage proceeds in high yield under mild conditions. Secondly, the aldimine is cyclised by treatment with strong acid; hydrolysis of the imine competes and reduces the efficiency of this step and for this reason trifluoroacetic acid with boron trifluoride is a useful reagent.[140]

The second step is similar to those in the Combes and Skraup syntheses, in that the acid initially proton-ates, causing elimination of ethanol and the production of a species that can attack the aromatic ring as an electrophile. Final elimination of a second mole of alcohol completes the process.

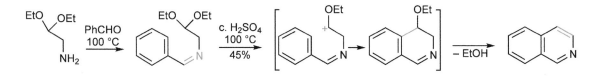

The electrophilic nature of the cyclisation step explains why the process works best for araldehyde-imines carrying electron-donating substituents (especially when these are oriented *para* to the point of closure leading to 7-substituted isoquinolines) and least well for systems deactivated by electron-withdrawing groups.

The problem of imine hydrolysis can be avoided by cyclising at a lower oxidation level, with tosyl on nitrogen for subsequent elimination as toluenesulfinic acid. The ring closure substrates can be obtained by reduction and tosylation of imine condensation products,[141] by benzylating the sodium salt of an

N-tosylaminoethanal-acetal,[142] or *via* a Mitsunobu reaction between a benzylic alcohol and an *N*-sulfonyl-aminoacetal.[143] Cyclisation of benzylaminoethanal-acetals using chlorosulfonic acid gives the aromatic isoquinoline directly.[144]

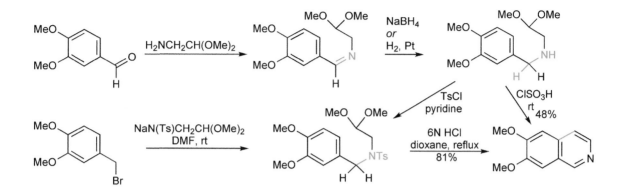

9.15.1.7 *Isoquinolines from 2-Aryl-Ethanamides or 2-Aryl-Ethamine-Imines*

The amide or imine from reaction of 2-aryl-ethanamines with an acid derivative or with an aldehyde, can be ring-closed to a 3,4-dihydro- or 1,2,3,4-tetrahydroisoquinoline respectively. Subsequent dehydrogenation can produce the aromatic heterocycle.

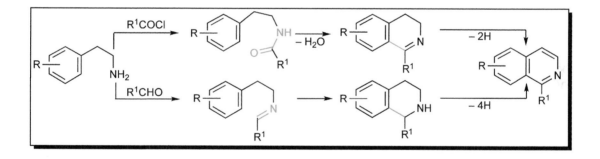

The Bischler–Napieralski Synthesis[145]

In the classical process, a 2-aryl-ethanamine reacts with a carboxylic acid chloride or anhydride to form an amide, which can be cyclised, with loss of water, to a 3,4-dihydro-isoquinoline, then readily dehydrogenated to the isoquinoline using, for example, palladium, sulfur or diphenyl disulfide. Common cyclisation agents are phosphorus pentoxide, often with phosphoryl chloride,[146] and phosphorus pentachloride. The electrophilic intermediate is very probably an imino chloride,[147] or imino phosphate; the former have been isolated and treated with Lewis acids when they are converted into isonitrilium salts, which cyclise efficiently to 3,4-dihydroisoquinolines.[148]

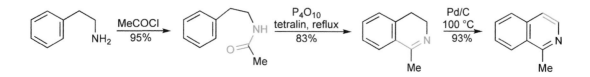

Here, once again, the cyclising step involves electrophilic attack on the aromatic ring so the method works best for activated rings, and *meta*-substituted-aryl ethanamides give exclusively 6-substituted isoquinolines.

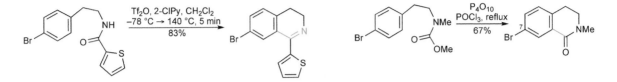

Carbamates can be cyclised to 1-isoquinolones using trifluoromethanesulfonic anhydride and 4-dimethylaminopyridine,[149] or phosphorus pentoxide and phosphoryl chloride.[150] One of the mildest combinations for ring closure is the use of trifluomethanesulfonic anhydride with 2-chloropyridine.[151]

The Pictet–Gams Modification

Conducting the Bischler–Napieralski sequence with a potentially unsaturated aryl-ethanamine, a fully aromatic isoquinoline can be obtained directly. The amide of a 2-methoxy- or 2-hydroxy-2-aryl-ethanamine is heated with the usual type of cyclisation catalyst. It is not clear whether dehydration to an unsaturated amide or to an oxazolidine[152] is an initial stage in the overall sequence.

9.15.1.8 Isoquinolines from Activated 2-Aryl-Ethanamines and Aldehydes (The Pictet–Spengler synthesis[153])

2-Aryl-ethanamines react with aldehydes easily and in good yields to give imines. 1,2,3,4-Tetrahydroisoquinolines result from their cyclisation with acid catalysis. Note that the lower oxidation level imine, versus amide, leads to a tetrahydro- not a dihydroisoquinoline. Routine dehydrogenation easily converts the tetrahydro-isoquinolines into fully aromatic species.

After protonation of the imine, a Mannich-type electrophile is generated; since these are intrinsically less electrophilic than the intermediates in Bischler–Napieralski closures, a strong activating substituent must normally be present on the benzene ring, and appropriately sited, for efficient ring closure. However the use of triflic acid, even with unactivated imines, brings about ring closure, probably *via* a dication,[154] and 2-aryl-ethanamine-carbamates can be converted in tetrahydro-isoquinolines by reaction with aldehydes using perfluorooctanesulfonic acid (PFOSA) with hexafluoropropan-2-ol (HFIP), and in water.[155]

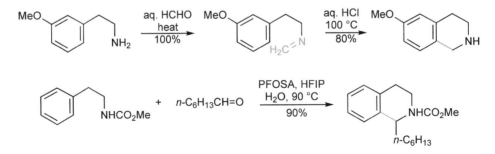

Highly activated, hydroxylated aromatic rings permit Pictet–Spengler ring closure under very mild, 'physiological' conditions.[156]

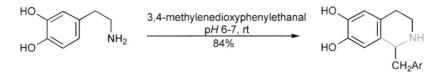

9.15.1.9 *Isoquinolines from* ortho-*Alkynyl-Araldehyde-Imines*

ortho-Iodo-araldehyde imines react directly with internal alkynes, using palladium(0) catalysis, generating isoquinolines in which the original nitrogen substituent has been lost as isobutene; the scheme shows one of many examples of this important process.[157]

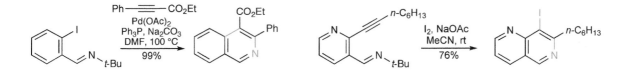

Electrophile-prompted closures of this same type of substrate are also successful, the electrophile ending at the isoquinoline 4-position; these processes are also successful with pyridine starting materials, in the example shown a 1,6-naphthyridine (pyrido[3,2-c]pyridine) is produced.[158]

If, instead of an imine, an oxime is used, the result is an isoquinoline *N*-oxide. Such closures can be effected with silver catalysis,[159] or with iodine, 4-iodoisoquinoline *N*-oxides being the products in the latter cases.[160]

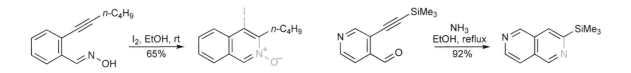

Even simpler, for the naphthyridines, all the isomeric *ortho*-alkynyl-pyridinyl aldehydes react with ammonia in refluxing ethanol to produce naphthyridines (or with hydroxylamine to give naphthyridine-*N*-oxides).[161]

9.15.1.10 *Other Methods for Isoquinolines*

ortho-Dialdehydes can be converted into isoquinoline-3-esters by condensation with a derivative of glycine, the process involving a Wittig–Horner reaction and imine formation.

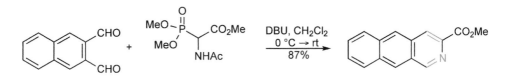

Usefully difunctionalised, 1,3-dichloro-isoquinolines are readily synthesized *via* Beckmann rearrangement of indanedione monoximes.[162]

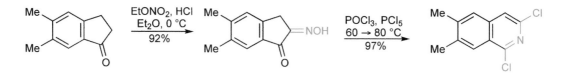

9.15.2 Examples of Notable Syntheses of Quinoline and Isoquinoline Compounds

9.15.2.1 Chloroquine
Chloroquine[163] is a synthetic antimalarial.

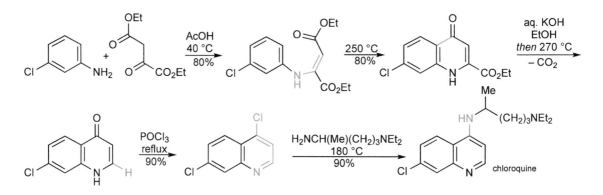

9.15.2.2 Papaverine
Papaverine[164] is an alkaloid from opium; it is a smooth-muscle relaxant and thus useful as a coronary vasodilator – the synthesis illustrates the Pictet–Gams variation of the Bischler–Napieralski sequence.

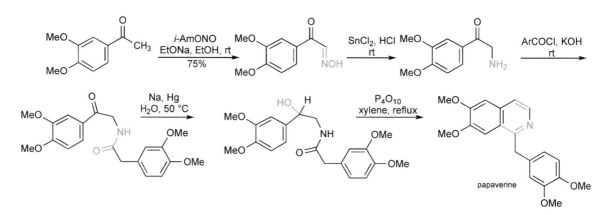

9.15.2.3 Methoxatin
Methoxatin[165] is an enzyme cofactor of bacteria which metabolises methanol and latterly was recognized,[166] 55 years after the 13th, to be a 14th vitamin, also now referred to as pyrroloquinoline quinone (PQQ). This total synthesis is a particularly instructive one since it includes an isatin synthesis (20.16.4), a quinoline synthesis and an indole synthesis.

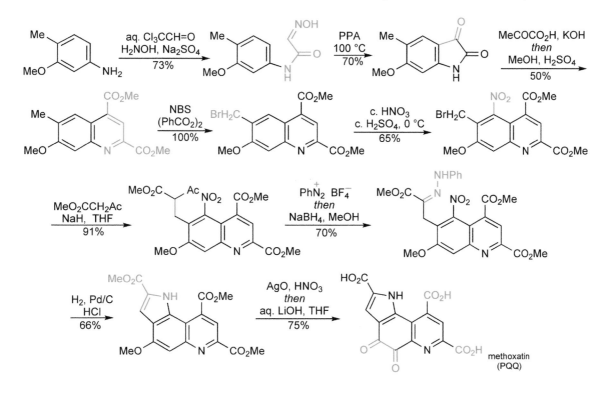

Exercises

Straightforward revision exercises (consult Chapters 7 and 9):

(a) At which positions do quinoline and isoquinoline undergo nitration? Why these positions?

(b) At which positions do quinoline and isoquinoline react most readily with nucleophiles? Why these positions?

(c) At which positions do quinoline and isoquinoline react most readily with radical reagents?

(d) How might one selectively reduce the heterocyclic ring of quinoline or isoquinoline?

(e) How could one convert 4-methylquinoline into 4-ethylquinoline?

(f) How could one convert: (i) isoquinoline into 2-methyl-1-isoquinolone; (ii) quinoline into 2-cyanoquinoline?

(g) What reactants would combine to produce 6-methoxy-2,4-diethylquinoline?

(h) What ring synthesis method would be suitable for converting 4-methoxyaniline into 6-methoxyquinoline?

(i) How could one prepare 2-ethyl-3-methylquinoline-4-carboxylic acid from aniline?

(j) What ring synthesis method would be suitable for the preparation of 6-methoxyisoquinoline?

(k) How could 2-(4-methoxyphenyl)ethanamine be converted into 7-methoxy-1-phenylisoquinoline?

More advanced exercises:

1. Predict the structures of the high yield mono-nitration products: (i) $C_{16}H_{12}N_2O_2$ from 1-benzylisoquinoline; (ii) $C_{10}H_8N_2O_3$ from 6-methoxyquinoline; (iii) $C_{10}H_8N_2O_3$ from 7-methoxyisoquinoline.

2. Write a sequence to rationalise the conversion of quinoline into 3-bromoquinoline by reaction with Br_2 in CCl_4/pyridine.

3. Suggest a structure for product $C_{16}H_{16}ClNO_4$ from 1,3-dichloroisoquinoline and $NaCH(CO_2Et)_2$

4. Deduce a structure for the product, $C_{15}H_{18}N_2OS$ formed on treatment of 2-t-BuCONH-quinoline successively with $3 \times n$-BuLi then dimethyl disulfide.

5. Write a sequence of mechanistic steps to explain the conversion of 2-methylisoquinolinium iodide into 2-methyl-1,2,3,4-tetrahydroisoquinoline with sodium borohydride in ethanol.

6. Draw the most stable tautomer of 3-oxyquinoline, and of 1-, 4- and 8-oxyisoquinolines.

7. Suggest a mechanistic sequence to rationalise the formation of methyl 2-methylquinoline-3-carboxylate from the reaction of aniline with methyl acetoacetate ($\rightarrow C_{11}H_{13}NO_2$) and then this with DMF/POCl$_3$.

8. Deduce the structure of the product quinolones: (i) $C_{12}H_{11}NO_4$ resulting from reaction of 2-methoxyaniline with dimethyl acetylenedicarboxylate then heating at 250 °C; (ii) $C_{10}H_6ClNO_3$ from 3-chloroaniline and diethyl ethoxymethylenemalonate (EtOCH=C(CO$_2$Et)$_2$) then heating at 250 °C, then heating with aq. NaOH.

9. Deduce structures for the quinolines produced from the following combinations: (i) $C_{16}H_{11}NO_2$ from isatin/NaOH then acetophenone; (ii) $C_{10}H_7NO_3$ from isatin/KOH then 3-chloropyruvic acid; (iii) $C_{10}H_7NO_3$ from N-acetylisatin and NaOH.

10. Deduce structures for the heterocyclic products from the following combinations: (i) $C_{11}H_7N_3O_2$ from 2-aminobenzaldehyde and barbituric acid; (ii) $C_{14}H_{11}NO_6$ from 4,5-methylenedioxy-2-aminobenzaldehyde and dimethyl acetylenedicarboxylate; (iii) $C_{14}H_{11}NS$ from 2-aminoacetophenone and 2-acetylthiophen; (iv) $C_{21}H_{19}NO$ from 2-aminobenzophenone and dimedone; (v) $C_{15}H_{12}N_2O_2S$ from 2-aminopyridine-3-aldehyde and 1-phenylsulfonylacetone; (vi) $C_{15}H_{11}N_3$ from 4-amino-pyrimidine-5-aldehyde and α-tetralone.

References

[1] Bressel, U., Katritzky, A. R. and Lea J. R., *J. Chem. Soc. (B)*, **1971**, 4.

[2] Austin, M. W. and Ridd, J. H., *J. Chem. Soc.*, **1963**, 4204; Moodie, R. B., Schofield, K., and Williamson, M. J., *Chem. Ind. (London)*, **1963**, 1283; Dewar, M. J. S. and Maitlis, P. M., *J. Chem. Soc.*, **1957**, 2521.

[3] Katritzky, A. R., Scriven, E. F. V., Majumder, S., Akhmedova, R. G., Vakulenko, A. V., Akhmedov, N. G., Murugan, R. and Abboud, K. A., *Org. Biomol. Chem.*, **2005**, *3*, 538.

[4] Gutteridge, C. E., Hoffman, M. M., Bhattacharjee, A. K. and Gerena, L., *J. Heterocycl. Chem.*, **2007**, *44*, 633.

[5] Beisler, J. A., *Tetrahedron*, **1970**, *26*, 1961.

[6] McCasland, G. E., *J. Org. Chem.*, **1946**, *11*, 277.

[7] Butler, J. C. and Gordon, M., *J. Heterocycl. Chem.*, **1975**, *12*, 1015

[8] Gordon, M. and Pearson, D. E., *J. Org. Chem.*, **1964**, *29*, 329.

[9] Brown, W. D. and Gouliaev, A. H., *Org. Synth.*, **2005**, *81*, 98.

[10] Kress, T. J. and Constantino, S. M., *J. Heterocycl. Chem.*, **1973**, *10*, 409.

[11] Okada, E., Sakaemura, T. and Shimomura, N., *Chem. Lett.*, **2000**, 50.

[12] Hoogerwerff, S. and van Dorp, W. A., *Chem. Ber.*, **1880**, *12*, 747.

[13] Quéguiner, G. and Pastour, P., *Bull. Soc. Chim. Fr.*, **1968**, 4117.

[14] O'Murcha, C., *Synthesis*, **1989**, 880.

[15] Cochran, J. C. and Little, W. F., *J. Org. Chem.*, **1961**, *26*, 808.

[16] Hoogerwerff, S. and van Dorp, W. A., *Recl. Trav. Chim. Pays-Bas*, **1885**, *4*, 285.

[17] Geissman, T. A., Schlatter, M. J., Webb, I. D. and Roberts, J. D., *J. Org. Chem.*, **1946**, *11*, 741.

[18] Makosza, M., Kinowski, A., Danikiewicz, W. and Mudryk, B., *Justus Liebigs Ann. Chem.*, **1986**, 69.

[19] Achmatowicz, M., Thiel, O. R., Gorins, G., Goldstein, C., Affouard, C., Jensen, R. and Larsen, R. D., *J. Org. Chem.*, **2008**, *73*, 6793.

[20] 'Oxidative amino-dehydrogenation of azines', van der Plas, H., *Adv. Heterocycl. Chem.*, **2004**, *86*, 42.

[21] Zoltewicz, J., Helmick, L. S., Oestreich, T. M., King, R. W. and Kandetzki, P. E., *J. Org. Chem.*, **1973**, *38*, 1947.

[22] Tondys, H., van der Plas, H. C. and Wozniak, M., *J. Heterocycl. Chem.*, **1985**, *22*, 353.

[23] Bergstrom, F. W., *Justus Liebigs Ann. Chem.*, **1935**, *515*, 34; Ewing, G. W. and Steck, E. A., *J. Am. Chem. Soc.*, **1946**, *68*, 2181.

[24] Wozniak, M., Baranski, A., Nowak, K. and van der Plas, H. C., *J. Org. Chem.*, **1987**, *52*, 5643; Wozniak, M., Baranski, A., Nowak, K. and Poradowska, H., *Justus Liebigs Ann. Chem.*, **1990**, 653; Wozniak, M. and Nowak, K., *Justus Liebigs Ann. Chem.*, **1994**, 355.

[25] Baik, W., Yun, S., Rhee, J. U. and Russell, G. A., *J. Chem. Soc., Perkin Trans. 1*, **1996**, 1777.

[26] Vanderwalle, J. J. M., de Ruiter, E., Reimlinger, H. and Lenaers, R. A., *Chem. Ber.*, **1975**, *108*, 3898.

[27] Simchen, G. and Krämer, W., *Chem. Ber.*, **1969**, *102*, 3666.

[28] Sanders, G. M., van Dijk, M. and den Hertog, H. J., *Recl. Trav. Chim. Pays-Bas*, **1974**, *93*, 198.

[29] Wolf, C., Tumambac, G. E. and Villalobos, C. N., *Synlett*, **2003**, 1801.

[30] Steinschifter, W. and Stadlbauer, W., *J. prakt. Chem.*, **1994**, *336*, 311.

[31] Verbeek, J., George, A. V. E., de Jong, R. C. P. and Brandsma, L., *J. Chem. Soc., Chem. Commun.*, **1984**, 257; Godard, A., Jaquelin, J.-M. and Quéguiner, G., *J. Organomet. Chem.*, **1988**, *354*, 273; Jacquelin, J. M., Robin, Y., Godard, A. and Quéguiner, G., *Can. J. Chem.*, **1988**, *66*, 1135.

[32] Gilman, H. and Beel, J. A., *J. Am. Chem. Soc.*, **1951**, *73*, 32.

[33] Marsais, F., Godard, A. and Quéguiner, G., *J. Heterocycl. Chem.*, **1989**, *26*, 1589.

[34] Rebstock, A.-S., Mongin, F., Trécourt, F. and Quéguiner, G., *Tetrahedron Lett.*, **2002**, *43*, 767.

[35] Alvarez, M., Salas, M., Rigat, L., de Veciana, A. and Joule, J. A., *J. Chem. Soc., Perkin Trans. 1*, **1992**, 351.

[36] Fernández, M., de la Cuesta, E. and Avendaño, C., *Synthesis*, **1995**, 1362.

[37] Gilman, H. and Soddy, T. S., *J. Org. Chem.*, **1957**, *22*, 565; *ibid.*, **1958**, *23*, 1584; Pinder, R. M. and Burger, A., *J. Med. Chem.*, **1968**, *11*, 267; Ishikura, M., Mano, T., Oda, I. and Terashima, M., *Heterocycles*, **1984**, *22*, 2471.

[38] Wommack, J. B., Barbee, T. G., Thoennes, D. J., McDonald, M. A. and Pearson, D. E., *J. Heterocycl. Chem.*, **1969**, *6*, 243.

[39] Baradarani, M. M., Dalton, L., Heatley, F. and Joule, J. A., *J. Chem. Soc., Perkin Trans. 1*, **1985**, 1503.

[40] Prasad, A. S., Stevenson, T. M., Citineni, J. R., Nyzam, V. and Knochel, P., *Tetrahedron*, **1997**, *53*, 7237.

[41] Dumouchel, S., Mongin, F., Trécourt, F. and Quéguiner, G., *Tetrahedron Lett.*, **2003**, *44*, 2033.

[42] Girard, G. R., Bondinelli, W. E., Hillegass, L. M., Holden, K. G., Pendleton, R. G. and Vzinskas, I. *J. Med. Chem.*, **1989**, *32*, 1566.

[43] Nose, A. and Kudo, T., *Chem. Pharm. Bull.*, **1984**, *32*, 2421.

[44] Ranu, B. C., Jana, U. and Sarkar, A., *Synth. Commun.*, **1998**, *28*, 485.

[45] Vierhapper, F. W. and Eliel, E. L., *J. Org. Chem.*, **1975**, *40*, 2729; Patrick, G. L., *J. Chem. Soc., Perkin Trans. 1*, **1995**, 1273; Skupinska, K. A., McEachern, E. J., Skerlj, R. T. and Bridger, G. J., *J. Org. Chem.*, **2002**, *67*, 7890; Koltunov, K. Yu., Surya Prakash, G. K., Rasul, G. and Olah, G. A., *J. Og. Chem.*, **2007**, *72*, 7394.

[46] Katayama, H., Ohkoshi, M. and Yasue, M., *Chem. Pharm. Bull.*, **1980**, *28*, 2226.

[47] Birch, A. J. and Lehman, P. G., *Tetrahedron Lett.*, **1974**, 2395.

[48] Hückel, W. and Graner, G., *Chem. Ber.*, **1957**, *90*, 2017.

[49] Braude, E. A., Hannah, J. and Linstead, R., *J. Chem. Soc.*, **1960**, 3249.

[50] Jackman, L. M. and Packham, D. I., *Chem. Ind. (London)*, **1955**, 360.

[51] '1,2-Dihydroisoquinolines', Dyke, S. F., *Adv. Heterocycl. Chem.*, **1972**, *14*, 279.

[52] Muren, J. F. and Weissman, A., *J. Med. Chem.*, **1971**, *14*, 49.

[53] Minter, D. E. and Stotter, P. L., *J. Org. Chem.* **1981**, *46*, 3965.

[54] Day, F. H., Bradsher, C. K. and Chen, T.-K., *J. Org. Chem.*, **1975**, *40*, 1195.

[55] Albini, A., Bettinetti, G. F. and Minoli, G., *Tetrahedron Lett.*, **1979**, 3761; *idem*, *Org. Synth.*, **1983**, *61*, 98.

[56] Buchardt, O., *Acta Chem. Scand.*, **1964**, *18*, 1389.

[57] Woodward, R. B. and Doering, W. E., *J. Am. Chem. Soc.*, **1945**, *67*, 860.

[58] 'The Quinolones', Ed. Andriole, V. T., Academic Press, London, **1988**.

[59] Pfister-Guillouzo, G., Guimon, C., Frank, J., Ellison, J. and Katritzky, A. R., *Justus Libeigs Ann. Chem.*, **1981**, 366.

[60] '3(2H)-Isoquinolones and their saturated derivatives', Hazai, L., *Adv. Heterocycl. Chem.*, **1991**, *52*, 155.

[61] Adams, A. and Hey, D. H., *J. Chem. Soc.*, **1949**, 255; Schofield, K. and Swain, T., *ibid.*, 1367.

[62] Kawazoe, Y. and Yoshioka, Y., *Chem. Pharm. Bull.*, **1968**, *16*, 715.

[63] Horning, D. E., Lacasse, G. and Muchowski, J. M., *Can. J. Chem.*, **1971**, *49*, 2785.

[64] Morel, A. F., Larghi, E. L. and Selvero, M. M., *Synlett*, **2005**, 2755.

[65] Gabriel, S. and Coman, J., *Chem. Ber.*, **1900**, *33*, 980.

[66] Kanishi, K., Onari, Y., Goto, S. and Takahashi, K., *Chem. Lett.*, **1975**, 717.

[67] Ogata, Y., Kawasaki, A. and Hirata, H., *J. Chem. Soc., Perkin Trans. 2*, **1972**, 1120.

[68] Kaslow, C. E. and Stayner, R. D., *J. Am. Chem. Soc.*, **1945**, *67*, 1716.

[69] Wislicenus, W. and Kleisinger, E., *Chem. Ber.*, **1909**, *42*, 1140.

[70] Burger, A. and Modlin, L. R., *J. Am. Chem. Soc.*, **1940**, *62*, 1079.

[71] Dyson, P. and Hammick, D. Ll., *J. Chem. Soc.*, **1937**, 1724.

[72] Quast, H. and Schmitt, E., *Justus Liebigs Ann. Chem.*, **1970**, *732*, 43.

[73] Gensler, W. J. and Shamasundar, K. T., *J. Org. Chem.*, **1975**, *40*, 123.

[74] For '-one' production by ferricyanide oxidation: Bunting, J. W., Lee-Young, P. A. and Norris, D. J., *J. Org. Chem.*, **1978**, *43*, 1132; by O_2 oxidation: Ruchirawat, S., Sunkul, S., Thebtaranonth, Y. and Thirasasna, N., *Tetrahedron Lett.*, **1977**, 2335.

[75] Loska, R., Majcher, M. and Makosza, M., *J. Org. Chem.*, **2007**, *72*, 5574.

[76] Schultz, O.-E. and Amschler, U., *Justus Liebigs Ann. Chem.*, **1970**, *740*, 192; see also Bunting, J. W. and Meathrel, W. G., *Can. J. Chem.*, **1974**, *52*, 303.

[77] Zincke, T. H. and Weisspfenning, G., *Justus Liebig's Ann. Chem.*, **1913**, *396*, 103.

[78] Barbier, D., Marazano, C., Das, B. C. and Potier, P., *J. Org. Chem.*, **1996**, *61*, 9596.

[79] Barbier, D., Marazano, C., Riche, C., Das, B. C. and Potier, P., *J. Org. Chem.*, **1998**, *63*, 1767.

[80] Weinstock, J. and Boekelheide, V., *Org. Synth., Coll. Vol. IV*, **1963**, 641.

[81] 'Reissert compounds', Popp, F. D., *Adv. Heterocycl. Chem.*, **1968**, *9*, 1; 'Developments in the chemistry of Reissert compounds (1968–1978)', *idem*, **1979**, *24*, 187.

[82] Ezquerra, J. and Alvarez-Builla, J., *J. Chem. Soc., Chem. Commun.*, **1984**, 54.

[83] Chênevert, R., Lemieux, E., and Voyer, N., *Synth. Commun.*, **1983**, *13*, 1095.

[84] Itoh, T., Nagata, K., Miyazaki, M., Kameoka, K. and Ohsawa, A., *Tetrahedron*, **2001**, *57*, 8827; Siek, O., Schaller, S., Grimme, S. and Liebscher, J., *Synlett*, **2003**, 337.

[85] Boger, D. L., Brotherton, C. E., Panek, J. S. and Yohannes, D., *J. Org. Chem.*, **1984**, *49*, 4056.

[86] Yamaguchi, R., Hatano, B., Nakayasu, T. and Kozima, S., *Tetrahedron Lett.*, **1997**, *38*, 403.

[87] Yamaguchi, R., Mochizuki, K., Kozima, S. and Takaya, H., *Chem. Lett.*, **1994**, 1809.

[88] Yamaguchi, R., Omoto, Y., Miyake, M. and Fujita, K., *Chem. Lett.*, **1998**, 547.

[89] Yadav, J. S., Reddy, B. V. S., Sreenivas, M. and Sathaiah, K., *Tetrahedron Lett.*, **2005**, *46*, 8905.

90 Nair, V., Devipriya, S. and Suresh, E., *Tetrahedron*, **2008**, *64*, 3567.
91 Yavari, I., Ghazanfarpour-Darjani, M., Sabbaghan, M. and Hossaini, Z., *Tetrahedron Lett.*, **2007**, *48*, 3749.
92 Nair, V., Sreekanth, A. R., Abhilash, N., Bhadbhade, M. M. and Gonnade, R. C., *Org. Lett.*, **2002**, *4*, 3575.
93 Ochiai, E., *J. Org. Chem.*, **1953**, *18*, 534; Yokoyama, A., Ohwada, T., Saito, S. and Shudo, K., *Chem. Pharm. Bull.*, **1997**, *45*, 279.
94 Ochiai, E. and Ikehara, M., *J. Pharm. Soc. Japan*, **1953**, *73*, 666.
95 Harusawa, S., Hamada, Y. and Shiorii, T., *Heterocycles*, **1981**, *15*, 981.
96 Miyashita, A., Matsuda, H., Iijima, C. and Higashino, T., *Heterocycles*, **1992**, *33*, 211.
97 Hayashida, M., Honda, H. and Hamana, M., *Heterocycles*, **1990**, *31*, 1325.
98 Fakhfakh, M. A., Franck, X., Fournet, A., Hocquemiller, R. and Figadère, B., *Tetrahedron Lett*, **2001**, *42*, 3847.
99 Diaz, J. L., Miguel, M. and Lavilla, R., *J. Org. Chem.*, **2004**, *69*, 3550.
100 Long, R. and Schofield, K., *J. Chem. Soc.*, **1953**, 3161; see also Roberts, E. and Turner, E. E., *J. Chem. Soc.*, **1927**, 1832.
101 'Advances in the chemistry of naphthyridines', Litvinov, V. P., *Adv. Heterocycl. Chem.*, **2006**, *91*, 189.
102 Nakatani, K., Sando, S., Toshida, K. and Saito, I., *Tetrahedron Lett.*, **1999**, *40*, 6029.
103 Lauer, W. M. and Kaslow, C. E., *Org. Synth., Coll. Vol. III*, 580; Reynolds, G. A. and Hauser, C. R., *ibid.*, 593.
104 Heindel, N. D., Brodof, T. A. and Kogelschatz, J. E., *J. Heterocycl. Chem.*, **1966**, *3*, 222.
105 Zewge, D., Chen., C.-y., Deer, C., Dormer, P. G. and Hughes, D. L., *J. Org. Chem.*, **2007**, *72*, 4276.
106 Price, C. C. and Roberts, R. M., *Org. Synth., Coll. Vol. III*, **1955**, 272; 'Aminomethylene malonates and their use in heterocyclic synthesis', Hermecz, I., Keresztúri, G. and Vasvári-Debreczy, L., *Adv. Heterocycl. Chem.*, **1992**, *54*, 1.
107 Lavergne, O., Leseur-Ginot, L., Rodas, F. P., Kasprzyk, P. G., Pommier, J., Demarquay, D., Pre'vost, G., Ulibarri, G., Rolland, A., Schiano-Liberatore, A., Harnett, J., Pons, D., Camera, J. and Bigg, D. C. H., *J. Med. Chem.*, **1998**, *41*, 5410.
108 Elban, M. A., Sun, W., Eisenhauer, B. M., Gao, R. and Hecht, S. M., *Org. Lett.*, **2006**, *8*, 3513.
109 Chen, B., Huang, X. and Wang, J., *Synthesis*, **1987**, 482.
110 'The Skraup synthesis of quinolines', Manske, R. H. F. and Kulka, M., *Org. React.*, **1953**, *7*, 59.
111 Clarke, H. T. and Davis, A. W., *Org. Synth., Coll. Vol. I*, **1932**, 478.
112 Mosher, H. S., Yanko, W. H. and Whitmore, F. C., *Org. Synth., Coll. Vol. III*, **1955**, 568.
113 Song, Z., Mertzman, M. and Hughes, D. L., *J. Heterocycl. Chem.*, **1993**, *30*, 17.
114 Tokuyama, H., Sato, M., Ueda, T. and Fukuyama, T., *Heterocycles*, **2001**, *54*, 105.
115 Matsugi, M., Tabusa, F. and Minamikawa, J.-i., *Tetrahedron Lett.*, **2000**, *41*, 8523.
116 Ranu, B. C., Hajra, A. and Jana, U., *Tetrahedron Lett.*, **2000**, *41*, 531.
117 Wu, Y.-C., Liu, L., Li, H.-J., Wang, D. and Chen, Y.-J., *J. Org. Chem.*, **2006**, *71*, 6592.
118 Zhang, X., Campo, M. A., Yao, T. and Larock, R. C., *Org. Lett.*, **2005**, *7*, 763.
119 'The Friedländer synthesis of quinolines', Cheng, C.-C. and Yan, S.-J., *Org. React.*, **1982**, *28*, 37.
120 For a useful synthesis of such compounds see Okabe, M. and Sun, R.-C., *Tetrahedron*, **1995**, *51*, 1861.
121 Fehnel, E. A., *J. Org. Chem.*, **1966**, *31*, 2899.
122 Hsiao, Y., Rivera, N. R., Yasuda, N., Hughes, D. L. and Reider, P. J., *Org. Lett.*, **2001**, *3*, 1101.
123 Dormer, P. G., Eng, K. K., Farr, R. N., Humphrey, G. R., McWilliams, J. C., Reider, P. J., Sager, J. W. and Volante, R. P., *J. Org. Chem.*, **2003**, *68*, 467.
124 Jia, C.-S., Zhang, Z., Tu, S.-J. and Wang, G.-W., *Org. Biomol. Chem.*, **2006**, *4*, 104.
125 Wu, J., Xia, H.-G. and Gao, K., *Org. Biomol. Chem.*, **2006**, *4*, 126
126 Ryabukhin, S. V., Volochnyuk, D. M., Plaskon, A. S., Naumchik, V. S. and Tolmachev, A.. A., *Synthesis*, **2007**, 1214.
127 Ghassamipour, S. and Sardarian, A. R., *Tetrahedron Lett.*, **2009**, *50*, 514.
128 Cho, C. S., Ren, W. X. and Shim, S. C., *Tetrahedron Lett.*, **2006**, *47*, 6781.
129 McNaughton, B. R. and Miller, B. L., *Org. Lett.*, **2003**, *5*, 4257.
130 Yasuda, N., Hsiao, Y., Jensen, M. S., Rivera, N. R., Yang, C., Wells, K. M., Yau, J., Palucki, M., Tan, L., Dormer, P. G., Volante, R. P., Hughes, D. L. and Reider, P. J., *J. Org. Chem.*, **2004**, *69*, 1959.
131 Calaway, P. K. and Henze, H. R., *J. Am. Chem. Soc.*, **1939**, *61*, 1355
132 Arcadi, A., Marinelli, F. and Rossi, E., *Tetrahedron*, **1999**, *55*, 13233; Abbiati, G., Arcadi, A., Marinelli, F., Rossi, E. and Verdecchia, *Synlett*, **2006**, 3218; Arcadi, A., Bianchi, G., Inesi, A., Marinelli, F. and Rossi, L., *Synlett*, **2007**, 1031.
133 Tois, J., Vahermo, M. and Koskinen, A., *TetrahedronLett.*, **2005**, *46*, 735.
134 Cho, I.-S., Gong, L. and Muchowski, J. M., *J. Org. Chem.*, **1991**, *56*, 7288.
135 Zhichkin, P., Beer, C. M. C., Rennells, W. M. and Fairfax, D. J., *Synlett*, **2006**, 379.
136 Gabriele, B., Mancuso, R., Salerno, G., Ruffolo, G. and Plastina, P., *J Org. Chem.*, **2007**, *72*, 6873.
137 Sandelier, M. J. and DeShong, P., *Org. Lett.*, **2007**, *9*, 3209.
138 Jones, C. P., Anderson, K. W. and Buchwald, S. L., *J. Org. Chem.*, **2007**, *72*, 7968.
139 'The synthesis of isoquinolines by the Pomeranz–Fritsch reaction', Gensler, W. J., *Org. React.*, **1951**, *6*, 191.
140 Kucznierz, R., Dickhaut, J., Leinert, H. and von der Saal, W., *Synth. Commun.*, **1999**, *29*, 1617.
141 Birch, A. J., Jackson, A. H. and Shannon, P. V. R., *J. Chem. Soc., Perkin Trans. 1*, **1974**, 2185.
142 Boger, D. L., Brotherton, C. E. and Kelley, M. D., *Tetrahedron*, **1981**, *37*, 3977.
143 García, A., Castedo, L. and Domínguez, D., *Synlett*, **1993**, 271; Larghi, E. L. and Kaufman, T. S., *Tetrahedron Lett.*, **1997**, *38*, 3159.
144 Kido, K. and Watanabe, Y., *Heterocycles*, **1980**, *14*, 1151.
145 'The preparation of 3,4-dihydroisoquinolines and related compounds by the Bischler–Napieralski reaction', Whaley, W. M. and Govindachari, *Org. React.*, **1951**, *6*, 74.
146 Okuda, K., Kotake, Y. and Ohta, S., *Biorg. Me. Chem. Lett.*, **2003**, *13*, 2853.
147 Nagubandi, S. and Fodor, G., *J. Heterocycl. Chem.*, **1980**, *17*, 1457.
148 Ban, Y., Wakamatsu, T. and Mori, M., *Heterocycles*, **1977**, *6*, 1711.

[149] Banwell, M. G., Bissett, B. D., Busato, S., Cowden, C. J., Hockless, D. C. R., Holman, J. W., Reed, R. W. and Wu, A. W., *J. Chem. Soc., Chem. Commun.*, **1995**, 2551.

[150] Wang, X., Tan, J. and Grozinger, K., *Tetrahedron Lett.*, **1998**, *39*, 6609.

[151] Movassaghi, M. and Hill, M. D., *Org. Lett.*, **2008**, *10*, 3485.

[152] Fitton, A. O., Frost, J. R., Zakaria, M. M. and Andrew, G., *J. Chem. Soc., Chem. Commun*, **1973**, 889.

[153] 'The Pictet–Spengler synthesis of tetrahydroisoquinolines and related compounds', Whaley, W. M. and Govindachari, T. R., *Org. React.*, **1951**, *6*, 151.

[154] Yokoyama, A., Ohwada, T. and Shudo, K., *J. Org. Chem.*, **1999**, *64*, 611.

[155] Saito, A., Takayama, M., Yamazaki, A., Numaguchi, J. and Hanzawa, Y., *Tetrahedron*, **2007**, *63*, 4039.

[156] Schöpf, C. and Salzer, W., *Justus Liebigs Ann. Chem.*, **1940**, *544*, 1.

[157] Roesch, K. R., Zhang, H. and Larock, R. C., *J. Org. Chem.*, **2001**, *66*, 8042.

[158] Huang, Q., Hunter, J. A. and Larock, R. C., *J. Org. Chem.*, **2002**, *67*, 3437.

[159] Yeom, H.-S., Kim, S. and Shin, S., *Synlett*, **2008**, 924.

[160] Huo, Z., Tomeba, H. and Yamamoto, Y., *Tetrahedron Lett.*, **2008**, *49*, 5531.

[161] Numata, A., Kondo, Y. and Sakamoto, T., *Synthesis*, **1999**, 306.

[162] Bartmann, W., Konz, E. and Rüger, W., *J. Heterocycl. Chem.* **1987**, *24*, 677.

[163] Surrey, A. R. and Hammer, H. F., *J. Am. Chem. Soc.*, **1946**, *68*, 113.

[164] Pictet A. and Gams, A., *Chem. Ber.*, **1909**, *42*, 2943.

[165] Gainor, J. A. and Weinreb, S. M., *J. Org. Chem.*, **1982**, *47*, 2833.

[166] Kasahara, T. and Kato, T., *Nature*, **2003**, *422*, 832.

10

Typical Reactivity of Pyrylium and Benzopyrylium Ions, Pyrones and Benzopyrones

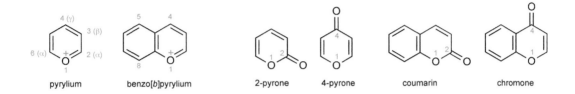

pyrylium   benzo[b]pyrylium   2-pyrone   4-pyrone   coumarin   chromone

The pyrylium cation presents an intriguing dichotomy – it is both 'aromatic', and therefore, one would be tempted to understand, 'stable', yet it is very reactive – the tropylium cation and the cyclopentadienyl anion can also be described in this way. However, all is relative, and that pyrylium cations react rapidly with nucleophiles to produce adducts that are not aromatic, is merely an expression of their relative stability – if they were not aromatic it is doubtful whether such cations could exist at all. Pyrylium perchlorate is actually surprisingly stable – it does not decompose below 275 °C, but, nonetheless, it will react with water, even at room temperature, producing a non-aromatic product. (**NOTE:** All perchlorates should be treated with caution – heating can cause explosive decomposition.)

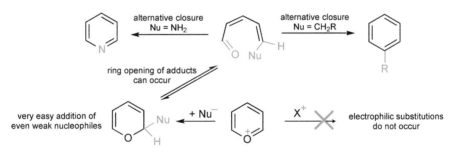

Typical reactions of pyrylium cations

The properties of pyrylium cations are best compared with those of pyridinium cations: the system does not undergo electrophilic substitution nor, indeed, are benzopyrylium cations substituted in the benzene ring. This is a contrast with the chemistry of quinolinium and isoquinolinium cations and is a comment on the stronger deactivating effect of the positively charged oxygen.

Heterocyclic Chemistry 5th Edition   John Joule and Keith Mills
© 2010 Blackwell Publishing Ltd

Pyrylium ions readily add nucleophilic reagents, at an α-position, generating 1,2-dihydro-pyrans, which then often ring open. Virtually all the known reactions of pyrylium salts fall into this general category. Pyrylium cations are more reactive in such nucleophilic additions than pyridinium cations – oxygen tolerates a positive charge less well than nitrogen. The analogy with carbonyl chemistry is obvious – the nucleophilic additions that characterise pyrylium systems are nothing more or less than those that are encountered frequently in acid-catalysed (*O*-protonated) chemistry of carbonyl groups.

The initial product of ring opening can take part in an alternative ring closure, generating a benzenoid aromatic system (if Nu contains active hydrogen attached to carbon) or a pyridine (if Nu is an amine nitrogen).

Comparison of nucleophilic addition to pyrylium systems with that to *O*-protonated aldehydes/ketones

In benzopyrylium systems, one finds exactly comparable behaviour – a readiness to add nucleophiles, adjacent to the positively charged oxygen, in the heterocyclic ring. The interaction of the two isomeric bicycles with ammonia is instructive: benzo[*c*]pyrylium can be converted into isoquinoline, but benzo[*b*] pyrylium cannot be converted into quinoline for, although in the last case the addition can and does take place, in the subsequent ring-opened species, no low-energy mechanism is available to allow the nitrogen to become attached to the benzene ring.

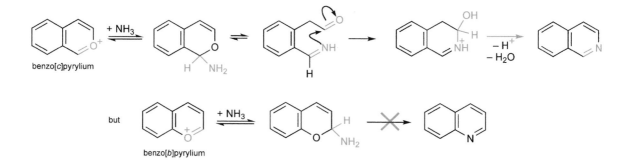

Pyrones, which are the ring-oxygen equivalents of pyridones, are simply α- and γ-hydroxy-pyrylium salts from which an *O*-proton has been removed. There is little to recommend that 2- and 4-pyrones be viewed as aromatic: they are perhaps best seen as cyclic unsaturated lactones and cyclic β-oxy-α,β-unsaturated-ketones, respectively, for example 2-pyrones are hydrolysed by alkali, just like simpler esters (lactones). It is instructive that, whereas the pyrones are converted into pyridones by reaction with amines or ammonia, the reverse is not the case – pyridones are not transformed into pyrones by water or hydroxide. Some electrophilic *C*-substitutions are known for pyrones and benzopyrones, the oxygen guiding the electrophile *ortho* or *para*, however there is a tendency for electrophilic addition to the C–C double bond of the heterocyclic ring, again reflecting their non-aromatic nature. Easy Diels–Alder additions to 2-pyrones are further evidence for diene, rather than aromatic, character.

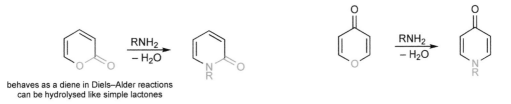

behaves as a diene in Diels–Alder reactions
can be hydrolysed like simple lactones

Some typical reactions of pyrones

The cyclisation of an unsaturated 1,5-dicarbonyl compound produces pyrylium salts, providing of course that a suitable acidic medium is chosen – it must not contain nucleophilic species that would add to the salt, once formed. Acid-catalysed ring closure of 1,3,5-triketones produces 4-pyrones, as shown; 2-pyrones are formed via the construction of an α,β-unsaturated ester that has a 5-carbonyl group.

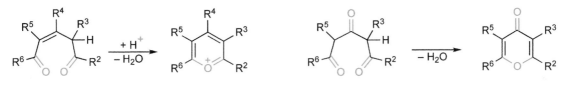

Typical pyrylium and 4-pyrone ring syntheses

Benzopyrylium salts are formed when phenols react with 1,3-dicarbonyl compounds under acidic, dehydrating conditions. The comparable use of 1,3-keto-esters with phenols leads to benzopyrones.

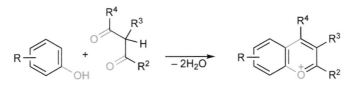

Typical benzo[*b*]pyrylium ring synthesis

11

Pyryliums, 2- and 4-Pyrones: Reactions and Synthesis

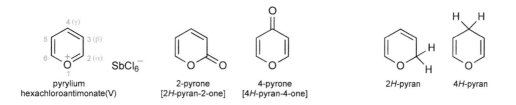

pyrylium
hexachloroantimonate(V)

2-pyrone
[2*H*-pyran-2-one]

4-pyrone
[4*H*-pyran-4-one]

2*H*-pyran

4*H*-pyran

Pyrylium salts, especially perchlorates, tetrafluoroborates and hexachloroantimonates(V), are stable, but reactive compounds. Perchlorates have been used extensively, since pyrylium perchlorates tend to be sparingly soluble, and thus relatively easily isolated, however all perchlorates should be treated with **CAUTION**: perchlorates, particularly dry perchlorates can decompose explosively. Wherever possible, other salts should be substituted, for example comparative preparations for 2,4,6-trimethylpyrylium salts have been described: perchlorate,[1] tetrafluoroborate[2] and trifluoromethanesulfonate,[3] the last having the advantage of better solubility in organic solvents. No pyrylium salts have been identified in living organisms, though the benzo[*b*]pyrylium system plays an important role in the flower pigments (see 32.5.6).

Almost all the known reactions of the pyrylium nucleus involve addition of a nucleophile, usually at an α-position, occasionally at C-4, as the first step. A feature of pyrylium chemistry is the ring opening of adducts produced by such additions, followed by cyclisation in a different manner, to give a new heterocyclic or homocyclic product (ANRORC processes).

Straightforward electrophilic or radical substitutions at ring positions are unknown. Controlled oxidations, like those of pyridinium salts to 2-pyridones, are likewise not known in pyrylium chemistry.

11.1 Reactions of Pyrylium Cations[4,5]
11.1.1 Reactions with Electrophilic Reagents
11.1.1.1 Proton Exchange
2,4,6-Triphenylpyrylium undergoes exchange at the 3- and 5-positions in hot deuterioacetic acid, but the process probably involves, not protonation of the pyrylium cation, but formation of an equilibrium concentration of an adduct, with acetate added to C-2, allowing enol ether protonation and thus exchange.[6]

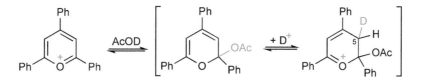

11.1.1.2 Nitration
Nitration of 2,4,6-triphenylpyrylium proceeds on the benzene rings;[7] no nitrations of pyrylium rings are known.

11.1.2 Addition Reactions with Nucleophilic Reagents
Pyrylium salts usually add nucleophiles at a carbon adjacent to the oxygen, and such reactions are analogous with those of *O*-protonated carbonyl compounds.

11.1.2.1 Water and Hydroxide Ion
The degree of susceptibility of pyrylium salts to nucleophilic attack varies widely: pyrylium cation itself is even attacked by water at 0 °C, whereas 2,4,6-trimethylpyrylium is stable in water at 100 °C. Hydroxide anion, however, adds very readily to C-2 in all cases.

The reaction of 2-methyl-4,6-diphenylpyrylium is typical:[8] the immediate 2-hydroxy-2*H*-pyran, which is a cyclic enol hemiacetal, is in equilibrium with a dominant concentration of the acyclic tautomer, reached probably *via* a proton-catalysed process, since methoxide adducts remain cyclic.[9] Treatment of such acyclic unsaturated diketones with acid regenerates the original pyrylium salt (11.3.1).

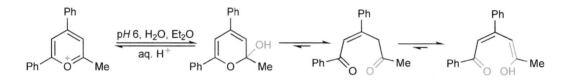

With pyryliums carrying α-alkyl groups, more vigorous alkaline treatment leads to an alternative closure, producing arenes, for example reaction of 2,4,6-trimethylpyrylium with warm alkali causes a subsequent cyclising aldol condensation of the acyclic intermediate to give 3,5-dimethylphenol.[10]

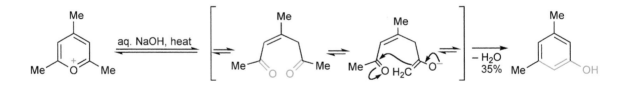

11.1.2.2 Ammonia and Primary and Secondary Amines
Ammonia and primary amines react with pyrylium salts to give pyridines and *N*-alkyl- or *N*-aryl-pyridinium salts, respectively.[11] The transformation represents a good method for preparing the nitrogen heterocycles, providing the pyrylium salt can be accessed in the first place. The initial adduct exists as one of a number of ring-opened tautomeric possibilities,[12] depending upon conditions; it is probably an amino-dienone that recloses to give the nitrogen heterocycle.

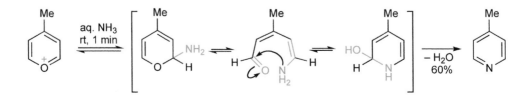

The reaction of a secondary amine cannot, of course, lead to a pyridine, however in pyryliums carrying an α-methyl, ring closure to an arene can occur, this time *via* an enamine.[6a]

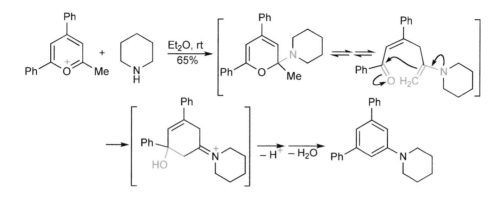

Other reactants containing a primary amino group will also convert pyryliums into *N*-substituted nitrogen heterocycles: *N*-amino azoles[13] are amongst several types of hydrazine derivatives to have been utilised: these give *N*-(heteroaryl)-pyridinium salts. Reaction of pyryliums with hydroxylamine comparably leads (predominantly) to the formation of pyridine *N*-oxides.[1,14]

11.1.2.3 Organometallic Addition
Organometallic addition takes place at an α-position, or occasionally at C-4 when the α-positions are substituted and C-4 is unsubstituted[15] or when organocuprates are used.[16] The initial *2H*-pyrans undergo electrocyclic ring opening (and more rapidly than the comparable cyclohexadiene/hexatriene transformation[17]) affording dienones or dienals with retention of geometrical integrity.

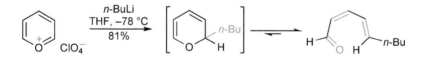

11.1.2.4 Other Carbanionic Additions
By processes comparable to organometallic addition, cyanide addition to 2,4,6-trimethylpyrylium leads to a ring-opened dienone.[18] Reactions with stabilised anions, such as those from nitromethane or 1,3-dicarbonyl compounds, proceed though a series of equilibria to recyclised, aromatic compounds.[4a]

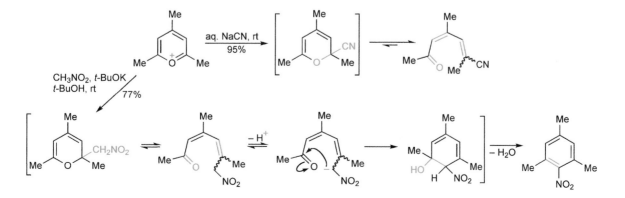

Following addition of triphenylphosphonium methylide, Wittig condensation, electrocyclic ring opening and double bond equilibration, all *trans*-7-substituted 2,4,6-heptatrienals can be accessed.[19]

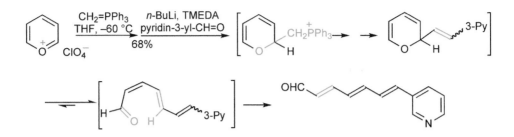

11.1.3 Substitution Reactions with Nucleophilic Reagents

There are a small number of pyrylium reactions that can be categorised as nucleophilic substitutions. 4-Pyrones react with acetic anhydride at carbonyl oxygen to produce 4-acetoxy-pyryliums, *in situ*, allowing nucleophilic substitution at C-4: the reaction of 2,6-dimethylpyrone with methyl cyanoacetate is typical.[20] Phosphorus pentachloride likewise converts 4-pyrones into 4-chloropyryliums.[21]

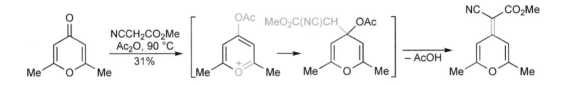

11.1.4 Reactions with Radicals

4-Alkylation of 2,6-disubstituted pyryliums has been achieved using tetraalkyltin compounds in the presence of UV light; the initial adducts are re-oxidised *in situ* to produce 4-substituted pyrylium salts.[22]

11.1.5 Reactions with Reducing Agents

The addition of hydride to pyryliums takes place mainly at an α-position, generating 2*H*-pyrans, which rapidly open to form the isolated products, dienones, best extracted immediately into an organic solvent to minimise further reaction; the minor products are the isomeric 4*H*-pyrans.[23] One-electron polarographic reduction generates radicals, which dimerise (cf. 8.5.3).[24]

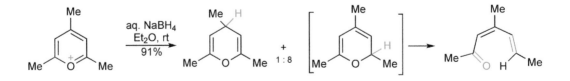

11.1.6 Photochemical Reactions

At first sight, the photochemistry of 4-hydroxy-pyryliums, i.e. of 4-pyrones in acid solution, seems extraordinary, in that they are converted into 2-pyrones, however a rationalisation involving, first, a bicyclic hydroxyallyl cation, second, a bicyclic epoxy-cyclopentenone, and then a second photoexcitation, makes the transformation clear; the sequence is shown below.[25] Irradiation at higher pH leads to a trapping of the first-formed photointermediate by solvent and thus the isolation of dihydroxy-cyclopentenones.[26]

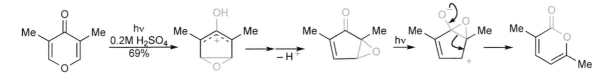

11.1.7 Reactions with Dipolarophiles; Cycloadditions

Pyrylium-3-olates,[27] formally 3-hydroxy-pyryliums rendered overall neutral by loss of the phenolic proton, though this is not usually the method for their formation, undergo cycloadditions across the 2,6-positions and in so doing parallel the reactivity of pyridinium-3-olates (8.7). Even unactivated alkenes will cycloadd when tethered and thus the process is intramolecular.[25] Usually, the pyrylium-3-olate is generated by elimination of acetic acid from a 6-acetoxy-2H-pyran-3(6H)-one (a 'pyranulose acetate'[28]) (see 18.2).

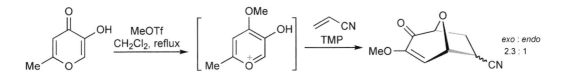

Borohydride reduction of the ketone carbonyl in such adducts, then ozonolysis, generates 2,5-*cis* disubstituted tetrahydrofurans.[29]

Another ingenious route for the generation of the dipolar species involves the carbonyl-*O*-alkylation[30] or *O*-silylation[31] of 3-oxygenated 4-pyrones. The example below shows *O*-methylation of a kojic acid derivative, then deprotonation of the 3-hydroxyl using a hindered base to trigger the dipolar cycloaddition.

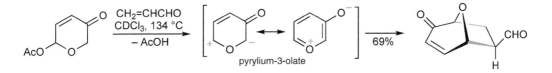

11.1.8 Alkyl-Pyryliums[32]

Hydrogens on alkyl groups at the α- and γ-positions of pyrylium salts are, as might be expected, quite acidic: reaction at a γ-methyl is somewhat faster than at an α-methyl.[33] Condensations with aromatic aldehydes (illustrated below),[34] triethyl orthoformate[35] and dimethylformamide[36] are all possible.

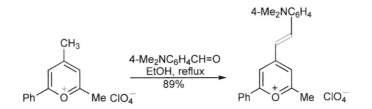

11.2 2-Pyrones and 4-Pyrones (2*H*-Pyran-2-ones and 4*H*-Pyran-4-ones; α- and γ-Pyrones)

11.2.1 Structure of Pyrones

2- and 4-Hydroxy-pyrylium salts are quite strongly acidic and are therefore much better known as their conjugate bases, the 2- and 4-pyrones. The simple 4-pyrones are quite stable crystalline substances, whereas the 2-pyrones are much less stable: 2-pyrone itself, which has the smell of fresh-mown hay, polymerises slowly on standing. There are relatively few simple pyrone natural products, in great contrast with the widespread occurrence and importance of their benzo derivatives, the coumarins and chromones (32.5.6), in nature.

11.2.2 Reactions of Pyrones

11.2.2.1 Electrophilic Addition and Substitution

4-Pyrone is a weak base, pK_{aH} −0.3 that is protonated on the carbonyl oxygen to afford 4-hydroxy-pyrylium salts, often crystalline. The reaction of 2,6-dimethyl-4-pyrone with *t*-butyl bromide in hot chloroform provides a neat way to form the corresponding 4-hydroxy-2,6-dimethylpyrylium bromide.[37] 2-Pyrones are much weaker bases and, though they are likewise protonated on carbonyl oxygen in solution in strong acids, salts cannot be isolated. This difference is mirrored in reactions with alkylating agents: the former give 4-methoxy-pyrylium salts with dimethyl sulfate,[38] whereas 2-pyrones require Meerwein salts, $Me_3O^+BF_4^-$, for carbonyl-*O*-methylation. Acid-catalysed exchange in 4-pyrone, presumably *via* C-protonation of a concentration of neutral molecule, takes place at the 3/5-positions.[39]

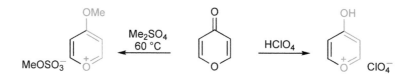

With bromine, 2-pyrone forms an unstable adduct that gives the substitution product 3-bromo-2-pyrone on warming,[40] however this can also be obtained satisfactorily by bromodecarboxylation of 2-pyrone-3-carboxylic acid; coumalic acid (2-pyrone-5-carboxylic acid) gives 5-bromo-2-pyrone[41] or 3,5-dibromo-2-pyrone[42] (See also 11.2.2.3 and 11.4.2 for syntheses of halo-2-pyrones).

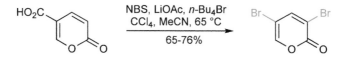

With nitronium tetrafluoroborate, the electrophile is assumed to attack first at carbonyl oxygen, leading subsequently to 5-nitro-2-pyrone.[43] Simple examples of electrophilic substitution of 4-pyrones are rare, however bis-dimethylaminomethylation of the parent heterocycle takes place under quite mild conditions.[44]

11.2.2.2 Attack by Nucleophilic Reagents

2-Pyrones are in many ways best viewed as unsaturated lactones, and as such they are easily hydrolysed by aqueous alkali; 4-pyrones, too, easily undergo ring-opening with base, though for these vinylogous lactones, initial attack is at C-2.[45]

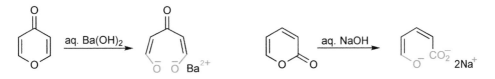

2-Pyrones can in principle add nucleophilic reactants at either C-2 (carbonyl carbon), C-4 or C-6: their reactions with cyanide anion,[46] and ammonia/amines are examples of the latter, whereas the addition of Grignard nucleophiles occurs at carbonyl carbon.

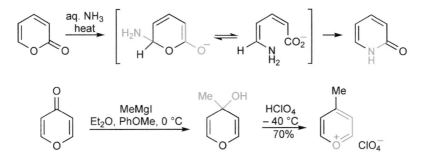

4-Pyrones also add Grignard nucleophiles at the carbonyl carbon, C-4; dehydration of the immediate tertiary alcohol product with mineral acid provides an important route to 4-mono-substituted pyrylium salts.[47] More vigorous conditions lead to the reaction of both 2- and 4-pyrones with two mole equivalents of organometallic reagent and the formation of 2,2-disubstituted-2*H*- and 4,4-disubstituted-4*H*-pyrans, respectively.[48] Perhaps surprisingly, hydride (lithium aluminium hydride) addition to 4,6-dimethyl-2-pyrone takes place, in contrast, at C-6.[49]

Ammonia and primary aliphatic and aromatic amines convert 4-pyrones into 4-pyridones:[50] this must involve attack at an α-position, then ring opening and reclosure; in some cases ring-opened products of reaction with two mole equivalents of the amine have been isolated, though such structures are not necessarily intermediates on the route to pyridones.[51] The transformation can also be achieved by, first, hydrolytic ring opening using barium hydroxide (see above), and then reaction of the barium salt with ammonium chloride.[52]

The reactions of 4-pyrones with hydrazines and hydroxylamine, can lead to recyclisations involving the second heteroatom of the attacking nucleophile, producing pyrazoles and isoxazoles, respectively, however in the simplest examples 4-pyrones react with hydroxylamine, giving either 1-hydroxy-4-pyridones or 4-hydroxy-amino-pyridine *N*-oxides;[53] again, prior hydrolytic ring opening using barium hydroxide has been employed.[43]

Nucleophilic displacement of leaving groups can also be carried out in suitable cases, for example, of the 4-methylthio in 3-cyano-2-pyrones.[54]

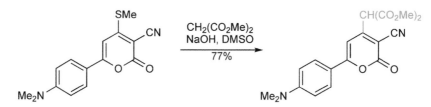

11.2.2.3 Organometallic Derivatives

3-Bromo-2-pyrone does not undergo exchange (or C-H-deprotonation) with *n*-butyllithium, however it has been transformed into a cuprate, albeit of singularly less nucleophilic character than typical cuprates.[55] 6-Substituted 5-iodo-2-pyrones, obtained by iodolactonisation, react with activated zinc, giving species that can be protonolysed or used to make 5,6-disubstituted 2-pyrones *via* Pd(0)-catalysed coupling.[56]

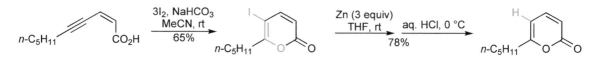

11.2.2.4 Cycloaddition Reactions[57]

2-Pyrone reacts readily as a diene in Diels–Alder additions, but the initial adduct often loses carbon dioxide, generating a second diene that then adds a second mole of the dienophile: reaction with maleic anhydride, shown below, is typical – a monoadduct can be isolated, which under more vigorous conditions loses carbon dioxide and undergoes a second addition.[58] When the dienophile is an alkyne, methyl propiolate for example, benzenoid products result from the expulsion of carbon dioxide.[59] Primary adducts, which have not lost carbon dioxide, can be obtained from reactions conducted at lower temperatures under very high pressure or in the presence of lanthanide catalysts.[60] A useful example is the reaction of 2-pyrone and substituted derivatives with alkynyl boronates leading to aryl boronates; 2-pyrone itself reacts in 86% yield with trimethylsilylethynyl boronate.[61]

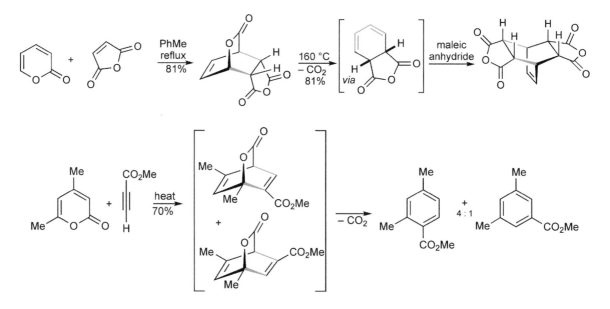

3-[62] and 5-Bromo[63] -2-pyrones present remarkable properties in their abilities to act as efficient dienes towards *both* electron-rich *and* electron-poor dienophiles (illustrated below); 3-(*para*-tolylthio)-2-pyrone also undergoes ready cycloadditions with electron-deficient alkenes.[64]

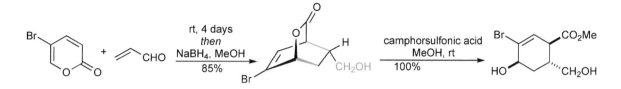

Under appropriate conditions, even unactivated alkenes will take part in intermolecular cycloadditions with 3- and 5-bromo-2-pyrones and with 3-methoxycarbonyl-2-pyrone.[65] Reactions can be conducted at 100 °C, or at room temperature under 10–12 kbar and with zinc chloride catalysis.

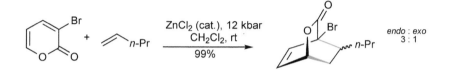

3,5-Dibromo-2-pyrone is a more reactive diene in both normal and inverse electron demand Diels–Alder cycloadditions: an example is shown below.[66]

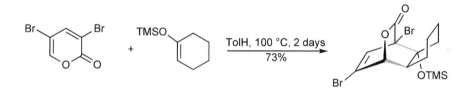

5-Alkenyl-2-pyrones, react as dienes, but in the alternative way indicated below.[67]

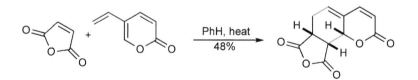

11.2.2.5 Photochemical Reactions

In addition to the photocatalysed rearrangement of 4-pyrones in acid solution (11.1.6) the other clear-cut photochemical reactions undergone are the transformation of 2-pyrone into a bicyclic β-lactone on irradiation in a non-hydroxylic solvent, and into an acyclic unsaturated ester-aldehyde on irradiation in the presence of methanol.[68]

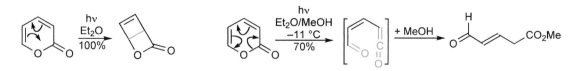

11.2.2.6 Side-Chain Reactions

4-Pyrones[69] and 2-pyrones[70] condense with aromatic aldehydes at 2- and 6-methyl groups and 2,6-dimethyl-4-pyrone can be lithiated at a methyl.[71]

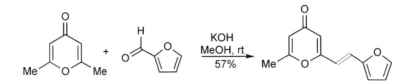

11.2.2.7 2,4-Dioxygenated Pyrones

2,4-Dioxygenated pyrones exist as the 4-hydroxy tautomers. Such molecules are easily substituted by electrophiles, at the position between the two oxygens (C-3)[72] and can be side-chain deprotonated using two mole equivalents of strong base.[73]

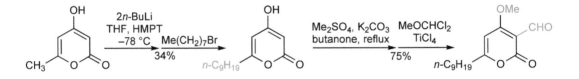

11.3 Synthesis of Pyryliums[1,7a]

Pyrylium rings are assembled by the cyclisation of a 1,5-dicarbonyl precursor, separately synthesised or generated *in situ*.

11.3.1 From 1,5-Dicarbonyl Compounds

1,5-Dicarbonyl compounds can be cyclised, with dehydration and in the presence of an oxidising agent.

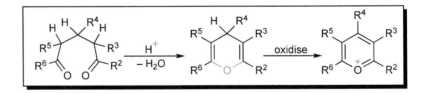

Mono-enolisation of a 1,5-diketone, then the formation of a cyclic hemiacetal, and its dehydration, produces 4H-pyrans, which require only hydride abstraction to arrive at the pyrylium oxidation level. The diketones are often prepared *in situ* by the reaction of an aldehyde with two moles of a ketone (compare Hantzsch synthesis, 8.14.1.2) or of a ketone with a previously prepared conjugated ketone – a 'chalcone' in the case of aromatic ketones/aldehydes. It is the excess chalcone that serves as the hydride acceptor in this approach.

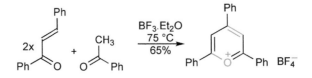

Early work utilised acetic anhydride as solvent with the incorporation of an oxidising agent, often iron(III) chloride (though it is believed that the acylium cation is the hydride acceptor); latterly the incorporation of 2,3-dichloro-5,6-dicyano-1,4-benzoquinone,[74] 2,6-dimethylpyrylium or, most often, the triphenylmethyl cation[75] have proved efficient. In some cases the 4*H*-pyran is isolated then oxidised in a separate step.[76]

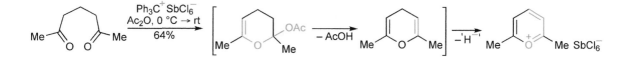

If an unsaturated dicarbonyl precursor is available, no oxidant needs to be added: a synthesis of the perchlorate of pyrylium itself, shown below, falls into this category: careful perchloric acid treatment of either glutaconaldehyde, or of its sodium salt, produces the parent salt.[12,77] (**CAUTION**: potentially explosive).

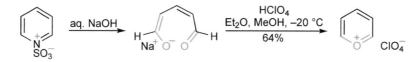

11.3.2 Alkene Acylation
Alkenes can be diacylated with an acid chloride or anhydride, generating an unsaturated 1,5-dicarbonyl compound, which then cyclises with loss of water.

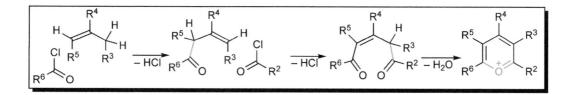

The aliphatic version of the classical aromatic Friedel–Crafts acylation process, produces, by loss of proton, a non-conjugated enone, which can then undergo a second acylation, thus generating an unsaturated 1,5-diketone. Clearly, if the alkene is not symmetrical, two isomeric diketones are formed.[78] Under the conditions of these acylations, the unsaturated diketone cyclises, loses water and forms a pyrylium salt. The formation of 2,6-di-*t*-butyl-4-methylpyrylium[79] illustrates the process – here a precursor alcohol generates the alkene *in situ*; halides that dehydrohalogenate can also be used.[80] A comparable sequence using acetic anhydride gives 2,4,6-trimethylpyrylium, best isolated as its much more stable and non-hygroscopic carboxymethanesulfonate.[81]

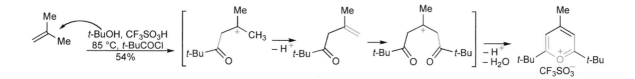

11.3.3 From 1,3-Dicarbonyl Compounds and Ketones

The acid-catalysed condensation of a ketone with a 1,3-dicarbonyl compound, with dehydration *in situ* produces pyrylium salts.

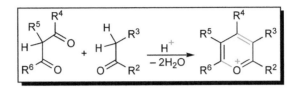

Aldol condensation between a 1,3-dicarbonyl component and a ketone with an α-methylene, under acidic, dehydrating conditions, produces pyrylium salts.[82] It is likely that the initial condensation is followed by a dehydration before the cyclic hemiacetal formation and loss of a second water molecule. The use of the bis-acetal of malondialdehyde, as a synthon for malondialdehyde, is one of the few ways available for preparing α-unsubstituted pyryliums.[1]

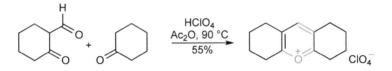

Variations on this theme include the use, as synthons for the 1,3-dicarbonyl component, of β-chloro-α,β-unsaturated ketones,[83] or of conjugated alkynyl aldehydes.[84]

11.4 Synthesis of 2-Pyrones

11.4.1 From 1,3-Keto(aldehydo)-Acids and Carbonyl Compounds

The classical general method for constructing 2-pyrones is that based on the cyclising condensation of a 1,3-keto(aldehydo)-acid with a second component that provides the other two ring carbons.

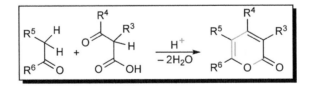

The long known synthesis of coumalic acid from treatment of malic acid with hot sulfuric acid illustrates this route: decarbonylation produces formylacetic acid, *in situ*, which serves as both the 1,3-aldehydo-acid component and the second component.[85] Decarboxylation of coumalic acid gives access to 2-pyrone itself.[86]

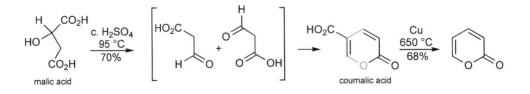

malic acid   coumalic acid

Conjugate additions of enolates to alkynyl-ketones[87] or to alkynyl-esters[88] are further variations on the synthetic theme.

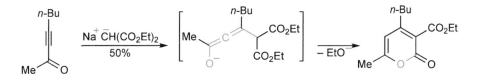

11.4.2   Other Methods

2-Pyrone itself can be prepared *via* Prins alkylation of but-3-enoic acid with subsequent lactonisation, giving 5,6-dihydro-2-pyrone, which, *via* allylic bromination and then dehydrobromination, is converted into 2-pyrone.[89] Alternative manipulation[90] of the dihydropyrone affords a convenient synthesis of a separable mixture of the important 3- and 5-bromo-2-pyrones (see 11.2.2.4).

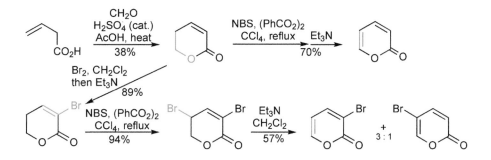

Phosphine-catalysed addition of ethyl allene carboxylate to aldehydes also involves the construction of the 5,6-bond.[91]

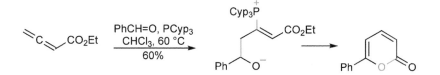

The esterification of a 1,3-ketoaldehyde enol with a diethoxyphosphinyl-alkanoic acid, forming the ester linkage of the final molecule first, allows ring closure involving C-3–C-4 bond formation *via* an intramolecular Horner–Emmons reaction.[92]

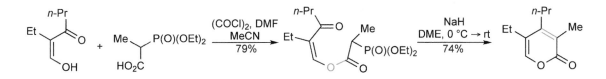

The palladium-catalysed coupling of alkynes with a 3-iodo-α,β-unsaturated ester, or with the enol triflate of a β-keto-ester as illustrated below, must surely be one of the shortest and most direct routes to 2-pyrones.[93]

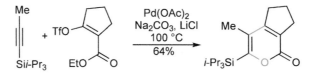

6-Chloro-2-pyrone is easily available by reaction of *trans*-glutaconic acid with phosphorus pentachloride.[94]

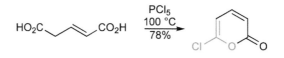

Acylation of the anion derived by deprotonating the methyl in a 2,2,6-trimethyl-1,3-dioxin-4-one, then thermolysis, provides a neat route to 2,4-dioxygenated pyrones.[95]

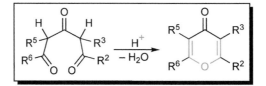

11.5 Synthesis of 4-Pyrones

4-Pyrones result from the acid-catalysed closure of 1,3,5-tricarbonyl precursors.

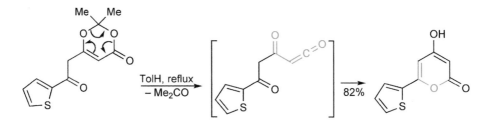

The construction of a 4-pyrone is essentially the construction of a 1,3,5-tricarbonyl compound, since such compounds easily form cyclic hemiacetals then requiring only dehydration. Strong acid has usually been used for this purpose, but where stereochemically sensitive centres are close, the reagent from triphenylphosphine and carbon tetrachloride can be employed.[96]

Several methods are available for the assembly of tricarbonyl precursors: the synthesis of chelidonic acid (4-pyrone-2,6-dicarboxylic acid)[97] represents the obvious approach of bringing about two Claisen condensations, one on each side of a ketone carbonyl group. Chelidonic acid can be decarboxylated to produce 4-pyrone itself.[98]

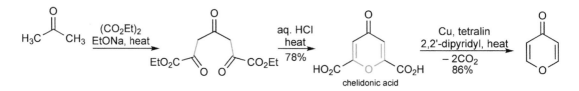

chelidonic acid

A variety of symmetrically substituted 4-pyrones can be made very simply by heating an alkanoic acid with polyphosphoric acid;[99] presumably a series of Claisen-type condensations, with a decarboxylation, lead to the assembly of the requisite acyclic, tricarbonyl precursor.

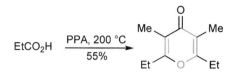

The Claisen condensation of a 1,3-diketone, *via* its dianion, with an ester,[100] or of a ketone enolate with an alkyne ester,[101] also give the desired tricarbonyl arrays. Applying this principle to 1,3-keto-esters leads through to 2,4-dioxygenated heterocycles.[102]

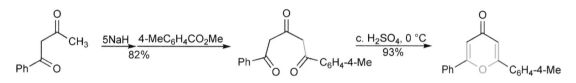

Another strategy to bring about acylation at the less acidic carbon of a β-keto ester, is to condense, firstly at the central methylene, with DMFDMA; this has the added advantage that the added carbon can then provide the fifth carbon of the target heterocycle.[103]

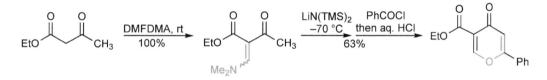

α-Unsubstituted 4-pyrones have similarly been constructed *via* acylation of 2-methoxyvinyl ketones.[104]

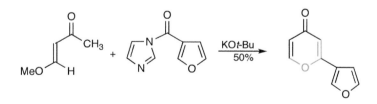

Dehydroacetic acid[105] was first synthesised in 1866;[106] it is formed very simply from ethyl acetoacetate by a Claisen condensation between two molecules, followed by the usual cyclisation and finally loss of ethanol. In a modern version, β-keto-acids can be self-condensed using carbonyl diimidazole as the condensing agent.[107]

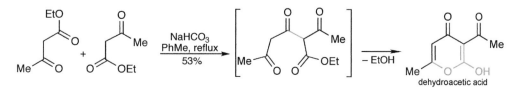
dehydroacetic acid

The acylation of the enamine of a cyclic ketone with diketene leads directly to bicyclic 4-pyrones, as indicated below.[108]

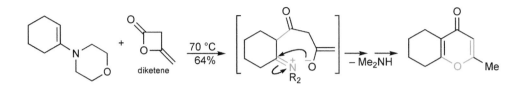

Exercises

Straightforward revision exercises (consult Chapters 10 and 11):

(a) Specify three nucleophiles that add easily to pyrylium salts, and draw the structures of the products produced thereby.

(b) Certain derivatives of six-membered oxygen heterocycles undergo 4+2 cycloaddition reactions: draw out three examples.

(c) Draw a mechanism for the transformation of 2-pyrone into 1-methyl-2-pyridone on reaction with methylamine.

(d) What steps must take place to achieve the conversion of a saturated 1,5-diketone into a pyrylium salt?

(e) Describe how 5,6-dihydro-2-pyrone can be utilised to prepare either 2-pyrone, or 3- and 5-bromo-2-pyrones.

(f) 1,3,5-Tricarbonyl compounds are easily converted into 4-pyrones. Describe two ways to produce a 1,3,5-trione or a synthon thereof.

More advanced exercises:

1. Write a sequence for the transformation of 2,4,6-trimethylpyrylium into 1-phenyl-2,4,6-trimethylpyridinium by reaction with aniline.

2. Devise a mechanism to explain the formation of 1,3,5-triphenylbenzene from reaction of 2,4,6-triphenylpyrylium perchlorate on reaction with 2 mole equivalents of $Ph_3P=CH_2$.

3. Suggest structures for the compounds in the following sequence: 2-methyl-5-hydroxy-4-pyrone reacted with MeOTf → $C_7H_9O_3^+$TfO$^-$ (a salt), then this with 2,2,6,6-tetramethylpiperidine (a hindered base) → $C_7H_8O_3$, a dipolar substance, and this then with acrylonitrile → $C_{10}H_{11}NO_3$.

4. Write out a mechanism for the conversion of 4-pyrone into 1-phenyl-4-pyridone by reaction with aniline. Write structures for the products you would expect from reaction of methyl coumalate (5-methoxycarbonyl-2-pyrone) with benzylamine.

5. Deduce structures for the pyrylium salts formed by the following sequences: (i) pinacolone (Me_3CCOMe) condensed with pivaldehyde ($Me_3CCH=O$) gave $C_{11}H_{20}O$, which was then reacted with pinacolone in the presence of NaNH$_2$, generating $C_{17}H_{32}O_2$ and this with $Ph_3C^+ClO_4^-$ in AcOH gave a pyrylium salt; (ii) cyclodecene and Ac$_2$O/HClO$_4$; (iii) PhCOMe and MeCOCH$_2$CHO with Ac$_2$O and HClO$_4$.

6. When dehydroacetic acid is heated with c. HCl 2,6-dimethyl-4-pyrone is formed in 97% yield – explain.

7. When ethyl acetoacetate is reacted with HCl, isodehydroacetic acid (ethyl 4,6-dimethyl-2-pyrone-5-carboxylate) is formed – explain.

8. Deduce structures for the pyrones formed by the following sequences: (i) PhCOCH$_3$ with PhC≡CCO$_2$Et in the presence of NaOEt; (ii) butanoic acid heated with PPA at 200 °C; (iii) n-BuCOCH$_2$CO$_2$H with carbonyl diimidazole; (iv) PhCOCH$_2$COCH$_3$ with excess NaH then methyl 4-chlorobenzoate; (v) CH$_3$COCH=CHOMe with KOt-Bu and PhCOCl.

References

[1] Bangert, K., Boekelheide, V., Hafner, K. and Kaiser, H., *Org. Synth., Coll. Vol. 5*, 1106.

[2] Balaban, A. T. and Boulton, A. J., *Org. Synth., Coll. Vol. 5*, 1112.

[3] Balaban, A. T. and Boulton, A. J., *Org. Synth., Coll. Vol. 5*, 1114.

[4] A great deal of the pioneering work on pyryliums, by Roumanian and Russian workers, is described in the Russian literature. This is well reviewed in 'Pyrylium salts. Part I. Syntheses', Balaban, A. T., Schroth, W. and Fischer, G., *Adv. Heterocycl. Chem.*, **1969**, *10*, 241; 'Pyrylium salts. Synthesis, reactions and physical properties', Balaban, A. T., Dinculescu, A., Dorofeenko, G. N., Fischer, G. W., Koblik, A. V., Mezheritskii, V. V. and Schroth, W., *Adv. Heterocycl. Chem., Suppl. 2*, **1982**.

[5] 'Cycloadditions and reactions of oxa-aromatics with nucleophiles', Ohkata, K. and Akiba, K.-Y., *Adv. Heterocycl. Chem.*, **1996**, *65*, 283.

[6] (a) Gârd, E., Vasilescu, A., Mateescu, G. D. and Balaban, A. T., *J. Labelled Cmpds.*, **1967**, *3*, 193; (b) Farcasiu, D., Vasilescu, A. and Balaban, A. T., *Tetrahedron*, **1971**, *27*, 681.

[7] Le Fèvre, C. G. and Le Fèvre, R. J. W., *J. Chem. Soc.*, **1932**, 2894.

[8] Williams, A., *J. Am. Chem. Soc.*, **1971**, *93*, 2733.

[9] Katritzky, A. R., Brownlee, R. T. C. and Musumarra, G., *Heterocycles*, **1979**, *12*, 775.

[10] 'Aromatic compounds from pyrylium salts', Dimroth, K. and Wolf, K. H., Newer Methods of Preparative Organic Chemistry, Vol. *3*, Academic Press, New York, **1964**, 357; Rajoharison, H. G., Soltani, H., Arnaud, M., Roussel, C. and Metzger, J., *Synth. Commun.*, **1980**, *10*, 195.

[11] 'Conversion of primary amino groups into other functionality mediated by pyrylium cations', Katritzky, A. R., *Tetrahedron*, **1980**, *36*, 679; Toma, C. and Balaban, A. T., *Tetrahedron Suppl.*, **1966**, *7*, 9.

[12] Toma, C. and Balaban, A. T., *Tetrahedron Suppl.*, **1966**, *7*, 1; Katritzky, A. R., Brownlee, R. T. C. and Musumarra, G., *Tetrahedron*, **1980**, *36*, 1643.

[13] Katritzky, A. R. and Suwinski, J. W., *Tetrahedron*, **1975**, *31*, 1549.

[14] Pedersen, C. L., Harrit, N. and Buchardt, O., *Acta Chem. Scand.*, **1970**, *24*, 3435; Balaban, A. T., *Tetrahedron*, **1968**, *24*, 5059; Schmitz, E., *Chem. Ber.*, **1958**, *91*, 1488.

[15] Dimroth, K. and Wolf, K. H., *Angew. Chem.*, **1960**, *72*, 777.

[16] Furber, M., Herbert, J. M. and Taylor, R. J. K., *J. Chem. Soc., Perkin Trans. 1*, **1989**, 683.

[17] Royer, J., Saffieddine, A. and Dreux, J., *Bull Chem. Soc. Fr.*, **1972**, 1646; Marvell, E. N., Chadwick, T., Caple, G., Gosink, G. and Zimmer, G., *J. Org. Chem.*, **1972**, *37*, 2992.

[18] Balaban, A. T. and Nenitzescu, C. D., *J. Chem. Soc.*, **1961**, 3566.

[19] Hemming, K. and R. J. K. Taylor, *J. Chem. Soc., Chem. Commun.*, **1993**, 1409.

[20] Reynolds, G. A., Van Allen, J. A. and Petropoulos, C. C., *J. Heterocycl. Chem.*, **1970**, *7*, 1061.

[21] Razus, A. C., Birzan, L., Pavel, C., Lehadus, O., Corbu, A. C. and Enache, C., *J. Heterocycl. Chem.*, **2006**, *43*, 963.

[22] Baciocchi, E., Doddi, G., Ioele, M. and Ercolani, G., *Tetrahedron*, **1993**, *49*, 3793.

[23] Safieddine, A., Royer, J. and Dreux, J., *Bull. Soc. Chim. Fr.*, **1972**, 2510; Marvel, E. N. and Gosink, T., *J. Org. Chem.*, **1972**, *37*, 3036.

[24] Farcasiu, D., Balaban, A. T. and Bologa, U. L., *Heterocycles*, **1994**, *37*, 1165.

[25] Barltrop, J. A., Barrett, J. C., Carder, R. W., Day, A. C., Harding, J. R., Long, W. E. and Samuel, C. J., *J. Am. Chem. Soc.*, **1979**, *101*, 7510.

[26] Pavlik, J. W., Kirincich, S. J. and Pires, R. M., *J. Heterocycl. Chem.*, **1991**, *28*, 537; Pavlik, J. W., Keil, E. B. and Sullivan, E. L., *J. Heterocycl. Chem.*, **1992**, *29*, 1829.

[27] Hendrickson, J. B. and Farina, J. S., *J. Org. Chem.*, **1980**, *45*, 3359; 'Recent studies on 3-oxidopyrylium and its derivatives', Sammes, P. G., *Gazz. Chim. Ital.*, **1986**, *116*, 109.

[28] Sammes, P. G. and Street, L. J., *J. Chem. Soc., Perkin Trans. 1*, **1983**, 1261.

[29] Fishwick, C. W. G., Mitchell, G. and Pang, P. F. W., *Synlett*, **2005**, 285.

[30] Wender, P. A. and Mascareñas, J. L., *J. Org. Chem.*, **1991**, *56*, 6267; idem, ibid., **1992**, *57*, 2115.

[31] Rumbo, A., Castedo, L., Mouriño, A. and Mascareñas, J. L., *J. Org. Chem.*, **1993**, *58*, 5585.

[32] 'Reactions of α- and γ-alkyl groups in pyrylium salts and some transformations reaction products', Mezheritskii, V. V., Wasserman, A. L. and Dorofeenko, G. N., *Heterocycles*, **1979**, *12*, 51

[33] Simalty, M., Strzelecka, H. and Khedija, H., *Tetrahedron*, **1971**, *27*, 3503.

[34] Dilthey, W. and Fischer, J., *Chem. Ber.*, **1924**, *57*, 1653; Kelemen, J. and Wizinger, A., *Helv. Chim. Acta*, **1962**, *45*, 1918.

[35] Kirner, H.-D. and Wizinger, R., *Helv. Chim. Acta*, **1961**, *44*, 1766.

[36] Van Allan, J. A., Reynolds, G. A., Maier, D. P. and Chang, S. C., *J. Heterocycl. Chem.*, **1972**, *9*, 1229.

[37] Cioffi, E. A. and Bailey, W. F., *Tetrahedron Lett.*, **1998**, *39*, 2679.

[38] Baeyer, A., *Chem. Ber.*, **1910**, *43*, 2337.

[39] Beak, P. and Carls, G. A., *J. Org. Chem.*, **1964**, *29*, 2678.

[40] Pirkle, W. H. and Dines, M., *J. Org. Chem.*, **1969**, *34*, 2239.

[41] Cho, C.-G., Park, J.-S., Jung, I.-H. and Lee, H., *Tetrahedron Lett.*, **2001**, *42*, 1065.

[42] Cho, C.-G., Kim, Y.-W., Lim, Y.-K., Park, J.-S., Lee, H. and Koo, S., *J. Org. Chem.*, **2002**, *67*, 290.

[43] Pirkle, W. H. and Dines, M., *J. Heterocycl. Chem.*, **1969**, *6*, 313.

[44] Eiden, F. and Herdeis, C., *Arch. Pharm. (Weinheim)*, **1976**, *309*, 764.

[45] Collie, J. N. and Wilsmore, N. T. M., *J. Chem. Soc.*, **1896**, 293.

[46] Vogel, G., *J. Org. Chem.*, **1965**, *30*, 203.

[47] Baeyer, A. and Piccard, J., *Justus Liebigs Ann. Chem.*, **1911**, *384*, 208; Köbrich, G., ibid., **1961**, *648*, 114.

[48] Gompper, R. and Christmann, O., *Chem. Ber.*, **1961**, *94*, 1784.

[49] Vogel, G., *Chem. Ind.*, **1962**, 268.

[50] Adams, R. and Johnson, J. L., *J. Am. Chem. Soc.*, **1949**, *71*, 705; Campbell, K. N., Ackerman, J. F. and Campbell, B. K., *J. Org. Chem.*, **1950**, *15*, 221; Hünig, S. and Köbrich, G., *Justus Liebigs Ann. Chem.*, **1958**, *617*, 181.

[51] Borsche, W. and Bonaacker, I., *Chem. Ber.*, **1921**, *54*, 2678; Van Allen, J. A., Reynolds, G. A., Alassi, J. T., Chang, S. C. and Joines, R. C., *J. Heterocycl. Chem.*, **1971**, *8*, 919.

[52] Watkins, W. J., Robinson, G. E., Hogan, P. J. and Smith, D., *Synth. Commun.*, **1994**, *24*, 1709.

[53] Parisi, F., Bovina, P. and Quilico, A., *Gazz. Chim. Ital.*, **1960**, *90*, 903; Yates, P., Jorgenson, M. J. and Roy, S. K., *Canad. J. Chem.*, **1962**, *40*, 2146.

[54] Mizuyama, N., Murakami, Y., Nagoaka, J., Kohra, S., Ueda, K., Hiraoka, K., Shigemitsu, Y. and Tominaga, Y., *Heterocycles*, **2006**, *68*, 1105.

[55] Posner, G. H., Harrison, W. and Wettlaufer, D. G., *J. Org. Chem.*, **1985**, *50*, 5041.

[56] Bellina, F., Biagetti, M., Carpita, A. and Rossi, R., *Tetrahedron lett.*, **2001**, *42*, 2859.

[57] 'Diels-Alder cycloadditions of 2-pyrones and 2-pyridones', Afarinkia, K., Vinader, V., Nelson, T. D. and Posner, G. H., *Tetrahedron*, **1992**, *48*, 9111.

[58] Diels, O. and Alder, K., *Justus Liebigs Ann. Chem.*, **1931**, *490*, 257; Goldstein, M. J. and Thayer, G. L., *J. Am. Chem. Soc.*, **1965**, *87*, 1925; Shimo, T., Kataoka, K., Maeda, A. and Somekawa, K., *J. Heterocycl. Chem.*, **1992**, *29*, 811.

[59] Salomon, R. G., and Burns, J. R. and Dominic, W. J., *J. Org. Chem.*, **1976**, *41*, 2918.

[60] Markó, I. E., Seres, P., Swarbrick, T. M., Staton, I. and Adams, H., *Tetrahedron Lett.*, **1992**, *33*, 5649; Markó, I. E., Evans, G. R., Seres, P., Chellé, I. and Janousek, Z., *Pure Appl. Chem*, **1996**, *68*, 113.

[61] Delaney, P. M., Moore, J. E. and Harrity, J. P. A., *Chem. Commun.*, **2006**, 3323.

[62] Posner, G. H., Nelson, T. D., Kinter, C. M. and Afarinkia, K., *Tetrahedron Lett.*, **1991**, *32*, 5295.

[63] Afarinkia, K. and Posner, G. H., *Tetrahedron Lett.*, **1992**, *51*, 7839.

[64] Posner, G. H., Nelson, T. D., Kinter, C. M. and Johnson, N., *J. Org. Chem.*, **1992**, *57*, 4083.

[65] Afarinka, K., Daly, N. T., Gomez-Farnos, S. and Joshi, S., *Tetrahedron Lett.*, **1997**, *38*, 2369; Posner, G., Hutchings, R. H. and Woodard, B. T., *Synlett*, **1997**, 432.

[66] Cho, C.-G., Kim, Y.-W. and Kim, W.-K., *Tetrahedron Lett.*, **2001**, *42*, 8193.

[67] Liu, Z. and Meinwald, J., *J. Org. Chem.*, **1996**, *61*, 6693.

[68] Corey, E. J. and Streith, J., *J. Am. Chem. Soc.*, **1964**, *86*, 950; Pirkle, W. H. and McKendry, L. H., *J. Am. Chem. Soc.*, **1969**, *91*, 1179; Chapman, O. L., McKintosh, C. L. and Pacansky, J., *J. Am. Chem. Soc.*, **1973**, *95*, 614.

[69] Woods, L. L., *J. Am. Chem. Soc.*, **1958**, *80*, 1440.

[70] Adam, W., Saha-Möller, C. R., Veit, M. and Welke, B., *Synthesis*, **1994**, 1133.

[71] West, F. G., Fisher, P. V. and Willoughby, C. A., *J. Org. Chem.*, **1990**, *55*, 5936; West, F. G., Amann, C. M. and Fisher, P. V. *Tetrahedron Lett.*, **1994**, *35*, 9653.

[72] De March, P., Moreno-Mañas, M., Pleixats, R. and Roca, J. C., *J. Heterocycl. Chem.*, **1984**, *21*, 1369.

[73] Groutas, W. C., Huang, T. L., Stanga, M. A., Brubaker, M. J. and Moi, M. K., *J. Heterocycl. Chem.*, **1985**, *22*, 433; Poulton, G. A. and Cyr, T. P., *Canad. J. Chem.*, **1980**, *58*, 2158.

[74] Carretto, J. and Simalty, M., *Tetrahedron Lett.*, **1973**, 3445.

[75] Rundel, W., *Chem. Ber.*, **1969**, *102*, 374; Farcasiu, D., Vasilescu, A. and Balaban, A. T., *Tetrahedron*, **1971**, *27*, 681; Farcasiu, D., *Tetrahedron*, **1969**, *25*, 1209.

[76] Undheim, K. and Ostensen, E. T., *Acta Chem. Scand.*, **1973**, *27*, 1385.

[77] Klager, F. and Träger, H., *Chem. Ber.*, **1953**, *86*, 1327.

[78] Balaban, A. T. and Nenitzescu, C. D., *Justus Liebigs Ann. Chem.*, **1959**, *625*, 74; *idem*, *J. Chem. Soc.*, **1961**, 3553; Praill, P. F. G. and Whitear, A. L., *ibid.*, 3573.

[79] Anderson, A. G. and Stang, P. J., *J. Org. Chem.*, **1976**, *41*, 3034.

[80] Balaban, A. T., *Org. Prep. Proced. Int.*, **1977**, *9*, 125.

[81] Dinculescu, A. and Balaban, A. T., *Org. Prep. Proedc. Int.*, **1982**, *14*, 39.

[82] Schroth, W. and Fischer, G. W., *Chem. Ber.*, **1969**, *102*, 1214; Dorofeenko, G. N., Shdanow, Ju. A., Shungijetu, G. I. and Kriwon, W. S. W., *Tetrahedron*, **1966**, *22*, 1821.

[83] Schroth, W., Fischer, G. W. and Rottmann, J., *Chem. Ber.*, **1969**, *102*, 1202.

[84] Stetter, H. and Reischl, A., *Chem. Ber.*, **1960**, *93*, 1253.

[85] Wiley, R. H. and Smith, N. R., *Org. Synth., Coll. Vol. IV*, **1963**, 201.

[86] Zimmerman, H. E., Grunewald, G. L. and Paufler, R. M., *Org. Synth., Coll. Vol. V*, **1973**, 982.

[87] Anker, R. M. and Cook, A. H., *J. Chem. Soc.*, **1945**, 311.

[88] El-Kholy, I., Rafla, F. K. and Soliman, G., *J. Chem. Soc.*, **1959**, 2588.

[89] Nakagawa, M., Saegusa, J., Tonozuka, M., Obi, M., Kiuchi, M., Hino, T. and Ban, Y., *Org. Synth., Coll. Vol. VI*, **1988**, 462.

[90] Posner, G., Afarinkia, K. and Dai, H. *Org. Synth.*, **1994**, *73*, 231.

[91] Zhu, X.-F., Schaffner, A. P., Li, R. C. and Kwon, O., *Org. Lett.*, **2005**, *7*, 2977.

[92] Stetter, H. and Kogelnik, H.-J., *Synthesis*, **1986**, 140.

[93] Larock, R. C., Han, X. and Doty, M. J., *Tetrahedron Lett.*, **1998**, *39*, 5713.

[94] Pirkle, W. H. and Dines, M., *J. Am. Chem. Soc.*, **1968**, *90*, 2318; Bellina, F., Biagetti, M., Carpita, A. and Rossi, R., *Tetrahedron Lett.*, **2003**, *44*, 607.

[95] Katritzky, A. R., Wang, Z., Wang, M., Hall, C. D. and Suzuki, K., *J. Org. Chem.*, **2005**, *70*, 4854.

[96] Arimoto, H., Nishiyama, S. and Yamamura, S., *Tetrahedron Lett.*, **1990**, *31*, 5491.

[97] Riegel, E. R. and Zwilgmeyer, F. Z., *Org. Synth., Coll. Vol. II*, **1943**, 126.

[98] De Souza, C., Hajikarimian, Y. and Sheldrake, P. W., *Synth. Commun.*, **1992**, *22*, 755.

[99] Mullock, E. B. and Suschitzky, H., *J. Chem. Soc., C*, **1967**, 828.

[100] Miles, M. L., Harris, T. M. and Hauser, C. R., *Org. Synth., Coll. Vol. V*, **1973**, 718; Miles, M. L. and Hauser, C. R., *ibid.*, 721.

[101] Soliman, G. and El-Kholy, I. E.-S., *J. Chem. Soc.*, **1954**, 1755.

[102] Schmidt, D., Conrad, J., Klaiber, I. and Beifuss, U., *Chem. Commun.*, **2006**, 4732.

[103] McCombie, S. W., Metz, W. A., Nazareno, D., Shankar, B. B. and Tagat, J. *J. Org, Chem.*, **1991**, *56*, 4963.

[104] Morgan, T. A. and Ganem, B., *Tetrahedron Lett.*, **1980**, *21*, 2773; Koreeda, M. and Akagi, H., *ibid.*, 1197.

[105] Arndt, F., *Org. Synth., Coll. Vol. III*, **1955**, 231; 'Dehydroacetic acid, triacetic acid lactone, and related pyrones', Moreno-Mañas, M. and Pleixats, R., *Adv. Heterocycl. Chem.*, **1993**, *53*, 1.

[106] Oppenheim, A. and Precht, H., *Chem. Ber.*, **1866**, *9*, 324.

[107] Ohta, S., Tsujimura, A. and Okamoto, M., *Chem. Pharm. Bull.*, **1981**, *29*, 2762.

[108] Hünig, S., Benzing, E. and Hübner, K., *Chem. Ber.*, **1961**, *94*, 486.

12

Benzopyryliums and Benzopyrones:
Reactions and Synthesis

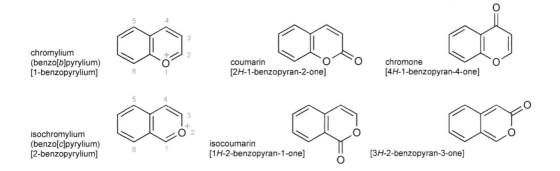

chromylium
(benzo[b]pyrylium)
[1-benzopyrylium]

coumarin
[2H-1-benzopyran-2-one]

chromone
[4H-1-benzopyran-4-one]

isochromylium
(benzo[c]pyrylium)
[2-benzopyrylium]

isocoumarin
[1H-2-benzopyran-1-one]

[3H-2-benzopyran-3-one]

1-Benzopyryliums, coumarins and chromones are very widely distributed throughout the plant kingdom, where many secondary metabolites contain them. Not the least of these are the 'flavonoids', which make up the majority of flower pigments (see 32.5.6). In addition, many flavone (2-arylchromone) and coumarin derivatives have marked toxic and other physiological properties in animals, though they play no part in the normal metabolism. The isomeric 2-benzopyrylium[1] system does not occur naturally and only a few isocoumarins[2] occur as natural products, and as a consequence much less work on these has been described.

Processes initiated by nucleophilic additions to the positively charged heterocyclic ring are the main, almost the only, types of reaction known for benzopyryliums. The absence of examples of electrophilic substitution in the benzene ring is to be contrasted with substitution in quinolinium and isoquinolinium salts, emphasising the greater electron-withdrawing, and thus deactivating, effect of positively charged oxygen.

Coumarins, chromones, and isocoumarins react with both nucleophiles and electrophiles in much the same way as do quinolones and isoquinolones.

12.1   Reactions of Benzopyryliums

Much more work has been done on 1-benzopyryliums than on 2-benzopyryliums, because of their relevance to the flavylium (2-phenyl-1-benzopyrylium) nucleus, which occurs widely in the anthocyanins, and much of that work has been conducted on flavylium itself. As with pyrylium salts, benzopyrylium salts usually add nucleophiles at the carbon adjacent to the oxygen.

12.1.1   Reactions with Electrophilic Reagents

No simple examples are known of electrophilic or radical substitution of either heterocyclic or homocyclic rings of benzopyrylium salts; flavylium[3] and 1-phenyl-2-benzopyrylium[5] salts nitrate in the substituent

Heterocyclic Chemistry 5th Edition   John Joule and Keith Mills
© 2010 Blackwell Publishing Ltd

benzene ring. Having said this, the cyclisation of coumarin-4-propanoic acid may represent Friedel–Crafts-type intramolecular attack on the carbonyl-*O*-protonated form, i.e. on a 2-hydroxy-1-benzopyrylium system, at C-3.[4]

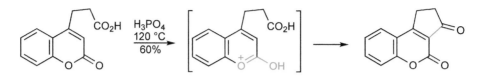

12.1.2 Reactions with Oxidising Agents

Oxidative general breakdown of flavylium salts was utilised in early structural work on the natural compounds. Baeyer–Villiger oxidation is such a process, whereby the two 'halves' of the molecule can be separately examined (after ester hydrolysis of the product).[5] Flavylium salts can be oxidised to flavones using thallium(III) nitrate,[6] and benzopyrylium itself can be converted into coumarin with manganese dioxide.[7]

12.1.3 Reactions with Nucleophilic Reagents

12.1.3.1 Water and Alcohols

Water and alcohols add readily at C-2, and sometimes at C-4, generating chromenols or chromenol ethers.[8] It is difficult to obtain 2*H*-chromenols pure, since they are always in equilibrium with ring-opened chalcones.[9]

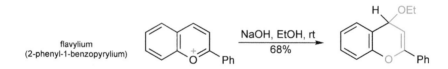

Controlled conditions are required for the production of simple adducts, for under more vigorous alkaline treatment, ring opening, then carbon–carbon bond cleavage *via* a retro-aldol mechanism takes place, and such processes, which are essentially the reverse of a route used for the synthesis of 1-benzopyryliums (12.3.1.2) were utilised in early structural work on anthocyanin flower pigments.

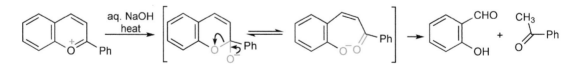

12.1.3.2 Ammonia and Amines

Ammonia and amines add to benzopyryliums, and simple adducts from secondary amines have been isolated.[10]

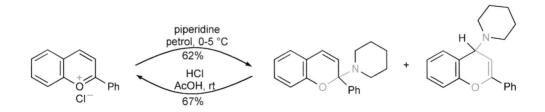

It is important to realise that 1-benzopyrylium salts cannot be converted into quinolines or quinolinium salts by reaction with ammonia or primary amines (cf. pyryliums to pyridines, 11.1.2.2), whereas 2-benzopyrylium salts are converted, efficiently, into isoquinolines or isoquinolinium salts, respectively.[11]

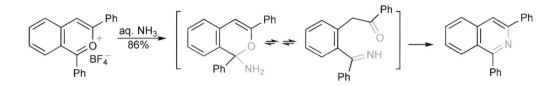

12.1.3.3 Carbon Nucleophiles

Organometallic carbon nucleophiles add to flavylium salts,[12] as do activated aromatics like phenol,[13] and enolates such as those from cyanoacetate, nitromethane[14] and dimedone,[15] all very efficiently, at C-4. Cyanide and azide add to 2-benzopyryliums at C-1.[16]

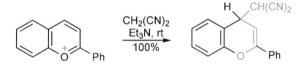

Silyl enol ethers, or allylsilanes will add at C-2 to 1-benzopyrylium salts generated by *O*-silylation of chromones; in the case of silyl ethers of α,β-unsaturated ketones, cyclisation of the initial adduct is observed (cf. 8.12.2).[17]

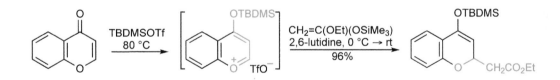

12.1.4 Reactions with Reducing Agents

Catalytic hydrogenation of flavylium salts is generally straightforward and results in the saturation of the heterocyclic ring. Lithium aluminium hydride reduces flavylium salts, generating 4*H*-chromenes,[18] unless there is a 3-methoxyl, when 2*H*-chromenes are the products.[19] 2-Benzopyryliums add hydride at C-1.[20]

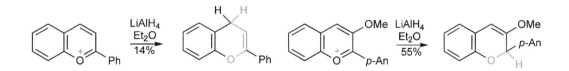

12.1.5 Alkyl-Benzopyryliums

Alkyl groups oriented α or γ to the positively charged oxygen in benzopyryliums have acidified hydrogens that allow aldol-type condensations.[1,21]

12.2 Benzopyrones (Chromones, Coumarins and Isocoumarins)
12.2.1 Reactions with Electrophilic Reagents
12.2.1.1 Addition to Carbonyl Oxygen

Addition of a proton to carbonyl oxygen produces a hydroxy-benzopyrylium salt; chromones undergo this protonation more easily than the coumarins, for example passage of hydrogen chloride through a mixture of chromone and coumarin in ether solution leads to the precipitation of only chromone hydrochloride (i.e. 4-hydroxy-1-benzopyrylium chloride).[22] *O*-Alkylation requires the more powerful alkylating agents;[1,23] *O*-silylation of benzopyrones is easy (12.1.3.3).

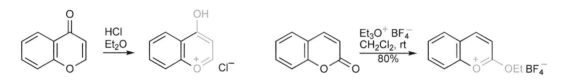

12.2.1.2 C-Substitution

C-Substitution of coumarins and chromones has been observed in both rings: in strongly acidic media, in which presumably it is a hydroxy-benzopyrylium cation that is attacked, substitution takes place at C-6, for example nitration.[24] This can be contrasted with the dimethylaminomethylation of chromone,[25] iodination of flavones[26] or the chloromethylation of coumarin[27] where hetero-ring substitution takes place, presumably *via* the non-protonated (non-complexed) heterocycle (**CAUTION**: CH₂O/HCl also produces some ClCH₂OCH₂Cl, a carcinogen).

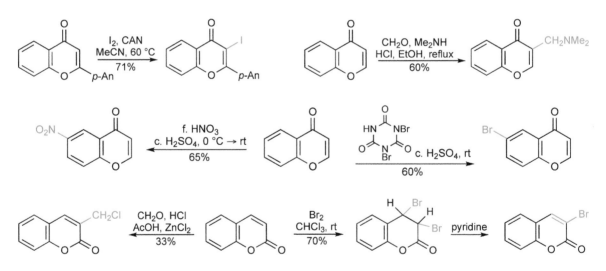

Reaction of coumarin or chromone with bromine results in simple addition to the heterocyclic ring double bond, subsequent elimination of hydrogen bromide giving 3-bromocoumarin[28] or 3-bromochromone.[29] Copper(II) halides with alumina in refluxing chlorobenzene is an alternative method for 3-halogenation of coumarins.[30] Bromine in the presence of an excess of aluminium chloride (the 'swamping catalyst' effect) converts coumarin into 6-bromocoumarin;[31] chromone can be efficiently brominated at C-6 using dibromoisocyanuric acid.[32]

12.2.2 Reactions with Oxidising Agents

Non-phenolic coumarins are relatively stable to oxidative conditions.[33] Various oxidative methods were used extensively in structure determinations of natural flavones.

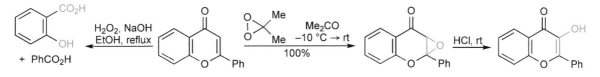

Flavones and isoflavones (3-aryl-chromones) are quantitatively converted into 2,3-epoxides by exposure to dimethyl dioxirane; flavone oxides are quantitatively converted by acid into 3-hydroxy-flavones, which are naturally occurring.[34]

12.2.3 Reactions with Nucleophilic Reagents

12.2.3.1 Hydroxide

Coumarins are quantitatively hydrolysed to give yellow solutions of the salts of the corresponding *cis*-cinnamic acids (coumarinic acids), which cannot be isolated, since acidification brings about immediate re-lactonisation; prolonged alkali treatment leads to isomerisation and the formation of the *trans*-acid (coumaric acid) salt.

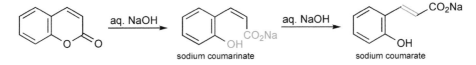

Cold sodium hydroxide comparably reversibly converts chromones into the salts of the corresponding ring-opened phenols, *via* initial attack at C-2, more vigorous alkaline treatment leading to reverse-Claisen degradation of the 1,3-dicarbonyl side-chain.

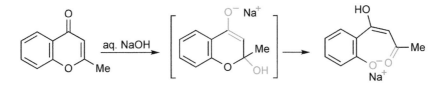

12.2.3.2 Ammonia, Amines and Hydrazines

Ammonia and amines do not convert coumarins into 2-quinolones, nor chromones into 4-quinolones, but isocoumarins do produce isoquinolones.[35] Ring-opened products from chromones and secondary amines can be obtained where the nucleophile has attacked at C-2.

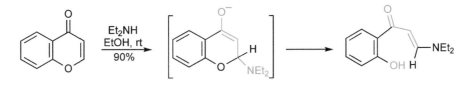

The interaction of 3-iodochromone with five-membered azoles, such as imidazole, leads to substitution at the 2-position, presumably *via* an addition/elimination sequence, as indicated.[36]

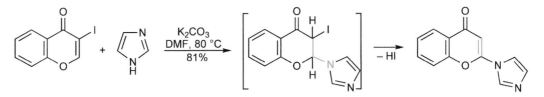

3-Formyl chromones[37,38] react with arylhydrazines to produce 4-acyl-pyrazoles.[39]

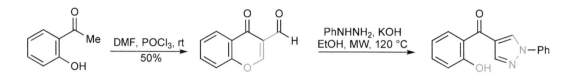

12.2.3.3 Carbon Nucleophiles

Grignard reagents react with chromones at the carbonyl carbon; the resulting chromenols can be converted by acid into the corresponding 4-substituted 1-benzopyrylium salts.[26]

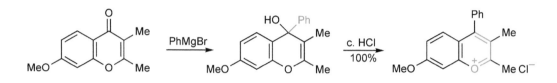

Coumarins and isocoumarins[16] react with Grignard reagents, often giving mixtures of products resulting from ring opening of the initial carbonyl adduct; the reaction of coumarin with methylmagnesium iodide illustrates this.[40]

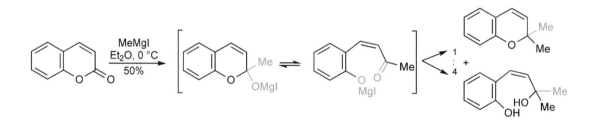

By conversion into a benzopyrylium salt with a leaving group, nucleophiles can be introduced at the chromone 4-position: treatment with acetic anhydride presumably forms a 4-acetoxy-benzopyrylium.[41]

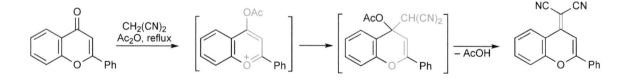

In efficient reactions, coumarin can be made to react with electron-rich aromatics using phosphoryl chloride, alone, or with zinc chloride.[42]

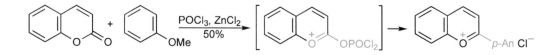

12.2.3.4 Organometallic Derivatives
Flavone can be lithiated at C-3.[43]

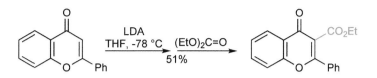

12.2.3.5 Reactions with Reducing Agents
Both coumarin and chromone are converted by diborane then alkaline hydrogen peroxide into 3-hydroxy-chroman.[44] Catalytic reduction of coumarin or chromone saturates the C–C double bond.[45] For both systems, hydride reagents can of course react either at carbonyl carbon or at the conjugate position and mixtures therefore tend to be produced. Zinc amalgam in acidic solution converts benzopyrones in 4-unsubstituted benzopyrylium salts.[46]

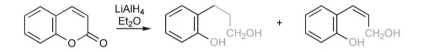

12.2.3.6 Reactions with Dienophiles; Cycloadditions
Coumarins, but not apparently chromones, serve as dienophiles in Diels–Alder reactions, though under relatively forcing conditions;[47] in water such additions take place at 150 °C and under high pressure, at 70 °C.[48]

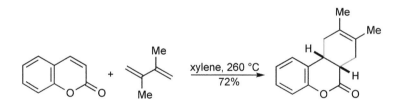

It can be taken as a measure of the low intrinsic aromaticity associated with fused pyrone rings, that 3-acyl-chromones undergo hetero-Diels–Alder additions with enol ethers,[49] and ketene acetals,[50] 3-formylchromone reacting the most readily.

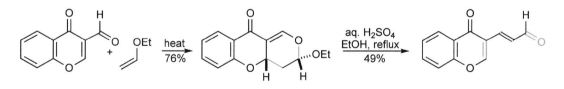

Chromone-3-esters, on the other hand, serve as dienophiles under Lewis acid catalysis.[51]

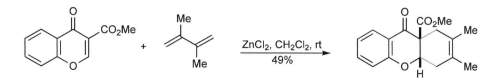

2-Benzopyran-3-ones, generated by cyclising dehydration of an *ortho*-formyl-arylacetic acids take part in intramolecular Diels–Alder additions, as shown below.[52]

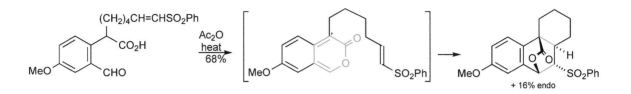

Decomposition of aryl-diazoketones, which have an alkene tethered *via* an ortho-ester, generates a 4-oxido-isochromylium salt for intramolecular cycloaddition to the alkene.[53]

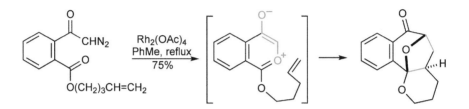

12.2.3.7 Photochemical Reactions

Coumarin has been studied extensively in this context; in the absence of a sensitiser, it gives a *syn* head-to-head dimer; in the presence of benzophenone as sensitiser, the *anti* isomer is formed;[54] the *syn* head-to-tail dimer is obtained by irradiation in acetic acid.[55] Cyclobutane-containing products are obtained in modest yields by sensitiser-promoted cycloadditions of coumarins or 3-acyl-oxycoumarins, with alkenes, ketene diethyl acetal or cyclopentene.[56]

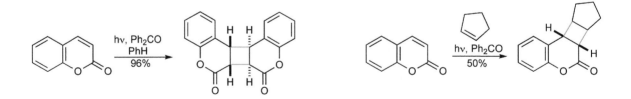

12.2.3.8 Alkyl-coumarins and Alkyl-chromones

Methyl groups at C-2, but not at C-3, of chromones undergo condensations with aldehydes, because only the former can be deprotonated to give conjugated enolates.[57]

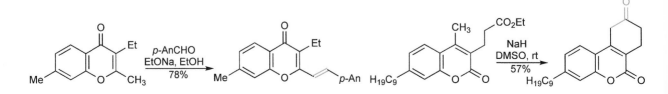

The 4-position of coumarins is the only one at which alkyl substituents have enhanced acidity in their hydrogens,[58] and this is considerably less than that of the methyl groups of 2-methyl-chromones.[59]

12.3 Synthesis of Benzopyryliums, Chromones, Coumarins and Isocoumarins

There are three important ways of putting together 1-benzopyryliums, coumarins and chromones; all begin with phenols. The isomeric 2-benzopyrylium and isocoumarin nuclei require the construction of an *ortho*-carboxy- or *ortho*-formyl-arylacetaldehyde (homophthalaldehyde).

Subject to the restrictions set out below, phenols react with 1,3-dicarbonyl compounds to produce 1-benzopyryliums or coumarins, depending on the oxidation level of the 1,3-dicarbonyl component.

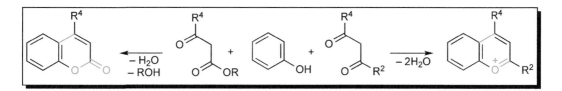

ortho-Hydroxy-benzaldehydes react with carbonyl compounds having an α-methylene, to give 1-benzopyryliums or coumarins, depending on the nature of the aliphatic unit.

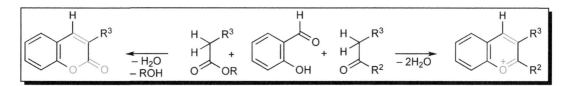

ortho-Hydroxyaryl alkyl ketones react with esters to give chromones.

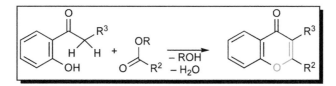

12.3.1 Ring Synthesis of 1-Benzopyryliums[1b]

12.3.1.1 From Phenols and 1,3-Dicarbonyl Compounds

The simplest reaction, that between a diketone and a phenol, works best with resorcinol, for the second hydroxyl facilitates the cyclising electrophilic attack. This synthesis can give mixtures with unsymmetrical diketones, and it is therefore well suited to the synthesis of 1-benzopyryliums with identical groups at C-2 and C-4,[60] however diketones in which the two carbonyl groups are appreciably different in reactivity can also produce high yields of single products.[61]

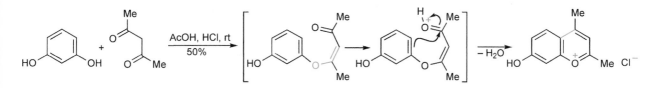

Acetylenic ketones, synthons for 1,3-keto-aldehydes, also take part regioselectively in condensations,[62] as do chalcones (a chalcone has the form ArCH=CHC(O)Ar), though of course an oxidant must be incorporated in this latter case.[63] Hexafluorophosphoric acid is recommended for the condensation of phloroglucinols and alkynyl-ketones.[64]

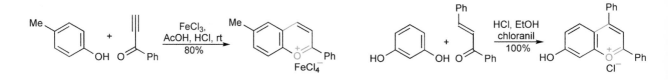

For hetero-ring-unsubstituted targets, the bis-acetal of malondialdehyde can be employed: in this variant a heterocyclic acetal-ether is first obtained, from which two mole equivalents of ethanol must then be eliminated.[65]

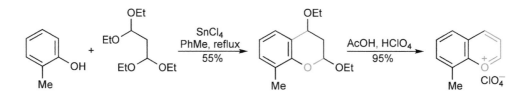

12.3.1.2 From ortho-*Hydroxy-araldehydes and ketones*

Salicylaldehydes can be condensed, by base or acid catalysis, with ketones that have an α-methylene. When base catalysis is used, the intermediate hydroxy-chalcones can be isolated,[8] but overall yields are often better when the whole sequence is carried out in one step, using acid.[66] It is important to note that because this route does not rely upon an electrophilic cyclisation onto the benzene ring, 1-benzopyryliums free from benzene ring (activating) substituents can be produced.

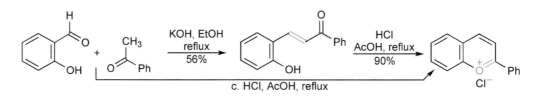

12.3.2 Ring Synthesis of Coumarins

12.3.2.1 From Phenols and 1,3-Keto-Esters

The Pechmann Synthesis[67]

Phenols react with β-keto-esters, including cyclic keto-esters,[68] to give coumarins under acid-catalysed conditions – concentrated sulfuric acid,[69] hydrogen fluoride,[70] a cation exchange resin,[71] indium(III) chloride[72] and sulfamic acid[73] (solvent free) are amongst those that have been used.

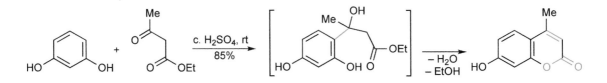

The Pechmann synthesis works best with the more nucleophilic aromatic compounds, such as resorcinols: electrophilic attack on the benzene ring *ortho* to phenolic oxygen by the protonated ketone carbonyl is the probable first step, though aryl-acetoacetates, prepared from a phenol and diketene, also undergo ring closure to give coumarins.[74] The greater electrophilicity of the ketonic carbonyl determines the orientation of combination. The production of hetero-ring-unsubstituted coumarins can be achieved by condensing with formylacetic acid, generated *in situ* by the decarbonylation of malic acid.

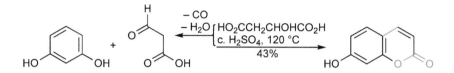

Coumarins can be obtained directly, in a one-pot procedure, from phenols and a propiolate, using palladium or platinum catalysis.[75,76]

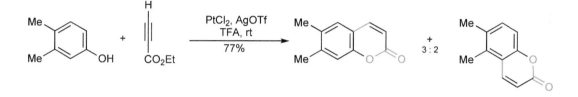

12.3.2.2 *From* ortho-*Hydroxy-Araldehydes and Anhydrides or Esters*

The simplest synthesis of coumarins is a special case of the Perkin condensation, i.e. the condensation of an aromatic aldehyde with an anhydride. *ortho*-Hydroxy-*trans*-cinnamic acids cannot be intermediates since they do not isomerise under the conditions of the reaction; nor can *O*-acetylsalicylaldehyde be the immediate precursor of the coumarin, since it is not cyclised by sodium acetate on its own.[77]

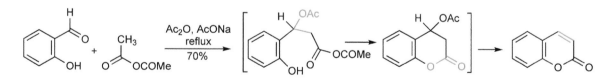

The general approach can be enlarged and conditions for condensation made milder by the use of further-activated esters, thus condensation with methyl nitroacetate produces 3-nitro-coumarins,[78] condensations with Wittig ylides[79] allow *ortho*-hydroxyaryl ketones to be used[80] and the use of diethyl malonate (or malonic acid[81]) (a 3-ester can be removed by hydrolysis and decarboxylation[82]), malononitrile, ethyl trifluoroacetoacetate, or substituted acetonitriles in a Knoevenagel condensation, produces coumarins with a 3-ester,[83] 3-trifluoroacetyl,[84] 3-cyano, or 3-alkyl or -aryl substituent.[85] Condensation with *N*-acetylglycine generates 3-acetylamino-coumarins.[86]

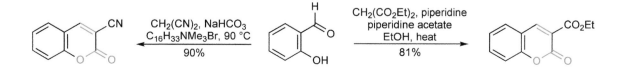

12.3.2.3   From ortho-*Hydroxyaraldehydes and Bis(methylthio)methylidene-Ketones*
In a route which certainly involves formation of the ester linkage as a first step, *ortho*-hydroxy-araldehydes react with bis(methylthio)methylidene-ketones (easily generated from methyl ketones by reaction with base, then carbon disulfide, then iodomethane), the ring closure taking place without further intervention.[87]

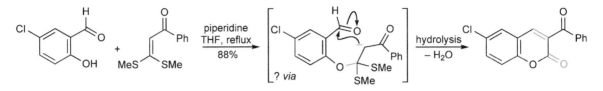

12.3.3   Ring Synthesis of Chromones
12.3.3.1   From ortho-*Hydroxyacyl-Arenes with Esters*
Most syntheses of chromones require the prior construction of a 1-(*ortho*-hydroxyaryl)-1,3-diketone, or equivalent, and it is in the manner by which this intermediate is generated that the methods differ.

Claisen condensation between an ester and the methylene adjacent to the carbonyl of the acylarene produces a 1-(*ortho*-hydroxyaryl)-1,3-diketone. The Claisen condensation can be conducted in the presence of the acidic phenolic hydroxyl by the use of excess strong base;[88] triethylamine as solvent and base can also be utilised.[89] Alternatively, the process is conducted in two steps: first, acylation of the phenolic hydroxyl, and secondly, an intramolecular[90] base-catalysed Claisen condensation, known as the Baker–Venkataraman rearrangement: a synthesis of flavone itself is illustrative.[91]

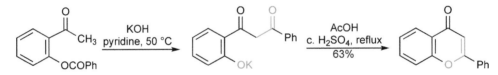

The use of diazabicycloundecene (DBU) allows the whole sequence to be conducted without isolation of intermediates, as shown in the example below.[92]

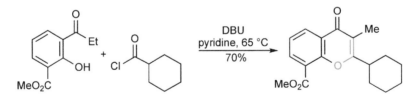

The production of a 2-unsubstituted chromone by this route requires the use of triethyl orthoformate.[93] Reaction with dimethylformamide dimethyl acetal, followed by an electrophile, bromine in the example, gives 3-substituted chromones.[94]

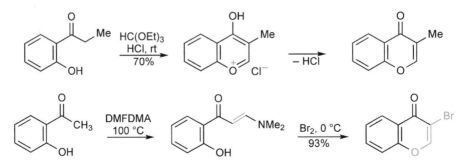

A variant of this route to 2-unsubstituted chromones employs oxalic acid half-ester half-acid chloride, which gives a 2-ethoxycarbonyl-chromone, hydrolysis and decarboxylation of which achieves the required result.[95] Diethyl carbonate as the ester gives rise to 2,4-dioxygenated heterocycles, which exist as 4-hydroxy-coumarins.[96] The condensation of a salicylate with an ester, using three mole equivalents of base also leads through to 4-hydroxy-coumarins, as illustrated below.[97]

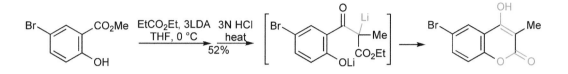

2-Aminobenzopyrones result from the ring closure of 1-(*ortho*-hydroxyaryl)-1,3-ketoamides.[98]

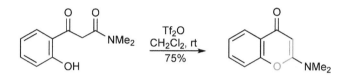

The variants on this route are many: for example condensation of *ortho*-hydroxyacetophenone with the Vilsmeier reagent produces 3-formylchromone (12.2.3.2).

Isoflavones (3-aryl-chromones) can also be prepared in this way: boron trifluoride-catalysed Friedel–Crafts acylation of a reactive phenol with an aryl acetic acid is followed by reaction with dimethylfor-mamide and phosphorus pentachloride.[99]

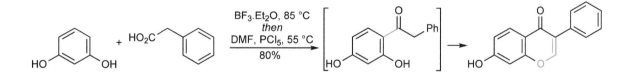

At a lower oxidation level, *ortho*-hydroxy-acyl-arenes undergo base-catalysed aldol condensations with aromatic aldehydes to give chalcones,[100] which can be cyclised to 2,3-dihydro-chromones *via* an intramo-lecular Michael process; the dihydro-chromones can in turn be dehydrogenated to produce chromones by a variety of methods, for example by bromination then dehydrobromination or by oxidation with the trityl cation, iodine, dimethyldioxirane or iodobenzene diacetate.[101]

Yet another variant uses *ortho*-fluorobenzoyl chloride in condensation with a 1,3-keto-ester;[102] the fluoride is displaced in an intramolecular sense by enolate oxygen, and the chromone obtained directly, as shown below.

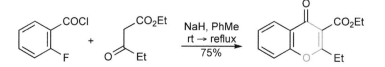

12.3.3.2 From ortho-*Hydroxyaryl Alkynyl Ketones*

ortho-Hydroxyaryl alkynyl ketones are intermediates in palladium(0)-catalysed coupling of *ortho*-hydroxyaryl iodides with terminal alkynes in the presence of carbon monoxide, ring closing to chromones *in situ*.[103]

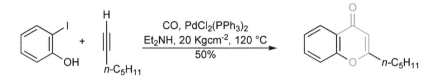

The alkynyl-ketones required for the 6-*endo-dig* cyclisation process[104] can be synthesised separately, and cyclise under mild conditions, either base catalysed or using iodine chloride, producing in the last case 3-iodo-chromones. Enamino-ketones also intervene in a very flexible sequence, in which the cyclisation precursor is produced by coupling an acetylene with an *ortho*-silyloxyaryl acid chloride; treatment of the resulting alkynone with a secondary amine leads to the chromone.[105]

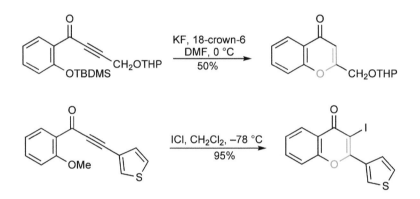

12.3.4 Ring Synthesis of 2-Benzopyryliums

The first synthesis[106] of the 2-benzopyrylium cation provided the pattern for subsequent routes, in which it is the aim to produce a homophthaldehyde, or diketo analogue, for acid-catalysed closure.

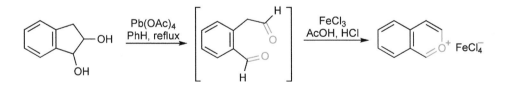

Most of the 2-benzopyrylium salts that have been synthesised subsequently are 1,3-disubstituted and their precursors prepared by Friedel–Crafts acylation of activated benzyl ketones.[6,107]

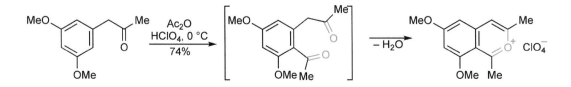

12.3.5 Ring Synthesis of Isocoumarins

One approach to isocoumarins is comparable to that above for 2-benzopyryliums, but replacing the aromatic aldehyde with an acid group.[108]

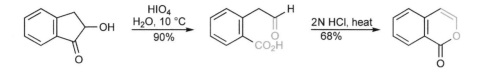

Benzoates carrying an *ortho*-acetylenic substituent, from Sonogashira couplings (4.2.5.1), can be ring closed using mercuric acetate, which also allows the introduction of an iodine at C-4,[109] or if that functionality is not required, simple acid-catalysed closure works well.[110] The use of iodine chloride similarly gives 4-iodo isocoumarins.[111]

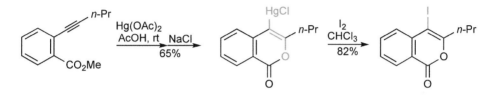

The most general route involves coupling an alkyne with *ortho*-iodobenzoic acid or with methyl *ortho*-iodobenzoate.[112] Both mono- (giving 3-substituted isocoumarins[113]) and disubstituted alkynes will serve, allowing considerable flexibility for the construction of substituted isocoumarins.

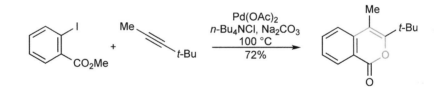

12.3.6 Notable Examples of Benzopyrylium and Benzopyrone Syntheses

12.3.6.1 Pelargonidin Chloride

The first synthesis of pelargonidin chloride used methyl ethers as protecting groups for the phenolic hydroxyls during the Grignard addition step.[114]

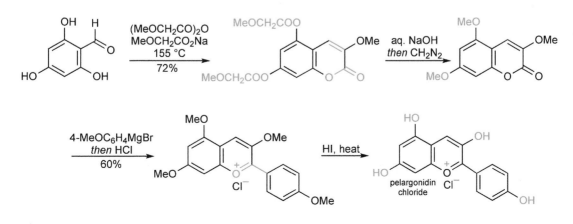

12.3.6.2 Apigenin

The scheme below shows two contrasting routes to apigenin. The modern use of excess of a very strong base, and the reaction of the resulting 'polyanion' obviated the need for phenolic protection in one synthesis.[115]

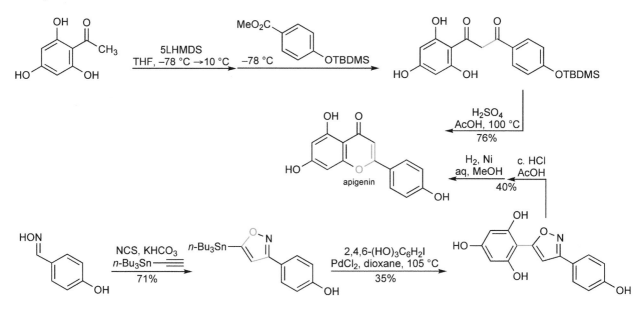

An elegant and flexible strategy for the assembly of a synthon for the *ortho*-hydroxyaryl-1,3-diketone required for a chromone synthesis depends on the use of an isoxazole as surrogate for the 1,3-diketone unit (25.7). An isoxazole was produced by the cycloaddition (25.12.1.2) of an aryl nitrile oxide to tri-*n*-butylstannylacetylene, the product coupled with 2,4,6-trihydroxyiodobenzene and then the N–O bond hydrogenolytically cleaved.[116]

Exercises
Straightforward revision exercises (consult Chapters 10, 11 and 12):
(a) At which position(s) do benzopyrylium ions react with nucleophiles, for example water?
(b) What is the typical structure of an anthocyanin flower pigment? What is the typical structure of a flavone flower pigment?
(c) At which atom do coumarins and chromones protonate?
(d) At which positions do coumarins and chromones undergo electrophilic substitution?
(e) Describe a cycloaddition reaction in which: (i) a coumarin and (ii) a chromone take part.
(f) How could one construct a 1-benzopyrylium salt from a phenol?
(g) How can *ortho*-hydroxyaryl-aldehydes be used to prepare coumarins?
(h) How can *ortho*-hydroxyaryl-ketones be used to prepare chromones?

More advanced exercises:
1. When salicylaldehyde and 2,3-dimethyl-1-benzopyrylium chloride are heated together in acid, a condensation product $C_{18}H_{15}O_2^+$ Cl^- is formed. Treatment of the salt with a weak base (pyridine) generates a neutral compound, $C_{18}H_{14}O_2$. Suggest structures for these two products.
2. When ethyl 2-methylchromone-3-carboxylate is treated with NaOH, then HCl, a product $C_{11}H_8O_4$ is produced that does not contain a carboxylic acid group, but does dissolve in dilute alkali: suggest a structure and the means whereby it could be formed.

3. Deduce the structures of intermediate and final product in the sequence: salicylaldehyde/ MeOCH$_2$CO$_2$Na/Ac$_2$O/heat → C$_{10}$H$_8$O$_3$, this then with 1 mole equivalent of PhMgBr → C$_{16}$H$_{14}$O$_3$ and finally this with HCl → C$_{16}$H$_{13}$O$_2$$^+Cl^-$.

4. Predict the structure of the major product from the interaction of resorcinol (1,3-dihydroxybenzene) and: (i) PhCOCH$_2$COMe in AcOH/HCl; (ii) methyl 2-oxocyclopentanecarboxylate/H$_2$SO$_4$.

References

1. A large proportion of work on benzo[c]pyryliums is by Russian and Hungarian workers and is described in relatively inaccessible journals, however it is well reviewed as: 'Benzo[c]pyrylium salts: syntheses, reactions and physical properties', Kuznetsov, E. V., Shcherbakova, I. V. and Balaban, A. T., *Adv. Heterocycl. Chem.*, **1990**, *50*, 157.
2. 'Isocoumarins. Developments since 1950', Barry, R. D., *Chem. Rev.*, **1964**, *64*, 229.
3. Le Fèvre, R. J. W., *J. Chem. Soc.*, **1929**, 2771.
4. Holker, J. S. E. and Underwood, J. G., *Chem. Ind. (London)*, **1964**, 1865.
5. Jurd, L., *Tetrahedron*, **1966**, *22*, 2913; ibid., **1968**, *24*, 4449.
6. Meyer-Dayan, M., Bodo, B., Deschamps-Valley, C. and Molho, D., *Tetrahedron Lett.*, **1978**, 3359.
7. Degani, I. and Fochi, R., *Ann. Chim. (Rome)*, **1968**, *58*, 251.
8. Hill, D. W. and Melhuish, R. R., *J. Chem. Soc.*, **1935**, 1161.
9. Jurd, L., *Tetrahedron*, **1969**, *25*, 2367.
10. Sutton, R., *J. Org. Chem.*, **1972**, *37*, 1069.
11. Dimroth, K. and Odenwälder, H., *Chem. Ber.*, **1971**, *104*, 2984.
12. Lowenbein, A., *Chem. Ber.*, **1924**, *57*, 1517.
13. Pomilio, A. B., Müller, O., Schilling, G. and Weinges, K., *Justus Liebigs Ann. Chem.*, **1977**, 597.
14. Kröhnke, F. and Dickoré, K., *Chem. Ber.*, **1959**, *92*, 46.
15. Jurd, L., *Tetrahedron*, **1965**, *21*, 3707.
16. Shcherbakova, I. V., Kuznetsov, E. V., Yudilevich, I. A., Kompan, O. E., Balaban, A. T., Abolin, A. H., Polyakov, A. V. and Struchkov, Yu. T., *Tetrahedron*, **1988**, *44*, 6217; Le Roux, J.-P., Desbene, P.-L. and Cherton, J.-C., *J. Heterocycl.Chem.*, **1981**, *18*, 847.
17. Lee, Y.-G., Ishimaru, K., Iwasaki, H., Okhata, K. and Akiba, K., *J. Org. Chem.*, **1991**, *56*, 2058.
18. Marathe, K. G., Philbin, E. M. and Wheeler, T. S., *Chem. Ind. (London)*, **1962**, 1793.
19. Clark-Lewis, J. W. and Baig, M. I., *Austr. J. Chem.*, **1971**, *24*, 2581.
20. Müller, A., Lempert-Stréter, M. and Karczag-Wilhelms, A., *J. Org. Chem.*, **1954**, *19*, 1533.
21. Heilbron, I. M. and Zaki, A., *J. Chem. Soc.*, **1926**, 1902.
22. Wittig, G., Baugert, F. and Richter, H. E., *Justus Liebigs Ann. Chem.*, **1925**, *446*, 155.
23. Meerwein, H., Hinz, G., Hofmann, P., Kroenigard, E. and Pfeil, E., *J. Prakt. Chem.*, **1937**, *147*, 257.
24. Clayton, A., *J. Chem. Soc.*, **1910**, 2106; Joshi, P. P., Ingle, T. R. and Bhide, B. V., *J. Ind. Chem. Soc.*, **1959**, *36*, 59.
25. Wiley, P. F., *J. Am. Chem. Soc.*, **1952**, *74*, 4326.
26. Zhang, F. J. and Li, Y. L., *Synthesis*, **1993**, 565.
27. Dean, F. M. and Murray, S., *J. Chem. Soc., Perkin Trans. 1*, **1975**, 1706.
28. Fuson, R. C., Kneisley, J. W. and Kaiser, E. W., *Org. Synth., Coll. Vol. III*, **1955**, 209; Perkin, W. H., *Justus Liebigs Ann. Chem.*, **1871**, *157*, 115.
29. Arndt, F., *Chem. Ber.*, **1925**, *58*, 1612.
30. Thapliyal, P. C., Singh, P. K. and Khauna, R. N., *Synth. Commun.*, **1993**, *23*, 2821.
31. Pearson, D. E., Stamper, W. E. and Suthers, B. R., *J. Org. Chem.*, **1963**, *28*, 3147.
32. Ellis, G. P. and Thomas, I. L., *J. Chem. Soc., Perkin Trans. 1*, **1973**, 2781.
33. For the use of singlet oxygen see Matsuura, T., *Tetrahedron*, **1977**, *33*, 2869.
34. Adam, W., Golsch, D., Hadjiarapoglou, L. and Patonay, T., *J. Org. Chem.*, **1991**, *56*, 7292; Adam, W., Hadjiarapoglou, L. and Levai, A., *Synthesis*, **1992**, 436.
35. Muller, E., *Chem. Ber.*, **1909**, *42*, 423.
36. Sugita, Y. and Yokoe, I., *Heterocycles*, **1996**, *43*, 2503.
37. Högberg, T., Vora, M., Drake, S. D., Mitscher, L. A. and Chu, D. T., *Acta Chem. Scand.*, **1984**, *B38*, 359.
38. Patonay, T., Attila Kiss-Szikszai, A., Silva, V. M. L., Silva, A. M. S., Pinto, D. C. G. A., Cavaleiro, J. A. S. and Jeko, J., *Eur. J. Org. Chem.*, **2008**, 1937.
39. Ulven, T., Receveur, J.-M., Grimstrup, M., Rist, Ø., Frimurer, T. M., Gerlach, L.-O., Mosolff, J., Mathiesen, M., Kostenis, E., Uller, L. and Högberg, T., *J Med. Chem.*, **2006**, *49*, 6638.
40. Shriner, R. C. and Sharp, A. G., *J. Org. Chem.*, **1939**, *4*, 575; Hepworth, J. D. and Livingstone, R., *J. Chem. Soc., C*, **1966**, 2013.
41. Reynolds, G. A., Van Allan, J. A. and Petropoulos, C. C., *J. Heterocycl. Chem.*, **1970**, *7*, 1061.
42. Goswami, M. and Chakravarty, A., *J. Ind. Chem. Soc.*, **1932**, *9*, 599; Michaelidis, Ch. and Wizinger, R., *Helv. Chim. Acta*, **1951**, *34*, 1761.
43. Costa, A. M. B. S. R. C. S., Dean, F. M., Jones, M. A. and Varma, R. S., *J. Chem. Soc., Perkin Trans. 1*, **1985**, 799.
44. Kirkiacharian, B. S. and Raulais, D., *Bull. Soc. Chim. Fr.*, **1970**, 1139.
45. Smith, L. I. and Denyes, R. O., *J. Am. Chem. Soc.*, **1936**, *58*, 304; John, W., Günther, P. and Schmeil, M., *Chem. Ber.*, **1938**, *71*, 2637.
46. Elhabiri, M., Figueiredo, P., Fougerousse, A. and Brouillard, R., *Tetrahedron Lett.*, **1995**, *36*, 4611.
47. Adams, R., McPhee, W. D., Carlin, R. B. and Wicks, Z. W., *J. Am. Chem. Soc.*, **1943**, *65*, 356; Adams, R. and Carlin, R. B., ibid., 360.
48. Girotti, R., Marrocchi, A., Minuti, L., Piermatti, O., Pizzo, F. and Vaccaro, L., *J. Org. Chem.*, **2006**, *71*, 70.
49. Ghosh, C. K., Tewari, N. and Bhattacharya, A., *Synthesis*, **1984**, 614; Coutts, S. J. and Wallace, T. W., *Tetrahedron*, **1994**, *50*, 11755.

[50] Wallace, T. W., Wardell, I., Li, K.-D., Leeming, P., Redhouse, A. D. and Challand, S. R., *J. Chem. Soc., Perkin Trans. 1*, **1995**, 2293.

[51] Ohkata, K., Kubo, T., Miyamoto, K., Ono, M., Yamamoto, J. and Akiba, K., *Heterocycles*, **1994**, *38*, 1483.

[52] Bush, E. J., Jones, D. W. and Ryder, T. C. L. M., *J. Chem. Soc., Perkin Trans. 1*, **1997**, 1929.

[53] Plüg, C., Friedrichsen, W. and Debaerdemaeker, T., *J. prakt. Chem.*, **1997**, *339*, 205.

[54] Schenk, G. O., von Willucki, I. and Krauch, C. H., *Chem. Ber.*, **1962**, *95*, 1409; Anet, R., *Canad. J. Chem.*, **1962**, *40*, 1249; Hammond, G. S., Stout, C. A. and Lamola, A. A., *J. Am. Chem. Soc.*, **1964**, *86*, 3103.

[55] Krauch, C. H., Farid, S. and Schenk, G. O., *Chem. Ber.*, **1966**, *99*, 625.

[56] Hanifen, J. W. and Cohen, E., *Tetrahedron Lett.*, **1966**, 1419; Kobayashi, K., Suzuki, M. and Suginome, H., *J. Chem. Soc., Perkin Trans. 1*, **1993**, 2837.

[57] Heilbron, I. M., Barnes, H. and Morton, R. A., *J. Chem. Soc.*, **1923**, 2559.

[58] Archer, R. A., Blanchard, W. B., Day, W. A., Johnson, D. W. Lavagnino, E. R., Ryan, C. W. and Baldwin, J. E., *J. Org. Chem.*, **1977**, *42*, 2277.

[59] Mahal, H. S. and Venkataraman, K., *J. Chem. Soc.*, **1933**, 616.

[60] Bülow, C. and Wagner, H., *Chem. Ber.*, **1901**, *34*, 1189.

[61] Bülow, C. and Wagner, H., *Chem. Ber.*, **1901**, *34*, 1782; Sweeny, J. G. and Iacobucci, G. A., *Tetrahedron*, **1981**, *37*, 1481.

[62] Johnson, A. W. and Melhuish, R. R., *J. Chem. Soc.*, **1947**, 346.

[63] Robinson, R. and Walker, J., *J. Chem. Soc.*, **1934**, 1435.

[64] Kueny-Stotz, M., Isorez, G., Chassaing, S. and Brouillard, R., *Synlett*, **2007**, 1067.

[65] Bigi, F., Casiraghi, G., Casnati, G. and Sartori, G., *J. Heterocycl. Chem.*, **1981**, *18*, 1325.

[66] Pratt, D. D. and Robinson, R., *J. Chem. Soc.*, **1923**, 745.

[67] 'The Pechmann reaction', Sethna, S. and Phadke, R., *Org. React.*, **1953**, *7*, 1.

[68] Sen, H. K. and Basu, U., *J. Ind. Chem. Soc.*, **1928**, *5*, 467.

[69] Russell, A. and Frye, J. R., *Org. Synth., Coll. Vol. III*, **1955**, 281.

[70] Dann, O. and Mylius, G., *Justus Liebigs Ann. Chem.*, **1954**, *587*, 1.

[71] John, E. V. O. and Israelstam, S. S., *J. Org. Chem.*, **1961**, *26*, 240.

[72] Bose, D. S., Rudradas, A. P. and Babu, M. H., *Tetrahedron Lett.*, **2002**, *43*, 9195.

[73] Singh, P. R., Singh, D. U. and Samant, S. D., *Synlett*, **2004**, 1909.

[74] Lacey, R. N., *J. Chem. Soc.*, **1954**, 854.

[75] Trost, B. M. and Toste, F. D., *J. Am. Chem. Soc.*, **1996**, *118*, 6305.

[76] Oyamada, J. and Kitamura, T., *Tetrahedron*, **2006**, *62*, 6918.

[77] Crawford, M. and Shaw, J. A. M., *J. Chem. Soc.*, **1953**, 3435.

[78] Dauzonne, D. and Royer, R., *Synthesis*, **1983**, 836.

[79] Harayama, T., Nakatsuka, K., Nishioka, H., Murakami, K., Hayashida, N. and Ishii, H., *Chem. Pharm. Bull.*, **1994**, *42*, 2170.

[80] Mali, R. S. and Yadav, V. J., *Synthesis*, **1977**, 464.

[81] Steck, W., *Canad. J. Chem.*, **1971**, *49*, 2297.

[82] Rouessac, F. and Leclerc, A., *Synth. Commun.*, **1993**, *23*, 1147.

[83] Horning, E. C., Horning, M. G. and Dimmig, D. A., *Org. Synth., Coll. Vol. III*, **1955**, 165.

[84] Chizhov, D. L., Sosnovskikh, V. Ya., Pryadeina, M. V., Burgart, Y. V., Saloutin, V. I. and Charushin, V. N., *Synlett*, **2008**, 281.

[85] Brufola, G., Friguelli, F., Piermatti, O. and Pizzo, F., *Heterocycles,*, **1996**, *43*, 1257.

[86] Kudale, A. A., Kendall, J., Warford, C. C., Wilkins, N. D. and Bodwell, G. J., *Tetrahedron Lett.*, **2007**, *48*, 5077.

[87] Rao, H. S. P. and Sivakumar, S., *J. Org. Chem.*, **2006**, *71*, 8715.

[88] Wiley, P. F., *J. Am. Chem. Soc.*, **1952**, *74*, 4329; Schmutz, J., Hirt, R. and Lauener, H., *Helv. Chim. Acta*, **1952**, *35*, 1168; Mozingo, R., *Org. Synth., Coll. Vol. III*, **1955**, 387; Hirao, I., Yamaguchi, M. and Hamada, M., *Synthesis*, **1984**, 1076; Banerji, H. and Goomer, N. C., *ibid.*, **1980**, 874.

[89] Looker, J. H., McMechan, J. H. and Mader, J. W., *J. Org. Chem.*, **1978**, *43*, 2344.

[90] Baker, W., *J. Chem. Soc.*, **1933**, 1381; Schmid, H. and Banholzer, K., *Helv. Chim. Acta*, **1954**, *37*, 1706.

[91] Wheeler, T. S., *Org. Synth., Coll. Vol. IV*, **1963**, 478.

[92] Riva, C., De Toma, C., Donadel, L., Boi, C., Peunini, R., Motta, G. and Leonardi, A., *Synthesis*, **1997**, 195.

[93] Becket, G. J. P., Ellis, G. P. and Trindade, M. I. U., *J. Chem. Res.*, **1978** (S) 47; (M) 0865.

[94] Gammill, R. B., *Synthesis*, **1979**, 901; Yokoe, I., Keiko Maruyama, K., Sugita, Y., Harashida, T. and Shirataki, Y., *Chem. Pharm. Bull.*, **1994**, *42*, 1697.

[95] Baker, W., Chadderton, J., Harborne, J. B. and Ollis, W. D., *J. Chem. Soc.*, **1953**, 1852.

[96] Boyd, J. and Robertson, A., *J. Chem. Soc.*, **1948**, 174.

[97] Davis, S. E., Church, A. C., Tummons, R. C. and Beam, C. F., *J. Heterocycl. Chem.*, **1997**, *34*, 1159.

[98] Morris, J., Wishka, D. G. and Fang, Y., *Synth. Commun.*, **1994**, *24*, 849.

[99] Balasubramanian, S. and Nair, M. G., *Synth. Commun.*, **2000**, *30*, 469.

[100] Geissman, T. A. and Clinton, R. O., *J. Am. Chem. Soc.*, **1946**, *68*, 697.

[101] Lorette, N. B., Gage, T. B. and Wender, S. H., *J. Org. Chem.*, **1951**, *16*, 930; Schönberg, A. and Schutz, G., *Chem. Ber.*, **1960**, *93*, 1466; Patonay, T., Cavaleiro, J. A. S., Lévai, A. and Silva, A. M. S., *Heterocycl. Commun.*, **1997**, *3*, 223; Bernini, R., Mincione, E., Sanetti, A., Bovicelli, P. and Lupatteli, P., *Tetrahedron Lett.*, **1997**, *38*, 4651; Prakash, O. and Tanwer, M. P., *J. Chem. Res (S)*, **1995**, *143*, (M), 1429; Litkei, G., Gulácsis, K., Antus, A. and Blaskó, G., *Justus Liebigs Ann. Chem.*, **1995**, 1711.

[102] Coppola, G. M. and Dodsworth, R. W., *Synthesis*, **1981**, 523; Cremins, P. J., Hayes, R. and Wallace, T. W., *Tetrahedron*, **1993**, *49*, 3211.

[103] Torii, S., Okumoto, H., Xu, L. H., Sadakane, M., Shostakovsky, M. V., Ponomaryov, A. B. and Kalinin, V. N., *Tetrahedron*, **1993**, *49*, 6773.

[104] Nakatani, K., Okamoto, A. and Saito, I., *Tetrahedron*, **1996**, *52*, 9427.

[105] Bhat, A. S., Whetstone, J. L. and Brueggemeier, R. W., *Tetrahedron Lett.*, **1999**, *40*, 2469.

[106] Blount, B. K. and Robinson, R., *J. Chem. Soc.*, **1933**, 555.

[107] Bringmann, G. and Jansen, J. R., *Justus Liebigs Ann. Chem.*, **1985**, 2116.

[108] Schöpf, C. and Kühne, R., *Chem. Ber.*, **1950**, *83*, 390.

[109] Larock, R. C. and Harrison, L. W., *J. Am. Chem. Soc.*, **1984**, *106*, 4218.

[110] Le Bras, G., Hamze, A., Messaoudi, S., Provot, O., Le Calvez, P.-B. and Brion, J.-D., *Synthesis*, **2008**, 1607.

[111] Yao, T. and Larock, R. C., *J. Org. Chem.*, **2003**, *68*, 5936.

[112] Liao, H.-Y. and Cheng, C.-H., *J. Org. Chem.*, **1995**, *60*, 3711; Larock, R. C., Yum, E. K., Doty, M. J. and Sham, K. K. C., *ibid.*, 3270.

[113] Subramanian, V., Batchu, V. R., Barange, D. and Pal, M., *J. Org. Chem.*, **2005**, *70*, 4778.

[114] Willstätter, R., Zechmeister, L. and Kindler, W., *Chem. Ber.*, **1924**, *47*, 1938.

[115] Nagarathnam, D. and Cushman, M., *J. Org. Chem.* **1991**, *56*, 4884.

[116] Gothelf, K., Thomsen, I. and Torssell, K. B. G., *Acta Chem. Scand.*, **1992**, *46*, 494; Gothelf, K. V. and Torssell, K. B. G., *ibid.*, **1994**, *48*, 165; Ellemose, S., Kure, N. and Torssell, K. B. G., *ibid.*, **1995**, *49*, 524.

13

Typical Reactivity of the Diazines: Pyridazine, Pyrimidine and Pyrazine

pyridazine   pyrimidine   pyrazine

The diazines – pyridazine, pyrimidine and pyrazine – contain two imine nitrogen atoms, so the lessons learnt with regard to pyridine (Chapters 7 and 8) are, in these heterocycles, exaggerated. Two heteroatoms withdraw electron density from the ring carbons even more than one in pyridine, so unsubstituted diazines are even more resistant to electrophilic substitution than is pyridine. A corollary is that this same increased electron deficiency at carbon makes the diazines more easily attacked by nucleophiles than pyridine. The availability of nitrogen lone pair(s) is also reduced: each of the diazines is appreciably less basic than pyridine, reflecting the destabilising influence of the second nitrogen on the *N*-protocation. Nevertheless, diazines will form salts and will react with alkyl halides and with peracids to give *N*-alkyl quaternary salts and *N*-oxides, respectively. Generally speaking, such electrophilic additions take place at one nitrogen only, because the presence of the positive charge in the products renders the second nitrogen extremely unreactive towards a second electrophilic addition.

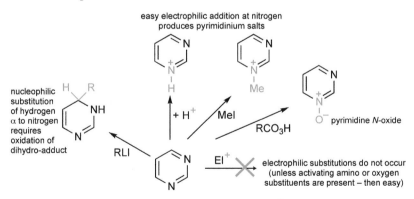

Typical reactions of a diazine illustrated with pyrimidine

A very characteristic feature of the chemistry of diazines, which is associated with their strongly electron-poor nature, is that they add nucleophilic reagents easily. Without halide to be displaced, such adducts require an oxidation to complete an overall substitution. However, halo-diazines, where the halide is α or γ to a nitrogen, undergo very easy nucleophilic displacements, the intermediates being particularly well stabilised. In line with their susceptibility to nucleophilic addition, diazines also undergo substitution by nucleophilic radicals, in acid solution, with ease.

Heterocyclic Chemistry 5th Edition   John Joule and Keith Mills
© 2010 Blackwell Publishing Ltd

All positions on each of the diazines, with the sole exception of the 5-position of a pyrimidine, are α and/or γ to an imine ring nitrogen and, in considering nucleophilic addition/substitution, it must be remembered that there is also an additional nitrogen that is withdrawing electron density. As a consequence, all the monohalo-diazines are more reactive than either 2- or 4-halo-pyridines. The 2- and 4-halo-pyrimidines are particularly reactive because the anionic intermediates (shown below for attack on a 2-halo-pyrimidine) derive direct mesomeric stabilisation from *both* nitrogen atoms.

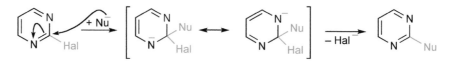

Delocalisation of negative charge for nucleophilic substitution of a 2-halo-pyrimidine

Despite this particularly strong propensity for nucleophilic addition, *C*-lithiation of diazines can be achieved by either metal–halogen exchange or, by deprotonation *ortho* to chloro- or alkoxyl substituents (DoM), though very low temperatures must be utilised in order to avoid nucleophilic addition of the reagent.

Considerable use has been made in diazine chemistry of palladium(0)-catalysed coupling processes, with the diazine as either a halide or as an organometallic derivative; one example chosen at random is shown below (see 4.2 for a detailed discussion).

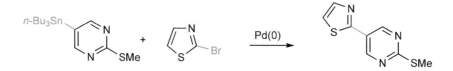

Further examples of the enhancement of those facets of pyridine chemistry associated with the imine electron withdrawal include a general stability towards oxidative degradation and, on the other hand, a tendency to undergo rather easy reduction of the ring.

Although there is always debate about quantitative measures of aromaticity, it is agreed that the diazines are less resonance stabilised than pyridines – they are 'less aromatic'. Thus, Diels–Alder additions are known for all three systems, with the heterocycle acting as an azadiene; initial adducts lose a small molecule – hydrogen cyanide in the pyrimidine example shown – to afford a final stable product.

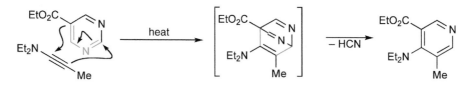

A pyrimidine acting as an azadiene in a Diels–Alder cycloaddition

N-Oxides, just as in the pyridine series, show a duality of effect – *both* electrophilic substitutions *and* nucleophilic displacements are enhanced. Just as in pyridine *N*-oxide chemistry, a very useful transformation is the introduction of halide α to a nitrogen on reaction with reagents such as phosphorus oxychloride. The importance of this transformation can be realised by noting that the *un*substituted heterocycle is converted, in the two steps, into a halo-diazine, with its potential for subsequent displacement reactions with nucleophiles.

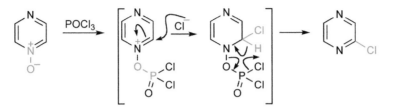

Conversion of pyrazine *N*-oxide into 2-chloropyrazine

The most studied diazine derivatives are the oxy- and amino-pyrimidines, since uracil, thymine and cytosine are found as bases in DNA and RNA. Carbonyl tautomers are the preferred forms. It is the enamide-like character of the double bonds in diazine diones that allows electrophilic substitution – uracil, for example, can be brominated at C-5. One amino-substituent permits electrophilic ring substitution and two amino, or one amino and one oxy, substituents, permit substitution with even weakly electrophilic reactants.

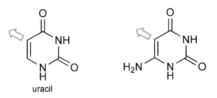

uracil

Electrophilic substitution of activated pyrimidines is easy

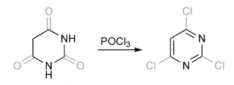

Diazinones can be converted into halo-diazines

Diazinones, like pyridones, react with phosphorus halides with overall conversion into halides. Anions produced by *N*-deprotonation of diazinones are ambident, with phenolate-like resonance contributors, but they generally react with electrophilic alkylating agents at nitrogen, rather than oxygen, giving *N*-alkyl diazinones.

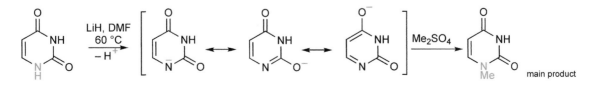

Diazine alkyl groups, with the exception of those at the 5-position of pyrimidine, can undergo condensation reactions that utilise a side-chain carbanion produced by removal of a proton. As in pyridine chemistry, formation of these anions is made possible by delocalisation of the charge onto one (or more) of the ring nitrogen atoms.

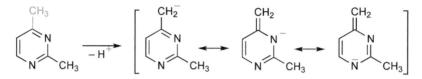

Stabilisation of side-chain carbanions

Each of the diazines can be constructed from an appropriate source of two nitrogens and a dicarbonyl compound. In the case of pyridazines, the nitrogen source is, of course, hydrazine and this in combination with 1,4-dicarbonyl compounds readily produces dihydro-pyridazines, which are very easily dehydrogenated to the aromatic heterocycle. Pyrimidines result from the interaction of a 1,3-dicarbonyl component and an amidine (as shown) or a urea (giving 2-pyrimidones) or a guanidine (giving 2-amino-pyrimidines), without the requirement for an oxidation step.

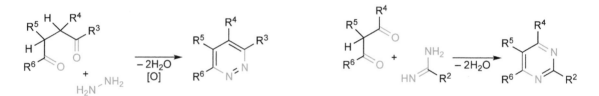

The two commonly used ring synthetic routes to pyridazines and pyrimidines

To access a pyrazine in this way one needs a 1,2-diamine and a 1,2-dicarbonyl compound, and a subsequent oxidation, but if both components are unsymmetrical, mixtures are formed. The dimerisation of 2-aminocarbonyl compounds also generates symmetrically substituted dihydro-pyrazines – perhaps the best known examples of such dimerisations involve the natural amino acids and their esters, which dimerise to give dihydropyrazine-2,5-diones – 'diketopiperazines'.

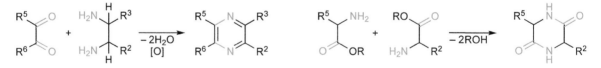

Two ring synthetic routes to the pyrazine ring system

14

The Diazines: Pyridazine, Pyrimidine, and Pyrazine: Reactions and Synthesis

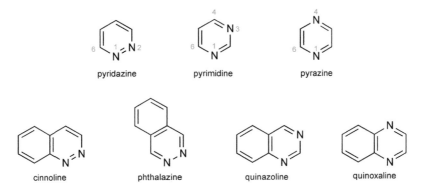

The three diazines, pyridazine,[1] pyrimidine,[2] and pyrazine[3] are stable, colourless compounds that are soluble in water. The three parent heterocycles, unlike pyridine, are expensive and not readily available and so are seldom used as starting materials for the synthesis of their derivatives. There are only four ways in which a benzene ring can be fused to a diazine: cinnoline, phthalazine, quinazoline and quinoxaline are the bicyclic systems thus generated.

One striking aspect of the physical properties of the diazine trio is the high boiling point of pyridazine (207 °C), 80–90 °C higher than that of pyrimidine (123 °C), pyrazine (118 °C), or indeed other azines, including 1,3,5-triazine, all of which also boil in the range 114–124 °C. The high boiling point of pyridazine is attributed to the polarisability of the N–N unit, which results in extensive dipolar association in the liquid.

The most important naturally occuring diazines are the pyrimidine bases uracil, thymine and cytosine, which are constituents of the nucleic acids (see 32.4). The nucleic acid pyrimidines are often drawn horizontally transposed from the representations used in this chapter, i.e. with N-3 to the 'north-west', mainly to draw attention to their structural similarity to the pyrimidine ring of the nucleic acid purines, which are traditionally drawn with the pyrimidine ring on the left. There are relatively few naturally occurring pyrazines or pyridazines.

14.1 Reactions with Electrophilic Reagents
14.1.1 Addition at Nitrogen[4]
14.1.1.1 Protonation
The diazines, pyridazine (pK_{aH} 2.3), pyrimidine (1.3), and pyrazine (0.65) are essentially mono-basic substances, and considerably weaker, as bases, than pyridine (5.2). This reduction in basicity is believed to be largely a consequence of destabilisation of the mono-protonated cations due to a combination of inductive and mesomeric withdrawal by the second nitrogen atom. Secondary effects, however, determine the order of basicity for the three systems: repulsion between the lone pairs on the two adjacent nitrogen atoms in

Heterocyclic Chemistry 5th Edition   John Joule and Keith Mills
© 2010 Blackwell Publishing Ltd

pyridazine means that protonation occurs more readily than if inductive effects, only, were operating. In the case of pyrazine, mesomeric interaction between the protonated and neutral nitrogen atoms probably destabilises the cation.

N,N'-Diprotonation is very much more difficult and has only been observed in very strongly acidic media. Of the trio, pyridazine ($pK_{aH(2)}$ −7.1) is the most difficult from which to generate a dication, probably due to the high energy associated with the juxtaposition of two immediately adjacent positively charged atoms, but pyrimidine ($pK_{aH(2)}$ −6.3) and pyrazine ($pK_{aH(2)}$ −6.6) are only marginally easier to doubly protonate.

Substituents can affect basicity (and nucleophilicity) both inductively and mesomerically, but care is needed in the interpretation of pK_{aH} changes, for example it is important to be sure which of the two nitrogens of the substituted azine is protonated (see also 14.1.1.2).

14.1.1.2 Alkylation

The diazines react with alkyl halides to give mono-quaternary salts, though somewhat less readily than comparable pyridines. Dialkylation cannot be achieved with simple alkyl halides, however the more reactive trialkyloxonium tetrafluoroborates do convert all three systems into di-quaternary salts.[5]

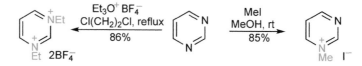

Pyridazine is the most reactive in alkylation reactions and this again has its origin in the lone-pair/lone-pair interaction between the nitrogen atoms. This phenomenon is known as the 'α effect'[6] and is also responsible, for example, for the relatively higher reactivity of hydrogen peroxide as a nucleophile, compared with water.

Unsymmetrically substituted diazines can give rise to two isomeric quaternary salts. Substituents influence the orientation mainly by steric and inductive, rather than mesomeric effects. For example, 3-methylpyridazine alkylates mainly at N-1, even though N-2 is the more electron-rich site. Again, quaternisation of 3-methoxy-6-methylpyridazine takes place adjacent to the methyl substituent, at N-1, although mesomeric release would have been expected to favour attack at N-2.[7]

14.1.1.3 Oxidation

All three systems react with peracids,[8] giving *N*-oxides, but care must be taken with pyrimidines[9] due to the relative instability of the products under the acidic conditions. Pyrazines[10] form *N,N'*-dioxides the most easily, but pyridazine[10] requires forcing conditions, and pyrimidines, apart from some examples in which further activation is present, give poor yields.[11]

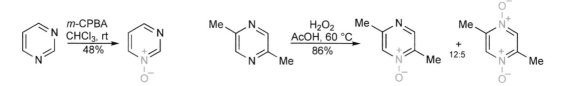

The regiochemistry of *N*-oxidation of substituted azines is governed by the same factors as alkylation (14.1.1.2), for example 3-methylpyridazine gives the 1-oxide as main (3:1) product,[12] but the pattern is not a simple one, for 4-methylpyrimidine *N*-oxidises principally (3.5:1) at the nitrogen adjacent to the methyl.[13] The acidity of the medium can also influence the regiochemistry of oxidation, for example 3-cyanopyridazine reacts at N-1 with peracetic acid, but under strongly acidic conditions, in which the heterocycle is mainly present as its N-1-protonic salt, oxidation, apparently involving attack on this salt, occurs at N-2.[14]

14.1.2 Substitution at Carbon

Recalling the resistance of pyridines to electrophilic substitution, it is not surprising to find that introduction of a second azomethine nitrogen, in any of the three possible orientations, greatly increases this resistance: no nitration or sulfonation of a diazine or simple alkyl-diazine has been reported, though some halogenations are known. It is to be noted that C-5 in pyrimidine is the only position, in all three diazines, which is not in an α- or γ-relationship to a ring nitrogen, and is therefore equivalent to a β-position in pyridine. Diazines carrying electron-releasing (activating) substituents undergo electrophilic substitution much more easily (14.9.2.1 and 14.10).

14.1.2.1 Halogenation

Chlorination of 2-methylpyrazine occurs under such mild conditions that it is almost certain that an addition/elimination sequence is involved, rather than a classical aromatic electrophilic substitution.[15] Halogenation of pyrimidines may well also involve such processes.[16]

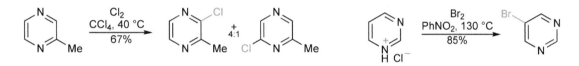

14.2 Reactions with Oxidising Agents

The diazines are generally resistant to oxidative attack at ring carbons, though alkaline oxidising agents can bring about degradation *via* intermediates produced by initial nucleophilic addition (14.3). Alkyl substituents[17] and fused aromatic rings[18] can be oxidised to carboxylic acid residues, leaving the heterocyclic ring untouched. An oxygen can be introduced into pyrimidines at vacant C-2 and/or C-4 positions using various bacteria.[19] Dimethyldioxirane converts *N,N*-dialkylated uracils into 5,6-diols probably *via* 5,6-epoxides.[20]

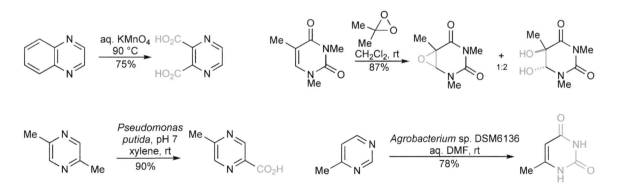

14.3 Reactions with Nucleophilic Reagents

The diazines are very susceptible to nucleophilic addition: pyrimidine, for example, is decomposed when heated with aqueous alkali by a process that involves hydroxide addition as a first step. It is converted into pyrazole by reaction with hot hydrazine.

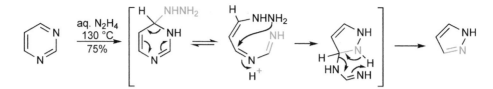

14.3.1 Nucleophilic Substitution with 'Hydride' Transfer

14.3.1.1 Alkylation and Arylation

The diazines readily add alkyl- and aryllithiums, and Grignard reagents, to give dihydro-adducts that can be aromatised by oxidation with reagents such as potassium permanganate or 2,3-dichloro-5,6-dicyano-1,4-benzoquinone (DDQ). In reactions with organolithiums, pyrimidines react at C-4,[13] and pyridazines at C-3, but Grignard reagents add to pyridazines at C-4.[21]

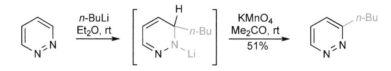

An important point is that in diazines carrying chlorine or methylthio substituents, attack does not take place at the halogen- or methylthio-bearing carbon; halogen-[22,23] and methylthio-containing[24] products are therefore obtained.

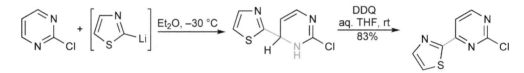

14.3.1.2 Amination

The Chichibabin reaction can be carried out under the usual conditions in a few cases,[25] but is much less general than for pyridines. This may be partly a consequence of the lower aromaticity of the diazines, for, although the initial addition is quite easy, the subsequent loss of hydride (re-aromatisation) is difficult. However, high yields of 4-aminopyridazine, 4-aminopyrimidine and 2-aminopyrazine can be obtained by oxidation of the dihydro-adduct *in situ* with potassium permanganate.[26]

14.3.2 Nucleophilic Substitution with Displacement of Good Leaving Groups

All the halo-diazines, apart from 5-halo-pyrimidines, react readily with 'soft' nucleophiles, such as amines, thiolates and malonate anions, with substitution of the halide. Even 5-bromopyrimidine can be brought into reaction with nucleophiles using microwave heating.[27] All cases are more reactive than 2-halo-pyridines: the relative reactivities can be summarised:

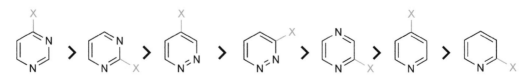

Pyrrolo[2,3-*b*]pyrazines can be produced *via* successive nucleophilic displacements of the halogens of 2,3-dichloropyrazine.[28] This result should be contrasted with the *tele*-substitution products obtained from this same substrate with a dithiane anion as the nucleophile.[29]

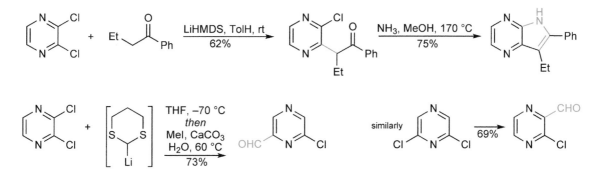

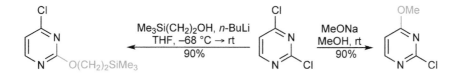

Nucleophilic displacement of halogen with ammonia[30] and amines[31] can be accelerated by carrying out the displacements in acid solution, when the protonated heterocycle is more reactive than the neutral heterocycle.[23,32] Halogen can also be easily removed hydrogenolytically, for example treatment of 2,4-dichloropyrimidine, readily available from uracil, with hydrogen, in the presence of palladium, or with hydrogen iodide, gives pyrimidine itself.[33]

The difference in reactivity between 2- and 4-halo-pyrimidines is relatively small and a discussion of the selectivity in nucleophilic displacement reactions of 2,4-dichloropyrimidine (an important synthetic intermediate) is instructive.

Reaction with sodium methoxide in methanol is highly selective for the 4-chlorosubstitutent,[34] whereas lithium 2-(trimethylsilyl)ethoxide is equally selective, but for the 2-chloro substituent.[35] The former is the normal situation for nucleophilic displacements[36] – 4-chloro > 2-chloro – the second case is the exception, where strong co-ordination of lithium in a non-polar solvent to the more basic nitrogen, N-1, leads to activation, and possibly also internal attack, at C-2. Under acidic conditions, an approximately 1:1 mixture of the two methoxy products is formed. Here, hydrogen bonding to the proton on N-1 provides the mechanism for encouraging attack at C-2. Selectivity with other nucleophiles is dependent on the nature of the nucleophile and on reaction conditions.

A more certain approach to selective 4- followed by 2-substitution involves the use of 4-chloro-2-methylthiopyrimidine with the first nucleophile, then oxidation to the sulfone, followed by reaction at C-2 with the second nucleophile displacing methylsulfone (as methanesulfinate anion).[37] The reaction of 2,4,6-trichloropyrimidine with the sodium salts of Boc-protected amines, generated with NaH in DMF, gives much better 4(6):2 selectivity than reaction with the corresponding amines.[38]

The displacement of fluoride from 2-fluoropyrimidine by aliphatic amines is about 100 times as fast as the displacements of the corresponding chloride or bromide, and reactions can be carried out at room temperature. Aryl-amines are unreactive under these conditions, but do react in the presence of trifluoro-acetic acid or boron trifluoride. These mild conditions allow the use of the fluoro-compound in solid phase synthesis (using excess of the fluoropyrimidine to ensure complete conversion).[39]

Nucleophilic displacement reactions are also sensitive to the presence of other substituents in the ring, either by electronic or steric effects and this sometimes leads to a reversal of the typical selectivity,[40] as can changes in the nucleophile, for example tri-*n*-butylstannyllithium attacks 2,4-dichloropyrimidine at C-2.[41]

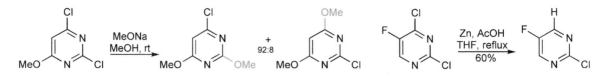

Halo-pyrimidines with other electron-donating substituents in the ring tend to be much less reactive to nucleophilic substitution: this can be overcome by the use of the very nucleophilic *O,N*-dimethylhydroxylamine, followed by hydrogenolysis to reveal the amine.[42]

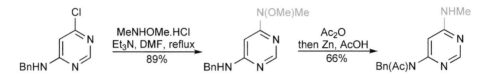

A device that is also used in pyridine and purine chemistry is the initial replacement of halogen with a tertiary amine; the resulting salt, now having a better leaving group, undergoes nucleophilic substitution more easily.[43]

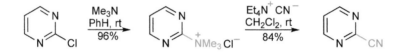

Alkylsulfonyl groups are also good leaving groups (as alkylsulfinate anion) in all of the diazines,[44] generally better than chloro, sometimes considerably so, for example 3-methanesulfonylpyridazine reacts 90 times faster with methoxide than does 3-chloropyridazine. Sulfinates can be used to catalyse displacements of chlorine *via* the intermediacy of the sulfone.[45]

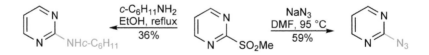

Even methoxy groups can be displaced by carbanions.[46]

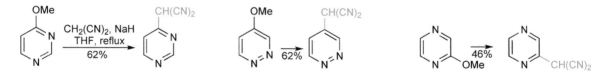

Monsubstitution of 3,6-diiodopyridazine is easy, further manipulation *via* various palladium-catalysed couplings (see also 4.2) providing a good route to 3,6-differently-substituted pyridazines.[47] 4-Methoxyben-zylamine, as a surrogate for ammonia *via* cleavage by subsequent acid hydrolysis, displaces both chlorines in 3,6-dichloropyridazine at 165 °C.[48]

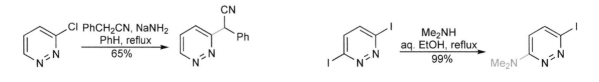

Stannanes can be prepared *via* nucleophilic displacements at low temperature.[49]

14.4 Metallation and Reactions of *C*-Metallated Diazines[50]

14.4.1 Direct Ring C–H Metallation

All three diazines undergo H/D exchange at all ring positions with MeONa/MeOD at 164 °C;[51] the transient carbanions that allow the exchange are formed somewhat faster than for pyridines, and again this is probably due to the acidifying, additional inductive withdrawal provided by the second nitrogen.

The three parent diazines have been lithiated adjacent to nitrogen (for pyrimidine at C-4, not C-2) using the non-nucleophilic lithium tetramethylpiperidide,[52] but the resulting heteroaryl-lithiums are very unstable, readily forming dimeric compounds by self addition. In some cases, the use of a somewhat higher temperature allows equilibration to a thermodynamic anion.[53]

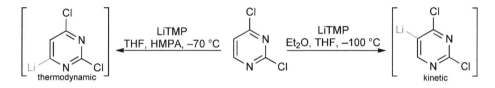

Moderate to good yields of trapped products can be obtained either by using very short lithiation times (pyridazine and pyrazine) or by *in situ* trapping, where the electrophile is added *before* the metallating agent.[54] 4-Lithiopyridazine can be prepared by transmetallation of the corresponding tri-*n*-butylstannane using *n*-butyllithium.

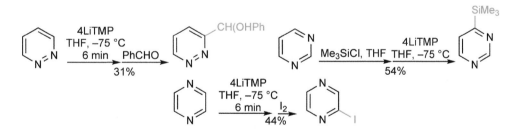

Direct zincation of each of the diazines can be achieved using a zinc diamide/lithium amide mixture, pyrazine and pyrimidine (at C-4) in THF at room temperature, pyridazine (at C-3) requiring reaction at reflux.[55] Pyridazine and pyrimidine can also be directly zincated using a phosphazene base (*t*-Bu-P4) with zinc iodide.[56]

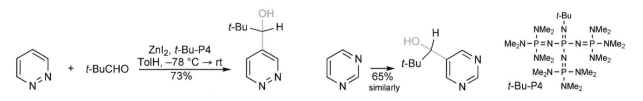

Lithiation of diazines with directing groups (methoxy, methylthio, chloro, fluoro, even iodo, and various carboxamides) is straightforward[57] and such derivatives are used widely. In contrast to the useful carboxamide directing group, thiocarboxamide is *not* an *ortho*-directing group; *N-t*-butylpyridazine-3-thiocarboxamide lithiates at C-5 and *N-t*-butylpyrazine-2-thiocarboxamide lithiates at C-5.[58] 2,4-Dimethoxypyrimidine and 2,4-bis(methylthio)pyrimidine, i.e. protected uracils, can be selectively metallated: the former is 5-lithiated with LiTMPMgCl.LiCl and TMP₂Mg.2LiCl brings about 6-magnesiation of both protected uracils.[59]

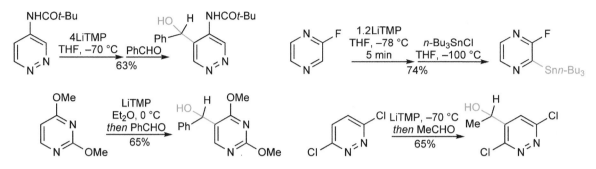

Studies of the positional lithiation of pyridin-2-yl-substituted diazines are instructive: each was subjected to excess lithium tetramethylpiperidide at −78 °C and then quenched with DCl; the structures indicate the percentage of deuteration at each position.[60]

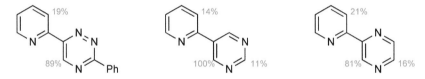

14.4.2 Metal–Halogen Exchange

Lithio-diazines are also accessible *via* halogen exchange with alkyl-lithiums, but very low temperatures must be used in order to avoid nucleophilic addition to the ring.[61] The examples below show how 5-bromopyrimidine can be lithiated at C-4, using LDA, or alternatively can be made to undergo exchange, using *n*-butyllithium.[62] Note, also, that in some cases, reactions are carried out by adding the electrophile *before* lithiation, a practice which incidentally illustrates that metal–halogen exchange with *n*-butyllithium is faster than the addition of *n*-butyllithium to a carbonyl compound.[63]

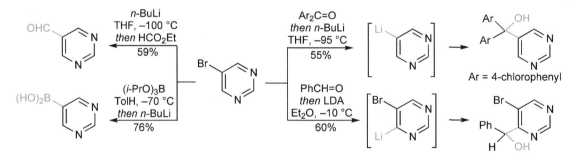

Lithio-pyrimidines, -pyrazines, and -pyridazines have been converted by exchange with zinc chloride into the more stable zinc compounds[64] for use in palladium-catalysed couplings. Diazine Grignard reagents, which can be prepared and used at 0 °C, or even, in some cases, at room temperature, are available *via* halogen exchange reactions using *i*-propylmagnesium chloride.[65] Cerium compounds (which give better results than lithio-pyrimidines in reactions with enolisable ketones) can be prepared from either bromo- or lithio-pyrimidine.[66]

14.5   Reactions with Reducing Agents

Due to their lower aromaticity, the diazines are more easily reduced than pyridines. Pyrazine and pyridazine can be reduced to hexahydro-derivatives with sodium in hot ethanol; under these conditions pyridazine has a tendency for subsequent reductive cleavage of the N–N bond.

Partial reductions of quaternary salts to dihydro-compounds can be achieved with borohydride, but such processes are much less well studied than in pyridinium salt chemistry (8.6).[67] 1,4-Dihydropyrazines have been produced with either silicon[68] or amide[69] protection at the nitrogen atoms, and all the diazines can be reduced to tetrahydro derivatives with carbamates on nitrogen, which aids in stabilisation and thus allows isolation.[70] 2-Amino-pyrimidines are reduced to 3,4,5,6-tetrahydro derivatives with triethylsilane in trifluoroacetic acid at room temperature, the products thus retaining a guanidine unit.[71]

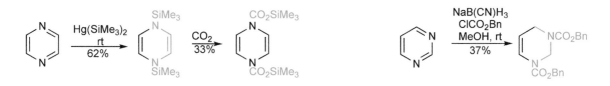

14.6   Reactions with Radicals

Nucleophilic radicals add readily to diazines under Minisci conditions.[72] Additions to pyrimidine often show little selectivity, C-2 *versus* C-4, but a selective Minisci reaction on 5-bromopyrimidine provided a convenient synthesis of the 4-benzoyl-derivative on a large scale.[73] Similar reactions on pyridazines shows selectivity for C-4,[74] even when C-3 is unsubstituted. Pyrazines[75] can, of course, substitute in only one type of position.

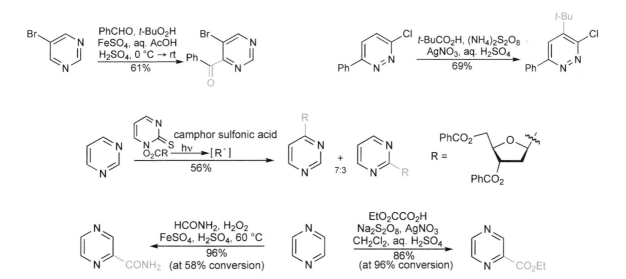

14.7   Electrocyclic Reactions

All the diazines, providing they also have electron-withdrawing substituents, undergo Inverse Electron Demand Diels–Alder (IEDDA) additions with dienophiles. Intramolecular reactions occur the most readily; these do not even require the presence of activating substituents. The immediate products of such process usually lose nitrogen (pyridazine adducts) or hydrogen cyanide (adducts from pyrimidines and pyrazines) to generate benzene and pyridine products,[76] respectively, as illustrated below.[77]

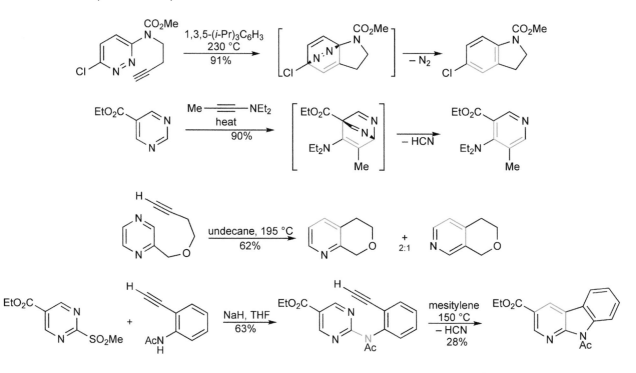

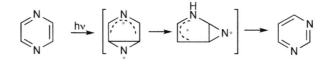

Pyrazine isomerises to pyrimidine, in the vapour phase, on exposure to UV light.[78]

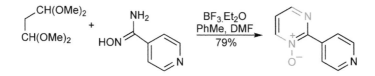

14.8 Diazine *N*-Oxides[79]

Although pyridazine and pyrazine *N*-oxides can be readily prepared by oxidation of the parent heterocycles, pyrimidine *N*-oxides are more difficult to obtain in this way, but they can conveniently be prepared by ring synthesis.[80]

Pyridazine and pyrazine *N*-oxides behave like their pyridine counterparts in electrophilic substitution.[81] Displacement of nitro β to the *N*-oxide function occurs about as readily as that of a γ-nitro group, but certainly, displacements on *N*-oxides proceed faster[82] than for the corresponding base.

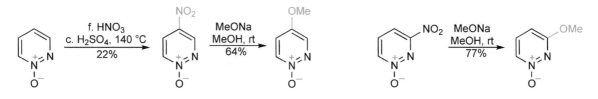

Nucleophilic substitution by halide, cyanide, carbon nucleophiles, such as enamines, and acetate (by reaction with acetic anhydride), with concomitant loss of the oxide function, occur smoothly in all three systems,[83] though the site of introduction of the nucleophile is not always that predicted by analogy with pyridine chemistry (α to the *N*-oxide), as illustrated by two of the examples below.[13]

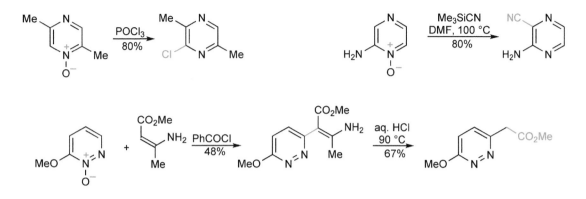

The *N*-oxide grouping can also serve as an activating substituent to allow regioselective lithiation[84] or for the further acidification (14.11) of side-chain methyl groups for condensations with, for example, aromatic aldehydes or amyl nitrite.[85]

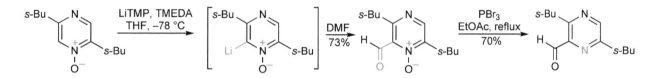

14.9 Oxy-Diazines

By far the most important naturally occurring diazines are the pyrimidinones uracil, thymine and cytosine, which, as the nucleosides uridine, thymidine and cytidine, are components of the nucleic acids, and as a consequence, a great deal of synthetic chemistry[86] has been directed towards these types of compound in the medicinal context (33.6.3 and 33.7). (In this chapter, the designations '4-pyrimidinone' and 'pyrimidin-4-one', etc, are both used, both are found widely in the literature, rather than the strictly correct '4(1*H*)-pyrimidinone', etc.)

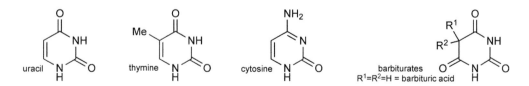

14.9.1 Structure of Oxy-Diazines

With the exception of 5-hydroxypyrimidine, which is analogous to 3-hydroxypyridine, all the mono-oxygenated diazines exist predominantly as carbonyl tautomers and are thus categorised as diazinones.

The dioxy-diazines present a more complicated picture, for, in some cases, where both oxygens are α or γ to a nitrogen, and both might be expected to exist in carbonyl form, one actually takes up the hydroxy form: a well-known example is 'maleic hydrazide'. One can rationalise the preference easily in this case, as resulting from the removal of the unfavourable interaction between two adjacent, partially positive nitrogen atoms in the dicarbonyl form. On the other hand, uracil exists as the dione and most of its reactions[87] can be interpreted on this basis. Barbituric acid adopts a tricarbonyl tautomeric form.

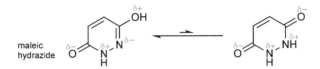

14.9.2 Reactions of Oxy-Diazines

Note, for many synthetic transformations, it is convenient to utilise halo- or alkoxy-diazines, in lieu of the (oxidation level) equivalent carbonyl compounds. Often this device facilitates solubility; a final hydrolysis converts to the carbonyl form.

14.9.2.1 Reactions with Electrophilic Reagents

The deactivating effect of two ring nitrogens cannot always be overcome by a single oxygen substituent: 3-pyridazinone can be neither nitrated nor halogenated, or again, of the singly oxygenated pyrimidines, only 2-pyrimidinone can be nitrated;[88] pyrazinones seem to be the most reactive towards electrophilic substitution.

5-Hydroxypyrimidine, the only phenolic diazine, is unstable even to dilute acid and no electrophilic substitutions have been reported.

Uracils undergo a range of electrophilic substitution reactions at carbon, such as halogenation, phenylsulfenylation,[89] mercuration,[90] nitration,[91] and hydroxy- and chloromethylation.[92] (**CAUTION**: *chloromethylations using formaldehyde and HCl produce the carcinogenic di-(chloromethyl) ether as a by-product.*) Bromination of uracils has been shown to proceed *via* the bromohydrin adduct (for fluorination see 31.1.1), and similarly of 2-pyrimidinone, *via* the bromohydrin-hydrate;[93] iodine with tetrabutylammonium peroxydisulfate allows iodination.[94] Isopropylidene uridine reacts with aromatic aldehydes in the presence of DABCO to give the 5-(arylmethanol). Both the isopropylidene group and the free 5′-hydroxyl are required for a successful reaction, leading to the conclusion that C-5 is activated by addition of the 5′-alkoxide to C-6.[95]

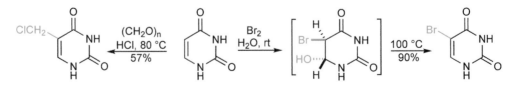

Uracil derivatives can be nitrated at C-5 under conditions that allow retention of a sugar residue at N-1.[96] Nitration at N-3 can also be achieved: N-3-nitro-compounds react with amines *via* an ANRORC mechanism, with displacement of nitramide and incorporation of the amine as a substituted N-3, as shown below.[97] This sequence has been utilised to prepare ^{15}N-3-labelled pyrimidines (31.2.2). An analogous ANRORC process takes 1-(2,4-dinitrophenyl)-uracils to 1-arylamino-uracils by reaction with arylamines at room temperature.[91]

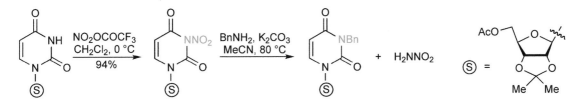

The presence of both an oxygen and an amino substituent (see also 14.10) can allow substitution with carbon electrophiles.[98,99]

14.9.2.2 Reactions with Nucleophilic Reagents

Diazinones are quite susceptible to nucleophilic attack, reaction taking place generally *via* Michael-type adducts rather than by attack at a carbonyl group. Grignard reagents add to give dihydro compounds and good leaving groups can be displaced.[100]

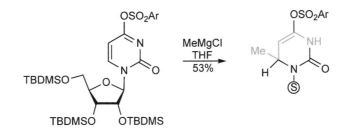

The reaction of cyanide with a protected 5-bromouridine[101] is instructive: under mild conditions a *cine*-substituted product is obtained *via* a Michael addition followed by β-elimination of bromide, but at higher temperatures, conversion of the 6- into the 5-cyano-isomer is observed, i.e. the product of apparent, direct displacement of bromide is obtained. The higher-temperature product arises *via* an isomerisation involving another Michael addition, then elimination of the 6-cyano group.

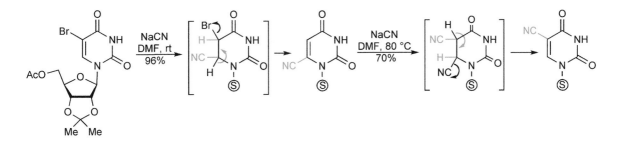

In a related reaction with the anion of phenylacetonitrile, the initial addition is followed by an internal alkylation, generating a cyclopropane-containing product.[102]

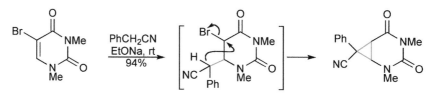

Ipso-substitution of 5-halo-uracils by amines can also be carried out – iodine under copper catalysis,[103] fluorine with ultraviolet irradiation[104] and bromine simply by heating.[105]

The conversion of 1,3-dimethyluracil into a mixture of *N,N'*-dimethylurea and the disodium salt of formylacetic acid begins with the addition of hydroxide at C-6.[106] The propensity for uracils to add nucleophiles can be put to synthetic use by reaction with double nucleophiles, such as ureas or guanidines, when a sequence of addition, ring opening and reclosure can achieve (at first sight) extraordinary transformations.[107]

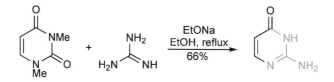

Products of 2- and 4-substitution of hydrogen are obtained by reaction of the sodium salt of imidazole with the phenyl pyrimidin-5-yl carbinol mesylate, with none of the product of direct substitution. Similar results are obtained with the sodium salts of pyrrole and indole, but other nucleophiles, such as amines, give complex product mixtures.[108]

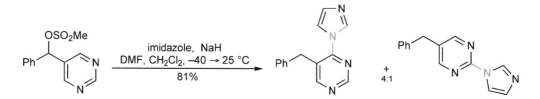

14.9.2.3 Deprotonation of N-Hydrogen and Other Reactions at Nitrogen

Like pyridones, oxy-diazines are readily deprotonated under mild conditions, to give ambident anions which can be alkylated conveniently by phase-transfer methods, alkylation usually occurring at nitrogen.[109] *N*-Arylations of uracils also proceed in this way with, for example, 1-fluoro-4-nitrobenzene.[110] 3-Pyridazinones alkylate cleanly on N-2 under phase-transfer conditions,[111] but the regiochemistry of uracil alkylation is sometimes difficult to control (see also below). Uracils are sufficiently acidic to take part in Mitsunobu reactions.[112]

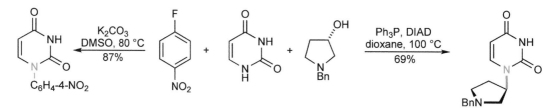

Carbon substitution can also be effected in some cases *via* delocalised *N*-anions, as in the reaction of 6-methyluracil with formaldehyde,[113] or with diazonium salts.[114]

O-Alkylation is also possible and is particularly important in ribosides, where it occurs intramolecularly and can be used to control the stereochemistry of substitution in the sugar residue, as illustrated in the following sequence for replacement of the 3′-hydroxyl with azide and with overall retention of configuration.[115]

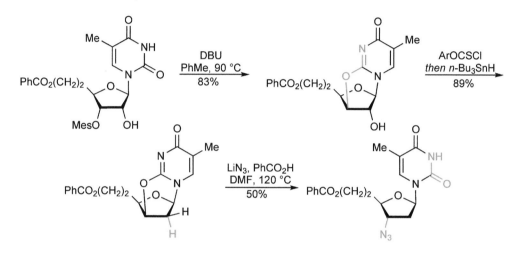

N-Alkylation of *O*-silylated derivatives,[116] for example with glycosyltrifluoro-acetimidates,[117] is an important method for unambiguous *N*-alkylation,[118] especially for ribosylation of uracils.[119]

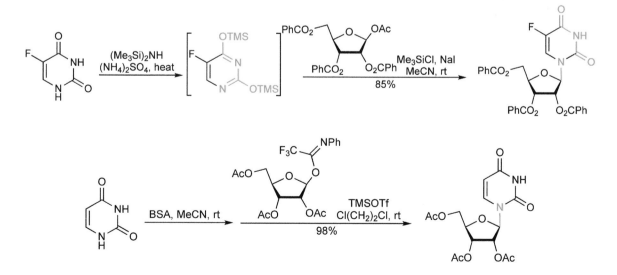

Stereospecific ribosylation of uracils and other pyrimidine bases can be carried out by attachment to the 5-hydroxymethyl substituent of the sugar, followed by internal delivery to C-2.[120]

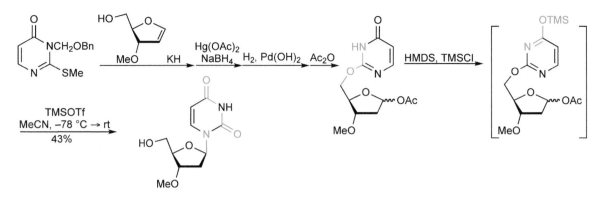

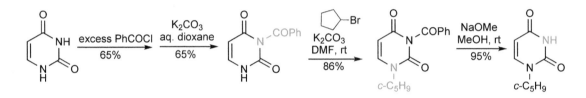

Selective alkylation of uracil at N-1 can be achieved *via* alkylation of the 3-benzoyl-derivative, itself obtained by selective hydrolysis of the dibenzoyl compound.[121]

14.9.2.4 *C-Metallation and Reactions of C-Metallated Diazinones*

C-Lithiation of uridine derivatives has been thoroughly studied as a means for the introduction of functional groups at C-5 and C-6. Chelating groups at C-5′ (hydroxyl or methoxymethoxy) favour 6-metallation,[122] as do equilibrating conditions, indicating that this is the most stable lithio derivative. Kinetic lithiation, at C-5, can be achieved when weakly chelating silyloxy-groups are used as protecting groups for the sugar.[123] It is remarkable that protection of the N-3–H group is not necessary and this is illustrated again in the 6-lithiation of uracil carrying an ethoxymethyl-substitutent on N-1.[124]

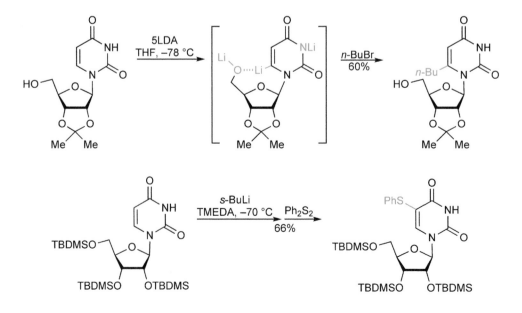

NH-Protection is also unnecessary for the side-chain metallation of 6-methylpyrimidin-2-one.[125]

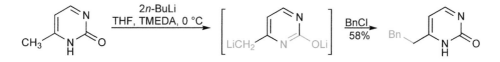

Zinc derivatives of uracils can be prepared directly by reaction of the appropriate halide with zinc dust. They react with a limited range of electrophiles, but are particularly useful for palladium-catalysed couplings[126] (see also 4.1).

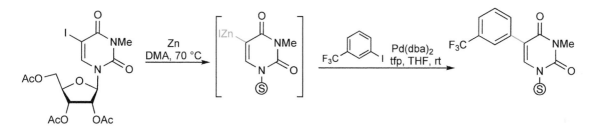

Halogen–magnesium exchange can be achieved with 5- or 6-iodouracils, again *without* masking the *N*-hydrogen groups.[127]

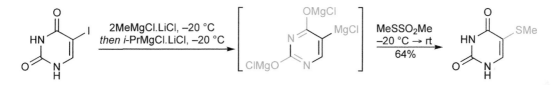

14.9.2.5 Replacement of Oxygen

Oxy-diazines, with the oxygen α to nitrogen, can be converted into halo-[30,128] and thio compounds[129] using the same reagents used for 2- and 4-pyridones, including *N*-bromosuccinimide with triphenylphosphine.[130] The reactions of *O*-silylated pyrazinones with phosphorus(III) bromide or phosphorus(V) chloride are also efficient.[131]

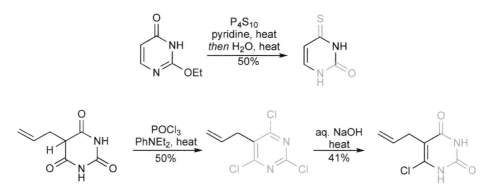

Diazinones can also be converted into amino-diazines, without the (classical) intermediacy of an isolated halo derivative, by various processes including the use of 1,2,4-triazole, as illustrated below.[132]

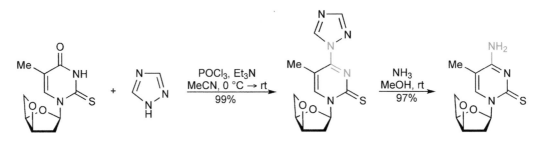

A direct replacement of the 4-oxygen of 1-substituted uracils by amines is possible by reaction in the presence of BOP and DBU.[133]

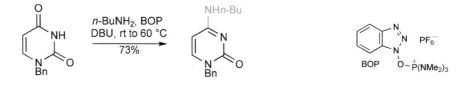

5-Hydroxypyrimidine-2,4-diones react as 5-ketones and undergo Wittig condensation, the double bond thus formed isomerising back into the more stable position in the ring.[134] Barbituric acid and C-5 derivatives can be converted into uracils by first forming a 6-mesylate and then catalytic hydrogenolysis.[135]

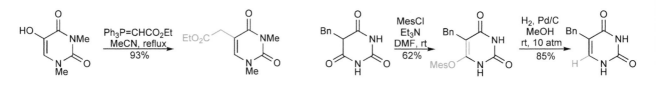

14.9.2.6 Electrocyclic Reactions
Mesoionic pyrazinium-3-olates undergo cycloadditions[136] similar to those known for pyridinium-3-olates (8.7) and pyrylium-3-olates (Section 11.1.7).

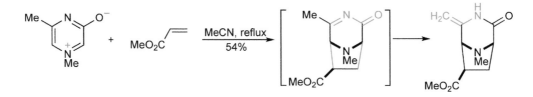

Heterodienophiles have also been studied: singlet oxygen across the 2,5-positions of pyrazinones.[137] The immediate cycloadduct is isolable when acryloyl cyanide is used as the heterodiene component in reaction with a pyrimidine-2,4-dione.[138]

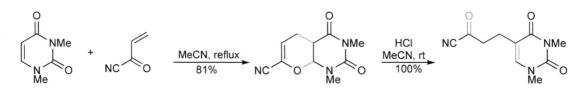

The reaction of deoxyuridine with nitrile oxides gives products of apparent electrophilic substitution, but these probably arise by ring opening of a cycloadduct.[139]

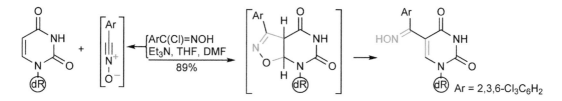

Extensive studies have been made of the versatile 3,5-dihalo-2(1H)-pyrazinones and their derivatives.[140] A remarkable ring synthesis[141] provides the dichloropyrazinones and these, and their derivatives, undergo cycloadditions.[142]

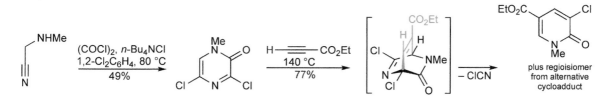

Because of possible relevance to mutagenesis, considerable effort has been devoted to study of the photochemical transformations of oxypyrimidines; uracil, for example, takes part in a [2 + 2] cycloaddition with itself,[143] or with vinylene carbonate (1,3-dioxol-2-one).[144] Uracils undergo radical additions;[145] these too are of possible relevance to mutagenesis mechanisms.

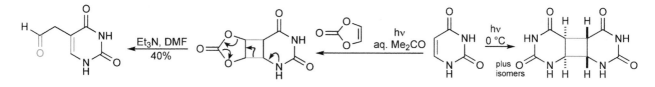

14.10 Amino-Diazines

Amino-diazines exist in the amino form. They are stronger bases than the corresponding unsubstituted systems and always protonate on one of the ring nitrogen atoms; where two isomeric cations are possible, the order of preference for protonation is of a ring nitrogen, which is $\gamma > \alpha > \beta$ to the amino group, as can be seen in the two examples below. A corollary of this is that those amino-diazines that contain a γ-amino-azine system are the strongest bases.

The alkali-promoted rearrangement of quaternary salts derived from 2-aminopyrimidine provides the simplest example of the Dimroth rearrangement.[146] The larger the substituent on the positively charged ring nitrogen, the more rapidly the rearrangement proceeds, no doubt as a result of the consequent relief in strain between the substituent and the adjacent amino group.

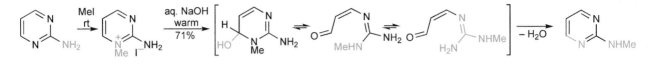

All of the amino-diazines react with nitrous acid to give the corresponding diazinones,[30] by way of highly reactive diazonium salts; even 5-aminopyrimidine does not give a stable diazonium salt, though a low yield of 2-chloropyrimidine can be obtained by diazotisation of 2-aminopyrimidine in concentrated hydrochloric acid.[147]

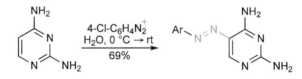

One amino group is sufficient in most cases to allow easy electrophilic substitution, halogenation[148] for example, and two amino groups activate the ring to attack even by weaker electrophilic reagents – for example by thiocyanogen.[149] Diamino-pyrimidines will couple with diazonium salts,[150] which provides a means for the introduction of a third nitrogen substituent.

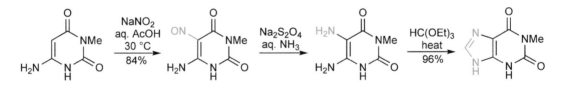

Amino-oxy-pyrimidines,[151] and amino-dioxy-pyrimidines[152] can be *C*-nitrosated, and such 5,6-dinitrogen-substituted pyrimidines, after reduction to 5,6-diamino-pyrimidines, are important intermediates for the synthesis of purines (an example is shown below; see also 27.11.1.1) and pteridines (14.14).

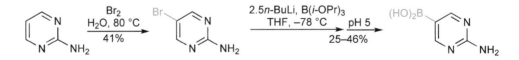

2-Amino-5-bromopyrimidine can be converted into a 5-boronic acid, without the need to mask the amino group.[153]

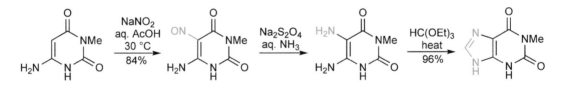

14.11 Alkyl-Diazines

All alkyl-diazines, with the exception of 5-alkyl-pyrimidines, undergo condensations that involve deprotonation of the alkyl group,[154] in the same way as α- and γ-picolines.[155] The intermediate anions are stabilised by mesomerism involving one, or in the case of 2- and 4-alkyl-pyrimidines, both nitrogens.

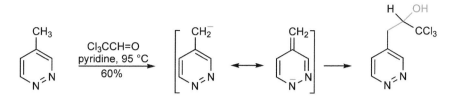

In pyrimidines, a 4-alkyl- is deprotonated more readily than a 2-alkyl-group;[156] here again one sees the greater stability associated with a γ-quinonoid resonating anion. Side-chain radical halogenation selects a pyrimidine-5-methyl over a pyrimidine-4-methyl; the reverse selectivity can be achieved by halogenation in acid solution – presumably an *N*-protonated, side-chain-deprotonated species, i.e. the enamine tautomer, is involved.[157]

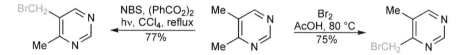

14.12 Quaternary Diazinium Salts

The already high susceptibility of the diazines to nucleophilic addition is greatly increased by quaternisation. Addition of organometallic reagents to *N*-acyl quaternary salts has been achieved in some cases, but is much more restricted than is the case with pyridines (8.12.2). Thus, allylstannanes[158] and -silanes[159] and silyl enol ethers have been added to diazine salts (hydride also traps such salts (14.5)). Pyridazines give good yields of mono-adducts with attack mainly α to the acylated nitrogen, but the regioselectivity of silyl ether addition[160] is, in some cases, sensitive to substituents. Pyrazine gives mainly double addition products[161] and pyrimidine produces only the double adduct. Reissert adducts (cf. 9.13) have been described for pyridazine and pyrimidine.[162]

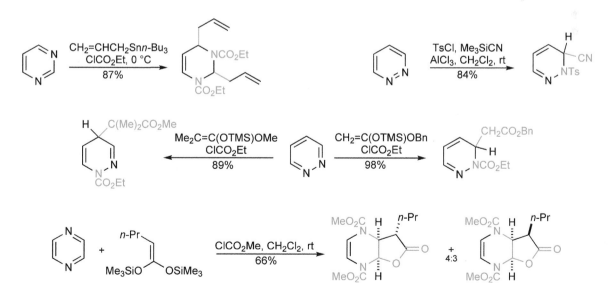

14.13 Synthesis of Diazines

Routes for the ring synthesis of the isomeric diazines are, as one would expect, quite different one from the other, and must therefore be dealt with separately.

14.13.1 Pyridazines

14.13.1.1   From a 1,4-Dicarbonyl Compound and a Hydrazine

A common method for the synthesis of pyridazines involves a 1,4-dicarbonyl compound reacting with hydrazine; unless the four-carbon component is unsaturated, a final oxidative step is needed to give an aromatic pyridazine.

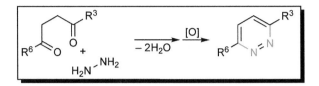

The most useful procedure utilises a 1,4-keto-ester giving a dihydro-pyridazinone, which can be easily dehydrogenated to the fully aromatic heterocycle, often by *C*-bromination then dehydrobromination;[163] alternatively, simple air oxidation can often suffice.[164] 6-Aryl-pyridazin-3-ones have been produced by this route in a number of ways: using an α-amino nitrile as a masked ketone in the four-carbon component,[165] or by reaction of an acetophenone with glyoxylic acid and then hydrazine.[166] Friedel–Crafts acylation using succinic anhydride is an alternative route to 1,4-keto-acids, reaction with hydrazine giving 6-aryl-pyridazinones.[167] Alkylation of an enamine with a phenacyl bromide produces 1-aryl-1,4-diketones, allowing synthesis of 3-aryl-pyridazines.[168]

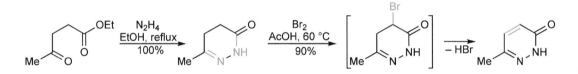

Maleic anhydride and hydrazine give the hydroxy-pyridazinone ('maleic hydrazide') directly,[169] the additional unsaturation in the 1,4-dicarbonyl component meaning that an oxidative step is not required; conversion of 3-hydroxypyridazin-6-one into 3,6-dichloropyridazine makes this useful intermediate very easily available. Mucohalo acids (18.1.1.4), synthons for 4-carboxy-aldehydes, are an oxidation level down and produce 1-aryl-pyridazin-3-ones on reaction with arylhydrazines.[170]

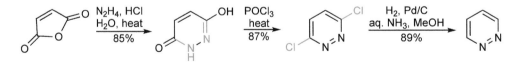

The use of saturated 1,4-diketones can suffer from the disadvantage that they can react with hydrazine in two ways, giving mixtures of the desired dihydro-pyridazine and an *N*-amino-pyrrole; this complication does not arise when unsaturated 1,4-diketones are employed.[171] There is also no structural ambiguity when aryl-hydrazines are reacted with pent-4-ynoic acid catalysed by zinc chloride, producing 2-aryl-3,4-dihydro-6-methyl-pyridazin-3-ones.[172] Synthons for unsaturated 1,4-diketones are available as cyclic acetals from furans (18.1.4), and react with hydrazines to give the fully aromatic pyridazines directly.[173]

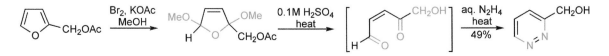

14.13.1.2 By Cycloaddition of a 1,2,4,5-Tetrazine with an Alkyne

Cycloaddition of a 1,2,4,5-tetrazine with an alkyne (or its equivalent), with elimination of nitrogen gives pyridazines.

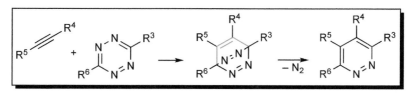

This process works best when the tetrazine has electron-withdrawing substituents, but 1,2,4,5-tetrazine itself will react with a range of simple alkynes, enamines and enol ethers, under quite moderate conditions.[174] A wide range of substituents can be incorporated *via* the acetylene, including nitro and trimethylsilyl, affording the means to access other substituted pyridazines.[175] The preparation of a boronic ester is another example.[176]

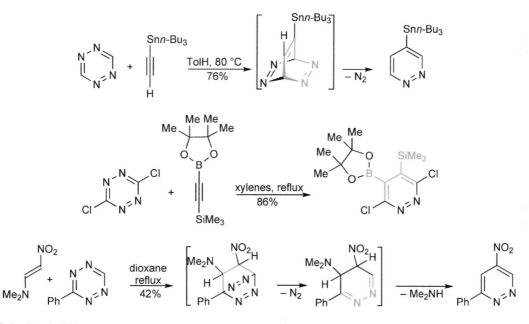

14.13.2 Pyrimidines

14.13.2.1 From a 1,3-Dicarbonyl Compound and an N–C–N Fragment

The most general pyrimidine ring synthesis involves the combination of a 1,3-dicarbonyl component with an N–C–N fragment such as a urea, an amidine or a guanidine.

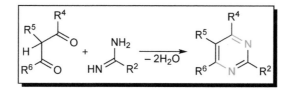

The choice of N–C–N component – amidine,[177] guanidine,[178] or a urea[179] (thiourea[180]) – governs the substitution at C-2 in the product heterocycle. Although not formally 'N–C–N' components, formamide,[181] or an orthoester, plus ammonia[182] can serve instead of urea in this type of approach. The dicarbonyl component can be generated *in situ*, for example formylacetic acid (by decarbonylation of malic acid), or a synthon used (1,1,3,3-tetramethoxypropane for malondialdehyde). When a nitrile serves as a carbonyl equivalent, the resulting heterocycle now carries an amino-substituent.[183] The use of 2-bromo-1,1,3,3-tetramethoxypropane provides a route to 5-bromopyrimidine,[184] methanetricarboxaldehyde reacts with amidines to give 5-formyl-pyrimidines,[185] and 2-aminomalondialdehyde leads to 5-amino-pyrimidines.[186] The examples below illustrate some of these.[187]

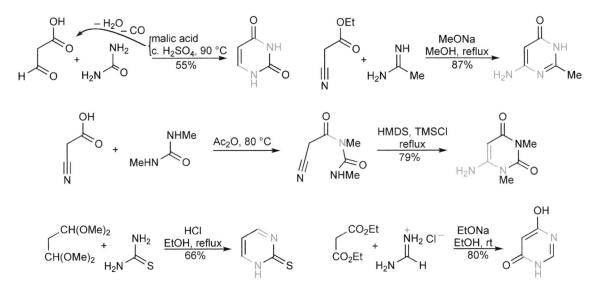

Other synthons for 1,3-dicarbonyl compounds that have been successfully applied include β-chloro-α,β-unsaturated ketones and aldehydes,[188] β-dimethylamino-α,β-unsaturated ketones (easily obtained from ketones by reaction with DMFDMA),[189] β-alkoxy-enones[190] and vinyl-amidinium salts.[191] Alkynyl-ketones react with *S*-alkyl-isothioureas, giving 2-alkylthio-pyrimidines[192] and propiolic acid reacts with urea to give uracil directly in about 50% yield.[193] 1,3-Keto-esters with formamidine produce 4-pyrimidinones[194] and *C*-substituted formamidines with ethyl cyanoacetate give 2-substituted-6-amino-4-pyrimidinones.[195] In analogy, pyrimidines fused to other rings, for example as in quinazolines, can be made from *ortho*-aminonitriles[196] and in general, from β-enamino esters.[197]

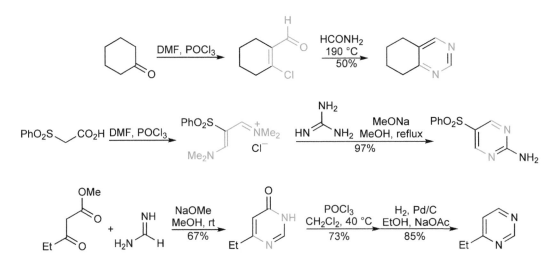

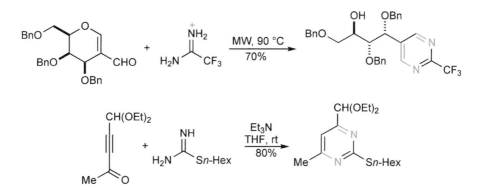

Barbituric acid and barbiturates can be synthesised by reacting a malonate with a urea,[198] or a bis primary amide of a substituted malonic acid with diethyl carbonate.[199]

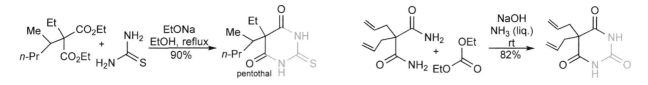

14.13.2.2 By Cycloadditions

Cycloaddition of a 1,3,5-triazine with an alkyne (or its equivalent) gives pyrimidines after loss of hydrogen cyanide.

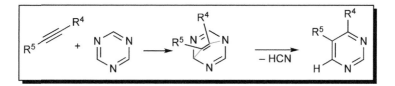

The formation of pyrimidines[200] *via* aza-Diels–Alder reactions is similar to the preparation of pyridazines from tetrazines (cf. 14.13.1.2).

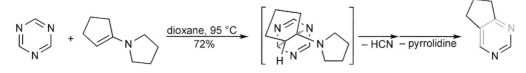

14.13.2.3 From 3-Ethoxyacryloyl Isocyanate and Primary Amines

Primary amines add to the isocyanate group in a 3-alkoxyacryloyl isocyanate; ring closure then gives pyrimidines *via* intramolecular displacement of the alkoxy-group.

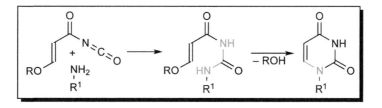

Uracils can be prepared *via* reaction of primary amines with 3-ethoxyacryloyl isocyanate;[201] this method is particularly suitable for complex amines and has found much use in recent years in the synthesis of, for example, carbocyclic nucleoside analogues as potential anti-viral agents.[202] The immediate product of amine/isocyanate interaction can be cyclised under either acidic or basic[203] conditions; the method can also be applied to thiouracil synthesis by the use of the corresponding isothiocyanate.

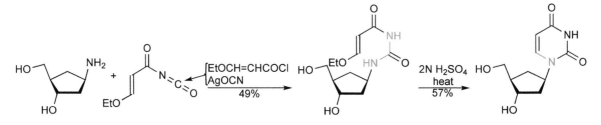

Though different bonds are made, it is useful to include here the enaminothioimidate shown below for the synthesis of pyrimidin-2-ones and -thiones, by reaction with isocyanates and isothiocyanates, respectively, proceeding *via* a cycloaddition between the azadiene and the imine unit of the –N=C=S, then loss of dimethylamine to aromatise.[204]

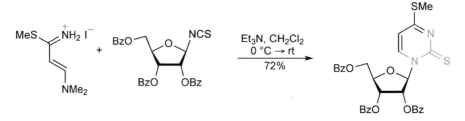

14.13.2.4 From an Aldehyde, a 1,3-Dicarbonyl-Compound and a Urea
The Biginelli Synthesis[205]
This very old synthesis[206] of 1,4-dihydropyrimidin-2-ones, which is analogous to the Hantzsch pyridine synthesis (8.14.1.2), is much used, particularly for library synthesis, and many variants of the reaction conditions have been described; most often the condensation is acid or Lewis-acid catalysed. The products are important in their own right, but can also be dehydrogenated to give pyrimidin-2-ones.[207] If guanidine is used instead of the urea component, 2-amino-1,4-dihydropyrimidines result.[208]

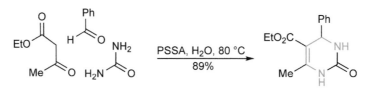

14.13.2.5 From Ketones
Condensation of ketones with two mole equivalents of a nitrile in the presence of trifluoromethanesulfonic acid anhydride is a useful method for the production of a limited range of pyrimidines, where the substituents at C-2 and C-4 are identical.[209, 210]

Methyl ketones give 4-monosubstituted-pyrimidines when reacted with formamide at high temperature.[211] Although the yields, apart from aryl-methyl-ketones, are not large, the method is extremely simple. Aryl higher-alkyl ketones also react in this way, giving 4-aryl-5-alkyl-pyrimidines.

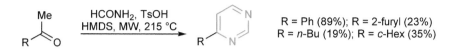

R = Ph (89%); R = 2-furyl (23%)
R = *n*-Bu (19%); R = *c*-Hex (35%)

14.13.3 Pyrazines

Pyrazine is not easily made in the laboratory. Commercially, the high temperature cyclodehydrogenation of precursors such as *N*-hydroxyethylethane-1,2-diamine is used.

14.13.3.1 From the Self-Condensation of a 2-Amino-Ketone

Symmetrical pyrazines result from the spontaneous self condensation of two mole equivalents of a 2-amino-ketone, or 2-amino-aldehyde, followed by an oxidation.

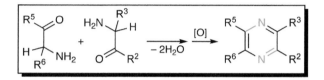

2-Amino-carbonyl compounds, which are stable only as their salts, are usually prepared *in situ* by the reduction of 2-diazo-, -oximino- or -azido-ketones. The dihydropyrazines produced by this strategy are very easily aromatised, for example by air oxidation, and often distillation alone is sufficient to bring about disproportionation.[212]

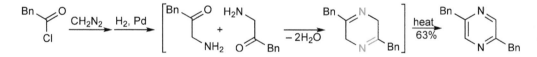

α-Amino esters are more stable than α-amino-ketones but nonetheless easily self-condense to give heterocycles, known as 2,5-diketopiperazines. These compounds are resistant to oxidation, but can be used to prepare aromatic pyrazines after first converting them into dichloro- or dialkoxy-dihydropyrazines.[213]

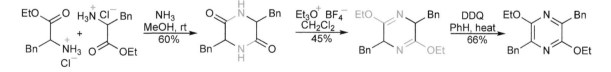

14.13.3.2 From 1,2-Dicarbonyl Compounds and 1,2-Diamines

1,2-Dicarbonyl compounds undergo double condensation with 1,2-diamines; an oxidation is then required.

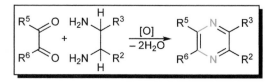

This method is well suited to the formation of symmetrical pyrazines,[214] but if both diketone and diamine are unsymmetrical, two isomeric pyrazines are formed. The dihydro-pyrazines can be dehydrogenated and they will also react with aldehydes and ketones, with introduction of another alkyl group at the same time as achieving the aromatic oxidation level.[215]

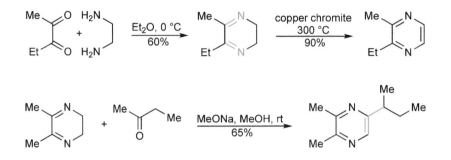

Other dinitrogen components that also carry unsaturation are α-amino acid amides,[216] from which pyrazinones can be formed; a special example is aminomalonamide and a pyrazinone synthesis using this unit is shown below.[217]

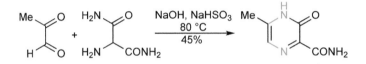

The direct synthesis of aromatic pyrazines using a 1,2-dicarbonyl compound requires a 1,2-diamino-*alkene*, but very few simple examples of such compounds are known; diaminomaleonitrile[218] is stable and serves in this context.

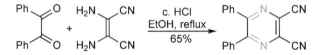

An ingenious modification of the general method uses 5,6-diaminouracil as a masked unsaturated 1,2-diamine: the products can be hydrolysed with cleavage of the pyrimidinone ring finally arriving at 2-amino-pyrazine-3-acids as products.[219]

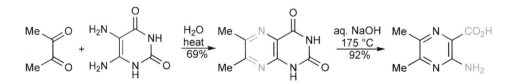

14.13.3.3 From 1,2-Diketone Mono-Oximes

Alkyl-pyrazines can be produced by an ingenious sequence involving an electrocyclic ring closure of a 1-hydroxy-1,4-diazatriene, aromatisation being completed by loss of the oxygen from the original oxime hydroxyl group.[220]

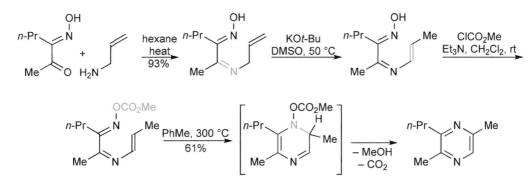

14.13.4 Notable Syntheses of Diazines

14.13.4.1 4,6-Diamino-5-thioformamido-2-methylpyrimidine

4,6-Diamino-5-thioformamido-2-methylpyrimidine can be converted into 2-methyladenine.

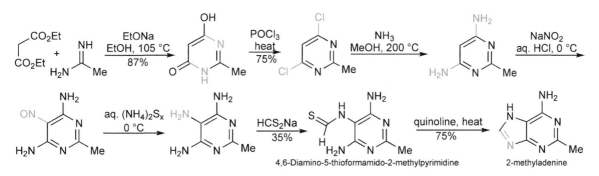

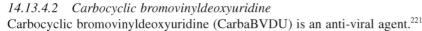

14.13.4.2 Carbocyclic bromovinyldeoxyuridine

Carbocyclic bromovinyldeoxyuridine (CarbaBVDU) is an anti-viral agent.[221]

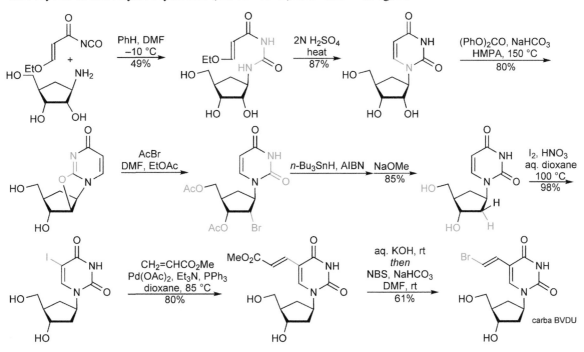

14.13.4.3 Coelenterazine

Coelenterazine, a bioluminescent compound from a jellyfish, with potential for use in bioassays, has been synthesised in an overall 25% yield from chloropyrazine.[222]

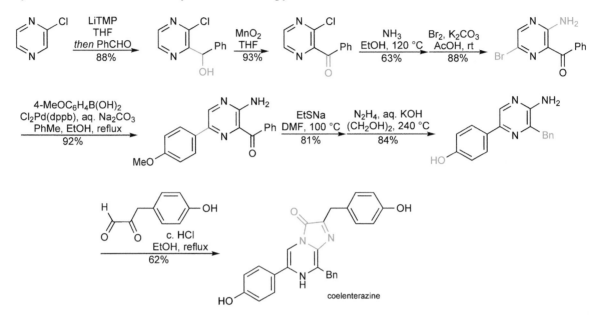

14.14 Pteridines

Pyrazino[2,3-*d*]pyrimidines are known as 'pteridines',[223] because the first examples of the ring system, as natural products, were found in pigments, like xanthopterin (yellow), in the wings of butterflies (*Lepidoptera*). The pteridine ring system has subsequently been found in coenzymes that use tetrahydrofolic acid (derived from the vitamin folic acid), and in the cofactor of the oxomolybdoenzymes[224] and comparable tungsten enzymes.

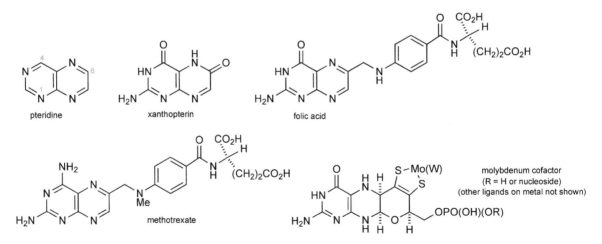

The synthesis of the pteridine ring system has been approached by two obvious routes: one is the fusion of the pyrazine ring onto a pre-formed 4,5-diamino-pyrimidine, and the second, the elaboration of the pyrimidine ring on a pre-formed pyrazine. The first of these, the *Isay synthesis*, suffers from the disadvantage that condensation of the heterocyclic 1,2-diamine with an unsymmetrical 1,2-dicarbonyl compound usually leads to a mixture of 6- or 7-substituted isomers.[225] It was to avoid this difficulty that the alternative strategy, the *Taylor synthesis*, now widely used, starting with a pyrazine, was developed.[226] This approach has the further advantage that because the pyrazine ring is pre-synthesised, using 2-cyanoglycinamide,[227] it eventually produces, regioselectively, 6-substituted pteridines – substitution at the 6-position is the common pattern in natural pteridines.

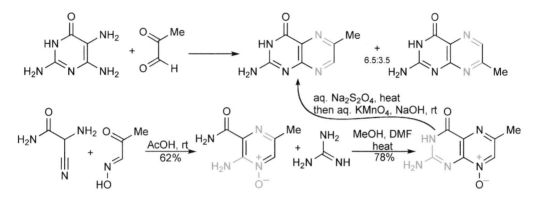

Exercises
Straightforward revision exercises (consult Chapters 13 and 14):
(a) Why is it difficult to form diprotonic salts from diazines?
(b) How do uridine, thymidine and cytidine differ?
(c) Are the diazines more or less reactive towards *C*-electrophilic substitution than pyridine?
(d) What factor assists and what factor mediates against nucleophilic displacement of hydrogen in diazines?
(e) Which is the only chlorodiazine that does not undergo easy nucleophilic displacement, and why?
(f) What precaution is usually necessary in order to lithiate a diazine?
(g) Write out one example each where a pyridazine, a pyrimidine and a pyrazine undergo a cycloaddition, acting as a diene or an azadiene.
(h) How could one convert an oxy-diazine, where the oxygen is α to a nitrogen: (i) into a corresponding chloro-diazine, (ii) efficiently into a corresponding *N*-methyl-diazinone, (iii) into a corresponding amino-diazine, but without involving a chloro-diazine.
(i) What is the product from hydrazine and a 1,4-keto-ester? How could it be converted into a pyridazinone?
(j) Given pentane-2,4-dione, how could one prepare: (i) 4,6-dimethylpyrimidine, (ii) 4,6-dimethyl-2-pyrimidone (iii) 2-amino-4,6-dimethylpyrimidine?
(k) What substitution pattern is the easiest to achieve in the ring synthesis of pyrazines?

More advanced exercises:
1. What compounds are produced at each stage in the following sequences: (i) pyridazin-3-one reacted with $POCl_3$ ($\rightarrow C_4H_3N_2Cl$) and this with NaOMe ($\rightarrow C_5H_6N_2O$); (ii) chloropyrazine with $BuNH_2$/120 °C ($\rightarrow C_8H_{13}N_3$).
2. What are the structures of the compounds formed: (i) $C_6H_9IN_2S$ from 3-methylthiopyridazine and $C_6H_8ClIN_2$ from 3-chloro-6-methylpyridazine, each with MeI; (ii) $C_5H_2Cl_2N_2O$ from treatment of 2,6-dichloropyrazine with LiTMP then HCO_2Et; (iii) $C_{14}H_{12}N_2O_2$ from 2,6-dimethoxypyrazine with LiTMP, then I_2 then PhC≡CH/Pd(0).
3. Decide the structures of the compounds produced by the following sequences: (i) $C_6H_9N_3$ from 2-aminopyrimidine first with $NaNO_2$/c. HCl/–15 °C and then the product with Me_2NH; (ii) $C_{18}H_{14}N_2$ from 3-methyl-6-phenylpyridazine with $PhCH=O/Ac_2O$/heat.

4. Write sequences and structures for intermediates and final products in the following ring syntheses: (i) chlorobenzene with succinic anhydride/AlCl$_3$ ($\rightarrow$ C$_{10}$H$_9$ClO$_3$), then this with N$_2$H$_4$ ($\rightarrow$ C$_{10}$H$_9$ClN$_2$O) and finally this with Br$_2$/AcOH ($\rightarrow$ C$_{10}$H$_7$ClN$_2$O); (ii) 2,5-dimethylfuran reacted with Br$_2$ in MeOH ($\rightarrow$ C$_8$H$_{14}$O$_3$), then this firstly with aqueous acid and then hydrazine ($\rightarrow$ C$_6$H$_8$N$_2$).

5. What would be the pyrimidine products from the following combinations: (i) 1,1-dimethoxybutan-3-one with guanidinium hydrogen carbonate ($\rightarrow$ C$_5$H$_7$N$_3$); (ii) ethyl cyanoacetate with guanidine/NaOEt ($\rightarrow$ C$_4$H$_6$N$_4$O); (iii) ethyl cyanoacetate with urea/EtONa ($\rightarrow$ C$_4$H$_5$N$_3$O$_2$); (iv) (EtO)$_2$CHCH$_2$CH(OEt)$_2$/HCl/urea ($\rightarrow$ C$_4$H$_4$N$_2$O),

6. Decide the structures of the final products and the intermediates (in part (i)) from the following combinations of reactants: (i) MeOCH$_2$COMe with EtO$_2$CH/Na ($\rightarrow$ C$_5$H$_8$O$_3$), then this with thiourea ($\rightarrow$ C$_6$H$_8$N$_2$OS), then this with H$_2$/Ni ($\rightarrow$ C$_6$H$_8$N$_2$O); (ii) PhCOCH$_2$CO$_2$Et with EtC(=NH)NH$_2$ ($\rightarrow$ C$_{12}$H$_{12}$N$_2$O); (iii) PhCOCHO with MeCH(NH$_2$)CONH$_2$ ($\rightarrow$ C$_{11}$H$_{10}$N$_2$O).

References

[1] 'Pyridazines', Tishler, M. and Stanovnik, B., *Adv. Heterocycl. Chem.*, **1968**, *9*, 211; 'Recent advances in pyridazine chemistry', *ibid.*, **1979**, *24*, 363; 'Advances in pyridazine chemistry', *ibid.*, **1990**, *49*, 385.

[2] 'Conversion of simple pyrimidines into derivatives with a carbon functional group', Sakamoto, T. and Yamanaka, H., *Heterocycles*, **1981**, *15*, 583.

[3] 'Recent advances in pyrazine chemistry', Cheeseman, G. W. H. and Werstiuk, E. S. G., *Adv. Heterocycl. Chem.*, **1972**, *14*, 99.

[4] 'Reactions of annular nitrogens of azines with electrophiles', Grimmett, M. R. and Keene, B. R. T., *Adv. Heterocycl. Chem.*, **1988**, *43*, 127.

[5] Curphey T. J. and Prasad, K. S., *J. Org. Chem.*, **1972**, *37*, 2259.

[6] Zoltewicz, J. A. and Deady, L. W., *J. Am. Chem. Soc.*, **1972**, *94*, 2765.

[7] Lund, H. and Lunde, P., *Acta Chem. Scand.*, **1967**, *21*, 1067.

[8] Koelsch, C. F. and Gumprecht, W. H., *J. Org. Chem.*, **1958**, *23*, 1603; Kubota, T. and Watanabe, H., *Bull. Chem. Soc. Jpn.*, **1963**, *36*, 1093.

[9] Yamanaka, H., Ogawa, S., and Sakamoto, T., *Heterocycles*, **1981**, *16*, 573.

[10] Suzuki, I., Nakadate, M. and Sueyoshi, S., *Tetrahedron Lett.*, **1968**, 1855.

[11] Brederick, H., Gompper, R. and Herlinger, H., *Chem. Ber.*, **1958**, *91*, 2832.

[12] Kano, H., Ogata, M., Watanabe, H. and Ishizuka, I., *Chem. Pharm. Bull.*, **1961**, *9*, 1017; Ogata, M. and Kano, H., *ibid.*, **1963**, *11*, 29.

[13] Ogata, M., Watanabe, H., Tori, K. and Kano, H., *Tetrahedron Lett.*, **1964**, 19.

[14] Sato, N., *J. Heterocycl. Chem.*, **1983**, *20*, 169.

[15] Gainer, H., Kokorudz, M. and Langdon, W. K., *J. Org. Chem.*, **1961**, *26*, 2360; Bourguignon, J., Lemarchand, M. and Quéquiner, G., *J. Heterocycl. Chem.*, **1980**, *17*, 257.

[16] Kress, R. J. and Moore, L. L., *J. Heterocycl. Chem.*, **1973**, *10*, 153; Pews, R. G., *Heterocycles*, **1990**, *31*, 109.

[17] Bowman, A., *J. Chem. Soc.*, **1937**, 494; Keiner, A., *Angew. Chem., Int. Ed. Engl.*, **1992**, *31*, 774.

[18] Jones, R. G. and McLaughlin, K. C., *Org. Synth., Coll. Vol. IV*, **1963**, 824.

[19] Gotor, V., Quirós, M., Liz, R., Frigda, J. and Fernandez, R., *Tetrahedron*, **1997**, *53*, 6421.

[20] Lupattelli, P., Saladino, R. and Mincione, E., *Tetrahedron Lett.*, **1993**, *34*, 6313.

[21] Letsinger, R. L. and Lasco, R., *J. Org. Chem.*, **1956**, *21*, 812.

[22] Strekowski, L., Harden, D. B., Grubb, W. B., Patterson, S. E., Czarny, A., Makrosz, M. J., Cegla, M. T. and Wydra, R. L., *J. Heterocycl. Chem.*, **1990**, *27*, 1393.

[23] Bursavich, M. G., Lombardi, S. and Gilbert, A. M., *Org. Lett.*, **2005**, *7*, 4113.

[24] Strewkowski, L., Harden, D. and Watson, R. A., *Synthesis*, **1988**, 70.

[25] Shreve, R. N. and Berg, L., *J. Am. Chem. Soc.*, **1947**, *69*, 2116.

[26] 'Oxidative amino-dehydrogenation of azines', van der Plas, H., *Adv. Heterocycl. Chem.*, **2004**, *86*, 42.

[27] Cherng, Y.-J., *Tetrahedron*, **2002**, *58*, 887.

[28] Chekmarev, D. S., Shorshnev, S. V., Stepanov, A. E. and Kasatkin, A. N., *Tetrahedron*, **2006**, *62*, 9919.

[29] Torr, J. E., Large, J. M., Horton, P. N., Hursthouse, M. B. and McDonald, E., *Tetrahedron Lett.*, **2006**, *47*, 31.

[30] Erickson A. E. and Spoerri, P. E., *J. Am. Chem. Soc.*, **1946**, *68*, 400.

[31] Overberger, C. G., Kogon, I. C. and Minin, R., *Org. Synth., Coll. Vol. IV*, **1963**, 336.

[32] Banks, C. K., *J. Am. Chem. Soc.*, **1944**, *66*, 1127; Chapman, N. B. and Rees, C. W., *J. Chem. Soc.*, **1954**, 1190.

[33] Whittaker, N., *J. Chem. Soc.*, **1953**, 1646; Brown, D. J. and Waring, P., *Aust. J. Chem.*, **1973**, *26*, 443.

[34] Katritzky, A. R., Baykut, G., Rachwal, S., Szafran, M., Caster, K. C. and Eyler, J., *J. Chem. Soc., Perkin Trans. 2*, **1989**, 1499.

[35] Acevedo, O. L., Andrews, R. S., Dunkel, M. and Cook, D. P., *J. Heterocycl. Chem.*, **1994**, *31*, 989.

[36] For another example see Delia, T. J., Stark, D. and Glenn, S. K., *J. Heterocycl. Chem.*, **1995**, *32*, 1177.

[37] Chen, J. J., Thakur, K. D., Clark, M. P., Laughlin, S. K., George, K. M., Bookland, R. G., Davis, J. R., Cabrera, E. J., Eswaran, V., De, B. and Zhang, Y. G., *Bioorg. Med. Chem. Lett.*, **2006**, *16*, 5633.

[38] Zanda, M., Talaga, P., Wagner, A. and Mioskowski, C., *Tetrahedron Lett.*, **2000**, *41*, 1757.

[39] Gibson, C. and Kessler, H., *Tetrahedron Lett.*, **2000**, *41*, 1725.

[40] Yukawa, M., Niiya, T., Goto, Y., Sakamoto, T., Yoshizawa, H., Watanabe, A. and Yamanaka, H., *Chem. Pharm. Bull.*, **1989**, *37*, 2892.

[41] Sandosham, J. and Undheim, K., *Tetrahedron*, **1994**, *50*, 275.

[42] Desaubry, L., Wermuth, C. G., Boehrer, A., Marescaux, C. and Bourguignon, J.-J., *Bioorg. Med. Chem. Lett.*, **1995**, *5*, 139.

[43] Klötzer, W., *Monatsh. Chem.*, **1956**, *87*, 131 and 526.

[44] Brown, D. J. and Ford, P. W., *J. Chem. Soc. (C)*, **1967**, 568; Barlin, G. B. and Brown, W. V., *J. Chem. Soc. (B)*, **1967**, 648; *idem*, *J. Chem. Soc. (C)*, **1969**, 921; Budesinsky, Z. and Vavrina, J., *Coll. Czech. Chem. Commun.*, **1972**, *37*, 1721; Forbes, I. T., Johnson, C. N. and Thompson, M., *Synth. Commun.*, **1993**, *23*, 715; Font, D., Linden, A., Heras, M. and Villalgordo, J. M., *Tetrahedron*, **2006**, *62*, 1433.

[45] Miyashita, A., Suzuki, Y., Ohta, K., Iwamoto, K. and Higashito, T., *Heterocycles*, **1998**, *47*, 407.

[46] Yamanake, H. and Ohba, S., *Heterocycles*, **1990**, *31*, 895.

[47] Draper, T. L. and Bailey, T. R., *J. Org. Chem.*, **1995**, *60*, 748.

[48] Xing, L., Petitjean, A., Schmidt, R. and Cuccia, L., *Synth. Commun.*, **2007**, *37*, 2349.

[49] Darabantu, M., Boully, L., Turck, A. and Plé, N., *Tetrahedron*, **2005**, *61*, 2897; Kamei, K., Maeda, N., Katsuragi-Ogino, R., Koyama, M., Nakajima, M., Tatsuoka, T., Ohno, T. and Inoue, T., *Biorg. Med. Chem. Lett.*, **2005**, *15*, 2990.

[50] 'Advances in the directed metallation of azines and diazines (pyridines, pyrimidines, pyrazines, pyridazines, quinolines, benzodiazines and carbolines). Part 2. Metallation of pyrimidines, pyrazines, pyridazines, and benzodiazines', Turck, A., Plé, N., Mongin, F., and Quéguiner, G., *Tetrahedron*, **2001**, *57*, 4489.

[51] Zoltewicz, J. A., Grahe, G. and Smith, C. L., *J. Am. Chem. Soc.*, **1969**, *91*, 5501.

[52] Wada, A., Yamamoto, J. and Kanatomo, S., *Heterocycles*, **1987**, *26*, 585; Turck, A., Majovic, L. and Quéguiner, G., *Synthesis*, **1988**, 881; Plé, N., Turck, A., Bardin, F. and Quéguiner, G., *J. Heterocycl. Chem.*, **1992**, *29*, 467; Turck, A., Plé, N., Trohay, D., Ndzi, B. and Quéguiner, G., *ibid.*, **1992**, *29*, 699; Mattson, R. J. and Sloan, C. P., *J. Org. Chem.*, **1990**, *55*, 3410; Plé, N., Turck, A., Heynderickx, A. and Quéguiner, G., *J. Heterocycl. Chem.*, **1994**, *31*, 1311; Turck, A., Plé, N., Mojovic, L., Ndzi, B., Quéguiner, G., Haider, N., Schuller, H. and Heinisch, G., *J. Heterocycl. Chem.*, **1995**, *32*, 841; Nakamura, H., Aizawa, M. and Murai, A., *Synlett*, **1996**, 1015.

[53] Turck, A., Plé, N., Mojovic, L. and Quéguiner, G., *J. Heterocycl. Chem.*, **1990**, *27*, 1377.

[54] Plé, N., Turck, A., Couture, K. and Quéguiner, G., *J. Org. Chem.*, **1995**, *60*, 3781.

[55] Seggio, A., Chevallier, F., Vaultier, M. and Mongin, F., *J. Org. Chem.*, **2007**, *72*, 6602.

[56] Imahori, T., Suzuwa, K. and Kondo, Y., *Heterocycles*, **2008**, *76*, 1057.

[57] 'Directed metallation of π-deficient azaaromatics: strategies of functionalisation of pyridines, quinolines, and diazines', Quéguiner, G., Marsais, F., Snieckus, V. and Epsztajn, J., *Adv. Heterocycl. Chem.*, **1991**, *52*, 187; Toudic, F., Heynderickx, A., Plé, N., Turck, A. and Quéguiner, G., *Tetrahedron*, **2003**, *59*, 6375.

[58] Fruit, C., Turck, A., Plé, N. and Quéguiner, G., *Heterocycles*, **1999**, *51*, 2349; Fruit, C., Turck, A., Plé, N., Mojovic, L. and Quéguiner, G., *Tetrahedron*, **2002**, *58*, 2743.

[59] Mosrin, M., Boudet, N. and Knochel, P., *Org. Biomol. Chem.*, **2008**, *6*, 3237.

[60] Berghian, C., Darabantu, M., Turck, A. and Plé, N., *Tetrahedron*, **2005**, *61*, 9637.

[61] Gronowitz, S. and Röe, J., *Acta Chem. Scand.*, **1965**, *19*, 1741; Frissen, A. E., Marcelis, A. T. M., Buurman, D. G., Pollmann, C. A. M. and van der Plas, H. C., *Tetrahedron*, **1989**, *45*, 5611.

[62] Kress, T. J., *J. Org. Chem.*, **1979**, *44*, 2081; Rho, T. and Abah, Y. F., *Synth. Commun.*, **1994**, *24*, 253; Li, W., Nelson, D. P., Jensen, M. S., Hoerrner, R. S., Cai, D., Larsen, R. D. and Reider, P. J., *J. Org. Chem.*, **2002**, *67*, 5394.

[63] Taylor, H. M., Jones, C. D., Davenport, J. D., Hirsch, K. S., Kress, T. J. and Weaver, D., *J. Med. Chem.*, **1987**, *30*, 1359.

[64] Turck, A., Plé, N., Lepretre-Graquere, A. and Quéguiner, G., *Heterocycles*, **1998**, *49*, 205.

[65] Leprêtre, A., Turck, A., Plé, N., Knochel, P. and Quéguiner, G., *Tetrahedron*, **2000**, *56*, 265.

[66] Zheng, J. and Undheim, K., *Acta Chem. Scand.*, **1989**, *43*, 816; Shimura, A., Momotake, A., Togo, H. and Yokoyama, M., *Synthesis*, **1999**, 495.

[67] 'Recent advances in the chemistry of dihydroazines', Weis, A. L., *Adv. Heterocycl. Chem.*, **1985**, *38*, 1.

[68] Becker, H. P. and Neumann, W. P., *J. Organomet. Chem.*, **1972**, *37*, 57; Bessenbacher, C., Kaim, W. and Stahl, T., *Chem. Ber.*, **1989**, *122*, 933.

[69] Gottlieb, R. and Pfleiderer, W., *Justus Liebigs Ann. Chem.*, **1981**, 1451.

[70] Russell, J. R., Garner, C. D. and Joule, J. A., *J. Chem. Soc., Perkin Trans. 1*, **1992**, 409.

[71] Baskaran, S., Hanan, E., Byun, D. and Shen, W., *Tetrahedron Lett.*, **2004**, *45*, 2107.

[72] 'Advances in the synthesis of substituted pyridazines *via* introduction of carbon functional groups into the parent heterocycle', Heinisch, G., *Heterocycles*, **1987**, *26*, 481.

[73] Phillips, O. A., Keshava Murthy, K. S., Fiakpui, C. Y. and Knaus, E. E., *Can. J. Chem.*, **1999**, *77*, 216.

[74] Samaritoni, J. G. and Babbitt, G., *J. Heterocycl. Chem.*, **1991**, *28*, 583.

[75] Houminer, Y., Southwick, E. W. and Williams, D. L., *J. Org. Chem.*, **1989**, *54*, 640; Minisci, F., Recupero, F., Punta, C., Gambarotti, C., Antonietti, F., Fontana, F. and Pedulli, G. F., *Chem. Commun.*, **2002**, 2496.

[76] 'Hetero Diels–Alder methodology in organic synthesis', Ch. 10, Boger, D. L. and Weinreb, S. M., Academic press, **1987**; Stolle, W. A. W., Frissen, A. E., Marcelis, A. T. M. and van der Plas, H. C., *J. Org. Chem.*, **1992**, *57*, 3000.

[77] Boger, D. L. and Coleman, R. S., *J. Org. Chem.*, **1984**, *49*, 2240; Neunhoffer, H. and Werner, G., *Justus Liebigs Ann. Chem.*, **1974**, 1190; Biedrzycki, M., de Bie, D. A. and van der Plas, H. C., *Tetrahedron*, **1989**, *45*, 6211.

[78] Pavlik, J. W. and Vongnakorn, T., *Tetrahedron Lett.*, **2007**, *48*, 7015.

[79] 'Pyrimidine N-oxides: synthesis, structure and chemical properties', Yamanaka, H., Sakamoto, T. and Niitsuma, S., *Heterocycles*, **1990**, *31*, 923.

[80] Kocevar, M., Mlakar, B., Perdih, M., Petric, A., Polanc, S. and Vercek, B., *Tetrahedron Lett.*, **1992**, *33*, 2195.

[81] Itai, T. and Natsume, S., *Chem. Pharm. Bull.*, **1963**, *11*, 83.

[82] Sako, S., *Chem. Pharm. Bull.*, **1963**, *11*, 261; Klein, B., O'Donnell, E. and Gordon, J. M., *J. Org. Chem.*, **1964**, *29*, 2623.

[83] Ogata, M., *Chem. Pharm. Bull.*, **1963**, *11*, 1522; Iwao, M. and Kuraishi, T., *J. Heterocycl. Chem.*, **1978**, *15*, 1425; Sato, N., *J. Heterocycl. Chem.*, **1989**, *26*, 817; Koelsch, C. F. and Gumprecht, W. H., *J. Org. Chem.*, **1958**, *23*, 1603.

[84] Aoyagi, Y., Maeda, A., Inoue, M., Shiraishi, M., Sakakibara, Y., Fukui, Y. and Ohta, A., *Heterocycles*, **1991**, *32*, 735.

[85] Itai, T., Sako, S. and Okusa, G., *Chem. Pharm. Bull.*, **1963**, *11*, 1146; Ogata, M., *ibid.*, 1517.

[86] 'Synthesis of nucleosides', Vorbrüggen, H. and Ruh-Pohlenz, C., *Org. React.*, **2000**, *55*, 1.

[87] 'Uracils: versatile starting materials in heterocyclic synthesis', Wamhoff, H., Dzenis, J. and Hirota, K., *Adv. Heterocycl. Chem.*, **1992**, *55*, 129.

[88] Johnson, C. D., Katritzky, A. R., Kingsland, M. and Scriven, E. F. V., *J. Chem. Soc., (B)*, **1971**, 1.

[89] Lee, C. H. and Kim, Y. H., *Tetrahedron Lett.*, **1991**, *32*, 2401.

[90] Bergstrom, D. E. and Ruth, J. L., *J. Carbohyd. Nucleosides, Nucleotides*, **1977**, *4*, 257.

[91] Gondela, A. and Walczak, K., *Tetrahedron*, **2007**, *63*, 2859.

[92] Skinner, W. A., Schelstraete, M. G. M. and Baker, B. R., *J. Org. Chem.*, **1960**, *25*, 149; Delia, T. J., Scovill, J. P., Munslow, W. D. and Burckhalter, J. H., *J. Med. Chem.*, **1976**, *19*, 344.

[93] Wang, S. Y., *J. Org. Chem.*, **1959**, *24*, 11; Tee, O. S. and Banerjee, S., *Can. J. Chem.*, **1974**, *52*, 451.

[94] Whang, J. P., Yang, S. G. and Kim, Y. H., *Chem. Commun.*, **1997**, 1355.

[95] Sajiki, H., Yamada, A., Yasunaga, K., Tsunoda, T., Amer, M. F. A. and Hirota, K., *Tetrahedron Lett.*, **2003**, *44*, 2179.

[96] Giziewicz, J., Wnuk, S. F. and Robins, M. J., *J. Org. Chem.*, **1999**, *64*, 2149.

[97] Ariza, X., Bou, V. and Vilarrasa, J., *J. Am. Chem. Sooc.*, **1995**, *117*, 3665.

[98] Quiroga, J., Trilleras, J., Gálvez, J., Insuasty, B., Abonía, R., Nogueras, M., Cobo, J. and Marchal, A., *Tetrahedron Lett.*, **2008**, *49*, 5672.

[99] Prukala, D., *Tetrahedron Lett.*, **2006**, *47*, 9045.

[100] Fateen, A. K., Moustafa, A. H., Kaddah, A. M. and Shams, N. A., *Synthesis*, **1980**, 457; Bischofberger, N., *Tetrahedron Lett.*, **1989**, *30*, 1621.

[101] Inoue, H. and Ueda, T., *Chem. Pharm. Bull.*, **1978**, *26*, 2657.

[102] Hirota, K., Sajiki, H., Maki, Y., Inoue, H. and Ueda, T., *J. Chem. Soc., Chem. Commun.*, **1989**, 1659.

[103] Arterburn, J. B., Pannala, M. and Gonzalez, A. M., *Tetrahedron Lett.*, **2001**, *42*, 1457.

[104] Kanciurzewska, A., Raczkowski, M., Ciszewski, K. and Celewicz, L., *Tetrahedron Lett.*, **2003**, *44*, 761.

[105] Barawkar, D. A. and Gaesh, K. N., *Bioorg. Med. Chem. Lett.*, **1993**, *3*, 347.

[106] Lovett, E. G. and Lipkin, D., *J. Org. Chem.*, **1977**, *42*, 2574.

[107] Hirota, K., Watanabe, K. A. and Fox, J. J., *J. Org. Chem.*, **1978**, *43*, 1193; Hirota, K., Kitade, Y., Sagiki, H. and Maki, Y., *Heterocycles*, **1984**, *22*, 2259.

[108] Sard, H., Gonzalez, M. D., Mahadevan, A. and McKew, J. *J. Org. Chem.*, **2000**, *65*, 9261.

[109] Heddayatulla, M., *J. Heterocycl. Chem.*, **1981**, *18*, 339; Tanabe, T., Yamauchi, K. and Kinoshita, M., *Bull. Chem. Soc. Jpn.*, **1977**, *50*, 3021.

[110] Gondela, A. and Walczak, K., *Tetrahedron Lett.*, **2006**, *47*, 4653.

[111] Yamada, T. and Ohki, M., *Synthesis*, **1981**, 631.

[112] Richichi, B., Cicchi, S., Chiacchio, U., Romeo, G. and Brandi, A., *Tetrahedron Lett.*, **2002**, *43*, 4013.

[113] Cline, R. E., Fink, R. M. and Fink, K., *J. Am. Chem. Soc.*, **1959**, *81*, 2521.

[114] Ottenheijm, H. C. J., van Nispen, S. P. J. M. and Sinnige, M. J., *Tetrahedron Lett.*, **1976**, 1899.

[115] Hiebl, J., Zbiral, E., Balzarini, J. and De Clercq, E., *J. Med. Chem.*, **1992**, *35*, 3016; Verheyden, J. P. H., Wagner, D. and Moffatt, J. G., *J. Org. Chem.*, **1971**, *36*, 250.

[116] Müller, C. E., *Tetrahedron Lett.*, **1991**, *32*, 6539.

[117] Liao, J., Sun, J. and Yu, B., *Tetrahedron Lett.*, **2008**, *49*, 5036.

[118] 'Adventures in silicon-organic chemistry', Vorbrüggen, H., *Acc. Chem. Res.*, **1995**, *28*, 509.

[119] 'Nucleoside synthesis, organosilicon methods', Lukevics, E. and Zablacka, A., Ellis Horwood, London, **1991**; Matsuda, A., Kurasawa, Y. and Watanabe, K., *Synthesis*, **1981**, 748.

[120] Jung, M. E. and Castro, C., *J. Org. Chem.*, **1993**, *58*, 807; Lipshutz, B. H., Hayakawa, H., Kato, K., Lowe, R. F. and Stevens, K. L., *Synthesis*, **1994**, 1476.

[121] Paryzek, Z. and Tabaczka, B., *Org. Prep. Proc. Int.*, **2001**, *33*, 400.

[122] Tanaka, H., Hayakawa, H., Iijima, S., Haraguchi, K. and Miyasaka, T., *Tetrahedron*, **1985**, *41*, 861.

[123] Tanaka, H., Nasu, I. and Miyasaka, T., *Tetrahedron Lett.*, **1979**, *21*, 4755; Hayakawa, H., Tanaka, H., Obi, K., Itoh, M. and Miyasaka, T., *ibid.*, **1987**, *28*, 87; Shimizu, M., Tanaka, H., Hayakawa, H. and Miyasaka, T., *ibid.*, **1990**, *31*, 1295; Armstrong, R. W., Gupta, S. and Whelihan, F., *ibid.*, **1989**, *30*, 2057.

[124] Benhida, R., Aubertin, A.-M., Grierson, D. S. and Monneret, C., *Tetrahedron Lett.*, **1996**, *37*, 1031.

[125] Murray, T. P., Hay, J. V., Portlock, D. E. and Wolfe, J. F., *J. Org. Chem.*, **1974**, *39*, 595.

[126] Stevenson, T. M., Prasad, A. S. B., Citineni, J. R. and Knochel, P., *Tetrahedron Lett.*, **1996**, *37*, 8375; Prasad, A. S. B. and Knochel, P., *Tetrahedron*, **1997**, *49*, 16711; Berillon, L., Wagner, R. and Knochel, P., *J. Org. Chem.*, **1998**, *63*, 9117.

[127] Kopp, F. and Knochel, P., *Org. Lett.*, **2007**, *9*, 1639.

[128] Coad, P., Coad, R. A., Clough, S., Hyepock, J., Salisbury, R. and Wilkins, C., *J. Org. Chem.*, **1963**, *28*, 218; Ishikawa, I., Khachatrian, V. E., Melik-Ohanjanian, R. G., Kawahara, N., Mizuno, Y. and Ogura, H., *Chem. Pharm. Bull.*, **1992**, *40*, 846; Goe, G. L., Huss, C. A., Keay, J. G. and Scriven, E. F. V., *Chem. Ind.*, **1987**, 694.

[129] Fox, J. J. and van Praag, D., *J. Am. Chem. Soc.*, **1960**, *82*, 486.

[130] Sugimoto, O., Mori, M. and Tanji, K., *Tetrahedron Lett.*, **1999**, *40*, 7477.

[131] Sato, N. and Narita, N., *J. Heterocycl. Chem.*, **1999**, *36*, 783.

[132] Reese, C. B. and Varaprasad, C. V. N. S., *J. Chem. Soc., Perkin Trans. 1*, **1994**, 189.

[133] Wan, Z.-K., Wacharasindhu, S., Binnun, E. and Mansour, T., *Org. Lett.*, **2006**, *8*, 2425.

[134] Hirota, K., Suematsu, M., Kuwabara, Y., Asao, T. and Senda, S., *J. Chem. Soc., Chem. Commun.*, **1981**, 623.

[135] Candiani, I., Cabri, W., Bedeschi, A., Martinengo, T. and Penco, S., *Heterocycles*, **1992**, *34*, 875.

[136] Yates, N. D., Peters, D. A., Beddoes, R. L., Scopes, D. I. C. and Joule, J. A., *Heterocycles*, **1995**, *40*, 331.

[137] Nishio, T., Nakajima, N. and Omote, Y., *Tetrahedron Lett.*, **1981**, *22*, 753.

[138] Zhuo, J.-C. and Wyler, H., *Helv. Chim. Acta*, **1993**, *76*, 1916.

[139] Kim, J. N. and Ryu, E. K., *J. Org. Chem.*, **1992**, *57*, 1088.

[140] '3,5-Dihalo-2(1*H*)-pyrazinones: versatile scaffolds in organic synthesis', Pawar, V. G. and De Borggraeve, W. M., *Synthesis*, **2006**, 2799.

[141] Vekemans, J., Pollers-Wieërs, C. and Hoornaert, G., *J. Heterocycl. Chem.*, **1983**, *20*, 919.

[142] Tutonda, M., Vanderzande, D., Hendrickx, M. and Hoornaert, G., *Tetrahedron*, **1990**, *46*, 5715.

[143] 'Chemical investigations of the molecular origin of biological radiation damage', Fahr, E., *Angew. Chem., Int. Ed. Engl.*, **1969**, *8*, 578.

[144] Bergstrom, D. E. and Agosta, W. C., *Tetrahedron Lett.*, **1974**, 1087.

[145] Itahara, T. and Ide, N., *Bull. Chem. Soc. Jpn.*, **1992**, *65*, 2045.

[146] Brown, D. J., Hoerger, E. and Mason, S. F., *J. Chem. Soc.*, **1955**, 4035; Perrin, D. D. and Pitman, I. H., *ibid.*, **1965**, 7071.

[147] Kogon, I. C., Minin, R. and Overberger, C. G., *Org. Synth., Coll. Vol. IV*, **1963**, 182.

[148] English, J. P., Clark, J. H., Clapp, J. W., Seeger, D. and Ebel, R. H., *J. Am. Chem. Soc.*, **1946**, *68*, 453; Sato, N. and Takeuchi, R., *Synthesis*, **1990**, 659.

[149] Maggiolo, A. and Hitchings, G. H., *J. Am. Chem. Soc.*, **1951**, *73*, 4226.

[150] Lythgoe, B., Todd, A. R. and Topham, A., *J. Chem. Soc.*, **1944**, 315.

[151] Bergmann, F., Kalmus, A., Ungar-Waron, H. and Kwietny-Govrin, H., *J. Chem. Soc.*, **1963**, 3729.

[152] Müller, C. E., *Synthesis*, **1993**, 125.

[153] Clapham, K. M., Smith, A. E., Batsanov, A. S., McIntyre, L., Pountney, A., Bryce, M. R. and Tarbit, B., *Eur. J. Org. Chem.*, **2007**, 5712.

[154] e.g. Mizzoni, R. H. and Spoerri, P. E., *J. Am. Chem. Soc.*, **1954**, *76*, 2201.

[155] 'Methylpyridines and other methylazines as precursors of bicycles and polycycles', Abu-Shanab, F. A., Wakefield, B. J. and Elnagdi, M. H., *Adv. Heterocycl. Chem.*, **1997**, *68*, 181.

[156] Yamanaka, H., Abe, H. and Sakamoto, T., *Chem. Pharm. Bull.*, **1977**, *25*, 3334.

[157] Strekowski, L., Wydra, R. L., Janda, L. and Harden, D. B., *J. Org. Chem.*, **1991**, *56*, 5610.

[158] Itoh, T., Hasegawa, H., Nagata, K., Matsuya, Y. and Ohsawa, A., *Heterocycles*, **1994**, *37*, 709; Itoh, T., Hasegawa, H., Nagata, K., Matsuya, Y., Okada, M. and Ohsawa, A., *Chem. Pharm. Bull.*, **1994**, *42*, 1768.

[159] Itoh, T., Miyazaki, M., Nagata, K. and Ohsawa, A., *Heterocycles*, **1998**, *49*, 67.

[160] Itoh, T., Miyazaki, M., Nagata, K. and Ohsawa, A., *Heterocycles*, **1997**, *46*, 83.

[161] Rudler, H., Denise, B., Xu, Y. and Vaissermann, J., *Tetrahedron Lett.*, **2005**, *46*, 3449; Rotzoll, S., Ullah, E., Fischer, C., Michalik, D., Spannenberg, A. and Langer, P., *Tetrahedron*, **2006**, *62*, 12084.

[162] Veeraraghavan, S., Bhattacharjee, D. and Popp, F. D., *J. Heterocycl. Chem.*, **1981**, *18*, 443; Dostal, W. and Heinish, G., *Heterocycles,*, **1986**, *24*, 793.

[163] Overend, W. G. and Wiggins, L. F., *J. Chem. Soc.*, **1947**, 239.

[164] Padwa, A., Rodriguez, A., Tohidi, M. and Fukunaga, T., *J. Am. Chem. Soc.*, **1983**, *105*, 933; Ho, T.-L. and Chang, M.-H., *J. Chem. Soc., Perkin Trans. 1*, **1999**, 2479.

[165] Albright, J. D., McEvoy, F. J. and Moran, D. B., *J. Heterocycl. Chem.*, **1978**, *15*, 881.

[166] Coates, W. J. and McKillop, A., *Synthesis*, **1993**, 334.

[167] Steck, E. A., Brundage, R. P. and Fletcher, L. T., *J. Am. Chem. Soc.*, **1953**, *75*, 1117.

[168] Altomare, C., Cellamare, S., Summo, L., Catto, M., Carotti, A., Thull, U., Carrupt, P.-A., Testa, B. and Stoeckli-Evans, H., *J. Med. Chem.*, **1998**, *41*, 3812.

[169] Mizzoni, R. H. and Spoerri, P. E., *J. Am. Chem. Soc.*, **1951**, *73*, 1873; Horning, R. H. and Amstutz, E. D., *J. Org. Chem.*, **1955**, *20*, 707; Atkinson, C. M. and Sharpe, C. J., *J. Chem. Soc.*, **1959**, 3040.

[170] Zhang, J., Morton, H. E. and Ji, J., *Tetrahedron Lett.*, **2006**, *47*, 8733.

[171] Lutz, R. E. and King, S. M., *J. Org. Chem.*, **1952**, *17*, 1519.

[172] Alex, K., Tillack, A., Schwarz, N. and Beller, M., *Tetrahedron Lett.*, **2008**, *49*, 4607.

[173] Clauson-Kaas, N. and Limborg, F., *Acta Chem. Scand.*, **1947**, *1*, 619.

[174] Sauer, J., Heldmann, D. K., Hetzenegger, J., Krauthan, J., Sichert, H. and Schuster, J., *Eur. J. Org. Chem.*, **1998**, 2885.

[175] Marcelis, A. T. M. and van der Plas, H. C., *Heterocycles*, **1985**, *23*, 683; Birkofer, L. and Hänsel, E., *Chem. Ber.*, **1981**, *114*, 3154; Boger, D. L. and Patel, M., *J. Org. Chem.*, **1988**, *52*, 1405; Sakamoto, T., Funami, N., Kondo, Y. and Yamanaka, H., *Heterocycles*, **1991**, *32*, 1387.

[176] Helm, M. D., Plant, A. and Harrity, J. P. A., *Chem. Commun.*, **2006**, 4278.

[177] Kenner, G. W., Lythgoe, B., Todd, A. R. and Topham, A., *J. Chem. Soc.*, **1943**, 388.

[178] Burgess, D. M., *J. Org. Chem.*, **1956**, *21*, 97; Van Allan, J. A., *Org. Synth., Coll. Vol. IV*, **1963**, 245.

[179] Sherman, W. R. and Taylor, E. C., *Org. Synth., Coll. Vol. IV*, **1963**, 247.

[180] Foster, H. M. and Snyder, H. R., *Org. Synth., Coll. Vol. IV*, **1963**, 638; Crosby, D. G., Berthold, R. V. and Johnson, H. E., *ibid., Vol. V*, **1973**, 703.

[181] Bredereck, H., Gompper, R. and Morlock, G., *Chem. Ber.*, **1957**, *90*, 942; Bredereck, H., Gompper, R. and Herlinger, H., *Chem. Ber.*, **1958**, *91*, 2832.

[182] Papet, A.-L. and Marsura, A., *Synthesis*, **1993**, 478.

[183] Fülle, F. and Müller, C. E., *Heterocycles*, **2000**, *53*, 347.

[184] Bredereck, H., Effenberger, F. and Schweizer, E. H., *Chem. Ber.*, **1962**, *95*, 803.

[185] Takagi, K., Bajnati, A. and Hubert-Habert, M., *Bull. Soc. Chim. Fr.*, **1990**, 660.

[186] Reichardt, C. and Schagerer, K., *Justus Liebigs Ann. Chem.*, **1982**, 530.

[187] Davidson, D. and Baudisch, O., *J. Am. Chem. Soc.*, **1926**, *48*, 2379; Hunt, R. R., McOmie, J. F. W. and Sayer, E. R., *J. Chem. Soc.*, **1959**, 525; Maggiolo, A., Phillips, A. P. and Hitchings, G. H., *J. Am. Chem. Soc.*, **1951**, *73*, 106; Kenner, G. W., Lythgoe, B., Todd, A. R. and Topham, A., *J. Chem. Soc.*, **1943**, 388.

[188] Ziegenbein, W. and Franke, W., *Angew. Chem.*, **1959**, *71*, 628.

[189] Mosti, L., Menozzi, G. and Schenone, P., *J. Heterocycl. Chem.*, **1983**, *20*, 649; Wang, F. and Schwabacher, A. W., *Tetrahedron Lett.*, **1999**, *40*, 4779.

[190] Bellur, E. and Langer, P., *Tetrahedron*, **2006**, *62*, 5426.

[191] Gupton, J. T., Gall, J. E., Riesinger, S. W., Smith, S. Q., Bevirt, K. M., Sikorski, J. A., Dahl, M. L. and Arnold, Z., *J. Heterocycl. Chem.*, **1991**, *28*,1281.

[192] Verron, J., Malherbe, P., Prinssen, E., Thomas, A. W., Nock, N. and Masciadri, R., *Tetrahedron Lett.*, **2007**, *48*, 377.

[193] De Pasquale, R. J., *J. Org. Chem.*, **1977**, *42*, 2185.

[194] Butters, M., *J. Heterocycl. Chem.*, **1992**, *29*, 1369.

[195] Slee, D. H., Zhang, X., Moorjani, M., Lin, E., Lanier, M. C., Chen, Y., Rueter, J. K., Lechner, S. M., Markison, S., Malany, S., Joswig, T., Santos, M., Gross, R. S., Williams, J. P., Castro-Palomino, J. C., Crespo, M. I., Prat, M., Gual, S., Diaz, J.-L., Wen, J., O'Brien, Z. and Saunders, J., *J. Med. Chem.*, **2008**, *51*, 400.

[196] Ch. II in '*o*-Aminonitriles', Taylor, E. C. and McKillop, A., *Adv. Org. Chem.*, **1970**, *7*, 79.

[197] 'Heterocyclic β-enaminoesters, versatile synthons in heterocyclic synthesis', Wamhoff, H., *Adv. Heterocycl. Chem.*, **1985**, *38*, 299.

[198] Dickey, J. B. and Gray, A. R., *Org. Synth., Coll. Vol. II*, **1943**, 60.

[199] Shimo, K. and Wakamatsu, S., *J. Org. Chem.*, **1959**, *24*, 19.

[200] Boger, D. L., Schumacher, J., Mullican, M. D., Patel, M. and Panek, J. S., *J. Org. Chem.*, **1982**, *47*, 2673; Boger, D. L. and Menezes, R. F., *ibid.*, **1992**, *57*, 4331.

[201] Shaw, G. and Warrener, R. N., *J. Chem. Soc.*, **1958**, 157.

[202] Hronowski, L. J. J. and Szarek, W. A., *Can. J. Chem.*, **1985**, *63*, 2787.

[203] Ueno, Y., Kato, T., Sato, K., Ito, Y., Yoshida, M., Inoue, T., Shibata, A., Ebihara, M. and Kitade, Y., *J. Org. Chem.*, **2005**, *70*, 7925.

[204] Pearson, M. S. M., Robin, A., Bourgougnon, N., Jean Claude Meslin, J. C. and Deniaud, D., *J. Org. Chem.*, **2003**, *68*, 8583; Robin, A., Julienne, K., Meslin, J.-C. and Deniaud, D., *Eur. J. Org. Chem.*, **2006**, 634.

[205] 'The Biginelli dihydropyrimidine synthesis', Kappe, C. O. and Stadler, A., *Org. React.*, **2004**, *63*, 1.

[206] Biginelli, P., *Ber. Deut. Chem. Gesel.*, **1881**, *24*, 2962.

[207] e.g. Shanmugam, P. and Perumal, P. T., *Tetrahedron*, **2006**, *62*, 9726.

[208] Wyatt, E. E., Galloway, W. R. J. D., Thomas, G. L., Welch, M., Loiseleur, O., Plowright, A. T. and Spring, D. R., *Chem. Commun.*, **2008**, 4962.

[209] Martínez, A. G., Fernández, A. H., Jiménez, F. M., Fraile, A. G., Subramanian, L. R. and Hanack, M., *J. Org. Chem.*, **1992**, *57*, 1627; Herrera, A., Martinez, R., González, B., Illescas, B., Martin, N. and Seoane, C., *Tetrahedron Lett.*, **1997**, *38*, 4873.

[210] Herrera, A., Martínez-Alvarez, R., Chioua, M., Chatt, R., Chioua, R., Sánchez, A. and Almy, J., *Tetrahedron*, **2006**, *62*, 2799.

[211] Tyagarajan, S. and Chakravarty, P. K., *Tetrahedron Lett.*, **2005**, *46*, 7889.

[212] Birkofer, L., *Chem. Ber.*, **1947**, *80*, 83.

[213] Blake, K. W., Porter, A. E. A. and Sammes, P. G., *J. Chem. Soc., Perkin Trans. 1*, **1972**, 2494.

[214] Flament, I. and Stoll, M., *Helv. Chim. Acta*, **1967**, *50*, 1754.

[215] Masuda, H., Tanaka, M., Akiyama, T. and Shibamoto, T., *J. Agric. Food Chem.*, **1980**, *28*, 244.

[216] Bradbury, R. H., Griffiths, D. and Rivett, J. E., *Heterocycles*, **1990**, *31*, 1647.

[217] Muehlmann, F. L. and Day, A. R., *J. Am. Chem. Soc.*, **1956**, *78*, 242.

[218] Rothkopf, H. W., Wöhrle, D., Müller, R. and Kossmehl, G., *Chem. Ber.*, **1975**, *108*, 875.

[219] Weijlard, J., Tishler, M. and Erickson, A. E., *J. Am. Chem. Soc.*, **1945**, *67*, 802.

[220] Büchi, G. and Galindo, J., *J. Org. Chem.*, **1991**, *56*, 2605.

[221] Herdewijn, P., De Clerq, E., Balzarini, J. and Vandehaeghe, H., *J. Med. Chem.*, **1985**, *28*, 550.

[222] Jones, K., Keenan, M. and Hibbert, F., *Synlett*, **1996**, 509; Keenan, M., Jones, K. and Hibbert, F., *Chem. Commun.*, **1997**, 323.

[223] 'Pteridine Chemistry', Ed. Pfleiderer, W. and Taylor, E. C., Pergammon Press, London, **1964**; 'Pteridines. Properties, reactivities and biological significance', Pfleiderer, W., *J. Heterocycl. Chem.*, **1992**, *29*, 583.

[224] For reviews see *J. Biol. Inorg. Chem.*, **1997**, *2*, 772, 773, 782, 786, 790, 797, 804, 810 and 817.

[225] Waring, P. and Armarego, W. L. F., *Aust. J. Chem.*, **1985**, *38*, 629.

[226] Taylor, E. C., Perlman, K. L., Sword, I. P., Séquin-Frey, M. and Jacobi, P. A., *J. Am. Chem. Soc.*, **1973**, *29*, 3610.

[227] Cook, A. H., Heilbron, I. and Smith, E., *J. Chem. Soc.*, **1949**, 1440.

15

Typical Reactivity of Pyrroles, Furans and Thiophenes

In this chapter are gathered the most important generalisations that can be made, and the general lessons that can be learned about the reactivity, and relative reactivities, one with the other, of the prototypical five-membered aromatic heterocycles: pyrroles, furans and thiophenes.

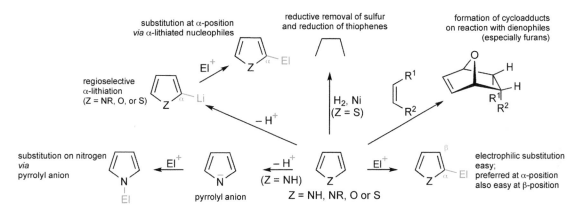

Typical reactions of pyrroles, furans and thiophenes

The chemistry of pyrrole and thiophene is dominated by a readiness to undergo electrophilic substitution, preferentially at an α-position but, with only slightly less alacrity, also at a β-position, should the α positions be blocked. It is worth re-emphasising the stark contrast between the five- and six-membered heterocycles – the five-membered systems considered in this chapter react much more readily with electrophiles than does benzene, but the azines react much less readily (cf. Chapters 7–9 and 13, 14).

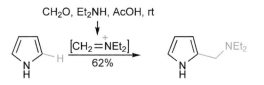

An example of easy α-electrophilic substitution of pyrrole with a weak electrophile

Heterocyclic Chemistry 5th Edition   John Joule and Keith Mills
© 2010 Blackwell Publishing Ltd

Positional selectivity in these five-membered systems, and their high reactivity to electrophilic attack, are well explained by a consideration of the Wheland intermediates (and by implication, the transition states that lead to them) for electrophilic substitution. Intermediate cations from both α- and β-attack are stabilised (shown for attack on pyrrole). The delocalisation, involving donation of electron density from the hetero-atom, is greater in the intermediate from α-attack, as illustrated by the number of low-energy resonance contributors. Note that the C–C double bond in the intermediate for β-attack is not and cannot become involved in delocalisation of the charge.

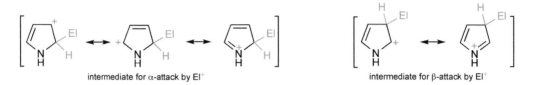

Intermediates for electrophilic substitution of pyrrole

There is a simple parallelism between the reaction of a pyrrole with an electrophile and the comparable reaction of an aniline, and indeed the reactivity of pyrrole towards electrophiles is in the same range as that of aniline.

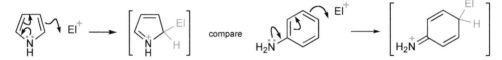

Electrophilic attack on electron-rich pyrrole compared with attack on electron-rich aniline

The five-membered heterocycles do not react with electrophiles at the heteroatom; perhaps this surprises the heterocyclic newcomer, most obviously with respect to pyrrole, for here, it might have been anticipated, the nitrogen lone pair would be easily donated to an incoming electrophile, as it would be in reactions of its saturated counterpart, pyrrolidine. The difference is that in pyrrole, electrophilic addition at the nitrogen would lead to a substantial loss of resonance stabilisation – the molecule would be converted into a cyclic butadiene, with an attached nitrogen carrying a positive charge *localised* on that nitrogen atom. The analogy with aniline falls down for, of course, anilines do react easily with simple electrophiles (e.g. protons) at nitrogen. The key difference is that, although some stabilisation in terms of overlap between the aniline nitrogen lone pair and benzenoid π-system is lost, the majority of the stabilisation energy, associated with the six-electron benzenoid π-system, is retained when aniline nitrogen donates its lone pair of electrons to a proton (electrophile).

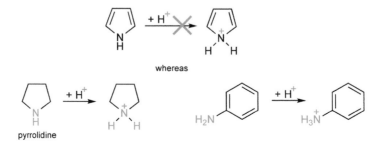

Of the trio – pyrrole, furan and thiophene – the first is by far the most susceptible to electrophilic attack: this susceptibility is linked to the greater electron-releasing ability of neutral trivalent nitrogen, and the concomitant greater stability of a positive charge on tetravalent nitrogen. This finds its simplest expression in the relative basicities of saturated amines, ethers and sulfides, respectively, which are seen to parallel nicely the relative order of reactivity of pyrrole, furan and thiophene towards electrophilic attack at carbon, but involving major assistance by donation from the heteroatom, i.e. the development of positive charge on the heteroatom.

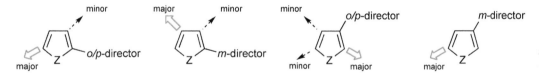

In qualitative terms, the much greater reactivity of pyrrole is illustrated by its rapid reaction with weak electrophiles like the benzenediazonium cation and nitrous acid, neither of which reacts with furan or thiophene. It is relevant to note that *N,N*-dimethylaniline reacts rapidly with these reactants, whereas anisole does not.

Substituents ranged on five-membered rings have directing effects comparable to those that they exert on a benzene (or pyridine) ring. Alkyl groups, for example, direct *ortho* and *para*, and nitro groups direct *meta* although, strictly, the terms *ortho/meta/para* cannot be applied to the five-membered situation. The very strong tendency for α-electrophilic substitution is, however, the dominating influence in most instances, and products resulting from attack following guidance from the substituent are generally minor products in mixtures where the dominant substitution is at an available α-position. The influence of substituents is felt least in furans.

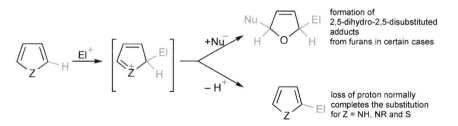

Effect of substitutents on regioselectivity of electrophilic substitution in five-membered heterocycles

A significant aspect of the chemistry of furans is the occurrence of 2,5-additions initiated by electrophilic attack: a Wheland intermediate is formed normally, but then adds a nucleophile, when a sufficiently reactive one is present, instead of then losing a proton. Conditions can, however, usually be chosen to allow the formation of a 'normal' α-substitution product. The occurrence of such processes in the case of furan is generally considered to be associated with its lower aromatic resonance stabilisation energy – there is less to regain by loss of a proton and the consequent return to an aromatic furan.

Formation of adducts from furans

The lower aromaticity of furans also manifests itself in a much greater tendency to undergo cycloadditions, as a 4-π, diene component in Diels–Alder reactions. That is to say, furans are much more like dienes, and less like six-electron aromatic systems, than are pyrroles and thiophenes. However, the last two systems can be made to undergo cycloadditions by carrying out high-pressure reactions or, in the case of pyrroles, by 'reducing the aromaticity' by the device of inserting an electron-withdrawing group onto the nitrogen.

In direct contrast with electron-deficient heterocycles like pyridines and the diazines, the five-membered systems *do not undergo nucleophilic substitutions*, except in situations (especially in furan and thiophene chemistry) where halide is situated *ortho* or *para* to a nitro group. In the manipulation of the five-membered heterocycles, extensive use has been made of the various palladium(0)-catalysed couplings regimes, as illustrated with one example (see 4.2 for a detailed discussion).

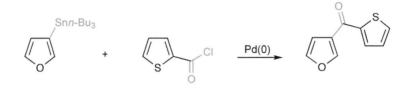

Deprotonations are extremely important: furan and thiophene are *C*-deprotonated by strong bases, such as *n*-butyllithium or lithium diisopropylamide, at their α-positions, because here the heteroatom can exert its greatest acidifying influence by inductive withdrawal of electron density, to give anions that can then be made to react with the whole range of electrophiles, affording α-substituted furans and thiophenes. This methodology compliments the use of electrophilic substitutions to introduce groups, also regioselectively α, but has the advantage that even weak electrophiles, such as aldehydes and ketones, can be utilised. The employment of metallated *N*-substituted (blocked) pyrroles is an equally valid strategy for producing α-substituted pyrroles. Pyrroles that have an *N*-hydrogen are deprotonated at the nitrogen, and the pyrryl anion thus generated is nucleophilic at the heteroatom, providing a means for the introduction of groups on nitrogen.

The potential for interaction of the heteroatom (electron donation) with positive charge on a side-chain, especially at an α-position, has a number of effects: amongst the most important is the enhanced reactivity of side-chain derivatives carrying leaving groups. Similarly, carbonyl groups attached to five-membered heterocycles have somewhat reduced reactivity, as implied by the resonance contributor shown.

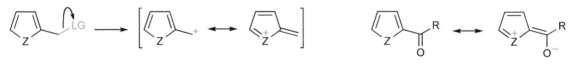

Side-chain reactivity

Generally speaking, the five-membered heterocycles are far less stable to oxidative conditions than benzenes or pyridines, with thiophenes bearing the closest similarity – in many ways thiophenes, of the trio, are the most like carboaromatic compounds. Hydrogenation of thiophenes, particularly over nickel as catalyst, leads to saturation and removal of the heteroatom. Some controlled chemical reductions of pyrroles and furans are known, which give dihydro-products.

The ring synthesis of five-membered heterocycles has been extensively investigated, and many and subtle methods have been devised. Each of these three heterocyclic systems can be prepared from 1,4-dicarbonyl-compounds, for furans by acid-catalysed cyclising dehydration, and for pyrroles and thiophenes by interaction with ammonia or a primary amine, or a source of sulfur, respectively.

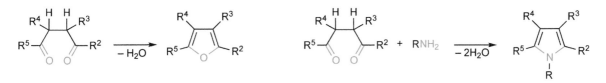

Ring synthesis of five-membered heterocycles from 1,4-dicarbonyl compounds

As illustrations of the variety of methods available, the three processes below show: (i) the addition of isonitrile anions to α,β-unsaturated nitro-compounds, with loss of nitrous acid to bring about aromatisation, (ii) the interaction of thioglycolates with 1,3-dicarbonyl-compounds, for the synthesis of thiophene 2-esters, and (iii) the cycloaddition/cycloreversion preparation of furans from oxazoles.

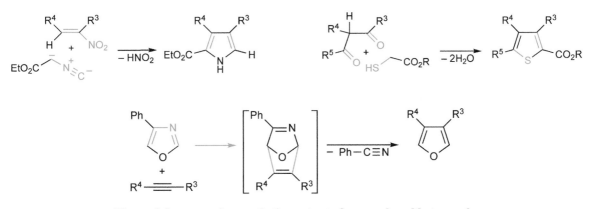

Three of the many ring synthetic routes to five-membered heterocycles

16

Pyrroles: Reactions and Synthesis

pyrrole
1H-pyrrole

Pyrrole[1] and the simple alkyl-pyrroles are colourless liquids, with relatively weak odours rather like that of aniline, which, also like the anilines, darken by autoxidation. Pyrrole itself is readily available commercially, and is manufactured by alumina-catalysed gas-phase interaction of furan and ammonia. Pyrrole was first isolated from coal tar in 1834 and then in 1857 from the pyrolysate of bone, the chemistry of which is similar to an early laboratory method for the preparation of pyrrole – the pyrolysis of the ammonium salt of the sugar acid, mucic acid. The word pyrrole is derived from the Greek for red, which refers to the bright red colour which pyrrole imparts to a pinewood shaving moistened with concentrated hydrochloric acid.

The early impetus for the study of pyrroles came from degradative work relating to the structures of two pigments central to life processes, the blood respiratory pigment haem, and chlorophyll, the green photosynthetic pigment of plants (32.3).[2] Chlorophyll and haem are synthesised in the living cell from porphobilinogen, the only aromatic pyrrole to play a role – a vitally important role – in fundamental metabolism.[3,4]

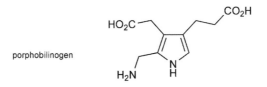

porphobilinogen

16.1 Reactions with Electrophilic Reagents[5]

Whereas pyrroles are resistant to nucleophilic addition and substitution, they are very susceptible to attack by electrophilic reagents and undergo easy *C*-substitution. Pyrrole itself, *N*- and *C*-monoalkyl- and to a lesser extent *C,C'*-dialkyl-pyrroles, are polymerised by strong acids, so that many of the electrophilic reagents useful in benzene chemistry cannot be used. However, the presence of an electron-withdrawing substituent, such as an ester, prevents polymerisation and allows the use of the strongly acidic, nitrating and sulfonating agents.

Heterocyclic Chemistry 5th Edition   John Joule and Keith Mills
© 2010 Blackwell Publishing Ltd

16.1.1 Substitution at Carbon

16.1.1.1 Protonation

In solution, reversible proton addition occurs at all positions, being by far the fastest at the nitrogen, and about twice as fast at C-2 as at C-3.[6] In the gas phase, mild acids like $C_4H_9^+$ and NH_4^+ protonate pyrrole *only* on carbon and with a larger proton affinity at C-2 than at C-3.[7] Thermodynamically, the stablest cation is the 2*H*-pyrrolium ion, formed by protonation at C-2 and observed pK_{aH} values for pyrroles are for these 2-protonated species. The weak *N*-basicity of pyrroles is the consequence of the absence of mesomeric delocalisation of charge in the 1*H*-pyrrolium cation.

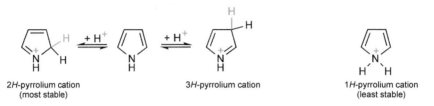

2*H*-pyrrolium cation   3*H*-pyrrolium cation   1*H*-pyrrolium cation
(most stable)   (least stable)

The pK_{aH} values of a wide range of pyrroles have been determined:[8] pyrrole itself is an extremely weak base with a pK_{aH} value of −3.8; this, as a 0.1 molar solution in 1N acid, corresponds to only one protonated molecule to about 5000 unprotonated. However, basicity increases very rapidly with increasing alkyl substitution, so that 2,3,4,5-tetramethylpyrrole, with a pK_{aH} of +3.7, is almost completely protonated on carbon as a 0.1 molar solution in 1N acid (this can be compared with aniline, which has a pK_{aH} of +4.6). Thus alkyl groups have a striking stabilising effect on cations – isolable, crystalline salts can be obtained from pyrroles carrying *t*-butyl groups.[9]

Reactions of Protonated Pyrroles

The 2*H*- and 3*H*-pyrrolium cations are essentially iminium ions and as such are electrophilic: they play the key role in polymerisation (see 16.1.8) and reduction (16.7) of pyrroles in acid. In the reaction of pyrroles with hydroxylamine hydrochloride, which produces ring-opened 1,4-dioximes, it is probably the more reactive 3*H*-pyrrolium cation that is the starter.[10] Primary amines, RNH_2, can be protected, by conversion into 1-R-2,5-dimethylpyrroles (16.16.1.1), recovery of the amine being by way of this reaction with hydroxylamine.[11,12]

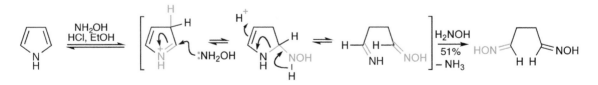

16.1.1.2 Nitration

Nitrating mixtures suitable for benzenoid compounds cause complete decomposition of pyrrole, but reaction occurs smoothly with acetyl nitrate at low temperature, giving mainly 2-nitropyrrole. This nitrating agent is formed by mixing fuming nitric acid with acetic anhydride to form acetyl nitrate and acetic acid, thus removing the strong mineral acid. In the nitration of pyrrole with this reagent, it has been shown that C-2 is 1.3×10^5 and C-3 is 3×10^4 times more reactive than benzene.[13] A combination of PPh_3, $AgNO_3$ and Br_2 also produces a comparable mixture of nitro-pyrroles.[14]

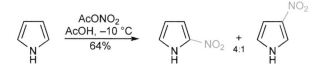

N-Substitution of pyrroles gives rise to increased proportions of β-nitration, even an *N*-methyl producing a β:α ratio of 1:3, and the much larger *t*-butyl actually reverses the relative positional reactivities, with a β:α ratio of 4:1.[15] The intrinsic α-reactivity can be effectively completely blocked with a very large substituent such as a triisopropylsilyl (TIPS) group, especially useful since it can be subsequently easily removed.[16]

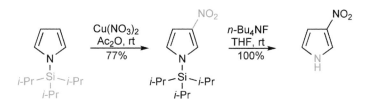

16.1.1.3 Sulfonation and Reactions with Other Sulfur Electrophiles

For sulfonation, a mild reagent of low acidity must be used: the pyridine–sulfur trioxide compound smoothly converts pyrrole into a sulfonate initially believed to be the 2-isomer,[17] but subsequently shown to be pyrrole-3-sulfonic acid.[18] It seems likely that this isomer results from reversibility of the sulfonation, and the eventual formation of the more stable acid. Chlorosulfonation of 1-phenylsulfonylpyrrole is clean and an efficient route to pyrrole 3-sulfonic-acid derivatives.[19]

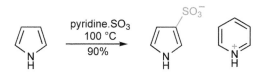

Sulfenylation of pyrrole[20] and thiocyanation of pyrrole[21] or of 1-phenylsulfonylpyrrole[22] also provide means for the electrophilic introduction of sulfur groups, at lower oxidation levels, and in contrast to the sulfonations, at the pyrrole α-position.

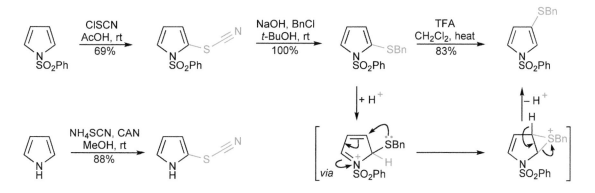

Acid catalyses rearrangement of sulfur substituents from the α-position to give β-substituted pyrroles[23] (see also 16.1.1.5 and 16.12), perhaps initiated by protonation at the sulfur-bearing α-carbon.

16.1.1.4   Halogenation

Pyrrole reacts with halogens so readily that unless controlled conditions are used, tetrahalo-pyrroles are the only isolable products, and these are stable.[24] Attempts to mono-halogenate simple alkyl-pyrroles fail, probably because of side-chain halogenation and the generation of extremely reactive pyrryl-alkyl halides (16.11).

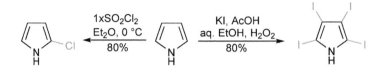

Although unstable compounds, 2-bromo- and 2-chloropyrrole (also using SO_2Cl_2) can be prepared by direct halogenation of pyrrole with the *N*-halo-succinimides;[25] 2-bromopyrrole can be conveniently prepared using 1,3-dibromo-4,4-dimethylhydantoin and can be stabilised by conversion into its *N-t*-butoxycarbonyl derivative.[26] Formation of *N*-tosyl derivatives[27] is also recommended for stabilising 2-bromopyrrole.

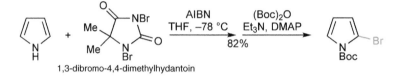

1,3-dibromo-4,4-dimethylhydantoin

N-Triisopropylsilylpyrrole monobrominates and monoiodinates cleanly and nearly exclusively at C-3, and with two mole equivalents of *N*-bromosuccinimide it dibrominates, at C-3 and C-4[17,28] *N*-Tosylpyrrole 3,4-dibrominates with bromine in hot acetic acid,[29] whereas *N*-Boc-pyrrole gives the 2,5-dibromo derivative using NBS at 0 °C.[30] Selective replacement of trimethylsilyl groups of *N*-blocked 3,4-bis(trimethylsilyl)-pyrroles with iodine requires the halogen with CF_3CO_2Ag at −78 °C.[31]

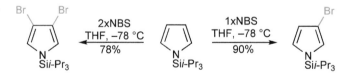

16.1.1.5   Acylation

Direct acetylation of pyrrole with acetic anhydride at 200 °C leads to 2-acetylpyrrole as main product, together with some 3-acetylpyrrole, but no *N*-acetylpyrrole.[32] *N*-Acetylpyrrole can be obtained in high yield by heating pyrrole with *N*-acetylimidazole.[33] Alkyl substitution facilitates *C*-acylation, so that 2,3,4-trimethylpyrrole yields the 5-acetyl-derivative, even on refluxing in acetic acid. The more reactive trifluoro-roacetic anhydride and trichloroacetyl chloride react with pyrrole efficiently, even at room temperature, to give 2-substituted products, alcoholysis or hydrolysis of which provides a clean route to pyrrole-2-esters or -acids.[34] Nitration of 2-(chloroacetyl)pyrrole occurs at C-4 and this regioselectivity applies also to acyla-tion[35] of pyrroles with electron-withdrawing groups at C-2.

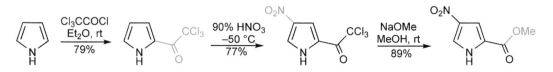

N-Acyl benzotriazoles very effectively acylate pyrrole at C-2 using TiCl₄ catalysis; TIPS-pyrrole is acylated with this reagent at C-3.[36]

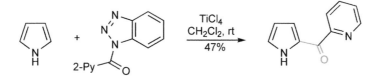

Vilsmeier[37,38] acylation of pyrroles, formylation with dimethylformamide/phosphoryl chloride in particular, is a generally applicable process.[39] The actual electrophilic species is an *N,N*-dialkyl-chloromethyleneiminium cation (the chloride is available commercially as a solid).[40] Here again, the presence of a large pyrrole-*N*-substituent perturbs the intrinsic α-selectivity, formylation of *N*-tritylpyrrole favouring the β-position by 2.8 : 1 and trifluoroacetylation of this pyrrole giving only the 3-ketone;[41] the use of bulky *N*-silyl substituents allows β-acylation with the possibility of subsequent removal of the *N*-substituent.[42] The final intermediate in a Vilsmeier reaction is an iminium salt requiring hydrolysis to produce the isolated product aldehyde. When a secondary lactam is used, hydrolysis does not take place and a cyclic imine is obtained.[43]

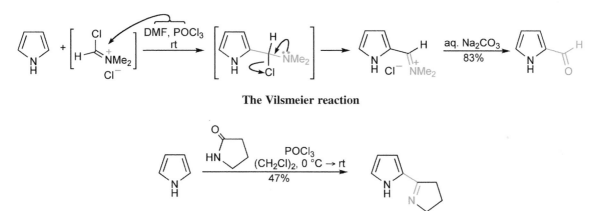

The Vilsmeier reaction

Acylation of 1-phenylsulfonylpyrrole, with its deactivating *N*-substituent, requires more forcing conditions in the form of a Lewis acid as catalyst, the regioselectivity of attack depending both on the choice of catalyst and on the particular acylating agent, as illustrated.[44] However, no Lewis acid is required if the mixed anhydride of the required acid and trifluoroacetic acid are employed to acylate *N*-tosyl-pyrroles, this proceeding exclusively at the 2-position.[45]

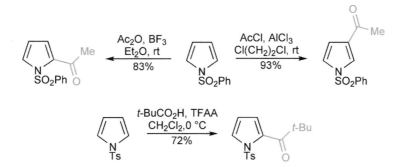

Regioselection in Friedel–Crafts acylations, depending on the Lewis acid employed, also applies to pyrroles with electron-withdrawing/stabilising groups, like esters, on carbon.[46] Lewis-acid catalysed acylation of 3-acyl-pyrroles, easily obtained by hydrolysis of 1-phenylsulfonyl-3-acyl-pyrroles, proceeds smoothly to give 2,4-diacyl-pyrroles, substitution *meta* to the acyl group and also at the remaining pyrrole α-position;[47] Vilsmeier formylation of methyl pyrrolyl-2-carboxylate takes place at C-5, the α-selectivity being dominant in this case.[48]

16.1.1.6 Alkylation

Mono-*C*-alkylation of pyrroles cannot be achieved by direct reaction with simple alkyl halides, either alone or with a Lewis-acid catalyst, for example pyrrole does not react with methyl iodide below 100 °C; above about 150 °C, a series of reactions occurs leading to a complex mixture made up mostly of polymeric material together with some poly-methylated pyrroles. The more reactive allyl bromide reacts with pyrrole at room temperature, but mixtures of mono- to tetra-allyl-pyrroles together with oligomers and polymers are obtained.

Providing an appropriate acidic catalyst is chosen – one that will not cause polymerisation of pyrrole – reaction with alkenes carrying an electron-withdrawing group can be achieved. Examples include nitro-alkenes using sulfamic acid[49] and conjugated ketones using InCl₃.[50] The use of the triflate salt of an optically active amine as catalyst induces alkylations using $R^1CH=CHCOR^2$ with high enantioselectivity.[51]

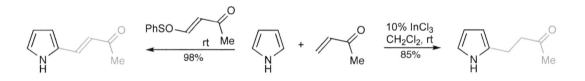

Alkylations with conjugated enones carrying a leaving group at the β-position proceed smoothly, producing mono-alkenylated pyrroles.[52]

16.1.1.7 Condensation with Aldehydes and Ketones

Condensations of pyrroles with aldehydes and ketones occur easily by acid catalysis, but the resulting pyrrolyl-carbinols cannot usually be isolated, for under the reaction conditions proton-catalysed loss of water produces 2-alkylidene-pyrrolium cations that are themselves reactive electrophiles. Thus, in the case of pyrrole itself, reaction with aliphatic aldehydes in acid inevitably leads to resins, probably linear polymers. Reductive trapping of these cationic intermediates, producing alkylated pyrroles, can be synthetically useful, however all free positions react; acyl and alkoxycarbonyl-substituents are unaffected.[53]

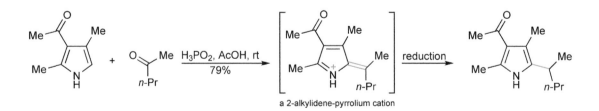

a 2-alkylidene-pyrrolium cation

Syntheses of dipyrromethanes have usually involved pyrroles with electron-withdrawing substituents and only one free α-position, the dipyrromethane resulting from attack by the electrophilic 2-alkylidene-pyrrolium intermediate on a second mole equivalent of the pyrrole.[54]

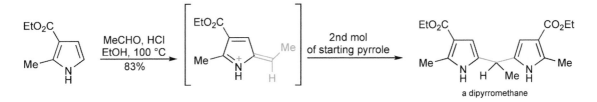

By careful choice of conditions, the simplest dipyrromethane, bis(pyrrol-2-yl)methane, can be obtained directly from pyrrole with aqueous formalin in acetic acid;[55] reaction in the presence of potassium carbonate allows 2,5-bis-hydroxymethylpyrrole to be isolated.[56] This diol reacts with pyrrole in dilute acid to give tripyrrane and from this, reaction with 2,5-bis(hydroxymethyl)pyrrole gives porphyrinogen, which can be oxidised with chloranil (2,3,5,6-tetrachloro-*p*-benzoquinone) to porphine, the simplest porphyrin.

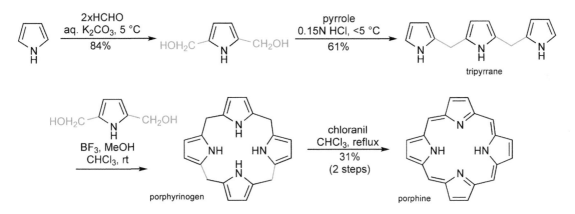

Acetone, reacting in a comparable manner, gives a cyclic tetramer directly and in high yield, perhaps because the geminal methyl groups tend to force the pyrrole rings into a coplanar conformation, greatly increasing the chances of cyclisation of the linear tetrapyrrolic precursor.[57]

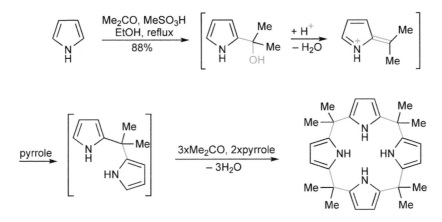

Condensations with aromatic aldehydes carrying appropriate electron-releasing substituents produce cations that are sufficiently stabilised by mesomerism to be isolated. Such cations are coloured: the reaction with *p*-dimethylaminobenzaldehyde is the basis for the classical Ehrlich test, deep red/violet colours being produced by pyrroles (and also by furans and indoles) that have a free nuclear position. Under appropriate conditions one can combine four mole equivalents of pyrrole and four of an aldehyde to produce a

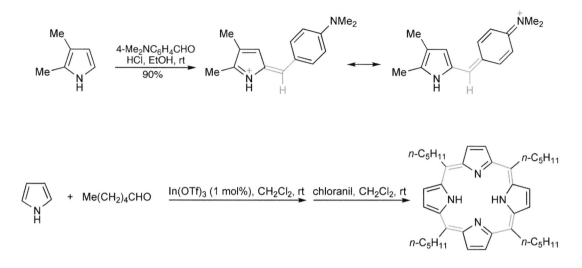

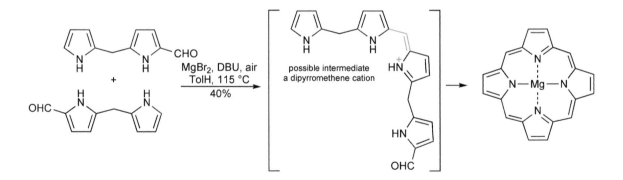

tetrasubstituted porphyrinogen in one pot,[58] but, usually, immediate oxidation is employed to proceed to the *meso*-tetra-substituted porphyrin.[59]

Analogous condensations, but with a pyrrole aldehyde lead to mesomeric dipyrromethene cations, which play an important part in porphyrin synthesis. Thus, using formyldipyrromethane as the aldehyde and a second mole as the pyrrole component, with air as oxidant, porphine is formed directly, as its magnesium derivative, possibly *via* a dipyrromethene cationic intermediate.[60]

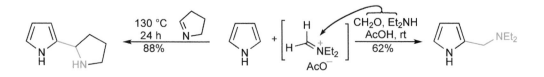

16.1.1.8 Condensation with Imines and Iminium Ions

The imine and iminium functional groupings are, of course, the nitrogen equivalents of carbonyl and *O*-protonated carbonyl groups, and their reactivity is analogous. The Mannich reaction of pyrrole produces dialkylaminomethyl derivatives, the iminium electrophile being generated *in situ* from formaldehyde, dialkylamine and acetic acid.[61] There are only a few examples of the reactions of imines themselves with pyrroles; the condensation of 1-pyrroline with pyrrole as reactant and solvent is one such example.[62] *N*-Tosyl-imines react with pyrrole with $Cu(OTf)_2$ as catalyst.[63]

The mineral-acid-catalysed polymerisation of pyrrole involves a series of Mannich reactions, but under controlled conditions, pyrrole can be converted into an isolable trimer, which is probably an intermediate in the polymerisation. The key to understanding the formation of the observed trimer is that the less stable, therefore more reactive, β-protonated pyrrolium cation is the electrophile that initiates the sequence, attacking a second mole equivalent of the heterocycle. The 'dimer', an enamine, is too reactive to be isolable, however 'pyrrole trimer', relatively protected as its salt, reacts further only slowly.[64]

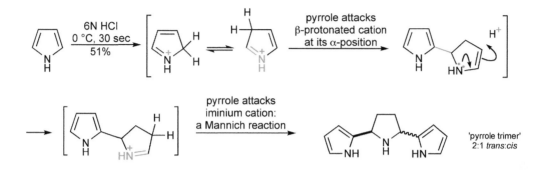

16.2 Reactions with Oxidising Agents[65]

Simple pyrroles are generally easily attacked by strong oxidising agents, frequently with complete breakdown. When the ring does survive, maleimide derivatives are the commonest products, even when there was originally a 2- or 5-alkyl substituent. This kind of oxidative degradation played an important part in early porphyrin structure determination, in which chromium trioxide in aqueous sulfuric acid or fuming nitric acid were usually used as oxidising agents. Hydrogen peroxide is a more selective reagent and can convert pyrrole itself into a tautomeric mixture of pyrrolin-2-ones in good yield (16.15.1.2).

Pyrroles which have a ketone or ester substituent are more resistant to ring degradation and high-yielding side-chain oxidation can be achieved using cerium(IV) ammonium nitrate, with selectivity for an α-alkyl.[66]

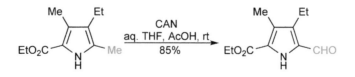

Pyrrole can be 2,2′-dimerised with the hypervalent iodine(III) reagent phenyliodine bis(trifluoroacetate) *via* what may be an SET process.[67]

16.3 Reactions with Nucleophilic Reagents

Pyrrole and its derivatives do not react with nucleophilic reagents by addition or by substitution, except in the same type of situation that allows nucleophilic substitution in benzene chemistry, i.e. where the leaving group is *ortho* or *para* to an electron-withdrawing group: the two examples below are illustrative.[68]

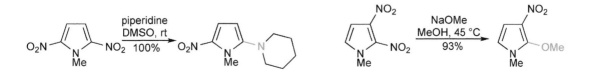

A key step in a synthesis of ketorolac, an analgesic and anti-inflammatory agent, involves an intramolecular nucleophilic displacement of a methanesulfonyl group activated by a 5-ketone.[69]

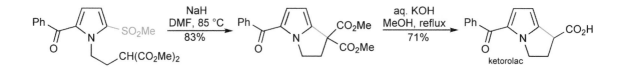

16.4 Reactions with Bases

16.4.1 Deprotonation of *N*-Hydrogen and Reactions of Pyrryl Anions

Pyrrole *N*-hydrogen is much more acidic (pK_a 17.5) than that of a comparable saturated amine, say pyrrolidine ($pK_a \sim 44$), or aniline (pK_a 30.7), and of the same order as that of 2,4-dinitroaniline. Any very strong base will effect complete conversion of an *N*-unsubstituted pyrrole into the corresponding pyrryl anion, perhaps the most convenient being commercial *n*-butyllithium solution. The pyrryl anion is nucleophilic at nitrogen (however, see 16.4.2) and thus provides the means for the introduction of groups onto pyrrole nitrogen, for example using alkyl halides. Typical reactions of the pyrryl anion are exemplified below by the preparation of 1-triisopropylsilylpyrrole. However, reactions at nitrogen can proceed *via* smaller, equilibrium concentrations of pyrryl anion, as in the formation of 1-chloropyrrole (in solution) by treatment with sodium hypochlorite[70] or a preparation of 1-*t*-butoxycarbonylpyrrole shown below.[71]

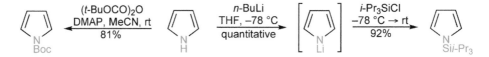

16.4.2 Lithium, Sodium, Potassium and Magnesium Derivatives

N-Salts of pyrroles can react with electrophiles to give either *N*- or *C*-substituted pyrroles: generally speaking, the more ionic the metal–nitrogen bond and/or the better the solvating power of the solvent, the greater is the percentage of attack at nitrogen.[72] Based on these principles, several methods are available for efficient *N*-alkylation of pyrroles, including the use of potassium hydroxide in dimethylsulfoxide,[73] or in benzene with 18-crown-6,[74] phase-transfer methodology,[75] K_2CO_3 in an ionic liquid[76] or of course by reaction of the pyrryl anion generated using *n*-butyllithium. *N*-Amination, for example, using chloramine (NH_2Cl) can be achieved efficiently, either by quantitative conversion into the anion using sodium hydride,[77] or using phase-transfer conditions.[78] *N*-Arylation of pyrroles requires palladium catalysis (4.2).

Pyrryl Grignard reagents, obtained by treating an *N*-unsubstituted pyrrole with a Grignard reagent, tend to react at carbon with alkylating and acylating agents, but sometimes give mixtures of 2- and 3-substituted products with the former predominating,[79] *via* neutral, non-aromatic intermediates. Clean α-substitution can be achieved for example with bromoacetates[80] as exemplified below, or using 2-acylthio-pyridines as acylating agents.[81]

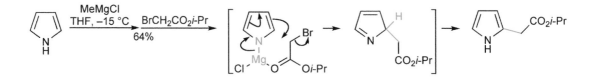

16.5 C-Metallation and Reactions of *C*-Metallated Pyrroles
16.5.1 Direct Ring C–H Metallation

The *C*-lithiation of pyrroles requires the absence of the acidic *N*-hydrogen, i.e. the presence of an *N*-substituent, either alkyl[82] or, if required, a removable group[83] like phenylsulfonyl,[84] carboxylate,[85] trimethylsilylethoxymethyl,[86] *t*-butylaminocarbonyl,[87] diethoxymethyl[88] or *t*-butoxycarbonyl. Even in the absence of chelation assistance to lithiation, which is certainly an additional feature in each of the latter examples, *metallation proceeds at the α-position.* *N*-Methylpyrrole, rather amazingly, can be converted into a dilithio-derivative, either 2,4- or 2,5-dilithio-1-methylpyrrole depending on the exact conditions.[89] Lithiation of 1-*t*-butoxycarbonyl-3-*n*-hexylpyrrole occurs at C-5, avoiding both steric and electronic discouragement to the alternative C-2 deprotonation.[90]

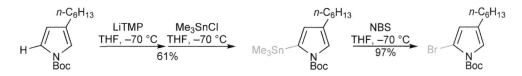

Reactions of the species produced by the α-lithiation of *N*-substituted-pyrroles are widely used for the introduction of groups, either by reaction with electrophiles or by coupling processes based on palladium chemistry (4.2). Some examples where removable *N*-blocking groups have been used in the synthesis of 2-substituted pyrroles, *via* lithiation, are shown below.[86,91]

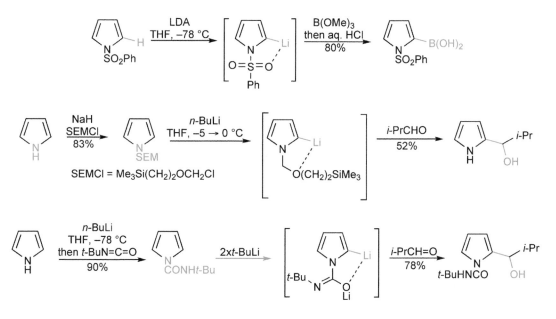

16.5.2 Metal–Halogen Exchange

Metal–halogen exchange on *N*-protected-pyrroles can provide access to either 2- or 3-lithio-pyrroles. Thus, for example, 2-bromo-1-*t*-butoxycarbonylpyrrole and its 2,5-dibromo-counterpart give monolithiated reagents and from the latter, even a dilithiated species can be generated.[30]

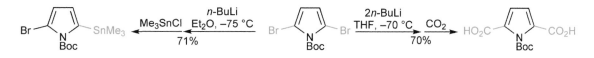

Metal–halogen exchange using 3-halo-*N*-triisopropylsilyl-pyrroles[92] allows the introduction of groups to the pyrrole β-position and can complement direct electrophilic substitution of *N*-triisopropylsilylpyrrole, which is β-selective (see 16.1.2 and 16.1.4). Sequential mono-lithiations of 1-tosyl-3,4-dibromopyrrole allows selective functionalisation.[29]

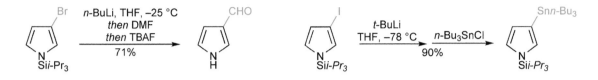

16.6 Reactions with Radicals

Pyrrole itself tends to give tars under radical conditions. A 2-toluensulfonyl-substituent can be displaced by radicals. Electrophilic radical substitution of 1-phenylsulfonylpyrrole[93] occurs at an α-position; the formation of a pyrrol-2-ylacetic acid is typical.[94] 3-Substituted pyrroles are attacked by radicals at C-2.[95]

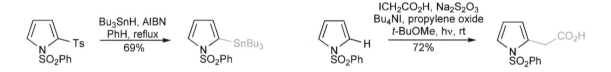

Pyrroles can be dimerised, regiospecifically, *via* a radical cation produced from the pyrrole by reaction with phenyliodine(III) bis(trifluoroacetate).[96]

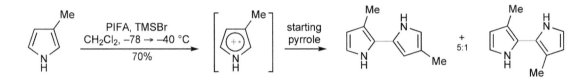

16.7 Reactions with Reducing Agents

Simple pyrroles are not reduced by hydride reducing agents or diborane, but are reduced in acidic media, in which the species under attack is the protonated pyrrole. The products are 2,5-dihydropyrroles, accompanied by some of the pyrrolidine as by-product.[97] Reduction[98] of pyrroles to pyrrolidines can be effected catalytically over a range of catalysts, is especially easy if the nitrogen carries an electron-withdrawing group, and is not complicated by carbon–heteroatom hydrogenolysis and ring opening, as is the case for furans. Reduction of 2-acyl-pyrroles to pyrrolidine alcohols, over 5% Rh/Al₂O₃, proceeds with high stereoselectivity, with the control probably arising from the alcohol produced by initial reduction of the carbonyl group.

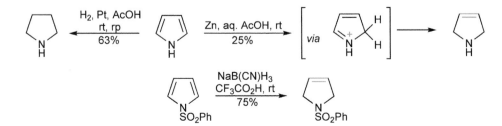

Birch reduction of pyrrole carboxylic esters and tertiary amides gives dihydro-derivatives; the presence of an electron-withdrawing group on the nitrogen serves both to remove the acidic *N*-hydrogen and also to reduce the electron density on the ring. Quenching the immediate reduced species – an enolate – with an alkyl halide produces alkylated dihydropyrroles.[99]

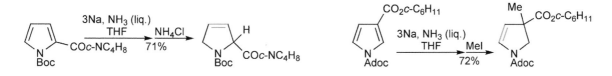

16.8 Electrocyclic Reactions (Ground State)

Simple pyrroles do not react as 4π components in Diels–Alder cycloadditions: exposure of pyrrole to benzyne, for example, leads only to 2-phenylpyrrole, in low yield.[100] However *N*-substitution, particularly with an electron-withdrawing group, does allow such reactions to occur,[101] for example adducts with arynes are obtained using 1-trimethylsilylpyrrole.[102] Whereas pyrrole itself reacts with dimethyl acetylenedicarboxylate only by α-substitution, even at 15 kbar,[103] *N*-acetyl- and *N*-alkoxycarbonyl-pyrroles give cycloadducts,[104] addition being much accelerated by high pressure or by aluminium chloride catalysis.[105] The most popular *N*-substituted pyrrole in this context has been *N*-Boc-pyrrole,[106] with benzyne (from diazotization of anthranilic acid) for example, a 60% yield of the cycloadduct is obtained.[107]

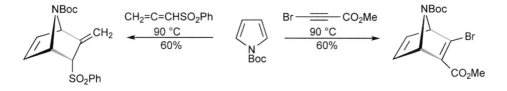

A process that has proved valuable in synthesis is the addition of singlet oxygen to *N*-alkyl- and especially *N*-acyl-pyrroles[108] producing 2,3-dioxa-7-aza-bicyclo[2.2.1]heptanes, which react with nucleophiles, such as silyl enol ethers, mediated by tin(II) chloride, generating 2-substituted-pyrroles that can be used, as the example shows, for the synthesis of indoles *via* intramolecular electrophilic attack by the carbonyl group at the pyrrole β-position.

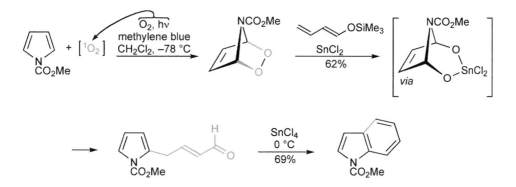

Intermolecular examples of pyrroles serving as 2π components in cycloadditions are rare but examples include *N*-tosyl-2- and -3-nitropyrrole[109] and *N*-tosyl-2,4-diacyl-pyrroles,[110] however in an intramolecular sense tricyclic 6-azaindoles are produced where the 4π component is a 1,2,4-triazine (29.2.1).[111]

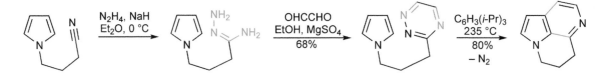

Vinyl-pyrroles will take part in Diels–Alder processes as 4π components,[112] providing the aromaticity of the ring has been reduced by the presence of a phenylsulfonyl group on the pyrrole nitrogen, the presumed initial product easily isomerising in the reaction conditions to reform an aromatic pyrrole.[113]

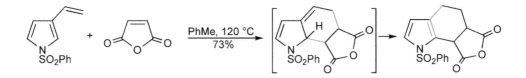

16.9 Reactions with Carbenes and Carbenoids

The reaction of pyrrole with dichlorocarbene proceeds in part *via* a dichlorocyclopropane intermediate, ring expansion of which leads to 3-chloropyridine.[114,115] *N*-Methylpyrrole with ethoxycarbonylcarbene gives only substitution products.[116]

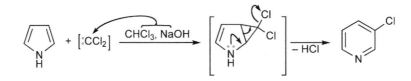

Isolable cyclopropane-containing adducts can be obtained from *N*-Boc-pyrrole,[117] and ozonolysis of the mono adduct is a means for the synthesis of heavily functionalised cyclopropanes.[118]

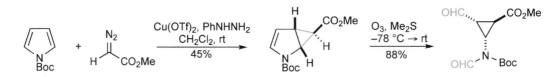

Rhodium-catalysed addition of a vinyl carbene produces a cyclopropanated intermediate that undergoes a Cope rearrangement, neatly producing an 8-azabicyclo[3.2.1]octadiene – the ring skeleton of cocaine.[119]

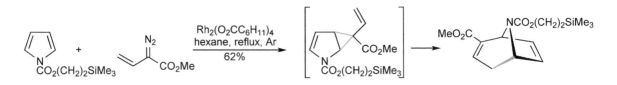

16.10 Photochemical Reactions[120]

The photo-catalysed rearrangement of 2- to 3-cyanopyrrole is considered to involve a 1,3-shift in an initially formed bicyclic aziridine.[121]

16.11 Pyrryl-C-X Compounds

Pyrroles of this type, where X is halogen, alcohol, alkoxy or amine, and especially protonated alcohol or alkoxy, or quaternised amine, easily lose X, generating reactive electrophilic species. Thus ketones can be reduced to alkane, *via* the loss of oxygen from the initially formed alcohol (cf. 16.12), and quaternary ammonium salts, typified by 2-dimethylaminomethylpyrrole metho-salts, react with nucleophiles by loss of trimethylamine in an elimination/addition sequence of considerable synthetic utility.[122]

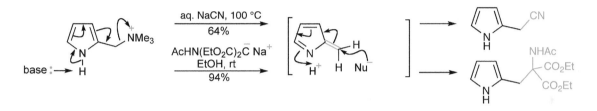

16.12 Pyrrole Aldehydes and Ketones

These are stable compounds which do not polymerise or autoxidise. For the most part, pyrrole-aldehydes and -ketones are typical aryl-ketones, though less reactive – such ketones can be viewed as vinylogous amides. They can be reduced to alkyl-pyrroles by the Wolff–Kishner method, or by sodium borohydride *via* elimination from the initial alcoholic product (cf. 16.11).[123] Treatment of acyl-1-phenylsulfonyl-pyrroles with *t*-butylamine-borane also effects conversion to the corresponding alkyl derivatives.[124]

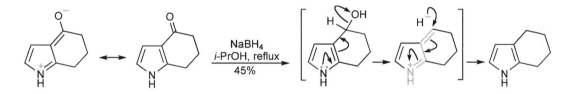

β- and α-Acyl-pyrroles can be equilibrated, one with the other, using acid; for *N*-alkyl-*C*-acyl-pyrroles, the equilibrium lies completely on the side of the β-isomer.[125]

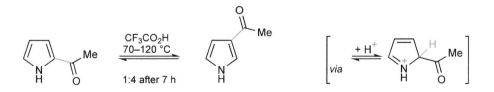

16.13 Pyrrole Carboxylic Acids

The main feature within this group is the ease with which loss of the carboxyl group occurs. Simply heating[126] pyrrole acids causes loss of carbon dioxide in what is essentially *ipso*-displacement of carbon dioxide by proton.[127] This facility is of considerable relevance to pyrrole synthesis since several of the ring-forming routes (e.g. see 16.16.1.2 and 16.16.1.6) produce pyrrole esters, in which the ester function may not be required ultimately.

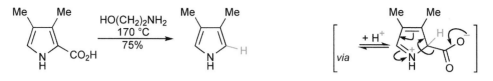

Ipso displacement of carboxyl groups by electrophiles, such as halogens,[128] or under nitrating conditions, or with aryl-diazonium cations, occurs more readily than substitution of hydrogen.

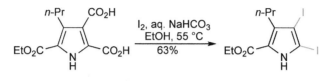

16.14 Pyrrole Carboxylic Acid Esters

The electrophilic substitution of these stable compounds has been much studied; the *meta*-directing effect of a 2-ester overcomes the normally dominant pyrrole tendency for α-substitution.[129]

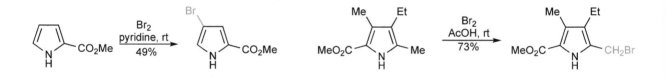

An ester group can also activate a side-chain alkyl for halogenation, and such pyrrolyl-alkyl halides have been used extensively in synthesis.[130] Cerium(IV) triflate in methanol can be used for the analogous introduction of methoxide onto an alkyl side-chain.[131] The rates of alkaline hydrolysis of α- and β-esters are markedly different, the former being faster than the latter.[132]

16.15 Oxy- and Amino-Pyrroles
16.15.1 2-Oxy-Pyrroles

2-Oxypyrroles exist in the hydroxyl form, if at all, only as a minor component of the tautomeric mixture that favours 3-pyrrolin-2-one over 4-pyrrolin-2-one by 9:1.[133]

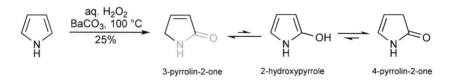

3-pyrrolin-2-one   2-hydroxypyrrole   4-pyrrolin-2-one

After *N*-protection, silylation produces 2-silyloxy-pyrroles, which react with aldehydes to give 5-substituted 3-pyrrolin-2-ones.[134]

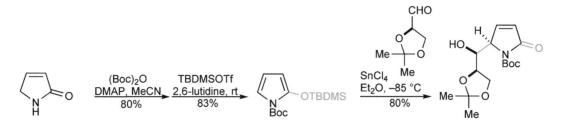

16.15.2   3-Oxy-Pyrroles

3-Oxy-pyrroles exist largely in the carbonyl form, unless flanked by an ester group at C-2 that favours the hydroxyl-tautomer by intramolecular hydrogen bonding.[135]

16.15.3   Amino-Pyrroles

Amino-pyrroles have been very little studied because they are relatively unstable and difficult to prepare.[136] Simple 2-amino-pyrroles can be prepared, but must be stored in acidic solution.[137] An alternative is to reduce nitro-pyrroles over Pd/C in the presence of anhydrides, which produces amides of the otherwise unstable amino-pyrroles.[138] Another trapping procedure employs a 1,4-dione and thus engenders another pyrrole ring (cf. 16.16.1.1).[139]

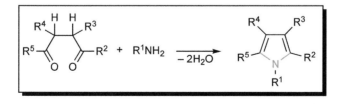

16.16   Synthesis of Pyrroles[5,140]

16.16.1   Ring Synthesis

16.16.1.1   From 1,4-Dicarbonyl Compounds and Ammonia or Primary Amines[141]

1,4-Dicarbonyl compounds react with ammonia or primary amines to give pyrroles.

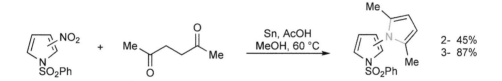

Paal–Knorr Synthesis[142]

Pyrroles are formed by the reaction of ammonia or a primary amine with a 1,4-dicarbonyl compound[143] (see also 17.12.1.1 and 18.13.1.1). Successive nucleophilic additions of the amine nitrogen to each of the two carbonyl carbon atoms and the loss of two mole equivalents of water represent the net course of the synthesis; a reasonable sequence[144] for this is shown below, using the synthesis of 2,5-dimethylpyrrole[145] as an example.

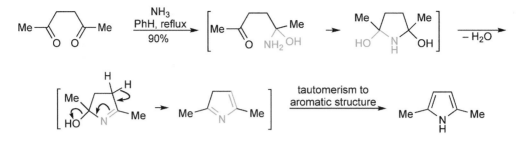

Several variations have been shown to improve the efficiency: microwave irradiation makes the process very rapid[146] and the use of iodine on a clay support[147] are just two of these. An alternative to the use of ammonia for the synthesis of *N*-unsubstituted-pyrroles employs hexamethyldisilazide with alumina,[148] or a solution of ammonia can be generated *in situ* conveniently, using the reaction of magnesium nitride, Mg_3N_2, with methanol.[149]

The best synthon for unstable succindialdehyde, for the ring synthesis of *C*-unsubstituted pyrroles, is 2,5-dimethoxytetrahydrofuran (18.1.1.4),[150] or 1,4-dichloro-1,4-dimethoxybutane obtainable from it.[151] 2,5-Dimethoxytetrahydrofuran will react with aliphatic and aromatic amines,[12,152] amino esters, aryl-sulfonamides,[153] trimethylsilylethoxycarbonylhydrazine[154] or primary amides to give the corresponding *N*-substituted-*C*-unsubstituted-pyrroles.[155]

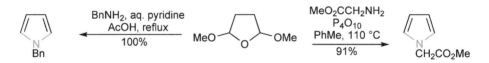

16.16.1.2 From α-Aminocarbonyl-Compounds and Activated Ketones

α-Amino-ketones react with carbonyl compounds that have an α-methylene grouping, preferably further activated, for example by ester, as in the illustration.

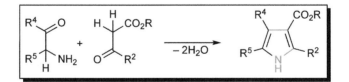

Knorr Synthesis

This widely used general approach to pyrroles utilizes two components: one, the α-aminocarbonyl component, supplies the nitrogen and C-2 and C-3, and the second component supplies the remaining two carbons and must possess a methylene group α to a carbonyl. The Knorr synthesis works well only if the methylene group of the second component is further acidified (e.g. as in acetoacetic ester, i.e. it is a 1,3-dicarbonyl compound, or equivalent) to enable the desired condensation leading to pyrrole to compete effectively with the self-condensation of the α-aminocarbonyl component. The synthesis of 4-methylpyrrole-3-carboxylic acid and therefrom, 3-methylpyrrole, illustrates the process.

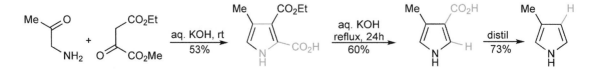

Since free α-aminocarbonyl compounds self-condense very readily producing dihydropyrazines (14.13.3.1), they have traditionally been prepared and used in the form of their salts, to be liberated for the condensation reaction by the base present in the reaction mixture. Alternatively, carbonyl-protected amines, such as aminoacetal ($H_2NCH_2CH(OEt)_2$), have been used. In one such case, the enol ether of a 1,3-keto-aldehyde was the synthon for the activated carbonyl component.[156]

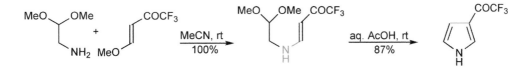

A way of avoiding the difficulty of handling α-aminocarbonyl-compounds is to prepare them in the presence of the second component, with which they are to react. Zinc–acetic acid or sodium dithionite[157] can be used to reduce oximino-groups to amino, while leaving ketone and ester groups untouched. In the classical synthesis, which gives this route its name, the α-aminocarbonyl component is simply an amino derivative of the other carbonyl component, and it is even possible to generate the oximino precursor of the amine *in situ*.[158]

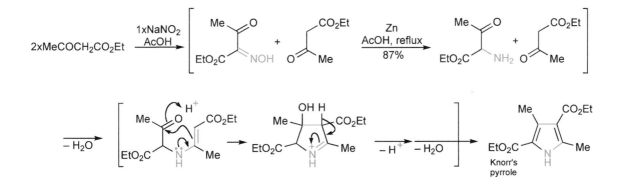

It is believed that in the mechanism, shown for Knorr's pyrrole, an N–C-2 bond is the first formed, which implies that the nitrogen becomes attached to the more electrophilic of the two carbonyl groups of the other component. Similarly, the C-3–C-4 bond is made to the more electrophilic carbonyl group of the original α-aminocarbonyl-component, where there is a choice.

Alternatives for the assembly of the α-aminocarbonyl-component in protected form include the reaction of a 2-bromo-ketone with sodium diformamide producing an α-formamido-ketone,[159] and the reaction of a Weinreb amide of an *N*-Cbz α-amino acid with a Grignard reagent, the release of the *N*-protection in the presence of the second component, produces the pyrrole.[160]

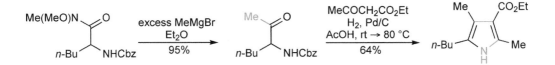

A final example in this category involves the enamines produced by addition of an α-amino acid ester to dimethyl acetylenedicarboxylate leading to 3-hydroxypyrroles by Claisen-type ring closure.[161]

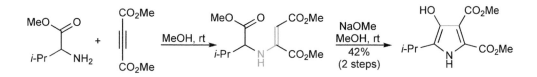

16.16.1.3 From Tosylmethyl Isocyanide and α,β-Unsaturated Esters or Ketones and from Isocyano-Acetates and α,β-Unsaturated Nitro-Compounds

Tosylmethyl isocyanide anion reacts with α,β-unsaturated esters, ketones or sulfones with loss of toluene-sulfinate. Isocyano-acetates react with α,β-unsaturated nitro-compounds with loss of nitrous acid.

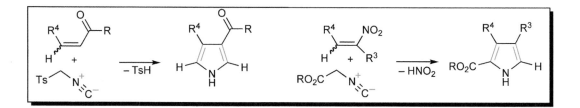

The van Leusen Synthesis

The stabilised anion of tosylmethyl isocyanide[162] (TosMIC) (or of benzotriazol-1-ylmethyl isocyanide – BetMIC[163]) adds in Michael fashion to unsaturated ketones and esters, with subsequent closure onto isocyanide carbon, generating the ring. Proton transfer, then elimination of toluenesulfinate generates a 3H-pyrrole that tautomerises to an aromatic pyrrole that is unsubstituted at both α-positions.[164] Addition of the TosMIC anion to unsaturated nitro-compounds gives rise to 2,5-unsubstituted-3-nitropyrroles.[165]

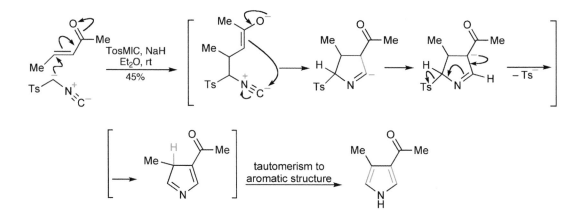

tautomerism to aromatic structure

The Barton–Zard Synthesis[166]

In this approach, conjugate addition of the anion from an isocyano-acetate to an α,β-unsaturated nitro-compound with eventual loss of nitrous acid, produces 5-unsubstituted pyrrole-2-esters.[167] The example[168] below shows a mechanistic sequence that can be seen to parallel that in the van Leusen synthesis. The most useful route to the α,β-unsaturated nitro-compound involves the base-catalysed condensation of an alde-hyde with a nitroalkane giving an α-hydroxy-nitroalkane; it can alternatively be generated *in situ*, in the presence of the isonitrile, using diazabicycloundecane (DBU) as base on the *O*-acetate of the α-hydroxy-nitroalkane[169] (for an example see 16.16.2.1). The process works even when the unsaturated nitro unit is a component of a polycyclic aromatic compound.[170]

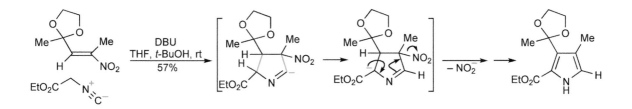

Extrapolations and improvements to this approach continue to enlarge its usefulness – α,β-unsaturated sulfones react with isocyano-acetates and isocyano-nitriles to give pyrroles.[171] Potassium carbonate can be used as the base,[172] vinyl arenes and hetarenes react at the side-chain double bond to give 3-aryl(hetaryl)-pyrroles,[173] and acetylenic-esters produce pyrrole-2,4-dicarboxylates, methyl *t*-butyl ether as solvent to avoid peroxides.[174]

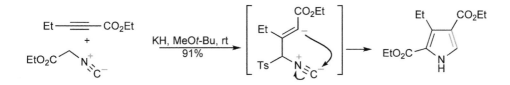

16.16.1.4 From Azines (RCH₂CH=N)₂

The Piloty–Robinson Synthesis

Improvements in the long-known Piloty–Robinson synthesis make it a useful approach, not only for sym-metrically 3,4-disubstituted pyrroles, but also for more complex systems. In its simplest form, an aldehyde azine (RCH₂CH=NN=CHCH₂R) reacts with an acid chloride to generate an *N*-acyl-pyrrole carrying identi-cal groups R at the 3- and 4-positions; microwave heating facilitates the process, which involves, as its key step, a 3,3-sigmatropic rearrangement.[175] The use of *N*,*N*′-di-Boc-hydrazine and two successive copper-catalysed couplings to different iodo-alkenes produces unsymmetrical hydrazine bis-enamines, leading to unsymmmetrically substituted *N*-Boc-pyrroles.[176]

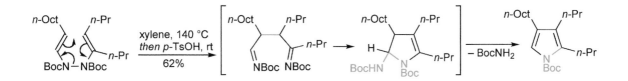

16.16.1.5 From Oximes and Alkynes
The Trofimov Synthesis[177]

This approach, though less well known than it perhaps deserves, involves simply heating a ketoxime and acetylene in the presence of an alkali metal hydroxide, generally in DMSO. The pyrrole products are 2,3-unsubstituted. The process is simple, though it does require handling acetylene at high temperature. Conditions are available to produce either the pyrrole or, directly, an *N*-vinyl-pyrrole, complete with a protecting group on nitrogen. The scheme suggests a probable mechanism.

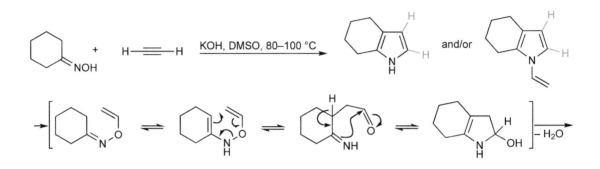

16.16.1.6 From Precursors with Four Carbons and a Nitrogen; From C₄N Units

There are a number of routes to pyrroles in which a precursor is assembled that has the four carbons and the nitrogen destined to be those of the final pyrrole, and requiring only the making of one final bond and rearrangement of double bonds into the aromatic configuration. The nature of that bond-making ring-closing step differs from one example to another.

For example, condensation of a 1,3-dicarbonyl compound with ethyl glycinate, using triethylamine as base, produces an enamino-ketone, which can then be ring closed, the step in this case being an aldol condensation.[178]

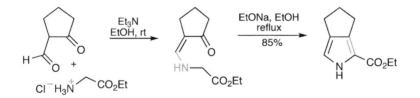

Another example is the 5-*endo-dig* closure of 4-tosylamino-alkynes, which initially generates dihydro-pyrroles, the elimination of toluenesulfinate from which produces the aromatic system.[179]

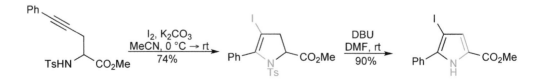

γ-Chloro-enones react with primary amines to generate γ-amino-enones which, adsorbed on silica, cyclise on microwave irradation.[180]

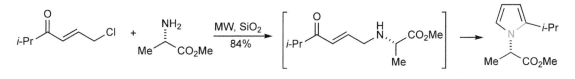

Copper-assisted cycloisomerisation of conjugated alkynyl-imines[181] gives pyrroles, even when the imine is actually the imine 'double bond' of a pyridine.

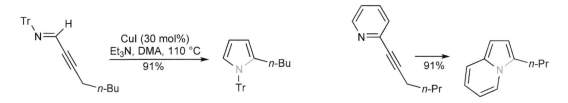

16.16.1.7 From Alkynes and Oxido-Oxazoliums[182]
Dipolar cycloaddition of alkynes to mesoionic oxido-oxazoliums, followed by expulsion of carbon dioxide, yields pyrroles.

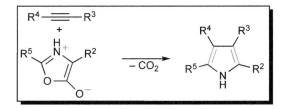

Dehydration of *N*-acylamino-acids generates azlactones; these are in equilibrium with mesoionic species, which can be trapped by reaction with alkynes, final loss of carbon dioxide giving the aromatic pyrrole.

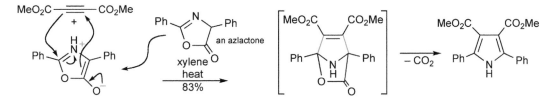

16.16.2 Some Notable Syntheses of Pyrroles
16.16.2.1 Octaethylporphyrin[183]
This synthesis of octaethylporphyrin, widely used as a model compound, uses a Barton–Zard sequence and leads to a pyrrole-2-ester which is then hydrolysed and decarboxylated.

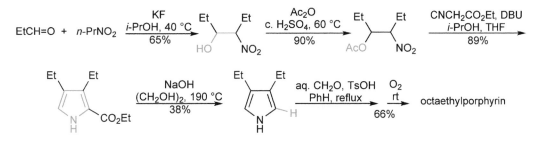

16.16.2.2 Octaethylhemiporphycene[184,185]

All of the non-natural isomers (porphycenes) of the porphyrin ring system comprising permutations of four pyrrole rings, four methines and having an 18 π-electron main conjugation pathway, have been synthesised. The scheme below shows the use of a MacDonald condensation[186] to assemble a tetrapyrrole and then the use of the McMurray reaction to construct the macrocycle.[187]

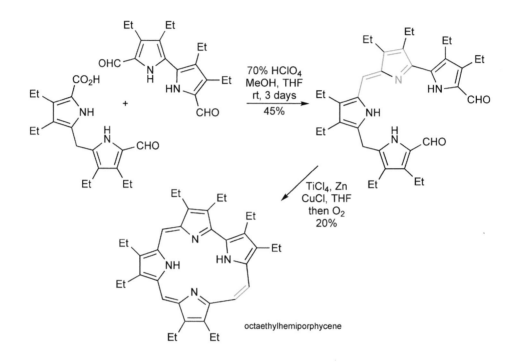

octaethylhemiporphycene

16.16.2.3 Benzo[1,2-b:4,3-b′]dipyrroles

Several ingenious approaches[188] have been described for the elaboration of the pyrrolo-indole unit (strictly a benzo[1,2-b:4,3-b′]dipyrrole) three of which are present in the potent anti-tumour compound CC-1065;[189] the approach shown here uses the van Leusen approach (16.16.1.3).[190]

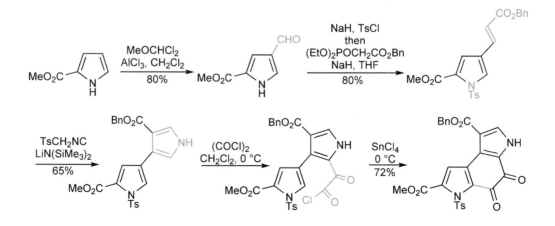

Exercises
Straightforward revision exercises (consult Chapters 15 and 16):
(a) Why does pyrrole not form salts by protonation on nitrogen?

(b) Starting from pyrrole, how would one prepare, cleanly, 2-bromo-pyrrole, 3-bromo-pyrrole, 2-formyl-pyrrole, 3-nitro-pyrrole? (More than one step necessary in some cases.)

(c) What would be the structures of the products from the following reactions: (i) pyrrole with CH_2O/pyrrolidine/AcOH; (ii) pyrrole with NaH/MeI; (iii) 1-tri-*i*-propylsilylpyrrole with LDA then $Me_3CCH=O$?

(d) How could one produce a 3-lithiated pyrrole?

(e) How could a pyrrole system be encouraged to react as a diene in Diels–Alder-type processes.

(f) How could pyrrole be converted into pyrrol-2-yl-CH_2CN in two steps?

(g) By what mechanism are pyrrole carboxylic acids readily decarboxylated on heating?

(h) Which ring synthesis method and what reactants would be appropriate for the synthesis of a pyrrole, unsubstituted on the ring carbons, but carrying $CH(Me)(CO_2Me)$ on nitrogen?

(i) With what reactant would ethyl acetoacetate ($MeCOCH_2CO_2Et$) need to be reacted to produce ethyl 2-methyl-4,5-diphenylpyrrole-3-carboxylate?

(j) With what reactant would TosMIC ($TsCH_2NC$) need to be reacted to produce methyl 4-ethylpyrrol-3-carboxylate?

(k) With what reactants would 3-nitrohex-3-ene need to be treated to produce ethyl 3,4-diethylpyrrole-2-carboxylate?

More advanced exercises:
1. Two isomeric mono-nitro derivatives, $C_5H_6N_2O_2$, are formed in a ratio of $6:1$, by treating 2-methylpyrrole with Ac_2O/HNO_3. What are their structures and which would you predict to be the major product?

2. Write structures for the products of the following sequences: (i) pyrrole treated with Cl_3CCOCl, then the product with Br_2, then this product with MeONa/MeOH $\rightarrow C_6H_6BrNO_2$, (ii) pyrrole treated with DMF/$POCl_3$, then with $MeCOCl$/$AlCl_3$, then finally with aq. NaOH $\rightarrow C_7H_7NO_2$, (iii) 2-chloropyrrole treated with DMF/$POCl_3$, then aq. NaOH, then the product with $LiAlH_4 \rightarrow C_5H_6ClN$.

3. Write structures for the products formed by the reaction of pyrrole with $POCl_3$ in combination with: (i) *N,N*-dimethylbenzamide; (ii) pyrrole-2-carboxylic acid *N,N*-dimethylamide; (iii) 2-pyrrolidone $\rightarrow C_8H_{10}N_2$, in each case followed by aq. NaOH.

4. Treatment of 2-methylpyrrole with HCl produces a dimer, not a trimer as does pyrrole itself. Suggest a structure for the dimer, $C_{10}H_{14}N_2$, and explain the non-formation of a trimer.

5. Treatment of 2,5-dimethylpyrrole with Zn/HCl gave a mixture of two isomeric products $C_6H_{11}N$: suggest structures.

6. (i) Heating 1-methoxycarbonylpyrrole with diethyl acetylenedicarboxylate at 160 °C produced diethyl 1-methoxycarbonylpyrrole-3,4-dicarboxylate; suggest a mechanism and a key intermediate; (ii) deduce the structure of the product, $C_{11}H_{12}N_2O_2$, resulting from successive treatment of 1-methoxycarbonylpyrrole with singlet oxygen then a mixture of 1-methylpyrrole and $SnCl_2$.

7. Deduce structures for the products formed at each stage by treating pyrrole successively with: (i) Me_2NH/HCHO/AcOH, (ii) CH_3I, (iii) piperidine in hot EtOH $\rightarrow C_{10}H_{16}N_2$.

8. From a precursor that does not contain a pyrrole ring, how might one synthesise: (i) 1-*n*-propylpyrrole; (ii) 1-(thien-2-yl)pyrrole; (iii) 1-phenylsulfonylpyrrole?

9. Reaction of $MeCOCH_2CO_2Et$ with HNO_2, then a combination of Zn/AcOH and pentane-2,4-dione gave a pyrrole, $C_{11}H_{15}NO_3$. Deduce the structure of the pyrrole, write out a sequence for its formation, and suggest a route whereby it could then be converted into 2,4-dimethyl-3-ethylpyrrole.

10. How might one prepare: (i) diethyl 4-methylpyrrole-2,3-dicarboxylate, (ii) ethyl 2,4,5-trimethylpyrrole-3-carboxylate; (iii) ethyl 4-amino-2-methylpyrrole-3-carboxylate; (iv) ethyl 3,4,5-trimethylpyrrole-2-carboxylate?

References

[1] 'The synthesis of highly functionalized pyrroles: a challenge in regioselectivity and chemical reactivity', *Synthesis*, **2007**, 3095; 'Pyrrole. From Dippel to Du Pont', Anderson, H. J., *J. Chem. Ed.*, **1995**, *72*, 875; 'The chemistry of pyrroles', Jones, R. A. and Bean, G. P., Academic Press, New York, **1977**; 'Physicochemical properties of pyrroles', Jones, R. A., *Adv. Heterocycl. Chem.*, **1970**, *11*, 383.

[2] 'The pyrrole pigments', Smith, K. M. In 'Rodd's Chemistry of Carbon Compounds', **1977**, Vol. *IVB*, and supplement, **1997**, Ch. 12.

[3] 'Nature's pathways to the pigments of life', Battersby, A. R., *Nat. Prod. Rep.*, **1987**, *4*, 77.

[4] 'The colours of life. An introduction to the chemistry of porphyrins and related compounds', Milgram, L. R., Oxford University Press, Oxford, **1997**.

[5] 'The synthesis of 3-substituted pyrroles from pyrrole', Anderson, H. J. and Loader, C. E., *Synthesis*, **1985**, 353.

[6] Bean, G. P., *J. Chem. Soc., Chem. Commun.*, **1971**, 421; Muir, D. M. and Whiting, M. C., *J. Chem. Soc., Perkin Trans. 2*, **1975**, 1316.

[7] Nguyen, V. Q. and Turecek, F., *J. Mass Spectrom.*, **1996**, *31*, 1173.

[8] Chiang, Y., Hinman, R. L., Theodoropulos, S. and Whipple, E. B., *Tetrahedron*, **1967**, *23*, 745.

[9] Gassner, R., Krumbholz, E. and Steuber, F. W., *Justus Liebigs Ann. Chem.*, **1981**, 789.

[10] Findlay, S. P., *J. Org. Chem.*, **1956**, *21*, 644; Garrido, D. O. A., Buldain, G. and Frydman, B., *J. Org. Chem.*, **1984**, *49*, 2619.

[11] Breukelman, S. P., Leach, S. E., Meakins, G. D. and Tirel, M. D., *J. Chem. Soc., Perkin Trans. 1*, **1984**, 2801.

[12] Ragan, J. A., Jones, B. P., Castaldi, M. J., Hill, P. D. and Makowski, T. W., *Org. Synth.*, **2002**, *78*, 63.

[13] Cooksey, A. R., Morgan, K. J. and Morrey, D. P., *Tetrahedron*, **1970**, *26*, 5101.

[14] Iranpoor, N., Firouzabadi, H., Nowrouzi, N. and Firouzabadi, D., *Tetrahedron Lett.*, **2006**, *47*, 6879.

[15] Doddi, G., Mencarelli, P., Razzini, A. and Stegel, F., *J. Org. Chem.*, **1979**, *44*, 2321.

[16] Muchowski, J. M. and Naef, R., *Helv. Chim. Acta*, **1984**, *67*, 1168; Bray, B. L., Mathies, P. H., Naef, R., Solas, D. R., Tidwell, T. T., Artis, D. R. and Muchowski, J. M., *J. Org. Chem.*, **1990**, *55*, 6317.

[17] Terentyev, A. P., Yanovski, L.A. and Yashunskjy, V. G., *J. Gen Chem. USSR (US translation)*, **1950**, *20*, 539.

[18] Mizuno, A., Kan, Y., Fukami, H., Kamei, T., Miyazaki, K., Matsuki, S. and Oyama, Y., *Tetrahedron Lett.*, **2000**, *41*, 6605.

[19] Janosik, T., Shirani, H., Wahlström, N., Malky, I., Stensland, D. and Bergman, J., *Tetrahedron*, **2006**, *62*, 1699.

[20] Carmona, O., Greenhouse, R., Landeros, R. and Muchowski, J. M., *J. Org. Chem.*, **1980**, *45*, 5336.

[21] Nair, V., George, T. G., Nair, L. G. and Panicker, S. B., *Tetrahedron Lett.*, **1999**, *40*, 1195.

[22] Kakushima, M. and Frenette, R., *J. Org. Chem.*, **1984**, *49*, 2025.

[23] Ortiz, C. and Greenhouse, R., *Tetrahedron Lett.*, **1985**, *26*, 2831.

[24] Treibs, A. and Kolm, H. G., *Justus Liebigs Ann. Chem.*, **1958**, *614*, 176.

[25] Cordell, G. A., *J. Org. Chem.*, **1975**, *40*, 3161; Gilow, H. M. and Burton, D. E., *ibid.*, **1981**, *46*, 2221.

[26] Chen, W. and Cava, M. P., *Tetrahedron Lett.*, **1987**, *28*, 6025; Chen, W., Stephenson, E. K., Cava, M. P. and Jackson, Y. A., *Org. Synth.*, **1992**, *70*, 151.

[27] Knight, L. W., Huffman, J. W. and Isherwood, M. L., *Synlett*, **2003**, 1993.

[28] Kozikowski, A. P. and Cheng, X.-M., *J. Org. Chem.*, **1984**, *49*, 3239; Shum, P. W. and Kozikowski, A. P., *Tetrahedron Lett.*, **1990**, *31*, 6785.

[29] Zonta, C., Fabris, F. and De Lucchi, O., *Org. Lett.*, **2005**, *7*, 1003.

[30] Fürstner, A., Krause, H. and Thiel, O. R., *Tetrahedron*, **2002**, *58*, 6373; Knight, L. W., Huffman, J. W. and Isherwood, M. L., *Synlett*, **2003**, 1993.

[31] Liu, J.-H., Chan, W.-W. and Wong, H. N. C., *J. Org. Chem.*, **2000**, *56*, 3274.

[32] Anderson, A. G. and Exner, M. M., *J. Org. Chem.*, **1977**, *42*, 3952.

[33] Reddy, G. S., *Chem. Ind.*, **1965**, 1426.

[34] Harbuck, J. W. and Rapoport, H., *J. Org. Chem.*, **1972**, *37*, 3618; Bailey, D. M., Johnson, R. E. and Albertson, N. F., *Org. Synth., Coll. Vol. VI*, **1988**, 618; Chadwick, D. J., Meakins, G. D. and Rhodes, C. A., *J. Chem. Res., (S)*, **1980**, 42; Jaramillo, D., Liu, Q., Aldrich-Wright, J. and Tor, Y., *J. Org. Chem.*, **2004**, *69*, 8151.

[35] Bélanger, P. *Tetrahedron Lett.*, **1979**, 2505.

[36] Katritzky, A. R., Suzuki, K., Singh, S. K. and He, H.-Y., *J. Org.Chem.*, **2003**, *68*, 5720.

[37] Vilsmeier, A. and Haack, A., *Chem. Ber.*, **1927**, *60*, 119.

[38] 'The Vilsmeier reaction of fully conjugated carbocycles and heterocycles', Jones, G. and Stanforth, S. P., *Org. React.*, **1997**, *49*, 1.

[39] Smith, G. F., *J. Chem. Soc.*, **1954**, 3842; Silverstein, R. M., Ryskiewicz, E. E. and Willard, C., *Org. Synth., Coll. Vol. IV*, **1963**, 831; de Groot, J. A., Gorter-La Roy, G. M., van Koeveringe, J. A. and Lugtenburg, J., *Org. Prep. Proc. Int.*, **1981**, *13*, 97.

[40] Jugie, G., Smith, J. A. S. and Martin, G. J., *J. Chem. Soc., Perkin Trans. 2*, **1975**, 925.

[41] Chadwick, D. J. and Hodgson, S. T., *J. Chem. Soc., Perkin Trans. 1*, **1983**, 93.

[42] Simchen, G. and Majchrzak, M. W., *Tetrahedron Lett.*, **1985**, *26*, 5035.

[43] Rapoport, H. and Castagnoli, N., *J. Am. Chem. Soc.*, **1962**, *84*, 2178; Oishi, T., Hirama, M., Sita, L. R. and Masamune, S., *Synthesis*, **1991**, 789.

[44] Kimbaris, A. and Varvounis, G., *Tetrahedron*, **2000**, *56*, 9675; Gracia, S., Schulz, J., Pellet-Rostaing, S. and Lemaire, M., *Synlett*, **2008**, 1852.

[45] Song, C., Knight, D. W. and Whatton, M. A., *TetraheronLett.*, **2004**, *45*, 9573.

[46] Tani, M., Ariyasu, T., Nishiysama, C., Hagiwara, H., Watanabe, T., Yokoyama, Y. and Murakami, Y., *Chem. Pharm. Bull.*, **1996**, *44*, 48.

[47] Cadamuro, S., Degani, I., Dughera, S., Fochi, R., Gatti, A. and Piscopo, L., *J. Chem. Soc., Perkin Trans. 1*, **1993**, 273.

[48] Hong, F., Zaidi, J., Pang, Y.-P., Cusack, B. and Richelson, E., *J. Chem. Soc., Perkin Trans. 1*, **1997**, 2997.

[49] An, L.-T., Zou, J.-P., Zhang, L.-L. and Zhang, Y., *Tetrahedron Lett.*, **2007**, *48*, 4297.

[50] Yadav, J. S., Abraham, S., Reddy, B. V. S. and Sabitha, G., *Tetrahedron Lett.*, **2001**, *42*, 8063.

[51] Paras, N. A. and MacMillan, W. C., *J. Am. Chem. Soc.*, **2001**, *123*, 4370.

[52] Hayakawa, K., Yodo, M., Ohsuki, S. and Kanematsu, K., *J. Am. Chem. Soc.*, **1984**, *106*, 6735.

[53] Gregorovich, B. V., Liang, K. S. Y., Clugston, D. M. and MacDonald, S. F., *Can. J. Chem.*, **1968**, *46*, 3291.

[54] Fischer, H. and Schubert, F., *Hoppe-Seyler's Z. Physiol. Chem.*, **1926**, *155*, 88.

[55] Wang, Q. M. and Bruce, D. W., *Synlett*, **1995**, 1267.

[56] Taniguchi, S., Hasegawa, H., Nishimura, M. and Takahashi, M., *Synlett*, **1999**, 73.

57 Rothemund, P. and Gage, C. L., *J. Am. Chem. Soc.*, **1955**, *77*, 3340; Corwin, A. H., Chivvis, A. B. and Storm, C. B., *J. Org. Chem.*, **1964**, *29*, 3702.

58 Gonsalvez, A. M. d'A. R., Varejão, J. M. T. B. and Pereira, M. M., *J. Heterocycl. Chem.*, **1991**, *28*, 635.

59 Smith, B. M., Kean, S. D., Wyatt, M. F. and Graham, A. E., *Synlett*, **2008**, 1953; Lucas, R., Vergnaud, J., Teste, K., Zerrouki, R., Sol, V. and Krausz, P., *Tetrahedron Lett.*, **2008**, *49*, 5537.

60 Dogutan, D. K., Ptaszek, M. and Lindsey, J. S., *J. Org. Chem.*, **2007**, *72*, 5008.

61 Hanck, A. and Kutscher, W., *Z. Physiol. Chem.*, **1964**, *338*, 272.

62 Fuhlhage, D. W. and VanderWerf, C. A. *J. Am. Chem. Soc.*, **1958**, *80*, 6249.

63 Temelli, B. and Unaleroglu, C., *Tetrahedron Lett.*, **2005**, *46*, 7941.

64 'The acid-catalysed polymerisation of pyrroles and indoles', Smith, G. F., *Adv. Heterocycl. Chem.*, **1963**, *2*, 287; Zhao, Y., Beddoes, R. L. and Joule, J. A., *J. Chem. Res.*, (*S*), **1997**, 42–43; (*M*), **1997**, 0401–0429.

65 'The oxidation of monocyclic pyrroles', Gardini, G. P., *Adv. Heterocycl. Chem.*, **1973**, *15*, 67.

66 Thyrann, T. and Lightner, D. A., *Tetrahedron Lett.* **1995**, *36*, 4345; *ibid.*, **1996**, *37*, 315; Moreno-Vargas, A. J., Robina, I., Fernández-Bolaños, J. G. and Fuentes, J., *ibid.*, **1998**, *39*, 9271.

67 Dohi, T., Morimoto, K., Maruyama, A. and Kita, Y., *Org. Lett.*, **2006**, *8*, 2007.

68 Doddi, G., Mercarelli, P. and Stegel, F., *J. Chem. Soc., Chem. Commun.*, **1975**, 273; Di Lorenzo, A. D., Mercarelli, P. and Stegel, F., *J. Org. Chem.*, **1986**, *51*, 2125.

69 Franco, F., Greenhouse, R. and Muchowski, J. M., *J. Org. Chem.*, **1982**, *47*, 1682.

70 De Rosa, M., *J. Org. Chem.*, **1982**, *47*, 1008.

71 Grehn, L. and Ragnarsson, U., *Angew. Chem., Int. Ed. Engl.*, **1984**, *23*, 296.

72 Hobbs, C. F., McMillin, C. K., Papadopoulos, E. P. and VanderWerf, C. A., *J. Am. Chem. Soc.*, **1962**, *84*, 43.

73 Heaney, H. and Ley, S. V., *J. Chem. Soc., Perkin Trans. 1*, **1973**, 499.

74 Santaniello, E., Farachi, C., and Ponti, F., *Synthesis*, **1979**, 617.

75 Jonczyk, A. and Makosza, M., *Rocz. Chem.*, **1975**, *49*, 1203; Wang, N.-C., Teo, K.-E. and Anderson, H. J., *Can. J. Chem.*, **1977**, *55*, 4112; Hamaide, T., *Synth. Commun.*, **1990**, 2913.

76 Jorapur, Y. R., Jeong, J.M. and Chi, D. Y., *Tetrahedron Lett.*, **2006**, *47*, 2435.

77 Hynes, J., Doubleday, W. W., Dyckman, A. J., Godfrey, J. D., Grosso, J. A., Kiau, S. and Leftheris, K., *J. Org. Chem.*, **2004**, *69*, 1368.

78 Bhattacharya, A., Patel, N. C., Plata, R. E., Peddicord, M., Ye, Q., Parlanti, L., Palaniswamy, V. A. and Grosso, J. A., *Tetrahedron Lett.*, **2006**, *47*, 5341.

79 Skell, P. S. and Bean, G. P., *J. Am. Chem. Soc.*, **1962**, *84*, 4655; Bean, G. P., *J. Heterocycl. Chem.*, **1965**, *2*, 473.

80 Schloemer, G. C., Greenhouse, R. and Muchowski, J. M., *J. Org. Chem.*, **1994**, *59*, 5230.

81 Nicolau, C., Claremon, D. A. and Papahatjis, D. P., *Tetrahedron Lett.*, **1981**, *22*, 4647.

82 Brittain, J. M., Jones, R. A., Arques, J. S. and Saliente, T. A., *Synth. Commun.*, **1982**, 231.

83 'Pyrrole protection', Jolicoeur, B., Chapman, E. E., Thompson, A. and Lubell, W. D., *Tetrahedron*, **2006**, *62*, 11531.

84 Hasan, I, Marinelli, E. R., Lin, L.-C. C., Fowler, F. W. and Levy, A. B., *J. Org. Chem.*, **1981**, *46*, 157; Grieb, J. G. and Ketcha, D. M., *Synth. Commun.*, **1995**, *25*, 2145.

85 Katritzky, A. R. and Akutagawa, K., *Org. Prep. Proc. Int.*, **1988**, *20*, 585.

86 Edwards, M. P., Doherty, A. M., Ley, S. V. and Organ, H. M., *Tetrahedron*, **1986**, *42*, 3723; Muchowski, J. M. and Solas, D. R., *J. Org. Chem.*, **1984**, *49*, 203.

87 Gharpure, M., Stoller, A., Bellamy, F., Firnau, G. and Snieckus, V., *Synthesis*, **1991**, 1079.

88 Bergauer, M. and Gmeiner, P., *Synthesis*, **2001**, 2281.

89 Chadwick, D. J., *J. Chem. Soc., Chem. Commun.*, **1974**, 790; Chadwick, D. J. and Willbe, C., *J. Chem. Soc., Perkin Trans. 1*, **1977**, 887; Chadwick, D. J., McKnight, M. V. and Ngochindo, R., *ibid.*, **1982**, 1343.

90 Groenedaal, L., M. E. Van Loo, M. E., Vekemans, J. A. J. M. and Meijer, E. W., *Synth. Commun.*, **1995**, *25*, 1589.

91 Grieb, J. G. and Ketcha, D. M., *Synth. Commun.*, **1995**, *25*, 2145.

92 Alvarez, A., Guzmán, A., Ruiz, A., Velarde, E. and Muchowski, J. M., *J. Org. Chem.*, **1992**, *57*, 1653.

93 Aboutayab, K., Cadick, S., Jenkins, K., Joshi, S. and Khan, S., *Tetrahedron*, **1996**, *52*, 11329.

94 Byers, J. H., Duff, M. P. and Woo, G. W., *Tetrahedron Lett.*, **2003**, *44*, 6853.

95 Guadarrama-Morales, O., Méndez, F. and Miranda, L. D., *Tetrahedron Lett.*, **2007**, *48*, 4515.

96 Dohi, T., Morimoto, K., Ito, M. and Kita, Y., *Synthesis*, **2007**, 2913.

97 Hudson, C. B. and Robertson, A. V., *Tetrahedron Lett.*, **1967**, 4015; Ketcha, D. M., Carpenter, K. P. and Zhou, Q., *J. Org. Chem.*, **1991**, *56*, 1318.

98 Andrews, L. H. and McElvain, S. M., *J. Am. Chem. Soc.*, **1929**, *51*, 887; Kaiser, H.-P. and Muchowski, J. M., *J. Org. Chem.*, **1984**, *49*, 4203; Jefford, C. W., Sienkiewicz, K. and Thornton, S. R., *Helv. Chim. Acta*, **1995**, *78*, 1511; Hext, N. M., Hansen, J., Blake, A. J., Hibbs, D. E., Hursthouse, M. B., Shishkin, O. V. and Mascal, M., *J. Org. Chem.*, **1998**, *63*, 6016.

99 Donohoe, T. J. and Guyo, P. M., *J. Org. Chem.*, **1996**, *61*, 7664; Donohoe, T. J., Guyo, P. M., Beddoes, R. L. and Helliwell, M., *J. Chem. Soc., Perkin Trans. 1*, **1998**, 667; Donohoe, T. J., Guyo, P. M. and Helliwell, M., *Tetrahedron Lett.*, **1999**, *40*, 435; Donohoe, T. J. and House, D., *J. Org. Chem.*, **2002**, *67*, 5015.

100 Wittig, G. and Reichel, B., *Chem. Ber.*, **1963**, *96*, 2851.

101 'Chemistry of 7-azabicyclo[2.2.1]hepta-2,5-dienes, 7-azabicyclo[2.2.1]hept-2-enes and 7-azabicyclo[2.2.1]heptanes', Chen, Z. and Trudell, M. L., *Chem. Rev.*, **1996**, *96*, 1179.

102 Anderson, P. S., Christy, M. E., Lundell, G. F. and Ponticello, G. S., *Tetrahedron Lett.*, **1975**, 2553.

103 Kotsuki, H., Mori, Y., Nishizawa, H., Ochi, M. and Matsuoka, K., *Heterocycles*, **1982**, *19*, 1915.

104 Kitzing, R., Fuchs, R., Joyeux, M. and Prinzbach, H., *Helv. Chim. Acta*, **1968**, *51*, 888.

105 Bansal, R. C., McCulloch, A. W. and McInnes, A. G., *Can. J. Chem.*, **1969**, *47*, 2391; *ibid.* **1970**, *48*, 1472.

106 Zhang, C. and Trudell, M. L., *J. Org. Chem.*, **1996**, *61*, 7189; Pavri, N. P. and Trudell, M. L., *Tetrahedron Lett.*, **1997**, *38*, 7993.

107 Lautens, M., Fagnou, K. and Zunic, V., *Org. Lett.*, **2002**, *4*, 3465.

[108] Lightner, D. A., Bisacchi, G. S. and Norris, R. D., *J. Am. Chem. Soc.*, **1976**, *98*, 802; Natsume, M. and Muratake, H., *Tetrahedron Lett.*, **1979**, 3477.

[109] Rosa, C. D., Kneetman, M. and Mancini, P., *Tetrahedron Lett.*, **2007**, *48*, 1435.

[110] Chrétien, A., Chataigner, I. and Piettre, S. R., *Tetrahedron*, **2005**, *61*, 7907.

[111] Li, J.-H. and Snyder, J. K., *J. Org. Chem.*, **1993**, *58*, 516.

[112] 'Cycloaddition reactions with vinyl heterocycles', Sepúlveda-Arques, J., Abarca-González, B. and Medio-Simón, *Adv. Heterocycl. Chem.*, **1995**, *63*, 339.

[113] Xiao, D. and Ketcha, D. M., *J. Heterocycl. Chem.*, **1995**, *32*, 499.

[114] Jones, R. C. and Rees, C. W., *J. Chem. Soc. (C)*, **1969**, 2249; Gambacorta, Nicoletti, R., Cerrini, S., Fedeli, W. and Gavuzzo, E., *Tetrahedron Lett.*, *1978*, 2439.

[115] de Angelis, F., Gambacorta, A. and Nicoletti, R., *Synthesis*, **1976**, 798.

[116] Maryanoff, B. E., *J. Org. Chem.*, **1979**, *44*, 4410; Pomeranz, M. and Rooney, P., *J. Org. Chem.*, **1988**, *53*, 4374.

[117] Hedley, S. J., Ventura, D. L., Dominiak, P. M., Nygren, C. L. and Davies, H. M. L., *J. Org. Chem.*, **2006**, *71*, 5349.

[118] Böhm, C., Schinnerl, M., Bubert, C., Zabel, M., Labahn, T., Parisini, E. and Reiser, O., *Eur. J. Org. Chem.*, **2000**, 2955.

[119] H. M. L. Davies, E. Saikali and W. B. Young, *J. Org. Chem.*, **1991**, *56*, 5696; Davies, H. M. L., Julius, J., Matasi, J. J., Hodges, L. M., Huby, N. J. S., Thornley, C., Kong, N. and Houser, J. H., *J. Org. Chem.*, **1997**, *62*, 1095.

[120] 'Photochemical reactions involving pyrroles, Parts I and II', D'Auria, M., *Heterocycles*, **1996**, *43*, 1305 & 1529.

[121] Hiraoka, H., *J. Chem. Soc., Chem. Commun.*, **1970**, 1306; Barltrop, J. A., Day, A. C. and Ward, R. W., *ibid.*, **1978**, 131.

[122] Herz, W. and Tocker, S., *J. Am. Chem. Soc.*, **1955**, *77*, 6353; Herz, W., Dittmer, K. and Cristol, S. J., *ibid.*, **1948**, *70*, 504.

[123] Greenhouse, R., Ramirez, C. and Muchowski, J. M., *J. Org. Chem.*, **1985**, *50*, 2961; Schumacher, D. P. and Hall, S. S., *ibid.*, **1981**, *46*, 5060.

[124] Ketcha, D. M., Carpenter, K. P., Atkinson, S. T. and Rajagopalan, H. R., *Synth. Commun.*, **1990**, *20*, 1647.

[125] Carson, J. R. and Davis, N. M., *J. Org. Chem.*, **1981**, *46*, 839.

[126] e.g., Badger, G. M., Harris, R. L. N. and Jones, R. A., *Aust. J. Chem.*, **1964**, *17*, 1022 and Lancaster, R. E. and VanderWerf, C. A., *J. Org. Chem.*, **1958**, *23*, 1208.

[127] Dunn, G. E. and Lee, G. K. J., *Can. J. Chem.*, **1971**, *49*, 1032.

[128] Nikitin, E. B., Nelson, M. J. and Lightner, D. A., *J. Heterocycl. Chem.*, **2007**, *44*, 739.

[129] Anderson, H. J. and Lee, S.-F., *Can. J. Chem.*, **1965**, *43*, 409.

[130] Norris, R. D. and Lightner, D. A., *J. Heterocycl. Chem.*, **1979**, *16*, 263.

[131] Thyrann, T. and Lightner, D. A., *Tetrahedron Lett.*, **1996**, *37*, 315.

[132] Khan, M. K. A. and Morgan, K. J., *Tetrahedron*, **1965**, *21*, 2197; Williams, A. and Salvadori, G., *J. Chem. Soc., Perkin Trans. 2*, **1972**, 883.

[133] Bocchi, V., Chierichi, L., Gardini, G. P. and Mondelli, R., *Tetrahedron*, **1970**, *26*, 4073.

[134] Casiraghi, G., Rassu, G., Spanu, P. and Pinna, L., *J. Org. Chem.*, **1992**, *57*, 3760; *idem, Tetrahedron Lett.*, **1994**, *35*, 2423; 'Furan-, pyrrole-, and thiophene-based siloxydienes for synthesis of densely functionalised homochiral compounds', Casiraghi, G. and Rassu, G., *Synthesis*, **1995**, 607.

[135] Momose, T., Tanaka, T., Yokota, T., Nagamoto, N. and Yamada, K., *Chem. Pharm. Bull.*, **1979**, *27*, 1448.

[136] 'Synthesis of amino derivatives of five-membered heterocycles by Thorpe-Ziegler cyclisation', Granik, V. G., Kadushkin, A. V. and Liebscher, J., *Adv. Heterocycl. Chem.*, **1998**, *72*, 79.

[137] De Rosa, M., Issac, R. P. and Houghton, G., *Tetrahedron Lett.*, **1995**, *36*, 9261.

[138] Fu, L. and Gribble, G. W., *Tetrahedron Lett.*, **2007**, *48*, 9155.

[139] Fu, L. and Gribble, G. W., *Tetrahedron Lett.*, **2008**, *49*, 3545.

[140] 'Recent synthetic methods for pyrroles and pyrrolenines (2H- or 3H-pyrroles)', Patterson, J. M., *Synthesis*, **1976**, 281.

[141] Bishop, W. S., *J. Am. Chem. Soc.*, **1945**, *67*, 2261.

[142] Knorr, L., *Chem. Ber.*, **1884**, *17*, 1635.

[143] For one of several methods for the synthesis of 1,4-dicarbonyl compounds see: Wedler, C. and Schick, H., *Synthesis*, **1992**, 543.

[144] Amarnath, V., Anthony, D. C., Amarnath, K., Valentine, W. M., Wetteran, L. A. and Graham, D. G., *J. Org. Chem.*, **1991**, *56*, 6924.

[145] Young, D. M. and Allen, C. F. H., *Org. Synth., Coll. Vol. II*, **1943**, 219.

[146] Danks, T. N., *Tetrahedron Lett.*, **1999**, *40*, 3957.

[147] Banik, B. K., Samajdar, S. and Banik, I., *J. Org.Chem.*, **2004**, *69*, 213.

[148] Rousseau, B., Nydegger, E., Gossauer, A., Bennau-Skalmowski, B. and Vorbrüggen, H., *Synthesis*, **1996**, 1336.

[149] Veitch, G. E., Bridgwood, K. L., Rands-Trevor, K. and Ley, S. V., *Synlett*, **2008**, 2597.

[150] Elming, N. and Clauson-Kaas, N., *Acta Chem. Scand.*, **1952**, *6*, 867.

[151] Lee, S. D., Brook, M. A. and Chan, T. H., *Tetrahedron Lett.*, **1983**, 1569; Chan T. H. and Lee, S. D., *J. Org. Chem.*, **1983**, *48*, 3059.

[152] Lee, C. K., Jun, J. H. and Yu, J. S., *J. Heterocycl. Chem.*, **2000**, *37*, 15; Fürstner, A., Manne, U., Seidel, G. and Laurich, D., *Org. Synth.*, **2006**, *83*, 103.

[153] Abid, M., Teixeira, L. and Török, B., *Tetrahedron Lett.*, **2007**, *48*, 4047.

[154] McLeod, M., Boudreault, N. and Leblanc, Y., *J. Org. Chem.*, **1996**, *61*, 1180.

[155] Josey, A. D., *Org. Synth., Coll. Vol. V*, **1973**, 716; Jefford, C. W., Thornton, S. R. and Sienkiewicz, K., *Tetrahedron Lett.* **1994**, *35*, 3905; Fang, V., Leysend, D. and Ottenheijm, H. C. J., *Synth. Commun.*, **1995**, *25*, 1857.

[156] Okada, E., Masuda, R., Hojo, M. and Yoshida, R., *Heterocycles*, **1992**, *34*, 1435.

[157] Treibs, A., Schmidt, R. and Zinsmeister, R., *Chem. Ber.*, **1957**, *90*, 79.

[158] Fischer, H., *Org. Synth., Coll. Vol. II*, **1943**, 202.

[159] Yinglin, H. and Hongwer, H., *Synthesis*, **1990**, 615.

[160] Hamby, J. M. and Hodges, J. C., *Heterocycles*, **1993**, *35*, 843.

[161] Kolar, P. and Tisler, M., *Synth. Commun.*, **1994**, *24*, 1887.

[162] 'Synthetic uses of tosylmethyl isocyanide (TosMIC)', van Leusen, D. and van Leusen, A. M., *Org. React.*, **2001**, *57*, 417.

[163] Katritzky, A. R., Cheng, D. and Musgrave, R. P., *Heterocycles*, **1997**, *44*, 67.

[164] Hoppe, D., *Angew. Chem., Int. Ed. Engl.*, **1974**, *13*, 789; van Leusen, A. M., Siderius, H., Hoogenboom, B. E. and van Leusen, D., *Tetrahedron Lett.*, **1972**, 5337; Possel, O. and van Leusen, A. M., *Heterocycles*, **1977**, *7*, 77; Parvi, N. P. and Trudell, M. L., *J. Org. Chem.*, **1997**, *62*, 2649; for a related process see Houwing H. A. and van Leusen, A. M., *J. Heterocycl. Chem.*, **1981**, *18*, 1127.

[165] Ono, T., Muratani, E. and Ogawa, T., *J. Heterocycl. Chem.*, **1991**, *28*, 2053.

[166] 'Barton-Zard pyrrole synthesis and its application to synthesis of porphyrins, polypyrroles and dipyrromethene dyes', Ono, N., *Heterocycles*, **2008**, *75*, 243.

[167] Barton, D. H. R., Kervagoret, J. and Zard, S. Z., *Tetrahedron*, **1990**, *46*, 7587.

[168] Boëlle, J., Schneider, R., Gérardin, P. and Loubinoux, B., *Synthesis*, **1997**, 1451.

[169] Lash, T. D., Belletini, J. R., Bastian, J. A. and Couch, K. B., *Synthesis*, **1994**, 170.

[170] Ono, N., Hironaga, H., Ono, K., Kaneko, S., Murashima, T., Ueda, T., Tsukamura, C. and Ogawa, T., *J. Chem. Soc., Perkin Trans. 1*, **1996**, 417.

[171] Abel, Y., Haake, E., Haake, G., Schmidt, W., Struve, D., Walter, A. and Montforts, F.-P., *Helv. Chim. Acta*, **1998**, *81*, 1978.

[172] Bobál, P. and Lightner, D. A., *J. Heterocycl. Chem.*, **2001**, *38*, 527

[173] Smith, N. D., Huang, D. and Cosford, N. D. P., *Org. Lett.*, **2002**, *4*, 3537.

[174] Bhattacharya, A., Cheruki, S., Plata, R. E., Patel, N., Tamez, V., Grosso, J. A., Peddicord, M. and Palaniswamy, V. A., *Tetrahedron Lett.*, **2006**, *47*, 5481.

[175] Milgram, B. C., Eskildsen, K., Richter, S. M., Scheidt, W. R. and Scheidt, K. A., *J. Org. Chem.*, **2007**, *72*, 3941.

[176] Rivero, M. R. and Buchwald, S. L., *Org. Lett.*, **2007**, *9*, 973.

[177] Trofimov, B. A., *Adv. Heterocycl. Chem.*, **1990**, *51*, 177.

[178] Mataka, S., Takahashi, K., Tsuda, Y. and Tashiro, M., *Synthesis*, **1982**, 157; Walizei G. H. and Breitmaier, E., *ibid.*, **1989**, 337; Hombrecher H. K. and Horter, G., *ibid.*, **1990**, 389.

[179] Knight, D. W., Redfern, A. L. and Gilmore, J., *Chem. Commun.*, **1998**, 2207.

[180] Aydogan, F. and Demir, A. S., *Tetrahedron*, **2005**, *61*, 3019.

[181] Kel'in, A. V., Sromek, A. W. and Gevorgyan, V., *J. Am. Chem. Soc.*, **2001**, *123*, 2074.

[182] Bayer, H. O., Gotthard, H. and Huisgen, R., *Chem. Ber.*, **1970**, *103*, 2356; Huisgen, R., Gotthard, H., Bayer, H. O. and Schafer, F. C., *ibid.*, 2611; Padwa, A., Burgess, E. M., Gingrich, H. L. and Roush, D. M., *J. Org. Chem.*, **1982**, *47*, 786.

[183] Sessler, J. L., Mozaffari, A. and Johnson, M. R., *Org. Synth.*, **1992**, *70*, 68.

[184] 'Novel porphyrinoid macrocycles and their metal complexes', Vogel, E., *J. Heterocycl. Chem.*, **1996**, *33*, 1461.

[185] Sessler, J. L. and Weghorn, S. J., 'Expanded, contracted, and isomeric porphyrins', Pergamon, Oxford, **1997**.

[186] G. P. Arsenault, E. Bullock and S. F. MacDonald, *J. Am. Chem. Soc.*, **1960**, *82*, 4384.

[187] E. Vogel, M. Bröring, S. J. Weghorn, P. Scholz, R. Deponte, J. Lex, H. Schmickler, K. Schaffner, S. E. Braslavsky, M. Müller, S. Pörting, C. J. Fowler and J. C. Sessler, *Angew. Chem., Int. Ed. Engl.*, **1997**, *36*, 1651.

[188] 'CC-1065 and the duocarmycins: synthetic studies', Boger, D. C., Boyce, C. W., Garbaccio, R. M. and Goldberg, J. A., *Chem. Rev.*, **1997**, *97*, 787.

[189] 'The chemistry, mode of action and biological properties of CC 1065', Reynolds, V. L., McGovern, J. P. and Hurley, L. H., *J. Antiobiotics*, **1986**, *39*, 319.

[190] Carter, P., Fitzjohn, S., Halazy, S. and Magnus, P., *J. Am. Chem., Soc.*, **1987**, *109*, 2711.

17

Thiophenes: Reactions and Synthesis

thiophene

The simple thiophenes[1] are stable liquids that closely resemble the corresponding benzene compounds in boiling points and even in smell. They occur in coal-tar distillates – the discovery of thiophene in coal-tar benzene provides one of the classic anecdotes of organic chemistry. In the early days, colour reactions were of great value in diagnosis: an important one for benzene involved the production of a blue colour on heating with isatin (see 17.1.1.7) and concentrated sulfuric acid. In 1882, during a lecture-demonstration by Viktor Meyer before an undergraduate audience, this test failed, no doubt to the delight of everybody except the professor, and especially except the professor's lecture assistant. An inquiry revealed that the lecture assistant had run out of commercial benzene and had provided a sample of benzene that he had prepared by decarboxylation of pure benzoic acid. It was thus clear that commercial benzene contained an impurity and that it was this, not benzene, that was responsible for the colour reaction. In subsequent investigations, Meyer isolated the impurity via its sulfonic-acid derivative and showed it to be the first representative of a then new ring system, which was named thiophene from theion, the Greek word for sulfur, and another Greek word phaino which means shining, a root first used in phenic acid (phenol) because of its occurrence in coal tar, a by-product of the manufacture of 'illuminating gas'.

17.1 Reactions with Electrophilic Reagents
17.1.1 Substitution at Carbon
17.1.1.1 Protonation
Thiophene is stable to all but very strongly acidic conditions, so many reagent combinations that would lead to acid-catalysed decomposition or polymerisation of furans and pyrroles can be applied successfully to thiophenes.

Measurements of acid-catalysed exchange, or of protonolysis of other groups, for example silicon,[2] or mercury,[3] show the rate of proton attack at C-2 to be about 1000 times faster than at C-3.[4] The pK_{aH} for 2,5-di-*t*-butylthiophene forming a salt by protonation at C-2, is -10.2.[5]

Reactions of Protonated Thiophenes
The action of hot phosphoric acid on thiophene leads to a trimer;[6] its structure suggests that, in contrast with pyrrole (16.1.1.8), the electrophile involved in the first C–C bonding step is the α-protonated cation.

Heterocyclic Chemistry 5th Edition   John Joule and Keith Mills
© 2010 Blackwell Publishing Ltd

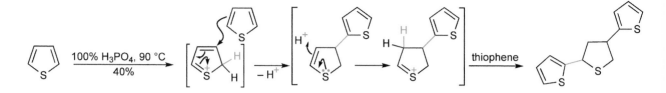

17.1.1.2 Nitration

Nitration of thiophene needs to be conducted in the absence of nitrous acid, which can lead to an explosive reaction;[7] the use of acetyl nitrate[8] or nitronium tetrafluoroborate[9] is satisfactory. The major 2-nitro-product is accompanied by approximately 10% of the 3-isomer.[10] Further nitration of either 2- or 3-nitrothiophenes[11] also leads to mixtures: equal amounts of 2,4- and 2,5-dinitrothiophenes from the 2-isomer, and mainly 2,4-dinitrothiophene from 3-nitrothiophene.[12] Similar, predictable isomer mixtures are produced in other nitrations of substituted thiophenes, for example 2-methylthiophene gives rise to 2-methyl-5- and 2-methyl-3-nitrothiophenes,[13] and 3-methylthiophene gives 4-methyl-2-nitro- and 3-methyl-2-nitrothiophenes,[14] in each case in ratios of 4:1.

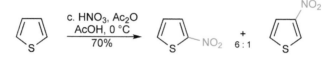

17.1.1.3 Sulfonation

As discussed in the introduction, the production of thiophene-2-sulfonic acid by sulfuric acid sulfonation of the heterocycle has been long known;[15] use of the pyridine–sulfur-trioxide complex is probably the best method.[16] 2-Chlorosulfonation[17] and 2-thiocyanation[18] are similarly efficient.

17.1.1.4 Halogenation

Halogenation of thiophene occurs very readily at room temperature and is rapid even at −30 °C in the dark; tetrasubstitution occurs easily.[19] The rate of halogenation of thiophene at 25 °C is about 10^8 times that of benzene.[20] 2-Bromo-, 2-chloro-[21] and 2-iodothiophenes[22] and 2,5-dibromo- and 2,5-dichlorothiophenes[23] can be produced cleanly under various controlled conditions. Controlled bromination of 3-bromothiophene produces 2,3-dibromothiophene.[24]

2,3,5-Tribromination of thiophene goes smoothly in 48% hydrobromic acid solution.[25] Since it has long been known that treatment of polyhalogeno-thiophenes with zinc and acid brings about selective removal of α-halogen, this compound can be used to access 3-bromothiophene[26] just as 3,4-dibromothiophene can be obtained by reduction of the tetrabromide.[27] One interpretation of the selective reductive removal is that it involves, first, electron transfer to the bromine, then transient 'anions', thus halogen can be selectively removed from that position where such an anion is best stabilised – normally an α-position (17.4.1).

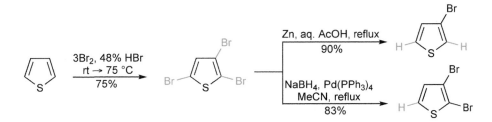

Monoiodination of 2-substituted thiophenes, whether the substituent is activating or deactivating, proceeds efficiently at the remaining α-position using iodine with iodobenzene diacetate.[28] 3-Alkyl-thiophenes can be monobrominated or monoiodinated at C-2 using *N*-bromosuccinimide,[29] or iodine with mercury(II) oxide,[30] respectively.

17.1.1.5 *Acylation*

The Friedel–Crafts acylation of thiophenes is a much-used reaction and generally gives good yields under controlled conditions, despite the fact that aluminium chloride reacts with thiophene to generate tars; this problem can be avoided by using tin tetrachloride and adding the catalyst to a mixture of the thiophene and the acylating agent.[31] Acylation with anhydrides, catalysed by phosphoric acid[32] is an efficient method. Reaction with acetyl *p*-toluenesulfonate, in the absence of any catalyst produces 2-acetylthiophene in high yield.[33] Vilmseier formylation of thiophene leads efficiently to 2-formylthiophene,[34] comparable substitution of 3-phenylthiophene gives 2-formyl-3-phenylthiophene,[35] the regioselectivity echoed in the Vilsmeier 2-formylation of 3-methylthiophene using *N*-formylpyrrolidine[36] (2-formyl-4-methylthiophene can be produced by lithiation (17.4.1) then reaction with dimethylformamide[37]).

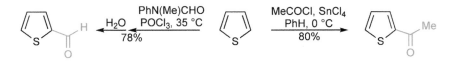

In acylations, almost exclusive α-substitution is observed, but where both α-positions are substituted, β-substitution occurs easily. This is nicely illustrated by the two ketones produced in the classic sequence shown below.[38,39]

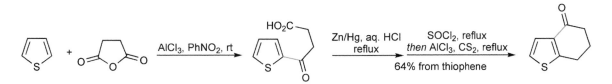

17.1.1.6 *Alkylation*

Alkylation occurs readily, but is rarely of preparative use, an exception being the efficient 2,5-bis-*t*-butylation of thiophene.[40]

17.1.1.7 Condensation with Aldehydes and Ketones

Acid-catalysed reaction of thiophene with aldehydes and ketones is not a viable route to hydroxyalkyl-thiophenes, for these are unstable under the reaction conditions. Chloromethylation can, however, be achieved,[41] and with the use of zinc chloride, even thiophenes carrying electron-withdrawing groups react[42] (**CAUTION**. *Bis(chloromethyl) ether, a carcinogen, is formed as a by-product*). Care is needed in choosing conditions; there is a tendency for formation of either di-2-thienyl-methanes[43] or 2,5-bis(chloromethyl)-thiophenes.[44]

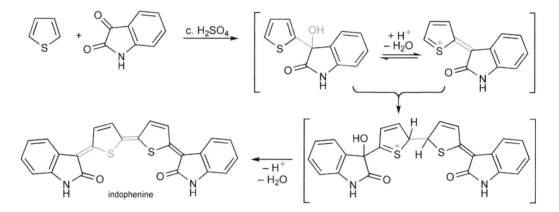

A reaction of special historical interest, mentioned in the introduction to this chapter, is the condensation of thiophene with isatin in concentrated sulfuric acid, to give the deep blue indophenine[45] as a mixture of geometrical isomers.[46]

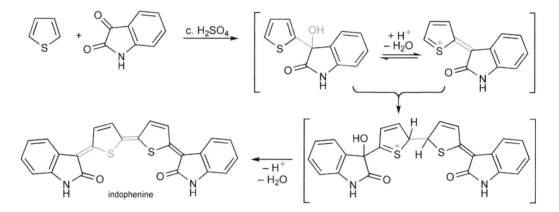

Hydroxyalkylation at the 5-position of 2-formylthiophene results from exposure of the thiophene alde-hyde and a second aldehyde, to samarium(II) iodide; in the example shown below, the other aldehyde is 1-methylpyrrole-2-carboxaldehyde.[47]

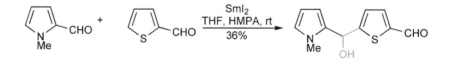

17.1.1.8 Condensation with Imines and Iminium Ions

Aminomethylation of thiophene[48] was reported long before the more common Mannich reaction – dimethylaminomethylation – which, although it can be achieved under routine conditions with methoxy-thiophenes,[49] requires the use of $Me_2N^+=CH_2\ Cl^-$ ('Eschenmoser's salt' is the iodide) for thiophene and alkyl thiophenes.[50]

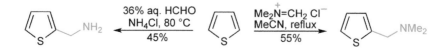

Another device for bringing thiophenes into reaction with Mannich intermediates is to utilise thiophene boronic acids – the Petasis reaction; primary aromatic amines can also be used as the amine component.[51]

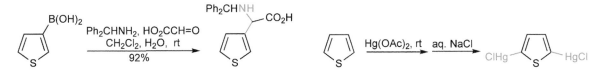

17.1.1.9 Mercuration
Mercuration of thiophenes occurs with great ease; mercuric acetate is more reactive than the chloride;[52] tetrasubstitution and easy replacement of the metal with halogen can also be achieved straightforwardly.[53]

17.1.2 Addition at Sulfur
In reactions not possible with the second-row-element-containing pyrrole and furan, thiophene sulfur can add electrophilic species. Thiophenium salts[54] though not formed efficiently from thiophene itself, are produced in high yields with polyalkyl-substituted thiophenes.[55] The sulfur in such salts is probably tetrahedral,[56] i.e. the sulfur is sp^3 hybridised (**CAUTION**: *Methyl fluorosulfonate is highly toxic*).

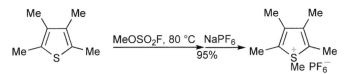

Even thiophene itself will react with carbenes, at sulfur, to produce isolable thiophenium ylides, and in these, the sulfur is definitely tetrahedral.[57] The rearrangement[58] of thiophenium bis(ethoxycarbonyl) methylide to the 2-substituted thiophene provides a rationalisation for the reaction of thiophene with ethyl diazoacetate,[59] which produces what appears to be the product of carbene addition to the 2,3-double bond; perhaps this proceeds *via* initial attack at sulfur followed by S → C-2 rearrangement, then collapse to the cyclopropane. Acid catalyses conversion of the cyclopropanated compound into a thiophene-3-acetic ester.[60] 2,5-Dichlorothiophenium bis(methoxycarbonyl)methylide has been used as an efficient source of the carbene: simply heating it in an alkene results in the transfer of (MeO$_2$C)$_2$C to the alkene.[61]

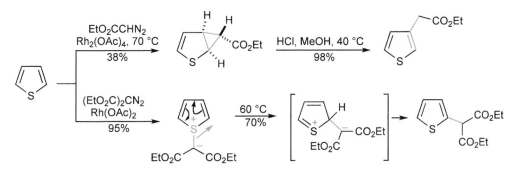

Uncontrolled *S*-oxidation of a thiophene leads to *S,S*-dioxides; that from thiophene itself has been isolated, but above −40 °C it dimerises giving eventual products depending on concentration,[62] but with substituted thiophenes the dioxides can be isolated. Peracids[63] or dimethyldioxirane[64] can be used, but do not succeed if there are electron-withdrawing substituents. A solution of fluorine in water (hypofluorous acid) will, however, achieve the objective, even with thiophenes carrying electron-withdrawing groups.[65] The *S,S*-dioxides are no longer aromatic thiophenes and react as dienes in Diels–Alder reactions; generally, sulfur dioxide is extruded from the initial adduct, leading to further reaction[66] – eventual aromatisation in the example shown.[67] Thiophene *S,S*-dioxides that also carry two strong electron-withdrawing groups behave as dienophiles.[68]

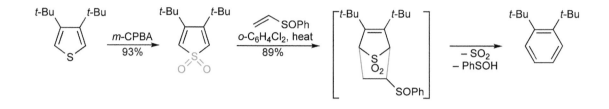

17.2 Reactions with Oxidising Agents

Apart from the *S*-oxidations discussed above, the thiophene ring system, unless carrying electron-releasing substituents, is relatively stable to oxidants; side-chains can be oxidised to carboxylic acid groups, though not usually in synthetically useful yields.

17.3 Reactions with Nucleophilic Reagents

Nitro substituents activate the displacement of leaving groups like halide, as in benzene chemistry, and extensive use of this has been made in thiophene work. Such nucleophilic displacements proceed at least 10^2 times faster than for benzenoid counterparts, and this may be accounted for by participation of the sulfur in the delocalisation of charge in the Meisenheimer intermediate.[69] Nitrogroups also permit the operation of VNS processes (3.3.3), as illustrated below.[70]

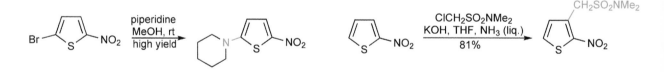

Activation provided by the sulfur may also account for the extremely easy displacement of iodine from the thiophene 2-position, using alkyl- or aryl-thiols as nucleophiles.[71]

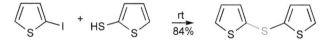

Copper and copper(I) salts have been used extensively in thiophene chemistry to catalyse displacement of bromine and iodine, but not chlorine, in simpler halo-thiophenes.[72,73]

17.4 Metallation and Reactions of C-Metallated Thiophenes

17.4.1 Direct Ring C–H Metallation

Monolithiation of thiophene takes place at C-2; two mole equivalents of lithiating agent easily produces 2,5-dilithiothiophene.[74] 2-Lithiated thiophene can be put to many uses, for example with *N*-tosylaziridine to introduce a 2-tosylaminoethyl side-chain.[75]

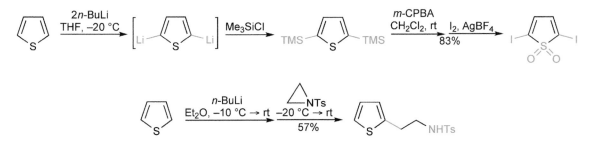

Lithiation at a thiophene β-position, in the presence of a free α-position, can be achieved with the assistance of an *ortho*-directing substituent at C-2.[76] Thiophene-2-carboxylic acid lithiates at C-3, *via ortho* assistance, using *n*-butyllithium,[77] but at C-5 using lithium diisopropylamide.[78] 3-(Hydroxyalkyl)thiophenes, again with *ortho* assistance, are lithiated at C-2.[79] The lithiation of 2-chloro-5-methoxythiophene at C-4 and C-3, in a ratio of 2:1, is instructive,[80] as is the deprotonation of 2- and 3-bromothiophenes at 5- and 2-positions, respectively, with lithium diisopropylamide.[81] The conversion of 3-isopropylthiophene into the 2-aldehyde by Vilsmeier formylation, but into the 5-aldehyde *via* lithiation, presents a nice contrast.[82] Lithiation of 3-hexyloxythiophene can be carried out with thermodynamic control at C-2 using *n*-butyllithium and with kinetic control using lithium diisopropylamide at C-5.[83]

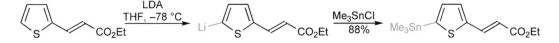

The formation of arynes has often been achieved by base-induced dehydrohalogenation, but for the formation of 3,4-didehydrothiophene, a fluoride-induced process can be used, following *ipso* electrophilic displacement of one of the silicons from 3,4-bis(trimethylsilyl)thiophene to generate the appropriate precursor.[84]

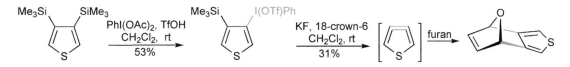

Direct magnesiation of thiophenes at C-2 can be achieved with lithium tri-*n*-butylmagnesate (Bu$_3$MgLi), at room temperature.[85]

17.4.2 Metal–Halogen Exchange

2-Bromo- and 2-iodothiophenes readily form thienyl Grignard reagents,[86] though 3-iodothiophene requires the use of Rieke magnesium.[87] Bromine and iodine at either α- or β-positions undergo exchange with alkyllithiums giving lithiated thiophenes. The reaction of 2,3-dibromothiophene with *n*-butyllithium produces 3-bromothien-2-yllithium.[88] 3-Lithiothiophene is unstable with respect to the 2-isomer at temperatures > –25 °C in ether solution, but is regiostable in hexane,[89] however it can be utilised straightforwardly at low temperature. The corresponding 3-zinc and 3-magnesium derivatives retain regio-integrity, even at room temperature.[90]

The use of thienyl Grignard reagents and lithiated thiophenes has been extensive and can be illustrated by citing formation of oxy-thiophenes, either by reaction of the former with *t*-butyl perbenzoate leading to thiophenone,[91] or the latter directly with bis(trimethylsilyl) peroxide,[92] the synthesis of thiophene carboxylic acids by reaction of the organometallic with carbon dioxide,[93] the synthesis of ketones by reaction with a nitrile,[94] or alcohols by reaction with aldehydes.[87] Syntheses of thiophene-3-boronic acid[95] and of 2-[96] and -3-stannanes,[97] and longer sequences leading to thieno[3,2-*b*]thiophene,[98,99] dithieno[3,2-*b*:2',3'-*d*] thiophene,[100] (another route to this tricycle utilises 3,4-dibromo-2,5-dilithiothiophene generated from the tetrabromide[101]), and dithieno[2,3-*b*:3',2'-*d*]thiophene[102] all involve lithiated thiophenes. Some of these are illustrated below.

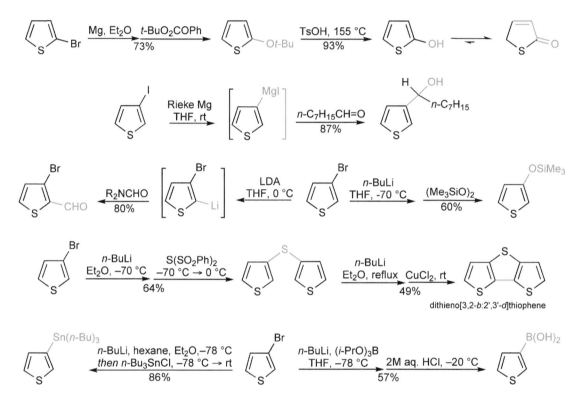

There are two complications that can arise in the formation and the use of lithiated thiophenes: the occurence of a 'Base Catalysed Halogen Dance',[103] and the isomerisation or ring opening of 3-lithiated thiophenes. As an example of the first of these, and one in which the phenomenon is put to good use, consider the transformation of 2-bromothiophene into 3-bromothiophene by reaction with sodamide in ammonia.[104] The final result is governed, in a set of equilibrations, by the stability of the final anion: the system settles to an anion in which the charge is both adjacent to halogen *and* at an α-position.

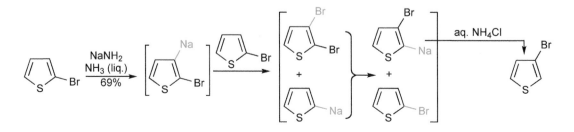

17.5 Reactions with Radicals

Aryl radicals generated by a variety of methods,[105] the most effective of which are aprotic diazotisation[106] and photolysis of iodo-arenes (particularly iodo-hetarenes[107]), attack to produce 2-aryl-thiophenes. However, this has been superseded as a route to aryl-thiophenes by palladium-catalysed coupling. Radicals generated in various ways have been utilised in elaborating thiophenes and in ring-closing reactions; examples are shown below.[108,109]

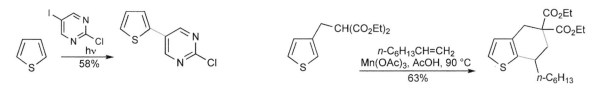

17.6 Reactions with Reducing Agents

Catalytic reductions of the thiophene ring, or of substituents attached to it, are complicated by two factors: poisoning of the catalyst, and the possibility of competing hydrogenolysis – reductive removal of sulfur, particularly with Raney nickel – indeed the use of thiophenes as templates on which to elaborate a structure, followed finally by desulfurisation, is an important synthetic strategy. This has been developed extensively for thiophene acids, where the desulfurisation can be achieved very simply by adding Raney alloy to an alkaline aqueous solution of the acid,[110] and for long-chain hydrocarbons[111] and large-ring ketones.[112]

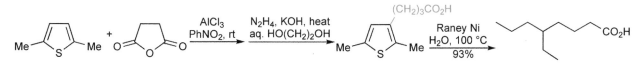

Sodium/ammonia[113] treatment also causes disruption of the ring in thiophene and simple thiophenes, however thiophene-2-carboxylic acid and 2-acyl-thiophenes can be converted into the 2,5-dihydro derivatives using lithium in ammonia, followed by protonation or trapping with an alkyl halide.[114] Side-chain reductions can be carried out with metal hydrides, which do not affect the ring.

Simple saturation of the ring can be achieved using 'ionic hydrogenation',[115] i.e. a combination of a trialkylsilane and acid, usually trifluoroacetic; the reduction proceeds *via* a sequence of proton then 'hydride' additions[116] and consequently requires electron-releasing substituents to facilitate the first step. 2,5-Dihydro-products accompany tetrahydrothiophenes from reductions with zinc and trifluoroacetic acid.[117]

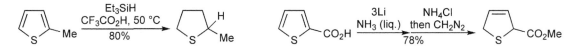

17.7 Electrocyclic Reactions (Ground State)[118]

Unactivated thiophenes show little tendency to react as 4π components in a Diels–Alder sense; however, maleic anhydride *will* react with thiophene to produce an adduct in high yield, under extreme conditions.[119] Electrophilic alkynes will react with thiophenes under vigorous conditions,[120] though the initial adduct extrudes sulfur, and substituted benzenes are obtained as products.

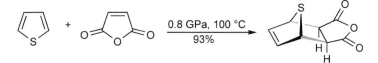

Thus, both α- and β-methoxy-substituted thiophenes react with dimethyl acetylenedicarboxylate in xylene to give modest yields of phthalates resulting from sulfur extrusion from initial adducts; in acetic acid as solvent, only substitution products are obtained.[121]

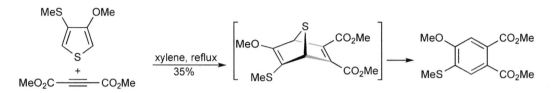

The strong tendency for thiophene *S,S*-dioxides (17.1.2) to undergo cycloaddition processes is echoed by thiophene *S*-oxides. Thus, when thiophenes are oxidised with *meta*-chloroperbenzoic acid and boron trifluoride (without which *S,S*-dioxides are formed), in the presence of a dienophile, adducts from 2 + 4 addition can be isolated.[122] Thiophenes that are 2,5- or 3,4-disubstituted with bulky groups can be converted into isolable *S*-oxides,[123] which undergo cycloadditions *syn* to the oxide, as exemplified below.[124,125]

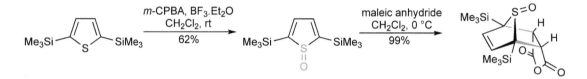

17.8 Photochemical Reactions

The classic photochemical reaction involving thiophenes is the isomerisation of 2-aryl-thiophenes to 3-aryl-thiophenes;[126] the aromatic substituent remains attached to the same carbon and the net effect involves interchange of C-2 and C-3, with C-4 and C-5 remaining in the same relative positions; scrambling of deuterium labelling is, however, observed, complicating interpretation of the detailed mechanism.

There are an appreciable number of examples in which photochemical ring closure of a 1-thienyl-2-aryl-(or heteroaryl-) ethene, carried out in the presence of an oxidant (often oxygen) to trap/aromatise a cyclised intermediate, leads to polycyclic products; an example[127,128] is shown below.

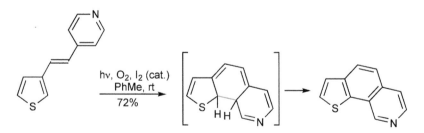

17.9 Thiophene-C–X Compounds: Thenyl Derivatives

The unit – thiophene linked to a carbon – is termed thenyl, hence thenyl chloride is the product of chloro-methylation (17.1.1.7); thenyl bromides are usually made by side-chain radical substitution,[129] substitution at an α-methyl being preferred over a β-methyl.[130]

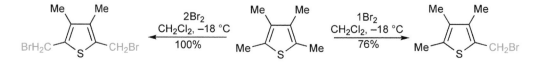

Relatively straightforward benzene analogue reactivity is found with thenyl halides, alcohols (conveniently preparable by reducing aldehydes) and amines, from, for example, reduction of oximes. One exception is that 2-thenyl Grignard reagents usually react to give 3-substituted derivatives, presumably *via* a non-aromatic intermediate.[131]

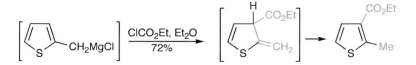

17.10 Thiophene Aldehydes and Ketones, and Carboxylic Acids and Esters

Here, the parallels with benzenoid counterparts continue, for these compounds have no special properties – their reactivities are those typical of benzenoid aldehydes, ketones, acids and esters. For example, in contrast to the easy decarboxylation of α-acids observed for pyrrole and furan, thiophene-2-acids do not easily lose carbon dioxide; nevertheless, high-temperature decarboxylations are of preparative value (see also 17.12.1.2).[132]

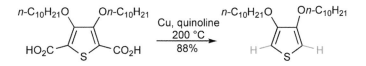

Just as in benzene chemistry, Wolff–Kischner or Clemmensen reduction of ketones is a much-used route to alkyl-thiophenes, hypochlorite oxidation of acetyl-thiophenes a good route to thiophene acids, Beckmann rearrangement of thiophene oximes is a useful route to acylamino-thiophenes and hence amino-thiophenes, and esters and acids are interconvertible without complications.

17.11 Oxy- and Amino-Thiophenes
17.11.1 Oxy-Thiophenes

These compounds are much more difficult to handle and much less accessible than phenols. Neither 2-hydroxythiophene nor its 4-thiolen-2-one tautomer are detectable, the compound existing as the conjugated enone isomer, 3-thiolen-2-one,[133] which can be obtained directly by oxidation of thiophene.[134]

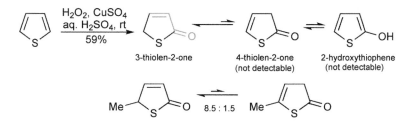

The presence of alkyl, or other groups, both stabilise the oxy compounds and the double bond to which they are attached. In these more stable compounds alternative tautomers are found, thus 5-methyl-2-hydroxythiophene exists as a mixture (actually separable by fractional distillation!) of the two enone tautomers[135] and ethyl 5-hydroxythiophene-2-carboxylate is in the hydroxy-form to the extent of 85%.[133]

3-Hydroxy-thiophenes are even more unstable than 2-hydroxy-thiophenes; 3-hydroxy-2-methylthiophene exists as a mixture of hydroxyl and carbonyl tautomeric forms, with the former predominating.[136] The parent can be prepared by gas-phase pyrolysis of a Meldrum's acid precursor; it exists as a 2.9:1 mixture of keto and hydroxy tautomers and gradually dimerises.[137]

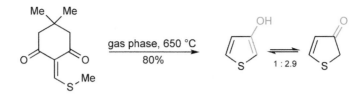

The acidities of the thiolenones are comparable with those of phenols, with pK_as of about 10. Oxy-thiophene anions can react at oxygen or carbon and products from reaction of electrophiles at both centres can be obtained.[138] Silylation generates 2-silyloxy derivatives which react at C-5 with aldehydes, in the presence of boron trifluoride.[139]

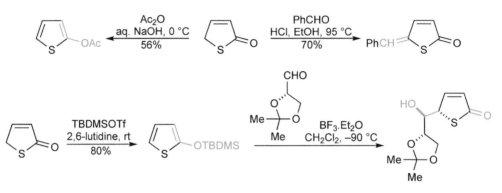

17.11.2 Amino-Thiophenes

Here again, these thiophene derivatives are much less stable than their benzenoid counterparts, unless the ring is provided with other substitution.[140] The unsubstituted amino-thiophenes (thiophenamines) can be obtained by reduction of the nitro-thiophenes,[141] but in such a way as to isolate them as salts – usually hexachlorostannates – or *via* Beckmann rearrangements[142] or Hofmann degradation,[143] as acyl-derivatives, which are stable. 3,4-Dinitration of 2,5-dibromothiophene, then reduction, produces 3,4-diaminothiophene.[144] Many substituted amines have been prepared by nucleophilic displacement of halogen in nitro-halo-thiophenes. In so far as it can be studied, in simple cases, and certainly in substituted amino-thiophenes the amino form is the only detectable tautomer.[145]

17.12 Synthesis of Thiophenes[146]

Thiophene is manufactured by the gas-phase interaction of C_4 hydrocarbons and elemental sulfur at 600 °C. Using *n*-butane, the sulfur first effects dehydrogenation and then interacts with the unsaturated hydrocarbon by addition, further dehydrogenation generating the aromatic system.

17.12.1 Ring Synthesis
17.12.1.1 From 1,4-Dicarbonyl Compounds and a Source of Sulfide

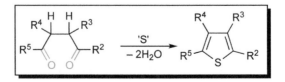

The reaction of a 1,4-dicarbonyl compound (cf. 16.16.1.1 and 18.13.1.1) with a source of sulfide, traditionally phosphorus sulfides, latterly Lawesson's reagent (LR),[147] or bis(trimethylsilyl)sulfide,[148] gives thiophenes, presumably, but not necessarily, *via* the bis(thioketone).

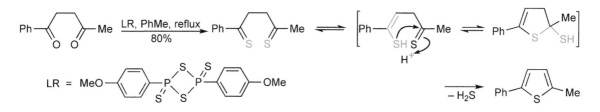

When the process is applied to 1,4-dicarboxylic acids, a reduction must occur at some stage, for thiophenes, and not 2- or 5-oxygenated thiophenes result.[149]

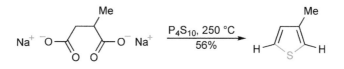

Much use has been made of conjugated diynes, which are also at the oxidation level of 1,4-dicarbonyl-compounds, which react smoothly with hydrosulfide or sulfide, under mild conditions, to give 3,4-unsubstituted thiophenes. Unsymmetrical 2,5-disubstituted thiophenes can be produced in this way too.[150] Since nearly all naturally occurring thiophenes are found in plant genera, and co-occur with polyynes, this laboratory ring synthesis may be mechanistically related to their biosynthesis.

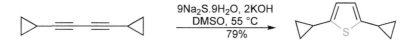

Finally in this category, the efficient synthesis of 3,4-dimethoxythiophene from 2,3-dimethoxy-1,3-butadiene on reaction with sulfur dichloride is notable; it was easily transformed into 'EDOT' (31.6.6.1).[151] Here the sulfur source is electrophilic in character.

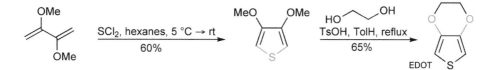

17.12.1.2 From Thiodiacetates and 1,2-Dicarbonyl Compounds

1,2-Dicarbonyl compounds condense with thio-diacetates (or thiobis(methyleneketones)) to give thiophene-2,5-diacids (-diketones).

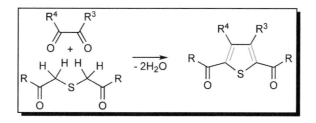

The Hinsberg Synthesis

Two consecutive aldol condensations between a 1,2-dicarbonyl compound and diethyl thiodiacetates give thiophenes. The immediate product is an ester-acid, produced[152] by a Stobbe-type mechanism, but the reactions are often worked up *via* hydrolysis to afford a diacid as the isolated product. The use of diethyl oxalate as the 1,2-dicarbonyl-component leads to 3,4-dihydroxy-thiophenes.[153]

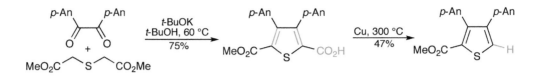

17.12.1.3   *From Thioglycolates and 1,3-Dicarbonyl Compounds*

Thioglycolates react with 1,3-dicarbonyl compounds (or equivalents) to give thiophene-2-carboxylic acid esters.

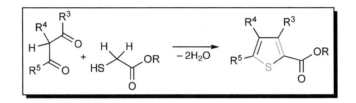

In most of the examples of this approach, thioglycolates, as donors of an S–C unit, have been reacted with 1,3-keto-aldehydes, to give intermediates that can be ring closed to give thiophenes, as exemplified below.[154]

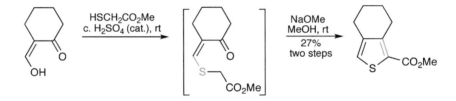

Alkynyl-ketones react with thioglycolates to generate comparable intermediates by conjugate addition to the triple bond.[155]

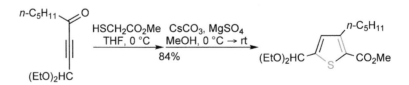

Vinamidinium salts used as 2-substituted malondialdehyde synthons, produce 3,5-unsubstituted thiophenes.[156]

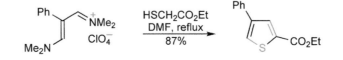

17.12.1.4 From α-Thio-Carbonyl Compounds

2-Keto-thiols add to alkenyl-phosphonium ions, affording ylides, which then ring close by Wittig reaction and give 2,5-dihydrothiophenes, which can be dehydrogenated.[157] Thiophene-2-esters can be comparably produced, without the need for dehydrogenation, by reaction of the 2-keto-thiol with methyl 3-methoxyacrylate.[158]

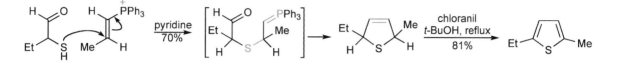

Microwave irradiation in the presence of triethylamine was used for the synthesis of 4,5-unsubstituted 2-amino-thiophenes using thioacetaldehyde dimer.[159]

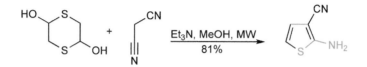

17.12.1.5 From Thio-Diketones

A route[160] in which the 3,4-bond is made by an intramolecular pinacol reaction is nicely illustrated[161] by the formation of a tricyclic thiophene with two cyclobutane fused rings. In this example, it was necessary to force the double dehydration required for aromatisation, because of the strain in the system. Starting materials for this route are easily obtained from sodium sulfide and two mole equivalents of a 2-bromo-ketone.

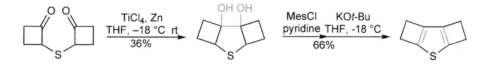

17.12.1.6 Using Carbon Disulfide

The addition of a carbanion to carbon disulfide with a subsequent *S*-alkylation provides a route to 2-alkylthio-thiophenes.[162] In the example below, the carbanion is the enolate of a cyclic 1,3-diketone.

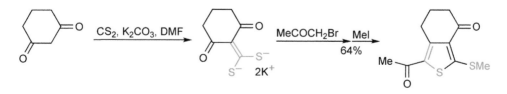

When the enolate is that derived from malononitrile,[163] 3-amino-4-cyano-thiophenes are the result.[164]

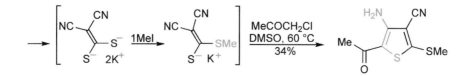

A truly delightful exploitation of this idea is a synthesis of thieno[2,3-*b*]thiophene, in which a diyne is lithiated to give a lithio-allene, which reacts with carbon disulfide.[165]

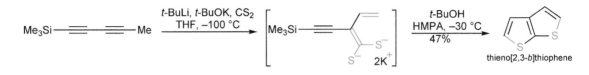

17.12.1.7   From Thiazoles

The cycloaddition/cycloreversion sequence that ensues when thiazoles (the best in this context is 4-phenylthiazole) are heated strongly with an alkyne, generates 2,5-unsubstituted thiophenes. Though the conditions are vigorous, excellent yields can be obtained.[84]

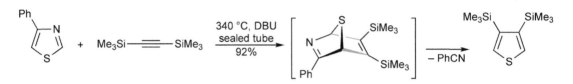

17.12.1.8   From Thio-Nitroacetamides

The *S*-alkylation of thio-nitroacetamides with 2-bromo-ketones produces 2-amino-3-nitro-thiophenes. The scheme below shows how the 3,4-bond making involves the intramolecular interaction of the introduced ketone carbonyl with an enamine/thioenol β-carbon.[166]

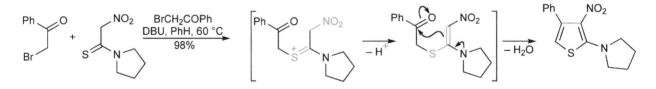

17.12.1.9   From Sulfur, α-Methylene-Ketones and Malononitrile or Cyanoacetate
The Gewald Synthesis[167]

2-Amino-3-cyano-thiophenes or 2-amino-thiophene-3-esters result from this route, generally conducted as a one-pot process, involving a ketone that has an α-methylene, ethyl cyanoacetate or malononitrile, sulfur, and morpholine. Various improvements to the original procedure include using microwave irradiation on solid support,[168] or with potassium fluoride on alumina as the base,[169] or solvent-free,[170] and using morpholinium acetate in excess morpholine for aryl alkyl ketones.[171]

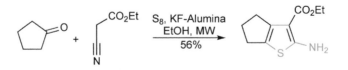

17.12.2   Examples of Notable Syntheses of Thiophene Compounds

17.12.2.1   Thieno[3,4-b]thiophene

Thieno[3,4-*b*]thiophene was prepared from 3,4-dibromothiophene utilising the two halogens in separate steps: palladium-catalysed coupling and lithiation by transmetallation, followed by introduction of sulfur and intramolecular addition to the alkyne.[172]

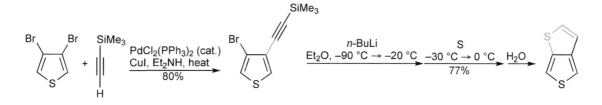

17.12.2.2 *2,2′:5′,3″-Terthiophene and Tetrakis(2-thienyl)methane*

This sequence, for the regioselective synthesis of 2,2′:5′,3″-terthiophene uses the reaction of a diyne with sodium sulfide to make the central ring.[173]

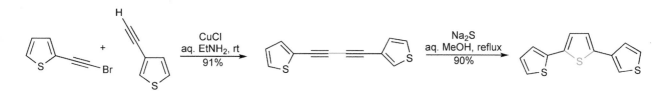

The synthesis of tetrakis(2-thienyl)methane also depends on this synthesis method in its final stage.[174]

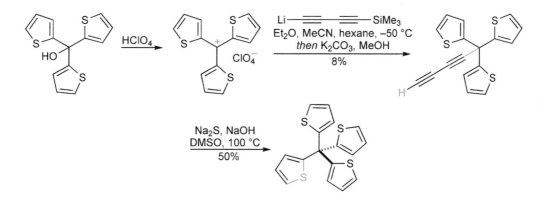

17.12.2.3 *Thieno[2,3-f:5,4-f′]bis[1]benzothiophene*[175]

This synthesis of a molecule with alternating thiophene and benzene rings depends on bromine-to-lithium exchange processes, the final ring closures involving intramolecular electrophilic attacks on the central thiophene ring by pronotated aldehyde groups.

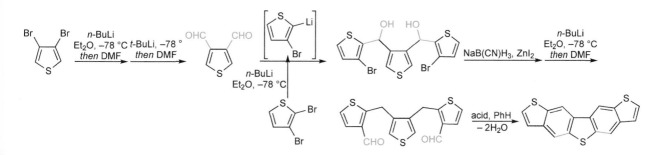

Exercises

Straightforward revision exercises (consult Chapters 15 and 17):

(a) How could one prepare 2-bromo-, 3-bromo- and 3,4-dibromothiophenes?

(b) What would be the products of carying out Vilsmeier reactions with 2-methyl- and 3-methylthiophenes?

(c) How could one convert 2,5-dimethylthiophene into: (i) its *S*-oxide and (ii) its *S,S*-dioxide?

(d) What routes could one use to convert thiophene into derivatives carrying at the 2-position: (i) $CH(OH)t$-Bu; (ii) $(CH_2)_2OH$; (iii) Ph?

(e) How could one prepare *n*-decane from thiophene?

(f) Draw the structures of the thiophenes that would be produced from the following reactant combinations: (i) octane-3,6-dione and Lawesson's reagent; (ii) dimethyl thiodiacetate $[S(CH_2CO_2Me)_2]$, cyclopentane-1,2-dione and base; (iii) pentane-2,4-dione, methyl thioglycolate $[HSCH_2CO_2Me]$ and base.

More advanced exercises:

1. Deduce the structure of the compound, $C_4H_3NO_2S$, produced from thiophene by the following sequence: $ClSO_3H$, then f. HNO_3, then H_2O/heat; the product is isomeric with that obtained by reacting thiophene with acetyl nitrate.

2. Suggest structures for the major and minor isomeric products, $C_5H_5NO_3S$, from 2-methoxythiophene with HNO_3/AcOH at −20 °C.

3. What compounds would be formed by the reaction of: (i) thiophene with propionic anhydride/H_3PO_4; (ii) 3-*t*-butylthiophene with PhN(Me)CHO/POCl$_3$, then aq. NaOH; (iii) thiophene with Tl(O$_2$CCF$_3$)$_3$, then aq. KI → C_4H_3IS.

4. Predict the principle site of deprotonation on treatment of 2- and 3-methoxythiophenes with *n*-BuLi.

5. Deduce structures for the compounds, C_4HBr_3S and $C_4H_2Br_2S$, produced successively by treating 2,3,4,5-tetrabromothiophene with Mg then H_2O and then the product again with Mg then H_2O.

6. Deduce the structure of the compound, $C_9H_6OS_2$, produced by the sequence: thiophene with BuLi, then CO_2 → $C_5H_4O_2S$, then this with thiophene in the presence of P_4O_{10}.

7. Deduce the structure of the thiophenes: (i) $C_6H_4N_4S$, produced by reacting $(NC)_2C=C(CN)_2$ with H_2S; (ii) $C_8H_8O_6S$ from diethyl oxalate, $(EtO_2CCH_2)_2S$/NaOEt, aq. NaOH, then Me_2SO_4; (iii) $C_{11}H_{16}S$ from 3-acetylcyclononanone with P_4S_{10}.

References

[1] 'Recent advances in the chemistry of thiophenes', Gronowitz, S., *Adv. Heterocycl. Chem.*, **1963**, *1*, 1.

[2] Deans, F. B. and Eaborn, C., *J. Chem. Soc.*, **1959**, 2303; Taylor, R., *J. Chem. Soc. (B)*, **1970**, 1364.

[3] Steinkopf, W. and Köhler, W., *Justus Liebigs Ann. Chem.*, **1937**, *532*, 250; Schreiner, H., *Monatsh. Chem.*, **1951**, *82*, 702.

[4] Schwetlick, K., Unverferth, K. and Mayer, R., *Z. Chem.*, **1967**, *7*, 58; Olsson, S., *Arkiv. Kemi*, **1970**, *32*, 89; Baker, R., Eaborn, C. and Taylor, R., *J. Chem. Soc., Perkin Trans. 2*, **1972**, 97.

[5] Carmody, M. P., Cook, M. J., Dassanayake, N. C., Katritzky, A. R., Linda, P. and Tack, R. D., *Tetrahedron*, **1976**, *32*, 1767.

[6] Curtis, R. F., Jones, D. M. and Thomas, W. A., *J. Chem. Soc. (C)*, **1971**, 234.

[7] Butler, A. R. and Hendry, J. B., *J. Chem. Soc. (B)*, **1971**, 102.

[8] Babasinian, V. S., *Org. Synth., Coll. Vol. II*, **1943**, 466.

[9] Olah, G. A., Kuhn, S. and Mlinko, A., *J. Chem. Soc.*, **1956**, 4257.

[10] Östman, B., *Acta Chem. Scand.*, **1968**, *22*, 1687.

[11] Blatt, A. H., Bach, S. and Kresch, L. W., *J. Org. Chem.*, **1957**, *22*, 1693.

[12] Östman, B., *Acta Chem. Scand.*, **1968**, *22*, 2754.

[13] Gronowitz, S. and Gjos, N., *Acta Chem. Scand.*, **1967**, *21*, 2823.

[14] Gronowitz, S. and Ander, I., *Chem. Scr.*, **1980**, *15*, 20.

[15] Steinkopf, W. and Ohse, W., *Justus Liebigs Ann. Chem.*, **1924**, *437*, 14.

[16] Terentev, A. P. and Kadatskii, G. M., *J. Gen. Chem., USSR*, **1952**, 189.

[17] Maccarone, E., Musumarra, G. and Tomaselli, G. A., *J. Org. Chem.*, **1974**, *39*, 3286.

[18] Söderbäck, E., *Acta Chem. Scand.*, **1954**, *8*, 1851.

[19] Steinkopf, W., Jacob, H. and Penz, H., *Justus Liebigs Ann. Chem.*, **1934**, *512*, 136.

[20] Marino, G., *Tetrahedron*, **1965**, *21*, 843.

[21] Goldberg, Y. and Alper, H., *J. Org. Chem.*, **1993**, *58*, 3072.

[22] Minnis, W., *Org. Synth., Coll. Vol. II*, **1943**, 357; Lew, H. Y. and Noller, C. R., *ibid., Coll. Vol. IV*, **1963**, 545.

[23] Muathen, H. A., *Tetrahedron*, **1996**, *52*, 8863.

[24] Antolini, L., Goldini, F., Iarossi, D., Mucci, A. and Schenetti, L., *J. Chem. Soc., Perkin Trans. 1*, **1997**, 1957.

[25] Brandsma, L. and Verkruijsse, H. D., *Synth. Commun.*, **1988**, *18*, 1763.

[26] Gronowitz, S. and Raznikiewicz, T., *Org. Synth., Coll Vol. V*, **1973**, 149; Hallberg, A., Liljefors, S. and Pedaja, P., *Synth. Commun.*, **1981**, *11*, 25.

[27] Gronowitz, S., *Acta Chem. Scand.*, **1959**, *13*, 1045.

[28] D'Auria, M. and Mauriello, G., *Tetrahedron Lett.*, **1995**, *36*, 4883.

[29] Hoffmann, K. J. and Carlsen, P. H. J., *Synth. Commun.*, **1999**, *29*, 1607.

[30] Pearson, D. L. and Tour, J. M., *J. Org. Chem.*, **1997**, *62*, 1376.

[31] Johnson, J. R. and May, G. E., *Org. Synth., Coll. Vol. II*, **1943**, 8; Minnis, W., *ibid.*, 520.

[32] Hartough, H. D. and Kosak, A. I., *J. Am. Chem. Soc.*, **1947**, *69*, 3093.

[33] Pennanen, S. I., *Heterocycles*, **1976**, *4*, 1021.

[34] Weston, A. W. and Michaels, R. J., *Org. Synth., Coll. Vol. IV*, **1963**, 915; Downie, I. M., Earle, M. J., Heaney, H. and Shuhaibar, K. F., *Tetrahedron*, **1993**, *49*, 4015.

[35] Finch, H., Reece, D. H. and Sharp, J. T., *J. Chem. Soc., Perkin Trans. 1*, **1994**, 1193.

[36] Meth-Cohn, O. and Ashton, M., *Tetrahedron Lett.*, **2000**, *41*, 2749.

[37] Okuyama, T., Tani, Y., Miyake, K. and Yokoyama, Y., *J. Org. Chem.*, **2007**, *72*, 1634; Smith, K. and Barratt, M. L., *ibid.*, 1031.

[38] Fieser, L. F. and Kennelly, R. G., *J. Am. Chem. Soc.*, **1935**, *57*, 1611.

[39] Yadav, P. P., Gupta, P., Chaturvedi, A. K., Shukla, P. K. and Maurya, R., *Bioorg. Med. Chem.*, **2005**, *13*, 1497.

[40] Kamitori, Y., Hojo, M., Masuda, R., Izumi, T. and Tsukamoto, S., *J. Org. Chem.*, **1984**, *49*, 4161.

[41] Wiberg, K. B. and Shane, H. F., *Org. Synth., Coll. Vol. III*, **1955**, 197; Emerson, W. S. and Patrick, T. M., *ibid., Coll. Vol. IV*, **1963**, 980.

[42] Janda, M., *Coll. Czech. Chem. Commun.*, **1961**, *26*, 1889.

[43] Blicke, F. F. and Burckhalter, J. F., *J. Am. Chem. Soc.*, **1942**, *64*, 477.

[44] Griffing, J. M. and Salisbury, L. F., *J. Am. Chem. Soc.*, **1948**, *70*, 3416.

[45] Ballantine, J. A. and Fenwick, R. G., *J. Chem. Soc. (C)*, **1970**, 2264.

[46] Tormos, G. V., Belmore, K. A. and Cava, M. P., *J. Am. Chem. Soc.*, **1993**, *115*, 11512.

[47] Yang, S.-M. and Fang, J.-M., *J. Chem. Soc., Perkin Trans. 1*, **1995**, 2669.

[48] Hartough, H. D. and Meisel, S. L., *J. Am. Chem. Soc.*, **1948**, *70*, 4018.

[49] Barker, J. M., Huddleston, P. R. and Wood, M. L., *Synth. Commun.*, **1975**, 59.

[50] Dowle, M. D., Hayes, R., Judd, D. B. and Williams, C. N., *Synthesis*, **1983**, 73.

[51] Harwood, L. M., Currie, G. S., Drew, M. G. B. and Luke, R. W. A., *Chem. Commun.*, **1996**, 1953; Petasis, N. A., Goodman, A., and Zavialov, I. A., *Tetrahedron*, **1997**, *53*, 16463.

[52] Briscoe, H. V. A., Peel, J. B. and Young, G. W., *J. Chem. Soc.*, **1929**, 2589.

[53] Steinkopf, W. and Köhler, W., *Justus Liebigs Ann. Chem.*, **1937**, *532*, 250.

[54] 'The chemistry of thiophenium salts and thiophenium ylides', Porter, A. E. A., *Adv. Heterocycl. Chem.*, **1989**, *45*, 151.

[55] Acheson, R. M. and Harrison, D. R., *J. Chem. Soc. (C)*, **1970**, 1764; Heldeweg, R. F. and Hogeveen, H., *Tetrahedron Lett.*, **1974**, 75.

[56] Hashmall, J. A., Horak, V., Khoo, L. E., Quicksall, C. O. and Sun, M. K., *J. Am. Chem. Soc.*, **1981**, *103*, 289.

[57] Gillespie, R. J., Murray-Rust, J., Murray-Rust, P. and Porter, A. E. A., *J. Chem. Soc., Chem. Commun.*, **1978**, 83.

[58] Gillespie, R. J., Porter, A. E. A. and Willmott, W. E., *J. Chem. Soc., Chem. Commun.*, **1978**, 85.

[59] Tranmer, G. K. and Capretta, A., *Tetrahedron*, **1998**, *54*, 15499; Monn, J. A. *et al.*, *J. Med. Chem.*, **1999**, *42*, 1027.

[60] Schenk, G. O. and Steinmetz, R., *Justus Liebigs Ann. Chem.*, **1963**, *668*, 19.

[61] Cuffe, J., Gillespie, R. J. and Porter, A. E. A., *J. Chem. Soc., Chem. Commun.*, **1978**, 641.

[62] Nakayama, J., Nagasawa, H., Sugihara, Y. and Ishii, A., *J. Am. Chem. Soc.*, **1997**, *119*, 9077.

[63] Melles, J. L. and Backer, H. J., *Recl. Trav. Chim. Pays-Bas*, **1953**, *72*, 314; van Tilborg, W. J. M., *Synth. Commun.*, **1976**, *6*, 583; McKillop, A. and Kemp, D., *Tetrahedron*, **1989**, *45*, 3299.

[64] Nakayama, J., Nagasawa, H., Sugihara, Y. and Ishii, A., *Heterocycles*, **2000**, *52*, 365.

[65] Rozen, S. and Bareket, Y., *J. Org. Chem.*, **1997**, *62*, 1457.

[66] Melles, J. L., *Recl. Trav. Chim. Pays-Bas*, **1952**, *71*, 869.

[67] Nakayama, J. and Hirashima, A., *J. Am. Chem. Soc.*, **1990**, *112*, 7648.

[68] Moiseev, A. M., Tyurin, D. D., Balenkova, E. S. and Nenajdenko, V. G., *Tetrahedron*, **2006**, *62*, 4139.

[69] Consiglio, G., Spinelli, D., Gronowitz, S., Hörnfeldt, A.-B., Maltesson, B. and Noto, R., *J. Chem. Soc., Perkin Trans. 1*, **1982**, 625.

[70] Makosza, M. and Kwast, E., *Tetrahedron*, **1995**, *51*, 8339.

[71] Lee, S. B. and Hong, J.-I., *Tetrahedron Lett.*, **1995**, *36*, 8439.

[72] Keegstra, M. A., Peters, T. H. A. and Brandsma, L., *Synth. Commun.*, **1990**, *20*, 213.

[73] Punidha, S., Sinha, J., Kumar, A. and Ravikanth, M., *J. Org. Chem.*, **2008**, *73*, 323.

[74] Chadwick, D. J. and Willbe, C., *J. Chem. Soc., Perkin Trans. 1*, **1977**, 887; Feringa, B. L., Hulst, R., Rikers, R. and Brandsma, L., *Synthesis*, **1988**, 316; Furukawa, N., Hoshino, H., Shibutani, T., Higaki, M., Iwasaki, F. and Fujihara, H., *Heterocycles*, **1992**, *34*, 1085.

[75] Fikentscher, R., Brückmann, R. and Betz, R, *Justus Liebigs Ann. Chem.*, **1990**, 113.

[76] Chadwick, D. J., McKnight, M. V. and Ngochindo, R., *J. Chem. Soc., Perkin Trans. 1*, **1982**, 1343; Chadwick, D. J. and Ennis, D. S., *Tetrahedron*, **1991**, *47*, 9901.

[77] Carpenter, A. J. and Chadwick, D. J., *Tetrahedron Lett.*, **1985**, *26*, 1777.

[78] Knight, D. W. and Nott, A. P., *J. Chem. Soc., Perkin Trans. 1*, **1983**, 791.

[79] Bures, E., Spinazzé, P. G., Beese, G., Hunt, I. R., Rogers, C. and Keay, B. A., *J. Org. Chem.*, **1997**, *62*, 8741; DuPriest, M. T., Zincke, P. W., Conrow, R. E., Kuzmich, D., Dantanarayana, A. P. and Sproull, S. J., *ibid.*, 9372.

[80] Gronowitz, S., Hallberg, A. and Frejd, T., *Chem. Scr.*, **1980**, *15*, 1.

[81] Hallberg, A. and Gronowitz, S., *Chem. Scr.*, **1980**, *16*, 42.

[82] Detty, M. R. and Hays, D. S., *Heterocycles*, **1995**, *40*, 925.

[83] Zöllner, M. J., Jahn, U., Becker, E., Kowalsky, W. and Johannes, H.-H., *Chem. Commun.*, **2009**, 565.

[84] Ye, X.-S. and Wong, H. N. C., *J. Org. Chem*, **1997**, *62*, 1940.

[85] Bayh, O., Awad, H., Mongin, F., Hoarau, C., Trécourt, F., Quéguiner, G., Marsais, F., Blanco, F., Abarca, B. and Ballesteros, R., *Tetrahedron*, **2005**, *61*, 4779.

[86] Goldberg, Yu., Sturkovich, R. and Lukevics, E., *Synth. Commun.*, **1993**, *23*, 1235.

[87] Wu, X. and Rieke, R. D., *J. Org. Chem.*, **1995**, *60*, 6658.

[88] Spagnolo, P. and Zanirato, P., *J. Chem. Soc., Perkin Trans. 1*, **1996**, 963.

[89] Wu, X., Chen, T. A., Zhu, L. and Rieke, R. D., *Tetrahedron Lett.*, **1994**, *35*, 3673.

[90] Rieke, R. D., Kim, S.-H. and Wu, X., *J. Org. Chem.*, **1997**, *62*, 6921.

[91] Frisell, C. and Lawesson, S.-O., *Org. Synth., Coll. Vol. V*, **1973**, 642.

[92] Camici, L., Ricci, A. and Taddei, M., *Tetrahedron Lett.*, **1986**, *27*, 5155.

[93] Acheson, R. M., MacPhee, K. E., Philpott, R. G. and Barltrop, J. A., *J. Chem. Soc.*, **1956**, 698.

[94] Alvarez, M., Bosch, J., Granados, R. and López, F., *J. Heterocycl. Chem.*, **1978**, *15*, 193.

[95] de Bettencourt-Dias, A., Viswanathan, S. and Rollett, A., *J. Am. Chem. Soc.*, **2007**, *129*, 15436.

[96] Raposo, M. M. M., Fonseca, A. M. C. and Kirsch, G., *Tetrahedron*, **2004**, *60*, 4071.

[97] Arnswald, M. and Neumann, W. P., *J. Org. Chem.*, **1993**, *58*, 7022.

[98] Fuller, L. S., Iddon, B. and Smith, K. A., *J. Chem. Soc., Perkin Trans. 1*, **1997**, 3465.

[99] 'The chemistry of thienothiophenes', Litvinov, V. P., *Adv. Heterocycl. Chem.*, **2006**, *90*, 125.

[100] Li, X.-C., Sirringhaus, H., Garnier, F., Holmes, A. B., Moratti, S. C., Feeder, N., Clegg, W., Teat, S. J. and Friend, R. H., *J. Am. Chem. Soc.*, **1998**, *120*, 2206.

[101] Frey, J., Proemmel, S., Armitage, M. A. and Holmes, A. B., *Org. Synth.*, **2006**, *83*, 209.

[102] Nenajdenko, V. G., Gribkov, D. V., Sumerin, V. V. and Balenkova, E. S., *Synthesis*, **2003**, 124.

[103] 'The base-catalysed halogen dance, and other reactions of aryl halides', Bunnett, J. F., *Acc. Chem. Res.*, **1972**, *5*, 139.

[104] Brandsma, L. and de Jong, R. L. P., *Synth. Commun.*, **1990**, *20*, 1697.

[105] Camazzi, C. M., Leardini, R., Tundo, A. and Tiecco, M., *J. Chem. Soc., Perkin Trans. 1*, **1974**, 271; Bartle, M., Gore, S. T., Mackie, R. K. and Tedder, J. M., *J. Chem. Soc., Perkin Trans. 1*, **1976**, 1636.

[106] Camaggi, C.-M., Leardini, R., Tiecco, M. and Tundo, A., *J. Chem. Soc., B*, **1970**, 1683; Vernin, G., Metzger, J. and Párkányi, C., *J. Org. Chem.*, **1975**, *40*, 3183.

[107] Ryang, H. S. and Sakurai, H., *J. Chem. Soc., Chem. Commun.*, **1972**, 594; Allen, D. W., Buckland, D. J., Hutley, B. G., Oades, A. C. and Turner, J. B., *J. Chem. Soc., Perkin Trans. 1*, **1977**, 621; D'Auria, M., De Luca, E., Mauriello, G. and Racioppi, R., *Synth. Commun.*, **1999**, *29*, 35.

[108] Chuang, C.-P. and Wang, S.-F., *Synth. Commun.*, **1994**, *24*, 1493; *idem*, *Synlett*, **1995**, 763.

[109] Araneo, S., Arrigoni, R., Bjorsvik, H.-R., Fontana, F., Minisci, F. and Recupero, F., *Tetrahedron Lett.*, **1996**, *37*, 7425.

[110] Hansen, S., *Acta Chem. Scand.*, **1954**, *8*, 695.

[111] Wynberg, H. and Logothetis, A., *J. Am. Chem. Soc.*, **1956**, *78*, 1958.

[112] Gol'dfarb, Ya. L., Taits, S. Z. and Belen'kii, L. I., *Tetrahedron*, **1963**, *19*, 1851.

[113] Birch, S. F. and McAllan, D. T., *J. Chem. Soc.*, **1951**, 2556.

[114] Blenderman, W. G., Joullié, M. M. and Preti, G., *Tetrahedron Lett.*, **1979**, 4985; Kosugi, K., Anisimov, A. V., Yamamoto, H., Yamashiro, R., Shirai, K. and Kumamoto, T., *Chem. Lett.*, **1981**, 1341; Altenbach, H.-J., Brauer, D. J. and Merhof, G. F., *Tetrahedron*, **1997**, *53*, 6019.

[115] 'Applications of ionic hydrogenation to organic synthesis', Kursanov, D. N., Parnes, Z. N. and Loim, N. M., *Synthesis*, **1974**, 633.

[116] Kursanov, D. N., Parnes, Z. N., Bolestova, G. I. and Belen'kii, L. I., Tetrahedron, **1975**, *31*, 311.

[117] Lyakhovetsky, Yu., Kalinkin, M., Parnes, Z., Latypova, F. and Kursanov, D., *J. Chem. Soc., Chem. Commun.*, **1980**, 766.

[118] 'Cycloaddition, ring-opening, and other novel reactions of thiophenes', Iddon, B., *Heterocycles*, **1983**, *20*, 1127; 'Cycloaddition reactions with vinyl heterocycles', Sepúlveda-Arques, J., Abarca-González, B. and Medio-Simón, M., *Adv. Heterocycl. Chem.*, **1995**, *63*, 339.

[119] Kumamoto, K., Fukuda, I. and Kotsuki, H., *Angew. Chem. Int. Ed.*, **2004**, *43*, 2015.

[120] Helder, R. and Wynberg, H., *Tetrahedron Lett.*, **1972**, 605; Kuhn, H. J. and Gollnick, K., *Chem. Ber.*, **1973**, *106*, 674.

[121] Corral, C., Lissavetzky, J. and Manzanares, I., *Synthesis*, **1997**, 29.

[122] Li, Y., Thiemann, T., Sawada, T., Mataka, S. and Tashiro, M., *J. Org. Chem.*, **1997**, *62*, 7926.

[123] Furukawa, N., Zhang, S-Z., Sato, S. and Higaki, M., *Heterocycles*, **1997**, *44*, 61.

[124] Furukawa, N., Zhang, S.-Z., Horn, E., Takahashi, O. and Sato, S., *Heterocycles*, **1998**, *47*, 793.

[125] Takayama, J., Sugihara, Y., Takayanagi, T. and Nakayama, J., *Tetrahedron Lett.*, **2005**, *46*, 4165.

[126] Wynberg, H., *Acc. Chem. Res.*, **1971**, *4*, 65.

[127] Marzinzik, A. L. and Rademacher, P., *Synthesis*, **1995**, 1131.

[128] Sato, K., Arai, S. and Yamagishi, T., *J. Heterocycl. Chem.*, **1996**, *33*, 57.

[129] Campaigne, E. and Tullar, B. F., *Org. Synth., Coll. Vol. IV*, **1963**, 921; Clarke, J. A. and Meth-Cohn, O., *Tetrahedron Lett.*, **1975**, 4705.

[130] Nakayama, J., Kawamura, T., Kuroda, K. and Fujita, A., *Tetrahedron Lett.*, **1993**, *34*, 5725.

[131] Gaertner, R., *J. Am. Chem. Soc.*, **1951**, *79*, 3934.

[132] Merz, A. and Rehm, C., *J. Prakt. Chem.*, **1996**, *338*, 672; Coffey, M., McKellar, B. R., Reinhardt, B. A., Nijakowski, T. and Feld, W. A., *Synth. Commun.*, **1996**, *26*, 2205.

[133] Jakobsen, H. J., Larsen, E. H. and Lawesson, S.-O., *Tetrahedron*, **1963**, *19*, 1867.

[134] Allen, D. W., Clench, M. R., Hewson, A. T. and Sokmen, M., *J. Chem. Res. (S)*, **1996**, 242.

[135] Gronowitz, S. and Hoffman, R. A., *Ark. Kemi*, **1960**, *15*, 499; Hörnfeldt, A.-B., *ibid.*, **1964**, *22*, 211.

[136] Thorstad, O., Undheim, K., Cederlund, B. and Hörnfeldt, A.-B., *Acta Chem. Scand.*, **1975**, *B29*, 647.

[137] Hunter, G. A. and McNab, H., *J. Chem. Soc., Chem. Commun.*, **1990**, 375.

[138] Lantz, R. and Hörnfeldt, A.-B., *Chem. Scr.*, **1976**, *10*, 126; Hurd, C. D. and Kreuz, K. L. *J. Am. Chem. Soc.*, **1950**, *72*, 5543.

[139] Rassu, G., Spanu, P., Pinna, L., Zanardi, F. and Casiraghi, G., *Tetrahedron Lett.*, **1995**, *36*, 1941; 'Furan-, pyrrole-, and thiophene-based siloxydienes for synthesis of densely functionalised homochiral compounds', Casiraghi, G. and Rassu, G., *Synthesis*, **1995**, 607.

[140] 'Synthesis of amino derivatives of five-membered heterocycles by Thorpe-Ziegler cyclisation', Granik, V. G., Kadushkin, A. V. and Liebscher, J., *Adv. Heterocycl. Chem.*, **1998**, *72*, 79.

[141] Steinkopf, W., *Justus Liebigs Ann. Chem.*, **1914**, *403*, 17; Steinkopf, W. and Höpner, T., *ibid.*, **1933**, *501*, 174.

[142] Meth-Cohn, O. and Narine, B., *Synthesis*, **1980**, 133.

[143] Campaigne, E. and Monroe, P. A., *J. Am. Chem. Soc.*, **1954**, *76*, 2447.

[144] Kenning, D. D., Mitchell, K. A., Calhoun, T. R., Funfar, M. R., Sattler, D. J. and Rasmussen, S. C. *J. Org. Chem.*, **2002**, *67*, 9073.

[145] Brunett, E. W., Altwein, D. M. and McCarthy, W. C., *J. Heterocycl. Chem.*, **1973**, *10*, 1067.

[146] 'The preparation of thiophens and tetrahydrothiophens', Wolf, D. E. and Folkers, K., *Org. Reactions*, **1951**, *6*, 410.

[147] Shridar, D. R., Jogibhukta, M., Shanthon Rao, P. and Handa, V. K., *Synthesis*, **1982**, 1061; Jones, R. A. and Civcir, P. U., *Tetrahedron*, **1997**, *53*, 11529.

[148] Freeman, F., Lee, M. Y., Lu, H., Wang, X. and Rodriguez, E., *J. Org. Chem.*, **1994**, *59*, 3695.

[149] Feldkamp, R. F. and Tullar, B. F., *Org. Synth., Coll. Vol. IV*, **1963**, 671.

[150] Schulte, K. E., Reisch, J. and Hörner, L., *Chem. Ber.*, **1962**, *95*, 1943; Kozhushkov, S., Hanmann, T., Boese, R., Knieriem, B., Scheib, S., Bäuerle, P. and de Meijere, A., *Angew. Chem., Int. Ed. Engl.*, **1995**, *35*, 781; Alzeer, J. and Vasella, A., *Helv. Chim. Acta*, **1995**, *78*, 177.

[151] Von Kieseritzky, F., Allared, F., Dahlsdtedt, E. and Hellberg, J., *Tetrahedron Lett.*, **2004**, *45*, 6049.

[152] Wynberg, H. and Kooreman, H. J., *J. Am. Chem. Soc.*, **1965**, *87*, 1739.

[153] Agarwal, N., Hung, C.-H. and Ravikanth, M., *Tetrahedron*, **2004**, *60*, 10671.

[154] Taylor, E. C. and Dowling, J. E., *J. Org. Chem.*, **1997**, *62*, 1599.

[155] Obrecht, D., Gerber, F., Sprenger, D. and Masquelin, T., *Helv. Chim. Acta*, **1997**, *80*, 531.

[156] Clemens, R. T. and Smith, S. Q., *Tetrahedron Lett.*, **2005**, *46*, 1319.

[157] McIntosh. J. M. and Khalil, H., *Can. J. Chem.*, **1975**, *53*, 209.

[158] Fevig, T. L., Phillips, W. G. and Lau, P. H., *J. Org. Chem.*, **2001**, *66*, 2493.

[159] Hesse, S., Perspicace, E. and Kirsch, G., *Tetrahedron Lett.*, **2007**,*48*, 5261.

[160] Nakayama, J., Machida, H., Saito, R. and Hoshino, M., *Tetrahedron Lett.*, **1985**, *26*, 1983.

[161] Nakayama, J. and Kuroda, K., *J. Am. Chem. Soc.*, **1993**, *115*, 4612.

[162] Prim, D. and Kirsch, G., *Synth. Commun.*, **1995**, *25*, 2449.

[163] Gewald, K., Rennent, S., Schindler, R. and Schäfer, H., *J. Prakt. Chem.*, **1995**, *337*, 472.

[164] Rehwald, M., Gewald, K. and Böttcher, G., *Heterocycles*, **1997**, *45*, 493.

[165] De Jong, R. L. P. and Brandsma, L., *J. Chem. Soc., Chem. Commun.*, **1983**, 1056; Otsubo, T., Kono, Y., Hozo, N., Miyamoto, H., Aso, Y., Ogura, F., Tanaka, T. and Sawada, M., *Bull. Chem. Soc. Jpn.*, **1993**, *66*, 2033.

[166] Reddy, K. V. and Rajappa, S., *Heterocycles*, **1994**, *37*, 347.

[167] Gewald, K., Schinke, E. and Böttcher, H., *Chem. Ber.*, **1966**, *99*, 94.

[168] Hoener, A. P. F., Henkel, B. and Gauvin, J.-C., *Synlett*, **2003**, 63

[169] Sridhar, M., Rao, R. M., Baba, N. H. K. and Kumbhare, R. M., *Tetrahedron Lett.*, **2007**, *48*, 3171.

[170] Huang, W., Li, J., Tang, J., Liu, H., Shen, J. and Jiang, H., *Synth. Commun.*, **2005**, *35*, 1351.

[171] Tormyshev, V. M., Trukhin, D. V., Rogozhnikova, O. Yu., Mikhalina, T. V., Troitskaya, T. I. and Flinn, A., *Synlett*, **2006**, 2559.

[172] Brandsma, L. and Verkruijsse, H. D., *Synth. Commun.*, **1990**, *20*, 2275.

[173] Kagan, J., Arora, S. K., Prakesh, I. and Üstunol, A., *Heterocycles*, **1983**, *20*, 1341.

[174] Matsumoto, K., Nakaminami, H., Sogabe, M., Kurata, H. and Oda, M., *Tetrahedron Lett.*, **2002**, *43*, 3049.

[175] Wex, B., Kaafarani, B. R., Kirschbaum, K. and Neckers, D. C., *J. Org. Chem.*, **2005**, *60*, 4502.

18

Furans: Reactions and Synthesis

furan

Furans[1] are volatile, fairly stable compounds with pleasant odours. Furan itself is slightly soluble in water. It is readily available, and its commercial importance is mainly due to its role as the precursor of the very widely used solvent tetrahydrofuran (THF). Furan is produced by the gas-phase decarbonylation of furfural (2-formylfuran, furan-2-carboxaldehyde), which in turn is prepared in very large quantities by the action of acids on vegetable residues, mainly from the manufacture of porridge oats and cornflakes. Furfural was first prepared in this way as far back as 1831 and its name is derived from *furfur*, which is the Latin word for bran; in due course, in 1870, the word furan was coined from the same root. Hydrogenation of furfural produces 2-methyltetrahydrofuran, commonly known as MeTHF, a solvent with significant advantages[2] over THF, for example it is only partially miscible with water, making isolation of products easier; it does not freeze until −136 °C, so it is suitable for reactions at very low temperature. 5-Hydroxymethylfurfural (and hence 2,5-diformylfuran by oxidation or 2,5-dimethylfuran by reduction[3]) can be produced from fructose[4] or glucose[5] by dehydration. 2,5-Dimethylfuran has potential as a bio-fuel – it has a higher energy density than ethanol.[6]

18.1 Reactions with Electrophilic Reagents

Of the three five-membered systems with one heteroatom considered in this book, furan is the 'least aromatic' and as such has the greatest tendency to react in such a way as to give addition products – this is true in the context of its interaction with the usual electrophilic substitution reagents, considered in this section, as well as in Diels–Alder-type processes (see 18.7).

18.1.1 Substitution at Carbon

18.1.1.1 Protonation
Furan and the simple alkyl-furans are relatively stable to aqueous mineral acids, though furan is instantly decomposed by concentrated sulfuric acid or by Lewis acids, such as aluminium chloride. Furan reacts only slowly with hydrogen chloride, either as the concentrated aqueous acid or in a non-hydroxylic organic solvent. Hot, dilute aqueous mineral acids cause hydrolytic ring opening.

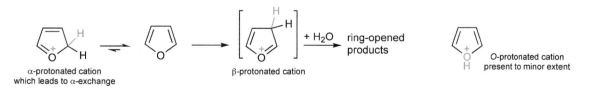

α-protonated cation which leads to α-exchange

β-protonated cation

+ H₂O → ring-opened products

O-protonated cation present to minor extent

Heterocyclic Chemistry 5th Edition   John Joule and Keith Mills
© 2010 Blackwell Publishing Ltd

No pK_{aH} value is available for O-protonation of furan, but it is probably much less basic at oxygen than an aliphatic ether. Acid-catalysed deuteration occurs at an α-position;[7] 3/4-deuterio-furans are not obtained because, although β-protonation probably occurs, the cation produced is more susceptible to water, leading to hydrolytic ring opening. An estimate of pK_{aH} −10.0 was made for the 2-protonation of 2,5-di-*t*-butylfuran, which implies a value of about −13 for furan itself.[8] An α-protonated cation, stable in solution, is produced on treatment of 2,5-di-*t*-butylfuran with concentrated sulfuric acid.[8,9]

Reactions of Protonated Furans

The hydrolysis (or alcoholysis) of furans involves nucleophilic addition of water (or an alcohol) to an initially formed cation, giving rise to open-chain 1,4-dicarbonyl-compounds or derivatives thereof. This is in effect the reverse of one of the general methods for the construction of furan rings (18.13.1.1). Succindialdehyde cannot be obtained from furan itself, presumably because this dialdehyde is too reactive under conditions for hydrolysis, but some alkyl-furans can be converted into 1,4-dicarbonyl products quite efficiently, and this can be viewed as a good method for their synthesis, and of cyclopentenones derived from them.[10] Other routes from furans to 1,4-dicarbonyl compounds are the hydrolysis of 2,5-dialkoxy-tetrahydro-furans (18.1.1.4) and by various oxidative procedures (18.2).

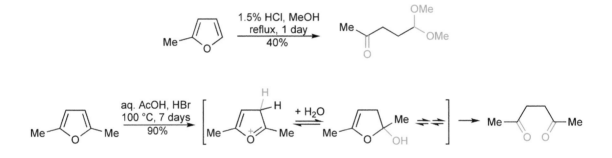

18.1.1.2 Nitration

Sensitivity precludes the use of concentrated acid nitrating mixtures. Reaction of furan, or substituted furans[11] with acetyl nitrate produces non-aromatic adducts, in which progress to a substitution product has been interrupted by nucleophilic addition of acetate to the cationic intermediate, usually[12] at C-5.[13] Aromatisation, by loss of acetic acid, to give the nitro-substitution product, will take place under solvolytic conditions, but is better effected by treatment with a weak base, like pyridine.[14] Further nitration of 2-nitrofuran gives 2,5-dinitrofuran as the main product.[15]

18.1.1.3 Sulfonation

Furan and its simple alkyl-derivatives are decomposed by the usual strong acid reagents, but the pyridine–sulfur-trioxide complex can be used, disubstitution of furan occurring even at room temperature.[16]

18.1.1.4 Halogenation

Furan reacts vigorously with chlorine and bromine at room temperature to give polyhalogenated products, but does not react at all with iodine.[17] Controlled conditions – bromine in dimethylformamide at room temperature – smoothly produce 2-bromo- or 2,5-dibromo-furans.[18] The bromination probably proceeds *via* a 2,5-dibromo-2,5-dihydro-adduct, indeed such species have been observed at low temperature using ^{1}H NMR spectroscopy.[19]

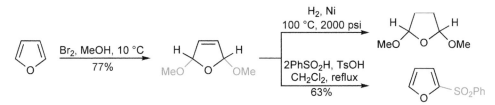

If the bromination is conducted in an alcohol, trapping of the intermediate by C-5 addition of the alcohol, then alcoholysis of C-2-bromide, produces 2,5-dialkoxy-2,5-dihydrofurans, as mixtures of *cis*- and *trans*-isomers;[20] hydrogenation of these species affords 2,5-dialkoxy-tetrahydrofurans, extremely useful as 1,4-dicarbonyl synthons – the unsubstituted example is equivalent to succindialdehyde.[21] The 2,5-dialkoxy-2,5-dihydrofurans are also useful for the synthesis of 2-substituted furans, for example with benzenethiol or phenylsulfinic acid, 2-sulfur-substituted furans are formed[22] and with enol ethers, 2-(2,2-dialkoxyethyl)-furans are formed.[23] Acid-catalysed hydrolysis produces butenolides[24] (see also 18.12.1).

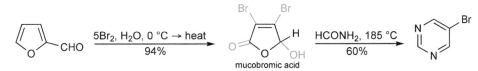

The intrinsically high reactivity of the furan nucleus is further exemplified by the reaction of furfural with excess halogen to produce 'mucohalic acids'; incidentally, mucobromic acid reacts with formamide to provide a useful synthesis of 5-bromopyrimidine.[25] On the other hand, with control, methyl furoate can be cleanly converted into its 5-monobromo or 4,5-dibromo derivatives; hydrolysis and decarboxylation of the latter then affording 2,3-dibromofuran;[26] bromination of 3-furoic acid produces 5-bromofuran-3-carboxylic acid.[27]

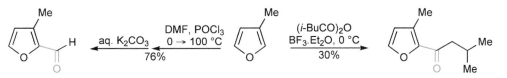

mucobromic acid

18.1.1.5 Acylation

Carboxylic acid anhydrides or halides normally require the presence of a Lewis acid (often boron trifluoride) for Friedel–Crafts acylation of furans, though trifluoroacetic anhydride will react alone. Aluminium-chloride-catalysed acetylation of furan proceeds 7×10^4 times faster at the α-position than at the β-position.[28] 3-Alkyl-furans substitute mainly at C-2;[29] 2,5-dialkyl-furans can be acylated at a β-position, but generally with more difficulty. 3-Bromofuran is efficiently acetylated at C-2 using aluminium chloride catalysis.[30]

Me — aq. K₂CO₃ — 76% ← DMF, POCl₃ 0 → 100 °C — Me — (*i*-BuCO)₂O BF₃.Et₂O, 0 °C 30% → Me

Vilsmeier formylation of furans is a good route to α-formyl-furans,[31] though the ready availability of furfural as a starting material, and methods involving lithiated furans (18.4), are important. Formylation of substituted furans follows the rule that the strong tendency for α-substitution overrides other factors, thus both 2-methylfuran[32] and methyl furan-3-carboxylate[33] give the 5-aldehyde; 3-methylfuran gives mainly the 2-aldehyde.[34]

18.1.1.6 Alkylation and Alkenylation

Traditional Friedel–Crafts alkylation is not generally practicable in the furan series, partly because of catalyst-induced polymerisation and partly because of polyalkylation. Instances of preparatively useful reactions include: production of 2,5-di-*t*-butylfuran[35] from furan or furoic acid[36] and the isopropylation of methyl furoate with double substitution, at the 3- and 4-positions.[34]

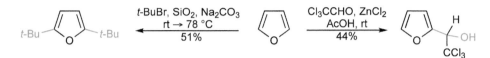

Intramolecular alkenylation at a furan α- or β-position by an alkyne occurs, with the formation of bicyclic derivatives, when promoted by mercury(II) acetate (or Hg(OAc)(OTf), generated *in situ* from mercuric acetate and scandium triflate).[37] In the case of closure onto a β-position, a spirocyclic intermediate from preferred attack at the α-position, may be involved, as shown.

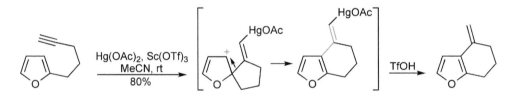

18.1.1.7 Condensation with Aldehydes and Ketones

This occurs by acid catalysis, but generally the immediate product, a furfuryl alcohol, reacts further; 2-(2,2,2-trichloro-1-hydroxy)ethylfuran can, however, be isolated.[38] Macrocycles known as tetraoxaquaterenes can be obtained by condensations[39,40] with dialkyl ketones or cyclohexanone *via* sequences, giving products exactly comparable to those described for pyrrole (16.1.1.7).

18.1.1.8 Condensation with Imines and Iminium Ions

Mono-alkyl-furans undergo Mannich substitution under normal conditions,[41] but furan itself requires a preformed iminium salt for 2-substitution.[42] *N*-Tosyl-imines, generated *in situ* from *N*-sulfinyl-*p*-toluenesulfonamide and aldehydes, bring about tosylaminoalkylation at C-2.[43] The use of furan boronic acids allows Mannich substitutions at both α- and β-positions, with primary or secondary amine components.[44]

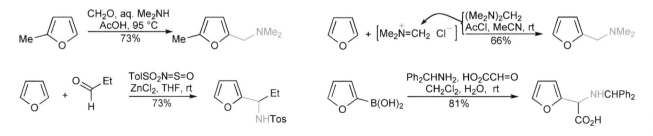

18.1.1.9 Mercuration
Mercuration takes place very readily with replacement of hydrogen, or carbon dioxide from an acid.[45]

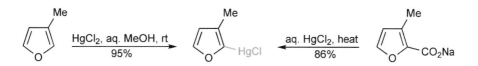

18.2 Reactions with Oxidising Agents
The bromine/methanol (18.1.1.4) oxidation of furans to give 2,5-dialkoxy-2,5-dihydrofurans and the cyclo-addition of singlet oxygen (18.7) are discussed elsewhere in this chapter. 2,5-Dialkoxy-2,5-dihydrofurans can also be obtained by electrochemical oxidation in alcohol solvents[20,46] or conveniently by oxidation with magnesium monoperoxyphthalate in methanol.[47] Reaction of furan with lead(IV) carboxylates produces 2,5-diacyloxy-2,5-dihydrofurans in useful yields.[48]

In related chemistry, ring-opened, Δ-2-unsaturated 1,4-diones can be obtained in *E*- or *Z*-form using reagents such as bromine in aqueous acetone or *meta*-chloroperbenzoic acid;[49] an example is given below.[36] Even but-2-en-1,4-dial (malealdehyde) itself can be produced by oxidation with dimethyldioxirane.[50] The urea/hydrogen peroxide adduct, with catalytic methyltrioxorhenium(VII), oxidises a range of furans to *cis*-ene-diones;[51] magnesium monoperoxyphthalate can also be used for this purpose.[52] The combination, cumyl hydroperoxide with molybdenum hexacarbonyl, can be used to access either *E*- or *Z*-isomers.[53] Oxidation of furans to produce 5-hydroxy-butenolides can be achieved smoothly with buffered sodium chlorite[54] (see also 18.12.1).

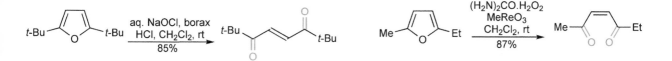

The oxidation of furyl-2-carbinols can produce 6-hydroxy-2*H*-pyran-3(6*H*)-ones (the Achmatowicz rearrangement), which have several synthetic uses,[55] the most important in the heterocyclic context being for the formation therefrom of pyrylium-3-olate species (11.1.7). The oxidation can be conducted with *meta*-chloroperbenzoic acid,[56,57] vanadium(III) acetylacetonate with *t*-butyl peroxide,[58] or with singlet oxygen[59] (18.7).

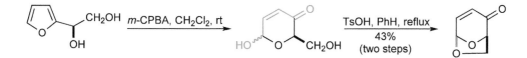

Comparable oxidation of 2-tosyaminomethyl-furans leads to 6-hydroxy-1-tosyl-1,6-dihydro-2*H*-pyridin-3-ones.[60]

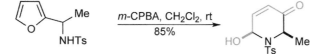

18.3   Reactions with Nucleophilic Reagents

Simple furans do not react with nucleophiles by addition or by substitution. Nitro substituents activate the displacement of halogen, as in benzene chemistry, and VNS methodology (3.3.3) can also be applied to nitro-furans.[61]

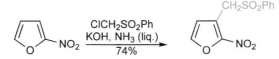

18.4   Metallation and Reactions of C-Metallated Furans
18.4.1   Direct Ring C–H Metallation

Metallation with alkyllithiums proceeds selectively at an α-position, indeed lithiation of furan is one of the earliest examples[62] of the now familiar practice of aromatic ring-metallation. The lithiation can be achieved in refluxing ether or indeed at low temperature.[63] More forcing conditions can bring about 2,5-dilithiation of furan.[64] Magnesiation at an α-position can also be achieved at room temperature, with lithium tri-*n*-butylmagnesate.[65] The preference for α-deprotonation is nicely illustrated by the demonstration that 3-lithiofuran, produced from 3-bromofuran by metal–halogen exchange at −78 °C, equilibrates to the more stable 2-lithiofuran if the temperature rises to > −40 °C,[66] however 3-lithiofuran can be utilised; for example it reacts with bis(trimethylsilyl)peroxide to provide the trimethylsilyl ether of 3-hydroxyfuran.[67] 2-Lithiofuran has been reacted with many electrophiles, for example the 2-boronic acid,[68] 2-tri-*n*-butylstannylfuran,[69] 2-bromofuran[70] and 2-iodofuran[71] (also from the 2-magnesate[65]) can be prepared efficiently in this way.

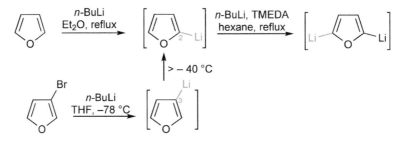

Lithium diisopropylamide can effect C-2-deprotonation of 3-halo-furans.[72] With furoic acid and two equivalents of lithium diisopropylamide, selective formation of the 5-lithio lithium 2-carboxylate takes place,[73] whereas *n*-butyllithium, *via ortho*-assistance, produces the 3-lithio lithium 2-carboxylate.[74]

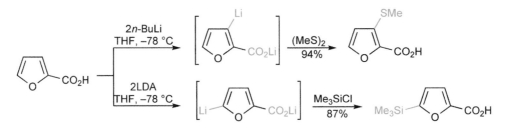

Ortho-direction of metallation to C-3 by 2-bis(dimethylamino)phosphate,[75] 2-oxazoline,[76] or 2-diethylaminocarbonyl[77] groups, and to C-2 by 3-hydroxymethyl[78] or 3-*t*-butoxycarbonylamino[79] also occur. 5-Lithiation of furans with non-directing groups at C-2 provides a route to 2,5-disubstituted furans, but appropriate choice of lithiating conditions can outweigh *ortho*-directing effects, as illustrated.[80]

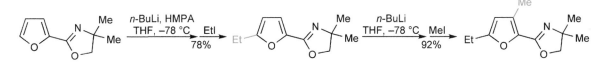

A synthetically useful regioselective 5-lithiation of 3-formylfuran[81] can be achieved by first adding lithium morpholide to the aldehyde and then lithiation at C-5, resulting in 2-substituted 4-formyl-furans, following loss of the amine during work-up.[82]

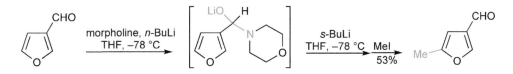

18.4.2 Metal–Halogen Exchange

Metallation at C-3 can be achieved *via* metal–halogen exchange.[83] The greater stability of a carbanion at an α-position shows up again in a mono-exchange of 2,3-dibromofuran with selective replacement of the α-bromine.[34,84]

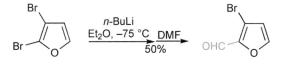

Oxidation of a 2-boronate ester is a means for the synthesis of butenolides (18.12.1).[85] Furan-3-boronic acid and 3-tri-*n*-butylstannylfuran can be made by reaction of the lithiated species with tri-*iso*-propyl borate[68] and tri-*n*-butylchlorostannane[69] respectively.

18.5 Reactions with Radicals

Reactions of furans with radical species as synthetically useful processes have been little developed; arylation[86] and alkylation[87] are selective for α-positions. Exposed to dibenzoyl peroxide, furan produces a stereoisomeric mixture of 1,4-addition products.

18.6 Reactions with Reducing Agents

The best way to reduce a furan to a tetrahydrofuran is using Raney nickel catalysis, though ring opening, *via* hydrogenolysis of C–O bonds can be a complication. Most furans are not reduced simply by metal/ammonia combinations, however furoic acids[88] and furoic acid tertiary amides[89] give dihydro derivatives.

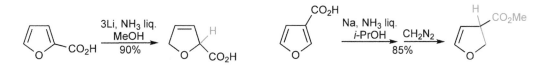

18.7 Electrocyclic Reactions (Ground State)

The 4 + 2 cycloaddition of furan to reactive dienophiles such as maleic anhydride[90] was one of the earliest described examples of the Diels–Alder reaction;[91] the isolated product is the *exo*-isomer,[92] though this is the thermodynamic product, the *endo*-isomer being the kinetic product and the cycloaddition being easily reversible.[93]

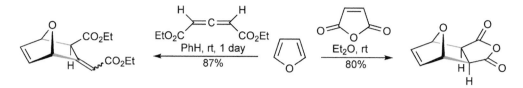

Furan also undergoes cycloadditions with allenes,[94] with benzyne[95] and with simpler dienophiles, like acrylonitrile and acrylate; various Lewis acidic catalysts can assist[96] in some cases, zinc iodide[97] is one such, hafnium tetrachloride another,[98] and improved *endo*:*exo* ratios are obtained in an ionic liquid as reaction solvent.[99] Maleate and fumarate esters react if the addition is conducted under high pressure.[100] This device can also be used to increase markedly the reactivity of 2-methoxyfuran and 2-acetoxyfuran towards dienophiles.[101] At higher reaction temperatures alkynes[102] and even electron-rich alkenes[103] will add to furan. 3- or 5-Halo-furans react faster in these cycloadditions.[104]

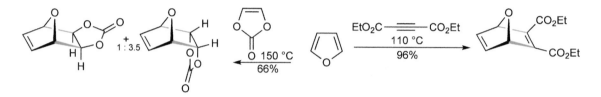

Although, as one would anticipate for the electron-rich component of a normal Diels–Alder pairing, 2-formylfuran is a poor diene; its dimethylhydrazone is a good one, though only ring-opened benzenoid products, derived subsequently from the adducts, are isolated.[105]

Examples of the exploitation of furan Diels–Alder cycloadditions for the construction of complex systems are many;[106] one delightful example is shown below. The residual dienophilic double bond of the Diels–Alder adduct between one of the two furan rings and dimethyl acetylenedicarboxylate then enters into cycloaddition with the second furan ring.[107]

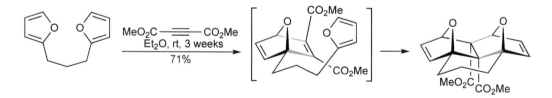

The dipolar cycloaddition of 2-oxyallyl cations[108] is also a process that has been exploited for the synthesis of substituted furans and polycyclic materials,[109] for example it can be made the means for the introduction of acylmethyl groups at the furan 2-position.[110]

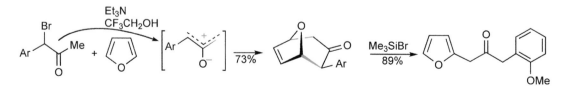

Many examples of furans participating in intramolecular Diels–Alder additions have been described;[111] the example below illustrates the mildness of conditions required in favourable cases.[112] Even unactivated alkenes will cycloadd to furans, in an intramolecular sense.[113]

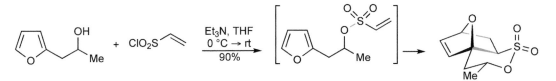

Furans also undergo cycloaddition with singlet oxygen.[114] This has been the basis for several routes to highly oxygenated compounds, for example in syntheses of 5-hydroxy-2(5*H*)-furanones[115] (4-hydroxybut-2-enolides, see 18.12.1), a structural unit which occurs in several natural products. Addition to a 3-substituted furan in the presence of a hindered base[116] or addition[117] to 2-trialkylsilyl-4-substituted furans[118] leads through, as shown, to 4-substituted 5-hydroxy-2(5*H*)-furanones. 5-Substituted furfurals also give 5-hydroxy-2(5*H*)-furanones with loss of the aldehyde carbon.[119] A particularly neat example is the reaction of 2-furoic acid which is converted in quantitative yield, *via* decarboxylation, into malaldehydic acid (the cyclic hemiacetal of *Z*-4-oxobut-2-enoic acid).[120] Addition of singlet oxygen to 2-methoxy-5-alkyl-furans followed by acid-catalysed *Z* to *E* isomerisation produces γ-keto-enoates.[121]

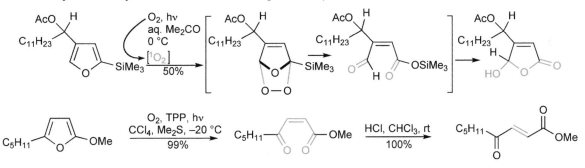

The few examples in which vinyl-furans take part as 4π components[122] in intramolecular cycloadditions include that shown below.[123] In simpler, intermolecular cases yields are generally poor and the extra-annular mode must compete with the more usual intra-annular mode; the inclusion of a bulky group at an α-position assists this differentiation.[124]

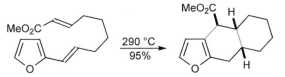

Decarboxylative Claisen rearrangement of furfuryl alcohol esters can be used to introduce 3-substitutents; comparable rearrangements take place with 2-thienyl and 2-pyrrolyl esters.[125]

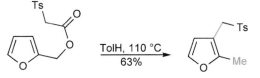

The 4,5-double bond of 2-methoxyfuran will participate as a dienophile.[126]

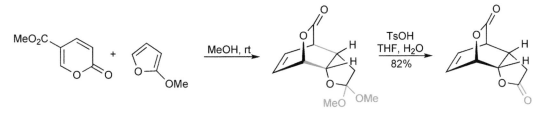

18.8 Reactions with Carbenes and Carbenoids

Dirhodium tetracetate[127] or copper complexed by trispyrazolylborate ligands[128] bring about the carbenoid addition of :CHCO₂Et from ethyl diazoacetate to furans. Generally, mixtures of cyclopropane and ring opened dienones/als are produced.

18.9 Photochemical Reactions

The cycloaddition of diaryl ketones and some aldehydes across the furan 2,3-double bond proceeds regi-oselectively to afford oxetano-dihydrofurans, proton-catalysed cleavage of the acetal linkage in which produces 3-substituted furans.[129]

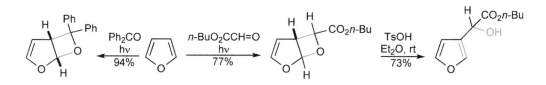

18.10 Furyl-C–X Compounds; Side-Chain Properties

The nucleophilic displacement of halide from furfuryl halides often produces mixtures of products resulting from straightforward displacement on the one hand, and displacement with nucleophilic addition to C-5 on the other;[130] the second mode proceeds through a non-aromatic intermediate, which then isomerises to aromatic product.

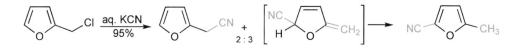

18.11 Furan Carboxylic Acids and Esters and Aldehydes

Save for their easy decarboxylation, furan acids (and their esters) are unexceptional. Carbon dioxide is easily lost[131] from either α- or β-acids and presumably involves ring-protonated intermediates and a decarboxylation analogous to that of β-keto-acids, at least in those examples where copper is not utilised.

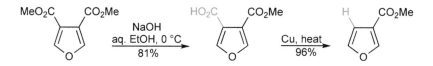

Nitration of 3-furoic acid takes place normally, and at C-5;[132] α-acids sometimes undergo *ipso*-substitution with decarboxylation,[133] for example 2-furoic acid gives the 5-nitro-2-furoic acid, accompanied by some 2,5-dinitrofuran.[134]

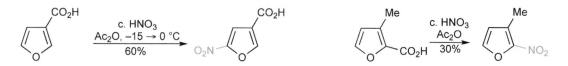

The reaction of furfural with anilines has been known since 1850; a controlled process is now available for its reaction with amines, catalysed by lanthanum or scandium triflate, forming *trans*-4,5-diamino-2-pentenones.[135]

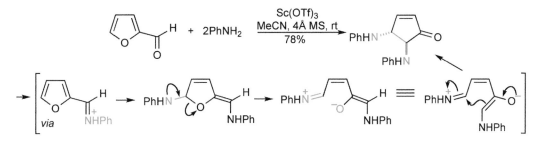

Isomaltol, which can be made from α-D-lactose, also takes part in an ANRORC sequence leading to 4-pyridones.[136]

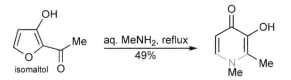

18.12 Oxy- and Amino-Furans
18.12.1 Oxy-Furans

2-Hydroxy-furans exist, if at all, at undetectably low concentrations in tautomeric equilibria involving 2(5*H*)-furanone[137] and 2(3*H*)-furanone forms, for example the angelica lactones can be equilibrated *via* treatment with an organic base, the more stable being the β-isomer; the chemistry of 2-oxy-furans, then, is that of unsaturated lactones. Less is known of 3-hydroxy-furans save again that the carbonyl tautomeric form (3(2*H*)-furanone) predominates.

Many natural products[138] and natural aroma components[139] contain 2-furanone units and considerable synthetic work has thereby been engendered.[140] In the context of these natural products, the name 'buteno-lide' is generally employed and compounds are therefore numbered as derivatives of 4-hydroxybutenoic acid and not as a furan, for example a tetronic acid is a 3-hydroxybut-2-enolide. Butenolides can be converted into furans by partial reduction of the lactone, then dehydration.[141]

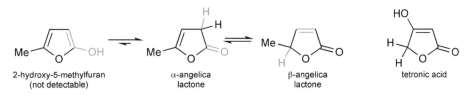

Strategies that have been developed for butenolide construction include the use of 2-trimethylsilyloxyfuran,[142] which reacts with electrophiles at furan C-5,[143] and, complimentarily, furans carrying a 2-oxy-tin (or 2-oxy-boron) substituent, which, *via* chelation control, react with electrophiles at C-3.[144] 2,5-Dimethoxy-2,5-dihydrofuran can generate 2-trimethylsilyloxyfuran *in situ*.[145]

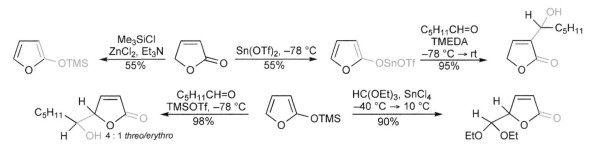

2-*t*-Butoxyfuran, available from the reaction of 2-lithiofuran with *t*-butyl perbenzoate,[146] can be lithiated at C-5, reaction with a carbonyl component, then hydrolysis with dehydration, furnishing alkylidene-butenolides.[147]

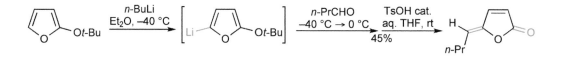

2-Trimethylsilyl-furans are converted into the butenolide by oxidation with peracid.[148]

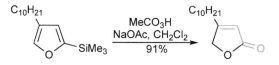

2-Methoxy- and 2-acetoxy-furans are available from 2,5-dimethoxy- and 2,5-diacetoxy-2,5-dihydro-furans (18.1.1.4) *via* acid-catalysed elimination.[149] They undergo Diels–Alder cycloadditions; the adducts can be further transformed into benzenoid compounds by acid-catalysed opening. 3,4-Dihydroxyfuran is undetectable in tautomeric equilibria between mono-enol and dicarbonyl forms; the dimethyl ether behaves as a normal furan, undergoing easy α-electrophilic substitution, mono- or dilithiation at the α-position(s),[150] and Diels–Alder cycloadditions.[151] 2,5-Bis(trimethylsilyloxy)furan is synthesised from succinic anhydride; it too undergoes Diels–Alder additions readily.[152] Both furan-2- and -3-thiols can be obtained by reaction of lithiated furans with sulfur; in each case the predominant tautomer is the thiol form.[153]

18.12.2   Amino-Furans[154]

So little has been described of the chemistry of amino-furans that general comment on their reactivity is difficult to make; it seems likely that simple amino-furans are too unstable to be isolable,[155] though 2-acylamino-furans have been described, for example Boc-masked 2-aminofuran is obtained *via* a Schmidt degradation of the carbonyl azide in *t*-butanol.[156] Heavily substituted amino-furans, in particular those with electron-withdrawing substituents on the ring or on the nitrogen are known.[157] For example methyl 2-aminofuran-5-carboxylate is relatively stable; it undergoes Diels–Alder cycloadditions in the usual manner (cf. 18.7).[158]

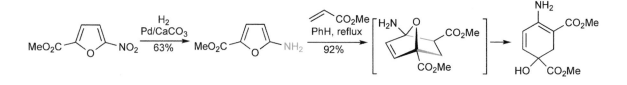

18.13   Synthesis of Furans

Furfural and thence furan, by vapour-phase decarbonylation, are available in bulk and represent the starting points for many furan syntheses. The aldehyde is manufactured[159] from xylose, obtained in turn from pentosans, which are polysaccharides extracted from many plants, e.g. corn cobs and rice husks. Acid catalyses the overall loss of three mole equivalents of water in very good yield. The precise order of events in the multi-step process is not known for certain, however a reasonable sequence[160] is shown below.

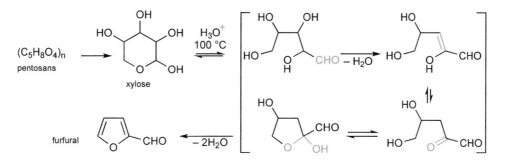

More controlled processes can be carried out using milder catalysts such as indium(III) chloride when enantiopure furan alcohols can be obtained.[161]

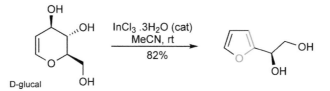

18.13.1 Ring Syntheses

Many routes to furans have been described, but the majority are variants on the first general method – the dehydrating ring closure of a 1,4-dicarbonyl substrate.

18.13.1.1 From 1,4-Dicarbonyl Compounds

1,4-Dicarbonyl compounds can be dehydrated, with acids, to form furans (cf. 16.16.1.1 and 17.12.1.1).

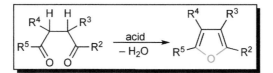

The Paal–Knorr Synthesis

The most widely used approach to furans is the cyclising dehydration of 1,4-dicarbonyl compounds, which provide all of the carbon atoms and the oxygen necessary for the nucleus. Usually, non-aqueous acidic conditions[162] are employed to encourage the loss of water. The process involves addition of the enol oxygen of one carbonyl group to the other carbonyl group, then elimination of water.[163]

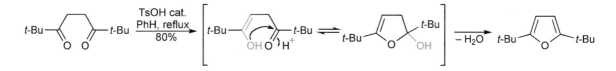

Access to a 1,4-dicarbonyl substrate can be realised in many diverse ways.[164] As just one example, the 1,4-dialdehyde (as a mono-acetal) necessary for a synthesis of diethyl furan-3,4-dicarboxylate was obtained *via* two Claisen condensations with ethyl formate, starting from diethyl succinate.[165] Isomerisation of alk-2-yne 1,4-diols gives 1,4-diketones which proceed in one pot to the 2,5-disubstituted furan, using a ruthenium catalyst with Xantphos and acid.[166] In another one-pot sequence either 2-butene-1,4-diones or 2-butyne-1,4-diones give furans after being first reduced to the saturated 1,4-dione by the formic acid with palladium-on-carbon reaction medium.[167]

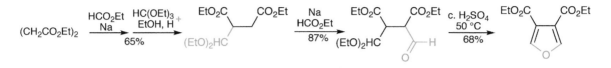

Ring closure of 4-keto-butanamides with trifluoroacetic anhydride or trifluoromethansulfonic anhydride produce (protected) 2-amino-furans (cf. 18.13.2).[168]

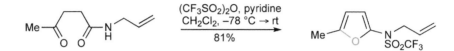

18.13.1.2 From α-Halo-Carbonyl and 1,3-Dicarbonyl Compounds

α-Halo-carbonyl compounds react with 1,3-dicarbonyl compounds in the presence of a base (not ammonia) to give furans.

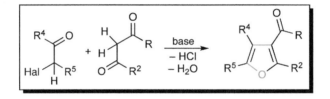

The Feist–Benary Synthesis

This classical synthesis rests on an initial aldol condensation at the carbonyl carbon of a 2-halo-carbonyl-component; ring closure is achieved *via* intramolecular displacement of halide by enolate oxygen; intermediates supporting this mechanistic sequence have been isolated in some cases.[169]

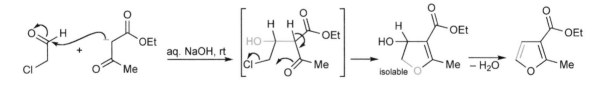

It is important to distinguish this synthesis from the alkylation of a 1,3-dicarbonyl enolate with a 2-halo-ketone, with displacement of halide, producing a 1,4-dicarbonyl unit for subsequent ring closure;[170] presumably the difference lies in the greater reactivity of the carbonyl group (aldehyde in the example) in the Feist–Benary sequence.

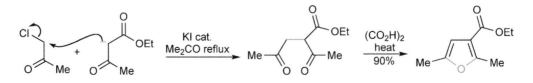

18.13.1.3 From C₄O Compounds

There is a range of furan syntheses that have one aspect in common – the precursor of the aromatic furan has: (i) four carbons, (ii) an oxygen at a terminus and (iii) two degrees of unsaturation located somewhere in the five-atom sequence. Treatment (often acid) of the precursor generates the furan, by a sequence of isomerisations, a ring closure and an elimination (often of water).

The simplest example here is the oxidation of *cis*-but-2-ene-1,4-diol, which gives furan *via* the hydroxy-aldehyde – the two degrees of unsaturation being the carbonyl and carbon–carbon double bonds.[171]

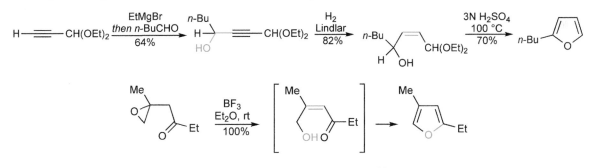

More elaborate 4-hydroxy-enals and -enones have been generated in a variety of ways, for example *via* alkynes[172] or often *via* epoxides,[173] it being sometimes unnecessary to isolate the hydroxy-enone,[174] or via Hörner–Wadsworth–Emmons condensation of β-ketophosphonates with α-acetoxyketones.[175] Acetal,[176] thioenolether[177] or terminal alkyne[178] can be employed as surrogate for the carbonyl group. 1,2,3-Trienyl-4-ols cyclise to give furans.[179] Some of these are exemplified below.

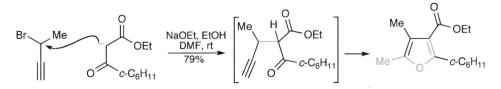

4-Pentynones can be closed to furans using potassium *t*-butoxide,[180] benzyl trimethylammonium methoxide[181] or mercury(II) triflate,[182] and 3-pentynones produce furans with zinc chloride,[183] or with, for example, *N*-bromosuccinimide, give 3-halo-furans.[184] The base-catalysed 2-alkylation of 1,3-dicarbonyl-compounds with propargyl halides, is followed *in situ* by 5-*exo*-dig ring closure.[185] Sundry other isomerisations have utilised gold to catalyse ring closure to furans.[186]

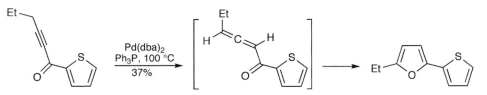

Allenyl-ketones pre-synthesised,[187] or produced *in situ* from palladium(0)- or copper(I)-catalysed isomerisation of conjugated[188,189] or non-conjugated[190] alkynyl-ketones, can be cyclised to furans. The ring closure can be effected with palladium[190] or gold[191] catalysis. Acylation of silylallenes leads to a furan directly.[192]

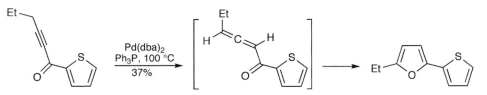

Reaction of aldehydes with methyl 3-nitropropanoate in a Henry reaction, leads through to butenolides.[193] A 3(2*H*)-furanone can be neatly produced by the intramolecular displacement of bromide from a 1-bromo-2,4-dione.[194]

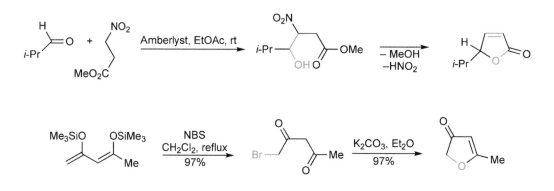

In what is formally a 5-*endo*-dig cyclisation, alk-3-yne-1,2-diols close under the influence of iodine to produce 3-iodo-furans.[195]

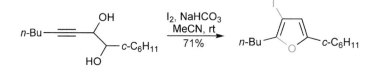

18.13.1.4 Miscellaneous Methods

Considerable ingenuity has been exercised in the development of routes to furans and many are available. This section includes a selection of such one-off routes.

Acyloins react with 'acetylene-transfer' reagents,[196] in one of the few furan syntheses that begins with formation of the ether unit; the cyclising step is a Wittig reaction.

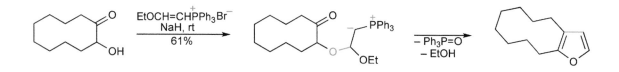

3-Aminofuran-2-carboxylates are formed from the interaction of ethyl glyoxylate with 2-cyano-ketones under Mitsunobu conditions to produce a vinyl ether, which is then ring closed with base.[197]

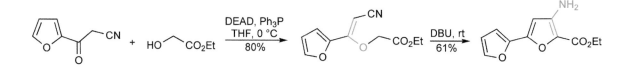

The useful 3,4-bis(tri-*n*-butylstannyl)-[198] and 3,4-bis(trimethylsilyl)-furans[199] are available *via* cycload-dition/cycloreversion steps using 4-phenyloxazole (cf. 17.12.1.7).

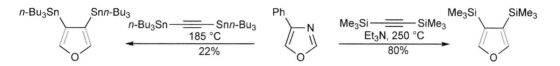

18.13.2 Examples of Notable Syntheses of Furans

18.13.2.1 Tris(furanyl)-18-Crown-6

Tris(furanyl)-18-crown-6 was prepared utilising the reactivity of furfuryl alcohols and chlorides.[200]

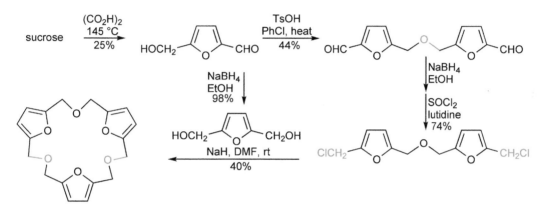

18.13.2.2 Furaneol

Furaneol is a natural flavour principle, isolated from pineapple and strawberry, and used in the food and beverage industries.[201]

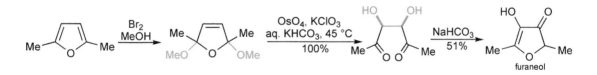

18.13.2.3 Ranitidine

Ranitidine is one of the most commercially successful medicines ever developed; it is used for the treatment of stomach ulcers and has been synthesised from furfuryl alcohol.

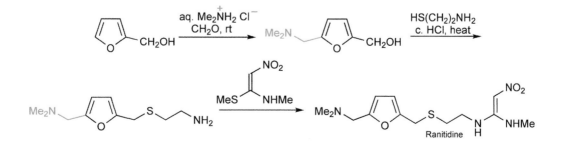

Exercises

Straightforward revision exercises (consult Chapters 15 and 18):

(a) Describe three distinctly different reactions of furans that confirm the relatively smaller aromatic resonance stabilisation of furan compared with thiophene and pyrrole (reactions which lead to non-aromatic products).

(b) At what position does furan undergo lithiation? How could one prepare the alternative lithio-isomer?

(c) With what type of dienophile do furans react most readily?

(d) What types of product can be obtained from the interaction of furans with singlet oxygen ($^1O_2$)?

(e) Do 2-hydroxy-furans exist?

(f) What is the most common method for the ring synthesis of furans? Write a mechanistic sequence for the ring-closing process.

(g) How could one synthesise 3,4-bis(tri-*n*-butylstannyl)furan?

More advanced exercises:

1. Hydrolysis of 2-methoxyfuran with aqueous acid produces 4-hydroxybut-2-enoic acid lactone and $MeO_2C(CH_2)_2CH=O$; write sequences involving protonation and reaction with water to rationalise formation of each of these.

2. Suggest structures for the products: (i) $C_{11}H_8O_2$ produced by treating 2-phenylfuran with the combination DMF/POCl$_3$ then aqueous base; (ii) $C_9H_{10}O_4$ from ethyl furoate/Ac$_2$O/SnCl$_4$; (iii) $C_5H_2N_2O_3$ from 3-cyanofuran with Ac$_2$O/HNO$_3$; (iv) $C_{14}H_{11}Cl_3O_6$ from methyl furoate, CCl$_3$CHO/H$_2$SO$_4$.

3. Electrochemical oxidation of 5-methylfurfuryl alcohol in methanol solvent afforded $C_8H_{14}O_4$, hydrogenation of which produced $C_8H_{16}O_4$, acid treatment of this gave a cyclic 1,2-dione, $C_6H_8O_2$. What are the structures of these compounds?

4. Trace the course of the following synthesis by writing structures for all intermediates: ethyl 2-methylfuran-3-carboxylate with LiAlH$_4$, then SOCl$_2$, then LiAlH$_4$ → C_6H_8O, treatment of which with Br$_2$/MeOH, then H$_2$O/60 °C, then aq. NaBH$_4$ gave $C_6H_{12}O_2$.

5. Write structures for the products of reacting 2-lithiofuran with: (i) cyclohexanone, (ii) Br(CH$_2$)$_7$Cl.

6. Suggest structures for the (main) product from the following combinations: (i) 3-methylfuran/DMF/POCl$_3$ then aq. NaOH; (ii) 2,3-dibromofuran/*n*-BuLi then H$_2$O; (iii) 3-bromofuran/LDA, then CH$_2$O → $C_5H_5BrO_2$; (iv) furfural with EtOH/H$^+$ → $C_9H_{14}O_3$, then this with BuLi followed by B(OBu)$_3$ and aqueous acid → $C_5H_5BO_4$; (v) 3-bromofuran/BuLi/–78 °C, then Bu$_3$SnCl → $C_{16}H_{30}OSn$ and this with MeCOCl/PdCl$_2$ → $C_6H_6O_2$.

7. Write structures for the products of reaction of: (i) furfuryl alcohol with H$_2$C=C=CHCN → $C_9H_9NO_2$; (ii) 2,5-dimethylfuran with CH$_2$=CHCOMe/15 kbar; (iii) furan with 2-chlorocyclopentanone/Et$_3$N/LiClO$_4$ → $C_9H_{10}O_2$.

8. (i) How could one prepare 2-trimethylsilyloxyfuran? (ii) What product, $C_6H_5NO_2$, would be formed from this with ICH$_2$CN/AgOCOCF$_3$?

9. What is the product, $C_{11}H_{10}O_3$, formed from the following sequence: 2-*t*-BuO-furan/*n*-BuLi, then PhCH=O, then TsOH?

10. Decide the structures of the furans produced by the ring syntheses summarised as follows: (i) CH$_2$=CHCH$_2$MgBr/EtCH=O then *m*-CPBA, then CrO$_3$/pyridine then BF$_3$; (ii) CH$_2$=C(Me)CH$_2$MgCl/ HC(OEt)$_3$, then *m*-CPBA, then aq. H$^+$; (iii) (MeO)$_2$CHCH$_2$COMe/ClCH$_2$CO$_2$Me/NaOMe then heat.

11. For the synthesis of tetronic acid summarised as follows, suggest structures for the intermediates: methylamine was added to dimethyl acetylenedicarboxylate (DMAD) → $C_7H_{11}NO_4$, selective reduction with LiAlH$_4$ then giving $C_6H_{11}NO_3$, which with acid cyclised → $C_5H_7NO_2$, aqueous acidic hydrolysis of which produced tetronic acid.

References

[1] 'The Furans', Dunlop, A. P. and Peters, F. N., Reinhold, New York, **1953**; 'The development of the chemistry of furans, 1952–1963', Bosshard, P. and Eugster, C. H., *Adv. Heterocycl. Chem.*, **1966**, *7*, 377; 'Recent advances in furan chemistry, Parts 1 and 2', Dean, F. M., *ibid.*, **1982**, *30*, 167 and *31*, 237; 'Regioselective syntheses of substituted furans', Hou, X. L., Cheung, H. Y., Hon, T. Y., Kwan, P. L., Lo, T. H., Tong, S. Y. and Wong, H. N. C., *Tetrahedron*, **1998**, *54*, 1955.

[2] Aycock, D. F., *Org. Process Res. Dev.*, **2007**, *11*, 156.

[3] Román-Leshkov, Y., Barrett, C. J., Liu, Z. Y. and Dumesic, J. A., *Nature*, **2007**, *447*, 982.

[4] Halliday, G. A., Young, R. J. and Grushin, V. V., *Org. Lett.*, **2003**, *5*, 2003.

[5] Zhao, H., Holladay, J. E., Brown, H. and Zhang, Z. C., *Science*, **2007**, *316*, 1597.

[6] *Chem. World.*, **2007**, *July*, 23.

[7] Unverferth, K. and Schwetlick, K., *J. Prakt. Chem.*, **1970**, *312*, 882; Salomaa, P., Kankaanperä, A., Nikander, E., Kaipainen, K. and Aaltonen, R., *Acta Chem. Scand.*, **1973**, *27*, 153.

[8] Carmody, M. P., Cook, M. J., Dassanayake, N. L., Katritzky, A. R., Linda, P. and Tack, R. D., *Tetrahedron*, **1976**, *32*, 1767.

[9] Wiersum, U. E. and Wynberg, H., *Tetrahedron Lett.*, **1967**, 2951.

[10] 'Synthesis of 1,4-dicarbonyl compounds and cyclopentenones from furans', Piancatelli, G., D'Auria, M. and D'Onofrio, F., *Synthesis*, **1994**, 867.

[11] Michels, J. G. and Hayes, K. J., *J. Am. Chem. Soc.*, **1958**, *80*, 1114.

[12] See however Kolb, V. M., Darling, S. D., Koster, D. F. and Meyers, C. Y., *J. Org. Chem.*, **1984**, *49*, 1636.

[13] Clauson-Kaas, N. and Faklstorp, J., *Acta Chem. Scand., Ser. B*, **1947**, *1*, 210; Balina, G., Kesler, P., Petre, J., Pham, D. and Vollmar, A., *J. Org. Chem.*, **1986**, *51*, 3811.

[14] Rinkes, I. J., *Recl. Trav. Chim. Pays-Bas*, **1930**, *49*, 1169.

[15] Doddi, G., Stegel, F. and Tanasi, M. T., *J. Org. Chem.*, **1978**, *43*, 4303.

[16] Skully, J. F. and Brown, E. V., *J. Org. Chem.*, **1954**, *19*, 894.

[17] Terent'ev, A. P., Belen'kii, L. I. and Yanovskaya, L. A., *Zhur. Obschei Khim.*, **1954**, *24*, 1265 (*Chem. Abstr.*, **1955**, *49*, 12327).

[18] Keegstra, M. A., Klomp, A. J. A. and Brandsma, L., *Synth. Commun.*, **1990**, *20*, 3371.

[19] Baciocchi, E., Clementi, S. and Sebastiani, G. V., *J. Chem. Soc., Chem. Commun.*, **1975**, 875.

[20] Ross, S. D., Finkelstein, M. and Uebel, J., *J. Org. Chem.*, **1969**, *34*, 1018.

[21] Burness, D. M., *Org. Synth., Coll. Vol. V*, **1973**, 403.

[22] Malanga, C., Mannucci, S. and Lardicci, L., *Tetrahedron Lett.*, **1998**, *39*, 5615; Malanga, C., Aronica, L. A. and Lardicci, L., *Synth. Commun.*, **1996**, *26*, 2317.

[23] Malanga, C. and Mannucci, S., *Tetrahedron Lett.*, **2001**, *42*, 2023.

[24] Ceñal, J. P., Carreras, C. R., Tonn, C. E., Padrón, J. I., Ramírez, M. A., Díaz, D. D., García-Tellado, F. and Martín, V. S., *Synlett*, **2005**, 1575.

[25] Kress, T. J. and Szymanski, E., *J. Heterocycl. Chem.*, **1983**, *20*, 1721.

[26] Chadwick, D. J., Chambers, J., Meakins, G. D. and Snowden, R. L., *J. Chem. Soc., Perkin Trans. 1*, **1973**, 1766.

[27] Wang, E. S., Choy, Y. M. and Wong, H. N. C., *Tetrahedron*, **1996**, *52*, 12137.

[28] Ciranni, G. and Clementi, S., *Tetrahedron Lett.*, **1971**, 3833.

[29] Finan, P. A. and Fothergill, G. A., *J. Chem. Soc.*, **1963**, 2723.

[30] Pelly, S. C., Parkinson, C. J., Van Otterlo, W. A. L. and De Koning, C. B., *J. Org. Chem.*, **2005**, *70*, 10474.

[31] Traynelis, V. J., Miskel, J. J. and Sowa, J. R., *J. Org. Chem.*, **1957**, *22*, 1269; Downie, I. M., Earle, M. J., Heaney, H. and Shuhaibar, K. F., *Tetrahedron*, **1993**, *49*, 4015.

[32] Taylor, D. A. H., *J. Chem. Soc.*, **1959**, 2767.

[33] Zwicky, G., Waser, P. G. and Eugster, C. H., *Helv. Chim. Acta*, **1959**, *42*, 1177.

[34] Chadwick, D. J., Chambers, J., Hargreaves, H. E., Meakins, G. D. and Snowden, R. C., *J. Chem. Soc., Perkin Trans. 1*, **1973**, 2327.

[35] Kamitori, Y., Hojo, M., Masuda, R., Izumi, T. and Tsukamoto, S., *J. Org. Chem.*, **1984**, *49*, 4161.

[36] Jurczak, J. and Pikul, S., *Tetrahedron Lett.*, **1985**, *26*, 3039; Fitzpatrick, J. E., Milner, D. J. and White, R., *Synth. Commun.*, **1982**, *12*, 489; Williams, P. D. and Le Goff, E., *J. Org. Chem.*, **1981**, *46*, 4143.

[37] Yamamoto, H., Sasaki, I., Imagawa, H. and Nishizawa, M., *Org. Lett.*, **2007**, *9*, 1399.

[38] Willard, J. R. and Hamilton, C. S., *J. Am. Chem. Soc.*, **1951**, *73*, 4805.

[39] Tanaka, S. and Tomokuni, H., *J. Heterocycl. Chem.*, **1991**, *28*, 991.

[40] Pajewski, R., Ostaszewski, R., Ziach, K., Kulesza, A. and Jurczak, J., *Synthesis*, **2004**, 865.

[41] Gill, E. W. and Ing, H. R., *J. Chem. Soc.*, **1958**, 4728; Eliel, E. L. and Peckham, P. A., *J. Am. Chem. Soc.*, **1950**, *72*, 1209.

[42] Heaney, H., Papageorgiou, G. and Wilkins, R. F., *Tetrahedron Lett.*, **1988**, *29*, 2377.

[43] Padwa, A., Zanka, A., Cassidy, M. P. and Harris, J. M., *Tetrahedron*, **2003**, *59*, 4939.

[44] Harwood, L. M., Currie, G. S., Drew, M. G. B. and Luke, R. W. A., *Chem. Commun.*, **1996**, 1953; Petasis, N. A., Goodman, A. and Zavialov, I. A., *Tetrahedron*, **1997**, *53*, 16463.

[45] Kutney, J. P., Hanssen, H. W. and Nair, G. V., *Tetrahedron*, **1971**, *27*, 3323; Büchi, G., Kovats, E. Sz., Enggist, P. and Uhde, G., *J. Org. Chem.*, **1968**, *33*, 1227.

[46] 'Electrochemical oxidation of organic compounds', Weinberg, N. L. and Weinberg, H. R., *Chem. Rev.*, **1968**, *68*, 449.

[47] D'Annibale, A. and Scettri, A., *Tetrahedron Lett.*, **1995**, *36*, 4659.

[48] Elming, N. and Clauson-Kaas, N., *Acta Chem. Scand.*, **1952**, *6*, 535; Trost, B. M. and Shi, Z., *J. Am. Chem. Soc.*, **1996**, *118*, 3037.

[49] Williams, P. D. and LeGoff, E., *J. Org. Chem.*, **1981**, *46*, 4143.

[50] Adger, B. M., Barrett, C., Brennan, J., McKervey, M. A. and Murray, R. W., *J. Chem. Soc., Chem. Commun.*, **1991**, 1553.

[51] Finlay, J., McKervey, M. A. and Gunaratne, H. Q. N., *Tetrahedron Lett.*, **1998**, *39*, 5651.

[52] Kim, G., Jung, S.-d., Lee, E.-J. and Kim, N., *J. Org. Chem.*, **2003**, *68*, 5395.

[53] Massa, A., Acocella, M. R., De Rosa, M., Soriente, A., Villano, R. and Scettri, A., *Tetrahedron Lett.*, **2003**, *44*, 835.

[54] Clive, D. L. J., Minaruzzaman and Ou, L., *J. Org. Chem.*, **2005**, *70*, 3318.

[55] 'Oxidative rearrangement of furyl carbinols to 6-hydroxy-2*H*-pyran-3(6*H*)-ones, a useful synthon for the preparation of a variety of heterocyclic compounds. A review', Georgiadis, M. P., Albizati, K. F. and Georgiadis, T. M., *Org. Prep. Proc. Int.*, **1992**, *24*, 95.

[56] Taniguchi, T., Nakamura, K. and Ogasawara, K., *Synlett*, **1996**, 971.

[57] Ohmori, N., Miyazaki, T., Kojima, S. and Ohkata, K., *Chem. Lett.*, **2001**, 906.

[58] Harding, M., Hodgson, R., Majid, T., McDowell, K. J. and Nelson, A., *Org. Biomol. Chem.*, **2003**, *1*, 338.

[59] Berberich, S. M., Cherney, R. J., Colucci, J., Courillon, C., Geraci, L. S., Kirkland, T. A., Marx, M. A., Schneider, M. F. and Martin, S. F., *Tetrahedron*, **2003**, *59*, 6819.

[60] Leverett, C. A., Cassidy, M. P. and Padwa, A., *J. Org. Chem.*, **2006**, *71*, 8591.

[61] Makosza, M. and Kwast, E., *Tetrahedron*, **1995**, *51*, 8339.

[62] Gilman, H. and Breur, F., *J. Am. Chem. Soc.*, **1934**, *56*, 1123; Ramanathan, V. and Levine, R., *J. Org. Chem.*, **1962**, *27*, 1216.

[63] Zeni, G., Alves, D., Braga, A. L., Stefani, H. A. and Nogueira, C. W., *Tetrahedron Lett.*, **2004**, *45*, 4823.

[64] Chadwick, D. J. and Willbe, C., *J. Chem. Soc., Perkin Trans. 1*, **1977**, 887.

[65] Mongin, F., Bucher, A., Bazureau, J. P., Bayh, O., Awad, H. and Trécourt, F., *Tetrahedron Lett.*, **2005**, *46*, 7989.

[66] Bock, I., Bornowski, H., Ranft, A. and Theis, H., *Tetrahedron*, **1990**, *46*, 1199.

[67] Camici, L., Ricci, A. and Taddei, M., *Tetrahedron Lett.*, **1986**, *27*, 5155.

[68] Thompson, W. J. and Gaudino, J., *J. Org. Chem.*, **1984**, *49*, 5237.

[69] Pinhey, J. T. and Roche, E. G., *J. Chem. Soc., Perkin Trans. 1*, **1988**, 2415.

[70] Verkruijsse, H. D., Keegstra, M. A. and Brandsma, L., *Synth. Commun.*, **1989**, *19*, 1047.

[71] Carman, C. S. and Koser, G. F., *J. Org. Chem.*, **1983**, *48*, 2534.

[72] Ly, N. D. and Schlosser, M., *Helv. Chim. Acta*, **1977**, *60*, 2085.

[73] Knight, D. W. and Nott, A. P., *J. Chem. Soc., Perkin Trans. 1*, **1981**, 1125.

[74] Carpenter, A. J. and Chadwick, D. J., *Tetrahedron Lett.*, **1985**, *26*, 1777.

[75] Näsman, J. H., Kopola, N. and Pensar, G., *Tetrahedron Lett.*, **1986**, *27*, 1391.

[76] Chadwick, D. J., McKnight, M. V. and Ngochindo, R., *J. Chem. Soc., Perkin Trans. 1*, **1982**, 1343.

[77] Grimaldi, T., Romero, M. and Pujol, M. D., *Synlett*, **2000**, 1788.

[78] Bures, E. J. and Keay, B. A., *Tetrahedron Lett.*, **1988**, *29*, 1247.

[79] Stanetty, P., Kolodziejczyk, K., Roiban, G.-D. and Mihoviloc, M. D., *Synlett*, **2006**, 789.

[80] Lenoir, J.-Y., Ribéreau, P. and Quéguiner, G., *J. Chem. Soc., Perkin Trans. 1*, **1994**, 2943.

[81] Hiroya, K. and Ogasawara, K., *Synlett*, **1995**, 175.

[82] Lee, G. C. M., Holmes, J. D., Harcourt, D. A. and Garst, M. E., *J. Org. Chem.*, **1992**, *57*, 3126.

[83] Fukuyama, Y., Kawashima, Y., Miwa, T. and Tokoroyama, T., *Synthesis*, **1974**, 443.

[84] Sornay, R., Meunier, J.-M. and Fournari, P., *Bull. Soc. Chim. Fr.*, **1971**, 990; Decroix, B., Morel, J., Paulmier, C. and Pastor, P., *ibid.*, **1972**, 1848.

[85] Pelter, A. and Rowlands, M., *Tetrahedron Lett.*, **1987**, *28*, 1203.

[86] Ayres, D. C. and Smith, J. R., *J. Chem. Soc., C*, **1968**, 2737; Camaggi, C. M., Leardini, R., Tiecco, M. and Tundo, A., *J. Chem. Soc., B*, **1969**, 1251; Maggiani, A., Tubul, A. and Brun, P., *Synthesis*, **1997**, 631.

[87] Janda, M., Srogl, J., Stibor, I., Nemec, M. and Vopatrná, P., *Tetrahedron Lett.*, **1973**, 637.

[88] Birch, A. J. and Slobbe, J., *Tetrahedron Lett.*, **1975**, 627; Divanford, H. R. and Jouillié, M. M., *Org. Prep. Proc. Int.*, **1978**, *10*, 94; Kinoshita, T., Miyano, K. and Miwa, T., *Bull. Chem. Soc. Jpn.*, **1975**, *48*, 1865; Beddoes, R. L., Lewis, M. L., Gilbert, P., Quayle, P., Thompson, S. P., Wang, S. and Mills, K., *Tetrahedron Lett.*, **1996**, *37*, 9119.

[89] Kinoshita, T., Ichinari, D. and Sinya, J., *J. Heterocycl. Chem.*, **1996**, *33*, 1313.

[90] Stockmann, H., *J. Org. Chem.*, **1961**, *26*, 2025.

[91] Diels, O. and Alder, K., *Chem. Ber.*, **1929**, *62*, 554.

[92] Woodward, R. B. and Baer, H., *J. Am. Chem. Soc.*, **1948**, *70*, 1161.

[93] Lee, M. W. and Herndon, W. C., *J. Org. Chem.*, **1978**, *43*, 518.

[94] Kurtz, P., Gold, H. and Disselnköter, H., *Justus Liebigs Ann. Chem.*, **1959**, *624*, 1; Kozikowski, A. P., Floyd, W. C. and Kuniak, M. P., *J. Chem. Soc., Chem. Commun.*, **1977**, 582.

[95] Kitamura, T., Todaka, M. and Fujiwara, Y., *Org. Synth.*, **2002**, *78*, 104.

[96] Lazlo, P. and Lucchetti, J., *Tetrahedron Lett.*, **1984**, *25*, 4387.

[97] Kienzle, F., *Helv. Chim. Acta*, **1975**, *58*, 1180; Brion, F., *Tetrahedron Lett.*, **1982**, *23*, 5299; Campbell, M. M., Kaye, A. D., Sainsbury, M. and Yavarzadeh, R., *Tetrahedron*, **1984**, *40*, 2461.

[98] Hayashi, Y., Nakamura, M., Nakao, S., Inoue, T. and Shoji, M., *Angew. Chem. Int. Ed.*, **2002**, *41*, 4079.

[99] Hemeon, I., DeAmicis, C., Jenkins, H., Scammells, P. and Singer, R. D., *Synlett*, **2002**, 1815.

[100] Dauben, W. G., Gerdes, J. M. and Smith, D. B., *J. Org. Chem.*, **1985**, *50*, 2576; Rimmelin, J., Jenner, G. and Rimmelin, P., *Bull. Soc. Chim. Fr.*, **1978**, *II*, 461.

[101] Kotsuki, H., Mori, Y., Ohtsuka, T., Nishizawa, H., Ochi, M. and Matsuoka, K., *Heterocycles*, **1987**, *26*, 2347.

[102] Eberbach, W., Penroud-Argüelles, M., Achenbach, H., Druckrey, E. and Prinzbach, H., *Helv. Chim. Acta*, **1971**, *54*, 2579; Weis, C. D., *J. Org. Chem.*, **1962**, *27*, 3520.

[103] Aljarilla, A., Murcia, M. C. and Plumet, J., *Synlett*, **2006**, 831.

[104] Padwa, A., Crawford, K. R., Straub, C. S., Pieniazek, S. N. and Houk, K. N., *J. Org. Chem.*, **2006**, *71*, 5432.

[105] Potts, K. T. and Walsh, E. B., *J. Org. Chem.*, **1988**, *53*, 1199.

[106] 'Synthetic applications of furan Diels-Alder chemistry', Kappe, C. O., Murphree, S. S. and Padwa, A., *Tetrahedron*, **1997**, *53*, 14179.

[107] Lautens, M. and Fillion, E., *J. Org. Chem.*, **1997**, *62*, 4418.

[108] Lautens, M. and Bouchain, G., *Org. Synth.*, **2002**, *79*, 251; Vidal-Pascual, M., Martinez-Lamenca, C. and Hoffmann, H. M. R., *Org. Synth.*, **2006**, *83*, 61.

[109] Jin, S.-j., Choi, J.-R., Oh, J., Lee, D. and Cha, J. K., *J. Am. Chem. Soc.*, **1995**, *117*, 10914; Montaña, A. M., Ribes, S., Grima, P. M., García, F., Solans, Y. and Font-Bardia, M., *Tetrahedron*, **1997**, *53*, 11669.

[110] Mann, J., Wilde, P. D. and Finch, M. W., *Tetrahedron*, **1987**, *45*, 5431.

[111] Gschwend, H. W., Hillman, M. J., Kisis, B. and Rodebaugh, R. K., *J. Org. Chem.*, **1976**, *41*, 104; Parker, K. A. and Adamchuk, M. R., *Tetrahedron Lett.*, **1978**, 1689; Harwood, L. M., Ishikawa, T., Phillips, H. and Watkin, D., *J. Chem. Soc., Chem. Commun.*, **1991**, 527.

[112] Metz, P., Meiners, U., Cramer, E., Fröhlich, R. and Wibbeling, B. W., *Chem. Commun.*, **1996**, 431; Metz, P., Seng, D., Fröhlich, R. and Wibbeling, B., *Synlett*, **1996**, 741.

[113] Choony, N., Dadabhoy, A. and Sammes, P. G., *Chem. Commun.*, **1997**, 512.

[114] 'Photo-oxidation of furans', Fering, B. C., *Recl. Trav. Chim. Pays Bas*, **1987**, *106*, 469; Gorman, A. A., Lovering, G. and Rodgers, M. A. J., *J. Am. Chem. Soc.*, **1979**, *101*, 3050; Iesce, M. R., Cermola, F., Graziano, M. L. and Scarpati, R., *Synthesis*, **1994**, 944.

[115] Patil, S. and Liu, F., *Org. Lett.*, **2007**, *9*, 195.

[116] Kernan, M. R. and Faulkner, D. J., *J. Org. Chem.*, **1988**, *53*, 2773.

[117] Lee, G. C. M., Syage, E. T., Harcourt, D. A., Holmes, J. M. and Garst, M. E., *J. Org. Chem.*, **1991**, *56*, 7007.

[118] Lee, G. C. M., Holmes, J. M., Harcourt, D. A. and Garst, M. E., *J. Org. Chem.*, **1992**, *57*, 3126.

[119] Cottier, L., Descotes, G., Eymard, L. and Rapp, K., *Synthesis*, **1995**, 303.

[120] White, J. D., Carter, J. P. and Kezar, H. S. J., *J. Org. Chem.*, **1982**, *47*, 929.

[121] Maras, A., Altay, A. and Ballini, R., *Synth. Commun.*, **2008**, *38*, 212.

[122] 'Cycloaddition reactions with vinyl heterocycles', Sepúlveda-Arques, J., Abarca-González, B. and Medio-Simón, M., *Adv. Heterocycl. Chem.*, **1995**, *63*, 339.

[123] Cornwall, P., Dell, C. P. and Knight, D. W., *J. Chem. Soc., Perkin Trans. 1*, **1993**, 2395.

[124] Avalos, L. S., Benítez, A., Muchowski, J. M., Romero, M. and Talamás, F. X., *Heterocycles*, **1997**, *45*, 1795.

[125] Craig, D., King, N. P., Kley, J. T. and Mountford, D. M., *Synthesis*, **2005**, 3279.

[126] Chen, C.-H. and Liao, C.-C., *Org. Lett.*, **2000**, *2*, 2049.

[127] Wenkert, E., Khatuya, H. and Klein, P. S., *Tetrahedron Lett*, **1999**, *40*, 5171.

[128] Caballero, A., Díaz-Requejo, M. M., Trofimenko, S., Belderrain, T. R. and Pérez, P. J., *J. Org. Chem.*, **2005**, *70*, 6101.

[129] Nakano, T., Rivas, C., Perez, C. and Tori, K., *J. Chem. Soc., Perkin Trans. 1*, **1973**, 2322; Rivas, C., Bolivar, R. A. and Cucarella, M., *J. Heterocycl. Chem.*, **1982**, *19*, 529; Zamojski, A. and Kozluk, T., *J. Org. Chem.*, **1977**, *42*, 1089; Jarosz, S. and Zamojski, A., *J. Org. Chem.*, **1979**, *44*, 3720.

[130] Yamamoto, F., Hiroyuki, M. and Oae, S., *Heterocycles*, **1975**, *3*, 1; Divald, S., Chun, M. C. and Joullié, M. M., *J. Org. Chem.*, **1976**, *41*, 2835.

[131] Wilson, W. C., *Org. Synth., Coll. Vol. I*, **1932**, 274; Boyd, M. R., Harris, T. M. and Wilson, B. J., *Synthesis*, **1971**, 545.

[132] Ferraz, J. P. and do Amaral, L., *J. Org. Chem.*, **1976**, *41*, 2350.

[133] Rinkes, I. J., *Recl. Trav. Chim. Pays-Bas*, **1930**, *49*, 1118.

[134] Hill, H. B. and White, G. R., *J. Am. Chem. Soc.*, **1902**, *27*, 193.

[135] Li, S.-W. and Batey, R. A., *Chem. Commun.*, **2007**, 3759.

[136] Fox, R. C. and Taylor, P. D., *Synth. Commun.*, **1999**, *29*, 989.

[137] 'The chemistry of 2(5*H*)-furanones', Hashem, A. and Kleinpeter, E., *Adv. Heterocycl. Chem.*, **2001**, *81*, 107.

[138] 'Natural 4-ylidenebutenolides and 4-ylidenetetronic acids', Pattenden, G., *Prog. Chem. Org. Nat. Prod.*, **1978**, *35*, 133.

[139] 'The role of heteroatomic substances in the aroma compounds of food stuffs', Ohlaff, G. and Flament, I., *Prog. Chem. Org. Nat. Prod.*, **1979**, *36*, 231.

[140] 'Recent advances in the chemistry of unsaturated lactones', Rao, Y. S., *Chem. Rev.*, **1976**, *76*, 625.

[141] Grieco, P. A., Pogonowski, C. S. and Burke, S., *J. Org. Chem.*, **1975**, *40*, 542; McMurray, J. E. and Donovan, S. F., *Tetrahedron Lett.*, **1977**, 2869.

[142] Yoshii, E., Koizumi, T., Kitatsuji, E., Kawazoe, T. and Kaneko, T., *Heterocycles*, **1976**, *4*, 1663; 'Furan-, pyrrole-, and thiophene-based siloxydienes for synthesis of densely functionalised homochiral compounds', Casiraghi, G. and Rassu, G., *Synthesis*, **1995**, 607.

[143] Jefford, C. W., Jaggi, D. and Boukouvalas, J., *Tetrahedron Lett.*, **1987**, *28*, 4037; Jefford, C. W., Sledeski, A. W. and Boukouvalas, J., *J. Chem. Soc., Chem. Commun.*, **1988**, 364; Asaoka, M., Sugimura, N. and Takei, H., *Bull. Soc. Chem. Jpn.*, **1979**, *52*, 1953; Yadav, J. S., Reddy, B. V. S., Narasimhulu, G. and Satheesh, G., *Tetrahedron Lett.*, **2008**, *49*, 5683.

[144] Jefford, C. W., Jaggi, D. and Boukouvalas, J., *J. Chem. Soc., Chem. Commun.*, **1988**, 1595.

[145] Garzelli, R., Samaritani, S. and Malanga, C., *Tetrahedron*, **2008**, *64*, 4183.

[146] Sornay, R., Neurier, J.-M. and Fournari, P., *Bull. Soc. Chim. Fr.*, **1971**, 990.

[147] Kraus, G. A. and Sugimoto, H., *J. Chem. Soc., Chem. Commun.*, **1978**, 30.

[148] Tanis, S. P. and Head, D. B., *Tetrahedron Lett.*, **1984**, *40*, 4451.

[149] D'Aleilio, G. F., Williams, C. J. and Wilson, C. L., *J. Org. Chem.*, **1960**, *25*, 1028; Sherman, E. and Dunlop, A. P., *ibid.*, 1309.

[150] Iten, P. X., Hofmann, A. A. and Eugster, C. H., *Helv. Chim. Acta*, **1978**, *61*, 430.

[151] Iten, P. X., Hofmann, A. A. and Eugster, C. H., *Helv. Chim. Acta*, **1979**, *62*, 2202.

[152] Brownbridge, P. and Chan, T.-H., *Tetrahedron Lett.*, **1980**, *21*, 3423.

[153] Niwa, E., Aoki, H., Tanake, H., Munakata, K. and Namiki, M., *Chem. Ber.*, **1966**, *99*, 3215; Cederlund, B., Lantz, R., Hörnfeldt, A.-B., Thorstad, O. and Undheim, K., *Acta Chem. Scand.*, **1977**, *B31*, 198.

[154] '2-Aminofurans and 3-aminofurans', Ramsden, C. A. and Milata, V., *Adv. Heterocycl. Chem.*, **2006**, *92*, 1.

[155] Lythgoe, D. J., McClenaghan, I. and Ramsden, C. A., *J. Heterocycl. Chem.*, **1993**, *30*, 113.

[156] Padwa, A., Brodney, M. A. and Lynch, S. M., *Org. Synth.*, **2002**, *78*, 202.

[157] 'Synthesis of amino derivatives of five-membered heterocycles by Thorpe-Ziegler cyclisation', Granik, V. G., Kadushkin, A. V. and Liebscher, J., *Adv. Heterocycl. Chem.*, **1998**, *72*, 79.

[158] Padwa, A., Dimitroff, M., Waterson, A. G. and Wu, T., *J. Org. Chem.*, **1997**, *62*, 4088.

[159] Adams, R. and Voorhees, V., *Org. Synth., Coll. Vol. I*, **1932**, 280.

[160] Bonner, W. A. and Roth, M. R., *J. Am. Chem. Soc.*, **1959**, *81*, 5454; Moye, C. J. and Krzeminski, Z. A., *Austr. J. Chem.*, **1963**, *16*, 258; Feather, M. S., Harris, D. W. and Nichols, S. B., *J. Org. Chem.*, **1972**, *37*, 1606.

[161] Babu, B. S. and Balasubramanian, K. K., *J. Org. Chem.*, **2000**, *65*, 4198.

[162] Nowlin, G., *J. Am. Chem. Soc.*, **1950**, *72*, 5754; Traylelis, V. J., Hergennother, W. L., Hanson, H. T. and Valicenti, J. A., *J. Org. Chem.*, **1964**, *29*, 123; Scott, L. T. and Naples, J. O., *Synthesis*, **1973**, 209.

[163] Amarnath, V. and Amarnath, K., *J. Org. Chem.*, **1995**, *60*, 301.

[164] Hegedus, L. S. and Perry, R. J., *J. Org. Chem.*, **1985**, *50*, 4955; Mackay, D., Neeland, E. G. and Taylor, N. J., *ibid.*, **1986**, *51*, 2351.

[165] Kornfeld, E. C. and Jones, R. G., *J. Org. Chem.*, **1954**, *19*, 1671.

[166] Pridmore, S. J., Slatford, P. A. and Williams, J. M. J., *Tetrahedron Lett.*, **2007**, *48*, 5111.

[167] Rao, H. S. P. and Jothilingam, S., *J. Org. Chem.*, **2003**, *68*, 5392.

[168] Padwa, A., Rashatasakhon, P. and Rose, M., *J. Org. Chem.*, **2003**, *68*, 5139.

[169] Ranu, B. C., Adak, L. and Banerjee, S., *Tetrahedron Lett.*, **2008**, *49*, 4613.

[170] Dann, O., Distler, H. and Merkel, H., *Chem. Ber.*, **1952**, *85*, 457.

[171] Clauson-Kaas, N., *Acta Chem. Scand.*, **1961**, *15*, 1177.

[172] Seyferth, H. E., *Chem. Ber.*, **1968**, *101*, 619.

[173] Cormier, R. A., Grosshans, C. A. and Skibbe, S. L., *Synth. Commun.*, **1988**, *18*, 677.

[174] Cormier, R. A. and Francis, M. D., *Synth. Commun.*, **1981**, *11*, 365.

[175] Díaz-Cortés, R., Silva, A. L. and Maldonado, L. A., *Tetrahedron Lett.*, **1997**, *38*, 2207.

[176] Cornforth, J. W., *J. Chem. Soc.*, **1958**, 1310; Burness, D. M., *Org. Synth., Coll. Vol. IV*, **1963**, 649; Kotake, H., Inomata, K., Kinoshita, H., Aoyama, S. and Sakamoto, Y., *Heterocycles*, **1978**, *10*, 105.

[177] Garst, M. E. and Spencer, T. A., *J. Am. Chem. Soc.*, **1973**, *95*, 250.

[178] Miller, D., *J. Chem. Soc., C*, **1969**, 12; Schreurs, P. H. M., de Jong, A. J. and Brandsma, L., *Recl. Trav. Chim. Pays-Bas*, **1976**, *95*, 75.

[179] Marshall, J. E. and DuBay, W. J., *J. Org. Chem.*, **1991**, *56*, 1685.

[180] Arcadi, A. and Rossi, E., *Tetrahedron*, **1998**, *54*, 15253.

[181] MaGee, D. I. and Leach, J. D., *Tetrahedron Lett.*, **1997**, *38*, 8129.

[182] Imagawa, H., Kurisaki, T. and Nishizawa, M., *Org. Lett.*, **2004**, *6*, 3679.

[183] Sniady, A., Durham, A., Morreale, M. S., Wheeler, K. A. and Dembinski, R., *Org. Lett.*, **2007**, *9*, 1175.

[184] Sniady, A., Wheeler, K. A. and Dembinski, R., *Org. Lett.*, **2005**, *7*, 1769.

[185] Couffignal, R., *Synthesis*, **1978**, 581.

[186] e.g. Liu, Y., Song, F., Song, Z., Liu, M. and Yan, B., *Org. Lett.*, **2005**, *7*, 5409; Zhou, C.-Y., Chan, P. W. H. and Che, C.-M., *ibid.*, **2006**, *8*, 325; Zhang, C.-Y. and Schmalz, H.-G., *Angew. Chem. Int. Ed.*, **2006**, *45*, 6704.

[187] Marshall, J. A. and Wang, X., *J. Org. Chem.*, **1991**, *56*, 960.

[188] Sheng, H., Lin, S. and Huang, Y., *Tetrahedron Lett.*, **1986**, *27*, 4893.

[189] Kel'in, A. V. and Gevorgyan, V., *J. Org. Chem.*, **2002**, *67*, 95.

[190] Sheng, H., Lin, S. and Huang, Y., *Synthesis*, **1987**, 1022.

[191] Zhou, C.-Y., Chan, P. W. H. and Che, C.-M., *Org. Lett.*, **2006**, *8*, 325.

[192] Danheiser, R. L., Stoner, E. J., Koyama, H. and Yamashita, D. S., *J. Am. Chem. Soc.*, **1989**, *111*, 4407.

[193] Ma, J., Wang, S. H. and Tian, G. R., *Synth. Commun.*, **2006**, *36*, 1229.

[194] Shamshina, J. L. and Snowden, T. S., *Tetrahedron Lett.*, **2007**, *48*, 3767.

[195] Bew, S. P. and Knight, D. W., *Chem. Commun.*, **1996**, 1007.

[196] Garst, M. E. and Spencer, T. A., *J. Org. Chem.*, **1974**, *39*, 584.

[197] Redman, A. M., Dumas, J. and Scott, W. J., *Org. Lett.*, **2000**, *2*, 2061.

[198] Yang, Y. and Wong, H. N. C., *Tetrahedron*, **1994**, *50*, 9583.

[199] Song, Z. Z., Ho, M. S. and Wong, H. N. C., *J. Org. Chem.*, **1994**, *59*, 3917.

[200] Timko, J. M., Moore, S. S., Walba, D. M., Hiberty, P. C. and Cram, D. J., *J. Am. Chem. Soc.*, **1977**, *99*, 4207.

[201] Büchi, G., Demole, E. and Thomas, A. F., *J. Org. Chem.*, **1973**, *78*, 123.

19

Typical Reactivity of Indoles, Benzo[*b*] thiophenes, Benzo[*b*]furans, Isoindoles, Benzo[*c*]thiophenes and Isobenzofurans

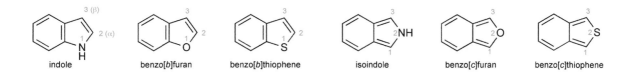

The fusion of a benzene ring to the 2,3-positions of a pyrrole generates one of the most important heterocyclic ring systems – indole. This chapter develops a description of the chemistry of indole, then discusses modifications necessary to rationalise the chemistry of the benzo[*b*]furan and benzo[*b*]thiophene analogues. Finally, the trio of heterocycles in which the benzene ring is fused at the five-membered ring 3,4-positions, isoindole, benzo[*c*]furan and benzo[*c*]thiophene are considered.

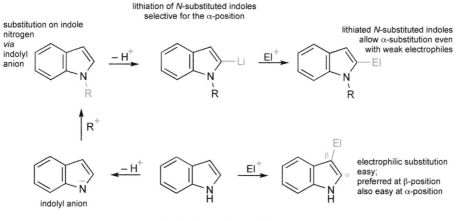

Typical reactions of indole

The chemistry of indoles is dominated by easy electrophilic substitution. Of the two rings, the heterocyclic ring is very electron-rich, by comparison with a benzene ring, so attack by electrophiles always takes place in the five-membered ring, except in special circumstances. Of the three positions on the heterocyclic ring, attack at nitrogen would destroy the aromaticity of the five-membered ring, and produce a localised cation, and so does not occur; both of the remaining positions can be readily attacked by electrophiles, leading to C-substituted products, but the β-position is preferred by a considerable margin. This contrasts

Heterocyclic Chemistry 5th Edition   John Joule and Keith Mills
© 2010 Blackwell Publishing Ltd

with the regiochemistry shown by pyrrole, but again can be well rationalised by a consideration of the Wheland intermediates for the two alternative sites of attack.

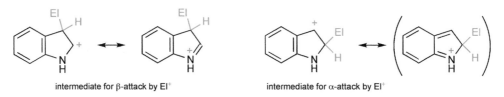

intermediate for β-attack by El⁺   intermediate for α-attack by El⁺

Intermediates for electrophilic substitution of indole

The intermediate for α-attack is stabilised – it is a benzylic cation – but it cannot derive assistance from the nitrogen without disrupting the benzenoid resonance (a resonance contributor, which makes a limited contribution, is shown in parenthesis). The more stable intermediate from β-attack, has charge located adjacent to nitrogen and able to derive the very considerable stabilisation attendant upon interaction with the nitrogen lone pair of electrons.

The facility with which indoles undergo substitution can be illustrated using the Mannich reaction: the electrophilic species in such reactions (C=N⁺R₂) is generally considered to be a 'weak' electrophile, yet substitution occurs easily under mild conditions.

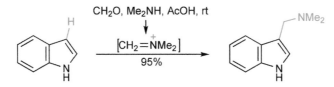

An example of easy β-electrophilic substitution of indole with a weak electrophile

There is a strong preference for attack at C-3, even when that position carries a substituent, and it is therefore important to consider, in detail, the 2-substitution of 3-substituted indoles. This could proceed in three ways: (i) initial attack at a 3-position followed by 1,2-migration to the 2-position; (ii) initial attack at the 3-position followed by reversal (when possible), then (iii); or (iii) direct attack at the 2-position. It has been definitely demonstrated, in the case of some irreversible substitutions, that the migration route operates, but equally it has been demonstrated that direct attack at an α-position can occur.

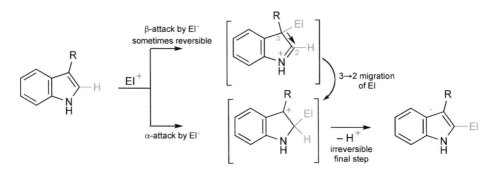

Possible mechanisms for the 2-substitution of 3-substituted indoles

Indoles react with strong bases losing the *N*-hydrogen and forming indolyl anions. When the counter ion is an alkali metal, these salts have considerable ionic character and react with electrophiles at the nitrogen, affording a practical route for *N*-alkylation (or acylation), for example in the synthesis of reversibly *N*-blocked indoles. These can be regioselectively lithiated at C-2, where the acidifying effect of the electronegative heteroatom is felt most strongly, often with additional chelation assistance from the *N*-substituent, thus providing a route to 2-substituted indoles.

As in all heterocyclic chemistry, the advent of palladium(0)-catalysed processes (see Section 4.2 for a detailed discussion) has revolutionised the manipulation of indoles, benzo[*b*]furans and benzo[*b*] thiophenes; the example below is typical.

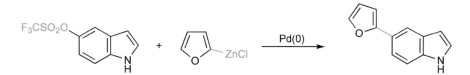

Palladium(0)-catalysed processes are very important in indole chemistry

The ready electron availability of indoles means that they are rather easily (aut)oxidised in the five-membered ring. Reductions can be made selective for either ring: in acid solution, dissolving metals attack the hetero-ring, and the benzenoid ring can be selectively reduced by Birch reduction conditions.

Indoles that carry leaving groups at benzylic positions, especially at C-3, undergo displacement processes extremely easily, encouraged by stabilisation of positive charge by the nitrogen or, alternatively, in basic conditions, by deprotonation of the indole hydrogen. One example of the latter is the lithium aluminium hydride reduction of 3-acyl-indoles that produces 3-alkyl-indoles. In a sense, the 3-ketones are behaving like vinylogous amides, and reduction intermediates are able to lose oxygen to give species that, on addition of a second hydride, produce the indolyl anion of the 3-alkyl-indole, converted into the indole during aqueous work-up.

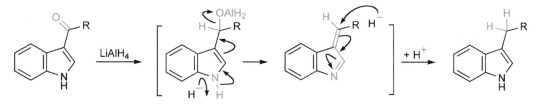

Reduction of 3-acylindoles to 3-alkylindoles

In comparison with indoles, benzo[*b*]furans and benzo[*b*]thiophenes have been studied much less fully, however similarities and some differences can be noted. Benzo[*b*]furans and benzo[*b*]thiophenes undergo electrophilic substitution, but the 3-regioselectivity is much lower than for indole, even to the extent that some attack takes place in the benzene ring of benzo[*b*]thiophene and that 2-substitution is favoured for benzo[*b*]furan. These changes are consequent upon the much poorer electron-donating ability of oxygen and sulfur – the nitrogen of indole is able to make a much bigger contribution to stabilising intermediates, particularly, as was shown above, for β-attack, and consequently to have a larger influence on regioselectivity. In the case of benzo[*b*]furan, it appears that simple benzylic resonance stabilisation in an intermediate from attack at C-2 outweighs the assistance that oxygen might provide to stabilise an adjacent positive charge.

Benzo[*b*]furans and benzo[*b*]thiophenes undergo lithiation at their 2-positions, consistent with the behaviour of furans, thiophenes, and of *N*-blocked pyrroles and indoles.

The chemical behaviour of isoindole, benzo[*c*]furan and benzo[*c*]thiophene is dominated by their lack of a 'complete' benzene ring: these three heterocycles undergo cycloaddition processes across the 1- and 3-positions with great facility, because the products do now have a regular benzene ring. Often, no attempt is made to isolate examples of these heterocycles, but they are simply generated in the presence of the dienophile with which it is desired that they react. As a result of this strong tendency, few of the classical electrophilic and nucleophilic processes have been much studied.

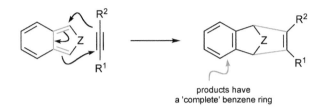

products have
a 'complete' benzene ring

Typical cycloaddition behaviour of isoindoles, benzo[*c*]furans and benzo[*c*]thiophenes

There has probably been more work carried out on the synthesis of indoles than on any other single heterocyclic system and consequently many routes are available; ring syntheses of benzo[*b*]furans and benzo[*b*]thiophenes have been much less studied. The Fischer indole synthesis, now more than 100 years old, is still widely used – an arylhydrazone is heated with an acid, a multi-step sequence ensues, ammonia is lost and an indole is formed.

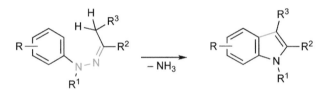

The Fischer indole synthesis

As an illustration of a modern and efficient route, 2,3-unsubstituted indoles are obtained from an *ortho*-nitrotoluene by heating with dimethylformamide dimethylacetal (DMFDMA), generating an enamine that, after reduction of the nitro group, closes with loss of dimethylamine, generating the aromatic heterocycle.

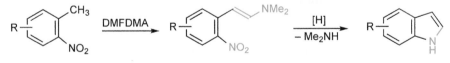

The Leimgruber–Batcho indole synthesis

Both benzo[*b*]furans and benzo[*b*]thiophenes can be obtained from the phenol or thiophenol respectively, by *O-/S*-alkylation with a bromoacetaldehyde acetal and then acid-catalysed ring closure involving intra-molecular electrophilic attack on the ring.

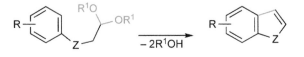

20

Indoles: Reactions and Synthesis

indole
[1*H*-indole]

indolenine
[3*H*-indole]

Indole[1] and the simple alkyl-indoles are colourless crystalline solids with a range of odours from naphthalene-like, in the case of indole itself, to faecal, in the case of skatole (3-methylindole). Many simple indoles are available commercially and all of these are produced by synthesis: indole, for example, is made by the high-temperature vapour-phase cyclising dehydrogenation of 2-ethylaniline. Most indoles are quite stable in air with the exception of those which carry a simple alkyl group at C-2: 2-methylindole autoxidises easily, even in a dark brown bottle.

The word indole is derived from the word India: a blue dye imported from India was known as indigo in the sixteenth century. Chemical degradation of the dye gave rise to oxygenated indoles (see 20.13), which were named indoxyl and oxindole; indole itself was first prepared in 1866 by zinc-dust distillation of oxindole.

For all practical purposes, indole exists entirely in the 1*H*-form, 3*H*-indole (indolenine) being present to the extent of only ca. 1 ppm. 3*H*-Indole can be generated in solution but tautomerises to 1*H*-indole within about 100 seconds at room temperature.[2]

20.1 Reactions with Electrophilic Reagents
20.1.1 Substitution at Carbon
20.1.1.1 Protonation
Indoles, like pyrroles, are very weak bases: typical pK_{aH} values are: indole, −3.5; 3-methylindole, −4.6; 2-methylindole. −0.3.[3] This means, for example, that in 6M sulfuric acid, two molecules of indole are protonated for every one unprotonated, whereas 2-methylindole is almost completely protonated under the same conditions. By NMR and UV examination, only the 3-protonated cation (3*H*-indolium cation) is detectable;[4] it is the thermodynamically stablest cation, retaining full benzene aromaticity (in contrast to the 2-protonated cation) with delocalisation of charge over the nitrogen and α-carbon. The spectroscopically undetectable *N*-protonated cation must be formed, and formed very rapidly, for acid-catalysed deuterium exchange at nitrogen is 400 times faster than at C-3,[5] indeed the *N*-hydrogen exchanges rapidly even at pH 7, when no exchange at C-3 occurs: clean conversion of indole into 3-deuterioindole can be achieved by successive deuterio-acid then water treatments.[6]

1*H*-indolium cation
(formed fastest)

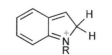

2*H*-indolium cation

3*H*-indolium cation
(stablest)

2-methyl stabilises cation

Heterocyclic Chemistry 5th Edition John Joule and Keith Mills
© 2010 Blackwell Publishing Ltd

That 2-methylindole is a stronger base than indole can be understood on the basis of stabilisation of the cation by electron release from the methyl group; 3-methylindole is a somewhat weaker base than indole.

Reactions of β-Protonated Indoles *(see also 20.1.1.9, 20.2 and 20.7)*

$3H$-Indolium cations are of course electrophilic species, in direct contrast with neutral indoles, and under favourable conditions will react as such. For example, the $3H$-indolium cation itself will add bisulfite at pH 4, under conditions that lead to the crystallisation of the product, the sodium salt of indoline-2-sulfonic acid (indoline is the widely used, trivial name for 2,3-dihydroindole). The salt reverts to indole on dissolution in water, however it can be N-acetylated and the resulting acetamide used for halogenation or nitration at C-5, final hydrolysis with loss of bisulfite affording the 5-substituted indole.[7]

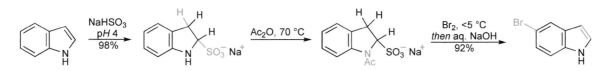

When N_b-acyl-tryptophans are exposed to strong acid, the indolium cation is trapped by cyclisation involving the side-chain nitrogen.[8] Comparable tricycles result from phenylselenylation of protected tryptophan[9] or reaction with 4-methyl-1,2,4-triazoline-3,5-dione,[10] or dimethyl(succinimido)sulfonium chloride (a CH_2SMe group ends up at the indole C-3).[11] If N-bromosuccinimide is employed, the initially formed 3-bromo-tricycle loses hydrogen bromide to produce an aromatic indole.[12]

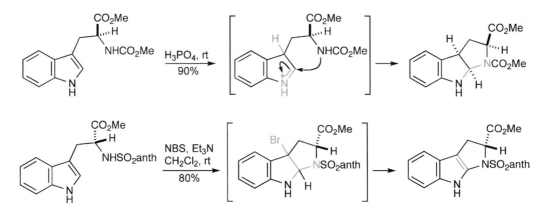

20.1.1.2 Nitration; Reactions with Other Nitrogen Electrophiles

Indole itself can be nitrated using benzoyl nitrate as a non-acidic nitrating agent; the usual mixed acid nitrating mixture leads to intractable products, probably because of acid-catalysed polymerisation. This can be avoided by carrying out the nitration using concentrated nitric acid and acetic anhydride at low temperature – under these conditions, N-alkylindoles, and indoles carrying electron-withdrawing N-substituents, but *not* indole itself, can be satisfactorily nitrated.[13]

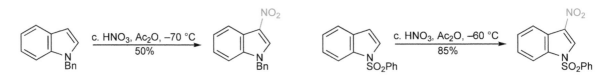

2-Methylindole gives a 3-nitro derivative with benzoyl nitrate,[14] but can also be nitrated successfully with concentrated nitric/sulfuric acids, but with attack at C-5. The absence of attack on the heterocyclic ring is explained by the complete protonation of 2-methylindole under these conditions; the regioselectivity of attack, *para* to the nitrogen, may mean that the actual moiety attacked is a hydrogensulfate adduct of the initial 3*H*-indolium cation, as shown in the scheme. 5-Nitration of 3*H*-indolium cations has been independently demonstrated using a 3,3-disubstituted 3*H*-indolium cation.[15] With an acetyl group at C-3, nitration with nitronium tetrafluoroborate in the presence of tin(IV) chloride takes place at either C-5 or C-6 depending on the temperature of reaction.[16]

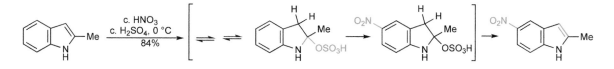

Indoles readily undergo electrophilic amination with bis(2,2,2-trichloroethyl) azodicarboxylate, the resulting acylated hydrazine being cleaved by zinc dust to give a 3-acetylamino-indole.[17]

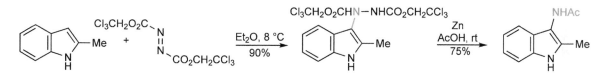

20.1.1.3 Sulfonation; Reactions with Other Sulfur Electrophiles

Sulfonation of indole,[18] at C-3, is achieved using the pyridine–sulfur trioxide complex in hot pyridine. Gramine is sulfonated in oleum to give 5- and 6-sulfonic acids, attack being on a diprotonated (C-3, side-chain-N) salt.[19] 1-Phenylsulfonylindoles are efficiently converted into 3-chlorosulfonyl-derivatives using chlorosulfonic acid at room temperature.[20] Sulfenylation of indole also occurs readily, at C-3, using a pre-formed sulfenyl chloride[21] or *N*-thioalkyl- or *N*-thioaryl-phthalimides with magnesium bromide,[22] or thiols activated with *N*-chlorosuccinimide[23] or Selectfluor™.[24] Thiocyanation of indole can be achieved in virtually quantitative yield with a combination of ammonium thiocyanate with cerium(IV) ammonium nitrate[25] or with iron(III) chloride.[26]

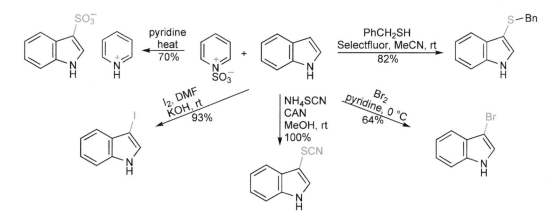

20.1.1.4 Halogenation

3-Halo-, and even more so, 2-halo-indoles are unstable and must be utilised as soon as they are prepared; *N*-acyl- or *N*-sulfonyl-haloindoles are much more stable. A variety of methods are available for the β-halogenation of indoles: bromine or iodine (the latter with potassium hydroxide) in dimethylformamide[27a] give very high yields; pyridinium tribromide[27b] works efficiently; iodination[27c] and chlorination[27d] tend to be carried out in alkaline solution or involve a preformed indolyl anion,[28] and general halogenation with copper(II) halides.[29] Initial *N*-chlorination, then rearrangement may be involved in some cases.

Reaction of 3-substituted indoles with halogens can be more complex; initial 3-halogenation occurs generating a 3-halo-3*H*-indole,[30] but the actual products obtained then depend upon the reaction conditions, solvent etc. Thus, nucleophiles can add at C-2 in the intermediate 3-halo-3*H*-indoles when, after loss of hydrogen halide, a 2-substituted indole is obtained as final product, for example in aqueous solvents, water addition produces oxindoles (20.13.1); comparable methanol addition gives 2-methoxyindoles. 2-Bromination of 3-substituted indoles can be carried out using *N*-bromosuccinimide in the absence of radical initiators.[31] 2-Bromo- and 2-iodo-indoles can be prepared very efficiently *via* α-lithiation (20.5.1).[32] 2-Halo-indoles are also available from the reaction of oxindoles with phosphoryl halides.[33] Some 2,3-diiodo-indoles can be obtained by iodination of the indol-2-ylcarboxylic acid.[34]

20.1.1.5 Acylation

Indole only reacts at an appreciable rate with acetic anhydride, alone, above 140 °C, giving 1,3-diacetylindole predominantly, together with smaller amounts of *N*- and 3-acetylindoles; 3-acetylindole is prepared by alkaline hydrolysis of product mixtures.[35] That β-attack occurs first is shown by the resistance of 1-acetylindole to *C*-acetylation, but the easy conversion of the 3-acetylindole into 1,3-diacetylindole. In contrast, acetylation in the presence of sodium acetate, or 4-dimethylaminopyridine,[36] affords exclusively *N*-acetylindole, probably *via* the indolyl anion (20.4.1). *N*-Acyl-indoles are much more readily hydrolysed than ordinary amides, aqueous sodium hydroxide at room temperature being sufficient. This lability is due in part to a much weaker mesomeric interaction of the nitrogen and carbonyl groups, making the latter more electrophilic, and in part to the relative stability of the indolyl anion, which makes it a better leaving group than amide anion. Trifluoroacetic anhydride, being much more reactive, acylates indole at room temperature, at C-3 in dimethylformamide (but at nitrogen in dichloromethane).[37]

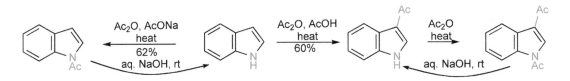

The use of a Lewis acid to catalyse Friedel–Crafts 3-acylation must be carried out with care, to avoid oligomerisation: the method involves adding tin(IV) chloride to the indole *first*, then adding the acid chloride or anhydride.[38]

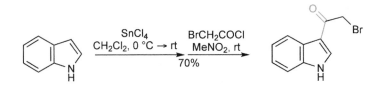

Simply heating indole with triethyl orthoformate at 160 °C leads to the alkylation of the indole nitrogen, introducing a diethoxymethyl group that can be used as a reversible *N*-blocking substituent – it allows 2-lithiation (cf. 20.5.1) and can be easily removed with dilute acid at room temperature.[39]

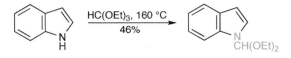

The Vilsmeier reaction is a very efficient method for the preparation of 3-formyl-indoles,[40] and for other 3-acyl-indoles using tertiary amides of other acids in place of dimethylformamide.[41] Even indoles carrying an electron-withdrawing group at the 2-position, for example ethyl indole-2-carboxylate, undergo smooth Vilsmeier 3-formylation.[42]

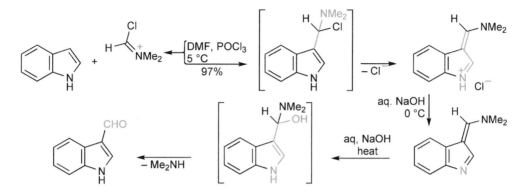

Isocyanides attack under the influence of aluminium chloride, thus introducing an imine unit directly.[43]

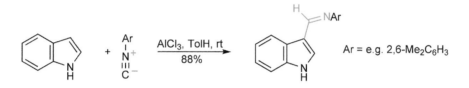

A particularly useful and high-yielding reaction is that between indoles and oxalyl chloride, which gives ketone-acid-chlorides convertible into a range of compounds, for example tryptamines; a synthesis of serotonin utilised this reaction.[44]

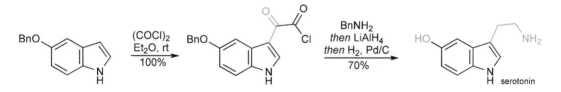

Acylation of 3-substituted indoles is more difficult, however 2-acetylation can be effected with the aid of boron trifluoride catalysis.[44] Indoles, with a carboxyl-containing side-chain acid at C-3, undergo intra-molecular acylation forming cyclic 2-acylindoles.[45] Intramolecular Vilsmeier processes, using tryptamine amides, have been used extensively for the synthesis of 3,4-dihydro-β-carbolines, a sub-structure found in many indole alkaloids (β-carboline is the widely used, trivial name for pyrido[3,4-b]indole). Note that it is the imine, rather than a ketone, that is the final product; the cyclic nature of the imine favours its reten-tion rather than hydrolysis to amine plus ketone as in the standard Vilsmeier sequence;[46] this ring closure is analogous to the Bischler–Napieralski synthesis of 3,4-dihydro-isoquinolines (9.15.1.7).

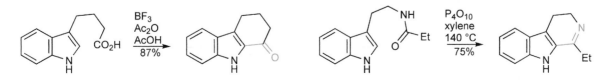

2-Acetylation of indol-3-ylacetic acid leads, *in situ*, to an enol-lactone: an indole fused to a 2-pyrone. This can be hydrolysed to the keto-acid, or the diene character of the 2-pyrone (11.2.2.4) can be utilised, as illustrated.[47]

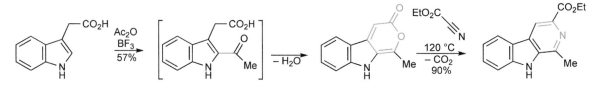

Deactivation of the pyrrole ring by electron-withdrawing substituents allows acylation in the six-membered ring. Lewis-acid-catalysed acylation of 3-trifluoroacetylindole takes place at C-5, and if such products are hydrolysed (a haloform reaction) to the 3-acids, decarboxylation then produces 5-acyl-indoles.[48]

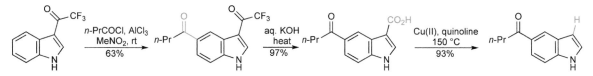

1-Pivaloylindole gives high yields of 6-substituted ketones on reaction with α-halo-acid-chlorides and aluminium chloride; simple acid chlorides react only at C-3.[49] The sequence below shows how a 1-pivaloyl-3-(indol-3-yl)propanoic acid undergoes Friedel–Crafts cyclisation to C-4, away from the deactivated heterocyclic ring.[50]

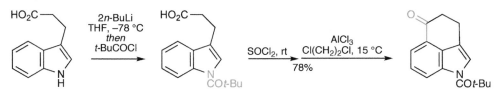

Acylation of 1-acetylindole in the presence of aluminium chloride goes cleanly at C-6, but with diacid chlorides (malonyl, succinyl), 2-substitution occurs. The former result is due to strong deactivation of C-5 by a 1-acetyl–Lewis-acid complex and the latter is probably due to complexation of one acid chloride to the 1-acetyl group, followed by intramolecular delivery of the other.[51]

20.1.1.6 Alkylation[52]

Indoles do not react with alkyl halides at room temperature. Indole itself begins to react with iodomethane in dimethylformamide at about 80 °C, when the main product is skatole. As the temperature is raised, further methylation occurs until eventually 1,2,3,3-tetramethyl-3*H*-indolium iodide is formed.

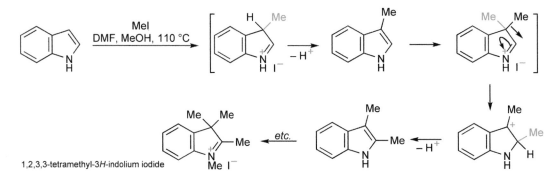

1,2,3,3-tetramethyl-3*H*-indolium iodide

The rearrangement of 3,3-dialkyl-3*H*-indolium ions by alkyl migration to give 2,3-dialkyl-indoles, as shown in the sequence above, is related mechanistically to the Wagner–Meerwein rearrangement, and is known as the Plancher rearrangement.[53] It is likely that most instances of 2-alkylation of 3-substituted-indoles by cationic reagents proceed by this route, and this was neatly verified in the formation of 1,2,3,4-tetrahydrocarbazole by boron-trifluoride-catalysed cyclisation of 4-(indol-3-yl)butan-1-ol. The experiment was conducted with material labelled at the benzylic carbon. The consequence of the rearrangement of the symmetrical spirocyclic intermediate, which results from attack at C-3, was the equal distribution of the label betweeen the C-1 and C-4 carbons of the product.[54] It is important to note that other experiments demonstrate that direct attack at C-2 can and does occur,[55] especially when this position is further activated by a 6-methoxyl group.[56]

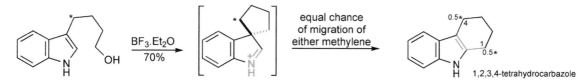

In another elegant experiment, the intervention of a 3,3-disubstituted 3*H*-indolium-intermediate in an overall α-substitution was proved by cyclisation of the mesylate of an optically active alcohol to give an optically *inactive* product, *via* an achiral, spirocyclic intermediate, from initial attack at the β-position.[57]

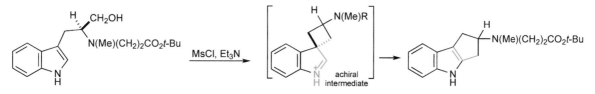

3-(2-Bromoethyl)indole, on treatment with potassium carbonate in refluxing acetonitrile, gives a stable spirocyclopropyl-indolenine, however this cyclisation is very slow when sodium bicarbonate is used as base and this allows efficient *N*-alkylation of piperidines with the bromide.[58] Nucleophiles, such as organo-lithiums and enolates, add to the indolenine without disrupting the cyclopropane ring.[59,60]

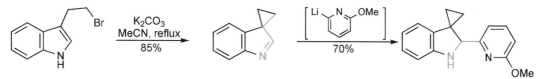

Reaction at C-3, with more electrophilic reagents, leading to allylated, benzylated and propargylated indoles, can be achieved under various mild conditions. All the following alkylate indoles at C-3, at room temperature: allyl halides with zinc triflate,[61] allyl and benzyl halides in aqueous acetone (with also ca. 20% attack at C-2),[62] allyl alcohols under ruthenium(IV) catalysis,[63] allylic and propargylic acetates with iodine[64] and propargyl tertiary alcohols with *p*-toluenesulfonic acid.[65] 1-Bromo-2-benzoylethyne directly alkynylates indole at room temperature on alumina.[66]

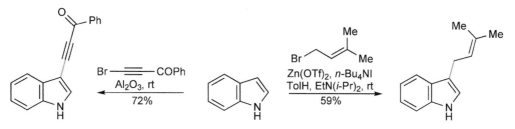

Tryptophans can be obtained directly from indoles by reaction of serine with the indole, in the presence of acetic anhydride.[67]

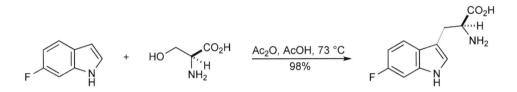

Indoles react with epoxides and aziridines in the presence of Lewis acids (see 20.4.1 for reaction of indolyl anions with such reactants) with opening of the three-membered ring and consequent 3-(2-hydroxyethylation) and 3-(2-aminoethylation) of the heterocycle. Both ytterbium triflate and phenylboronic acid are good catalysts for reaction with epoxides under high pressure;[68] silica gel is also an effective catalyst, but reactions are slow at normal pressure and temperature.[69] Reaction with aziridines can be catalysed by zinc triflate or boron trifluoride.[70]

Indoles react with homochiral aryloxiranes at the benzylic carbon in high optical yield, under very mild conditions (1% InBr₃, CH₂Cl₂, rt).[71] Reactions with *N*-Cbz aziridines are similarly catalysed by scandium triflate.[72]

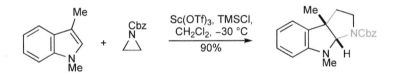

20.1.1.7 Reactions with α,β-Unsaturated Ketones, Nitriles and Nitro-Compounds

Such reactions are usually effected using acid (see below), or one of a number of mild Lewis acids, such as scandium iodide (with microwave heating),[73] indium bromide[74] or hafnium triflate,[75] and can be looked on as an extension of the reactions discussed in 20.1.1.6. In the simplest situation, indole reacts with methyl vinyl ketone in a conjugate fashion in acetic acid/acetic anhydride.[76]

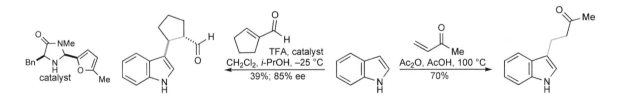

Analogous alkylations with unsaturated ketones can also be effected with silica-supported benzenesulfonic acid sodium salt[77] or, with some stereoselectivity, using a chiral imidazolidinone organo-catalyst.[78] Optical induction can also be achieved in the addition of indole to alkylidene malonates using bisoxazoline copper(II) complexes.[79]

The use of Montmorillonite clay[80] or ytterbium triflate,[81] allows α-alkylation of β-substituted indoles. This contrasts with the different, but very instructive, reaction pathway followed when mesityl oxide and 1,3-dimethylindole are combined in the presence of sulfuric acid – electrophilic attack at the already substituted β-position is followed by intramolecular nucleophilic addition of the enol of the side-chain ketone, to C-2.[82]

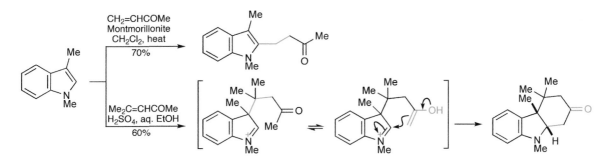

An extension of this methodology allows the synthesis of tryptophans by aluminium-chloride-catalysed alkylation with an iminoacrylate.[83]

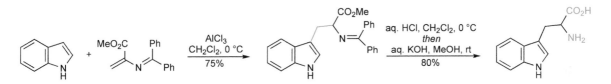

Nitroethene is sufficiently electrophilic to substitute indole without the need for acid catalysis.[84] Despite this, it has been shown that silica-gel-supported $CeCl_3.7H_2O/NaI$ brings about such reactions at room temperature under solvent-free conditions[85] or, to take another solvent extreme, the reaction occurs in water with a catalytic amount of a 'heteropoly acid' ($H_3PW_{12}O_{40}$).[86] The employment of 2-dimethylamino-1-nitroethene in trifluoroacetic acid leads to 2-(indol-3-yl)nitroethene – the reactive species is the protonated enamine and the process is similar to a Mannich condensation (20.1.1.9).[87] The use of 3-trimethylsilyl-indoles, with *ipso*-substitution of the silicon,[88] is an alternative means for effecting alkylation, avoiding the need for acid catalysis.

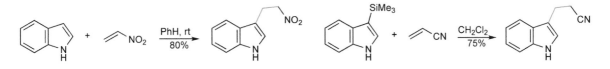

The formation of a nitroethene electrophile *in situ*, is believed to be involved in the reaction between an indole, paraformaldehyde and ethyl nitroacetate, giving precursors for tryptophans.[89]

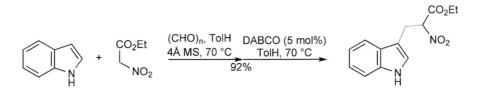

20.1.1.8 Reactions with Aldehydes and Ketones

Indoles react with aldehydes and ketones under acid catalysis – with simple carbonyl compounds, the initial products, indol-3-yl-carbinols are never isolated, for in the acidic conditions they dehydrate to 3-alkylidene-3*H*-indolium cations; those from aromatic aldehydes have been isolated in some cases;[90] reaction of 2-methylindole with acetone under anhydrous conditions gives the simplest isolable salt of this class.[91] Reaction with 4-dimethylaminobenzaldehyde (the Ehrlich reaction, see 16.1.1.7) gives a mesomeric and highly coloured cation. Only where dehydration is inhibited have 3-(hydroxyalkyl)-indoles been isolated, for example from reaction with diethyl mesoxalate[92] or ethyl glyoxylate.[93]

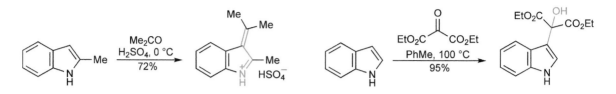

3-Alkylidene-3*H*-indolium cations are themselves electrophiles and can react with more of the indole, as illustrated for reaction with formaldehyde.[94]

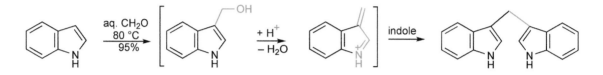

The introduction of a sugar moiety to the indole 3-position proceeds best with 2-substituted indoles; indium(III) chloride is used in combination with a glycosyl bromide.[95]

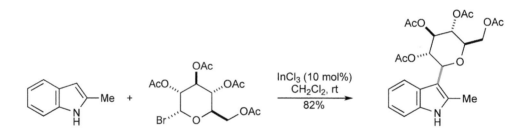

3-Alkylation of 2-alkyl- and 2-aryl-indoles can be achieved by trifluoracetic-acid-catalysed condensation with either aromatic aldehydes or aliphatic ketones in the presence of the triethylsilane, which reduces the intermediate 3-alkylidene-3*H*-indolium cations.[96]

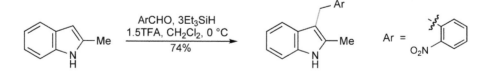

20.1.1.9   Reactions with Iminium Ions: Mannich Reactions[97]

Under neutral conditions and at 0 °C, indole reacts with a mixture of formaldehyde and dimethylamine by substitution at the indole nitrogen.[98] This *N*-substitution may involve a low equilibrium concentration of the indolyl anion (20.4.1) or may be the result of reversible kinetic attack followed by loss of proton. In neutral solution at higher temperature or in acetic acid, conversion into the thermodynamically more stable 3-dimethylaminomethylindole, gramine, takes place. Gramine is formed directly, smoothly and in high yield, by reaction in acetic acid.[99] The Mannich reaction is useful in synthesis because not only can the electrophilic iminium ion be varied widely, but also the product gramines are themselves intermediates for further manipulation (20.11).

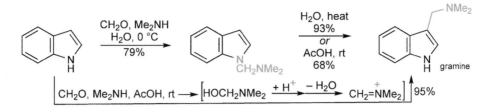

The iminium ion electrophile can also be prepared separately, as a crystalline solid known as 'Eschenmoser's salt' ($Me_2N^+=CH_2$ I^-)[100] and, with this, the reaction is normally carried out in a non-polar solvent. Examples that illustrate the variation in iminium ion structure that can be tolerated include the reaction of indole with quinolines, catalysed by indium(III) chloride,[101] with benzylidene derivatives of arylamines, catalysed by lanthanide triflates,[102] with ethyl glyoxylate imines[103] (no catalyst required) and with dihydro-1,4-oxazin-2-ones.[104]

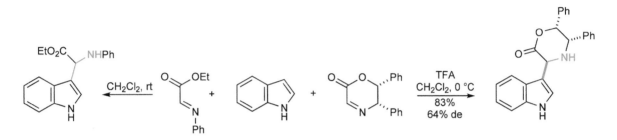

A related, and possibly more versatile, process can be carried out using an aldehyde and an arylsulfinic acid; the resulting sulfone can be displaced by a range of nucleophiles.[105]

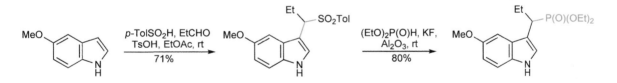

In the mineral-acid-catalysed 'dimerisation' of indole,[106] the indole is attacked by *protonated* indole, i.e. the iminium ion is protonated indole. In all manipulations of indoles it is necessary to be aware of their sensitivity to acidic conditions.

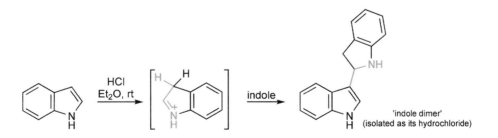

When protonated 3-bromoindole is employed as electrophile, a final elimination of hydrogen bromide gives rise to re-aromatised 2-substituted indoles; pyrrole (illustrated) or indoles will take part in this type of process.[107]

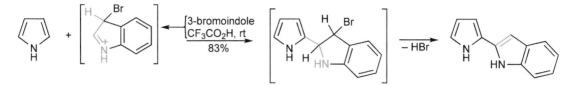

Conducted in an intramolecular sense, Mannich reactions have been much used for the construction of tetrahydro-β-carbolines.[108] Tryptamines carrying a 2-carboxylic acid group, which can be conveniently prepared (20.16.6.3), but are not easily decarboxylated as such, undergo cyclising Mannich condensation with aldehydes and ketones, with loss of the carbon dioxide in a final step.[109]

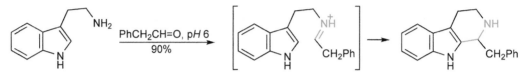

These cyclisations may proceed by direct electrophilic attack at the α-position, or by way of β-attack, then rearrangement. It may be significant that Mannich processes, as opposed to the alkylations discussed in Section 20.1.1.6, are reversible, which would allow a slower, direct α-substitution to provide the principal route to the α-substituted structure.

The cyclisation of nitrones derived from tryptamines is a similar process and can be carried out enantioselectively using a chiral Lewis acid.[110] A similar enantioselective intermolecular process is the copper-catalysed reaction of indoles with tosyl-imines of aromatic aldehydes.[111]

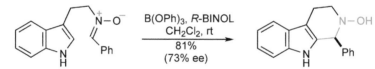

20.1.1.10 Diazo-Coupling and Nitrosation
The high reactivity of indole is shown up well by the ease with which it undergoes substitution with weakly electrophilic reagents, such as benzenediazonium chloride and nitrosating agents. Indoles react rapidly with nitrous acid; indole itself reacts in a complex manner, but 2-methylindole gives a 3-nitroso substitution product cleanly. This can also be obtained by a base-catalysed process using amyl nitrite as a source of the nitroso group; these basic conditions also allow 3-nitrosation of indole itself. 3-Nitroso-indoles exist predominantly in the oximino 3*H*-indole tautomeric form.[112]

20.1.1.11 Electrophilic Metallation
Mercuration *(CAUTION: mercury salts are highly toxic)*
Indole reacts readily with mercuric acetate at room temperature to give a 1,3-disubstituted product.[113] Even *N*-sulfonyl-indoles are substituted under mild conditions; the 3-mercurated compounds thus produced are useful in palladium-catalysed couplings[114] and can be used to prepare boronic acids (4.1.5.3).[115] 1-Phenylsulfonyl-3-substituted indoles mercurate at C-4, subsequent reaction with iodine giving 4-iodo-indoles.[116]

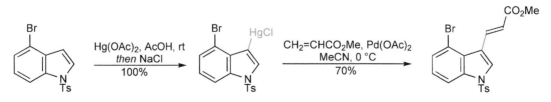

Thallation *(CAUTION: thallium salts are highly toxic)*

Thallium trifluoroacetate reacts rapidly with simple indoles, but well-defined products cannot be isolated. 3-Acyl-indoles, however, undergo a very selective substitution at C-4, due to chelation and protection of the heterocyclic ring by the electron-withdrawing 3-substituent.[117] The products are good intermediates for the preparation of 4-substituted indoles, for example 4-iodo- and thence 4-alkoxy-,[117b] 4-alkenyl-[118] and 4-methoxycarbonyl,[119] *via* palladium-mediated couplings. The regiochemistry is neatly complemented by thallation of *N*-acetylindoline, which goes to C-7, allowing introduction of substituents at this carbon[120] (cf. 20.5.1).

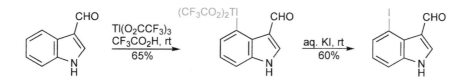

20.2   Reactions with Oxidising Agents

Autoxidation occurs readily with alkyl-indoles, thus, for example, 2,3-diethylindole gives an isolable 3-hydroperoxy-3*H*-indole. Generally, such processes give more complex product mixtures resulting from further breakdown of the initial hydroperoxide; singlet oxygen also produces hydroperoxides, but by a different mechanism. If the indole carries a side-chain capable of trapping the indolenine by intramolecular nucleophilic addition, then tricyclic hydroperoxides can be isolated.[121]

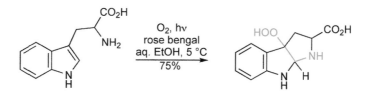

The reagent MoO$_5$.HMPA, known as 'MoOPH', brings about addition of the elements of methyl hydrogen peroxide to an *N*-acyl-indole, and these adducts in turn, can be utilised: one application is to induce loss of methanol, and thus the overall transformation of an indole into an indoxyl (20.13.2).[122]

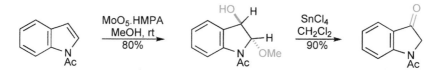

Oxidative cleavage of the indole 2,3-double bond can be achieved with ozone, sodium periodate,[123] potassium superoxide,[124] with oxygen in the presence of cuprous chloride[125] and with oxygen, photochemically in ethanolic solution.[126]

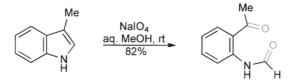

The conversion of 3-substituted indoles into their corresponding oxindoles can be brought about by reaction with dimethylsulfoxide in acid; the scheme below shows a reasonable mechanism for the process – once again a small concentration of β-protonated indole is the key.[127] In a comparable sequence using Swern conditions, Me₂S⁺ adds first to the indole β-position.[128] Dimethyldioxirane converts *N*-methoxycarbonyl-indoles (or -oxindoles) into 3-hydroxy-oxindoles.[129]

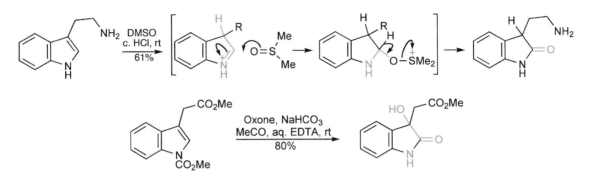

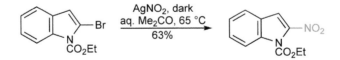

20.3 Reactions with Nucleophilic Reagents (see also 20.13.4)

As with pyrroles and furans, indoles undergo very few nucleophilic substitution processes. Most of those that are known involve special situations: *N*-substituted benzene-ring-nitro-indoles undergo vicarious nucleophilic substitutions (VNS) (3.3.3).[130] A related process involves addition of stabilised enol(ate)s *ortho* to a 5-sulfoxide, with loss of the oxygen from sulfur. The reaction is highly selective for C-4, even in the presence of 3-substituents.[131]

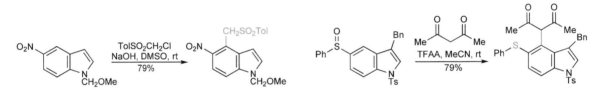

2-Iodo- and 2-bromo-*N*-protected-indoles undergo displacement by reaction with silver nitrite to give the corresponding 2-nitroindoles.[132]

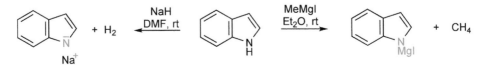

20.4 Reactions with Bases
20.4.1 Deprotonation of *N*-Hydrogen and Reactions of Indolyl Anions

As in pyrroles, the *N*-hydrogen in indoles is much more acidic (pK_a 16.2) than that of an aromatic amine (aniline has pK_a 30.7). Any very strong base will effect complete conversion of an *N*-unsubstituted indole into the corresponding indolyl anion, amongst the most convenient being sodium hydride, *n*-butyllithium or an alkyl Grignard reagent.

The indolyl anion has two main mesomeric structures showing the negative charge to reside mainly on nitrogen and the β-carbon. Electron-withdrawing substituents, particularly at the β-position, increase the acidity markedly, for example 3-formylindole is about five pK_a units more acidic than indole and 2-formylindole is some three units more acidic.[133]

In its reactions, the indolyl anion behaves as an ambident nucleophile; the ratio of *N*- to β-substitution with electrophiles depends on the associated metal, the polarity of the solvent, and the nature of the electrophile. Generally, the more ionic sodio and potassio derivatives tend to react at nitrogen, whereas magnesio derivatives have a greater tendency to react at C-3 (see also 20.1.1.4),[134] however, reaction of indolyl Grignards in HMPA leads to more attack at nitrogen.[135] Complimentarily, more reactive electrophiles show a greater tendency to react at nitrogen than less electrophilic species.

N-Alkylation of indoles can utilise indol-1-ylsodiums,[136] generated quantitatively with sodium hydride, or it can involve a small concentration of an indolyl anion, produced by phase-transfer methods.[137] Dimethyl carbonate with DABCO can be used for *N*-methylation, and the acidity of indoles, especially carbazoles, is sufficient for successful Mitsunobu alkylations.[138,139,140]

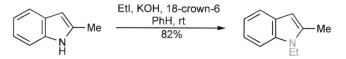

N-Aroyl-benzotriazoles react well with indolyl anions to give *N*-aroyl-indoles.[141] *N*-Acylation[142] and *N*-arylsulfonylation[143] can also be achieved efficiently using phase-transfer methodology.

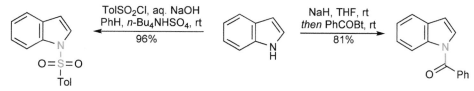

Indolyl *N*-Grignards,[144] or even better their zinc analogues,[145] undergo reaction predominantly at C-3 with a variety of carbon electrophiles such as aldehydes, ketones and acid halides, or reactive halo-heterocycles.[146] Including aluminium chloride in the zinc reactions produces high yields of 3-acyl-indoles.[147] The copper-catalysed reactions of indolyl-*N*-Grignards with *N*-t-butoxycarbonyl-aziridines also proceed well at C-3.[148]

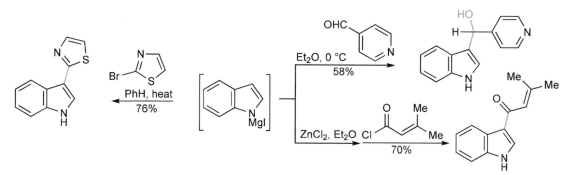

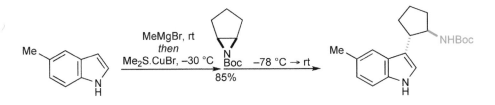

1-Lithio-indoles are equally useful; again, the position of attack depends on both solvent and the nature of the electrophile.[149] It is important to note that when an *N*-metallated 3-substituted indole alkylates at carbon, necessarily a 3,3-disubstituted-3*H*-indole is formed, which cannot re-aromatise to form an indole (see 20.1.1.6 for rearrangements of 3,3-disubstituted indolenines).

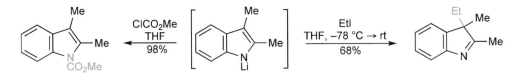

20.5 C-Metallation and Reactions of *C*-Metallated Indoles
20.5.1 Direct Ring C–H Metallation

C-Metallation of indoles has, in nearly all cases, been conducted in the absence of the much more acidic *N*-hydrogen i.e. the presence of an *N*-substituent like methyl,[156] or if required, a removable group: phenyl-sulfonyl,[157] lithium carboxylate[158] and *t*-butoxycarbonyl[159] have been used widely; also recommended are dialkylaminomethyl,[160] trimethylsilylethoxymethyl[161] and methoxymethoxy[162] (the *N*-substituent cannot be introduced into an indole – it requires a pre-formed 1-hydroxy-indole – but it is possible to reduce it off to leave an *N*-hydrogen-indole). Each of these removable substituents assists lithiation by intramolecular chelation and in some cases by electron withdrawal, reinforcing the intrinsic tendency for *metallation to proceed at the α-position*.

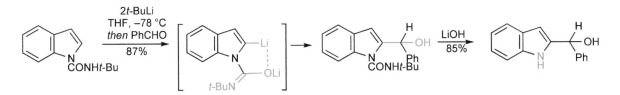

When *N*-acyl or *N*-sulfonyl groups are utilised as protecting groups during indole manipulations, it is important that there be efficient means for their final removal. Both types can be removed with hot base, providing the rest of the molecule is sturdy enough, but milder methods are available. Carbamates can be removed by reaction with hot aqueous methanolic potassium carbonate[150] or with *t*-butylamine in refluxing methanol,[151] and *N*-tosyl groups can be cleaved with thioglycolic acid at room temperature,[152] cesium carbonate in hot THF/methanol[153] or by photo-induced electron transfer from triethylamine.[154] Removal of *N*,*N*-dimethylaminosulfonyl groups can be achieved by electrolysis.[155]

Magnesiation at C-2 can be carried out at room temperature; as well as serving in the usual way as nucleophiles, magnesio-indoles can also be used directly for palladium-catalysed couplings.[163]

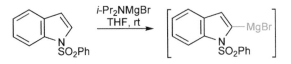

One of the most convenient *N*-protecting groups in indole α-lithiations is carbon dioxide[158] because the *N*-protecting group is installed *in situ* and, further, falls off during normal work-up. This technique has been used to prepared 2-halo-indoles[32] and to introduce a variety of substituents by reaction with appropriate electrophiles – aldehydes, ketones, chloroformates, etc.[158]

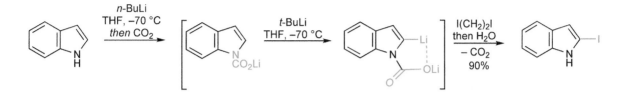

Given below are some α-substitutions achieved with various *N*-blocking/activating groups.[39,157,164,165]

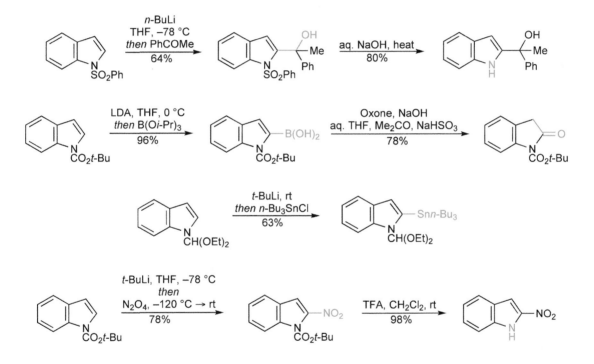

Direct 3-lithiation can be accomplished with *ortho*-assistance from a 2-(2-pyridyl)-[184] or a 2-carboxyl group.[166] Direct 3-lithiation even *without* a substituent at C-2 can be achieved with an *N*-di(*t*-butyl) fluorosilyl-,[167] *N*-tri-*i*-propylsilyl-[168] or *N*-(2,2-diethylbutanoyl)-substituent in place, the latter using *sec*-butyllithium in the presence of *N,N,N′,N″,N″*-pentamethyldiethylenetriamine.[169] Other directed metallation processes in the hetero-ring include: 2-lithiation of 1-substituted indole-3-carboxylic acids and amides,[170] and of 3-hydroxymethyl-1-phenylsulfonylindole.[171]

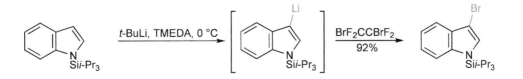

The dimethylamino group of gramine directs lithiation to C-4 when the indolic nitrogen is protected by the bulky TIPS group, but metallation occurs normally at C-2 when this nitrogen bears a simple methyl.[172] Comparable regioselectivity is found with 3-methoxymethyl-1-tri-*iso*-propylsilylindole.[173] 4-Lithiation of 5-(dimethylcarbamoyloxy)-1-(*t*-butyldimethylsilyl)indole and the 6-lithiation of 4-substituted-5-(dimethylcarbamoyloxy)-1-(*t*-butyldimethylsilyl)-indoles depend on *ortho*-directing effects.[174]

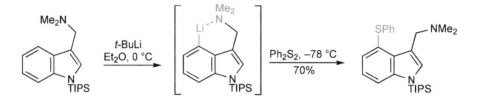

Metallation at C-7 can be achieved using a bulky *N*-2,2-diethylbutanoyl group, when a 3-substituent is also present to further discourage metallation at C-2.[175] An *N*-di-*t*-butylphosphinoyl group also directs metallation to C-7, but the *N*-substituent is difficult to remove, so a blocking strategy can be employed: the desired affect is achieved *via* a one-pot sequence from 1-(diethylaminocarbonyl)indole involving 2-lithiation, reaction with trimethylsilyl chloride (at C-2), and then *peri*-directed C-7-metallation.[176]

20.5.2 Metal–Halogen Exchange

The two halogens in 2,3-dibromo-1-methylindole can be exchanged selectively – first that at the α-position and then that at C-3[177] indeed 2,3-*di*lithio-1-methylindole can also be generated from 2,3-diiodo-1-methylindole using *t*-butyllithium at −100 °C. It is significant that 2,3-diiodo-1-phenylsulfonylindole undergoes comparable double metal–halogen exchange, but the dilithio-derivative ring opens, with the nitrogen anion acting as a leaving group (cf. 17.4.2), even at −100 °C, to lithium 2-(*N*-lithiophenylsulfonamido) phenylacetylide.

3-Lithio-indoles can be prepared by halogen exchange;[178,179,180,181] the *N*-*t*-butyldimethylsilyl-derivative is regiostable, even at 0 °C,[182] whereas 3-lithio-1-phenylsulfonylindole isomerises to the 2-isomer at temperatures above −100 °C, although hetero-ring opening and production of an alkyne is not a problem at that temperature.[183,184] The corresponding *N*-phenylsulfonyl-3-magnesium[185] and -3-zinc[186] species are regiostable, even at room temperaure; they can be prepared from the 3-iodoindole by reaction with ethylmagnesium bromide and lithium trimethylzincate respectively; the 2-zinc-reagent is comparably prepared.

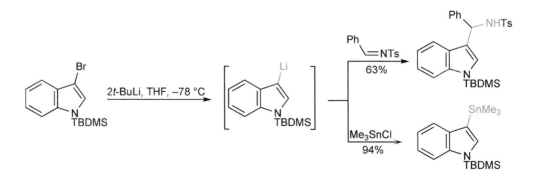

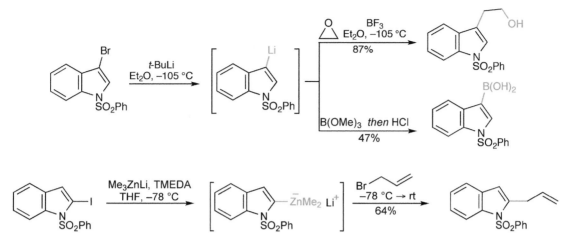

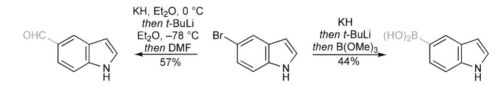

2-Iodoindole can be converted, using three equivalents of *n*-butyllithium, into the 1,2-dilithio compound, which reacts normally at C-2 with electrophiles.[187]

Lithium–bromine exchange can be achieved with each of the benzene-ring bromo-indoles after formation of the *N*-potassium salt, i.e. *N*-protection.[188]

20.6 Reactions with Radicals

Radicals such as benzyl and hydroxyl are unselective in their interaction with indoles, resulting in mixtures of products, so such reactions are of little synthetic use. On the other hand, benzoyloxylation of *N*-substituted indoles gives benzoates of indoxyl,[189] i.e. it effectively oxidises the indole heterocyclic ring, *via* β-attack by the strongly electrophilic benzoyloxy-radical. In contrast, the weakly electrophilic radical derived from malonate reacts selectively at C-2, *via* an atom-transfer mechanism.[190]

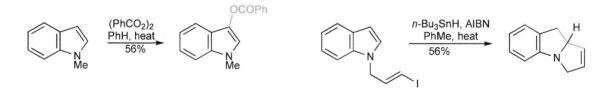

Some efficient oxidative[191] and reductive[192] intramolecular carbon-radical additions can be carried out. *Ipso*-replacement of toluenesulfonyl by tributylstannyl radical occurs readily at C-2[193] (but not at C-3) as does intramolecular replacement by carbon radicals.[194]

2-Indolyl radicals can be generated under standard conditions by reacting 2-bromoindole with tributyltin hydride.[195] 3-Methyl-1-tosylindole can be cyanated at C-2 in good yield by a mixture of TMSCN and PIFA *via* oxidation of the indole to a cation radical, then addition of cyanide anion.[196]

20.7 Reactions with Reducing Agents

The indole ring system is not reduced by nucleophilic reducing agents, such as lithium aluminium hydride or sodium borohydride; lithium/liquid ammonia does, however, reduce the benzene ring; 4,7-dihydroindole is the main product.[197]

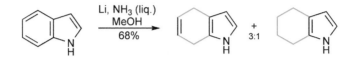

Reduction with lithium in the presence of trimethylsilyl chloride, followed by re-aromatisation, produces 4-trimethylsilylindole, an intermediate useful for the synthesis of 4-substituted indoles *via* electrophilic *ipso*-replacement of silicon.[198]

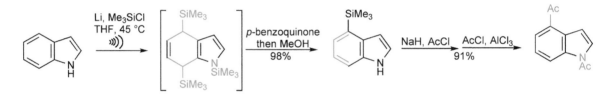

Reduction of the heterocyclic ring is readily achieved under acidic conditions; formerly, metal–acid combinations[199] were used, but now much milder conditions employ relatively acid-stable metal hydrides, such as sodium cyanoborohydride. Triethylsilane in trifluoroacetic acid is another convenient combination; 2,3-disubstituted indoles give 2,3-*cis*-indolines by this method.[200] Such reductions proceed by hydride attack on the β-protonated indole – the 3*H*-indolium cation.[201] Catalytic reduction of indole, again in acid solution, produces indoline initially, further slower reduction completing the saturation.[202] Rhodium-catalysed high-pressure hydrogenation of indoles with a *t*-butoxycarbonyl group on nitrogen proceeds smoothly to give 2,3-*cis*-indolines.[203]

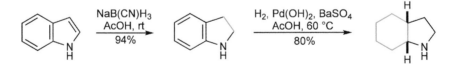

20.8 Reactions with Carbenes

No cyclopropane-containing products have been isolated from the interaction of an indole 2,3-double bond with carbenes (cf. 16.9). Methoxycarbonyl-substituted carbenes give rise only to a substitution product, at C-3 if available and at C-2 from 3-substituted indoles.[204,205]

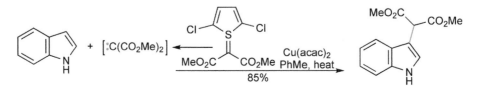

20.9 Electrocyclic and Photochemical Reactions

The heterocyclic double bond in simple indoles will take part in cycloaddition reactions with dipolar 4π components,[206] and with electron-deficient dienes (i.e. inverse electron demand), in most reported cases, held close using a tether;[207] a comparable effect is seen in the intermolecular cycloaddition of 2,3-cycloalkyl-indoles to *ortho*-quinone generating a 1,4-dioxane.[208] The introduction of electron-withdrawing substituents enhances the tendency for cycloaddition to electron-rich dienes: 3-acetyl-1-phenylsulfonylindole, for example, undergoes aluminium-chloride-catalysed cycloaddition with isoprene,[209] and 3-nitro-1-phenylsulfonylindole reacts with 1-acylamino-buta-1,3-dienes without the need for a catalyst.[210] Both 3- and 2-nitro-1-phenylsulfonyl-indoles undergo dipolar cycloadditions with azomethine ylides.[211]

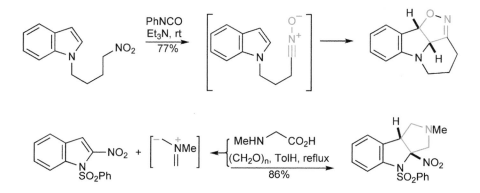

Both 2- and 3-vinylindoles can take part as 4π components in Diels–Alder cycloadditions;[212] mostly, but not always,[213] these employ *N*-acyl- or *N*-arylsulfonyl-indoles, in which the interaction between nitrogen lone pair and π-system has been reduced.[214] The example shows how this process can be utilised in the rapid construction of a complex pentacycle.[215]

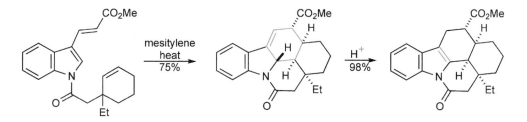

When tethered 1,2,4-triazines are used, their interaction with the indole 2,3-double bond generates carbolines. The tether can be incorporated into the product molecule,[216] or be designed to be broken *in situ*, as in the example below.[217] 1,2,4,5-Tetrazines react with the indole 2,3-bond in an intermolecular sense; the initial adduct loses nitrogen and then is oxidised to the aromatic level by a second mole equivalent of the tetrazine.[218]

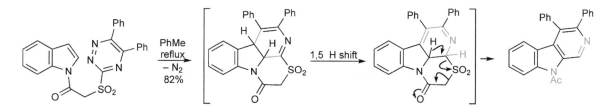

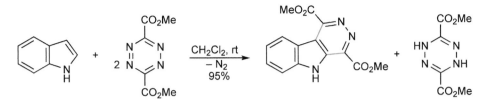

A 1-vinylamino-indole undergoes a 3,3-sigmatropic rearrangement giving the tricyclic ring system of the eseroline alkaloids.[219]

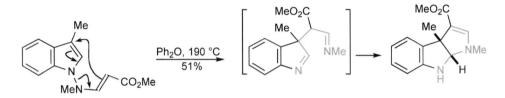

Claisen ortho-ester rearrangement of indol-3-yl-alkanols introduces the migrating group to the indole 2-position.[220]

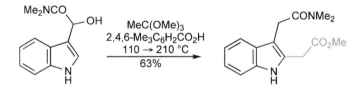

Under the influence of UV light, *N*-methylindoles add dimethyl acetylenedicarboxylate, generating cyclobuteno-fused products,[221] and even simple alkenes add in an apparent 2 + 2 fashion to *N*-acyl-indoles, but the mechanism probably involves radical intermediates.[222] Other photochemical additions to form *N*-benzoyl-indolines fused to four-membered rings include addition to the carbonyl group in benzophenone and the double bond in methyl acrylate.[223]

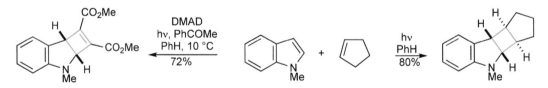

20.10 Alkyl-Indoles

Only alkyl groups at indole α-positions show any special reactions. Many related observations confirm that in a series of equilibria, β-protonation can lead to 2-alkylidene-indolines, and hence reactivity towards electrophiles at an α-, but not a β-alkyl group, for example in DCl at 100 °C 2,3-dimethylindole exchanges H for D only at the 2-methyl. This same phenomenon is seen in Mannich condensation[224] and trifluoro-acetylation[225] of 1,2,3-trimethylindole at the α-methyl.

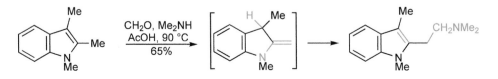

Side-chain lithiation is again specific for an α-substituent, *via* an *N*-lithium carboxylate,[226] or even without *N*-protection.[227]

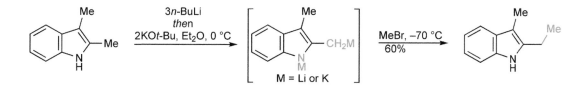

A quite different side-chain acylation can be achieved with aluminium chloride catalysis: here, association of the Lewis acid with the indole α-position is assumed to lead to a styrene intermediate, which is acylated.[228]

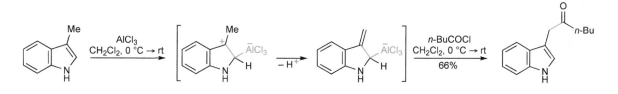

20.11 Reactions of Indolyl-C–X Compounds

Gramine and, especially, its quaternary salts are useful synthetic intermediates in that they are easily prepared and the dimethylamino group is easily displaced by nucleophiles – reactions with cyanide[229] and acetamidomalonate[230] anions, and boronic acids with rhodium(I) catalysis,[231] are typical.

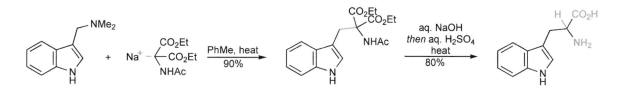

The easy displacement of the amine (ammonium) group proceeds by way of an elimination, involving loss of the indole hydrogen, and thus the intermediacy of a β-alkylidene-indolenine that then readily adds the nucleophile, regenerating the indole system. This mechanism has been verified by observing: (i) very much slower displacement with a corresponding 1-methyl-gramine, and (ii) racemisation on displacement using a substituted gramine in which the nitrogen-bearing carbon was a chiral centre.[232]

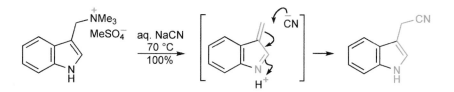

Utilising tri-*n*-butylphosphine to displace the dimethylamino group generates, *in situ*, a zwitterion which, by proton transfer, becomes a Wittig intermediate and thus can be utilised to prepare 3-vinyl-indoles.[233]

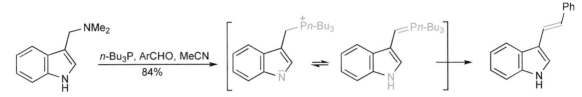

A related sequence is involved in the lithium aluminium hydride reduction of indol-3-yl-carbinols (which can be obtained from the corresponding ketones using milder reducing agents), with formation of the alkyl-indole. This constitutes a useful synthesis of 3-alkyl-indoles.[234] The one-pot conversion of 3-formylindole into 3-cyanomethylindole with a mixture of sodium cyanide and sodium borohydride probably involves a comparable elimination from the cyanohydrin, then reduction.[235]

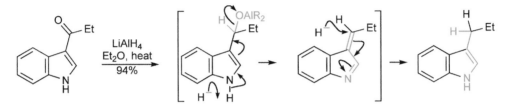

Yet another use for (*N*-protected) gramines is conversion into 3-bromo-indoles: this involves β-bromination and then retro-Mannich loss of the original substituent.[236] Combined with the directing ability of the original dimethylamino group (20.5.1) this provides a route to 4-substituted 3-bromo-indoles.

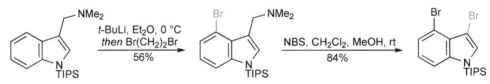

Although indolylic halides are generally unstable and not synthetically useful, *N*-acylated derivatives are much more stable, can be prepared by side-chain radical substitution, and can be utilised in nucleophilic substitution processes.[237]

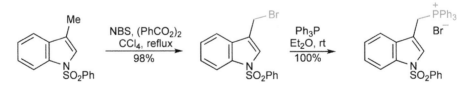

20.12 Indole Carboxylic Acids

Both indole-3-carboxylic[238] and indol-2-yl-acetic acids are easily decarboxylated in boiling water. In each case carbon dioxide is lost from a small concentration of β-protonated 3*H*-indolium cation, the loss being analogous to the decarboxylation of a β-keto-acid. Indole-1-carboxylic acid also decarboxylates very easily, but is sufficiently stable to allow isolation and use in acylation reactions.[239] Indole-2-carboxylic acids can only be decarboxylated by heating in mineral acid or in the presence of copper salts.[240]

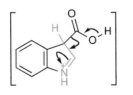

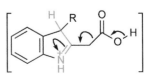

3-Trifluoroacetyl-indoles, very simply obtained from indoles by electrophilic substitution, are useful stable equivalents of indol-3-yl-carboxylic acid chlorides, giving amides or acids in reactions with lithium amides or aqueous base respectively. The reactivity of the *N*-hydrogen compounds is greater than of those with *N*-alkyl, indicating the intermediacy of a ketene in the reactions of the former.[241]

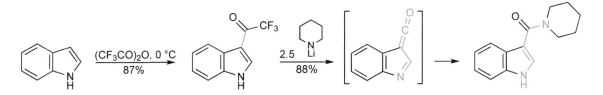

In a nice exemplification of the mesomeric interaction between indole nitrogen and a 3-carbonyl, which renders the 3-carbonyl somewhat amide-like (see also 20.11), 2,3-dicarboxylic acid anhydrides react with some nucleophiles selectively at the 2-carbonyl; inductive withdrawal by the ring nitrogen may also play a part in achieving this selectivity.[242]

20.13 Oxy-Indoles
Indoles with a hydroxyl group on the benzene ring behave like normal phenols; indoles with an oxygen at either of the heterocyclic ring positions are quite different.

20.13.1 Oxindole
Oxindole exists as the carbonyl-tautomer, the hydroxyl-tautomer ('2-hydroxyindole') being undetectable. There is nothing remarkable about the reactions of oxindole; for the most part it is a typical 5-membered lactam, except that deprotonation at the β-carbon ($pK_a \sim 18$) occurs more readily than with simple amides, because the resulting anion is stabilised by an aromatic indole resonance contributor. Such anions will react with electrophiles like alkyl halides and aldehydes[243,244] at the β-carbon, the last with dehydration and the production of aldol condensation products. Oxindoles can be oxidised to isatins (20.13.3) *via* easy 3,3-dibromination, then hydrolysis.[245] Bromination of oxindole with *N*-bromosuccinimide gives 5-bromooxindole.[193]

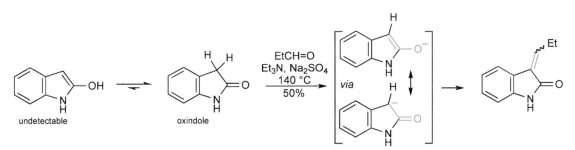

The interaction of oxindole with the Vilsmeier reagent produces 2-chloro-3-formylindole efficiently;[246] this difunctional indole has considerable potential for elaboration, for example nucleophilic displacement of the halogen, activated by the *ortho*-aldehyde, can produce indoles carrying a nitrogen substituent at C-2.[247]

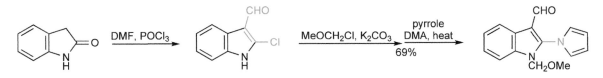

Of potential in palladium(0)-catalysed coupling processes to the indole 2-position is the 1-phenylsulfonylated 2-triflate readily obtained from 1-phenylsulfonyloxindole (see also 4.2.3).[248]

20.13.2 Indoxyl[249]

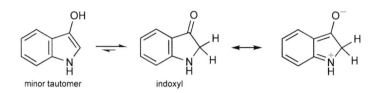

minor tautomer   indoxyl

3-Hydroxyindole certainly contributes in the tautomeric equilibrium with the carbonyl form, though it is the minor component. Indoxyl, pK_a 10.46,[250] is more acidic than oxindole, the anion produced is ambident; reactions with electrophiles at both oxygen and carbon are known.[251]

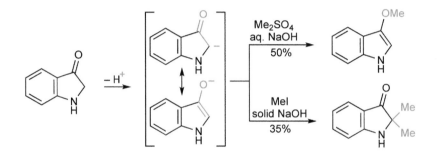

The indoxyl anion is particularly easily autoxidised, producing the ancient blue dye, indigo. The mechanism probably involves dimerisation of a radical formed by loss of an electron from the anion.

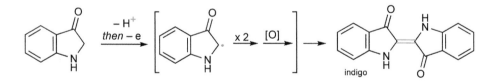

indigo

O-Acetylindoxyl[252] and *N*-acyl-indoxyls are more stable substances; the latter undergo normal ketone-carbonyl reactions, such as the Wittig reaction.[253]

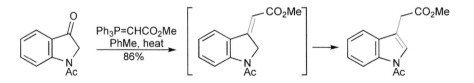

Mirroring oxindoles, aldol-type condensation at the 2-position in indoxyls can be accomplished either using the acetate of the enol form and base catalysis,[254] or with indoxyl itself, in either acid or basic conditions.[255] Borohydride reduction and dehydration allows these alkylidene condensation products to be converted into 2-substituted indoles.

Peroxide oxidation of *N*-phenylsulfonylindole-3-boronic acid gives *N*-phenylsulfonylindoxyl, which can be converted into the triflate of the 3-hydroxyindole tautomer.[256] The same *N*-protected indoxyl can be prepared by ring synthesis, as shown below.

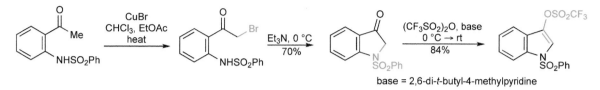

base = 2,6-di-*t*-butyl-4-methylpyridine

20.13.3 Isatin[257]

Isatin is a stable, bright orange solid that is commercially available in large quantities. Because it readily undergoes clean aromatic substitution reactions at C-5, *N*-alkylation *via* an anion, and ketonic reactions at the C-3-carbonyl group, for example enolate addition,[258] it is a very useful intermediate for the synthesis of indoles and other heterocycles.

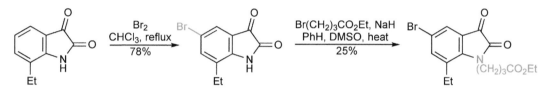

Conversion of isatins into oxindoles can be achieved by catalytic reduction in acid,[259] or by the Wolff–Kischner process.[260,261] 3-Substituted indoles result from Grignard addition at the ketone carbonyl, followed by lithium aluminium hydride reduction of the residual amide, then dehydration.[262] The reaction of isatin with triphenylphosphine provides an easy synthesis of 3-(triphenylphosphorylidene)oxindole, a Wittig reagent.[263]

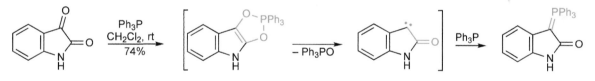

A process that produces pyrroles, from ketones and hydroxyproline, works well with isatins.[264]

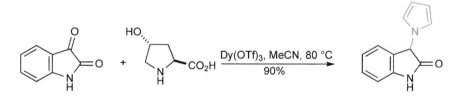

20.13.4 1-Hydroxyindole[265]

1-Hydroxyindole can be prepared in solution, but attempted purification leads to dimerisation *via* its nitrone tautomer; however, *O*-alkyl-derivatives can be formed easily and are stable.[266]

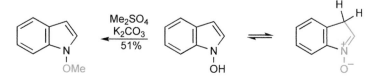

Lithiation of 1-methoxyindole takes place at C-2 and thus substituents can be introduced. Various nucleophilic substitutions, with departure of the 1-substituent can take place. One of the reactions below shows the introduction of a hydroxyl group onto the indole 5-position by aqueous acid treatment of a 1-hydroxyindole.[267] 1-Methoxy groups allow nucleophilic attack on the heterocyclic ring, as illustrated by the second example.[268]

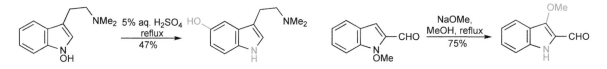

Cine-nucleophilic substitution of methoxy in a 1-methoxy-3-formyl-indole produces the 2-substituted product.[269]

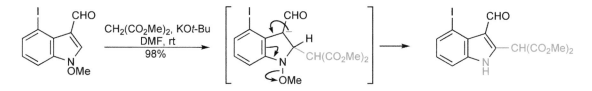

20.14   Amino-Indoles

2-Aminoindole exists mainly as the 3*H*-tautomer, presumably because of the energy advantage conveyed by amidine-type resonance. 3-Aminoindole is very unstable, and easily autoxidised.[270] Both 2- and 3-acylamino-indoles are stable and can be obtained by catalytic reduction of nitro-precursors in the presence of anhydrides.[271] 1-Amino-indoles can be prepared by direct amination of the indolyl anion.[272]

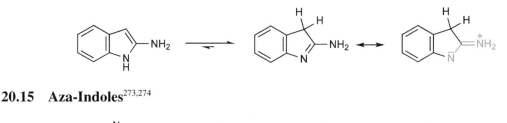

20.15   Aza-Indoles[273,274]

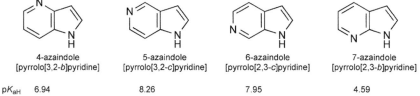

	4-azaindole [pyrrolo[3,2-*b*]pyridine]	5-azaindole [pyrrolo[3,2-*c*]pyridine]	6-azaindole [pyrrolo[2,3-*c*]pyridine]	7-azaindole [pyrrolo[2,3-*b*]pyridine]
pK_{aH}	6.94	8.26	7.95	4.59

The (mono)-aza-indoles, or more correctly pyrrolo-pyridines, where a carbon of the six-membered ring has been replaced by nitrogen, are of interest as bicyclic systems comprising an electron-rich ring fused to an electron-poor ring. Simple aza-indoles are not known in nature, but polycyclic products containing azaindole units have been isolated from sea sponges. Aza-indoles have elicited significant interest in medicinal chemistry as isosteres of indoles, particularly as components of azatryptamine analogues and even as di-deaza-purines.

Aza-indoles show the typical reactivity of both component heterocycles, but to a reduced and varying degree, with reduced electron density in the five-membered ring and increased electron density in the six-membered ring.

The pK_{aH}s for protonation on the pyridine nitrogen, of the four parent systems, demonstrate the degree of push–pull interaction between the two rings. For example, the pK_{aH}s of 5- and 7-azaindoles reflect, but to a greater degree, the pK_{aH}s of 4-aminopyridine (9.1) and 2-aminopyridine (7.2), respectively, and are partly explained by the more favourable γ-interaction between the donating and accepting groups in the former. This differential reactivity is exaggerated in mildly acidic solutions, such as are used in Mannich reactions, where the 5-azaindole is present predominantly in protonated form, while the 7-azaindole is mainly present in the form of its free base.

20.15.1   Electrophilic Substitution

Reactions with electrophilic reagents take place with substitution at C-3 or by addition at the pyridine nitrogen. All the aza-indoles are much more stable to acid than indole (cf. 20.1.1.9), no doubt due to the diversion of protonation onto the pyridine nitrogen, but the reactivity towards other electrophiles at C-3 is only slightly lower than that of indoles. Bromination and nitration occur cleanly in all four parent systems[275] and are more controllable than in the case of indole. Mannich and Vilsmeier reactions can be carried out in some cases, but when the latter fails, 3-aldehydes can be prepared by reaction with hexamine, possibly *via* the anion of the azaindole. Alkylation under neutral conditions results in quaternisation on the pyridine nitrogen and reaction with sodium salts allows N-1-alkylation. Acylation under mild conditions also occurs at N-1. The scheme below summarises these reactions for the most widely studied system – 7-azaindole. Acylation at C-3 in all four systems can be carried out at room temperature in the presence of excess aluminium chloride.[276]

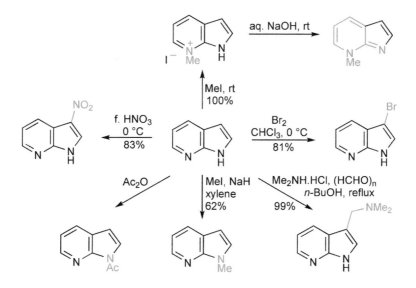

20.15.2   Nucleophilic Substitution

Only a few examples of nucleophilic substitution have been reported – displacement of halogen α and γ to the pyridine nitrogen can be carried out under vigorous conditions or with long reaction times. No Chichibabin substitutions have been reported. Reaction of 4-chloro-7-azaindole with a secondary amine results in normal substitution of the halogen, but reaction with primary amines gives 5-azaindole rearrangement products by the sequence shown below.[277]

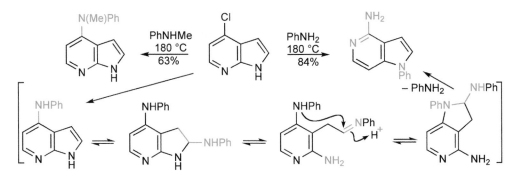

4-Chloro-7-azaindole undergoes nucleophilic displacement with cyanide; the halide is available *via* the *N*-oxide.[278]

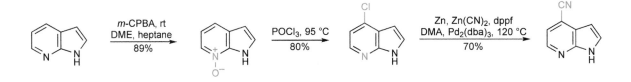

The reactivity of 4-chloro-1-methyl-5-azaindole, for which data is available, towards nucleophilic substitution of chlorine by piperidine[279] can be usefully compared with that of some related systems: it is significantly less reactive than the most closely related bicyclic systems, probably due to increased electron density in the six-membered ring resulting from donation from N-1.

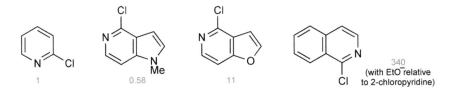

Relative rates for nucleophilic displacement with piperidine in MeO(CH₂)₂OH at 100 °C[280]

1-Phenylsulfonyl-7-azaindole is lithiated at C-2 by lithium diisopropylamide; subsequent reactions of the lithiated azaindole are normal, for example the preparation of a stannane.[281]

20.16   Synthesis of Indoles[282]
20.16.1   Ring Synthesis of Indoles
Indoles are usually prepared from non-heterocyclic precursors by cyclisation reactions on suitably substituted benzenes; they can also be prepared from pyrroles by construction of the homocyclic aromatic ring, and from indolines by dehydrogenation.

Because of the importance of indoles in natural product synthesis and pharmaceutical chemistry, reports of new routes to indoles and improvements to older reactions appear frequently. This section discusses the most important methods now available, often those that have been used most frequently and are the most adaptable.

20.16.1.1 From Aryl-hydrazones of Aldehydes and Ketones
Still the most widely used route, heating an arylhydrazone, usually with acid, sometimes in an inert solvent gives an indole with the loss of ammonia.

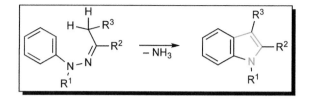

The Fischer Synthesis
The Fischer synthesis,[283] first discovered in 1883, involves the acid- or Lewis-acid-catalysed rearrangement of an arylhydrazone with the elimination of ammonia. The preparation of 2-phenylindole illustrates the process in its simplest form.[284]

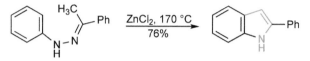

In many instances the reaction can be carried out simply by heating together the aldehyde or ketone and arylhydrazine in acetic acid;[285] the formation of the arylhydrazone and its subsequent rearrangement take place without the necessity for isolation of the arylhydrazone. Toluenesulfonic acid, cation-exchange resins, acidic clays and phosphorus trichloride have all been recommended for efficient cyclisations, sometimes even at or below room temperature.[286] Electron-releasing substituents on the benzene ring increase the rate of Fischer cyclisation, whereas electron-withdrawing substituents slow the process down,[287] though even arylhydrazones carrying nitro groups can be indolised satisfactorily with appropriate choice of acid and conditions, for example a two-phase mixture of toluene and phosphoric acid,[288] or boron trifluoride in acetic acid.[289] Electron-withdrawing substituents *meta* to the nitrogen give rise to roughly equal amounts of 4- and 6-substituted indoles; electron-releasing groups similarly oriented produce mainly the 6-substituted indole.[216] N_a-Aroyl-[290] and N_a-Boc-[291] arylhydrazones undergo normal Fischer closures, the latter with loss of the Boc group *in situ*. The Fischer process can be conducted on solid support in traceless mode by attaching the non-aromatic nitrogen to the support[292] (this is lost in the conversion – see below).

The mechanistic details of the multi-step Fischer sequence are best represented by the sequence shown below. Labelling studies proved the loss of the non-aromatic nitrogen as ammonia, and in some cases intermediates have been detected by ^{13}C and ^{15}N NMR spectroscopy.[293] The most important step – the one in which a carbon–carbon bond is made, marked **A** – is electrocyclic in character and analogous to the Claisen rearrangement of aryl allyl ethers.

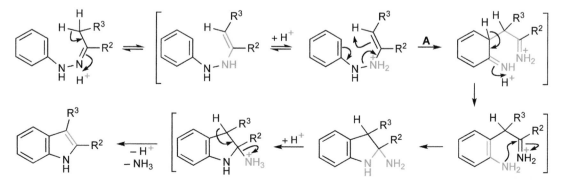

Support for this sequence also comes from the observation that, in many cases indolisation can be achieved thermally, at a temperature as low as 110 °C, in the special case of pre-formed ene-hydrazines, i.e. in which the first step of the normal sequence – acid-catalysed tautomerisation of imine to enamine – has already been accomplished.[294] The reaction does, however, still occur more rapidly in the presence of acid and this is interpreted as protonation of the nitrogen, as shown above, facilitating the electrocyclic step.

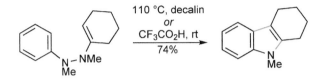

Fischer cyclisations can be achieved thermally, but generally much higher temperatures are required and proton transfer from solvent (typically a glycol) may be involved. However, using pre-formed *N*-trifluoroacetyl-ene-hydrazines allows thermal cyclisation at temperatures as low as 65 °C.[295] As the example below shows, in the case of derivatives of cyclopentanones, the 2-aminoindoline intermediate can be isolated at lower temperatures; subsequent elimination of trifluoroacetamide is easy and efficient.

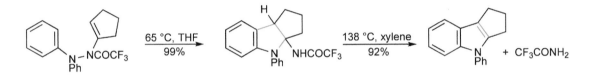

An extreme case of acid catalysis is the indolisation of phenylhydrazones of β-dicarbonyl-compounds in concentrated sulfuric acid;[296] in milder acid, only pyrazolones are produced from the interaction of β-keto-esters with hydrazines (25.12.1.1).

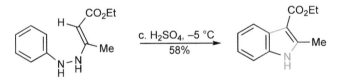

An aspect of the Fischer reaction that is of considerable practical importance is the ratio of the two possible indoles formed from unsymmetrical ketones; in many instances, mixtures result because ene-hydrazine formation occurs in both directions. It appears that strongly acidic conditions favour the least substituted ene-hydrazine.[297]

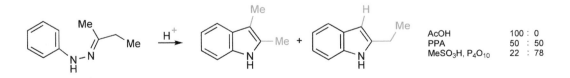

Cyclic enol ethers, enamides and related compounds are very useful stable aldehyde equivalents, as they can be precursors for tryptamine derivatives and analogues.[298,299]

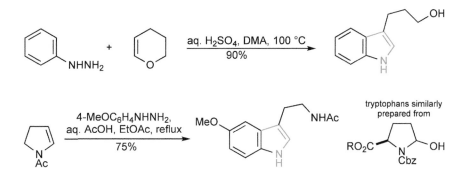

Indolenines (3*H*-indoles) are formed efficiently on Fischer cyclisation of the arylhydrazones of branched ketones; note, again, the use of a weaker acid medium to promote formation of the more substituted ene-hydrazine required for indolenine formation.[300] Subjected to higher temperatures of reaction, the aryl-hydrazones of branched ketones give rise to 2,3-disubstituted indoles, *via* a 2 → 3 migration (cf. 20.1.1.6) in the first-formed indolenine.[301]

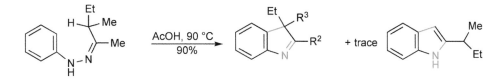

An important extension to the Fischer route is the ability to prepare arylhydrazones in ways other than from ketones/aldehydes. A generally applicable process is the palladium(0)-catalysed coupling of benzo-phenone hydrazone with aryl halides, which allows the convenient preparation of a wide range of aryl-hydrazones of benzophenone, then the benzophenone arylhydrazone can be either hydrolysed to the arylhydrazine, or even more conveniently, used directly in the Fischer cyclisation, where exchange occurs with the ketone.[302]

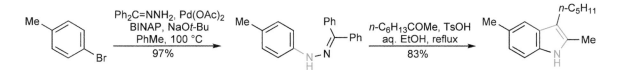

In another variant, *N*-Boc-hydrazine (*t*-BuOCONHNH₂) is coupled with an aryl halide *via* the carbamate nitrogen; the subsequent Fischer sequence can be conducted in the same pot after addition of acid, with loss of the Boc group.[303] Hydroamination of alkynes *via* various protocols[304] also produces arylhydrazones, ready for the Fischer process.

Transformations that are mechanistically analogous to the Fischer, and also produce indoles, use aryl-hydroxylamines instead of arylhydrazines, as shown below.[305]

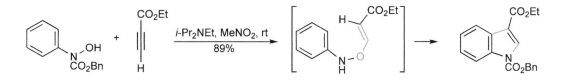

The Grandberg Synthesis

An exceptionally useful adaptation is the Grandberg synthesis of tryptamines from 4-halo-butanals, or more often in practice their acetals,[306] in which the nitrogen usually lost during the Fischer process is incorporated as the nitrogen of the aminoethyl side-chain.[307]

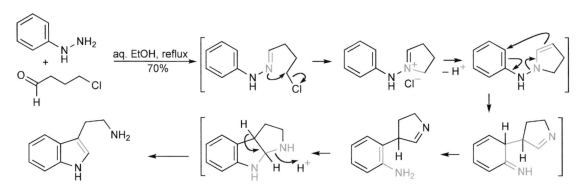

20.16.1.2   From ortho-(2-Oxoalkyl)-Anilines and Equivalents

Cyclisation of *ortho*-(2-oxoalkyl)-anilines by simple intramolecular condensation with loss of water, occurs spontaneously. There are several ways of generating the intermediate amino-ketone, or its equivalent; the prototype is the Reissert synthesis.

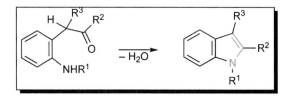

The Reissert Synthesis

The acidity of a methyl group *ortho* to nitro on a benzene ring is the means for condensation with oxalate in this route; the nitro group is then reduced to amino.[308]

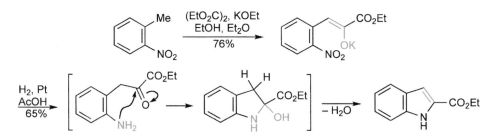

With the nitrogen already at the oxidation level of amine, but carrying a *t*-butoxycarbonyl group to assist the *ortho*-methyl (alkyl) lithiation, reaction with oxalate as in the classical sequence and final removal of the *N*-substituent with acid, again leads to an indole-2-ester.[309] The synthesis of 2-unsubstituted indoles is achieved by reaction of the *N,C*-dilithiated species with dimethylformamide.[310]

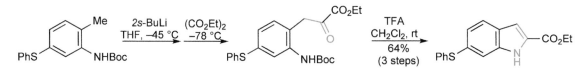

Various other routes produce Reissert-type ring-closure precursors. For example, the palladium-catalysed coupling, in the presence of a methoxyphenol additive, of *ortho*-halo-nitroarenes with methyl ketones, followed by titanium trichloride reduction of the products, leads directly to 3-unsubstituted indoles.[311] More obviously, *ortho*-halo trifluoroacetanilides can be coupled with β-keto esters or amides, the base incorporated in the mixture leading to hydrolysis and closure to the indole.[312]

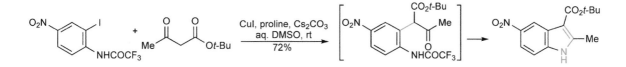

Coupling reactions using *N*-protected *ortho*-halo-anilines have been widely used to prepare *ortho*-(2-oxoalkyl)-anilines; in these instances no reductive step is required though the carbonyl-unit is sometimes incorporated in masked form, such as a 2-ethoxyvinyl-boronate, requiring deprotection.[313]

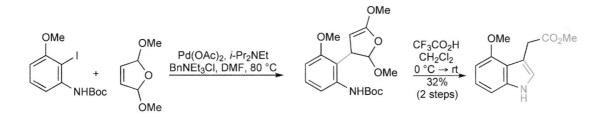

Aromatic nitro-compounds can be made to condense[314] with silyl-enol-ethers using tris(dimethylamino) sulfur (trimethylsilyl)difluoride (TASF); a non-aromatic nitronate intermediate is aromatised by reaction with bromine, without isolation, to provide a 2-(*ortho*-nitroaryl)-ketone and thence an indole after nitro group reduction.[315]

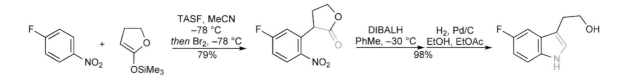

Leimgruber–Batcho Synthesis

The Leimgruber–Batcho synthesis[316] is one of the most widely used variations and also depends on the acidity of methyl groups *ortho* to aromatic nitro to allow introduction of the future indole α-carbon as an enamine and thence the synthesis of pyrrole-ring-unsubstituted indoles. Condensation with hot dimethyl-formamide dimethyl acetal (DMFDMA) (no added base being necessary) leads to an enamine; the condensation can be enhanced by microwave irradiation in the presence of a catalyst, such as ytterbium triflate.[317] Subsequent reduction of the nitro group, usually in acid conditions, leads directly to the hetero-ring-unsubstituted indole probably *via* a C-protonated amino-enamine. Mechanistically, this process is dependent on ionisation of the reagent producing methoxide (which deprotonates the aromatic methyl) and an electrophilic component, MeOCH=N$^+$Me$_2$, which combines with the side-chain-deprotonated aromatic. Both tris(piperidin-1-yl)methane and bis(dimethylamino)-*t*-butoxymethane (Bredereck's reagent) can function even better than DMFDMA.[318] A variety of benzene substituents are tolerated and the approach has been utilised for the syntheses of, amongst others, 4- and 7-indole-carboxylic esters.[319]

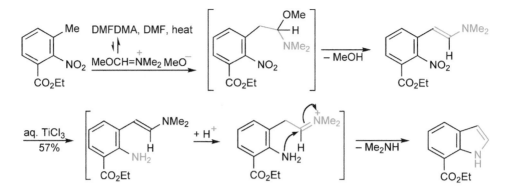

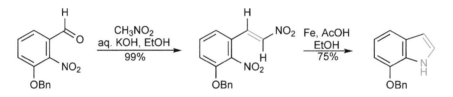

A Leimgruger–Batcho-type amino-enamine intermediate is likely to be involved following reduction of 2-(*ortho*-nitroaryl)-nitroethenes.[320] Reduction, traditionally with metal/acid combinations, but now with reagents such as palladium/carbon with ammonium formate and formic acid,[321] iron with acetic acid and silica gel,[322] or titanium(III) chloride,[323] gives the indole.

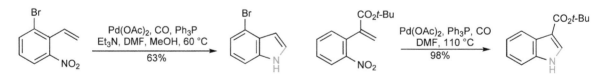

20.16.1.3 *From* ortho-*Nitro-Styrenes*

It is convenient to include here some ring closures involving *ortho*-nitro-styrenes, i.e. with the future C-2 at a lower starting oxidation level. *ortho*-Nitro-styrenes are readily available by a number routes: (i) reaction of an (*ortho*-bromomethyl)-nitroarene with a phosphine then Wittig condensation with an aldehyde; (ii) Wittig reaction employing an (*ortho*-nitro)-araldehyde as the carbonyl component; (iii) base-catalysed condensation of a methyl group *ortho* to an aromatic nitro group with an aldehyde and (iv) *ortho*-nitration of a styrene. Reduction of the nitro group over a catalyst, in the presence of carbon monoxide leads to the indole.[324]

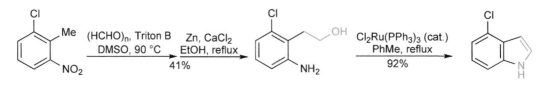

Methods involving ruthenium-catalysed condensations of arylamines with alcohols give indoles: the mechanism involves hydride transfer giving aldehyde intermediates. The process can be carried out intramolecularly[325] or intermolecularly, for example by the reaction of aniline with triethanolamine.[326] At a still lower oxidation level, 2-(*ortho*-aminoaryl)-ethanols can be converted[327] directly into indoles with a catalyst that can oxidise the alcohol to a carbonyl group, with expulsion of hydrogen.

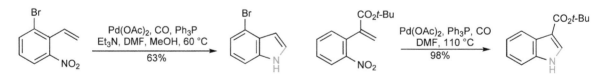

20.16.1.4 *From* ortho-*Alkynyl-Arylamines*

Cyclisation of *ortho*-alkynyl-arylamines can be achieved in various ways; Sonogashira reactions (4.2) provide the starting *ortho*-alkynyl-anilines.

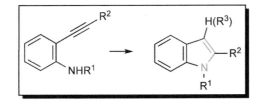

Sonogashira reactions allow easy access to arenes with an alkynyl substituent *ortho* to nitrogen, for example from *ortho*-iodo- and -bromo-nitrobenzenes,[328,329] or *ortho*-iodo- and -bromo-*N*-acyl- (or *N*-sulfonyl)-arylamines,[330] or even by coupling acetylenes with 2-iodoaniline itself.[331] Conversion of *ortho*-alkynyl-nitrobenzenes and -arylamines into indoles can be achieved in several ways, for example alkoxides add to the triple bond and form nitro-acetals, nitro-group reduction then acetal hydrolysis bringing ring closure.[329]

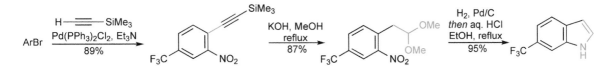

Direct cyclisation of *ortho*-alkynyl-anilines can be effected simply by treatment with tetra-*n*-butylammonium fluoride,[332] potassium *t*-butoxide or potassium hydride,[333] or simply with gold(III) chloride.[334] Treatment of 1-(2-arylethynyl)-2-nitroarenes with indium and aqueous hydrogen iodide produces 2-aryl-indoles, the reagent combination both reduces the nitro to amine and then the acid activates the alkyne for the ring closure.[335] Copper(II) salts[336] or diethylzinc in refluxing toluene[337] can be utilised with *N*-sulfonyl-*ortho*-alkynyl-anilines.

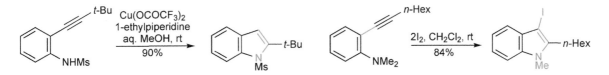

3-Iodo-indoles can be prepared simply by using the halogen (or bis(pyridine)iodonium(I) tetrafluoroborate[338]) to activate the alkyne for nucleophilic attack by the nitrogen.[339] Even *N*,*N*-dimethyl *ortho*-alkynyl-anilines take part in such closures, iodomethane being lost in the final stage, with formation of an *N*-methyl indole.[340]

When palladium is used to effect ring closure, the organopalladium intermediate can be either protonolysed, producing a 3-unsubstituted indole, or trapped out with consequent insertion of a substituent at the indole β-position.[341]

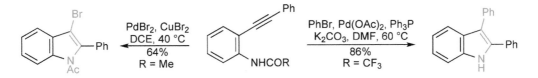

20.16.1.5   From ortho-*Halo Aryl-Amines and Alkynes*

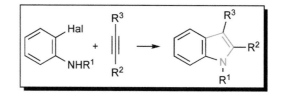

Some reaction conditions using terminal alkynes and *ortho*-halo-anilines lead directly through to indoles, but probably these involve initial alkynylation of the aromatic and then ring closure *in situ* (20.16.1.4).[342,343,344] Using titanium tetrachloride, the two steps *in situ* are hydroamination then Heck cyclisation.[345]

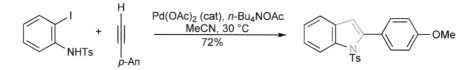

Disubstituted acetylenes can be utilised in a palladium-catalysed cyclo-condensation of *ortho*-halo-anilines giving the indole directly; the larger group (or hydroxyl-containing group) finishes at C-2.[346] A pyridin-2-yl-substituent tends to end at C-2, also, by coordinating the metal in the intermediate.[347]

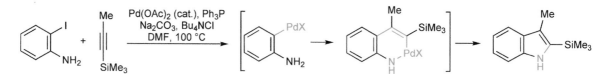

20.16.1.6   From ortho-*Toluidides*
Base-catalysed cyclo-condensation of an *ortho*-alkyl-anilide gives an indole.

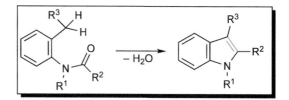

The Madelung Synthesis
In its original form, this route employed very harsh conditions (typically[348] sodium amide or potassium *t*-butoxide at 250–350 °C) to effect condensation between an unactivated aromatic methyl and an *ortho*-acylamino substituent, and was consequently limited to situations having no other sensitive groups. With the advent of the widespread use of alkyllithiums as bases, these cyclocondensations can now be brought about under much milder conditions.[349]

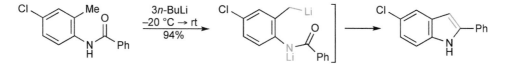

Modifications in which the benzylic hydrogens are acidified also allow the use of mild conditions; one example is the generation of a phosphonium ylide and then an intramolecular Wittig-like reaction, involving the amide carbonyl;[350] another variant uses a benzyl-silane.[351] The use of an amino-silane permits reaction at both nitrogen and benzylic carbon to take place in one pot.[352]

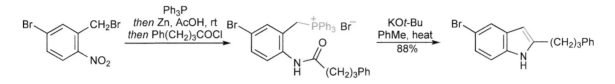

Formation of the 2–3 bond is also possible using the anion from an aryl-amino-nitrile adding intramolecularly to an unsaturated ketone or ester; the nitrile serves to acidify the future C-2-hydrogen and also to bring about aromatisation *via* final loss of hydrogen cyanide.[353]

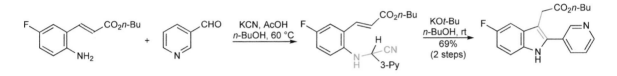

Finally, in this category there must be included cyclisations of the benzylic anions derived from *ortho*-isocyano-toluenes; the scheme shows this route in its simplest form. However, the synthesis is very flexible, for example the initial benzylic anion can be alkylated with halides or epoxides before the ring closure, thus providing 3-substituted indoles and, additionally, the final *N*-lithioindole can be *N*-alkylated by adding a suitable electrophile before work-up.[354]

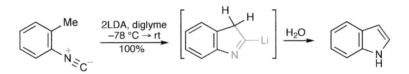

20.16.1.7 *From α-Arylamino-Carbonyl Compounds*
An α-arylamino-ketone is cyclised by electrophilic attack onto the aromatic ring.

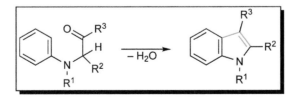

The Bischler Synthesis
In the original method, the Bischler synthesis, harsh acidic treatment of 2-arylamino-ketones (produced from a 2-halo-ketone and an arylamine) was used to bring about electrophilic cyclisation onto the aromatic ring; these conditions often result in mixtures of products *via* rearrangements.[355] However, *N*-acylated 2-arylamino-ketones and, particularly, acetals can be cyclised under much more controlled conditions, and in contrast to earlier work, this approach to indoles can even be used to produce hetero-ring-unsubstituted

indoles.[356] Lithium bromide, as a Lewis acid, will catalyse the ring closure of dimethoxyarylamino-ketones, without rearrangement, under essentially neutral conditions, indeed a one-pot procedure will take a dimethoxyaniline and the chloro-ketone through to the indole.[357]

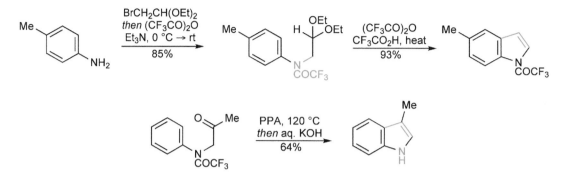

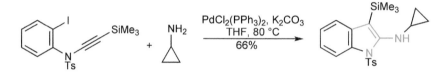

There is also a palladium(0) catalytic route involving formation of the 3–3a bond: assembly of an *N*-alkynyl-*ortho*-iodoaniline-tosylamide is followed by exposure to the catalyst and an amine, which becomes incorporated into the indole at C-2.[358]

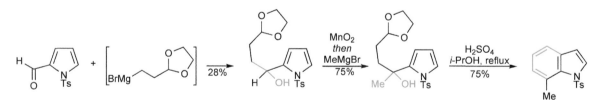

20.16.1.8 *From Pyrroles (see also 16.8)*

Several unrelated strategies have been utilised for the fusion of a benzene ring onto a pyrrole to generate an indole;[359] most follow a route in which a pyrrole, carrying a four-carbon side-chain at the α-carbon, is cyclised *via* an electrophilic attack at the adjacent pyrrole β-position; one of these is shown.[360] Another route involves the electrocyclisation of 2,3-divinyl-pyrroles.[361]

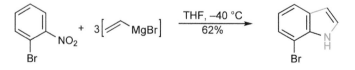

20.16.1.9 *From* ortho-*Substituted Nitro-Arenes*
Bartoli Synthesis

In the extraordinary, but nonetheless efficient and extremely practically simple process now known as the Bartoli synthesis, *ortho*-substituted nitrobenzenes[362] treated with three mole equivalents of vinylmagnesium bromide give 7-substituted indoles. The process works best when the 7-substitutent is large[363] (a bromine can be used as a removable 'large' group[364]) and it is thought that initial attack by the vinyl

Grignard is at the nitro group oxygen with subsequent elimination of magnesium enolate producing the nitroso equivalent of the original – it seems likely that this step is encouraged by non-planarity of the nitro group and the aromatic system, forced on the molecule by the large *ortho*-substituent. A second mole equivalent of vinyl Grignard then adds, again to oxygen generating an intermediate which undergoes a 3,3-sigmatropic rearrangement, much like that involved in the Fischer sequence, and finally hetero-ring closure occurs.[362,365]

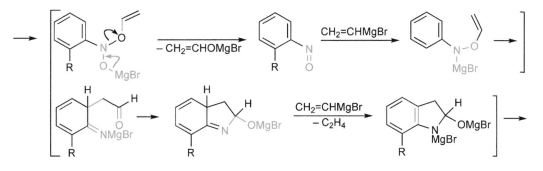

20.16.1.10 From N-Aryl Enamines
It is not clear whether the palladium-mediated cyclisations of anilino-acrylates and related systems[366] operate *via* a Heck sequence or *via* an electrophilic palladation of the enamine.

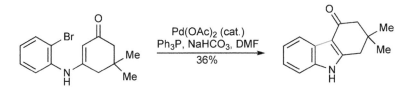

In a very useful modification, simple ketones with CH_2 adjacent to the carbonyl (cyclic ketones work much better than acyclic ketones) and *ortho*-iodo-arylamines react under palladium catalysis to give indoles directly. The use of dimethylformamide as solvent and DABCO as the base are crucial to the success of the route. Mechanistically, the sequence certainly proceeds through the enamine. As well as being conceptually and practically simple, this method tolerates functional groups that would be sensitive to the acid of the traditional Fischer sequence.[367] This method can also be applied to aldehydes, thus providing a direct route to 2-unsubstituted indoles, including side-chain-protected tryptophans.[368]

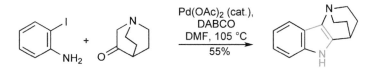

20.16.1.11 From Enamines and p-Benzoquinones
The Nenitzescu Synthesis
The Nenitzescu synthesis[369] is another process about which some of the mechanistic details remain unclear,[370] but which can be used for the efficient synthesis of certain 5-hydroxy-indoles.[371]

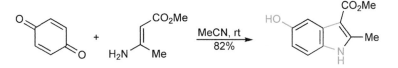

20.16.1.12 From Aryl-Amines
The Gassman Synthesis

The Gassman synthesis[372] produces sulfur-substituted indoles, but these can easily be hydrogenolysed if required.

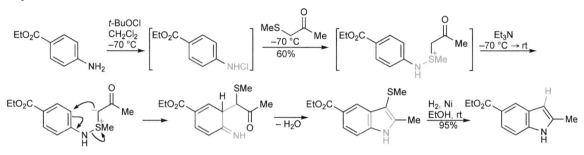

20.16.1.13 From ortho-Acyl-Anilides
The Fürstner Synthesis[373]

This flexible synthesis depends on the reductive cyclisation of *ortho*-acyl-anilides with low-valent titanium – the conditions used for the McMurray coupling of ketones. In the example below, the cyclisation precursor was built up *via* the acylation of 2-tri-*n*-butylstannylthiazole.[374]

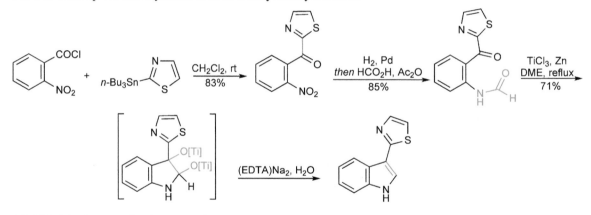

20.16.1.14 From ortho-Isocyano-Styrenes
The Fukuyama Synthesis[375]

ortho-Isocyano-styrenes, which are readily prepared by dehydration of the corresponding formamides, undergo tin-promoted radical cyclisation to give unstable 2-stannyl-indoles, which can either be hydrolysed to afford the corresponding 2-unsubstituted indole, or used without isolation for coupling with aryl halides using palladium(0) catalysis,[376] or converted into 2-iodoindoles *via ipso*-substitution with iodine.[377]

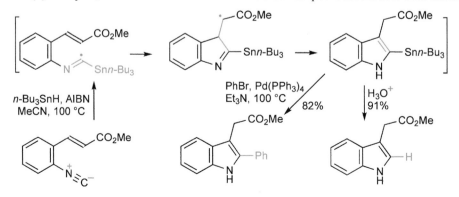

20.16.1.15 By Cyclisation of Nitrenes

Thermolysis of *ortho*-azido-styrenes gives nitrenes that insert into the side-chain to form indoles.[378] Similar nitrenes have been generated by reaction of nitro-compounds with trialkyl phosphites. The azide thermolysis method can be used to prepare 2-nitroindoles.[379]

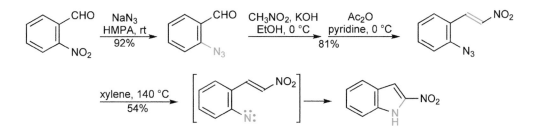

In a complementary sense, thermolysis of β-azido-styrenes also gives indoles, but here the intermediate may be an azirine;[380] this method, the *Hemetsberger–Knittel synthesis*, is particularly useful for the fusion of a pyrrole ring onto rings other than a benzene ring, as illustrated.[381]

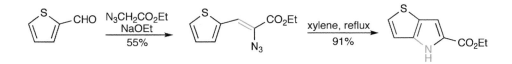

20.16.1.16 By Formation of the N–C-7a Bond

The formation of indoles by making the N–C-7a bond is relatively undeveloped. One method is to activate the nitrogen using phenyliodine bis(trifluoroacetate): a radical sequence is believed to operate.[382]

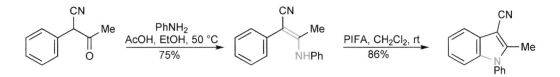

Intramolecular displacement of an *ortho*-halogen can be achieved at high temperature[383] or with copper(I)-catalysis.[384] Palladium-catalysed aminations of halide can be used to form either the N–C-2 or N–C-7a links, or both, for example in the double displacement shown below.[385]

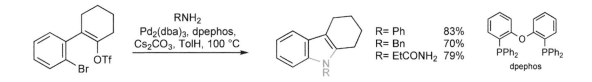

20.16.1.17 From Indolines

Indolines are useful intermediates for the synthesis of indoles with substituents in the carbocyclic ring. In electrophilic substitutions, they behave like anilines; the example shows *N*-acetylindoline undergoing regioselective 7-thallation. Nitration of indoline 2-carboxylic acid gives the 6-nitro-derivative; separation from the 5-nitro minor product is readily achieved by controlled pH extraction. The *N*-acetyl-derivative nitrates at C-5.[386]

Indolines can be obtained easily from indoles by reduction (see 20.7) and can be cleanly oxidised back to indoles using a variety of methods, including oxygen with cobalt catalysis (salcomine),[387] hypochlorite/dimethyl sulfide,[388] Mn(III)[389] and Au(III) compounds,[390] DDQ or manganese dioxide.[386]

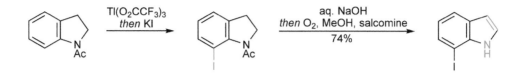

An attractive variant is to utilise certain products of reversible addition to 3*H*-indolium cations, such as the indole bisulfite adduct, or where there has been an intramolecular nucleophilic addition (20.1.1.2): such compounds, though they are indolines, are still at the oxidation level of indoles, needing only mild acid treatment to regenerate the aromatic system.[391]

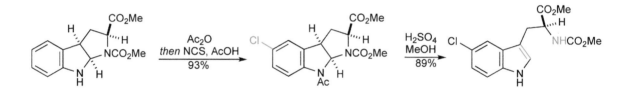

20.16.2   Ring Synthesis of Oxindoles[392]

The main synthesis of oxindoles is simple and direct and involves an intramolecular Friedel–Crafts alkylation reaction as the cyclising step.[393] Also straightforward in concept is the displacement of halogen from an *ortho*-halo-nitroarene with malonate, this leading to an oxindole after decarboxylation and reduction of the nitro group with spontaneous lactamisation.[261]

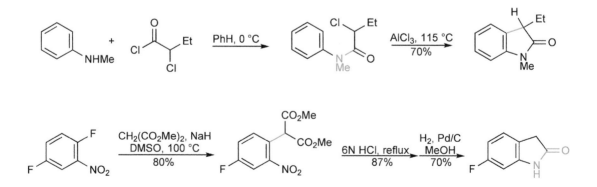

An alternative route to oxindoles depends on the intramolecular insertion of a rhodium carbenoid, derived from a 2-diazo-1,3-ketoamide, into an adjacent aromatic C–H bond.[394]

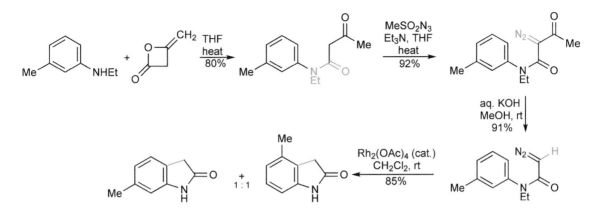

Oxindoles can also be prepared by palladium-catalysed enolate cyclisation of *ortho*-halo-anilides.[395]

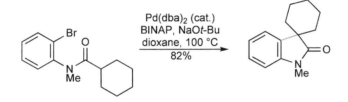

20.16.3 Ring Synthesis of Indoxyls

Indoxyls are normally prepared from anthranilic acids *via* alkylation with a haloacetic acid followed by a cyclising condensation with loss of carbon dioxide.[396,397] Indoxyl itself is best prepared by Friedel–Crafts type ring closure of *N*-phenylglycine activated with triphenylphosphine oxide/triflic anhydride in the presence of triethylamine at room temperature.[398]

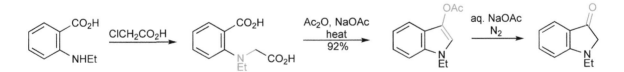

It is possible to directly chloroacylate an aniline using chloroacetonitrile and boron trifluoride, *ortho* to the nitrogen. After *N*-acylation, ring closure produces *N*-acetyl-indoxyls. The *Sugasawa synthesis* of indoles utilises these same *ortho*-chloroacetyl-anilines, *via* borohydride reduction and ring closure.[399]

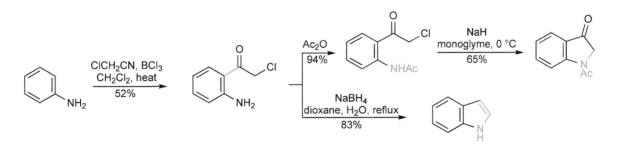

20.16.4 Ring Synthesis of Isatins

An isatin can be readily prepared *via* the reaction of an aniline with chloral, the resulting product converted into an oxime, and this cyclised in strong acid.[400] An alternative route to the oximinoacetanilide intermediates involves acylation of the aniline with 2,2-diacetoxyacetyl chloride and then reaction with hydroxylamine.[401]

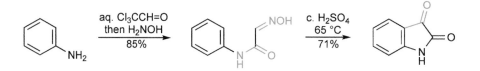

N,N-Dialkyl-anilines react with oxalyl chloride producing *N*-alkyl isatins.[402]

20.16.5 Synthesis of 1-Hydroxy-Indoles

The oxidation of indolines with sodium tungstate/hydrogen peroxide both aromatises and also oxidises the nitrogen, resulting in 1-hydroxy-indoles.[162] 1-Hydroxy-indoles can also be obtained *via* ring synthesis involving lead-promoted reductive cyclisation of *ortho*-nitrobenzyl-ketones (or -aldehydes).[403]

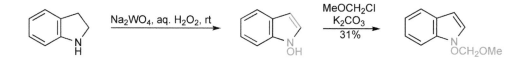

20.16.6 Examples of Notable Indole Syntheses

20.16.6.1 Ondansetron

Ondansetron is a selective, 5-hydroxytryptamine antagonist, used to prevent vomiting during cancer chemotherapy and radiotherapy.

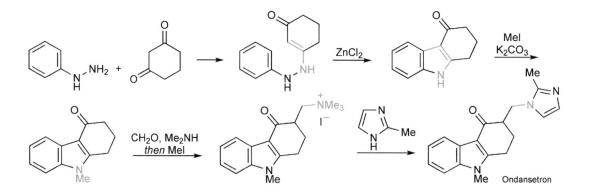

20.16.6.2 Staurosporine Aglycone[404]

Staurosporine and related molecules are under active investigation as potential anti-tumour agents. The synthesis illustrates several aspects of heterocyclic chemistry, including a 2-pyrone acting as a diene in an intramolecular Diels–Alder reaction, and the use of nitrene insertion for the formation of 5-membered nitrogen rings.

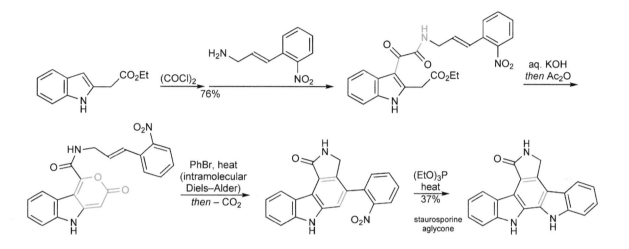

20.16.6.3 Serotonin

Serotonin has been synthesised by several routes; the method shown[405] relies on a Fischer indole synthesis, the requisite aryl-hydrazone being constructed by a process known as the *Japp–Klingemann reaction* in which the enol of a 1,3-dicarbonyl compound is reacted with an aryl-diazonium salt, with subsequent cleavage of the 1,3-dicarbonyl unit.

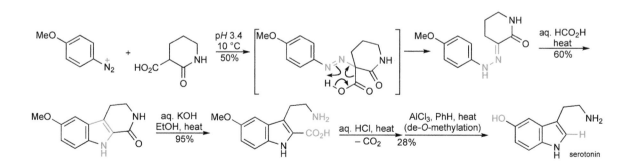

20.16.6.4 Chuangxinmycin[406]

This synthesis uses the approach of starting from a pyrrole: the cyclic ketone intermediate is in general a useful intermediate for the synthesis of 4-substituted indoles – in this case a sulfur substituent – it is already at the aromatic oxidation level needing only the loss of the 4-chlorophenylthiol.

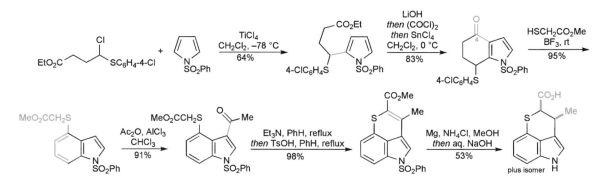

20.16.6.5 Dragmacidin D[407]

A synthesis of dragmacidin D, isolated from sea sponges, a selective inhibitor of threonine protein phosphatases, illustrates indole, pyrazine and imidazole chemistry. The final stages of the synthesis rest on palladium(0)-catalysed couplings: a bromo-iodo-pyrazine entered regioselectively into coupling with two indoles; the final stage, making the imidazole ring involved a 2-amino-ketone reacting with cyanamide.

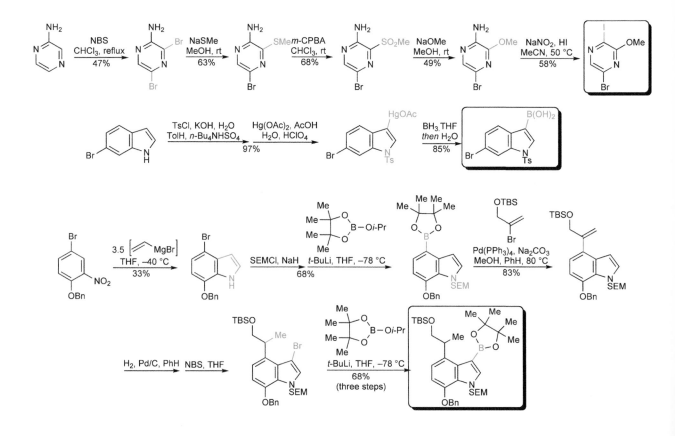

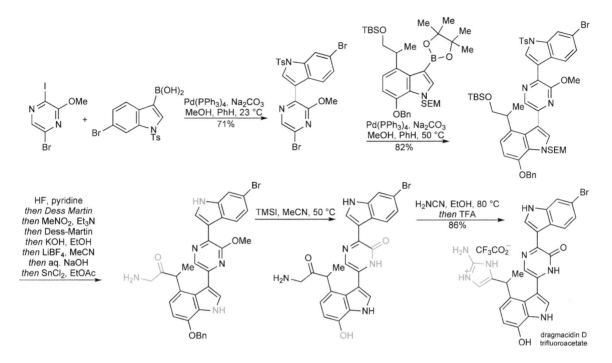

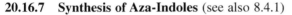

20.16.7 Synthesis of Aza-Indoles (see also 8.4.1)

Most syntheses of aza-indoles start from pyridines and parallel the standard indole syntheses discussed above. However, the Fischer reaction using pyridyl-hydrazones is much less consistent and useful than for arylhydrazones; the Madelung reaction is also not as useful, however the Bartoli route[408] (20.16.1.9) and the Gassman approach[409] (20.16.1.12) can be used to effect. The most successful methods involve palladium-catalysed coupling of acetylenes with amino-halo-pyridines either as one-[410] or two-step[333,411] processes. The starting amino-halo-pyridines are generally available *via* directed metallations.

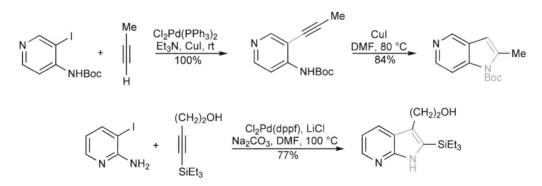

Syntheses utilising nitro-pyridines by Leimgruber–Batcho processes work well[412] and can be modified by the introduction of a further substituent at the enamine stage.[413]

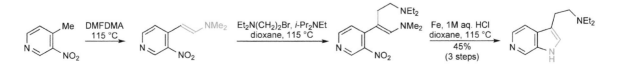

7-Aza-indoles can be prepared from 2,6-dichloropyridin-3-yl-epoxides by reaction with primary amines[414] and 5-, 6-, and 7-azaindole-2-esters can be made[415] *via* the Hemetsberger–Knittel route. Note that 4-aza-indoles cannot be made this way since cyclisation of the appropriate precursor takes place preferentially onto the ring nitrogen generating a pyrazolo[1,5-*a*]-pyridine.

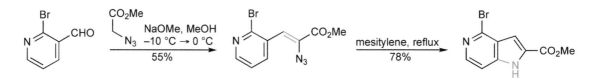

The sequence below shows the assembly of a ring-closure precursor using a VNS (3.3.3) reaction.[416]

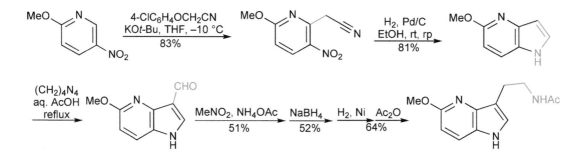

Synthesis from pyrroles is useful in particular cases.[412,417]

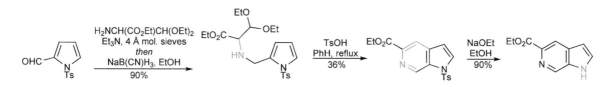

Exercises

Straightforward revision exercises (consult Chapters 19 and 20):

(a) What is the pK_{aH} of indole as a base and where does it protonate? What is the pK_a of indole as an acid?

(b) At what position is electrophilic substitution of indole fastest? Cite two examples.

(c) What are the structures of the intermediates and final product in the following sequence: indole with $(COCl)_2 \rightarrow C_{10}H_6ClNO_2$, then this with ammonia $\rightarrow C_{10}H_8N_2O_2$, then this with $LiAlH_4 \rightarrow C_{10}H_{12}N_2$? Explain the last transformation in mechanistic terms.

(d) How could one prepare from indole: (i) 3-formylindole; (ii) 3-(2-nitroethyl)indole; (iii) 3-dimethylaminomethylindole; (iv) 1-methylindole.

(e) At what position does strong base deprotonate an *N*-substituted indole? Name two groups that can be used to block the 1-position for such deprotonations and that could be removed later. How would these blocking groups be introduced onto the indole nitrogen?

(f) What is the mechanism of the conversion of 3-dimethylaminomethylindole into 3-cyanomethylindole on reaction with NaCN?

(g) Which phenyl-hydrazones would be required for the Fischer indole synthesis of: (i) 3-methylindole; (ii) 1,2,3,4-tetrahydrocarbazole; (iii) 2-ethyl-3-methylindole; (iv) 3-ethyl-2-phenylindole?

(h) How could one convert 2-bromoaniline into 2-phenylindole (more than one step is required)?

(i) What are the advantages of using an indoline (a 2,3-dihydroindole) as an intermediate for the synthesis of indoles?

More advanced exercises:

1. Indole reacts with a mixture of *N*-methyl-2-piperidone and POCl$_3$, followed by NaOH work-up to give C$_{14}$H$_{18}$N$_2$O. What is its structure?

2. Suggest a structure for the tetracyclic product, C$_{18}$H$_{19}$NO, formed when 3-methylindole is treated with 2-hydroxy-3,5-dimethylbenzyl chloride.

3. When indole dimer is subjected to acid treatment in the presence of indole, 'indole trimer', C$_{24}$H$_{21}$N$_3$, is produced. Suggest a structure for the 'trimer'. (Hint: consider which of the two reactants would be most easily protonated, and at which atom.)

4. Starting from indole, and using a common intermediate, how could one prepare: (i) indol-3-ylacetic acid and (ii) tryptamine?

5. What would be the products from the reactions of 5-bromo-3-iodo-1-phenylsulfonylindole with: (i) PhB(OH)$_2$/Pd(PPh$_3$)$_4$/aq. Na$_2$CO$_3$; (ii) ethyl acrylate/Pd(OAc)$_2$/Ph$_3$P/Et$_3$N?

6. Deduce a structure, and write out the mechanism for the conversion of 2-formylindole into a tricyclic compound, C$_{11}$H$_9$N, on treatment with a combination of NaH and Ph$_3$P$^+$CH = CH$_2$ Br$^-$.

7. When 3-ethyl-3-methyl-3*H*-indole is treated with acid, two products, each isomeric with the starting material, are formed – deduce their structures and explain the formation of two products.

8. Suggest a structure for the salt C$_{15}$H$_{13}$N$_2$$^+$ Br$^-$ formed by the following sequence: 2-(2-pyridyl)indole reacted first with *n*-BuLi then PhSO$_2$Cl ($\rightarrow$ C$_{19}$H$_{14}$N$_2$O$_2$S), then this sequentially with *t*-BuLi at -100 °C, then ethylene oxide ($\rightarrow$ C$_{21}$H$_{18}$N$_2$O$_3$S), aq. NaOH ($\rightarrow$ C$_{15}$H$_{14}$N$_2$O), and this finally reacted with PBr$_3$.

9. What are the products formed in the following sequence: indole/*n*-BuLi, then I$_2$, then LDA, then PhSO$_2$Cl $\rightarrow$ C$_{14}$H$_{10}$INO$_2$S, then this with LDA, then I$_2$ $\rightarrow$ C$_{14}$H$_9$I$_2$NO$_2$S.

10. When indol-3-yl-CH$_2$OH is heated with acid, di(indol-3-yl)methane is formed: suggest a mechanism for this transformation.

11. What product, C$_{10}$H$_{11}$NO, would be obtained from refluxing a mixture of phenylhydrazine and 2,3-dihydrofuran in acetic acid?

12. Draw structures for the aza-indoles resulting from treatment of 2-methyl-3-nitro- and 4-methyl-3-nitro-pyridines, respectively, with (EtO$_2$C)$_2$/EtONa, followed by H$_2$/Pd–C. Both products have the molecular formula C$_{10}$H$_{10}$N$_2$O$_2$.

13. Heating DMFDMA with the following aromatic compounds led to condensation products; subsequent reduction with the reagent shown gave indoles. Draw the structures of the condensation products and the indoles: (i) 2,6-dinitrotoluene then TiCl$_3$ gave C$_8$H$_8$N$_2$; (ii) 2-benzyloxy-6-nitrotoluene then H$_2$/Pt gave C$_{15}$H$_{13}$NO; (iii) 4-methoxy-2-nitrotoluene then H$_2$/Pd gave C$_9$H$_9$NO; (iv) 2,3-dinitro-1,4-dimethylbenzene then H$_2$/Pd gave C$_{10}$H$_8$N$_2$.

References

[1] 'The chemistry of indoles', Sundberg, R. J., Academic Press, New York, **1970**; 'Transition metals in the synthesis and functionalisation of indoles', Hegedus, L. S., *Angew. Chem., Int. Ed. Engl.*, **1988**, *27*, 1113; 'Indoles', Sundberg, R. J., Academic Press, London, **1996**.

[2] Gut, I. G. and Wirz, J., *Angew. Chem., Int. Ed. Engl.*, **1994**, *33*, 1153.

[3] Hinman, R. L. and Lang, J., *J. Am. Chem. Soc.*, **1964**, *86*, 3796.

[4] Hinman, R. L. and Whipple, E. B., *J. Am. Chem. Soc.*, **1962**, *84*, 2534.

[5] Challis, B. C. and Millar, E. M., *J. Chem. Soc., Perkin Trans. 2*, **1972**, 1111; Muir, D. M. and Whiting, M. C., *J. Chem. Soc., Perkin Trans. 2*, **1975**, 1316.

[6] Hinman, R. L. and Bauman, C. P., *J. Org. Chem.*, **1964**, *29*, 2437.

[7] Russell, H. F., Harris, B. J. and Hood, D. B., *Org. Prep. Proc. Int.*, **1985**, *17*, 391.

[8] Bourne, G. T., Crich, D., Davies, J. W. and Horwell, D. C., *J. Chem. Soc., Perkin Trans. 1*, **1991**, 1693.

[9] Ley, S. V., Cleator, E. and Hewitt, P. R., *Org. Biomol Chem*, **2003**, *1*, 3492.

[10] Baran, P., Guerrero, C. A. and Corey, E. J., *Org. Lett.*, **2003**, *5*, 1999.

[11] Kawahara, M., Nishida, A. and Nakagawa, M., *Org. Lett.*, **2000**, *2*, 671.

[12] Kamenecka, T. M. and Danishefsky, S. J., *Angew. Chem. Int. Ed.*, **1998**, *37*, 2993.

[13] Pelkey, E. T. and Gribble, G. W., *Synthesis*, **1997**, 1117.

[14] Berti, G., Da Settimo, A. and Nannipieri, E., *J. Chem. Soc., C*, **1968**, 2145.

[15] Brown, K. and Katritzky, A. R., *Tetrahedron Lett.*, **1964**, 803.

[16] Ottoni, O., Cruz, R. and Krammer, N. H., *Tetrahedron Lett.*, **1999**, *40*, 1117.

[17] Mitchell, H. and Leblanc, Y., *J. Org. Chem.*, **1994**, *59*, 682.

[18] Smith, G. F. and Taylor, D. A., *Tetrahedron*, **1973**, *29*, 669.

[19] Fatum, T. M., Anthoni, U., Christophersen, C. and Nielsen, P. H., *Heterocycles*, **1994**, *38*, 1619.

[20] Janosik, T., Shirani, H., Wahlström, N., Malky, I., Stensland, B. and Bergman, J., *Tetrahedron*, **2006**, *62*, 1699.

[21] Raban, M. and Chern, L.-J., *J. Org. Chem.*, **1980**, *45*, 1688; Gilow, H. M., Brown, C. S., Copeland, J. N. and Kelly, K. E., *J. Heterocycl. Chem.*, **1991**, *28*, 1025.

[22] Tudge, M., Tamiya, M., Savarin, C. and Humphrey, G. R., *Org. Lett.*, **2006**, *8*, 565.

[23] Schlosser, K. M., Krasutsky, A. P., Hamilton, H. W., Reed, J. E. and Sexton, K., *Org. Lett.*, **2004**, *6*, 819.

[24] Yadav, J. S., Reddy, B. V. S. and Reddy, Y. J., *Tetrahedron Lett.*, **2007**, *48*, 7034.

[25] Nair, V., George, T. G., Nair, L. G. and Panicker, S. B., *Tetrahedron Lett.*, **1999**, *40*, 1195.

[26] Yadav, J. S., Reddy, B. V. S., Krishna, A. D., Reddy, Ch. S. and Narsaiah, A. V., *Synthesis*, **2005**, 961.

[27] (a) Bocchi, V. and Palla, G., *Synthesis*, **1982**, 1096; (b) Piers, K., Meimaroglou, C., Jardine, R. V. and Brown, R. K., *Canad. J. Chem.*, **1963**, *41*, 2399; (c) Arnold, R. D., Nutter, W. M. and Stepp, W. L., *J. Org. Chem.*, **1959**, *24*, 117; (d) De Rosa, M. and Alonso, J. L. T., *J. Org. Chem.*, **1978**, *43*, 2639.

[28] Saulnier, M. G. and Gribble, G. W., *J. Org. Chem.*, **1982**, *47*, 757.

[29] Tang, S., Li, J.-H., Xie, Y.-X. and Wang, N.-X., *Synthesis*, **2007**, 1535.

[30] Dmitrienko, G. I., Gross, E. A. and Vice, S. F., *Canad. J. Chem.*, **1980**, *58*, 808.

[31] Zhang, P., Liu, R. and Cook, J. M., *Tetrahedron Lett.*, **1995**, *36*, 3103.

[32] Bergman, J. and Venemalm, L., *J. Org. Chem.*, **1992**, *57*, 2495.

[33] Brennan, M. R., Erickson, K. L., Szmalc, F. S., Tansey, M. J. and Thornton, J. M., *Heterocycles*, **1986**, *24*, 2879.

[34] Putey, A., Popowycz, F. and Joseph, B., *Synlett*, **2007**, 419.

[35] Saxton, J. E., *J. Chem. Soc.*, **1952**, 3592; Hart, G., Liljegren, D. R. and Potts, K. T., *ibid.*, **1961**, 4267.

[36] Nickisch, K., Klose, W. and Bohlmann, F., *Chem. Ber.*, **1980**, *113*, 2036.

[37] Cipiciani, A., Clementi, S., Linda, P., Savelli, G. and Sebastiani, G. V., *Tetrahedron*, **1976**, *32*, 2595.

[38] Ottoni, O., Neder, A. de V. F., Dias, A. K. B., Cruz, R. P. A. and Aquino, L. B., *Org. Lett.*, **2001**, *3*, 1005.

[39] Gmeiner, P., Kraxner, J. and Bollinger, B., *Synthesis*, **1996**, 1196.

[40] Smith, G. F., *J. Chem. Soc.*, **1954**, 3842; James, P. N. and Snyder, H. R., *Org. Synth., Coll. Vol. IV*, **1963**, 539.

[41] Anthony, W. C., *J. Org. Chem.*, *1960*, 25, **2049**.

[42] Nogrady, T. and Morris, L., *Canad. J. Chem.*, **1969**, *47*, 1999; Monge, A., Aldana, I., Lezamiz, I. and Fernandez-Alvarez, E., *Synthesis*, **1984**, 160.

[43] Tobisu, M., Yamaguchi, S. and Chatani, N., *Org. Lett.*, **2007**, *9*, 3351.

[44] Speeter, M. E. and Anthony, W. C., *J. Am. Chem. Soc.*, **1954**, *76*, 6208.

[45] Ishizumi, K., Shioiri, T. and Yamada, S., *Chem. Pharm. Bull.*, **1967**, *15*, 863.

[46] Späth, E. and Lederer, E., *Chem. Ber.*, **1930**, *63*, 2102.

[47] 'Indolo-2,3-quinodimethanes and stable analogues for regio- and stereocontrolled syntheses of [*b*]-anelated indoles', Pindur, U. and Erfanian-Abdoust, H., *Chem. Rev.*, **1989**, *89*, 1681.

[48] Li, J., Li, B., Chen, X. and Zhang, G., *Synlett*, **2003**, 1447.

[49] Nakatsuka, S., Teranishi, K. and Goto, T., *Tetrahedron Lett.*, **1994**, *35*, 2699; Teranishi, K., Nakatsuka, S. and Goto, T., *Synthesis*, **1994**, 1018.

[50] Teranishi, K., Hayashi, S., Nakatsuka, S. and Goto, T., *Tetrahedron Lett.*, **1994**, *35*, 8173.

[51] Cruz, R. P. A., Ottoni, O., Abella, C. A. M. and Aguino, L. B., *Tetrahedron Lett.*, **2001**, *42*, 1467; Jiang, Y. and Ma, D., *Tetrahedron Lett.*, **2002**, *43*, 7013.

[52] 'A journey across recent advances in catalytic and stereoselective alkylation of indoles', Bandini, M., Melloni, A., Tommasi, S. and Umani-Ronchi, A., *Synlett*, **2005**, 1199.

[53] Ciamician, G. and Plancher, G., *Chem. Ber.*, **1896**, *29*, 2475; Jackson, A. H. and Smith, P., *Tetrahedron*, **1968**, *24*, 2227.

[54] Jackson, A. H., Naidoo, B. and Smith, P., *Tetrahedron*, **1968**, *24*, 6119.

[55] Casnati, G., Dossena, A. and Pochini, A., *Tetrahedron Lett.*, **1972**, 5277.

[56] Iyer, R., Jackson, A. H., Shannon, P. V. R. and Naidoo, B., *J. Chem. Soc., Perkin Trans. 2*, **1973**, 872.

[57] Ganesan, A. and Heathcock, C. H., *Tetrahedron Lett.*, **1993**, *34*, 439.

[58] Johansen, J. E., Christie, B. D. and Rapoport, H., *J. Org. Chem.*, **1981**, *46*, 4914.

[59] Kraus, G. A. and Malpert, J. H., *Synlett*, **1997**, 107.

[60] Jiao, J., Zhang, Y. and Flowers, R. A., *Synlett*, **2006**, 3355.

[61] Zhu, X. and Ganesan, A., *J. Org. Chem.*, **2002**, *67*, 2705.

[62] Westermaier, M. and Mayr, H., *Org. Lett.*, **2006**, *8*, 4791.

[63] Zaitsev, A. B., Gruber, S. and Pregosin, P. S., *Chem. Commun.*, **2007**, 4692.

[64] Liu, Z., Liu, L., Shafiq, Z., Wu, Y.-C., Wang, D. and Chen, Y.-J., *Tetrahedron Lett.*, **2007**, *48*, 3963.

[65] Sanz, R., Miguel, D., Álvarez-Gutiérrez, J. M. and Rodríguez, F., *Synlett*, **2008**, 975.

[66] Sobenina, L. N., Demenev, A. P., Mikhaleva, A. I., Ushakov, I. A., Vasil'tsov, A. M., Ivanov, A. V. and Trofimov, B. A., *Tetrahedron Lett.*, **2006**, *47*, 7139.

[67] Blaser, G., Sanderson, J. M., Batsanov, A. S. and Howard, J. A. K., *Tetrahedron Lett.*, **2008**, *49*, 2795.

[68] Kotsuki, H., Teraguchi, M., Shimomoto, N. and Ochi, M. *Tetrahedron Lett.*, **1996**, *37*, 3727.

[69] Kotsuki, H., Hayashida, K., Shimanouchi, T. and Nishizawa, H., *J. Org. Chem.*, **1996**, *61*, 984.

[70] Sato, K. and Kozikowski, A. P., *Tetrahedron Lett.*, **1989**, *31*, 4073; Dubois, L., Mehta, A., Tourette, E. and Dodd, R. H., *J. Org. Chem.*, **1994**, *59*, 434.

[71] Bandini, M., Cozzi, P. G., Melchiorre, P. and Umani-Ronchi, A., *J. Org. Chem.*, **2002**, *67*, 5386.

[72] Nakagawa, M. and Kawahara, M., *Org. Lett.*, **2000**, *2*, 935.

[73] Zhan, Z.-P. and Lang, K., *Synlett*, **2005**, 1551.

[74] Agnusdei, M., Bandini, M., Melloni, A. and Umani-Ronchi, A., *J. Org. Chem.*, **2003**, *68*, 7126.

[75] Kawatsura, M., Aburatani, S. and Uenishi, J., *Synlett*, **2005**, 2492.

[76] Szmuszkovicz, J., *J. Am. Chem. Soc.*, **1957**, *79*, 2819.

[77] Gu, Y., Ogawa, C. and Kobayashi, S., *Org. Lett.*, **2007**, *9*, 175.

[78] King, H. D., Meng, Z., Denhart, D., Mattson, R., Kimura, R., Wu, D., Gao, Q. and Macor, J. E., *Org. Lett.*, **2005**, *7*, 3437.

[79] Zhou, J. and Tang, Y., *J. Am. Chem. Soc.*, **2002**, *124*, 9030.

[80] Iqbal, Z., Jackson, A. H. and Nagaraja Rao, K. R., *Tetrahedron Lett.*, **1988**, *29*, 2577.

[81] Harrington, P. E. and Kerr, M. A., *Synlett*, **1996**, 1047.

[82] Robinson, B. and Smith, G. F., *J. Chem. Soc.*, **1960**, 4574; Garnick, R. L., Levery, S. B. and Le Quesne, P. W., *J. Org. Chem.*, **1978**, *43*, 1226.

[83] Balsamini, C., Diamantini, G., Duranti, A., Spadoni, G. and Tontini, A., *Synthesis*, **1995**, 370.

[84] Ranganathan, D., Rao, C. B., Ranganathan, S., Mehrotra, A. K. and Iyengar, R., *J. Org. Chem.*, **1980**, *45*, 1185.

[85] Bartoli, G., Di Antonio, G., Giuli, S., Marcantoni, E., Marcolini, M. and Paoletti, M., *Synthesis*, **2008**, 320.

[86] Azizi, N., Arynasab, F. and Saidi, M. R., *Org. Biomol. Chem.*, **2006**, 4275.

[87] Büchi, G. and Mak, C.-P., *J. Org. Chem.*, **1977**, *42*, 1784.

[88] Majchrzak, M. W. and Simchen, G., *Synthesis*, **1986**, 956.

[89] Sui, Y., Liu, L., Zhao, J.-L., Wang, D. and Chen, Y.-J., *Tetrahedron Lett.*, **2007**, *48*, 3779.

[90] Burr, G. O. and Gortner, R. A., *J. Am. Chem. Soc.*, **1924**, *46*, 1224.

[91] Cook, A. H. and Majer, J. R., *J. Chem. Soc.*, **1944**, 486; Pindur, U. and Flo, C., *Monatsh. Chem.*, **1986**, *117*, 375.

[92] Pindur, U. and Kim, M.-H., *Tetrahedron*, **1989**, *45*, 6427.

[93] Zhuang, W. and Jorgensen, K. A., *Chem Commun.*, **2002**, 1336.

[94] Leete, E., *J. Am. Chem. Soc.*, **1959**, *81*, 6023; Thesing, J., *Chem. Ber.*, **1954**, *87*, 692.

[95] Mukherjee, D., Sarkar, S. K., Chowdhury, U. S. and Taneja, S. C., *Tetrahedron Lett.*, **2007**, *48*, 663.

[96] Appleton, J. E., Dack, K. N., Green, A. D. and Stelle, J., *Tetrahedron Lett.*, **1993**, *34*, 1529.

[97] 'Carbon-carbon alkylations with amines and ammonium salts', Brewster, J. H. and Eliel, E. L., *Org. Reactions*, **1953**, *7*, 99.

[98] Swaminathan, S. and Narisimhan, K., *Chem. Ber.*, **1966**, *99*, 889.

[99] Kühn, H. and Stein, O., *Chem. Ber.*, **1937**, *70*, 567.

[100] Kozikowski, A. P. and Ishida, H., *Heterocycles*, **1980**, *14*, 55.

[101] Yadav, J. S., Reddy, B. V. S., Vishnumurthy, P. and Premalatha, K., *Synthesis*, **2008**, 719.

[102] Xie, W., Bloomfield, K. M., Jin, Y., Dolney, N. Y. and Wang, G. P., *Synlett*, **1999**, 498.

[103] Jiang, B. and Huang, Z.-G., *Synthesis*, **2005**, 2198.

[104] Lei, F., Chen, Y.-J., Sui, Y., Liu, L. and Wang, D., *Synlett*, **2003**, 1160.

[105] Ballini, R., Palmieri, A., Petrini, M. and Torregiani, E., *Org. lett.*, **2006**, *8*, 4093; Petrini, M. and Shaikh, R. R., *Tetrahedron Lett.*, **2008**, *49*, 5645.

[106] 'The acid-catalysed polymerisation of pyrroles and indoles', Smith, G. F., *Adv. Heterocycl. Chem.*, **1963**, *2*, 287.

[107] Bocchi, V. and Palla, G., *Tetrahedron*, **1984**, *40*, 3251.

[108] Hahn, G. and Ludewig, H., *Chem. Ber.*, **1934**, *67*, **2031**.

[109] Narayanan, K. and Cook, J. M., *Tetrahedron Lett.*, **1990**, *31*, 3397.

[110] Kawate, T., Yamada, H., Matsumizu, M., Nishida, A. and Nakagawa, M., *Synlett*, **1997**, 761.

[111] Jia, Y.-X., Xie, J.-H., Duan, H.-F., Wang, L.-X. and Zhou, Q.-L., *Org. Lett.*, **2006**, *8*, 1621.

[112] Hodson, H. F. and Smith, G. F., *J. Chem. Soc.*, **1957**, 3546.

[113] Kirby, G. W. and Shah, S. W., *J. Chem. Soc., Chem. Commun.*, **1965**, 381.

[114] Harrington, P. J. and Hegedus, L. S., *J. Org. Chem.*, **1984**, *49*, 2657.

[115] Zheng, Q., Yang, Y. and Martin, A. R., *Tetrahedron Lett.*, **1993**, *34*, 2235.

[116] Brown, M. A. and Kerr, M. A., *Tetrahedron Lett.*, **2001**, *42*, 983.

[117] (a) Holins, R. A., Colnaga, L. A., Salim, V. M. and Seidl, M. C., *J. Heterocycl. Chem.*, **1979**, *16*, 993; (b) Somei, M., Yamada, F., Kunimoto, M. and Kaneko, C., *Heterocycles*, **1984**, *22*, 797.

[118] Somei, M., Hasegawa, T. and Kaneko, C., *Heterocycles*, **1983**, *20*, 1983; Somei, M. and Yamada, F., *Chem. Pharm. Bull.*, **1984**, *32*, 5064.

[119] Yamada, F. and Somei, M., *Heterocycles*, **1987**, *26*, 1173.

[120] Somei, M., Yamada, F., Hamada, H. and Kawasaki, T., *Heterocycles*, **1989**, *29*, 643.

[121] Nakagawa, M., Kato, S., Kataoka, S. and Hino, T., *J. Am. Chem. Soc.*, **1979**, *101*, 3136.

[122] Chien, C.-S., Hasegawa, A., Kawasaki, T. and Sakamoto, M., *Chem. Pharm. Bull.*, **1986**, *34*, 1493.

[123] Dolby, L. J. and Booth, D. L., *J. Am. Chem. Soc.*, **1966**, *88*, 1049.

[124] Balogh-Hergovich, E. and Speier, G., *Tetrahedron Lett.*, **1982**, *23*, 4473.

[125] Yukimasa, H., Sawai, H. and Takizawa, T., *Chem. Pharm. Bull.*, **1979**, *27*, 551.

[126] Mudry, C. A. and Frasca, A. R., *Tetrahedron*, **1973**, *29*, 603.

[127] Szabó-Pusztay, K. and Szabó, L., *Synthesis*, **1979**, 276.

[128] López-Alvarado, P., Steinhoff, J., Miranda, S., Avendaño, C. and Menéndez, J. C., *Synlett*, **2007**, 2792.

[129] Suárez-Castillo, O. R., Sánchez-Zavala, M., Meléndez-Rodríguez, M., Castelán-Duarte, L. E., Morales-Ríos, M. S. and Joseph-Nathan, P., *Tetrahedron*, **2006**, *62*, 3040.

[130] Wojciechowski, K. and Makosza, M., *Synthesis*, **1989**, 106.

[131] Akai, S., Kawashita, N., Wada, Y., Satoh, H., Alinejad, A.H., Kakiguchi, K., Kuriwaki, I. and Kita, Y., *Tetrahedron Lett.*, **2006**, *47*, 1881.

[132] Roy, S. and Gribble, G. W., *Tetrahedron Lett.*, **2005**, *46*, 1325.

[133] Yagil, G., *Tetrahedron*, **1967**, *23*, 2855; Scott, W. J., Bover, W. J., Bratin, K. and Zuman, P., *J. Org. Chem.*, **1976**, *41*, 1952.

[134] Nunomoto, S., Kawakami, Y., Yamashita, Y., Takeuchi, H. and Eguchi, S., *J. Chem. Soc., Perkin Trans. 1*, **1990**, 111.

[135] Reinecke, M. G., Sebastian, J. F., Johnson, H. W. and Pyun, C., *J. Org. Chem.*, **1972**, *37*, 3066.

[136] Rubottom, G. M. and Chabala, J. C., *Synthesis*, **1972**, 566.

[137] Barco, A., Benetti, S., Pollini, G. P. and Baraldi, P. G., *Synthesis*, **1976**, 124.

[138] Bombrun, A. and Casi, G., *Tetrahedron Lett.*, **2002**, *43*, 2187.

[139] Shieh, W.-C., Dell, S., Bach, A., Repic, O. and Blacklock, T. J., *J. Org. Chem.*, **2003**, *68*, 1954.

[140] Bremner, J. B., Samosorn, S. and Ambrus, J. I., *Synthesis*, **2004**, 2653.

[141] Katritzky, A. R., Khelashvili, L., Mohapatra, P. P. and Steel, P. J., *Synthesis*, **2007**, 3673.

[142] Illi, V. O., *Synthesis*, **1979**, 387; Santaniello, E., Farachi, C. and Ponti, F., *ibid.*, **1979**, 617.

[143] Illi, V. O., *Synthesis*, **1979**, 136.

[144] 'The indole Grignard reagents', Heacock, R. A. and Kaspárek, S., *Adv. Heterocycl. Chem.*, **1969**, *10*, 43.

[145] Bergman, J. and Venemalm, L., *Tetrahedron*, **1990**, *46*, 6061.

[146] Ayer, W. A., Craw, P. A., Ma, Y. and Miao, S., *Tetrahedron*, **1992**, *48*, 2919.

[147] Yang, C. X., Patel, H. H., Ku, Y.-Y., Shah, R. and Sawick, D., *Synth. Commun.*, **1997**, *27*, 2125.

[148] Ezquerra, J., Pedregal, C., Lamas, C., Pastor, A., Alvarez, P. and Vaquero, J. J., *Tetrahedron Lett.*, **1996**, *37*, 683.

[149] Fishwick, C. W. G., Jones, A. D. and Mitchell, M. B., *Heterocycles*, **1991**, *32*, 685; Onistschenko, A. and Stamm, H., *Chem. Ber.*, **1989**, *122*, 2397.

[150] Chakraborty, M., Kundu, T. and Harigaya, Y., *Synth. Commun.*, **2006**, *36*, 2069.

[151] Suárez-Castillo, O. R., Montiel-Ortega, L. A., Meléndez-Rodríguez, M. and Sánchez-Zavala, M., *Tetrahedron Lett.*, **2007**, *48*, 17.

[152] Haskins, C. M. and Knight, D. W., *Tetrahedron Lett.*, **2004**, *45*, 599.

[153] Bajwa, J. S., Chen, G.-P., Prasad, K., Repic, O. and Blacklock, T. J., *Tetrahedron Lett.*, **2006**, *47*, 6435.

[154] Hong, X., Mejía-Oneto, J. M., France, S. and Padwa, A., *Tetrahedron Lett.*, **2006**, *47*, 2409.

[155] Largeron, M., Farrell, B., Rousseau, J.-F., Fleury, M.-B., Potier, P. and Dodd, R. H., *Tetrahedron Lett.*, **2000**, *41*, 9403.

[156] Shirley, D. A. and Roussel, P. A., *J. Am. Chem. Soc.*, **1953**, *75*, 375.

[157] Sundberg, R. J. and Russell, H. F., *J. Org. Chem.*, **1973**, *38*, 3324.

[158] Katritzky, A. R. and Akutagawa, K., *Tetrahedron Lett.*, **1985**, *26*, 5935.

[159] Hasan, I., Marinelli, E. R., Lin, L.-C. C., Fowler, F. W. and Levy, A. B., *J. Org. Chem.*, **1981**, *46*, 157.

[160] Hlasla, D. J. and Bell, M. R., *Heterocycles*, **1989**, *29*, 849; Katritzky, A. R., Lue, P. and Chen, Y.-X., *J. Org. Chem.*, **1990**, *55*, 3688.

[161] Edwards, M. P., Doherty, A. M., Ley, S. V. and Organ, H. M., *Tetrahedron*, **1986**, *42*, 3723.

[162] Somei, M. and Kobayashi, T., *Heterocycles*, **1992**, *34*, 1295.

[163] Kondo, Y., Yoshida, A. and Sakamoto, T., *J. Chem. Soc., Perkin Trans. 1*, **1996**, 2331.

[164] Vazquez, E., Davies, I. W. and Payack, J. F., *J. Org. Chem.*, **2002**, *67*, 7551; Vazquez, E. and Payack, J. F., *Tetrahedron Lett.*, **2004**, *45*, 6549.

[165] Jiang, J. and Gribble, G. W., *Tetrahedron Lett.*, **2002**, *43*, 4115.

[166] Yokoyama, Y., Uchida, M. and Murakami, Y., *Heterocycles*, **1989**, *29*, 1661.

[167] Klingebiel, U., Luttke, W., Noltemeyer, M. and Schmidt, H. G., *J. Organometal. Chem.*, **1993**, *456*, 41.

[168] Matsuzono, M., Fukuda, T. and Iwao, M., *Tetrahedron Lett.*, **2001**, *42*, 7621.

[169] Fukuda, T., Mine, Y. and Iwao, M., *Tetrahedron*, **2001**, *57*, 975.

[170] Buttery, C. D., Jones, R. G. and Knight, D. W., *J. Chem. Soc., Perkin Trans. 1*, **1993**, 1425; Fisher, L. E., Labadie, S. S., Reuter, D. C. and Clark, R. D., *J. Org. Chem.*, **1995**, *60*, 6224.

[171] Saulnier, M. G. and Gribble, G. W., *Tetrahedron Lett.*, **1983**, *24*, 5435.

[172] Iwao, M., *Heterocycles*, **1993**, *36*, 29; Iwao, M. and Motoi, O., *Tetrahedron Lett.*, **1995**, *36*, 5929.

[173] Pérez-Serrano, L., Casarrubios, L., Domínguez, G., Freire, G. and Pérez-Castells, J., *Tetrahedron*, **2002**, *58*, 5407.

[174] Griffen, E. J., Roe, D. G. and Snieckus, V., *J. Org. Chem.*, **1995**, *60*, 1484.

[175] Fukuda, T., Maeda, R. and Iwao, M., *Tetrahedron*, **1999**, *55*, 9151.

[176] Hartung, C. G., Fecher, A., Chapell, B. and Snieckus, V., *Org. Lett.*, **2003**, *5*, 1899.

[177] Liu, Y. and Gribble, G. W., *Tetrahedron Lett.*, **2002**, *43*, 7135.

[178] Saulnier, M. G. and Gribble, G. W., *J. Org. Chem.*, **1982**, *47*, 757.

[179] Amat, M., Seffar, F., Llor, N. and Bosch, J., *Synthesis*, **2001**, 267.

[180] Wynne, J. H. and Stalick, W. M., *J. Org. Chem.*, **2002**, *67*, 5850.

[181] Conway, S. C. and Gribble, G. W., *Heterocycles*, **1990**, *30*, 627.

[182] Amat, M., Hadida, S., Sathyanarayana, S. and Bosch, J., *J. Org. Chem.*, **1994**, *59*, 10.

[183] Rubiralta, M., Casamitjana, N., Grierson, D. S. and Husson, H.-P., *Tetrahedron*, **1988**, *44*, 443.

[184] Johnson, D. A. and Gribble, G. W., *Heterocycles*, **1986**, *24*, 2127.

[185] Kondo, Y., Yoshida, A., Sato, S. and Sakamoto, T., *Heterocycles*, **1996**, *42*, 105.

[186] Kondo, Y., Takazawa, N., Yoshida, A. and Sakamoto, T., *J. Chem. Soc., Perkin Trans. 1*, **1995**, 1207.

[187] Herbert, J. M. and Maggiani, M., *Synth. Commun.*, **2001**, *31*, 947.

[188] Moyer, M. P., Shiurba, J. F. and Rapoport, H., *J. Org. Chem.*, **1986**, *51*, 5106.

[189] Kanaoka, T., Aiura, M. and Hariya, S., *J. Org. Chem.*, **1971**, *36*, 458.

[190] Byers, J. H., Campbell, J. E., Knapp, F. H. and Thissell, J. G., *Tetrahedron Lett.*, **1999**, *40*, 2677.

[191] Chuang, C.-P. and Wang, S.-F., *Synlett*, **1995**, 763.

[192] Ziegler, F. E. and Jeroncic, L. O., *J. Org. Chem.*, **1991**, *56*, 3479.

[193] Aboutayab, K., Caddick, S., Jenkins, K., Joshi, S. and Khan, S., *Tetrahedron*, **1996**, *52*, 11329.

[194] Caddick, S., Aboutayab, K., Jenkins, K. and West, R. I., *J. Chem. Soc., Perkin Trans. 1*, **1996**, 675.

[195] Dobbs, A. P., Jones, K. and Veal, K. T., *Tetrahedron Lett.*, **1995**, *36*, 4857.

[196] Dohi, T., Morimoto, K., Kiyono, Y., Tohma, H. and Kita, Y., *Org. Lett.*, **2003**, *7*, 537.

[197] O'Brien, S. and Smith, D. C. C., *J. Chem. Soc.*, **1960**, 4609; Remers, W. A., Gibs, G. J., Pidoocks, C. and Weiss, M. J., *J. Org. Chem.*, **1971**, *36*, 279; Ashmore, J. W. and Helmkamp, G. K., *Org. Prep. Proc. Int.*, **1976**, *8*, 223.

[198] Barrett, A. G. M., Dauzonne, D., O'Neil, I. A. and Renaud, A., *J. Org. Chem.*, **1984**, *49*, 4409.

[199] Dolby, L. J. and Gribble, G. W., *J. Heterocycl. Chem.*, **1966**, *3*, 124.

[200] Lanzilotti, A. E., Littell, R., Fanshawe, W. J., McKenzie, T. C. and Lovell, F. M., *J. Org. Chem.*, **1979**, *44*, 4809.

[201] Gribble, G. W. and Hoffman, J. H., *Synthesis*, **1977**, 859.

[202] Butula, I. and Kuhn, R., *Angew. Chem., Int. Ed. Engl.*, **1968**, *7*, 208.

[203] Coulton, S., Gilchrist, T. L. and Graham, K., *Tetrahedron*, **1997**, *53*, 791.

[204] Gillespie, R. J. and Porter, A. E. A., *J. Chem. Soc., Chem. Commun.*, **1979**, 50.

[205] Gibe, R. and Kerr, M. A., *J Org. Chem.*, **2002**, *67*, 6247.

[206] Dehaen, W. and Hassner, A., *J. Org. Chem.*, **1991**, *56*, 896.

[207] Benson, S. C., Li, J.-H. and Snyder, J. K., *J. Org. Chem.*, **1992**, *57*, 5285.

[208] Omote, Y., Harada, K., Tomotake, A. and Kashima, C., *J. Heterocycl. Chem.*, **1984**, *21*, 1841.

[209] Wenkert, E., Moeller, P. D. R. and Piettre, S. R., *J. Am. Chem. Soc.*, **1988**, *110*, 7188.

[210] Biolatto, B., Kneeteman, M. and Mancini, P., *Tetrahedron Lett.*, **1999**, *40*, 3343.

[211] Roy, S., Kishbaugh, T. L. S., Jasinski, J. P. and Gribble, G. W., *Tetrahedron Lett.*, **2007**, *48*, 1313.

[212] 'Cycloaddition reactions with vinyl heterocycles', Sepúlveda-Arques, J., Abarca-González, B. and Medio-Simón, M., *Adv. Heterocycl. Chem.*, **1995**, *63*, 339.

[213] Pindur, U. and Eitel, M., *Helv. Chim. Acta*, **1988**, *71*, 1060; Pindur, U., Eitel, M. and Abdoust-Houshang, E., *Heterocycles*, **1989**, *29*, 11.

[214] Saroja, B. and Srinivasan, P. C., *Synthesis*, **1986**, 748; Eberle, M. K., Shapiro, M. J. and Stucki, R., *J. Org. Chem.*, **1987**, *52*, 4661.

[215] Simoji, Y., Saito, F., Tomita, K. and Morisawa, Y., *Heterocycles*, **1991**, *32*, 2389.

[216] Benson, S. C., Lee, L. and Snyder, J. K., *Tetrahedron Lett.*, **1996**, *37*, 5061.

[217] Wan, Z.-K., and Snyder, J. K., *Tetrahedron Lett.*, **1998**, *39*, 2487.

[218] Benson, S. C., Palabrica, C. A. and Snyder, J. K., *J. Org. Chem.*, **1987**, *2 5*, 4610; Daly, K., Nomak, R. and Snyder, J. K., *Tetrahedon Lett.*, **1997**, *38*, 8611.

[219] Santos, P. F., Lobo, A. M. and Prabhakar, S., *Tetrahedron Lett.*, **1995**, *36*, 8099.

[220] Raucher, S. and Klein, P., *J. Org. Chem.*, **1986**, *51*, 123.

[221] Davis, P. D. and Neckers, D. C., *J. Org. Chem.*, **1980**, *45*, 456.

[222] Weedon, A. C. and Zhang, B., *Synthesis*, **1992**, 95.

[223] Julian, D. R. and Tringham, G. D., *J. Chem. Soc., Chem. Commun.*, **1973**, 13; Julian, D. R. and Foster, R., *J. Chem. Soc., Chem. Commun.*, **1973**, 311.

[224] Thesing, J. and Semler, G., *Justus Liebigs Ann. Chem.*, **1964**, *680*, 52.

[225] Bailey, A. S., Haxby, J. B., Hilton, A. N., Peach, J. M. and Vandrevala, M. H., *J. Chem. Soc., Perkin Trans. 1*, **1981**, 382.

[226] Katritzky, A. R. and Akutagawa, K., *J. Am. Chem. Soc.*, **1986**, *108*, 6808.

[227] Inagaki, S., Nishizawa, Y., Suguira, T. and Ishihara, H., *J. Chem. Soc., Perkin Trans. 1*, **1990**, 179; Naruse, Y., Ito, Y. and Inagaki, S., *J. Org. Chem.*, **1991**, *56*, 2256.

[228] Pal, M., Dakarapu, R. and Padakanti, S., *J. Org. Chem.*, **2004**, *69*, 2913.

[229] Thesing, J. and Schülde, F., *Chem. Ber.*, **1952**, *85*, 324.

[230] Howe, E. E., Zambito, A. J., Snyder, H. R. and Tishler, M., *J. Am. Chem. Soc.*, **1945**, *67*, 38.

[231] de la Herrán, G, Segura, A. and Csákÿ, A. G., *Org. Lett.*, **2007**, *9*, 961.

[232] Allbright, J. D. and Snyder, H. R., *J. Am. Chem. Soc.*, **1959**, *81*, 2239; Baciocchi, E. and Schiroli, A., *J. Chem. Soc (B)*, **1968**, 401.

[233] Low, K. H. and Magomedov, N. A., *Org. Lett.*, **2005**, *7*, 2003.

[234] Leete, E., *J. Am. Chem. Soc.*, **1959**, *81*, 6023; Littell, R. and Allen, G. R., *J. Org. Chem.*, **1973**, *38*, 1504.

[235] Yamada, F., Hashizumi, T. and Somei, M., *Heterocycles*, **1998**, *47*, 509.

[236] Chauder, B., Larkin, A. and Snieckus, V., *Org. Lett.*, **2002**, *4*, 815.

[237] Nagarathnam, D., *J. Heterocycl. Chem.*, **1992**, *29*, 953; *idem, Synthesis*, **1992**, 743; Zhang, P., Liu, R. and Cook, J. M., *Tetrahedron Lett.*, **1995**, *36*, 7411.

[238] Challis, B. C. and Rzepa, H. S., *J. Chem. Soc., Perkin Trans. 2*, **1977**, 281.

[239] Boger, D. and Patel, M., *J. Org. Chem.*, **1987**, *52*, 3934.

[240] Casini, G. and Goodman, L., *Canad. J. Chem.*, **1964**, *42*, 1235; Cairncross, A., Roland, J. R., Henderson, R. M. and Sheppard, W. A., *J. Am. Chem. Soc.*, **1970**, *92*, 3187.

[241] Hassinger, H. L., Soll, R. M. and Gribble, G. W., *Tetrahedron Lett.*, **1998**, *39*, 3095.

[242] Miki, Y., Tada, Y., Yanase, N., Hachiken, H. and Matsushita, K., *Tetrahedron Lett.*, **1996**, *37*, 7753; Miki, Y., Hachiken, H., Sugimoto, Y. and Yanase, N., *Heterocycles*, **1997**, *45*, 1759.

[243] Wieland, T. and Unger, O., *Chem. Ber.*, **1963**, *96*, 253; Elliott, I. W. and Rivers, P., *J. Org. Chem.*, **1964**, *29*, 2438; Nozoye, T., Nakai, T. and Kubo, A., *Chem. Pharm. Bull.*, **1977**, *25*, 196.

[244] Sun, L., Tran, N., Tang, F., App, H., Hirth, P., McMahon, G. and Tang, C., *J. Med. Chem.*, **1998**, *41*, 2588.

[245] Giovannini, E. and Portmann, P., *Helv. Chim. Acta*, **1948**, *31*, 1375.

[246] Schulte, K. E. Reisch, J. and Stoess, U., *Arch. Pharm (Weinheim)*, **1977**, *305*, 523.

[247] Comber, M. F. and Moody, C. J., *Synthesis*, **1992**, 731.

[248] Bourlot, A. S., Desarbre, E. and Mérour, J.-Y., *Synthesis*, **1994**, 411.

[249] 'Five membered monoheterocyclic compounds. The indigo group', Sainsbury, M., in Campbell, N., Rodd's Chemistry of Carbon Compounds, Vol IVB, Ch. 14, **1977** and Supplement, Ch. 14, **1985**.

[250] Capon, B. and Kwok, F.-C., *J. Am. Chem. Soc.*, **1989**, *111*, 5346.

[251] Étienne, A., *Bull Soc. Chim. France*, **1948**, 651.

[252] Arnold, R. D., Nutter, W. M. and Stepp, W. C., *J. Org. Chem.*, **1959**, *24*, 117.

[253] Kawasaki, T., Nonaka, Y., Uemura, M. and Sakamoto, M., *Synthesis*, **1991**, 701.

[254] Abramovitch, R. A. and Marko, A. M., *Canad. J. Chem.*, **1960**, *38*, 131; O'Sullivan, W. I. and Rothery, E. J., *Chem. Ind.*, **1972**, 849.

[255] Hooper, M. and Pitkethly, W. N., *J. Chem. Soc., Perkin Trans. 1*, **1972**, 1607.

[256] Conway, S. C. and Gribble, G. W., *Heterocycles*, **1990**, *30*, 627.

[257] 'The chemistry of isatin', Popp, F. D., *Adv. Heterocycl. Chem.*, **1975**, *18*, 1.

[258] Garden, S. J., Torres, J. C., Ferreira, A. A., Silva, R. B. and Pinto, A. C., *Tetrahedron Lett.*, **1997**, *39*, 1501.

[259] Muchowski, J. M., *Canad. J. Chem.*, **1969**, *47*, 857.

[260] Jackson, A. H., *Chem. Ind.*, **1965**, 1652.

[261] Gruda, I., *Canad. J. Chem.*, **1972**, *50*, 18.

[262] Bergman, J., *Acta. Chem. Scand.*, **1971**, *25*, 1277.

[263] Lathourakis, G. E. and Litinas, K. E., *J. Chem. Soc., Perkin Trans. 1*, **1996**, 491.

[264] Azizian, J., Karimi, A. R., Kazemizadeh, Z., Mohammadi, A. A. and Mohammadizadeh, M. R., *J. Org. Chem.*, **2005**, *70*, 1471; Yadav, J. S., Reddy, B. V. S., Jain, R. and Reddy, Ch. S., *Tetrahedron Lett.*, **2007**, *48*, 3295.

[265] '1-Hydroxyindoles', Somei, A., *Heterocycles*, **1999**, *50*, 1157.

[266] Kawasaki, T., Kodama, A., Nichida, T., Shimizu, K. and Somei, M., *Heterocycles*, **1991**, *32*, 221.

[267] Somei, M., Yamada, F. and Morikawa, H., *Heterocycles*, **1997**, *46*, 91.

[268] Somei, M., Kawasaki, T., Fukui, Y., Yamada, F., Kobayashi, T., Aoyama, H. and Shinmyo, D., *Heterocycles*, **1992**, *34*, 1877.

[269] Somei, M., Yamada, F. and Yamamura, G., *Chem. Pharm. Bull.*, **1998**, *46*, 191.

[270] Bird, C. W., *J. Chem. Soc.*, **1965**, 3490.

[271] Roy, S. and Gribble, G. W., *Tetrahedron Lett.*, **2008**, *49*, 1531.

[272] Sosnovsky, G. and Purgstaller, K., *Z. Naturforsch*, **1989**, *44b*, 582; Klein, J. T., Davis, L., Olsen, G. E., Wong, G. S., Huger, F. P., Smith, C. P., Petko, W. W., Cornfeldt, M., Wilker, J. C., Blitzer, R. D., Landau, E., Haroutunian, V., Matrin, L. L. and Effland, R. C., *J. Med. Chem.*, **1996**, *39*, 570.

[273] 'Synthesis and reactivity of 7-azaindole (1*H*-pyrrolo[2.3-*b*][pyridine)', Popowycz, F., Routier, S., Joseph, B. and Mérour, J.-Y., *Tetrahedron*, **2007**, *63*, 1031.

[274] 'Synthesis and reactivity of 4-, 5- and 6-azaindoles', Popowycz, F., Mérour, J.-Y. and Jospeh, B., *Tetrahedron*, **2007**, *63*, 8689.

[275] Prokopov, A. A. and Yakhontov, L. N., *Khim. Geterotsikl. Soedin.*, **1978**, 496; Yakhontov, L. N., Azimov, V. A. and Lapan, E. I., *Tetrahedron Lett.*, **1969**, 1909.

[276] Zhang, Z., Yang, Z., Wong, H., Zhu, J., Meanwell, N. A., Kadow, J. F. and Wang, T., *J. Org. Chem.*, **2002**, *67*, 6226.

[277] Girgis, N. S., Larson, S. B., Robins, R. K. and Cottam, H. B., *J. Heterocycl. Chem.*, **1989**, *26*, 317.

[278] Wang, X., Zhi, B., Baum, J., Chen, Y., Crockett, R., Huang, L., Heisenberg, S., Ng, J., Larsen, R., Martinelli, M. and Reider, P., *J. Org. Chem.*, **2006**, *71*, 4021.

[279] Bisagni, E., Legraverend, M. and Lhoste, J.-M., *J. Org. Chem.*, **1982**, *47*, 1500.

[280] Bourzat, J.-D. and Bisagni, E., *Bull. Soc. Chim. Fr.*, **1973**, *10*, 511.

[281] Desarbre, E., Coudret, S., Meheust, C. and Merour, J.-Y., *Tetrahedron*, **1997**, *53*, 3637.

[282] 'Practical methodologies for the synthesis of indoles', Humphrey, G. R. and Kuethe, J. T., *Chem. Rev.*, **2006**, *106*, 2875; 'Recent developments in indole ring synthesis-methodology and applications', Gribble, G. W., *J. Chem. Soc., Perkin Trans. 1*, **2000**, 1045.

[283] 'The Fischer indole synthesis', Robinson, B., John Wiley & Sons, Ltd, Chichester, New York, **1982**; 'Progress in the Fischer indole reaction', Hughes, D. L., *Org. Prep. Proc. Int.*, **1993**, *25*, 607.

[284] Shriner, R. L. Ashley, W. C. and Welch, E., *Org. Synth., Coll. Vol. III*, **1955**, 725.

[285] Rogers, C. V. and Corson, B. B., *Org. Synth., Coll. Vol. IV*, **1963**, 884.

[286] Murakami, Y., Yokoyama, Y., Miura, T., Hirasawa, H., Kamimura, Y. and Izaki, M., *Heterocycles*, **1984**, *22*, 1211; Baccolini, G. and Marotta, E., *Tetrahedron*, **1985**, *20*, 4615; Baccolini, G., Dalpozzo, R. and Todesco, P. E., *J. Chem. Soc., Perkin Trans. 1*, **1988**, 971; Dhakshinamoorthy, A. and Pitchumani, K., *Appl. Catal., A Gen.*, **2005**, *292*, 305.

[287] Przheval'skii, N. M., Kostromina, L. Yu. and Grandberg, I. I., *Khim. Geterotsikl. Soeden.*, **1988**, *24*, 188 (*Chem. Abstr.*, **1988**, *109*, 210837).

[288] Katritzky, A. R., Rachwal, S. and Bayyuk, S., *Org. Prep. Proc. Int.*, **1991**, *23*, 357.

[289] Ockenden, D. W. and Schofield, K., *J. Chem. Soc.*, **1957**, 3175.

[290] Rosenbaum, C., Röhrs, S., Müller, O. and Waldmann, H., *J. Med. Chem.*, **2005**, *48*, 1179.

[291] Lim, Y.-K. and Cho, C.-G., *Tetrahedron Lett.*, **2004**, *45*, 1857.

[292] Rosenbaum, C., Katzka, C., Marzinzik, A. and Waldmann, H., *Chem. Commun.*, **2003**, 1822.

[293] Douglas, A. W., *J. Am. Chem. Soc.*, **1978**, *100*, 6463; *ibid.*, **1979**, *101*, 5676.

[294] Posvic, H., Dombro, R., Ito, H. and Telinski, T., *J. Org. Chem.*, **1974**, *39*, 2575; Schiess, P. and Grieder, A., *Helv. Chim. Acta*, **1974**, *57*, 2643.

[295] Miyata, O., Takeda, N., Kimura, Y., Takemoto, Y., Tohnai, N., Miyata, M. and Naito, T., *Tetrahedron*, **2006**, *62*, 3629.

[296] Mills, K., Al Khawaja, I. K., Al-Saleh, F. S. and Joule, J. A., *J. Chem. Soc., Perkin Trans. 1*, **1981**, 636.

[297] Zhao, D., Hughes, D. L., Bender, D. R., De Marco, A. M. and Reider, P. J., *J. Org. Chem.*, **1991**, *56*, 3001.

[298] Campos, K. R., Woo, J. C. S., Lee, S. and Tillyer, R. D., *Org. Lett.*, **2004**, *6*, 79.

[299] Marais, W. and Holzapfel, C. W., *Synth. Commun.*, **1998**, *28*, 3691.

[300] Pausacker, K. H., *J. Chem. Soc.*, **1950**, 621.

[301] Liu, K. G., Robichaud, A. J., Lo, J. R., Mattes, J. F. and Cai, Y., *Org. Lett.*, **2006**, *8*, 5769.

[302] Wagaw, S., Yang, B. H. and Buchwald, S. L., *J. Am. Chem. Soc.*, **1998**, *120*, 6621.

[303] Chae, J. and Buchwald, S. L., *J. Org. Chem.*, **2004**, *69*, 3336.

[304] Cao, C., Shi, Y., and Odom, A. L., *Org. Lett.*, **2002**, *4*, 2853; Ahmed, M., Jackstell, R., Seayad, A. M., Klein, H. and Beller, M., *Tetrahedron Lett.*, **2004**, *45*, 869; Ackermann, L. and Born, R., *Tetrahedron Lett.*, **2004**, *45*, 9541; Khedkar, V., Tillack, A., Michalik, M. and Beller, M., *Tetrahedron*, **2005**, *61*, 7622.

[305] Martin, P., *Helv. Chim. Acta*, **1984**, *67*, 1647; Toyota, M. and Fukumoto, K., *J. Chem. Soc., Perkin Trans. 1*, **1992**, 547.

[306] Street, L. J., Baker, R., Castro, J. L., Chambers, M. S., Guiblin, A. R., Hobbs, S. C., Matassa, U. G., Reeve, A. J., Beer, M. S., Middlemiss, D. N., Noble, A. J., Stanton, J. A., Scholey, K. and Hargreaves, R. J., *J. Med. Chem.*, **1993**, *36*, 1529; Castro, J. L., Baker, R., Guiblin, A. R., Hobbs, S. C., Jenkins, M. R., Russell, M. G. N., Beer, M. S., Stanton, J. A., Scholey, K., Hargreaves, R. J., Graham, M. I. and Matassa, V. G., *J. Med. Chem.*, **1994**, *37*, 3023; Glenn, R. C., Martin, G. R., Hill, A. P., Hyde, R. M., Woollard, P. M., Salmon, J. A., Buckingham, J. and Robertson, A. D., *J. Med. Chem.*, **1995**, *38*, 3566.

[307] Grandberg, I. I., *Chem. Heterocycl. Compd. (Engl. Transl.)* **1974**, *10*, 501.

[308] Noland, W. E. and Baude, F. J., *Org. Synth., Coll. Vol. V*, **1973**, 567.

[309] Allen, D. A., *Synth. Commun.*, **1999**, *29*, 447.

[310] Clark, R. D., Muchowski, J. M., Fisher, L. E., Flippin, L. A., Repke, D. B. and Souchet, M., *Synthesis*, **1991**, 871.

[311] Rutherford, J. L., Rainka, M. P. and Buchwald, S. L., *J. Am. Chem. Soc.*, **2002**, *124*, 15168.

[312] Chen, Y., Xie, X. and Ma, D., *J. Org. Chem.*, **2007**, *72*, 9329.

[313] Satoh, M., Miyaura, N. and Suzuki, A., *Synthesis*, **1987**, 373; Beugelmans, R. and Roussi, G., *J. Chem. Soc., Chem. Commun.*, **1979**, 950; Bard, R. B. and Bunnett, J. F., *J. Org. Chem.*, **1980**, *45*, 1546; Suzuki, H., Thiruvikraman, S. V. and Osuka, A., *Synthesis*, **1984**, 616; Sakagami, H. and Ogasawara, K., *Heterocycles*, **1999**, *51*, 1131.

[314] RajanBabu, T. V., Reddy, G. S. and Fukunage, T., *J. Am. Chem. Soc.*, **1985**, *107*, 5473.

[315] RajanBabu, T. V., Chenard, B. L. and Petti, M. A., *J. Org. Chem.*, **1986**, *51*, 1704.

[316] 'Leimgruber-Batcho indole synthesis', Clark, R. D. and Repke, D. B., *Heterocycles*, **1984**, *22*, 195; Batcho, A. D. and Leimgruber, W., *Org. Synth., Coll. Vol. 7*, **1990**, 34.

[317] Sin, J., Baxendale, I. R. and Ley, S. V., *Org. Biomol. Chem.*, **2004**, *2*, 160.

[318] Lloyd, D. H. and Nichols, D. E., *Tetrahedron Lett.*, **1983**, *24*, 4561; Haefliger, W. and Knecht, H., *ibid.*, **1984**, *25*, 285.

[319] Ponticello, G. S. and Baldwin, J. J., *J. Org. Chem.*, **1979**, *44*, 4003; Kozikowski, A. P., Ishida, H. and Chen, Y.-Y., *J. Org. Chem.*, **1980**, *45*, 3350; Somei, M., Saida, Y. and Komura, N., *Chem. Pharm. Bull.*, **1986**, *34*, 4116.

[320] Rogers, C. B., Blum, C. A. and Murphy, B. P., *J. Heterocycl. Chem.*, **1987**, *24*, 941.

[321] Rajeswari, S., Drost, K. J. and Cava, M. P., *Heterocycles*, **1989**, *29*, 415.

[322] Sinhababu, A. K. and Borchardt, R. T., *J. Org. Chem.*, **1983**, *48*, 3347.

[323] Ijaz, A. S. and Parrick, J., *Sci. Int., (Lahore)* **1989**, *1*, 364.

[324] Söderberg, B. C. and Shriver, J. A., *J. Org. Chem.*, **1997**, *62*, 5838; Davies, I. W., Smitrovich, J. H., Sidler, R., Qu, C., Gresham, V. and Bazaral, C., *Tetrahedron*, **2005**, *61*, 6425; Söderberg, B. C. G., Banini, S. R., Turner, M. R., Minter, A. R. and Arrington, A. K., *Synthesis*, **2008**, 903; Scott, T. L., Burke, N., Carrero-Martínez, G. and Söderberg, B. C. G., *Tetrahedron*, **2007**, *63*, 1183.

[325] Tsuji, Y., Kotachi, S., Huh, K.-T. and Watanabe, Y., *J. Org. Chem.*, **1990**, *55*, 580.

[326] Cho, C. S., Kim, J. H., Kim, T.-J. and Shim, S. C., *Tetrahedron*, **2001**, *57*, 3321.

[327] Fujita, K.-i., Yamamoto, K. and Yamaguchi, R., *Org. Lett.*, **2002**, *4*, 2691.

[328] Sakamoto, T., Kondo, Y. and Yamanaka, H., *Heterocycles*, **1984**, *22*, 1347.

[329] Tischler, A. N. and Lanza, T. J., *Tetrahedron Lett.*, **1986**, *27*, 1653.

[330] Sakamoto, T., Kondo, Y., Iwashita, S., Nagano, T. and Yamanaka, H., *Chem. Pharm. Bull.*, **1988**, *36*, 1305; Rudisill, D. E. and Stille, J. K., *J. Org. Chem.*, **1989**, *54*, 5856.

[331] Arcadi, A., Cacchi, S. and Marinelli, F., *Tetrahedron Lett.*, **1986**, *27*, 6397.

[332] Yasuhara, A., Kanamori, Y., Kaneko, M., Numata, A., Kondo, Y. and Sakamoto, T., *J. Chem. Soc., Perkin Trans. 1*, **1999**, 529.

[333] Rodriguez, A. L., Koradin, C., Dohle, W. and Knochel, P., *Angew. Chem. Int. Ed.*, **2000**, *39*, 2488; Koradin, C., Dohle, W., Rodriguez, A. L., Schmid, B. and Knochel, P., *Tetrahedron*, **2003**, *59*, 1571.

[334] Zhang, Y., Donahue, J. P. and Li, C.-J., *Org. Lett.*, **2007**, *9*, 627.

[335] Kim, J. S., Han, J. H., Lee, J. J., Jun, Y. M., Lee, B. M. and Kim, B. H., *Tetrahedron Lett.*, **2008**, *49*, 3733.

[336] Hiroya, K., Itoh, S. and Sakamoto, T., *J. Org. Chem.*, **2004**, *69*, 1126; Hiroya, K., Itoh, S. and Sakamoto, T., *Tetrahedron*, **2005**, *61*, 10958.

[337] Yin, Y., Ma, W., Chai, Z. and Zhao, G., *J. Org. Chem.*, **2007**, *72*, 5731.

[338] Barluenga, J., Trincado, M., Rubio, E. and González, J. M., *Angew. Chem. Int. Ed.*, **2003**, *42*, 2406.

[339] Amjad, M. and Knight, D. W., *Tetrahedron Lett.*, **2004**, *45*, 539.

[340] Yue, D., Yao, T. and Larock, R. C., *J. Org. Chem.*, **2006**, *71*, 62.

[341] Kondo, Y., Sakamoto, T. and Yamanaka, H., *Heterocycles*, **1989**, *29*, 1013; 'The aminopalladation/reductive elimination domino reaction in the construction of functionalized indole rings', Battistuzzi, G., Cacchi, S. and Fabrizi, G., *Eur. J. Org. Chem.*, **2002**, 2671; Lu, B. Z., Zhao, W., Wei, H.-X., Dufour, M., Farina, V. and Senanayake, C. H., *Org. Lett.*, **2006**, *8*, 3271.

[342] Cacchi, S., Fabrizi, G. and Parisi, L. M., *Org. Lett.*, **2003**, *5*, 3843.

[343] Hong, K. B., Lee, C. W. and Yum, E. K., *Tetrahedron Lett.*, **2004**, *45*, 693.

[344] Palimkar, S. S., Kumar, P. H., Lahoti, R. J. and Srinivasan, K. V., *Tetrahedron*, **2006**, *62*, 5109.

[345] Ackermann, L., Kaspar, L. T. and Gschrei, C. J., *Chem. Commun.*, **2004**, 2824.

[346] Larock, R. C., Yum, E. K. and Refvik, M. D., *J. Org. Chem.*, **1998**, *63*, 7652.

[347] Roschangar, F., Liu, J., Estanove, E., Dufour, M., Rodríguez, S., Farina, V., Hickey, E., Hossain, A., Jones, P.-J., Lee, H., Lu, B. Z., Varsolona, R., Schröder, J., Beaulieu, P., Gillard, J. and Senanayake, C. H., *Tetrahedron Lett.*, **2008**, *49*, 363.

[348] Allen, C. F. H. and VanAllan, J., *Org. Synth., Coll. Vol. III*, **1955**, 597; Tyson, F. T., *ibid.*, 479.

[349] Houlihan, W. J., Parrino, V. A. and Uike, Y., *J. Org. Chem.*, **1981**, *46*, 4511.

350 Eitel, M. and Pindur, U., *Synthesis*, **1989**, 364; Prasitpan, N., Patel, J. N., De Croos, P. Z., Stockwell, B. L., Manavalan, P., Kar, L., Johnson, M. E. and Currie, B. L., *J. Heterocycl. Chem.*, **1992**, *29*, 335; Taira, S'i., Danjo, H. and Imamoto, T., *Tetrahedron Lett.*, **2002**, *43*, 2885.

351 Bartoli, G., Bosco, M., Dalpozzo, R. and Todesco, P. E., *J. Chem. Soc., Chem. Commun.*, **1988**, 807.

352 Smith, A. B., Visnick, M., Haseltine, J. N. and Sprengler, P. A., *Tetrahedron*, **1986**, *42*, 2957.

353 Opatz, T. and Ferenc, D., *Org. Lett.*, **2006**, *8*, 4473.

354 Ito, Y., Kobayashi, K. and Saegusa, T., *J. Am. Chem. Soc.*, **1977**, *99*, 3532; *idem, J. Org. Chem.*, **1979**, *44*, 2030; Ito, Y., Kobayashi, K., Seko, N. and Saegusa, T., *Bull. Chem. Soc. Jpn.*, **1984**, *57*, 73.

355 Campaigne, E. and Lake, R. D., *J. Org. Chem.*, **1959**, *24*, 478.

356 Nordlander, J. E., Catalane, D. B., Kotian, K. D., Stevens, R. M. and Haky, J. E., *J. Org. Chem.*, **1981**, *46*, 778; Sundberg, R. J. and Laurino, J. P., *ibid.*, **1984**, *49*, 249.

357 Pchalek, K., Jones, A. W., Wekking, M. M. T. and Black, D. StC., *Tetrahedron*, **2005**, *61*, 77.

358 Witulski, B., Alayrac, C. and Tevzadze-Saeftel, L, *Angew. Chem. Int. Ed.*, **2003**, *42*, 4257.

359 Kozikowski, A. P. and Cheng, X.-M., *Tetrahedron Lett.*, **1985**, *26*, 4047; Yokoyama, Y., Suzuki, H., Matsumoto, S., Sunaga, Y., Tani, M. and Murakami, Y., *Chem. Pharm. Bull.*, **1991**, *39*, 2830; Natsume, M. and Muratake, H., *Tetrahedron Lett.*, **1979**, 771; Muratake, H. and Natsume, M., *Tetrahedron Lett.*, **1987**, *28*, 2265; Yamashita, A., Scahill, T. A. and Toy, A., *ibid.*, **1985**, *26*, 2969; Muratake, H. and Natsume, M., *Heterocycles*, **1989**, *29*, 771.

360 Muratake, H. and Natsume, M., *Heterocycles*, **1989**, *29*, 783.

361 ten Have, R. and van Leusen, A. M., *Tetrahedron*, **1998**, *54*, 1913.

362 Pirrung, M. C., Wedel, M. and Zhao, Y., *Synlett*, **2002**, 143.

363 Bartoli, G., Palmieri, G., Bosco, M. and Dalpozzo, R., *Tetrahedron Lett.*, **1989**, *30*, 2129; Dodson, D., Todd, A. and Gilmore, J., *Synth. Commun.*, **1991**, *21*, 611.

364 Dobbs, A., *J. Org. Chem.*, **2001**, *66*, 638.

365 Bosco, M., Dalpozzo, R., Bartoli, G., Palmieri, G. and Petrini, M., *J. Chem. Soc., Perkin Trans. 2*, **1991**, 657; on solid support: Knepper, K. and Bräse, S., *Org. Lett.*, **2003**, *5*, 2829.

366 Iida, H., Yuasa, Y. and Kibayashi, C., *J. Org. Chem.*, **1980**, *45*, 2938; Sakamoto, T., Nagano, T., Kondo, Y. and Yamanaka, H., *Synthesis*, **1990**, 215.

367 Chen, C., Lieberman, D. R., Larsen, R. D., Verhoeven, T. R. and Reider, P. J., *J. Org. Chem.*, **1997**, *62*, 2676.

368 Jia, X. and Zhu, J., *J. Org. Chem.*, **2006**, *71*, 7826.

369 'The synthesis of 5-hydroxyindoles by the Nenitzescu reaction', Allen, G. R., *Org. React.*, **1973**, *20*, 337.

370 Kuckländer, U., *Tetrahedron*, **1972**, *28*, 5251.

371 Patrick, J. B. and Saunders, E. K., *Tetrahedron Lett.*, **1979**, 4009.

372 Gassman, P. G., Roos, J. J. and Lee, S. J., *J. Org. Chem.*, **1984**, *49*, 717; Gassman, P. G. and von Bergen, T. J., *Org. Synth., Coll. Vol. VI*, **1988**, 601.

373 'New developments in the chemistry of low-valent titanium', Fürstner, A. and Bogdanovic, B., *Angew. Chem., Int. Ed. Engl.*, **1996**, *35*, 2443.

374 Fürstner, A. and Ernst, A., *Tetrahedron*, **1995**, *51*, 773.

375 'Development of a novel indole synthesis and its application to natural products synthesis', Kobayashi, Y., and Fukuyama, T., *J. Heterocycl. Chem.*, **1998**, *35*, 1043.

376 Fukuyama, T., Chen, X. and Peng, G., *J. Am. Chem. Soc.*, **1994**, *116*, 3127.

377 Tokuyama, H., Kaburagi, Y., Chen, X. and Fukuyama, T., *Synthesis*, **2000**. 429.

378 Sundberg, R. J., Russell, H. F., Ligon,, W. F. and Lin, L.-S., *J. Org. Chem.*, **1972**, *37*, 719.

379 Pelkey, E. T. and Gribble, G. W., *Tetrahedron Lett.*, **1997**, *38*, 5603.

380 Knittel, D., *Synthesis*, **1985**, 186.

381 Galvez, J. E. and Garcia, F., *J. Heterocycl. Chem.*, **1984**, *21*, 215.

382 Du, Y., Liu, R., Linn, G. and Zhao, K., *Org. Lett.*, **2006**, *8*, 5919.

383 Schirok, H., *Synthesis*, **2008**, 1404.

384 Barberis, C., Gordon, T. D., Thomas, C., Zhang, X. and Cusack, K. P., *Tetrahedron Lett.*, **2005**, *46*, 8877.

385 Willis, M. C., Brace, G. N. and Holmes, I. P., *Angew. Chem. Int. Ed.*, **2005**, *44*, 403.

386 Lavrenov, S. N., Lakatosh, S. A., Lysenkova, L. N., Korolev, A. M. and Preobrazhenskaya, M. N., *Synthesis*, **2002**, 320.

387 Inada, A., Nakamura, Y. and Morita, Y., *Chem. Lett.*, **1980**, 1287; Somei, M. and Saida, Y., *Heterocycles*, **1985**, *23*, 3113.

388 Kawase, M., Miyake, Y. and Kikugawa, Y., *J. Chem. Soc., Perkin Trans. 1*, **1984**, 1401.

389 Ketcha, D. M., *Tetrahedron Lett.*, **1988**, *29*, 2151.

390 Kuehne, M. E. and Hall, T. C., *J. Org. Chem.*, **1976**, *41*, 2742.

391 Hino, T. and Taniguchi, M., *J. Am. Chem. Soc.*, **1978**, *100*, 5564.

392 Karp, G. M., *Org. Prep. Proced. Int.*, **1992**, *25*, 481.

393 Abramovitch, R. A. and Hey, D. H., *J. Chem. Soc.*, **1954**, 1697; Rutenberg, M. W. and Horning, E. C., *Org. Synth., Coll. Vol. IV*, **1963**, 620.

394 Doyle, M. P., Shanklin, M. S., Pho, H. Q. and Mahapatro, S. N., *J. Org. Chem.*, **1988**, *53*, 1017.

395 Shaughnessy, K. H., Hamann, B. C. and Hartwig, J. F., *J. Org. Chem.*, **1998**, *63*, 6546.

396 van Alphen, J., *Recl. Trav. Chim. Pays-Bas*, **1942**, *61*, 888; Su, H. C. F. and Tsou, K. C., *J. Am. Chem. Soc.*, **1960**, *82*, 1187.

397 Rodríguez-Domínguez, J. C., Balbuzano-Deus, A., López-López, M. A. and Kirsch, G., *J. Heterocycl. Chem.*, **2007**, *44*, 273.

398 Hendrickson, J. B. and Hussoin, Md. S., *J. Org. Chem.*, **1989**, *54*, 1144.

399 Sugasawa, T., Adachi, M., Sasakura, K. and Kitagawa, A., *J. Org. Chem.*, **1979**, *44*, 578; Nimtz, M. and Häflinger, G., *Liebig's Ann. Chem.*, **1987**, 765; Sasakura, K., Adachi, M. and Sugasawa, T., *Synth. Commun.*, **1988**, *18*, 265.

400 Marvel, C. S. and Hiers, G. S., *Org. Synth., Coll. Vol. I*, **1932**, 327; Kollmar, M., Parlitz, R., Oevers, S. R. and Helmchen, G., *Org. Synth.*, **2002**, *79*, 196.

401 Rewcastle, G. W., Sutherland, H. S., Weir, C. A., Blackburn, A. G. and Denny, W. A., *Tetrahedron Lett.*, **2005**, *46*, 8719.

[402] Cheng, Y., Zhan, Y.-H. and Meth-Cohn, O., *Synthesis*, **2002**, 34.

[403] Wong, A., Kuethe, J. T. and Davies, I. W., *J. Org. Chem.*, **2003**, *68*, 9865.

[404] Moody, C. J. and Rahimtoola, K. F., *J. Chem. Soc., Chem. Commun.*, **1990**, 1667.

[405] Abramovitch, R. A. and Shapiro, D., *J. Chem. Soc.*, **1956**, 4589; Henecka, H., Timmler, H., Lorenz, R. and Geiger, W., *Chem. Ber.*, **1957**, *90*, 1060.

[406] Ishibashi, H., Akamatsu, S., Iriyama, H., Hanaoka, K., Tabata, T. and Ikeda, M., *Chem. Pharm. Bull.*, **1994**, *42*, 271.

[407] Garg, N. K., Sarpong, R. and Stoltz, B. M., *J. Am. Chem. Soc.*, **2002**, *124*, 13179.

[408] Zhang, Z., Yang, Z., Meanwell, N. A., Kadow, J. F. and Wang, T., *J. Org. Chem.*, **2002**, *67*, 2345.

[409] Debenham, S. D., Chan, A., Liu, K., Price, K. and Wood, H. B., *Tetrahedron Lett.*, **2005**, *46*, 2283.

[410] Ujjainwalla, F. and Warner, D., *Tetrahedron Lett.*, **1998**, *39*, 5355; McLaughlin, M., Palucki, M. and Davies, I. W., *Org. Lett.*, **2006**, *8*, 3307.

[411] Xu, L., Lewis, I. R., Davidsen, S. K. and Summers, J. B., *Tetrahedron Lett.*, **1998**, *39*, 5159; Harcken, C., Ward, Y., Thomson, D. and Riether, D., *Synlett*, **2005**, 3121; de Mattos, M. C., Alatorre-Santamaría, S., Gotor-Fernández, V. and Gotor, V., *Synthesis*, **2007**, 2149; Sun, L.-P. and Wang, J.-X., *Synth. Commun.*, **2007**, *37*, 2187; Majumdar, K. C. and Mondal, S., *Tetrahedron Lett.*, **2007**, *48*, 6951.

[412] Mahadevan, I. and Rasmussen, M., *J. Heterocycl. Chem.*, **1992**, *29*, 359.

[413] Zhu, J., Wong, H., Zhang, Z., Yin, Z., Meanwell, N. A., Kadow, J. F. and Wang, T., *Tetrahedron Lett.*, **2006**, *47*, 5653.

[414] Schirok, H., *J. Org. Chem.*, **2006**, *71*, 5538.

[415] Roy, P. J., Dufresne, C., Lachance, N., Leclerc, J.-P., Boisvert, M. and Wang, Z., *Synthesis*, **2005**, 2751; Roy, P., Boisvert, M. and Leblanc, Y., *Org. Synth.*, **2007**, *84*, 262.

[416] Mazéas, D., Guillaumet, G. and Viaud, M.-C., *Heterocycles*, **1999**, *50*, 1065.

[417] Dekhane, M., Potier, P. and Dodd, R. H., *Tetrahedron*, **1993**, *49*, 8139.

21

Benzo[*b*]thiophenes and Benzo[*b*]furans: Reactions and Synthesis

benzothiophene
(thianaphthene)
[benzo[*b*]thiophene]

benzofuran
[benzo[*b*]furan]

Benzo[*b*]thiophene[1] and benzo[*b*]furan,[2] frequently (and in the rest of this chapter) referred to simply as benzothiophene and benzofuran, are respectively the sulfur and oxygen analogues of indole, but have been much less fully studied.

21.1 Reactions with Electrophilic Reagents

21.1.1 Substitution at Carbon

The electrophilic substitution of these systems is much less regioselective than that of indole, for which there is effectively complete selectivity of attack at C-3, even to the extent that the hetero-ring positions are only a little more reactive than some of the benzene ring positions. Measurements of detritiation of 2- and 3-tritiobenzothiophenes in trifluoroacetic acid showed rates which are effectively the same for both hetero-ring positions.[3] Nitric acid nitration of benzothiophene gives a mixture in which, although the main product is the 3-nitro derivative, lesser quantities of 2-nitro-, 4-nitro- 6-nitro- and 7-nitrobenzothiophenes are also all produced[4] however ceric ammonium nitrate in acetic anhydride at room temperature produces a high yield of 3-nitrobenzothiophene.[5] 2-Nitrobenzothiophene and 2-nitrobenzofuran can be obtained in good yields by the photo-promoted reaction of the corresponding 2-trimethylstannyl-heterocycles with dinitrogen tetroxide and tetranitromethane.[6] Friedel–Crafts acetylation[7] of benzothiophene gives a mixture or 3- and 2-acetyl-derivatives in a ratio of 7:3, however in other electrophilic substitutions the 3-isomer is the only product – iodination[8] falls into this category. Controlled reaction of benzothiophene with bromine produces 3-bromobenzothiophene in moderate yield,[9] however this compound is better prepared by room temperature, high-yielding 2,3-dibromination then regioselective metallation at C-2 and protonation.[10]

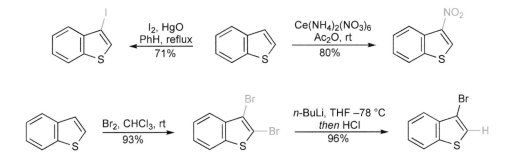

Heterocyclic Chemistry 5th Edition   John Joule and Keith Mills
© 2010 Blackwell Publishing Ltd

Benzofuran displays a lesser selectivity for 3-substitution: formylation of benzofuran gives only the 2-formyl-derivative,[11] and nitric acid nitration[12] produces 2-nitrobenzofuran, as does a combination of sodium nitrite and ceric ammonium nitrate (CAN)[13] or in high yield, CAN in acetic anhydride.[5] Dinitrogen tetroxide nitration produces a 5:2 mixture of 3- and 2-nitrobenzofurans, but an activating group on the benzene ring tips the balance and leads to benzene-ring substitution.[14] Treatment of benzofuran with halogens results in 2,3-addition products; reaction with the interhalogen BrCl gives 2-bromo-3-chloro adducts; from these addition products, by base-promoted hydrogen halide elimination, 3-monohalo-benzofurans can be obtained in high yields.[15] *N*-Bromosuccinimide smoothly 3-brominates 2-substituted benzofurans.[16] Friedel–Crafts substitution is complicated for hetero-ring-unsubstituted benzofurans because typical catalysts tend to cause resinification, but 3-acylations[17] are achieved using ferric chloride. Ytterbium-triflate-catalysed hydroxyalkylation by ethyl glyoxylate is efficient, and regioselective for C-3 for both benzothiophene and benzofuran.[18]

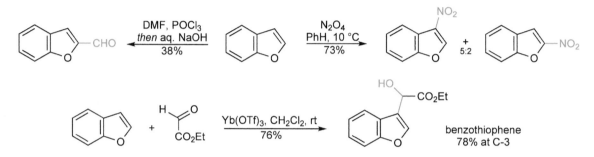

With substituents already present, the pattern of electrophilic substitution can be difficult to predict: some examples serve to illustrate this. Nitration of 2-bromobenzothiophene results in *ipso*-substitution and thus the formation of 2-nitrobenzothiophene, whereas 2-chlorobenzothiophene gives the 3-nitro-substitution product;[19] nitration of 3-bromobenzothiophene proceeds in moderate yield to give the 2-nitro derivative.[20] On the other hand, 3-carboxy- or 3-acyl-benzothiophenes nitrate mainly in the benzene ring.[21] Bromination[22] and Friedel–Crafts substitution[23] of 3-methyl- and 2-methylbenzothiophenes takes place cleanly at the vacant hetero-ring position; similarly 2-bromobenzothiophene undergoes Vilsmeier formylation at C-3.[24] 3-Methoxybenzothiophene gives the corresponding 2-aldehyde under Vilsmeier conditions.[25]

21.1.2 Addition to Sulfur in Benzothiophenes
Benzothiophenium salts[26] are produced by the reaction of the sulfur heterocycle with powerful alkylating reagents such as Meerwein salts; benzothiophenium salts can themselves act as powerful alkylating agents with fission of the C–S+ bond.[27]

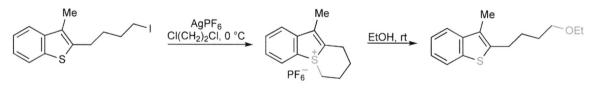

S-Oxidation produces 1,1-dioxides, which readily undergo cycloadditions as dienophiles,[28] or photodimerisation, the head-to-head dimer being the major product.[29] *S*-Oxidation of the sulfur using a microbiological method gives the *S*-oxide.[30]

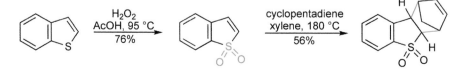

21.2 Reactions with Nucleophilic Reagents

Halogen at a benzothiophene 2-position is subject to displacement with amine nucleophiles,[31] and, more easily than halogen at the 3-position.

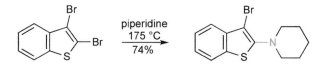

21.3 Metallation and Reactions of *C*-Metallated Benzothiophenes and Benzofurans

In some of the earliest uses of *n*-butyllithium, 2-lithiobenzofuran was obtained by metal–halogen exchange between the 2-bromo-heterocycle and *n*-butyllithium,[32] or by metallation of benzofuran.[33] The generation of 3-metallated benzofurans generally results in fragmentation with the production of 2-hydroxyphenylacetylene at room temperature,[34,98] though the 3-lithio derivative *can* be utilised by maintaining a very low temperature.[35] 2,3-Dibromobenzofuran lithiates at C-2.[36]

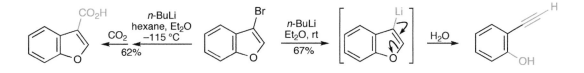

Sodium amide causes ring cleavage of benzothiophene to produce 2-ethynylphenyl thiol.[37] Ring opening in a rather different manner results from exposure of the heterocycle to lithium dimethylamide, followed by trapping with iodomethane, producing an enamine which must result from initial *addition* at C-2, perhaps by a minor pathway, but one which then leads to irreversible ring-opening elimination.[38]

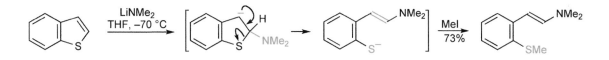

A ring opening can also be observed *via* butyllithium attack at sulfur.[39]

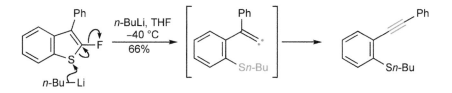

3-Lithio-benzothiophenes can be generated, and reacted with electrophiles, if the temperature is kept low.[40] Direct deprotonation of benzothiophenes follows the usual pattern for five-membered heterocycles and takes place adjacent to the heteroatom,[41] and in concord with this pattern, metal–halogen exchange processes favour a 2- over a 3-halogen; the sequence below shows how this can be utilised to develop substituted benzothiophenes.[42] 2-Lithiated reagents react with electrophiles: for example with *p*-toluenesulfonyl cyanide, 2-cyano derivatives are produced[43] and similarly, 2-trimethylstannylbenzofuran and -benzothiophene[6] and benzofuran-2-[44] and benzothiophene-2-boronic acids[45] can be prepared.

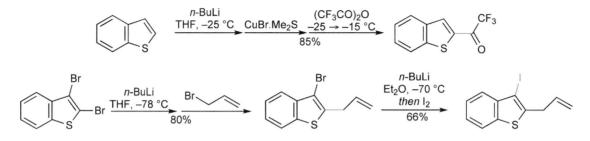

Metallation of thiophene 2- and 3-esters takes place adjacent to the ester functionality, i.e. at C-3 and C-2, respectively.[46]

21.4 Reactions with Radicals

There are few examples of radical substitution of benzofuran or benzothiophene: perfluoroalkylation of benzofuran is one such, as illustrated.[47] This process can also be applied to 2-substitution of thiophene, pyrrole, imidazole and indole.

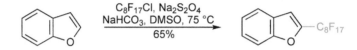

21.5 Reactions with Oxidising and Reducing Agents

Hydrodesulfurisation of benzothiophenes is conveniently achieved using Raney nickel.[48] Reduction of the hetero-rings of both benzofuran and benzothiophene giving 2,3-dihydro derivatives, notably with retention of the sulfur in the latter case, can be achieved using triethylsilane in acidic solution,[49] or with hydrogen over colloidal rhodium.[50] Reductive cleavage of benzofuran to 2-hydroxystyrene is caused by lithium with 4,4'-di-*t*-butylbiphenyl (DTBB).[51]

2,3-Dihydroxylation of benzofuran and benzothiophene giving *cis*-2,3-dihydro-2,3-dihydroxy derivatives can be achieved using *Pseudomonas putida*.[52] Benzofurans can be epoxidised at the hetero-ring double bond with dimethyl dioxirane, or alternatively converted into dioxetanes by reaction of that double bond with singlet oxygen. Both oxidised species are unstable and undergo a variety of complex further processes.[53]

21.6 Electrocyclic Reactions

2-[54] and 3-Vinyl-[55] benzofurans and benzothiophenes will serve as dienes in Diels–Alder cycloadditions, though under forcing conditions. The fusion of a pyridine ring onto benzothiophene can be achieved *via* a Staudinger reaction, using either 2- or 3-azides, to give phosphinimines, which undergo aza-Wittig condensations with unsaturated aldehydes, the ensuing electrocyclisation being followed by spontaneous dehydrogenation.[56]

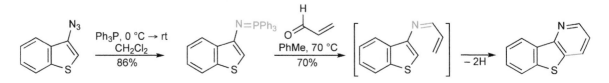

Rhodium-catalysed carbenoid addition to benzofuran using a chiral catalyst proceeds with high ee.[57]

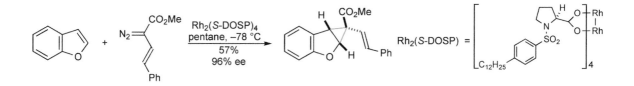

21.7 Oxy-[58] and Amino-Benzothiophenes and -Benzofurans

Benzothiophen-2-ones can be conveniently accessed by oxidation of 2-magnesio-benzothiophenes.[59] Benzothiophen-2-one will condense at the 3-position with aromatic aldehydes;[60] benzothiophen-3-one reacts comparably at its 2-position.[61]

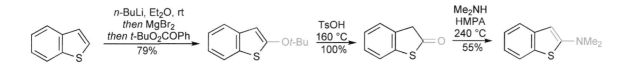

Both benzofuran-2-one, known trivially in the older literature as coumaranone, and best viewed as a lactone, and the isomeric benzofuran-3-one, form ambient anions by deprotonation at a methylene group, the former[62] requiring a stronger base than the latter.[63] Triflates suitable for metal-catalysed coupling processes are easily obtained from benzofuran-3-ones.[64]

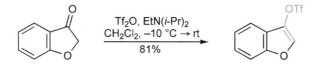

Little is known of simple 2- and 3-amino derivatives; 2-dialkylamino-benzothiophenes can be obtained by reaction of benzothiophene-2-thiol with secondary amines.[23a] In many ways 2-aminobenzothiophene behaves like a normal aromatic amine, but diazotisation leads directly to benzothiophen-2-one.[65]

21.8 Synthesis of Benzothiophenes and Benzofurans
21.8.1 Ring Synthesis
21.8.1.1 From 2-Arylthio- or 2-Aryloxy-Aldehydes, -Ketones or -Acids
Cyclisation of 2-arylthio- or 2-aryloxy-aldehydes, -ketones or -acids *via* intramolecular electrophilic attack on the aromatic ring, with loss of water, creates the heterocyclic ring; this route is a common method for the ring synthesis of benzothiophenes.

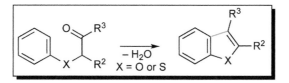

In order to produce hetero-ring unsubstituted benzothiophenes[66] an arylthioacetaldehyde acetal is generally employed, prepared in turn, from bromoacetaldehyde acetal and the thiophenol. An exactly parallel sequence produces 2,3-unsubstituted benzofurans.[67] 2-Aryloxy-acetates (or 1-aryloxy-acetones) heated with DMFDMA, produce enamines which cyclise on treatment with zinc chloride, giving 3-unsubstituted benzofuran-2-carboxylates (3-unsubstituted 2-acetyl-benzofurans).[68]

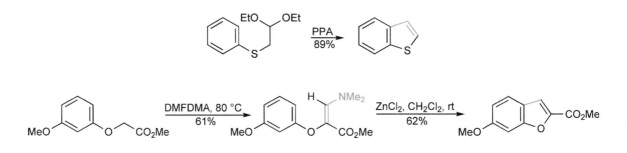

Comparable acid- (or Lewis acid) -catalysed ring closures of 2-arylthio-[69] and 2-aryloxy-[70] -ketones, and -2-arylthio-[71] and 2-aryloxyacetyl-[72] chlorides lead to 3-substituted heterocycles and 3-oxygenated heterocycles, respectively. It is possible to combine the preparation of the arylthio-ketone and the ring closure steps utilising two solid-supported reagents in a one-pot procedure, as illustrated.[73] Formation of 3-*aryl*-benzothiophenes by this route can be complicated by partial or complete isomerisation to the 2-aryl-heterocycle,[74] however using boron trifluoride as the Lewis acid produces *only* the 3-aryl-isomer.[75] 3-Tosylamino-benzofurans can be prepared from aryl glyoxal hydrates.[76]

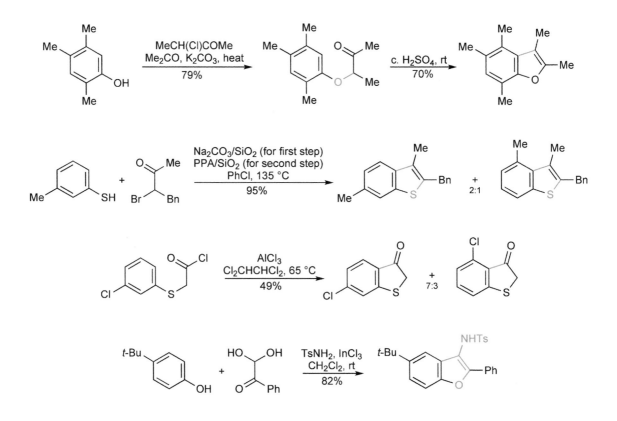

21.8.1.2   From 2-(ortho-Hydroxy(or Mercapto)aryl)-Acetaldehydes, -Ketones or -Acids
Cyclising dehydration of 2-(*ortho*-hydroxyaryl)-acetaldehydes, -ketones or -acids (and in some cases sulfur analogues) give the heterocycles; this route is important for benzofurans.

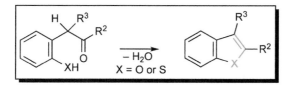

Claisen rearrangement of allyl phenolic ethers, followed by oxidation of the alkene generates *ortho*-hydroxy-arylacetaldehydes which close to give benzofurans under acid catalysis, the example showing the synthesis of 8-methoxypsoralen (33.8.1).[77] The formation of 2-substituted benzofurans from 2-(*ortho*-hydroxyaryl)-ketones is also very easy.[78]

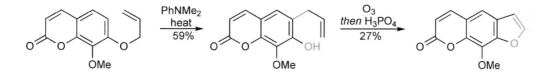

The employment of aryl 2-chloroprop-2-enyl sulfides (or ethers) as thio-Claisen rearrangement substrates neatly eliminates the necessity for an oxidative step, thus providing a route to 2-methyl-benzothiophenes (-benzofurans).[79]

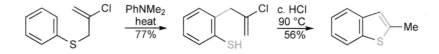

Propargyl aryl ethers undergo a Claisen rearrangement and then ring closure to produce 2-methyl-benzofurans directly.[80]

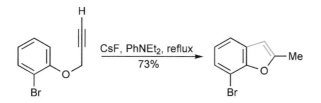

The electrocyclic rearrangement of *O*-aryl-ketoximes produces benzofurans. The acid-catalysed rearrangement[81] exactly parallels the rearrangement of phenylhydrazones, which gives indoles – the classical Fischer indole synthesis (20.16.1.1).

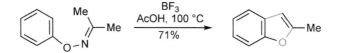

Another route to compounds of the same oxidation level involves Suzuki coupling of enol ether boronates[82] or Heck reaction with 2,5-dihydro-2,5-dimethoxyfuran, which leads to methyl benzofuran-3-acetate.[83]

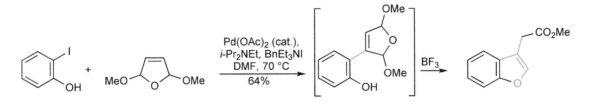

21.8.1.3 From 2-(ortho-Haloaryl)-Ketones or -Thioketones
Utilising various catalytic procedures, ring closure of ketones (thioketones), with replacement of an *ortho*-halogen, can be achieved.

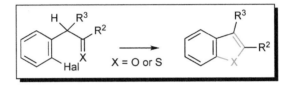

Copper(I)[84] and palladium(0)[85] catalysis can be used to ring close 2-(*ortho*-haloaryl)-ketones and the latter method has also been applied to thioketones.

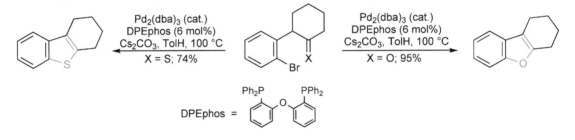

Furo[2,3-*b*]pyridin-4(7*H*)-ones can be generated from 2-pyridones carrying an alkynyl substituent at C-3.[86]

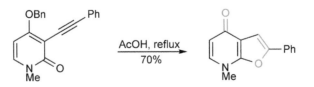

Conveniently included in this section is the ring closure (and subsequent decarboxylation if required) of 2-mercapto-3-arylprop-2-enoic acids with iodine and heating, best applied using microwaves.[87] This process probably involves *S*-iodination and then electrophilic cyclisation onto the benzene ring.

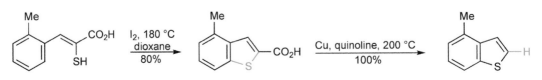

21.8.1.4 *From* ortho-*Acylaryloxy- or -Arylthioacetic Acids, Esters or Ketones*
Cyclising condensation of *ortho*-acylaryloxy- or -arylthioacetic acids (esters) or ketones gives the bicyclic heterocycles.

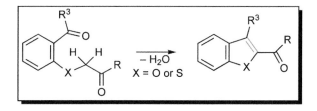

Intramolecular aldol/Perkin type condensation of *ortho*-formylaryloxyacetic acids and arylthioacetic esters produces benzofuran-[88] and benzothiophene-2-esters[89] respectively, as illustrated below. The synthesis can be performed in one pot, thus *ortho*-hydroxyaryl-aldehydes or -ketones, are *O*-alkylated with α-halo-ketones, then intramolecular aldol condensation *in situ* produces 2-acyl or 2-aroyl-benzofurans.[90,91] For benzothiophenes, the ring-closure substrates can also be obtained *via* methyl thioacetate displacement of fluoride from *ortho*-fluoro-araldehydes.[92]

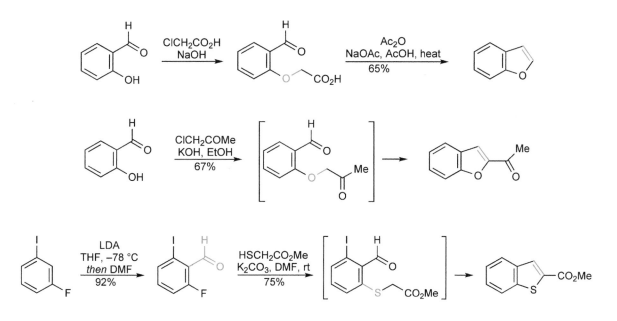

ortho-Haloaryl benzyl ethers react with *t*-butyllithium resulting in lithium–halogen exchange *and* metallation of the benzylic methylene group; addition of a carboxylic ester now leads to 3-hydroxy-2,3-dihydro-benzofurans requiring dehydration.[93]

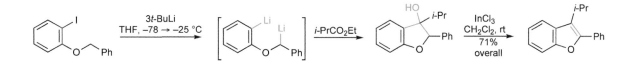

ortho-Formyl- or *ortho*-acylaryl benzyl ethers, can be comparably closed to produce 2-aryl-benzofurans, using potassium fluoride or caesium fluoride on alumina when the benzyl group carries an electron-withdrawing substituent,[94] or with simpler benzyl groups, using a phosphazene base.[95] Even the *S*-methyl groups of *ortho*-methylthio-arylcarboxamides can be deprotonated leading by cyclisation to benzothiophen-3-ones.[96]

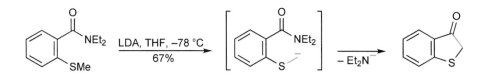

21.8.1.5 *From* ortho-*Halophenols; From* ortho-*Alkynyl-phenols*

Palladium(0)-catalysed coupling of an *ortho*-halophenolic ether (thioether) with a terminal alkyne (or with an alkynylboronic ester[97]) and ring closure promoted with an electrophile – iodine has been most often used – is an excellent method to make both benzothiophenes[98,99,100,101] and benzofurans.[102,103] *ortho*-Alkynyl-phenols can be comparably closed with palladium catalysis in the presence of copper(II) halides to give the corresponding 3-halo-benzofurans,[104] and *ortho*-alkynyl pyridin-2- and -3-yl acetates likewise ring close with iodine, generating furopyridines.[105]

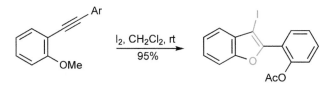

Variations on the theme include the use of gold(I) chloride to transform *ortho*-alkynylphenylthio-silanes into 3-trialkylsilyl-benzothiophenes,[106] carbonylative closures to produce 3-aroyl-benzofurans,[107] and cyclisation with hot lithium chloride.[108]

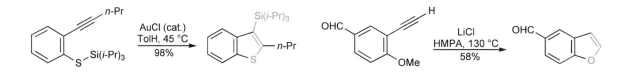

However, it is possible to produce the furan ring of a benzofuran *directly* by interaction between an *ortho*-iodo-phenol and an alkyne, the two carbon atoms of the triple bond providing C-2 and C-3 of the furan ring and the larger substituent of the alkyne (often SiR_3) ending up at the heterocyclic 2-position.[109]

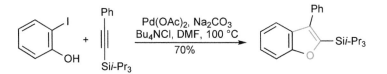

21.8.1.6 *From Partially Reduced Benzofurans and Benzothiophenes*

It can be an advantage for the introduction of benzene ring substitutents to operate with hetero-ring-reduced derivatives, the aromatic heterocycle being obtained by a final dehydrogenation. 2,3-Dihydrobenzothiophenes can be oxidised up with sulfuryl chloride or *N*-chlorosuccinimide;[110] 2,3-dichloro-5,6-dicyanobenzoquinone has been employed to dehydrogenate 2,3-dihydrobenzofurans.[111] In the example shown, a benzene ring substitution is followed by aromatisation *via* elimination of hydrogen iodide and isomerisation of the double bond into the aromatic position.[112]

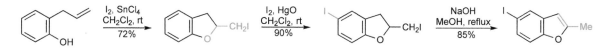

Exercises

Straightforward revision exercises (consult Chapters 19 and 21):

(a) In the electrophilic substitution of benzothiophene and benzofuran there is less selectivity than for comparable reactions of indole – why?

(b) What is the principal method for the efficient introduction of substituents to the 2-positions of benzofuran and benzothiophene?

(c) Beginning from a phenol carrying no substituents *ortho* to the hydroxyl, describe two methods for the synthesis of benzofurans.

(d) How can salicaldehydes be used for the synthesis of benzofurans?

More advanced exercises:

1. Suggest structures for the compounds formed at each stage in the following sequence: PhSH with ClCH$_2$COCH$_2$CO$_2$Et ($\rightarrow$ C$_{12}$H$_{14}$SO$_3$), then PPA/heat ($\rightarrow$ C$_{12}$H$_{12}$SO$_2$), then NH$_3$ ($\rightarrow$ C$_{10}$H$_9$NSO), then LiAlH$_4$ ($\rightarrow$ C$_{10}$H$_{11}$NS), then HCO$_2$H/heat, ($\rightarrow$ C$_{11}$H$_{11}$NSO), then POCl$_3$/heat giving finally a tricylic substance, C$_{11}$H$_9$NS.

2. Draw structures for the heterocycles formed from the following combinations: (i) C$_{13}$H$_{16}$O from 2,4,5-trimethylphenol with 3-chloro-2-butanone then the product with c. H$_2$SO$_4$; (ii) C$_{12}$H$_8$O$_4$ from 7-hydroxy-8-methoxycoumarin with CH$_2$=CHCH$_2$Br/K$_2$CO$_3$, then the product heated strongly giving an isomer, then reacted successively with O$_3$ then H$^+$; (iii) 4-trifluoromethylfluorobenzene with LDA then DMF ($\rightarrow$ C$_8$H$_4$F$_4$O), then with HSCH$_2$CO$_2$Me/NaH giving C$_{11}$H$_7$F$_3$O$_2$S; (iv) C$_9$H$_7$NO$_3$ from 4-fluoronitrobenzene with Me$_2$C=NONa then c. HCl/heat.

3. Deduce structures for the bi- and tetracyclic heterocycles formed in the following two steps respectively: 4-chlorophenylthioacetic acid with PCl$_3$ then AlCl$_3$ ($\rightarrow$ C$_8$H$_5$ClOS), then this with phenylhydrazine in hot AcOH $\rightarrow$ C$_{14}$H$_8$ClNS.

References

[1] 'Recent advances in the chemistry of benzo[*b*]thiophenes', Iddon, B. and Scrowston, R. M., *Adv. Heterocycl. Chem.*,**1970**, *11*, 177; 'Recent advances in the chemistry of benzo[*b*]thiophenes', Scrowston, R. M., *ibid.*, **1981**, *29*, 171.

[2] 'Recent advances in the chemistry of benzo[*b*]furan and its derivatives. Part I: Occurence and synthesis', Cagniant, P. and Cagniant, D., *Adv. Heterocycl. Chem.*, **1975**, *18*, 337.

[3] Eaborn, C. and Wright, G. J., *J. Chem. Soc., B*, **1971**, 2262.

[4] Armstrong, K. J., Martin-Smith, M., Brown, N. M. D., Brophy, G. C. and Sternhell, S., *J. Chem. Soc., C*, **1969**, 1766.

[5] Tanemura, K., Suzuki, T., Nishida, Y., Satsumabayashi, K. and Horaguchi, T., *J. Chem. Res. (S)*, **2003**, 497.

[6] Fargeas, V., Favresse, F., Mathieu, D., Beaudet, I., Charrue, P., Lebret, B., Piteau, M. and Quintard, J.-P., *Eur. J. Org. Chem.*, **2003**, 1711.

[7] Okuyama, T., Tani, Y., Miyake, K. and Yokoyama, Y., *J. Org. Chem.*, **2007** *72*, 1634.

[8] Van Zyl, G., Bredeweg, C. J., Rynbrandt, R. H. and Neckers, D. C., *Can. J. Chem.*, **1966**, *44*, 2283.

[9] Berens, U., Englert, U., Geyser, S., Runsink, J. and Salzer, A., *Eur. J. Org. Chem.*, **2006**, 2100.

[10] Heynderickx, A., Samat, A. and Guglielmetti, R., *Synthesis*, **2002**, 213.

[11] Krutosíková, A., Kováč, J., Dandárová, M. and Bobálová, M., *Coll. Czech. Chem. Commun.*, **1982**, *47*, 3288.

[12] v. Stoermer, R. and Richter, O., *Chem. Ber.*, **1897**, *30*, 2094; v. Stoermer, R. and Kahlert, B., *ibid.*, **1902**, *35*, 1633.

[13] Hwu, J. R., Chen, K.-L, Ananthan, S. and Patel, H. V., *Organometallics*, **1996**, *15*, 499.

[14] Bastian, G., Royer, R. and Cavier, R., *Eur. J. Med. Chem.*, **1983**, *18*, 365.

[15] Baciocchi, E., Sebastiani, G. V. and Ruzziconi, R., *J. Org. Chem.*, **1979**, *44*, 28; idem, *J. Am. Chem. Soc.*, **1983**, *105*, 6114.

[16] Yamaguchi, T. and Irie, M., *J. Org. Chem.*, **2005**, *70*, 10323.

[17] Campaigne, E., Weinberg, E. D., Carlson, G. and Neiss, E. S., *J. Med. Chem.*, **1965**, *8*, 136.

[18] Zhang, W. and Wang, P. G., *J. Org. Chem.*, **2000**, *65*, 4732.

[19] Dickinson, R. P., Iddon, B. and Sommerville, R. G., *Int. J. Sulfur Chem.*, **1973**, *8*, 233.

[20] Bachelet, J.-P., Royer, R. and Gatral, P., *Eur. J. Med. Chem. Chim. Ther.*, **1985**, *20*, 425.

[21] Brophy, G. C., Sternhell, S., Brown, N. M. D., Brown, I., Armstrong, K. J. and Martin-Smith, M., *J. Chem Soc., C*, **1970**, 933; Brown, I., Reid, S. T., Brown, N. M. D., Armstrong, K. J., Martin-Smith, M., Sneader, W. E., Brophy, G. C. and Sternhell, S., *J. Chem. Soc., C*, **1969**, 2755.

[22] Dickinson, R. P. and Iddon, B., *J. Chem. Soc., C*, **1971**, 182.

[23] (a) Sauter, F. and Golser, L., *Monatsh. Chem.*, **1967**, *98*, 2039; (b) Faller, P. and Cagniant, P., *Bull. Soc. Chim. Fr.*, **1962**, 30.

[24] Minh, T. Q., Thibaut, P., Christiaens, L. and Renson, M., *Tetrahedron*, **1972**, *28*, 5393.

[25] Ricci, A., Balucani, D. and Buu-Hoï, N. P., *J. Chem. Soc., C*, **1967**, 779.

[26] Acheson, R. M. and Harrison, D. R., *J. Chem. Soc., C*, **1970**, 1764.

[27] Cotruvo, J. A. and Degani, I., *J. Chem. Soc., Chem. Commun.*, **1971**, 436.

[28] Davies, W. and Porter, Q. N., *J. Chem. Soc.*, **1957**, 459.

[29] Harpp, D. N. and Heitner, C., *J. Org. Chem.*, **1970**, *35*, 3256; idem, *J. Am. Chem. Soc.*, **1972**, *94*, 8179.

[30] Boyd, D. R., Sharma, N. D., Haughey, S. A., Malone, J. F., McMurray, B. T., Sheldrake, G. N., Allen, C. C. R. and Dalton, H., *Chem. Commun.*, **1996**, 2363.

[31] Reinecke, M. G., Mohr, W. B., Adickes, H. W., de Bie, D. A., van de Plas, H. C. and Nijdam, J. *J. Org. Chem.*, **1973**, *38*, **1365**; Chippendale, K. E., Iddon, B., Suschitzky, H. and Taylor, D. S., *J. Chem. Soc., Perkin Trans. 1*, **1974**, 1168.

[32] Gilman, H. and Melstrom, D. S., *J. Am. Chem. Soc.*, **1948**, *70*, 1655.

[33] Shirley, D. A. and Cameron, M. D., *J. Am. Chem. Soc.*, **1950**, *72*, 2788.

[34] Reichstein, T. and Baud, J., *Helv. Chim. Acta*, **1937**, *20*, 892.

[35] Cugnan de Sevricourt, M. and Robba, M., *Bull. Soc. Chim. Fr.*, **1977**, 142.

[36] Bosold, F., Zulauf, P., Marsch, M., Harms, K., Lohrenz, J. and Boche, G., *Angew. Chem., Int. Ed. Engl.*, **1991**, *30*, 1455.

[37] Schroth, W., Jordan, H. and Spitzner, R., *Tetrahedron Lett.*, **1995**, *36*, 1421.

[38] Beyer, A. E. M. and Kloosterziel, H., *Recl. Trav. Chim. Pays Bas*, **1977**, *96*, 178.

[39] Belley, M., Douida, Z., Mancuso, J. and De Vleeschauwer, M., *Synlett*, **2005**, 247.

[40] Dore, G., Bonhomme, M. and Robba, M., *Tetrahedron*, **1972**, *28*, 2553; Dickinson, R. P. and Iddon, B., *J. Chem. Soc., C*, **1970**, 2592.

[41] Kerdesky, F. A. J. and Basha, A., *Tetrahedron Lett.*, **1991**, *32*, 2003.

[42] Dickinson, R. P. and Iddon, B., *J. Chem. Soc., C*, **1971**, 2504; Reinecke, M. G., Newsom, J. G. and Almqvist, K. A., *Synthesis*, **1980**, 327; Sura, T. P. and MacDowell, D. W. H., *J. Org. Chem.*, **1993**, *58*, 4360.

[43] Nagasaki, I., Suzuki, Y., Iwamoto, K., Higashino, T. and Miyashita, A., *Heterocycles*, **1997**, *46*, 443.

[44] Blettner, C. G. König, W. A., Tenzel, W. and Schotten, T., *Synlett*, **1998**, 295.

[45] Hedberg, M. H., Johansson, A. M., Fowler, C. J., Terenius, L. and Hacksell, U., *Biorg. Med. Chem. Lett.*, **1994**, *4*, 2527.

[46] Tietze, L. F., Lohmann, J. K. and Stadler, C., *Synlett*, **2004**, 1113.

[47] Huang, X.-T., Long, Z.-Y. and Chen, Q.-Y., *J. Fluorine Chem.*, **2001**, *111*, 107.

[48] Papa, D., Schwenk, E. and Ginsberg, H. F., *J. Org. Chem.*, **1949**, *14*, 723.

[49] Kursanov, D. N., Parnes, Z. N., Bolestova, G. I. and Belen'kii, L. I., *Tetrahedron*, **1975**, *31*, 311.

[50] Mévelle, V. and Roucoux, A., *Inorg. Chim. Acta*, **2004**, *357*, 3099.

[51] Yus, M., Foubelo, F., Ferrández, J. V. and Bachki, A., *Tetrahedron*, **2002**, *58*, 4907.

[52] Boyd, D. R., Sharma, N. D., Boyle, R., McMurray, B. T., Evans, T. A., Malone, J. F., Dalton, H., Chima, J. and Sheldrake, G. N., *J. Chem. Soc., Chem. Commun.*, **1993**, 49.

[53] Sauter, M. and Adam, W., *Acc. Chem. Res.*, **1995**, *28*, 289.

[54] Marrocchi, A., Minuti, L., Taticchi, A. and Scheeren, H. W., *Tetrahedron*, **2001**, *57*, 4959.

[55] Le Strat, F. and Maddaluno, J., *Org. Lett.*, **2002**, *4*, 2791.

[56] Degl'Innocenti, A., Funicello, M., Scafato, P., Spagnolo, P. and Zanirato, P., *J. Chem. Soc., Perkin Trans. 1*, **1996**, 2561.

[57] Davies, H. M. L., Kong, N. and Churchill, M. R., *J. Org. Chem.*, **1998**, *63*, 6586.

[58] 'Chemistry of benzo[*b*]thiophene-2,3-dione', Rajopadhye, M. and Popp, F. D., *Heterocycles*, **1988**, *27*, 1489.

[59] Vesterager, N. O., Pedersen, E. B. and Lawesson, S.-O., *Tetrahedron*, **1973**, *29*, 321.

[60] Conley, R. A. and Heindel, N. D., *J. Org. Chem.*, **1976**, *41*, 3743.

[61] Réamonn, L. S. S. and O'Sullivan, W. I., *J. Chem. Soc., Perkin Trans 1*, **1977**, 1009.

[62] Zaugg, H. E., Dunnigan, D. A., Michaels, R. J., Swett, R. J., Wang, T. S., Sommers, A. H. and DeNet, R. W., *J. Org. Chem.*, **1961**, *26*, 644.

[63] v. Auwers, K. and Schütte, H., *Chem. Ber.*, **1919**, *52*, 77.

[64] Morice, C., Garrido, F., Mann, A. and Suffert, J., *Synlett*, **2002**, 501.

[65] Stacy, G. W., Villaescusa, F. W. and Wollner, T. E., *J. Org. Chem.*, **1965**, *30*, 4074.

[66] Tilak, B. D., *Tetrahedron*, **1960**, *9*, 76.

[67] Spagnolo, P., Tiecco, M., Tundo, A. and Martelli, G., *J. Chem. Soc., Perkin Trans. 1*, **1972**, 556.

[68] del Carmen Cruz, M. and Tamariz, J., *Tetrahedron*, **2005**, *61*, 10061.

[69] Oliveira, M. M., Moustrou, C., Carvalho, L. M., Silva, J. A. C., Samat, A., Guglielmetti, R., Dubest, R., Aubard, J. and Oliveira-Campos, A. M. F., *Tetrahedron*, **2002**, *58*, 1709.

[70] Royer, R. and René, L., *Bull. Soc. Chim. Fr.*, **1970**, 1037.

[71] Werner, L. H., Schroeder, D. C. and Ricca, S., *J. Am. Chem. Soc.*, **1957**, *79*, 1675.

[72] Elvidge, J. A. and Foster, R. G., *J. Chem. Soc.*, **1964**, 981.

[73] Aoyama, T., Takido, T. and Kodomari, M., *Synlett*, **2005**, 2739.

[74] Banfield, J. E., Davies, W., Gamble, N. W. and Middleton, S., *J. Chem. Soc.*, **1956**, 4791.

[75] Kim, S., Yang, J. and DiNinno, F., *Tetrahedron Lett.*, **1999**, *40*, 2909.

[76] Chen, C.-X., Liu, L., Yang, D.-P., Wang, D. and Chen., Y.-J., *Synlett*, **2005**, 2047.

[77] de Souza, N. J., Nayak, P. V. and Secco, E., *J. Heterocycl. Chem.*, **1966**, *3*, 42.

[78] Tinsley, S. W., *J. Org. Chem.*, **1959**, *24*, 1197.

[79] Anderson, W. K., LaVoie, E. J. and Bottaro, J. C., *J. Chem. Soc., Perkin Trans. 1*, **1976**, 1.

[80] Ishii, H., Ohta, S., Nishioka, H., Hayashida, N. and Harayama, T., *Chem. Pharm. Bull.*, **1993**, *41*, 1166; Moghaddam, F. M., Sharifi, A. and Saidi, M. R., *J. Chem. Res. (S)*, **1996**, 338; Lingam, V. S. P. R., Vinodkumar, R., Mukkanti, K., Thomas, A. and Goplan, B., *Tetrahedron Lett.*, **2008**, *49*, 4260.

[81] Sheradsky, T., *Tetrahedron Lett.*, **1966**, 5225; Sheradsky, T. and Elgavi, A., *Isr. J. Chem.*, **1968**, *6*, 895; Kaminsky, D., Shavel, J. and Meltzer, R. I., *Tetrahedron Lett.*, **1967**, 859.

[82] Satoh, M., Miyaura, N. and Suzuki, A., *Synthesis*, **1987**, 373.

[83] Samizu, K. and Ogasawara, K., *Heterocycles*, **1994**, *38*, 1745.

[84] Chen, C.-y. and Dormer, P. G., *J. Org. Chem.*, **2005**, *70*, 6964.

[85] Willis, M. C., Taylor, D. and Gillmore, A. T., *Org. Lett.*, **2004**, *6*, 4755.

[86] Bossharth, E., Desbordes, P., Monteiro, N. and Balme, G., *Tetrahedron Lett.*, **2009**, *50*, 614.

[87] Allen, D., Callaghan, O., Cordier, F. L., Dobson, D. R., Harris, J. R., Hotten, T. M., Owton, W. M., Rathmell, R. E. and Wood, V. A., *Tetrahedron Lett.*, **2004**, *45*, 9645.

[88] Burgstahler, A. W. and Worden, L. R., *Org. Synth., Coll. Vol. V*, **1973**, 251.

[89] Bridges, A. J., Lee, A., Maduakor, E. C. and Schwartz, C. E., *Tetrahedron Lett.*, **1992**, *33*, 7499.

[90] Elliott, E. D., *J. Am. Chem. Soc.*, **1951**, *73*, 754.

[91] Rao, M. L. N., Awasthi, D. K. and Banerjee, D., *Tetrahedron Lett.*, **2007**, *48*, 431.

[92] Lee, S., Lee, H., Yi, K. Y., Lee, B. H., Yoo, S.-e., Lee, K. and Cho, N. S., *Bioorg. Med. Chem. Lett.*, **2005**, *15*, 2998.

[93] Sanz, R., Miguel, D., Martinez, A. and Pérez, A., *J. Org. Chem.*, **2006**, 4024.

[94] Hellwinkel, D. and Göke, K., *Synthesis*, **1995**, 1135.

[95] Kraus, G. A., Zhang, N., Verkade, J. G., Nagarajan, M. and Kisanga, P. B., *Org. Lett.*, **2000**, *2*, 2409.

[96] Mukherjee, C., Kamila, S. and De, A., *Tetrahedron*, **2003**, *59*, 4767.

[97] Colobert, F., Castanet, A.-S. and Abillard, O., *Eur. J. Org. Chem.*, **2005**, 3334.

[98] Flynn, B. L., Verdier-Pinard, P. and Hamel, E., *Org. Lett.*, **2001**, *3*, 651.

[99] Larock, R. C. and Yue, D., *Tetrahedron Lett.*, **2001**, *42*, 6011.

[100] Wang, C.-H., Hu, R.-R., Liang, S., Chen, J.-H., Yang, Z. and Pei, J., *Tetrahedron Lett.*, **2005**, *46*, 8153.

[101] Flynn, B. L., Verdier-Pinard, P. and Hamel, E., *Org. Lett.*, **2001**, *3*, 651.

[102] Yao, T., Yue, D. and Larock, R. C., *J. Org. Chem.*, **2005**, *70*, 9985.

[103] Yue, D., Yao, T. and Larock, R. C., *J. Org. Chem.*, **2005**, *70*, 10292.

[104] Liang, Y., Tang, S., Zhang, X.-D., Mao, L.-Q., Xie, Y.-X. and Li, J.-H., *Org. Lett.*, **2006**, *8*, 3017.

[105] Arcadi, A., Cacchi, S., Di Giuseppe, S., Fabrizi, G. and Marinelli, F., *Org. Lett.*, **2002**, *4*, 2409.

[106] Nakamura, I., Sato, T., Terada, M. and Yamamoto, Y., *Org. Lett.*, **2007**, *9*, 4081.

[107] Hu, Y., Zhang, Y., Yang, Z. and Fathi, R., *J. Org. Chem.*, **2002**, *67*, 2365.

[108] Hiroya, K., Hashimura, K. and Ogasawara, K., *Heterocycles*, **1994**, *38*, 2463.

[109] Larock, R. C., Yum, E. K., Doty, D. J. and Sham, K. K. C., *J. Org. Chem.*, **1995**, *60*, 3270; Bishop, B. C., Cottrell, I. F. and Hands, D., *Synthesis*, **1997**, 1315; Kundu, N. G., Pal, M., Mahanty, J. S. and De, M., *J. Chem. Soc., Perkin Trans. 1*, **1997**, 2815; Gill, G. S., Grobelny, D. W., Chaplin, J. H. and Flynn, B. L., *J. Org. Chem.*, **2008**, *73*, 1131.

[110] Tohma, H., Egi, M., Ohtsubo, M., Watanabe, H., Takizawa, S. and Kita, Y., *Chem. Commun.*, **1998**, 173.

[111] Büchi, G. and Chu, P.-S., *J. Org. Chem.*, **1978**, *43*, 3717; Stanetty, P. and Pürstinger, G., *J. Chem. Res.*, **1991**, *(S)* 78; *(M)* 0581.

[112] Onito, K., Hatakeyana, T., Takeo, M., Suginome, H. and Tokuda, M., *Synthesis*, **1997**, 23.

22

Isoindoles, Benzo[*c*]thiophenes and Isobenzofurans: Reactions and Synthesis

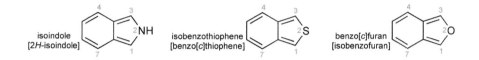

isoindole
[2*H*-isoindole]

isobenzothiophene
[benzo[*c*]thiophene]

benzo[*c*]furan
[isobenzofuran]

Isoindole,[1] benzo[*c*]thiophene[2] and isobenzofuran[3] are much less stable than their isomers, indole, and benzo[*b*]thiophene and benzo[*b*]furan. This is undoubtedly associated with their lower aromaticity, which can be appreciated qualitatively by noting that in these [*c*]-systems, the six-membered ring is not a complete benzenoid unit. Of the three unsubstituted heterocycles, benzo[*c*]thiophene is the most stable – it survives as a solid for a few days at −30 °C – but most chemistry has been carried out with substituted derivatives. The instability manifests itself in a strong tendency to add reagents so as to generate products that do have a complete benzene ring, in particular these heterocycles are susceptible to cycloaddition of dienophiles. In this context, then, it is not surprising that for isoindole, for which an alternative tautomer (1*H*-isoindole, sometimes called 'isoindolenine') is possible, which does have a complete benzenoid unit, an appreciable percentage of that alternative exists in equilibrium.[4] Indeed, some isoindoles exist largely as the tautomer with a C–N double bond – 1,3,4,7-tetramethylisoindole is an example[5] – but 1-phenylisoindole favours the 2*H*-tautomer to the extent of 91%.[6] Substituents on the benzenoid ring can also influence both the stability of the isoindole and the position of tautomeric equilibrium, for example 4,5,6,7-tetrabromoisoindole is a stable crystalline solid which exists wholly as the 2*H*-tautomer;[7] a pivaloyl substituent at C-5, though remote from the sensitive heterocyclic ring, can also stabilise an isoindole.[8] The position of such tautomeric equilibria can be altered by changing solvent – solvents such as dimethylsulfoxide tend to favour the *N*-hydrogen tautomer, where protic solvents like alcohols favour the imine tautomer.

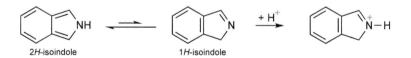

2*H*-isoindole   1*H*-isoindole

22.1   Reactions with Electrophilic Reagents

Isoindoles protonate to generate only one cation;[5] this electrophilic addition of protons sets the pattern for substitution in these systems, but there are relatively few clear cut examples, no doubt partly because of the instability of less substituted isoindoles, isobenzofurans and benzo[*c*]thiophenes. Detritiation studies showed the intrinsic reactivity of 2-methylisoindole in this electrophilic substitution to be 10^4 greater than that of 1-methylindole.[9]

Heterocyclic Chemistry 5th Edition   John Joule and Keith Mills
© 2010 Blackwell Publishing Ltd

2-t-Butylisoindole is much more stable than the unsubstituted heterocycle or other 2-substituted isoindoles, thus its reactions can be used as a measure of intrinsic reactivity, set aside from instability: even weak electrophiles such as aryl-diazonium ions attack it and it undergoes alkylation, in each case at the 1-position.[10]

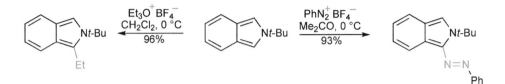

An interesting example of electrophilic substitution is the conversion of phthalimidine (2,3-dihydro-1*H*-isoindol-1-one) into 1-bromo-3-formylisoindole under Vilsmeier conditions (formation of 1-bromo-2*H*-isoindole may be the first step).[11] Mannich substitution of 2-methyl-1-phenylisoindole is another straightforward example.[12]

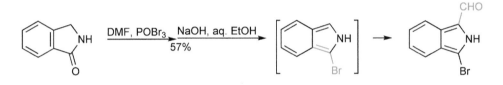

22.2 Electrocyclic Reactions

Each of the three systems has a strong tendency for cycloaddition with dienophiles across the 1- and 3-positions, thereby gaining the stabilising contribution of a complete benzene ring, isobenzofuran itself, for example, reacts instantly at 0 °C with maleic anhydride.[13] Isobenzofuran has been dubbed 'the most reactive isolable diene'. More typically, the synthesis of one of these non-isolable, or extremely reactive systems, is immediately followed by a trapping with a dienophile, so that discussion of synthesis must inevitably involve discussion of the product cycloaddition chemistry.

Reactions of 1,3-diphenylisobenzofuran, which is much more stable, are typical: it undergoes Diels–Alder cycloaddition with diethyl acetylenedicarboxylate[14] and adds singlet oxygen,[15] indeed this commercially available isobenzofuran has been very frequently used as a trapping reagent for alkenes and alkynes, both stable and transient. Less stable isobenzofurans are traditionally generated and reacted *in situ*: the reactions shown below of 1-methyl-3-phenylisobenzofuran[16] and 1-phenylisobenzofuran[17] are typical.

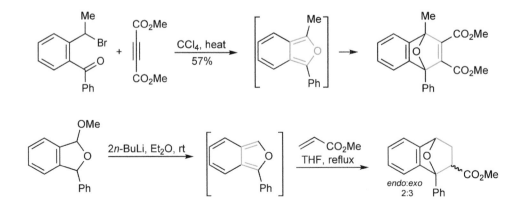

More modern methods for the production of isobenzofurans in solution have allowed detailed study of the rates of various substituted derivatives with *N*-methylmaleimide.[18] Diels–Alder additions are also known for benzo[*c*]thiophene[19,20] and isoindoles.[4]

22.3   Phthalocyanines[21]

The phthalocyanine macrocyclic system, formally derived from four isoindoles, is the basis for many blue dyestuffs. Metal derivatives have a cation complexed at the centre, much like the iron atom in heme (see 32.3). Phthalocyanine can be produced by the reductive cyclisation of 2-cyanobenzamide or, in a route which makes its relationship to isoindole more obvious, by the combination of four molecules of 1,3-diiminoisoindoline with the elimination of ammonia.[22]

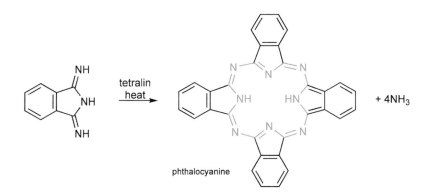

phthalocyanine

22.4   Synthesis of Isoindoles, Benzo[*c*]thiophenes and Isobenzofurans

22.4.1   Isoindoles

Isoindoles can be produced by eliminations from *N*-substituted isoindolines (1,3-dihydro-isoindoles), themselves readily produced by the reaction of a nitrogen nucleophile and a 1,2-bis(bromomethyl)-benzene:[23] examples are the pyrolytic elimination of acetic acid from the cyclic hydroxylamine acetate,[4] or, at a much lower temperature, of benzyl alcohol from an *N*-hydroxy-isoindoline benzyl ether,[24] or of methanesulfonic acid from a corresponding mesylate.[25] *N*-substituted isoindoles, too, have generally been made from an isoindoline by elimination processes, thus *N*-oxides can be made to lose water by pyrolysis,[26] or better, by treatment with acetic anhydride.[27]

A synthesis of 1-phenylisoindole represents a classical approach to the construction of a heterocycle: a precursor is assembled in which there is an amino group (initially protected in the form of a phthalimide) five atoms away from a carbonyl group with which it must interact and form a cyclic imine.[28]

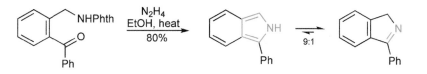

Several routes involve cycloreversions as final steps;[29] each of the starting materials shown below is available from the cycloadduct (cf. 16.8) of benzyne and 1-methoxycarbonylpyrrole.

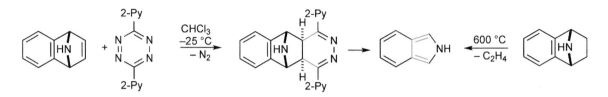

1,3-Diaryl-isoindoles can be constructed from 1,2-diaroyl-benzenes by reaction with an amine and a reducing agent.[30]

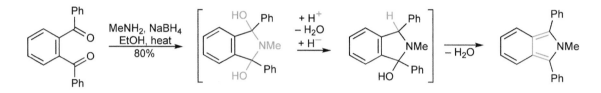

22.4.2   Benzo[*c*]thiophenes

Elimination from dihydrobenzo[*c*]thiophene *S*-oxides has been successfully applied, as for isoindoles, for the preparation of benzo[*c*]thiophenes, including the parent compound[20,31] and isolable 4,6,7-tri-*t*-butylbenzo[*c*]thiophene.[32]

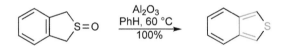

A general route to 1,3-diaryl-benzo[*c*]thiophenes starts with the synthesis of an aryl-substituted phthalide, reaction of this with an aryl (heteroaryl) Grignard reagent, then thionation using Lawesson's reagent.[33]

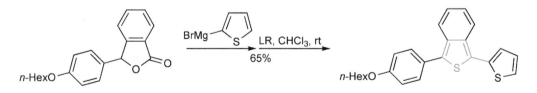

It is comparatively rare for the construction of a benzanellated heterocycle to involve formation of the benzene ring as a final step, however benzo[*c*]thiophenes can be made by this strategy, utilising a double Friedel–Crafts-type alkylation of a 2,5-disubstituted (to prevent attack at α-positions) thiophene with a 1,4-diketone.[34]

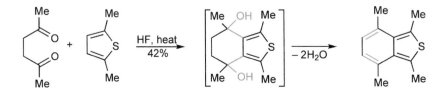

22.4.3 Isobenzofurans

Isobenzofuran can be isolated by trapping on a cold finger, following thermolysis of a suitable precursor such as 1,4-epoxy-1,2,3,4-tetrahydronaphthalene,[13,35] but although isolable, for trapping experiments it can be conveniently produced by either acid- or base-catalysed elimination of methanol from 1-methoxyphthalan in the presence of the intended dienophile.[36]

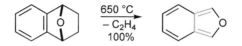

1-Methoxyphthalan is obtained by partial oxidation of 1,2-bis(hydroxymethyl)benzene with hypochlorite in methanol; treatment with lithium diisopropylamide gives isobenzofuran.[37] Conditions have been defined whereby this elimination can be run in such a way as to allow immediate ring lithiation; the lithiated isobenzofuran can then be further reacted to give alkylated isobenzofurans, which can be characterised by reaction with dimethyl acetylendicarboxylate.[38] This same starting material can be converted into isobenzofuran under neutral conditions, by reaction with a catalytic amount of Pd$_2$(dba)$_3$.CHCl$_3$ at 100 °C, again in the presence of dimethyl acetylenedicarboxylate.[39]

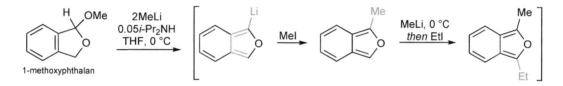

A spectacular demonstration of this approach is provided by the synthesis of the 'stretched' isobenzofuran shown below.[40] Note that the first cycloaddition is with a dienophile able to provide the *ortho*-related carbinol/aldehyde arrangement ready for the formation of another furan ring.

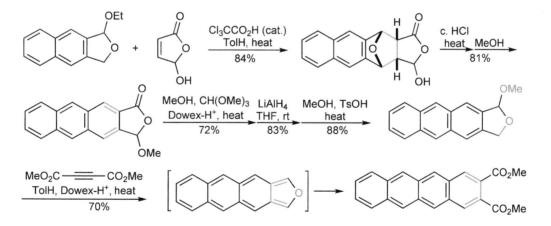

Most of the stable isobenzofurans are 1,3-diaryl-substituted, and are deep yellow. Such compounds are available *via* the partial reduction, and dehydrating cyclisation of 1,2-diaroyl-benzenes.[41] Both 1-mono- and 1,3-disubstituted isobenzofurans are available from phthalides by Grignard addition[42] and symmetrical 1,3-diaryl-isobenzofurans are efficiently formed from 3-methoxy-3*H*-isobenzofuran-1-one on reaction with two equivalents of aryl Grignard reagents, each process requiring final acid-catalysed elimination of water.[43]

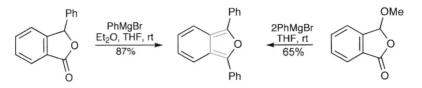

Exercises

Straightforward revision exercises (consult Chapters 19 and 22):

(a) The [*c*]-fused heterocycles considered in this chapter are much less stable than the [*b*]-fused isomers – why?

(b) What factors favour 1*H*-isoindoles over 2-*H*-isoindoles?

(c) What is the most characteristic reactivity of the [*c*]-fused heterocycles considered in this chapter? Give three examples of this typical reactivity.

(d) Describe one method each for the ring synthesis of isoindoles, benzo[*c*]thiophenes and isobenzofurans.

More advanced exercises:

1. Deduce structures for the compounds formed at each stage in the following sequence: (i) 1,2-bis(bromomethyl)-4-pivaloylbenzene with $H_2NH_2CC=CH/Et_3N \rightarrow C_{16}H_{19}NO$, which was then heated at 500 °C producing $C_{13}H_{15}NO$, which was trapped with *N*-phenylmaleimide $\rightarrow C_{23}H_{22}N_2O_3$ (what is the mechanism of the high temperature reaction?)

2. What products would be formed when: (i) phthalaldehyde is reacted, in sequence, with $2 \times NaHSO_3$, then $MeNH_2$, then $2 \times KCN \rightarrow C_{10}H_8N_2$; (ii) benzoic acid *N,N*-diethylamide is treated with *n*-BuLi, then PhCH=O then acid $\rightarrow C_{14}H_{10}O_2$, then this with PhMgBr then acid $\rightarrow C_{20}H_{14}O$ and finally this with O_2/methylene blue/hν/–50 °C $\rightarrow C_{20}H_{14}O_3$.

3. Phthalaldehyde reacts with $HO(CH_2)_2OH/CuSO_4$ to give $C_{10}H_{10}O_3$ – what is its structure? This product can be reduced with $NaBH_4$, and when followed by TsOH with $MeO_2CC=CCO_2Me$ in hot toluene, $C_{14}H_{12}O_5$ is formed – deduce a structure.

4. What would be the result of reacting benzo[*c*]thiophene with maleic anhydride, then hot NaOH, then acid – the product has the formula $C_{12}H_8O_4$?

References

[1] 'Isoindoles', White, J. D. and Mann, M. E., *Adv. Heterocycl. Chem.*, **1969**, *10*, 113; 'The chemistry of isoindoles', Bonnett, R. and North, S. A., *ibid.*, **1982**, *29*, 341.

[2] 'Benzo[*c*]thiophenes', Iddon, B., *Adv. Heterocycl. Chem.*, **1972**, *14*, 331.

[3] 'Benzo[*c*]furans', Friedrichsen, W., *Adv. Heterocycl. Chem.*, **1980**, *26*, 135; 'Isobenzofuran', Haddadin, M. J., *Heterocycles*, **1978**, *9*, 865; 'Progress in the chemistry of isobenzofurans; applications to the synthesis of natural products and polyaromatic hydrocarbons', Rodrigo, R., *Tetrahedron*, **1988**, *44*, 2093; 'Recent advances in the chemistry of benzo[*c*]furans and related compounds', Friedrichsen, W., *Adv. Heterocycl. Chem.*, **1999**, *73*, 1.

[4] Bonnett, R., Brown, R. F. C. and Smith, R. G., *J. Chem. Soc., Perkin Trans. 1*, **1973**, 1432.

[5] Bender, C. O. and Bonnett, R., *J. Chem. Soc., Chem. Commun.*, **1966**, 198.

[6] Veber, D. F. and Lwowski, W., *J. Am. Chem. Soc.*, **1964**, *86*, 4152.

[7] Kreher, R. and Herd, K. J., *Angew. Chem., Int. Ed. Engl.*, **1974**, *13*, 739.

[8] Kreher, R., Kohl, N. and Use, G., *Angew. Chem., Int. Ed. Engl.*, **1982**, *21*, 621.

[9] Laws, A. P. and Taylor, R., *J. Chem. Soc., Perkin Trans. 2*, **1987**, 591.

[10] Kreher, R. P. and Use, G., *Chem. Ber.*, **1989**, *122*, 337.

[11] von Dobeneck, H., Reinhard, H., Deubel, H. and Wolkenstein, D., *Chem. Ber.*, **1969**, *102*, 1357.

[12] Theilacker, W. and Kalenda, H., *Justus Liebigs Ann. Chem.*, **1953**, *584*, 87.

[13] Wiersum, U. E. and Mijs, W. J., *J. Chem. Soc., Chem. Commun.*, **1972**, 347.

[14] Berson, J. A., *J. Am. Chem. Soc.*, **1953**, *75*, 1240.

[15] Rio, G. and Scholl, M.-J., *J. Chem. Soc., Chem. Commun.*, **1975**, 474.

[16] Faragher, R. and Gilchrist, T. L., *J. Chem. Soc., Perkin Trans. 1*, **1976**, 336.

[17] Martin, C., Maillet, P., and Maddaluno, J., *Org. Lett.*, **2000**, *2*, 923.

[18] Tobia, D. and Rickborn, B., *J. Org. Chem.*, **1987**, *52*, 2611.

[19] Mayer, R., Kleinert, H., Richter, S. and Gewald, K., *Angew. Chem., Int. Ed. Engl.*, **1962**, *1*, 115.

[20] Cava, M. P., Pollack, N. M., Mamer, O. A. and Mitchell, M. J., *J. Org. Chem.*, **1971**, *36*, 3932.

[21] 'Advances in the chemistry of phthalocyanines', Lever, A. B. P., Hempstead, M. R., Leznoff, C. C., Liew, W., Melnik, M., Nevin, W. A. and Seymour, P., *Pure. Appl. Chem.*, **1986**, *58*, 1461.

[22] Elvidge, J. A. and Linstead, R. P., *J. Chem. Soc.*, **1955**, 3536.

[23] Bornstein, J. and Shields, J. E., *Org. Synth., Coll. Vol. V*, **1973**, 1064.

[24] Kreher, R. and Seubert, J., *Z. Naturforsch.*, **1966**, *20b*, 75.

[25] Kreher, R. and Herd, K. J., *Heterocycles*, **1978**, *11*, 409.

[26] Thesing, J., Schäfer, W. and Melchior, D., *Justus Liebigs Ann. Chem.*, **1964**, *671*, 119.

[27] Kreher, R. and Seubert, J., *Angew. Chem., Int. Ed. Engl.*, **1964**, *3*, 639; *ibid.*, **1966**, *5*, 967.

[28] Veber, D. F. and Lwowski, W., *J. Am. Chem. Soc.*, **1964**, *86*, 4152.

[29] Priestley, G. M. and Warrener, R. N., *Tetrahedron Lett.*, **1972**, 4295; Bornstein, J., Remy, D. E. and Shields, J. E., *J. Chem. Soc., Chem. Commun.*, **1972**, 1149.

[30] Haddadin, M. J. and Chelhot, N. C., *Tetrahedron Lett.*, **1973**, 5185.

[31] Holland, J. M. and Jones, D. W., *J. Chem. Soc., C*, **1970**, 536; Kreher, R. P. and Kalischko, J., *Chem.Ber.*, **1991**, 645.

[32] El-Shishtawy, R. M., Fukunishi, K. and Miki, S., *Tetrahedron Lett.*, **1995**, *36*, 3177.

[33] Mohanakrishnan, A. K. and Amaladass, P., *Tetrahedron Lett.*, **2005**, *46*, 4225.

[34] Dann, O., Kokorudz, M. and Gropper, R., *Chem. Ber.*, **1954**, *87*, 140.

[35] Warrener, R. N., *J. Am. Chem. Soc.*, **1971**, *93*, 2346.

[36] Mitchell, R. H., Iyer, V. S., Khalifa, N., Mahadevan, R., Venugopalan, S., Weerawarna, S. A. and Zhou, P., *J. Am. Chem. Soc.*, **1995**, *117* 1514.

[37] Naito, K. and Rickborn, B., *J. Org. Chem.*, **1980**, *45*, 4061.

[38] Crump, S. L. and Rickborn, B., *J. Org. Chem.*, **1984**, *49*, 304.

[39] Mikami, K. and Ohmura, H., *Org. Lett.*, **2002**, *4*, 3355.

[40] Tu, N. P. W., Yip, J. C. and Dibble, P. W., *Synthesis*, **1996**, 77.

[41] Zajec, W. W. and Pichler, D. E., *Can. J. Chem.*, **1966**, *44*, 833; Potts, K. T. and Elliott, A. J., *Org. Prep. Proc. Int.*, **1972**, *4*, 269.

[42] Newman, M. S., *J. Org. Chem.*, **1961**, *26*, 2630.

[43] Benderradji, F., Nechab, M., Einhorn, C. and Einhorn, J., *Synlett*, **2006**, 2035.

23

Typical Reactivity of 1,3- and 1,2-Azoles and Benzo-1,3- and -1,2-Azoles

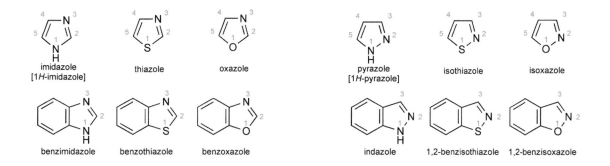

imidazole
[1*H*-imidazole]

thiazole

oxazole

pyrazole
[1*H*-pyrazole]

isothiazole

isoxazole

benzimidazole

benzothiazole

benzoxazole

indazole

1,2-benzisothiazole

1,2-benzisoxazole

The 1,3- and 1,2-azoles each contain one heteroatom in an environment analogous to that of the nitrogen in pyridine – an imine nitrogen – and one heteroatom like the nitrogen in pyrrole, or the sulfur in thiophene or the oxygen in furan. Consequently, the chemical reactions of the 1,2- and 1,3-azoles present a fascinating combination and mutual interaction of the types of reactivity that have been described earlier in this book for azines, on the one hand, and for pyrrole, thiophene and furan, on the other, with the variation in electronegativity of the five-membered-type heteroatom having a substantial differentiating effect.

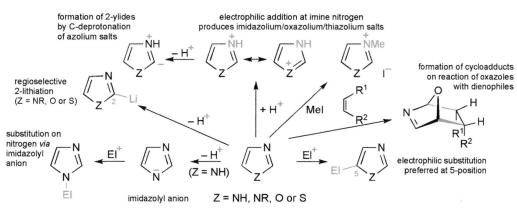

Typical reactions of 1,3-azoles

Many of the lessons to be learnt apply to both 1,3- and 1,2-azoles, though the direct linking of the two heteroatoms in the latter has a substantial inductive influence, altering properties in degree. Thus the 1,2-azoles tend to be less nucleophilic and less basic at the imine nitrogen than their 1,3-isomers. That such electrophilic additions occur readily, illustrates, as for pyridine, that an imine nitrogen lone pair is not involved in the aromatic sextet of electrons.

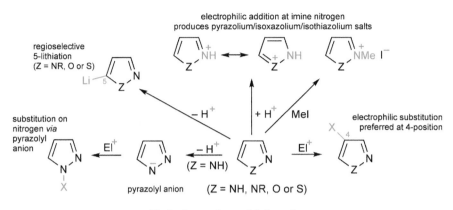

Typical reactions of 1,2-azoles

Electrophilic substitution in the azoles is intermediate in facility between pyridine on the one hand and pyrroles, thiophenes and furans on the other; the presence of the electron-withdrawing imine unit has an effect on the five-membered aromatic heterocycles just as it does when incorporated into a six-membered aromatic framework, i.e. the comparison is like that between benzene and pyridine (Chapter 7). The order of reactivity – pyrrole > furan > thiophene – is echoed in the azoles, though the presence of the basic nitrogen complicates such comparisons. The regiochemistry of electrophilic attack can be rationalised nicely by comparing the 'character' of the various ring positions – those that are activated in being five-membered in character and those that are deactivated by their similarity to α- and γ- positions in pyridine.

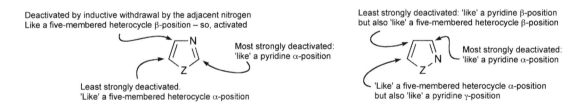

Influences on the positional reactivities of 1,3- and 1,2-azoles towards electrophilic substitution

The converse of electrophilic substitution following the five-membered pattern, is that nucleophilic substitution of halogen follows the pyridine pattern i.e. it is much faster at the 2-position of 1,3-azoles and at the 3-position of 1,2-azoles, than at other ring positions. Resonance contributors to the intermediates for such substitutions make the reason for this plain: the imine nitrogen can act as an electron sink for the attack, *only* at these positions.

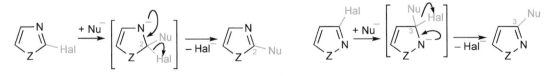

Intermediates for nucleophilic displacements on azoles

The utility of palladium(0)-catalysed processes for the elaboration of azoles has been developed extensively; one typical example serves to point up the importance of such process in azole chemistry (see Section 4.2 for a detailed discussion of palladium-catalysed processes).

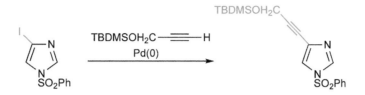

Continuing the analogy with pyridine reactivity, methyl groups at the 2-positions of 1,3-azoles and the 3-positions of 1,2-azoles carry acidified hydrogen atoms and can be deprotonated with strong bases. In further analogy with pyridines, quaternisation of the imine nitrogen makes such deprotonations even easier; the resulting enamines react with electrophiles at the side-chain carbon.

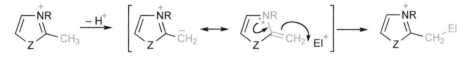

Deprotonation of 2-methyl-1,3-azolium salts

Ring deprotonation using a strong base is regioselective for the 2-position in the 1,3-azoles and for the 5-position in the 1,2-azoles.

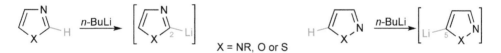

Preferred position of ring-deprotonation of azoles

The facility with which 1,3-azolium cations form ylides by 2-deprotonation is central to the biological activity of thiamine pyrophosphate (32.2.4). Such ylides have a neutral carbene resonance contributor.

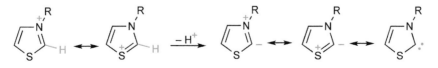

Thiazolium ylide resonance contributors

The benzo-1,3- and -1,2-azoles do not undergo electrophilic substitution on the hetero-ring; some examples of electrophilic halogenations are known in the benzene ring. These bicyclic systems are slightly weaker bases than the azoles, but react easily with alkyl halides to give salts. N-Deprotonation of indazoles or benzimidazoles allows substitution on nitrogen.

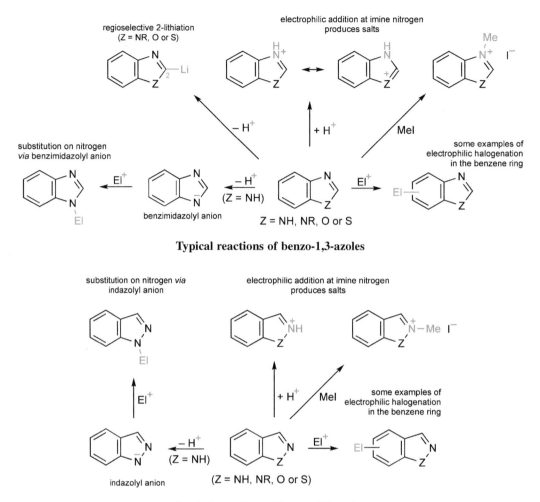

Typical reactions of benzo-1,3-azoles

Typical reactions of benzo-1,2-azoles

It has long been known that 1,3-azoles can be assembled from a component providing the two hetero-atoms, for example a thioamide or an amidine, and an α-haloketone. A more recent route employs the interaction between the anion of an isonitrile and an aldehyde, thioaldehyde or imine.

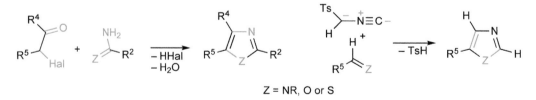

$Z = NR, O$ or S

Two routes for the ring synthesis of 1,3-azoles

To produce a 1,2-azole, a 1,3-dicarbonyl-compound needs to be condensed with a unit providing the two heteroatoms – a hydrazine or hydroxylamine. The 1,3-dipolar cycloaddition of nitrile oxides or nitrile imines to alkynes provides an important route to isoxazoles and pyrazoles.

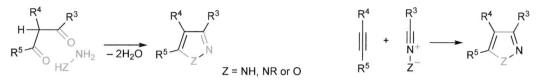

Two routes for the ring synthesis of 1,2-azoles

The benzo-1,3-azoles are commonly prepared from *ortho*-disubstituted benzenes by reaction with a carboxylic acid anhydride, or equivalent.

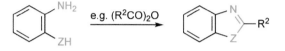

The usual ring synthesis route for benzo-1,3-azoles

Some of the routes to benzo-1,2-azoles require a heteroatom–heteroatom bond formation, but these are difficult to categorise generally, being of various mechanistic types. In one example, the generation of a sulfenyl chloride allows N–S bond formation.

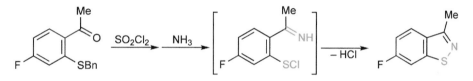

A benzisothiazole synthesis

24

1,3-Azoles: Imidazoles, Thiazoles and Oxazoles: Reactions and Synthesis

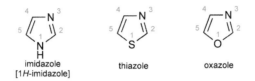

imidazole
[1*H*-imidazole]

thiazole

oxazole

The three 1,3-azoles, imidazole,[1] thiazole and oxazole,[2] are all very stable compounds that do not autoxidise. Oxazole and thiazole are water-miscible liquids with pyridine-like odours. Imidazole, which is a solid at room temperature, and 1-methylimidazole are also water soluble, but are odourless. They boil at much higher temperatures (256 °C and 199 °C) than oxazole (69 °C) and thiazole (117 °C); this can be attributed to stronger dipolar association resulting from the very marked permanent charge separation in imidazoles (the dipole moment of imidazole is 5.6 D; cf. oxazole, 1.4 D; thiazole, 1.6 D) and for imidazole itself, in addition, extensive intermolecular hydrogen bonding. The dihydro- and tetrahydro-1,3-azoles are named imidazoline/imidazolidine, thiazoline/thiazolidine and oxazoline/oxazolidine.

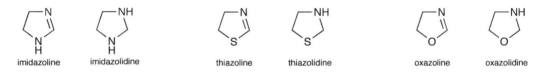

imidazoline   imidazolidine   thiazoline   thiazolidine   oxazoline   oxazolidine

24.1   Reactions with Electrophilic Reagents
24.1.1   Addition at Nitrogen
24.1.1.1   *Protonation.*
Imidazole, thiazole and alkyl-oxazoles, though not oxazole itself, form stable crystalline salts with strong acids, by protonation of the imine nitrogen, N-3, known as imidazolium, thiazolium and oxazolium salts.

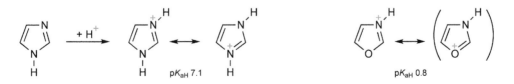

pK_{aH} 7.1

pK_{aH} 0.8

Imidazole, with a pK_{aH} of 7.1, is a very much stronger base than thiazole (2.5) or oxazole (0.8). That it is also stronger than pyridine (5.2) is due to the amidine-like resonance that allows both nitrogens to participate equally in carrying the charge. The particularly low basicity of oxazole can be understood as a

Heterocyclic Chemistry 5th Edition   John Joule and Keith Mills
© 2010 Blackwell Publishing Ltd

combination of inductive withdrawal by the oxygen and weaker mesomeric electron release from it. The 1,3-azoles are stable in hot, strong aqueous acid.

Hydrogen Bonding in Imidazoles

Imidazole, like water, is both a good donor and a good acceptor of hydrogen bonds; the imine nitrogen donates an electron pair and the *N*-hydrogen, being appreciably acidic (see 24.4), is an acceptor. This property is central to the mode of action of several enzymes that utilise the imidazole ring of a histidine (32.1) one of the 20 amino acids found in proteins.

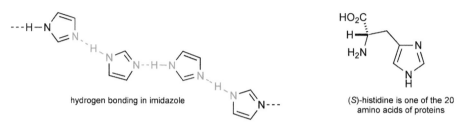

hydrogen bonding in imidazole

(*S*)-histidine is one of the 20 amino acids of proteins

Imidazoles with a ring *N*-hydrogen are subject to tautomerism, which becomes evident in unsymmetrically substituted compounds such as the methylimidazole shown. This special feature of imidazole chemistry means that to write simply '4-methylimidazole' would be misleading, for this molecule is in rapid tautomeric equilibrium with 5-methylimidazole. All such tautomeric pairs are inseparable and the convention used to cover this phenomenon is to write '4(5)-methylimidazole'. In some pairs, one tautomer predominates, for example 4(5)-nitroimidazole favours the 4-nitro-tautomer by 400:1.

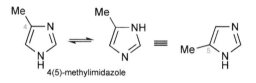

4(5)-methylimidazole

Tautomerism in imidazoles

24.1.1.2 Alkylation at Nitrogen

The 1,3-azoles are quaternised easily at the imine nitrogen with alkyl halides; the relative rates are: 1-methylimidazole:thiazole:oxazole, 900:15:1.[3] Microwave irradiation makes the process particularly rapid.[4] In the case of imidazoles that have an *N*-hydrogen, the immediate product is a protonated *N*-alkyl-imidazole; this can lose its *N*-hydrogen to unreacted imidazole (acting as a base) and react a second time, meaning that reactions with alkyl halides give mixtures of imidazolium, 1-alkyl-imidazolium and 1,3-dialkyl-imidazolium salts. The use of a limited amount of the alkylating agent, or reaction in basic solution,[5] when it is the imidazolyl anion (24.4.1) that is alkylated, can minimise these complications. A 4(5)-substituted imidazole can give two isomeric 1-alkyl-derivatives: generally the main product results from alkylation of that tautomer which minimizes steric interactions, i.e. 4(5)-substituted imidazoles give 1-alkyl-4-substituted imidazoles mainly. Formation of doubly *N*-alkylated derivatives can be achieved by reacting 1-trimethylsilylimidazole with an alkyl halide.[6]

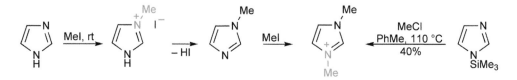

N-Alkylation of oxazoles,[7] or imidazoles carrying, for example, a phenylsulfonyl or acyl[8] group on nitrogen, is more difficult, requiring methyl triflate or a Meerwein salt (e.g. $Et_3O^+BF_4^-$) for smooth reaction. Subsequent simple alcoholysis of the imidazolium-sulfonamide (acylamide) releases the *N*-substituted imidazole.[9] Moreover, since acylation of 4(5)-substituted imidazoles gives the sterically less crowded 1-acyl-4-substituted imidazoles, subsequent alkylation, then hydrolytic removal of the acyl group, produces 1,5-disubstituted imidazoles.[10] Complementarily, the 1,4-disubstitution pattern can be achieved by alkylating 1-protected-5-substituted imidazoles (see 24.4.1) at N-3, then removing the *N*-protection.[11] Such protecting groups include trityl, removal of the triphenylmethyl group after alkylation requiring only simple acid treatment,[12] or 2-cyanoethyl, final elimination of acrylonitrile being the reverse of the Michael-type reaction used to introduce this group.[13]

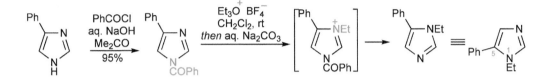

Another device to control the position of *N*-alkylation is applicable to histidine and histamine: a cyclic urea is first prepared by reaction with carbonyl dimidazole (24.1.1.3), forcing the alkylation onto the other nitrogen, methanolytic ring opening then providing the *N*-1-alkylated, carbamate-protected derivative.[14]

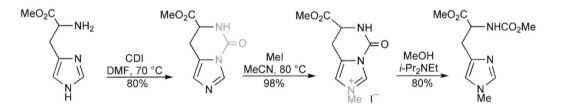

Exposure of imidazole to 'normal' Mannich conditions leads to *N*-(dimethylaminomethyl)imidazole, presumably *via* attack at the imine nitrogen, followed by loss of proton from the other nitrogen.[15]

N-Arylation or *N*-vinylation of imidazoles or benzimidazole can be achieved efficiently with metal catalysis[16] (see 4.2).

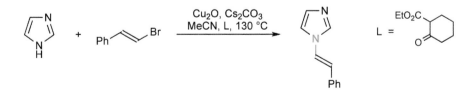

24.1.1.3 Acylation at Nitrogen

Acylation of imidazole produces *N*-acylimidazoles *via* loss of proton from the initially-formed *N*-3-acyl-imidazolium salt.[17] *N*-Acyl-imidazoles are even more easily hydrolysed than *N*-acyl-pyrroles; moist air is sufficient. The ready susceptibility to nucleophilic attack at carbonyl carbon has been capitalised upon: commercially available 1,1'-carbonyldiimidazole (CDI), prepared from imidazole and phosgene, can be used as a safe, phosgene synthon, and also in the activation of acids for formation of amides and esters *via* the *N*-acyl-imidazole.[18]

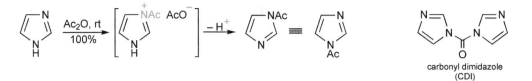

carbonyl dimidazole
(CDI)

In another application, *N*-acyl-imidazoles react with lithium aluminium hydride at 0 °C to give aldehydes, providing a route from the acid oxidation level.[19] In a similar way, 1-cyanoimidazole acts as a mild and efficient *N*-, *S*- and *C*-cyanating agent.[20]

24.1.2 Substitution at Carbon

24.1.2.1 Protonation

In acid solution, *via* a protonation-deprotonation sequence, hydrogen at the imidazole 5-position exchanges about twice as rapidly as at C-4 and >100 times faster than at C-2.[21] An altogether faster exchange, which takes place at room temperature in neutral or weakly basic solution, but not in acidic solution, brings about C-2–H exchange;[22] oxazole and thiazole also undergo this regioselective C-2–H exchange, the relative rates being in the order: imidazole > oxazole > thiazole.[23] The mechanism for this special process involves first, formation of protonic salt, then C-2–H deprotonation of the salt, producing a transient ylide, to which a carbene form is an important resonance contributor. It follows from this mechanism that quaternary salts of 1-alkyl-imidazoles, and of oxazole and thiazole will also undergo regioselective C-2–H exchange, and this is indeed the case. The relative rates of exchange, *via* the ylide mechanism, are in the order: oxazolium > thiazolium > *N*-methylimidazolium, in a ratio of about $10^5 : 10^3 : 1$.[24] Much attention[25] has been paid to thiazolium salts because of the involvement of exactly such an ylide in the mode of action of thiamin in its role as a component of a coenzyme in several biochemical processes[26] (32.2.4). '*N*-Heterocyclic carbenes', C-2-ylides from imidazolium salts, have become very important ligands in palladium catalysis (see 24.10 and 4.2.2).

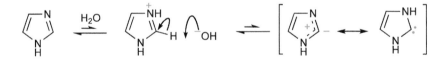

In the thermal decarboxylation of 1,3-azole-5-acids, the order of ease of loss of carbon dioxide is again oxazole- > thiazole- > *N*-methylimidazole-5-acids, but about 10^6 more slowly, than decarboxylation of the 2-acids.[27]

24.1.2.2 Nitration

Imidazole is much more reactive towards nitration than thiazole, substitution taking place *via* the salt,[28] as does nitration of alkyl-thiazoles.[29] Thiazole itself is untouched by nitric acid/oleum at 160 °C, but methyl-thiazoles are sufficiently activated to undergo substitution, the typical regioselectivity being for formation of more 5-nitro than 4-nitro derivatives;[30] the 2-position is not attacked: 4,5-dimethylimidazole is resistant to nitration. The much less reactive oxazoles do not undergo nitration.

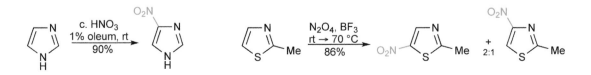

24.1.2.3 Sulfonation

Here again, thiazoles are much less reactive than imidazoles,[31] generally requiring high temperatures and mercury(II) sulfate as catalyst for any reaction to take place;[32] oxazole sulfonations are unknown.

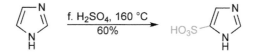

24.1.2.4 Halogenation

Imidazole,[33] and 1-alkyl-imidazoles,[34] are brominated with remarkable ease at all free nuclear positions. It is, at first sight, somewhat surprising that such relatively mild conditions allow bromination of imidazole at C-2, but it must be remembered that the neutral imidazole, not its protonic salt (cf. nitration and sulfonation), is available for attack, thus electrophilic addition of bromine to imine nitrogen, then addition of bromide at C-2 and finally elimination of hydrogen bromide may be the key to the 2-bromination.

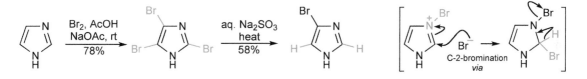

4(5)-Bromoimidazole can be obtained by reduction of tribromoimidazole with sodium sulfite,[35] *via* regioselective exchange of the 2- and 5-halogens using *n*-butyllithium, then water quenching,[36] or by bromination of imidazole with 4,4-dibromocyclohexa-2,5-dienone.[37] Chlorination with hypochlorite in alkaline solution effects substitution only at the 4- and 5-positions.[38] Iodination of imidazoles that have a free *N*-hydrogen, in alkaline solution and therefore *via* the imidazolyl anion, can also give fully halogenated products;[39] 4,5-diiodination of imidazole takes place in cold alkaline solution.[40]

Thiazole does not undergo bromination easily, though 2-methylthiazole brominates at C-5; when the 5-position is not free, no substitution occurs, thus 2,5-dimethylthiazole, despite its two activating substituents, is not attacked.[41] 2-Bromothiazole brominates to produce the 2,5-dibromo-isomer.[42] Halogenation of simple oxazoles has not been reported, however addition products, mainly 1,4-oriented and analogous to those obtained from furan (cf. 18.1.1.4), are obtained with substituted oxazoles.[43]

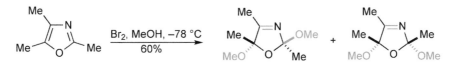

24.1.2.5 Acylation

Friedel–Crafts acylations are unknown for the azoles, clearly because of interaction between the basic nitrogen and the Lewis-acid catalyst. It is, however, possible to 2-aroylate 1-alkyl-imidazoles[44] or indeed imidazole itself[45] by reaction with the acid chloride in the presence of triethylamine, the substitution proceeding *via* an *N*-acyl-imidazolium ylide. Trifluoroacetylation of *N*-substituted imidazoles, and some oxazoles and thiazoles, also produces 2-substituted products in good yields.[46] In the reverse sense, 2-acyl-substituents can be cleaved by methanolysis, the mechanism again involving an imidazolium ylide.[47]

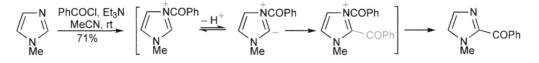

Another example of the utility of *N*-acyl-imidazolium ylides is a synthesis of 2-formylimidazole: the electrophile which attacks the ylide is in this case an *N*-benzoylimidazolium cation.[48]

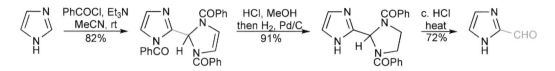

24.1.2.6   *Reactions with Aldehydes*

The discovery of *ipso*-displacement of silicon from the thiazole 2-position under mild conditions led to the development of this reaction as an essential component of a route to complex aldehydes. Subsequent quaternisation, saturation of the heterocyclic ring using sodium borohydride, and then mercury(II)- or copper(II)-catalysed treatment leads to the destruction of the thiazolidine and the formation of a new, homologous aldehyde; an example is shown below.[49]

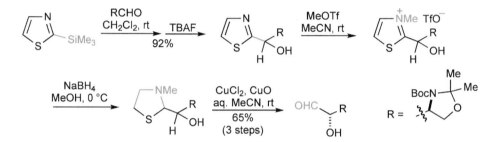

24.1.2.7   *Reactions with Iminium Ions*

The standard, acidic Mannich conditions do not allow simple *C*-substitutions of the imidazole, (cf. 24.1.1.2 for *N*-substitution), thiazole or oxazole systems.

24.2   Reactions with Oxidising Agents

Resistance to oxidative breakdown falls off in the order thiazoles > imidazoles > oxazoles. 2-Substituted thiazoles can be converted into *N*-oxides,[50] however peracids bring about degradation of imidazoles; oxazole *N*-oxides can only be prepared by ring synthesis. Substituted oxazoles are converted into imides, with loss of C-5.[51]

24.3   Reactions with Nucleophilic Reagents
24.3.1   With Replacement of Hydrogen

Generally speaking, the 1,3-azoles do not show the pyridine-type reactions in which hydrogen is displaced on reaction with a strong nucleophile.

24.3.2   With Replacement of Halogen

There are many examples of halogen at a 2-position undergoing nucleophilic displacement; examples chosen at random are 2-halo-thiazoles with sulfur nucleophiles[52] (indeed, more rapidly than for 2-halo-pyridines), 2-halo-1-substituted imidazoles,[53] and 2-chloro-oxazoles[54] with nitrogen nucleophiles.

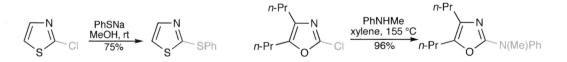

24.4 Reactions with Bases

24.4.1 Deprotonation of Imidazole *N*-Hydrogen and Reactions of Imidazolyl Anions

The pK_a for loss of the *N*-hydrogen of imidazole is 14.2; it is thus an appreciably stronger acid than pyrrole (pK_a 17.5) because of the enhanced delocalisation of charge, involving both nitrogens in the imidazolyl anion. Salts of imidazoles can be alkylated or acylated on nitrogen. One convenient method is to use the dry sodium/potassium salt obtained by evaporation of an aqueous alkaline solution;[55] sodium hydride in dimethylformamide also serves very well for this purpose.

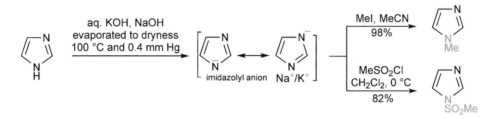

When there is a route for the entering group to be lost again, as in the addition to a cyano-conjugated alkene, a 4(5)-substituted imidazole will give the less hindered 1,4-disubstituted product rather than the 1,5-isomer.[56]

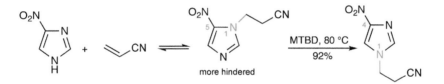

24.5 C-Metallation and Reactions of C-Metallated 1,3-Azoles[57]

24.5.1 Direct Ring C–H Metallation

The specific deprotonation of a C-2–hydrogen in the azoles, in neutral solution, *via* an ylide, has already been discussed (24.1.2.1). Preparative strong-base lithiation of oxazoles,[58] thiazoles[59] and *N*-methylimidazole[60] takes place preferentially at C-2, or at C-5 if the former position is substituted.[61] A variety of removable *N*-protecting groups have been used to achieve metallations for the eventual synthesis of *N*-unsubstituted imidazoles, including phenylsulfonyl,[62] dimethylaminosulfonyl,[63] dimethylaminomethyl,[15] trimethylsilylethoxymethyl,[64] diethoxymethyl,[65] 1-ethoxyethyl[66] and trityl.[67]

The intrinsic tendency to lithiate at C-2, then C-5, taken with metal–halogen exchange processes for the 4-position, are a powerful combination for elaborations of the 1,3-azoles. For example, the sequence shown below produces SEM-protected 5-substituted imidazoles,[63,68] with retention of a 2-silyl-substituent if required.[69]

Although oxazoles follow the pattern and lithiate at C-2, 4-substituted products are produced with some electrophiles; this is explained by a ring opening of the anion, to produce an enolate, which after *C*-electrophilic attack, recloses. An estimate by NMR spectroscopy showed the ring-cleaved tautomer to dominate the equilibrium.[70] Some electrophiles produce good yields of oxazole products: reaction of lithiated oxazole with hexachloroethane produces 2-chlorooxazole in good yield.[71] The open enolates can be trapped by reaction with chlorotrimethylsilane; the open, enol trimethylsilyl ether will undergo a thermal rearrangement to form 2-trimethylsilyloxazole.[72] More usefully, reaction with silyl triflates gives exclusively the *C*-silylated oxazole. 2-Triisopropylsilyloxazole, obtained in this way, is stable and can be lithiated at C-5.[73]

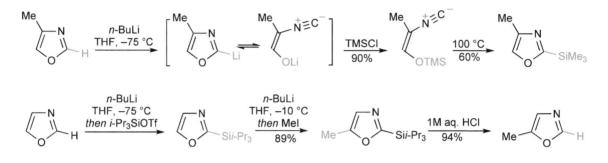

The ring-opened products can also be avoided by transmetallation to a zinc compound,[74] or by metallation with lithium tri-*n*-butylmagnesate[75] or *iso*-propylmagnesium chloride.[76]

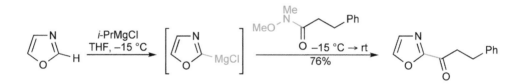

24.5.2 Metal–Halogen Exchange

The regioselectivity of metal–halogen exchange processes mirrors the C–H metallations discussed above: thus 2,4-dibromothiazole reacts at C-2 firstly.[77]

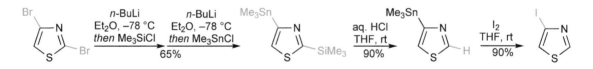

Metal–halogen exchange of 4(5)-bromoimidazole using *n*-butyllithium,[78] or of 4(5)-iodoimidazole, using methylmagnesium chloride with lithium chloride,[79] is possible without *N*-protection.

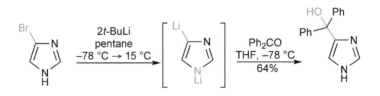

24.6 Reactions with Radicals

The preferred site for Minisci substitution of imidazoles in acid solution is C-2,[80] however intramolecular alkylation of a 4-formylimidazole in neutral solution occurs at C-5.[81] Intramolecular displacement of tosyl as a C-2 substitutent, has also been demonstrated.[82]

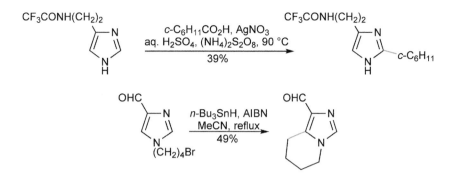

24.7 Reactions with Reducing Agents

Of the 1,3-azoles, oxazoles are the most easily reduced, catalytic sequences also bringing about C–O bond cleavage. 1,3-Azolium salts are easily attacked by hydride reducing agents: thiazolium salts produce tetra-hydro derivatives (24.1.2.6).[83]

24.8 Electrocyclic Reactions

Oxazoles readily undergo Diels–Alder type cycloaddition across the 2,5-positions, in parallel with the behaviour of furans (18.7). Thiazole and imidazole do not show this mode of reactivity, however they do react with highly electrophilic alkynes *via* initial electrophilic addition to the nitrogen, then nucleophilic intramolecular cyclising addition (cf. the comparable reactivity of quinolines, 9.13).[84]

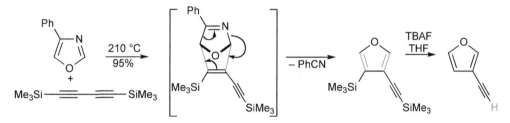

Oxazole cycloadditions have been reported with alkyne dienophiles[85] (tandem Diels–Alder addition and retro Diels–Alder loss of a nitrile leads on to furans), benzyne (the primary adduct can be isolated),[86] and with typical alkene dienophiles and the adducts can be transformed[87] into pyridines (8.14.1.4).

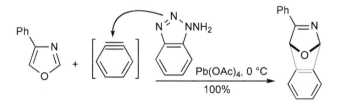

Electron-releasing substituents on the oxazoles increase the rate of reaction: 5-alkoxy-oxazoles are comparable in reactivity to typical all-carbon dienes. 4-(Trimethylsilyloxy)-oxazoles react smoothly to produce furans at modest temperatures.[88]

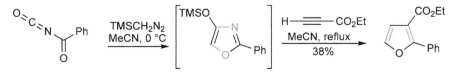

Particularly useful dienophiles are *N*-acyl-oxazolones – the products are protected *cis*-1,2-amino-alcohols.[89]

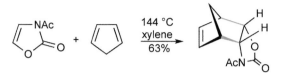

24.9 Alkyl-1,3-Azoles

Protons on methyl groups at 1,3-azole 2-positions are sufficiently acidic for strong-base deprotonation,[90] and are more acidic than methyl groups at other positions; an *ortho*-tertiary carboxamide can overcome this intrinsic tendency.[91] The presence of a 5-nitro group allows much milder base-catalysed condensations to occur by assisting in the stabilisation of the carbanion.[29] Condensation at the 2-methyl of thiazoles proceeds in organic acid solution.[92] *N*-Acylation also increases the acidity of 2-methyl groups, allowing *C*-acylation *via* a non-isolable enamide.[93]

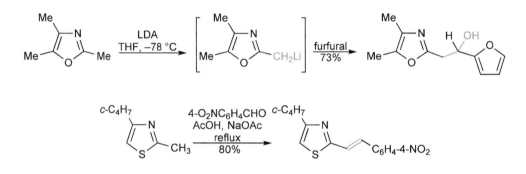

24.10 Quaternary 1,3-Azolium Salts

1,3-Dialkyl-imidazolium salts, for example 1-*n*-Butyl-3-methylimidazolium hexafluorophosphate (or tetra-fluoroborate) are a major group of ionic liquids (for a fuller discussion, see 31.5).

N-Alkoxycarbonyl-1,3-azolium salts, generated *in situ* by reaction with chloroformates, will react with allylstannanes,[94] or allylsilanes[95] by addition of the equivalent of an allyl anion. In the same way, silyl enol ethers add the equivalent of an enolate to give 2,3-dihydro-2-substituted imidazoles and thiazoles.[96]

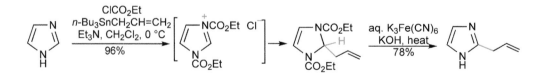

Azolium salts are readily attacked by hydroxide, addition at C-2 being followed by ring opening.[97]

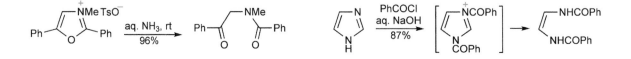

The C-2-exchange of azolium salts *via* an ylide mechanism was discussed in Section 24.1.2.1. Thiamin pyrophosphate acts as a coenzyme in several biochemical processes and in these, its mode of action depends on the intermediacy of a 2-deprotonated species (32.2.4). In the laboratory, thiazolium salts (3-benzyl-5-(2-hydroxyethyl)-4-methylthiazolium chloride is commercially available) will act as catalysts for the benzoin condensation, and in contrast to cyanide, the classical catalyst, allow such reactions to proceed with alkanals, as opposed to araldehydes; the key steps in thiazolium ion catalysis for the synthesis of 2-hydroxy-ketones are shown below and depend on the formation and nucleophilic reactivity of the C-2-ylide. Such catalysis provides acyl-anion equivalents.

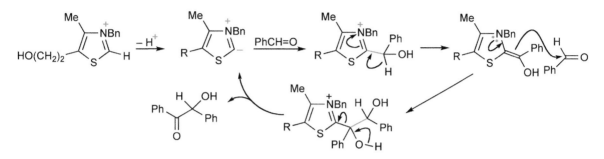

Significantly, replacing a thiazolium ring with an oxazolium ring gives a thiamin analogue in which there is no catalytic activity;[98] similarly 3,4-dimethyloxazolium iodide does not catalyse a benzoin condensation.[99] Nature has selected the heterocyclic system with the correct balance – oxazolium ylides are formed faster, but because of the greater stability that this reflects, do not then add to carbonyl groups as is required for catalytic activity. In keeping with the carbenoid character of thiazolium ylides, they dimerise; the dimers, either in their own right, or by reversion to monomer, are also catalysts for the benzoin condensation.[100]

What have become known as *N*-heterocyclic carbenes (NHCs) are important ligands for transition metals (see 4.2), especially in catalytic processes.[101] Isolable, crystalline carbenes can be derived even from a salt as simple as 1,3,4,5-tetramethylimidazolium. The NHC which has been most utilized in stabilizing both high- and low-valency states of metals, is the 1,3-bis(2,4,6-trimethylphenyl)imidazole carbene.[102] The pK_{aH}s for NHCs, in the range of 22–24, show them to be amongst the most basic neutral compounds; they are strongly nucleophilic.

24.11   Oxy-[103,104] and Amino-[105]1,3-Azoles

Amino-1,3-azoles exist as the amino tautomers. 2-Amino-1,3-azoles tend to be more stable than other isomers. All amino-1,3-azoles protonate on the ring nitrogen. 2-Aminothiazole has a pK_{aH} of 5.39, which compares with the value for 2-aminoimidazole of 8.46, reflecting, in part, the symmetry of the resonating guanidinium type system in the latter.

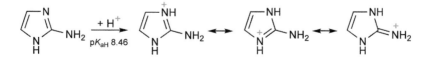

The amino-1,3-azoles behave as normal arylamines, for example undergoing carbonyl condensation reactions, easy electrophilic substitutions[106] and diazotisation,[107] though 2-amino-oxazoles cannot be diazotised,[108] presumably due to the greater electron withdrawal by the oxygen.

The oxygen-substituted 1,3-azoles exist in their carbonyl tautomeric forms. The bromination of thiazol-2-one, at C-5, is a nice demonstration of relative reactivity: here the double bond carries both sulfur and nitrogen, and it is the latter, i.e. the enamide rather than the enethiol ester character, that dictates the site of electrophilic attack.[109]

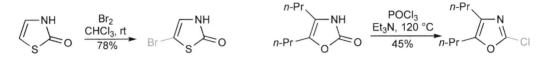

1,3-Azol-2-ones can be converted into the 2-halo-azoles by reaction with phosphorus halides.[53] Thiazolidine-2,4-dione is converted efficiently into 2,4-dibromothiazole by heating with phosphoryl bromide.[42] When accompanied by dimethylformamide, i.e. under Vilsmeier conditions, ring formylation also occurs and, after hydrogenolytic removal of halogen, the overall sequence can be seen to be a means for the preparation of 5-hydroxymethylthiazole.[110]

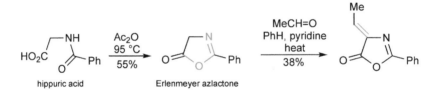

The 5-ones condense in an aldol fashion at C-4.[111] Alkylation of the 1,3-azolones can take place either on the oxygen, giving alkyloxy-1,3-azoles, or on nitrogen; for example thiazol-2-one reacts with diazomethane giving 2-methoxythiazole, but with methyl iodide/methoxide, to give 3-methylthiazol-2-one.[112]

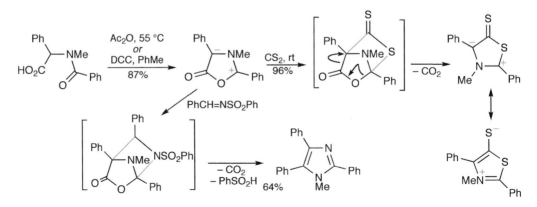

hippuric acid   Erlenmeyer azlactone

4(5)-Oxazolones are simply cyclic anhydrides of *N*-acyl-α-amino acids, and are constructed in the way that this implies. If the nitrogen also carries an alkyl group, cyclisation[113] can only lead to an overall neutral product by adopting a zwitterionic structure, for which no neutral canonical form can be written – a mesoionic structure. Mesoionic oxazolones (named 'münchnones' by Huisgen after their discovery at the University of München, Germany) undergo ready dipolar cycloadditions,[114] with loss of carbon dioxide from the initial adduct; the examples[115] show the conversion of a münchnone into a mesoionic thiazolone and into an imidazole.

24.12 1,3-Azole *N*-Oxides

The chemistry of azole-*N*-oxides is relatively underdeveloped compared, for example, with that of pyridine *N*-oxides, largely because of difficulty in their preparation from the azoles themselves.[116] Some ring synthetic methods can be used, for example the reaction of 1,2-dicarbonyl-mono-oximes with imines as shown.[117]

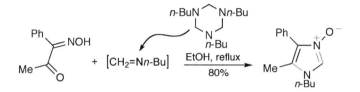

1-Benzyloxyimidazole undergoes useful 2-lithiations; hydrogenolysis produces 2-substituted 1-hydroxy-imidazoles and these, in turn, can be converted into the 2-substituted imidazoles by reduction with titanium(III) chloride.[118]

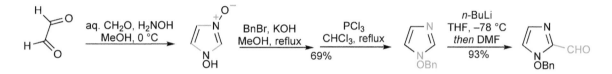

24.13 Synthesis of 1,3-Azoles[119,120,121]
24.13.1 Ring Synthesis
Considerable parallelism emerges from an examination of the major methods for the construction of oxazole, thiazole and imidazole ring systems.

24.13.1.1 From an α-Halo-Carbonyl-Component (or an Equivalent) and a Three-Atom Unit Supplying C-2 and Both of the Heteroatoms
This route is particularly important for thiazoles and imidazoles.

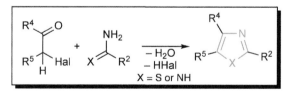

Simple examples of this strategy, which for the synthesis of thiazoles is known as the *Hantzsch synthesis*, are shown below (note there is an important pyridine ring synthesis, also named a Hantzsch synthesis (8.14.1.2)): the syntheses of 2,4-dimethylthiazole, where the heteroatoms are provided by thioacetamide,[122] and 2-aminothiazole, in which 1,2-dichloroethyl ethyl ether is utilised as a synthon for chloroacetaldehyde and the heteroatoms derive from thiourea.[123] The use of thioureas as the sulfur component with 2-chloro-acetamides as the second unit gives rise to 2,4-diamino-thiazoles,[124] and *N*-acyl-thioureas with 2-bromo-ketones produce 2-(*N*-acylamino)-thiazoles.[125] The first step in such ring syntheses is *S*-alkylation.[126] Useful variants include the use of an α-diazo-ketone in place of the α-halocarbonyl-component[127] and the conversion of 1,3-diketones into their 2-phenyl-iodonium derivatives and reaction of these with thioureas producing 2-amino-5-acyl-thiazoles.[128]

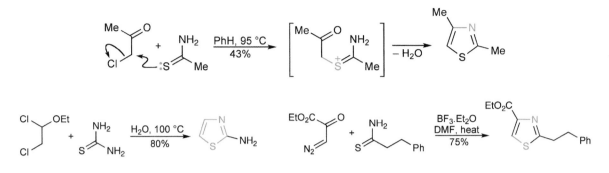

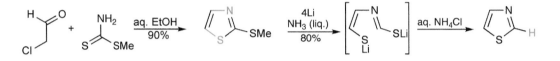

The interaction of ammonia with carbon disulfide produces ammonium dithiocarbamate in solution, which reacts with 2-halo-ketones to produce thiazole-2-thiones;[129] similarly, methyl dithiocarbamate serves as a component for the construction of 2-methylthiothiazole, reducable to thiazole itself by hydrogenolysis, thus providing a good route to the unsubstituted heterocycle.[130]

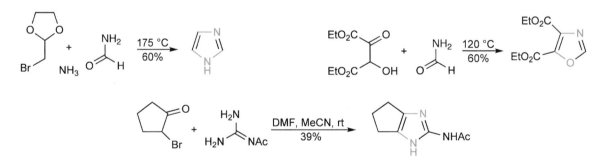

Imidazole itself can be prepared efficiently from bromoacetaldehyde ethylene acetal, formamide and ammonia; it is likely that displacement of halogen by ammonia occurs at an early stage.[131] A mono-enol ether of bromomalonaldehyde reacts with amidines giving 5-formyl-imidazoles.[132] Amidines react with 2-bromo-ketones giving 2-substituted imidazoles[133] and 2-acetylamino-imidazoles are formed efficiently from the interaction of 2-bromo-ketones and *N*-acetylguanidine.[134] The 2-hydroxy-aldehyde unit in sugars can be utilized in the synthesis of 4(5)-polyhydroxyalkyl-imidazoles, using ammonia and formamide: the reactions are carried out by heating at high pressure.[135]

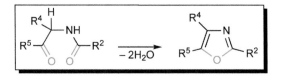

24.13.1.2 By Cyclising Dehydration of α-Acylamino-Carbonyl-Compounds
Cyclising dehydration of an α-acylamino-carbonyl-compound is particularly important for oxazoles, and can be adapted for thiazole or imidazole formation.

The classical method for making oxazoles, the *Robinson–Gabriel synthesis*, which is formally analogous to the cyclising dehydration of 1,4-dicarbonyl compounds to furans (18.13.1.1), is the acid-catalysed closure of α-acylamino-carbonyl-compounds and a simple example is shown below.[136]

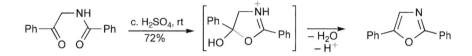

Amino acid-derived aldehydes can be converted into 5-unsubstituted oxazoles[137] and in a related sense, imidazoles can be produced by introducing ammonia to an 2-acylamino-ketone.[138]

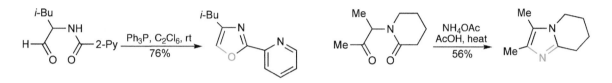

A route to all three ring systems depends on a rhodium-carbene insertion into the N–H of an amide, then direct dehydrative ring closure to an oxazole, or production of a thiazole with Lawesson's reagent, or of an imidazole with ammonia or a primary amine.[139]

The construction of an amide using aminomalononitrile and a carboxylic acid under typical peptide coupling conditions is accompanied by ring closure and thus the production of 5-amino-oxazoles.[140]

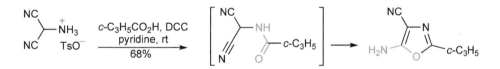

A synthesis shown below of ethyl oxazole-4-carboxylate illustrates a sophisticated use of this strategy, in which an imino ether and the potassium enolate of an aldehyde are involved;[141] iminoethers on reaction with aminoacetal give amidines, which close in acid to give 2-substituted imidazoles.[142,143] The process can be conducted without isolation of the intermediate.[144]

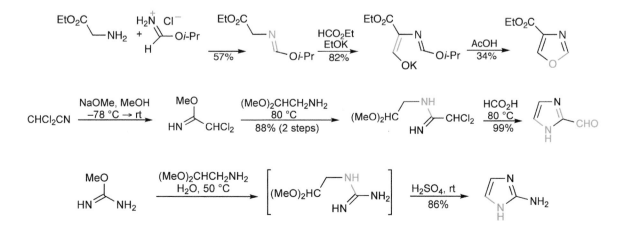

α-Acylthioketones react with ammonia to give thiazoles.[145]

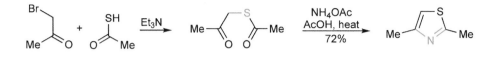

24.13.1.3   From Isocyanides
Tosylmethylisocyanide (TosMIC), can be used for the synthesis of all three 1,3-azole types.

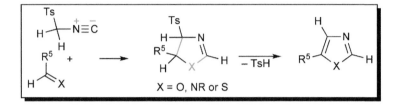

Tosylmethylisocyanide reacts with aldehydes, affording adducts that lose toluenesulfinate on heating, giving oxazoles;[146] with carbon disulfide it produces 4-tosyl-5-alkylthio-thiazoles (following a subsequent S-alkylation)[147] and, in analogy to its interaction with aldehydes, it adds to N-alkyl-imines[148,149] or N-dimethylaminosulfamyl-imines[150] when, following elimination of toluenesulfinate, imidazoles are formed. The analogous benzotriazolylmethyl isocyanide (BetMIC) can serve in the same way and has advantages in some situations.[151]

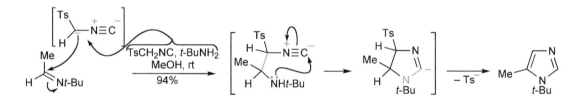

Anions derived from other isocyanides can be acylated (or thioformylated[152]), the products spontaneously closing to oxazoles[153] (thiazoles).

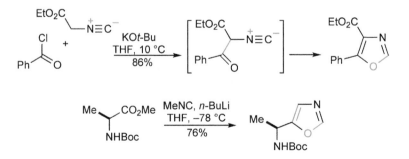

24.13.1.4   Oxazoles from α-Diazocarbonyl Compounds[154]
The carbene or carbenoid derived from an α-diazocarbonyl-compound cycloadds to nitriles to produce oxazoles with the nitrile R group at C-2.

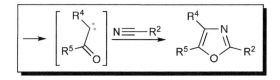

The generation of a carbene (or when using a metal catalyst, a carbenoid) from an α-diazocarbonyl-compound, in the presence of a nitrile, results in overall cycloaddition and the formation of an oxazole. Both α-diazo-ketones and α-diazo-esters have been used, the examples in the sequence below showing that the result in the latter situation is the formation of a 5-alkoxy-oxazole.[155] The exact sequence of events is not certain, but may involve a nitrile ylide, the result of electrophilic addition of the carbene to the nitrile nitrogen.

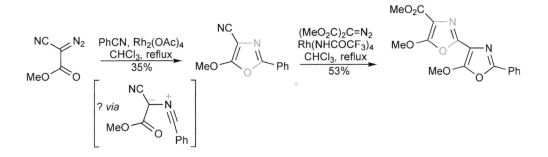

24.13.1.5 From 1,2-Diketones, Aldehydes and Ammonia
The microwave assisted interaction of a 1,2-diketone, an aldehyde and ammonia produces trisubstituted imidazoles.[156]

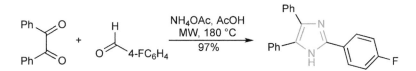

24.13.1.6 By Dehydrogenation
The ring synthesis of the tetrahydro-1,3-azoles is simply the formation of *N,N-, N,O-* or *N,S*-analogues of aldehyde cyclic acetals; the ring synthesis of the 4,5-dihydro-heterocycles requires reaction with an acid oxidation level component or the inclusion of iodine in a reacting mixture of ethylene diamine or amino-ethanol and an aldehyde. A good route to the aromatic systems is therefore the dehydrogenation of these reduced and partially reduced systems. Nickel peroxide,[157] manganese(IV) oxide,[158] copper(II) bromide/base[159] and bromotrichloromethane/diazabicycloundecane[160] have been used. The example shown uses cysteine methyl ester with a chiral aldehyde to form the tetrahydrothiazole.

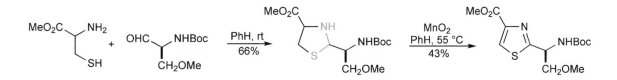

24.13.2 Examples of Notable Syntheses Involving 1,3-Azoles

24.13.2.1 4(5)-Fluorohistamine

4(5)-Fluorohistamine was synthesised[161] *via* nucleophilic displacement of a side-chain leaving group (cf. pyrroles, 16.11).

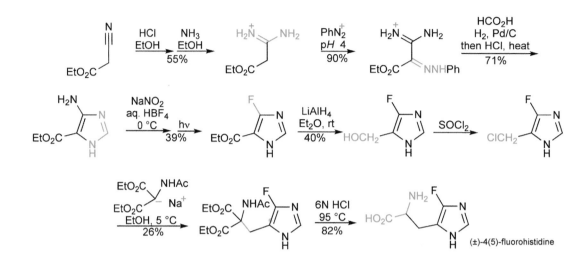

24.13.2.2 Thieno[2,3-d]imidazole[162]

The synthesis of thieno[2,3-*d*]imidazole illustrates again the selectivity in halogen–metal exchange processes in imidazoles. In this sequence a vinyl was used as *N*-protecting group, and it includes a nucleophilic displacement of bromine from the 4-position, activated by the 5-aldehyde.

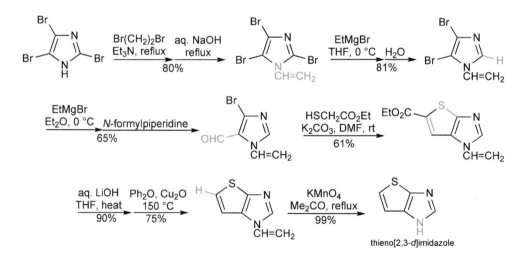

24.13.2.3 Inhibitor of Transforming Growth Factor β1, Type 1 Receptor

This sequence illustrates the interaction of an amidine and a 2-bromo-ketone with subsequent ring halogenation and palladium-catalysed coupling.[163]

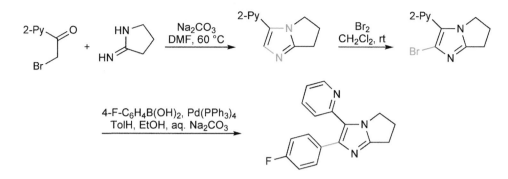

Exercises

Straightforward revision exercises (consult Chapters 23 and 24):

(a) Place the three 1,3-azoles in order of basicity and explain this order.

(b) What problems are associated with the *N*-alkylation of imidazoles in particular?

(c) What is carbonyl diimidazole (CDI) used for and why does it function well in that role?

(d) What factors must be considered in a discussion of electrophilic substitution of 1,3-azoles? How do they compare in this regard with pyrrole, thiophene and furan, respectively?

(e) What is the positional order of selectivity for deprotonation of 1,3-azoles? How, then, would you make 5-methylthiazole from thiazole?

(f) What is the most typical electrocyclic process undergone by the 1,3-azoles and which of them show this tendency to the highest degree?

(g) Describe one typical synthesis for: (i) an imidazole, (ii) a thiazole and (iii) an oxazole.

More advanced exercises:

1. Suggest structures for the halo-compounds formed in the following ways: (i) imidazole with NaOCl → $C_3H_2Cl_2N$; (ii) 1-methylimidazole with excess Br_2 in AcOH → $C_4H_3Br_3N_2$, then this with EtMgBr followed by water → $C_4H_4Br_2N_2$ and this in turn with *n*-BuLi, then $(MeO)_2CO$ gave $C_6H_7BrN_2O_2$.

2. Draw structures for the intermediates and final products that are formed when: (i) 4-phenyloxazole is heated with but-1-yn-3-one → $C_6H_6O_2$; (ii) 5-ethoxyoxazole is heated with dimethyl acetylenedicarboxylate → $C_{10}H_{12}O_6$.

3. Deduce structures for the products formed at each stage of the following syntheses: 1,2-dimethyl-5-nitroimidazole heated with $Me_2NCH(Ot\text{-}Bu)_2$ → $C_8H_{12}N_4O_2$; this then heated with Ac_2O → $C_{10}H_{14}N_4O_3$. This product reacted: (i) with guanidine $[H_2NC(NH_2)=NH]$ → $C_9H_{10}N_6O_2$ and (ii) with $MeNHNH_2$ → $C_9H_{11}N_5O_2$.

4. Deduce structures for the products formed in the following sequences: 1-methylimidazole/*n*-BuLi/ –30 °C, then TMSCl → $C_7H_{14}N_2Si$, then *n*-BuLi/–30 °C, then TMSCl → $C_{10}H_{22}N_2Si_2$, then this with MeOH/rt → $C_7H_{14}N_2Si$, which was different to the first product.

5. Explain the following: 4-bromo-1-methylimidazole treated with *n*-BuLi/–78 °C, then DMF gave $C_5H_6N_2O$. Carrying out the same sequence, but allowing the solution to warm to 0 °C before addition of DMF gave an isomeric product.

6. Thiazole-2-thione reacted with $Br(CH_2)_3Br$ to give, mainly, a salt $C_6H_8NS_2^+$ Br^-; suggest a structure and a mechanism for its formation.

7. Deduce structures for the 1,3-azoles that are produced from the following reactant combinations: (i) 1-chlorobutan-2-one and thiourea; (ii) thiobenzamide and chloroacetaldehyde; (iii) thioformamide and ethyl bromoacetate.

8. Write structures for the intermediates in the following synthesis of 3,4-bis(acetoxymethyl)furan: phenacyl bromide/NH_4^+ HCO_2^- → C_9N_7NO; this then heated with $AcOH_2CC≡CCH_2OAc$.

9. What imidazoles would be formed from the following reactant combination: (i) MeN≡C/*n*-BuLi and PhC≡N; (ii) 2-amino-1,2-diphenylethanone and H₂NC≡N?

References

[1] 'Advances in imidazole chemistry', Grimmett, M. R., *Adv. Heterocycl. Chem.*, **1970**, *12*, 103; 'Advances in imidazole chemistry', *ibid.*, **1980**, *27*, 241; 'The Azoles', Schofield, K., Grimmett, M. R. and Keene, B. R. T., Cambridge University Press, **1976**; 'Imidazole and Benzimidazole Synthesis', Grimmett, M. R., Academic Press, London, **1997**.

[2] 'Advances in oxazole chemistry', Lakhan, R. and Ternai, B., *Adv. Heterocycl. Chem.*, **1974**, *17*, 99; 'The chemistry of oxazoles', Turchi, I. J. and Dewar, M. J. S., *Chem. Rev.*, **1975**, *75*, 389; 'New chemistry of oxazoles', Hassner, A. and Fischer, B., *Heterocycles*, **1993**, *35*, 1441.

[3] Deady, L. W., *Aust. J. Chem.*, **1973**, *26*, 1949.

[4] Bogdal, D., Pielichowski, J. and Jaskot, K., *Heterocycles*, **1997**, *45*, 715.

[5] Baxter, R. A. and Spring, F. S., *J. Chem. Soc.*, **1945**, 232; Roe, A. M., *J. Chem. Soc.*, **1963**, 2195.

[6] Harlow, K. J., Hill, A. F. and Welton, T., *Synthesis*, **1996**, 697.

[7] Hayes, F. N., Rogers, B. S. and Ott, D. G., *J. Am. Chem. Soc.*, **1955**, *77*, 1850; Ott, D. G., Hayes, F. N. and Kerr, V. N., *ibid.*, **1956**, *78*, 1941.

[8] Ulibarri, G., Choret, N. and Bigg, D. C. H., *Synthesis*, **1996**, 1287.

[9] O'Connell, J. F. and Rapoport, H., *J. Org. Chem.*, **1992**, *57*, 4775.

[10] Olofson, R. A. and Kendall, R. V., *J. Org. Chem.*, **1970**, *35*, 2246.

[11] Shapiro, G. and Gomez-Lor, B., *Heterocycles*, **1995**, *41*, 215.

[12] Daninos-Zeghal, S., Mourabit, A. A., Ahond, A., Poupat, C. and Potier, P., *Tetrahedron*, **1997**, *53*, 7604.

[13] Collman, J. P., Bröning, M., Fu, L., Rapta, M. and Schwenninger, R., *J. Org. Chem.*, **1998**, *63*, 8084.

[14] Jain, R. and Cohen, L. A., *Tetrahedron*, **1996**, *52*, 5363.

[15] Katritzky, A. R., Rewcastle, G. W. and Fan, W.-Q., *J. Org. Chem.*, **1988**, *53*, 5685.

[16] Shen, G., Lv, X., Qian, W. and Bao, W., *Tetrahedron Lett.*, **2008**, *49*, 4556.

[17] Caplow, M. and Jencks, W. P., *Biochemistry*, **1962**, *1*, 883; Reddy, G. S., Mandell, L. and Goldstein, J. H., *J. Chem. Soc.*, **1963**, 1414.

[18] Morton, R. C., Mangroo, D. and Gerber, G. E., *Canad. J. Chem.*, **1988**, *66*, 1701.

[19] 'Syntheses using heterocyclic amides (azolides)', *Newer Methods of Prep. Org. Chem.*, **1968**, *5*, 61.

[20] Wu, Y.-q., Limburg, D. C., Wilkinson, D. E. and Hamilton, G. S., *Org. Lett.*, **2000**, *2*, 795.

[21] Wong, J. L. and Keck, J. H., *J. Org. Chem.*, **1974**, *39*, 2398.

[22] Staab, H. A., Irngartinger, H., Mannschreck, A. and Wu, M. T., *Justus Liebigs Ann. Chem.*, **1966**, *695*, 55.

[23] Staab, H. A., Wu, M. T., Mannschreck, A. and Schwalbach, G., *Tetrahedron Lett.*, **1964**, *15*, 845; Vaughn, J. D., Mughrabi, Z. and Wu, E. C., *J. Org. Chem.*, **1970**, *35*, 1141; Takeuchi, Y., Yeh, H. J. C., Kirk, K. L. and Cohen, L. A., *J. Org. Chem.*, **1978**, *43*, 3565.

[24] Haake, P., Bausher, L. P. and Miller, W. B., *J. Am. Chem. Soc.*, **1969**, *91*, 1113.

[25] Olofson, R. A. and Landesberg, J. M., *J. Am. Chem. Soc.*, **1966**, *88*, 4263; Coburn, R. A., Landesberg, J. M., Kemp, D. S. and Olofson, R. A., *Tetrahedron*, **1970**, *26*, 685.

[26] Breslow, R., *J. Am. Chem. Soc.*, **1958**, *80*, 3719; Breslow, R. and McNelis, E., *ibid.*, **1959**, *81*, 3080; Kluger, R., Karimian, K. and Kitamura, K. *ibid.*, **1987**, *109*, 6368

[27] Haake, P., Bausher, L. P. and McNeal, J. P., *J. Am. Chem. Soc.*, **1971**, *93*, 7045.

[28] Austin, M. W., Blackborow, J. R., Ridd, J. H. and Smith, B. V., *J. Chem. Soc.*, **1965**, 1051.

[29] Katritzky, A. R., Ögretir, C., Tarhan, H. O., Dou, H. M. and Metzger, J. V., *J. Chem. Soc., Perkin Trans. 2*, **1975**, 1614.

[30] Asato, G., *J. Org. Chem.*, **1968**, *33*, 2544.

[31] Barnes, G. R. and Pyman, F. L., *J. Chem. Soc.*, **1927**, 2711.

[32] Erlenmeyer, H. and Kiefer, H., *Helv. Chim. Acta*, **1945**, *28*, 985.

[33] Balaban, I. E. and Pyman, F. L., *J. Chem. Soc.*, **1922**, 947; Stensiö, K.-E., Wahlberg, K. and Wahren, R., *Acta Scand.*, **1973**, *27*, 2179; Bahnous, M., Mouats, C., Fort, Y. and Gros, P. C., *Tetrahedron Lett.*, **2006**, *47*, 1949.

[34] O'Connell, J. F., Parquette, J., Yelle, W. E., Wang, W. and Rapoport, H., *Synthesis*, **1988**, 767.

[35] Balaban, I. E. and Pyman, F. L., *J. Chem. Soc.*, **1924**, 1564.

[36] Iddon, B. and Khan, N., *J. Chem. Soc., Perkin Trans. 1*, **1987**, 1453.

[37] Caló, V., Ciminale, F., Lopez, L., Naso, F. and Todesco, P. E., *J. Chem. Soc., Perkin Trans. 1*, **1972**, 2567.

[38] Lutz, A. W. and DeLorenzo, S., *J. Heterocycl. Chem.*, **1967**, *4*, 399.

[39] Pauly, H. and Arauner, E., *J. Prakt. Chem.*, **1928**, *118*, 33; Groziak, M. P. and Wei, L., *J. Org. Chem.*, **1991**, *56*, 4296.

[40] Naidu, M. S. R. and Bensusan, H. B., *J. Org. Chem.*, **1968**, *33*, 1307.

[41] Nagasawa, F., *J. Pharm. Soc. Jpn.*, **1940**, *60*, 433 (*Chem. Abstr.*, **1940**, *34*, 5450); Ganapathi, K. and Kulkarni, K. D., *Current Sci.*, **1952**, *21*, 314 (*Chem. Abstr.*, **1951**, *45*, 5150).

[42] Stanetty, P., Schnürch, M. and Mihovilovic, M. D., *J. Org. Chem.*, **2006**, *71*, 3754.

[43] Hassner, A. and Fischer, B., *Tetrahedron*, **1989**, *45*, 6249.

[44] Regel, E. and Büchel, K.-H., *Justus Liebigs Ann. Chem.*, **1977**, 145.

[45] Bastiaansen, L. A. M. and Godefroi, E. F., *Synthesis*, **1978**, 675.

[46] Khodakovskiy, P. V., Voluchnyuk, D. M., Panov, D. M., Pervak, I. I., Zarudnitskii, E. V., Shishkin, O. V., Yurchenko, A. A., Shivanyuk, A. and Tolmachev, A. A., *Synthesis*, **2008**, 948.

[47] Antonini, I., Cristalli, G., Franchetti, P., Grifantini, M., Gulini, U. and Martelli, S., *J. Heterocycl. Chem.*, **1978**, *15*, 1201.

[48] Bastiaansen L. A. M., Van Lier, P. M. and Godefroi, E. F., *Org. Synth., Coll. Vol. VII*, **1990**, 287.

[49] 'The thiazole aldehyde synthesis', Dondoni, A. *Synthesis*, **1998**, 1681.

[50] Ochiai, E. and Hayashi, E., *J. Pharm. Soc. Jpn.*, **1947**, *67*, 34 (*Chem. Abstr.*, **1951**, *45*, 9533).

[51] Evans, D. A., Nagorny, P. and Xu, R., *Org. Lett.*, **2006**, *8*, 5669.

[52] Bosco, M., Forlani, L., Liturri, V., Riccio, P. and Todesco, P. E., *J. Chem. Soc. (B)*, **1971**, 1373; Bosco, M., Liturri, V., Troisi, L., Forlani, L. and Todesco, P. E., *J. Chem. Soc., Perkin Trans. 2*, **1974**, 508.

[53] de Bie, D. A., van der Plas, H. C. and Guersten, G., *Recl. Trav. Chim. Pays-Bas*, **1971**, *90*, 594.

[54] Gompper, R. and Effenberger, F., *Chem. Ber.*, **1959**, *92*, 1928.

[55] Begtrup, M. and Larsen, P., *Acta Scand.*, **1990**, *44*, 1050.

[56] Bhujanga Rao, A. K. S., Rao, C. G. and Singh, B. B., *Synth. Commun.*, **1991**, *21*, 427; idem, *J. Org. Chem.*, **1990**, *55*, 3702.

[57] 'Synthesis and reactions of lithiated monocyclic azoles containing two or more hetero-atoms. Part II: Oxazoles', Iddon, B., *Heterocycles*. **1994**, *37*, 1321; 'Part IV: Imidazoles', Iddon, B. and Ngochindo, R. I., ibid., **1994**, *38*, 2487; 'Part V: Isothiazoles and thiazoles', Iddon, B., ibid., **1995**, *41*, 533.

[58] Hodges, J. C., Patt, W. C. and Conolly, C. J., *J. Org. Chem.*, **1991**, *56*, 449.

[59] Dondoni, A., Mestellani, A. R., Medici, A., Negrini, E. and Pedrini, P., *Synthesis*, **1986**, 757; Dondoni, A., Fantin, G., Fagagnolo, M., Medici, A. and Pedrini, P., *J. Org. Chem.*, **1988**, *53*, 1748.

[60] 'Metallation and metal-halogen exchange reactions of imidazoles', Iddon, B., *Heterocycles*, **1985**, *23*, 417; Ohta, S., Matsukawa, M., Ohashi, N. and Nagayama, K., *Synthesis*, **1990**, 78.

[61] Ngochindo, R. I., *J. Chem. Soc., Perkin Trans. 1*, **1990**, 1645.

[62] Sundberg, R. J., *J. Heterocycl. Chem.*, **1977**, *14*, 517.

[63] Bell, A. S., Roberts, D. A. and Ruddock, K. S., *Tetrahedron Lett.*, **1988**, *29*, 5013; Bhagavatula, L., Premchandran, R. H., Plata, D. J., King, S. A. and Morton, H. E., *Heterocycles*, **2000**, *53*, 729.

[64] Lipshutz, B. H., Huff, B. and Hagen, W., *Tetrahedron Lett.*, **1988**, *29*, 3411.

[65] Curtis, N. J. and Brown, R. S., *J. Org. Chem.*, **1980**, *45*, 4038.

[66] Manoharan, T. S. and Brown, R. S., *J. Org. Chem.*, **1988**, *53*, 1107.

[67] Kirk, K. L., *J. Org. Chem.*, **1978**, *43*, 4381.

[68] Shapiro, G. and Gomez-Lor, B., *Heterocycles*, **1995**, *41*, 215.

[69] Vollinga, R. C., Menge, W. M. P. B. and Timmerman, H., *Recl. Trav. Chim. Pays-Bas*, **1993**, *112*, 123.

[70] Crowe, E., Hossner, F. and Hughes, M. J., *Tetrahedron*, **1995**, *52*, 8889.

[71] Atkins, J. M. and Vedejs, E., *Org. Lett.*, **2005**, *7*, 3351.

[72] Dondoni, A., Fantin, G., Fogagnolo, M., Medici, A. and Pedrini, P., *J. Org. Chem.*, **1987**, *52*, 3413.

[73] Miller, R. A., Smith, R. M. and Marcune, B., *J Org. Chem.*, **2005**, *70*, 9074.

[74] Harn, N. K., Gramer, C. J. and Anderson, B. A., *Tetrahedron Lett.*, **1995**, *36*, 9453.

[75] Bayh, O., Awad, H., Mongin, F., Hoarau, C., Bischoff, L., Trécourt, F., Quéguiner, G., Marsais, F., Blanco, F., Abarca, B. and Ballesteros, R., *J. Org. Chem.*, **2005**, *70*, 5190.

[76] Pippel, D. J., Mapes, C. M. and Mani, N. S., *J. Org. Chem.*, **2007**, *72*, 5828.

[77] Kelly, T. R., Jagoe, C. T. and Gu, Z, *Tetrahedron Lett.*, **1991**, *32*, 4263.

[78] Katritzky, A. R., Slawinski, J. J., Brunner, F. and Gorun, S., *J. Chem. Soc., Perkin Trans. 1*, **1989**, 1139.

[79] Kopp, F. and Knochel, P., *Synlett*, **2007**, 980.

[80] Jain, R., Cohen, L. A. and King, M. M., *Tetrahedron*, **1997**, *53*, 4539.

[81] Aldabbagh, F., Bowman, W. R. and Mann, E., *Tetrahedron Lett.*, **1997**, *38*, 7937.

[82] Aldabbagh, F. and Bowman, W. R., *Tetrahedron Lett.*, **1997**, *38*, 3793; idem, *Tetrahedron*, **1999**, *55*, 4109.

[83] Clark, G. M. and Sykes, P., *J. Chem. Soc., (C)*, **1967**, 1269 and 1411.

[84] Abbott, P. J., Acheson, R. M., Eisner, U., Watkin, D. J. and Carruthers, J. R., *J. Chem. Soc., Perkin Trans. 1*, **1976**, 1269; Huisgen, R., Giese, B. and Huber, H., *Tetrahedron Lett.*, **1967**, 1883; Acheson, R. M. and Vernon, J. M., *J. Chem. Soc.*, **1962**, 1148.

[85] Ohlsen, S. R. and Turner, S., *J. Chem. Soc., (C)*, **1971**, 1632; Hutton, J., Potts, B. and Southern, P. F., *Synth. Commun.*, **1979**, *9*, 789; König, H., Graf, F. and Weberndörfer, *Justus Liebigs Ann. Chem.*, **1981**, 668; Liotta, D., Saindane, M. and Ott, W., *Tetrahedron Lett.*, **1983**, *24*, 2473; Kawada, K., Kitagawa, O. and Kobayashi, Y., *Chem. Pharm. Bull.*, **1985**, *33*, 3670.

[86] Whitney, S. E. and Rickborn, B., *J. Org. Chem.*, **1988**, *53*, 5595.

[87] Condensation of oxazoles with dienophiles – a new method for synthesising pyridine bases', Karpeiskii, M. Ya. and Florent'ev, V. L., *Russ. Chem. Rev.*, **1969**, *38*, 540 (*Chem. Abstr.*, **1969**, *71*, 91346).

[88] Hari, Y., Iguchi, T. and Aoyama, T., *Synthesis*, **2004**, 1359.

[89] Deyrup, J. A. and Gingrich, H. L., *Tetrahedron Lett.*, **1977**, 3115; Ishizuka, T., Osaki, M., Ishihara, H. and Kunieda, T., *Heterocycles*, **1993**, *35*, 901.

[90] Noyce, D. S., Stowe, G. T. and Wong, W., *J. Org. Chem.*, **1974**, *39*, 2301; Lipshutz, B. H. and Hungate, R. W., *J. Org. Chem.*, **1981**, *46*, 1410; Knaus, G. N. and Meyers, A. I., *J. Org. Chem.*, **1974**, *39*, 1189.

[91] Cornwall, P., Dell, C. P., and Knight, D. W., *Tetrahedron Lett.*, **1987**, *28*, 3585.

[92] Van Arnum, S. D., Ramig, K., Stepsus, N. A., Dong, Y. and Outten, R. A., *Tetrahedron Lett.*, **1996**, *37*, 8659.

[93] Macco, A. A., Godefroi, E. F. and Drouen, J. J. M., *J. Org. Chem.*, **1975**, *40*, 252.

[94] Itoh, T., Hasegawa, H., Nagata, K. and Ohsawa, A., *J. Org. Chem.*, **1994**, *59*, 1319.

[95] Itoh, T., Miyazaki, M., Nagata, K. and Ohsawa, A., *Heterocycles*, **1998**, *49*, 67.

[96] Itoh, T., Miyazaki, M., Nagata, K., Matsuya, Y. and Ohsawa, A., *Tetrahedron*, **2000**, *56*, 4383.

[97] Ott, D. G., Hayes, F. N. and Kerr, V. N., *J. Am. Chem. Soc.*, **1956**, *78*, 1941; Ruggli, P., Ratti, R. and Henzi, E., *Helv. Chim. Acta*, **1929**, *12*, 332.

[98] Yount, R. G. and Metzler, D. E., *J. Biol. Chem.*, **1959**, *234*, 738.

[99] Hafferl, W., Lundin, R. and Ingraham, L. L., *Biochemistry*, **1963**, *2*, 1298.

[100] Castells, J., López-Calahorra, F., Geijo, F., Pérez-Dolz, R. and Bassedas, M., *J. Heterocycl. Chem.*, **1986**, *23*, 715; Teles. J. H., Melder, J.-P., Ebel, K., Schneider, R., Gehrer, E., Harder, W., Brode, S., Enders, D., Breuer, K. and Raabe, G., *Helv. Chim. Acta*, **1996**, *79*, 61.

101 'From the reactivity of N-heterocyclic carbenes to new chemistry in ionic liquids', Canal, J. P., Ramnial, T., Dickie, D. A. and Clyburne, J. A. C., *Chem. Commun.*, **2006**, 1809.

102 Arduengo, A. J, Dias, H. V. R., Harlow, R. L. and Kline, M., *J. Am. Chem. Soc.*, **1992**, *114*, 5530.

103 'Recent advances in oxazolone chemistry', Filler, R., *Adv. Heterocycl. Chem.*, **1965**, *4*, 75.

104 'The chemistry of 1,3-thiazolinone/hydroxy-1,3-thiazole systems', Barrett, G. C., *Tetrahedron*, **1980**, *36*, 2023.

105 '4-Unsubstituted, 5-amino and 5-unsubstituted, 4-aminoimidazoles', Lythgoe, D. J. and Ramsden, C. A., *Adv. Heterocycl. Chem.*, **1994**, *61*, 1.

106 Erlenmeyer, H. and Kiefer, H., *Helv. Chim. Acta*, **1945**, *28*, 985.

107 McLean, J. and Muir, G. D., *J. Chem. Soc.*, **1942**, 383; Kirk, K. L. and Cohen, L. A., *J. Am. Chem. Soc.*, **1973**, *95*, 4619; Vernin, G. and Metzger, J., *Bull. Soc. Chim. Fr.*, **1963**, 2504.

108 Gompper, R. and Christmann, O., *Chem. Ber.*, **1959**, *92*, 1944.

109 Klein, G. and Prijs, B., *Helv. Chim. Acta*, **1954**, *37*, 2057.

110 Kerdesky, F. A. and Seif, L. S., *Synth. Commun.*, **1995**, *25*, 2639.

111 Crawford, M. and Little, W. T., *J. Chem. Soc.*, **1959**, 729.

112 Zigeuner, G. and Rauter, W., *Monatsh. Chem.*, **1966**, *97*, 33.

113 Bayer, H. O., Huisgen, R., Knorr, R. and Schafer, F. C., *Chem. Ber.*, **1970**, *103*, 2581.

114 'Cycloaddition chemistry of anhydro-4-hydroxy-1,3-thiazolium hydroxides' (thioisomunchnones) for the synthesis of heterocycles', Padwa, A., Harring, S. R., Hertzog, D. L. and Nadler, W. R., *Synthesis*, **1994**, 993.

115 Huisgen, R., Funke, E., Gotthardt, H. and Panke, H.-L., *Chem. Ber.*, **1971**, *104*, 1532; Consonni, R., Croce, P. D., Ferraccioli, R. and La Rosa, C., *J. Chem. Res., (S)*, **1991**, 188.

116 'Thiazole and thiadiazole *S*-oxides', Clerici, F., *Adv. Heterocycl. Chem.*, **2001**, *81*, 107.

117 Mloston, G., Gendek, T. and Heimgartner, H., *Helv. Chim. Acta*, **1998**, *81*, 1585.

118 Eriksen, B. L., Vedso, P., Morel, S. and Begtrup, M., *J. Org. Chem.*, **1998**, *63*, 12.

119 'The preparation of thiazoles', Wiley, R. H., England, D. C. and Behr, L. C., *Org. React.*, **1951**, *6*, 367.

120 'Imidazole and Benzimidazole Synthesis', Grimmett, M. R., Academic Press, **1997**.

121 'Synthesis and biological activity of vicinal diaryl-substituted 1*H*-imidazoles', Bellina, F., Cauteruccio, S., and Rossi, R., *Tetrahedron*, **2007**, *63*, 4571.

122 Schwarz, G., *Org. Synth., Coll. Vol. III*, **1955**, 332.

123 Vogel's Textbook of Practical Organic Chemistry, 4th Edn., 929.

124 Flaig, R. and Hartmann, H., *Heterocycles*, **1997**, *45*, 875.

125 Bandarage, U. K., Come, J. H. and Green, J., *Tetrahedron Lett.*, **2006**, *47*, 8079.

126 Babadjamian, A., Gallo, R., Metzger, J. and Chanon, M., *J. Heterocycl. Chem.*, **1976**, *13*, 1205.

127 Kim, H.-S., Kwon, I.-C. and Kim, O.-H., *J. Heterocycl. Chem.*, **1995**, *32*, 937.

128 Moriarty, R. M., Vaid, B. K., Duncan, M. P., Ley, S. G., Prakash, O. and Goyal, S., *Synthesis*, **1992**,845; Kamproudi, H., Spyroudis, S. and Tarantili, P., *J. Heterocycl. Chem.*, **1996**, *33*, 575.

129 Buchman, E. R., Reims, A. O. and Sarjent, H., *J. Org. Chem.*, **1941**, *6*, 764.

130 Brandsma, L., de Jong, R. L. P. and VerKruijsse, H. D., *Synthesis*, **1985**, 948.

131 Brederick, H., Gompper, R., Bangert, R. and Herlinger, H., *Angew. Chem.*, **1958**, *70*, 269.

132 Shilcrat, S. C., Makhallati, M. K., Fortunak, J. M. D. and Pridgen, L. N., *J. Org. Chem.*, **1997**, *62*, 8449.

133 Liverton, N. J., Butcher, J. W., Claibourne, C. F., Claremon, D. A., Libby, B. E., Nguyen, K. T., Pitzenberger, S. M., Selnick, H. G., Smith, G. R., Tebben, A., Vacca, J. P., Varga, S. L., Agarwal, L., Dabcheck, K., Forsyth, A. J., Fletcher, D. S., Frantz, B., Hanlon, W. A., Harper, C. F., Hofsess, S. J., Kostura, M., Lin, J., Luell, S., O'Neill, E. A.,Orevillo, C. J., pang, M., parsons, J., Rolando, A., Sahly, Y., Visco, D. M. and O'Keefe, S. J., *J. Med. Chem.*, **1999**, *42*, 2180.

134 Litle, T. L. and Webber, S. E., *J. Org. Chem.*, **1994**, *59*, 7299.

135 Streith, J., Boiron, A., Frankowski, A., Le Nouen, D., Rudyk, H. and Tschamber, T., *Synthesis*, **1995**, 944; Siendt, H., Tschamber, T. and Streith, J., *Tetrahedron Lett.*, **1999**, *40*, 5191.

136 Wasserman, H. H. and Vinick, F. J., *J. Org. Chem.*, **1973**, *38*, 2407.

137 Morwick, T., Hrapchak, M., DeTuri, M. and Campbell, S., *Org. Lett.*, **2002**, *4*, 2665.

138 Suenaga, K., Shimogawa, H., Nakagawa, S. and Uemura, D., *Tetrahedron Lett.*, **2001**, *42*, 7079.

139 Davies, J. R., Kane, P D. and Moody, C. J., *Tetrahedron*, **2004**, *60*, 3967.

140 Freeman, F., Chen, T. van der Linden, J. B., *Synthesis*, **1997**, 861.

141 Cornforth, J. W. and Cornforth, R., *J. Chem. Soc.*, **1947**, 96.

142 Galeazzi, E., Guzmán, A., Nava, J. L., Liu, Y., Maddox, M. C. and Muchowski, J. M., *J. Org. Chem.*, **1995**, *60*, 1090.

143 Weinmann, H., Harre, M., Koenig, K., Merten, E. and Tilstam, U., *Tetrahedron Lett.*, **2002**, *43*, 593.

144 Voss, M. E., Beer, C. M., Mitchell, S. A., Blomgren, P. A. and Zhichkin, P. E., *Tetrahedron*, **2008**, *64*, 645.

145 Dubs, P. and Stuessi, R., *Synthesis*, **1976**, 696.

146 van Leusen, A. M., Hoogenboon, B. E. and Siderius, H., *Tetrahedron Lett.*, **1972**, 2369; van Leusen, A. M. and Oldenziel, O. H., *ibid.*, 2373; Kulkarni, B. A. and Ganesan, A., *Tetrahedron Lett.*, **1999**, *40*, 5637.

147 van Leusen, A. M. and Wildeman, J., *Synthesis*, **1977**, 501.

148 van Leusen, A. M., Wildeman, J. and Oldenziel, O. H., *J. Org. Chem.*, **1977**, *42*, 1153.

149 Gracias, V., Gasiecki, A. F. and Djuric, S. W., *Org. Lett.*, **2005**, *7*, 3183.

150 ten Have, R., Huisman, M., Meetsma, A. and van Leusen, A. M., *Tetrahedron*, **1997**, *53*, 11355.

151 Katritzky, A. R., Cheng, D. and Musgrave, R. P., *Heterocycles*, **1997**, *44*, 67.

152 Hartman, G. D. and Weinstock, L. M., *Org. Synth., Coll. Vol. VI*, **1988**, 620.

153 Schröder, R., Schöllkopf, U., Blume, E. and Hoppe, I., *Justus Liebigs Ann. Chem.*, **1975**, 533; Hamada, Y., Morita, S. and Shioiri, T., *Heterocycles*, **1982**, *17*, 321; Ohba, M., Kubo, H., Fujii, I. Ishibashi, H., Sargent, M. V. and Arbain, D., *Tetrahedron Lett.*, **1997**, *38*, 6697.

[154] Moody, C. J. and Doyle, K. J., *Progr. Heterocycl. Chem.*, **1997**, *9*, 1.

[155] Doyle, K. J. and Moody, C. J., *Tetrahedron*, **1994**, *50*, 3761.

[156] Wolkenberg, S. E., Wisnoski, D. D., Leister, W. H., Wang, Y., Zhao, Z. and Lindsley, C. W., *Org. Lett.*, **2004**, *6*, 1453.

[157] Evans, D. L., Minster, D. K., Jordis, U., Hecht, M., Mazzu, A. L. and Meyers, A. I., *J. Org. Chem.*, **1979**, *44*, 497.

[158] Ninomiya, K., Satoh, H., Sugiyama, T., Shinomiya, M. and Kuroda, R., *Chem. Commun.*, **1996**, 1825; Sowinski, J. A. and Toogood, P. T., *J. Org. Chem.*, **1996**, *61*, 7671.

[159] Barrish, J. C., Singh, J., Spergel, S. H., Han, W.-C., Kissick, T. P., Kronenthal, D. R. and Mueller, R. H., *J. Org. Chem.*, **1993**, *58*, 4494.

[160] Williams, D. R., Lowder, P. D., Gu, Y.-G. and Brooks, D. A., *Tetrahedron Lett.*, **1997**, *38*, 331.

[161] Montgomery, J. A., Hewson, K., Struck, R. F. and Sheely, Y. F., *J. Org. Chem.*, **1959**, *24*, 256; Kirk, K. L. and Cohen, L. A., *J. Am. Chem. Soc.*, **1973**, *95*, 4619.

[162] Hartley, D. J. and Iddon, B., *Tetrahedron Lett.*, **1997**, *38*, 4647.

[163] Callahan, J. F., Burgess, J. L., Fornwald, J. A., Gaster, L. M., Harling, J. D., Harrington, F. P., Heer, J., Kwon, C., Lehr, R., Mathur, A., Olson, B. A., Weinstock, J. and Laping, N. J., *J. Med. Chem.*, **2002**, *45*, 999.

25

1,2-Azoles: Pyrazoles, Isothiazoles, Isoxazoles: Reactions and Synthesis

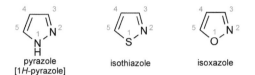

pyrazole
[1*H*-pyrazole]

isothiazole

isoxazole

The physical properties of the three 1,2-azoles, pyrazole,[1] isothiazole[2] and isoxazole[3] can be usefully compared and contrasted with those of their 1,3-isomeric counterparts. Echoing the higher boiling point of imidazole, pyrazole, which is the only one of the trio to be solid at room temperature, also has a much higher boiling point (187 °C) than isothiazole or isoxazole (114 °C and 95 °C), again reflecting the inter-molecular hydrogen bonding available only to pyrazole. This association probably takes the form of dimers, trimers and oligomers; dimeric forms are of course not available to imidazole. Each 1,2-azole has a pyridine-like odour, but is only partially soluble in water. The dihydro- and tetrahydro-heterocycles are named pyrazoline/pyrazolidine, isothiazoline/isothiazolidine and isoxazoline/isoxazolidine.

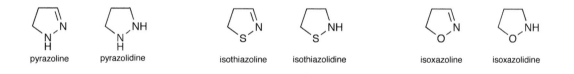

pyrazoline   pyrazolidine   isothiazoline   isothiazolidine   isoxazoline   isoxazolidine

Rapid tautomerism, involving switching of hydrogen from one nitrogen to the other, as in imidazoles (see 24.1.1.1), means that substituted pyrazoles are inevitably mixtures, and a nomenclature analogous to that used for imidazoles is employed to signify this: 3(5)-methylpyrazole, for example. In some cases, one tautomer is predominant.[4]

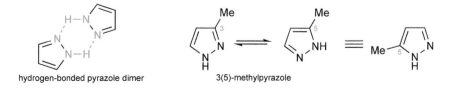

hydrogen-bonded pyrazole dimer   3(5)-methylpyrazole

Heterocyclic Chemistry 5th Edition   John Joule and Keith Mills
© 2010 Blackwell Publishing Ltd

25.1 Reactions with Electrophilic Reagents
25.1.1 Addition at Nitrogen
25.1.1.1 Protonation
Direct linking of two heteroatoms has a very marked base-weakening effect, as one can see by comparing ammonia with hydrazine and hydroxylamine (pK_{aH}s: NH_3, 9.3; H_2NNH_2, 7.9; $HONH_2$, 5.8), and this is mirrored in the 1,2-azoles: pyrazole with a pK_{aH} of 2.5 is some 4.5 pK_a units weaker than imidazole; isothiazole (−0.5) and isoxazole (−3.0) are some three pK_a units weaker than their 1,3-isomers. The higher basicity of pyrazole reflects the symmetry of the cation, with its two equivalent contributing resonance structures. Clearly, again, oxygen has a larger electron-withdrawing effect than sulfur.

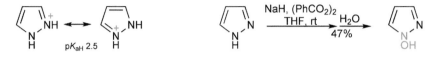

25.1.1.2 Oxidation at Nitrogen
The preparation of 1-hydroxy-pyrazoles can employ peracidic conditions (i.e. imine *N*-oxidation then loss of *N*-hydrogen)[5] or basic conditions,[6] when it is the pyrazolyl *N*-anion (25.4.1) that reacts with the oxidising agent, dibenzoyl peroxide.

25.1.1.3 Alkylation at Nitrogen
The 1,2-azoles are more difficult to quaternise than their 1,3-analogues: isothiazoles, for example, require reactive reagents such as benzyl halides or Meerwein salts (e.g. $Et_3O^+BF_4^-$).[7] Additionally, isoxazolium salts are particularly susceptible to ring cleavage (see 25.10). 3(5)-Substituted pyrazoles that have an *N*-hydrogen, can, in principle, give rise to two isomeric *N*-alkyl-pyrazoles after loss of proton from nitrogen, and there is the further complication that this initial product can, in principle, undergo a second alkylation, producing an 1,2-disubstituted quaternary salt.[8] However, the quaternisation of an already 1-substituted pyrazole generally requires more vigorous conditions, no doubt because of steric impediment to reaction due to the substituent on the adjacent nitrogen. Thus, in the *N*-methylation of pyrazole using dimethyl carbonate at 140 °C, a high yield of the mono-methylated product is obtained.[9] Microwave irradiation improves the rate of *N*-alkylation with phenacyl bromides, mono-alkylated salts being formed in high yields.[10]

Exposure of pyrazole to Mannich conditions produces an *N*-protected pyrazole, presumably *via* attack at the imine nitrogen, followed by loss of proton from the other nitrogen [11] and an *N*-tetrahydropyranyl pyrazole can be similarly prepared.[12]

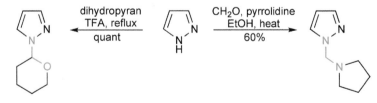

25.1.1.4 Acylation at Nitrogen
The introduction of an acyl[13] or phenylsulfonyl[14] group onto a pyrazole nitrogen is usually achieved in the presence of a weak base, such as pyridine; such processes proceed *via* imine N-2 acylation, then N-1$^+$–H deprotonation. Since acylation, unlike alkylation, is reversible, the more stable product is obtained.

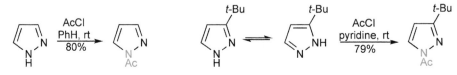

N-Acyl-pyrazoles are useful acylating agents: although they are ca. 20 times less reactive than *N*-acyl-imidazoles and this can be an advantage in the greater stability of the *N*-acyl-pyrazoles as reagents.[15]

Pyrazole reacts with cyanamide very efficiently to produce an *N*-derivative that can be utilised, by reaction with primary or secondary amines, to synthesise guanidines;[16] conversion to a doubly *t*-butoxycarbonyl-protected derivative allows this to be used for the direct synthesis of protected guanidines, as illustrated.[17]

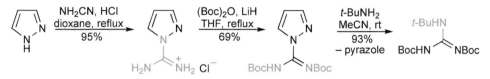

25.1.2 Substitution at Carbon

25.1.2.1 Nitration

Pyrazole,[18] isothiazole[19] and isoxazole[20] undergo straightforward nitration, at C-4. With acetyl nitrate or dinitrogen tetraoxide/ozone,[21] 1-nitropyrazole is formed, but this can be rearranged to 4-nitropyrazole in acid at low temperature.[22]

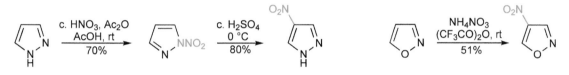

25.1.2.2 Sulfonation

Electrophilic sulfonation of isoxazole is of no preparative value; the substitution of only the phenyl-substituent of 5-phenylisoxazole with chlorosulfonic acid makes the same point.[23] Both isothiazole[24] and pyrazole[25] can be satisfactorily sulfonated at C-4.

25.1.2.3 Halogenation

Halogenation of pyrazole gives 4-monohalo-pyrazoles, for example 4-iodo-,[26] or 4-bromopyrazole[27] under controlled conditions. Poor yields are obtained on reaction of isothiazole[28] and isoxazole[29] with bromine, again with attack at C-4, but with stabilising groups present, halogenation proceeds better[30] and efficient 4-bromination of isoxazoles and pyrazoles is achieved using *N*-bromosuccinimide and microwave irradiation.[31] 3,4,5-Tribromopyrazole is formed efficiently in alkaline solution, presumably the pyrazolyl anion (25.4.1) is the reacting species.[32]

25.1.2.4 Acylation

Only for pyrazole, of the trio, have any useful electrophilic substitutions involving carbon electrophiles been described,[14,33] and, even here, only *N*-substituted pyrazoles react well.

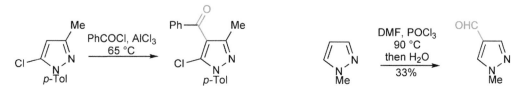

25.2 Reactions with Oxidising Agents

The 1,2-azole ring systems are relatively stable to oxidative conditions, even allowing substituent alkyl- or, more efficiently, acyl-groups to be oxidised up to carboxylic acid.[34] Ozone cleaves the isoxazole ring.[35]

25.3 Reactions with Nucleophilic Reagents

The 1,2-azoles do not generally react with nucleophiles with replacement of hydrogen. In favourable situations, an *N*-nitro group in 1-nitro-pyrazoles can act as a leaving group to allow substitution of hydrogen.[36]

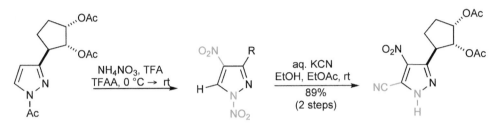

There is a limited number of examples of displacements of leaving groups from the isothiazole 5-position[37] when it is activated by a 4-keto or similar group, but 3-halo-groups are less easily displaced; 4-halides behave like halo-benzenes.

25.4 Reactions with Bases

25.4.1 Deprotonation of Pyrazole *N*-Hydrogen and Reactions of Pyrazolyl Anions

The pK_a for loss of the *N*-hydrogen of pyrazole is 14.2, thus it is an appreciably stronger acid than pyrrole (pK_a 17.5), and the same as that of imidazole; the anion has, like the imidazolyl anion, two, equally-contributing resonance forms.

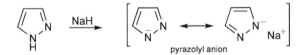

N-Alkylations can be conducted in strongly basic,[38] or phase-transfer conditions[39] or in the presence of 4-dimethylaminopyridine,[40] and it seems likely that under these conditions it is the pyrazolyl anion that is alkylated. The use of sodium hydrogen carbonate, without solvent, but with microwave heating, is also recommended.[41]

3(5)-Substituted pyrazoles sometimes give a product isomeric with that which is obtained by reaction in neutral solution.[8]

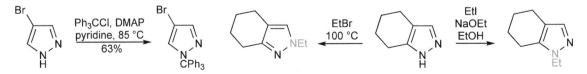

N-Arylation of pyrazoles can be achieved efficiently with metal catalysis (see 4.2)

25.5 C-Metallation and Reactions of C-Metallated 1,2-Azoles[42]

25.5.1 Direct Ring C–H Metallation

The C-5-lithiation of pyrazoles requires the absence of the *N*-hydrogen: removable *N*-protecting groups that have been used include phenylsulfonyl,[43] trimethylsilylethoxymethyl,[44] hydroxymethyl,[45] methylsulfonyl,[46] dimethylaminosulfonyl,[47] tetrahydropyran-2-yl,[12] and pyrrolidin-1-ylmethyl.[11] Lithiation of 1-benzyloxypyrazole[48] or 3-benzyloxyisothiazole[49] takes place at C-5 and of 1-substituted pyrazole *N*-oxides at C-3.[50]

Attempted *C*-lithiation of isoxazoles with hydrogen at C-3 leads inevitably to ring opening, with the oxygen as anionic leaving group,[51] indeed this type of cleavage was first recognised as long ago as 1891, when Claisen found that 5-phenylisoxazole was cleaved by sodium ethoxide[52] (see 25.10 for ring cleavage of isoxazolium salts). Comparable cleavages of isothiazoles that have a 3-hydrogen can also be a problem.

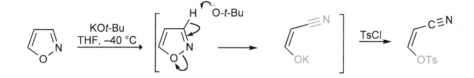

An alternative ring cleavage results from 5-lithiation of 3-phenylisoxazole and leads on to azetidinone-containing structures.[53]

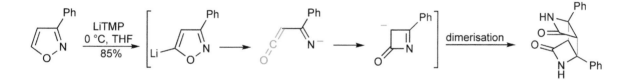

3-Methoxymethoxy-5-phenylisoxazole lithiates at C-4,[54] as does 3-*t*-butoxycarbonylamino-5-methylisoxazole (two equivalents of *n*-butyllithium required).[55]

The reactions of 5-lithiated isothiazoles and of 5-lithiated-1-substituted pyrazoles allow the introduction of substituents at that position by reaction with a range of electrophiles; three examples are shown below.[11,56,57,58]

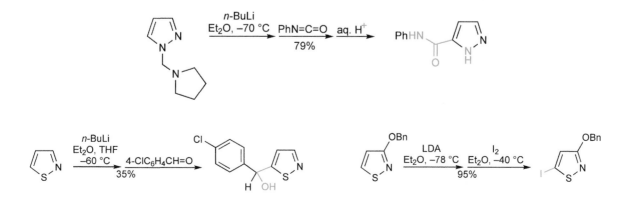

25.5.2 Metal–Halogen Exchange

It is significant that treatment of 4-bromo-1-phenylsulfonylpyrazole with *n*-butyllithium results in 5-lithiation and not metal–halogen exchange,[43] however 4-bromo-1-triphenylmethylpyrazole undergoes normal exchange and in this way, for example, a tin derivative can be obtained.[40]

Metal–halogen exchange can be achieved in the formation of 3-lithio-1-methylpyrazole from the bromo-pyrazole,[59] and reaction of 4-bromopyrazole with two equivalents of *n*-butyllithium produces a 1,4-dilithio-pyrazole that reacts with electrophiles at C-4.[60] 4-Iodoisothiazole can be converted into a magnesium compound that shows normal nucleophilic Grignard properties.[61]

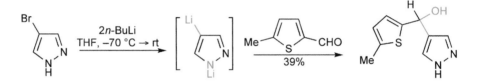

25.6 Reactions with Radicals

The interaction of 1,2-azoles with radical reagents is an area in which little is known so far. Displacement of tosyl from the 5-position of a protected pyrazole shows that there is potential for further development.[62]

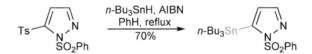

25.7 Reactions with Reducing Agents

Pyrazoles are relatively stable to catalytic and chemical reductive conditions, particularly when there is no substituent on nitrogen, though catalytic reduction can be achieved in acid solution.[63] Isothiazoles are reductively desulfurised using Raney nickel, with loss of the ring.[64] Catalytic hydrogenolysis of the N–O bond in isoxazoles takes place readily over noble-metal catalysts,[65] and this process is central to the stratagem in which isoxazoles are employed as masked 1,3-dicarbonyl compounds. The immediate products of N–O hydrogenolysis, β-amino-enones, can often be isolated as such, or further processed. The use of this ring cleavage to provide routes to pyrimidinones,[66] and to pyrazoles, as illustrated.[67]

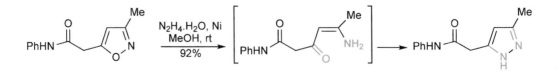

5-Alkoxy-isoxazoles undergo ring contraction with iron(II) chloride, producing azirine esters.[68]

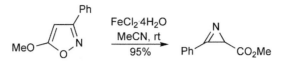

25.8 Electrocyclic and Photochemical Reactions

There are examples of 1,2-azoles being converted into their 1,3-isomers by irradiation, though such processes are of limited preparative value.[69]

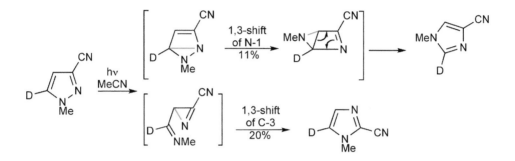

In a similar way, irradiation converts many simpler pyrazoles into imidazoles,[70] phenylisothiazoles[71] and methylisothiazoles[72] partially into the corresponding thiazoles, and 3,5-diarylisoxazoles into 2,5-disubstituted oxazoles.[73] The transformation of 1,2-azoles carrying, at C-3, a side-chain of three atoms terminating in a doubly bonded heteroatom, into isomeric systems with a new five-membered ring is a general process,[74] though there is no definitive view as to the details of its mechanism.

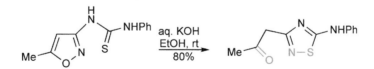

1,2-Azoles do not act as 1-azadienes in cycloadditions. 4-Nitro-isoxazoles react with dienes across the 4,5-bond[75] and in processes useful for the synthesis of purine analogues, 3(5)-amino-pyrazoles add to electron-deficient 1,3,5-triazines, across the pyrazole 4,5-bond, subsequent eliminations giving the final aromatic product.[76]

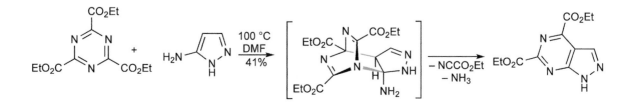

25.9 Alkyl-1,2-Azoles

4-Methyl-isothiazoles are not especially acidic, but it is rather surprising that 3-methyl-isothiazoles are also not reactive, whereas 5-methyl substituents will undergo condensation reactions.[77] This same effect is also found in isoxazoles: with the 3-position blocked to prevent ring degradation (25.5.1), 3,5-dimethylisoxazole exchanges, with methoxide in methanol, 280 times faster at the 5- than at the 3-methyl group and preparative deprotonations proceed exclusively at the 5-methyl substituent.[78] By working at low temperature, even 5-methylisoxazole can be deprotonated at the methyl without ring degradation.[79] Conversion to *N*-oxide[80] activates adjacent methyl groups, for example subsequent reaction with trimethylsilyl iodide permits side-chain iodination.[81]

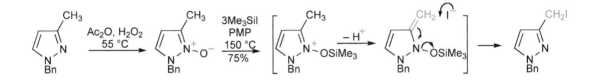

On subjection of 3-methyl-5-phenylisothiazole or 3-methyl-5-phenylisoxazole to lithiation conditions, competitive side-chain and C-4 deprotonation is observed, except when lithium *i*-propyl(cyclohexyl)amide (LICA) is used – this allows exclusive side-chain lithiation.[82]

25.10 Quaternary 1,2-Azolium Salts

The base-catalysed degradation of the ring of isoxazolium salts is particularly easy, requiring only alkali metal carboxylates to achieve it. The mechanism,[83] illustrated for the acetate-initiated degradation of 2-methyl-5-phenylisoxazolium iodide, involves initial 3-deprotonation with cleavage of the N–O bond; subsequent rearrangements lead to an enol-acetate which rearranges to a final keto-imide.

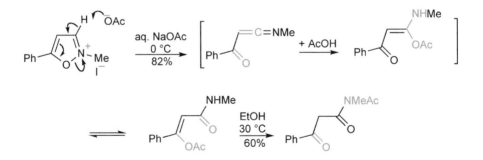

Isothiazolium salts[84] are often prepared by ring synthesis involving oxidative nitrogen-to-sulfur bond formation, however these quaternary salts can be prepared in the normal way from an isothiazole with a reactive electrophile. They are cleanly cleaved by reducing agents, for example sodium borohydride, producing β-enamino-thioketones.[85]

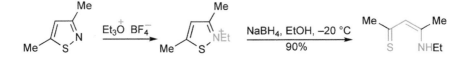

25.11 Oxy- and Amino-1,2-azoles

Only 4-hydroxy-1,2-azoles can be regarded as being phenol-like.[86] 3- and 5-Hydroxy-1,2-azoles exist mainly in carbonyl tautomeric forms, encouraged by resonance involving donation from a ring heteroatom, and are therefore known as pyrazolones,[87] isothiazolones and isoxazolones, though for all three systems, and depending on the nature of other substituents, an appreciable percentage of hydroxy tautomer exists in solution.

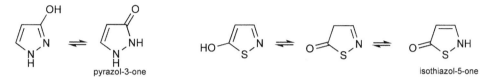

The reactivity of the 3- and 5-azolones centres mainly on their ability to react with electrophiles, such as halogens,[88] (giving 4,4-dihalo derivatives with excess reagent – 4,4-dibromo-3-methylpyrazol-5-one is a *para*-selective brominating agent for phenols and anilines[89]), or to nitrate,[90] or undergo Vilsmeier formylation;[91] the example shown below is the formylation of 'antipyrine' once used as an analgesic. Many dyestuffs have been synthesised *via* coupling of aryldiazonium cations with 5-pyrazolones at C-4 – tartrazine is such an example.

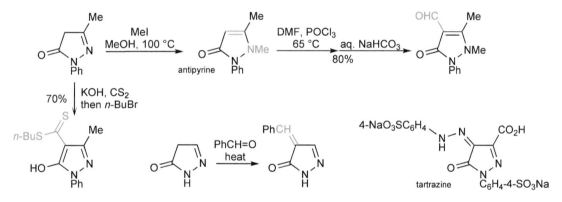

Pyrazolones also condense with aldehydes[92] in aldol-type processes, or react with other electrophiles, such as carbon disulfide,[93] in each case reaction presumably proceeding *via* the enol tautomer, or its anion. In basic solution isoxazol-3-ones alkylate either on oxygen or nitrogen, and the choice of base can influence the ratio.[52]

Amino-1,2-azoles[94] exist as the amino tautomers. Amino-pyrazoles and amino-isothiazoles are relatively well-behaved aromatic amines, for example 3(5)-aminopyrazole undergoes substituent-*N*-acetylation and easy electrophilic bromination at C-4.[95] Diazotisation and a subsequent Sandmeyer reaction provides routes to halo-isothiazoles[61,19] and azido-pyrazoles.[96]

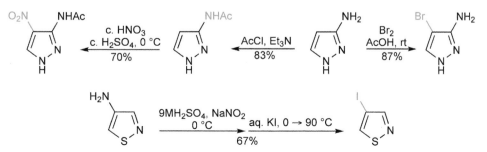

Diazotisation of 4-amino-pyrazoles, then deprotonation yields isolable diazo-pyrazoles.[97]

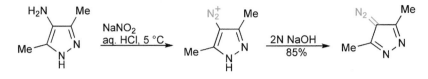

25.12 Synthesis of 1,2-Azoles[98]
25.12.1 Ring Synthesis

There are parallels, but also methods unique to particular 1,2-azoles, in the principal methods available for the construction of pyrazoles, isothiazoles and isoxazoles: neither the reaction of propene with sulfur dioxide and ammonia at 350 °C, which gives isothiazole itself[99] in 65% yield, nor a synthesis[100] from propargyl aldehyde and thiosulfate (shown below) have direct counterparts for the other 1,2-azoles.

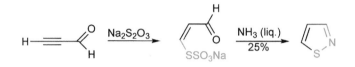

25.12.1.1 From 1,3-Dicarbonyl Compounds and Hydrazines or Hydroxylamine

Pyrazoles and isoxazoles can be made from a 1,3-dicarbonyl component and a hydrazine or hydroxylamine, respectively.

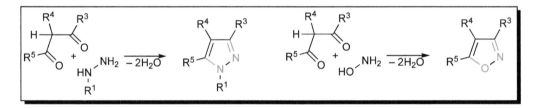

This, the most widely used route to pyrazoles and isoxazoles rests on the doubly nucleophilic character of hydrazines and hydroxylamine, allowing them to react in turn with each carbonyl group of a 1,3-diketone[101] or 1,3-keto-aldehyde, often with one of the carbonyl groups (especially when aldehyde) masked as enol ether,[102] acetal, imine[103] or enamine,[104,105] or other synthon for one of these.

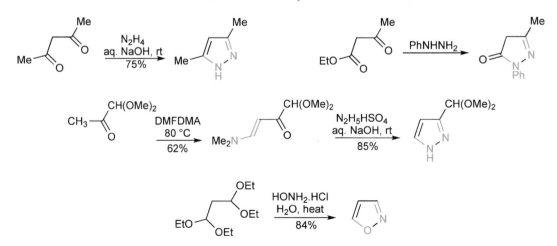

When β-keto-esters are used, the products are pyrazolones[106] or isoxazolones;[107] similarly, β-keto-nitriles with hydrazines give 3(5)-amino-pyrazoles.[108] 3(5)-Aminopyrazole itself is prepared *via* a dihydro precursor formed by addition of hydrazine to acrylonitrile then cyclisation;[109] hydrolysis of the first cyclic intermediate in this sequence and dehydrogenation *via* elimination of *p*-toluenesulfinate allows preparation of 3(5)-pyrazolone.[110]

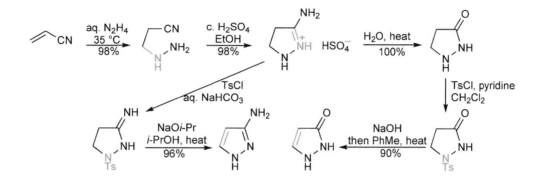

Generally speaking, unsymmetrical 1,3-dicarbonyl components produce mixtures of 1,2-azole products.[86] This difficulty can be circumvented in a number of ways: by the use of acetylenic aldehydes or ketones (as synthons), where a hydrazone or oxime can be formed first by reaction at the carbonyl group, and this can then be cyclised in a separate, second step.[111,112] A nice example of this is the formation of 5-silyl-pyrazoles and -isoxazoles using silyl-alkynones.[113] The regioselective reaction of arylhydrazines with 1,3-diketones shows what can sometimes be achieved by careful choice of reaction conditions and solvent.[114]

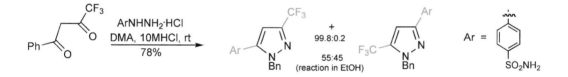

Pyrazole itself can be formed by the reaction of hydrazine with propargyl aldehyde.[10] Using β-chloro-,[115] β-alkoxy-[116] β-amino-[117] -enones as 1,3-dicarbonyl synthons are other ways to influence the regiochemistry of reaction.[118] 2-(Dimethoxymethyl)-ketones react with arylhydrazines at the ketonic carbonyl, thus reversing the intrinsic reactivity of a β-keto-aldehyde.[119]

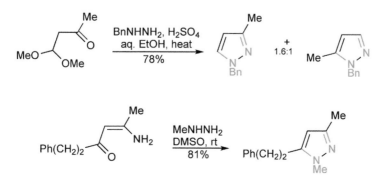

25.12.1.2 Dipolar Cycloadditions of Nitrile Oxides and Nitrile Imines

Isoxazoles are produced by the dipolar cycloaddition of nitrile oxides to alkynes; pyrazoles result from the comparable interaction of alkynes with nitrile imines.

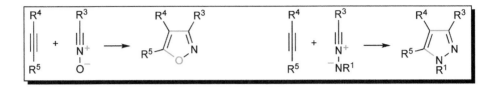

Nitrile oxides (R–C≡N⁺–O⁻),[120] which can be generated by base-catalysed elimination of hydrogen halide from halo-oximes (RC(Hal)=NOH), or by dehydration of nitro compounds[121] (RCH$_2$NO$_2$), readily add to alkenes and to alkynes, generating five-membered heterocycles. Addition to an alkene produces an isoxazoline, unless the alkene also incorporates a group capable of being eliminated in a step after the cycloaddition, as shown below.[122] However, isoxazolines can also be dehydrogenated to the aromatic system.[123,124]

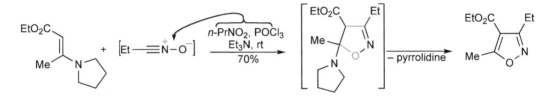

Cycloaddition of a nitrile oxide to an alkyne generates an aromatic isoxazole directly, but mixtures are sometimes obtained.[125,126]

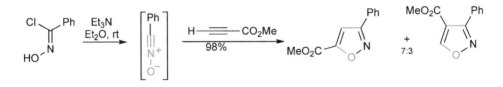

A useful route to 3-bromo-isoxazoles rests on the cycloaddition of bromonitrile oxide.[127]

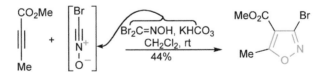

Nitrile imines can be generated in a similar way: the dehydrohalogenation of hydrazonyl halides (from *N*-halosuccinimide and a hydrazone), or, as in the sequence below, elimination of benzenesulfonate.[128]

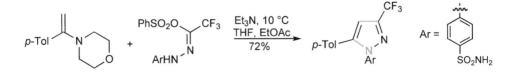

The cycloaddition of diazoalkanes to alkynes gives pyrazoles; the use of stannyl alkynes[129,130] produces tin derivatives of the heterocycle (cf. 4.1.6), for use in subsquent electrophilic *ipso*-displacements, or in palladium(0)-catalysed couplings. Aryl-diazomethanes can be generated *in situ*, from tosylhydrazones, for formation of pyrazoles.[131]

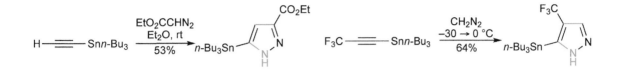

25.12.1.3 *From Oximes and Hydrazones*

Exposure of ketone oximes that have an α-hydrogen to two mole equivalents of *n*-butyllithium leads to *O*- and *C*-lithiation (*syn* to the oxygen); reaction with dimethylformamide as electrophile then allows *C*-formylation and ring closure *in situ* to a 5-unsubstituted isoxazole.[132] Similarly, reaction of the dianion to an acylating agent (an ester[133] or a Weinreb amide[134]) leads through to 5-substituted isoxazoles, or with diethyl oxalate, to isoxazole-5-esters.[135]

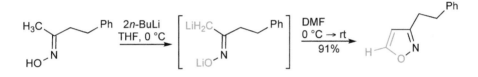

Displacement of the halogen of an α-bromo-ketone oxime with an alkyne leads to an intermediate, which closes to an isoxazole simply on treatment with mild base.[136]

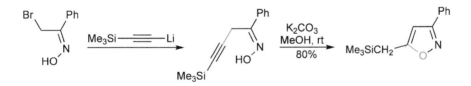

An elegant route to pyrazoles, involves forming a ring closure precursor by Horner–Emmons condensation of a tosylhydrazone-phosphonate with an aldehyde, which becomes the 5-substituent; intramolecular Michael addition and then loss of toluenesulfinate as the final aromatising step, completes the sequence.[137]

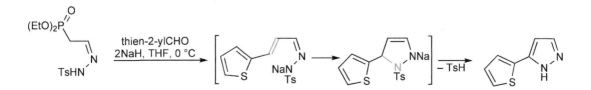

Exercises
Straightforward revision exercises (consult Chapters 23 and 25):
(a) Compare 1,2- with 1,3-azoles in pairs – which is the more basic? Why?

(b) What is incorrect about the name: '3-methylpyrazole'?

(c) Name some groups that can be used to mask the *N*-hydrogen in pyrazoles during *C*-lithiation.

(d) For what functionality are isoxazoles synthons if the *N–O* bond is cleaved? How could one cleave the *N–O* bond?

(e) How are 1,3-dicarbonyl compounds used for the synthesis of isoxazoles and pyrazoles?

(f) Describe a method involving an electrocyclic process for the ring synthesis of an isoxazole.

(g) Describe a method for the utilisation of the oxime of a dialkyl ketone, to make an isoxazole.

More advanced exercises:
1. Suggest structures for the isomeric products, $C_9H_7N_3O_2$ formed when 1-phenylpyrazole is reacted with: (i) c. H_2SO_4/c. HNO_3 or (ii) Ac_2O/HNO_3. Explain the formation of different products under the two conditions.

2. Draw structures for the products obtained by reacting 3,5-dimethylisoxazole with $NaNH_2$, then: (i) *n*-PrBr; (ii) CO_2; or (iii) $PhCO_2Me$.

3. Deduce structures for the products obtained by treating 5-methylisoxazole with $SO_2Cl_2 \rightarrow C_4H_4ClNO$, and this with aqueous sodium hydroxide $\rightarrow C_4H_4ClNO$ (which contains no rings).

4. Draw the structures of the products that would be formed from the reaction of $BnNHNH_2$ with $MeCOCH_2COCO_2Me$.

5. Deduce structures for the products formed in the following sequence: pyrazole/Me_2NSO_2Cl/$Et_3N \rightarrow C_5H_9N_3O_2S$, then this with *n*-BuLi/–70 °C, then TMSCl $\rightarrow C_8H_{17}N_3O_2SSi$, then this with PhCH=O/CsF $\rightarrow C_{12}H_{15}N_3SO_3$.

6. Draw the structures of the two products that are formed when hydroxylamine reacts with $PhCOCH_2CH=O$; suggest an unambiguous route for the preparation of 5-phenylisoxazole.

7. Deduce the structures of the heterocyclic substances produced: (i) C_7H_9NO, from cyclohexanone oxime with 2 mole equivalents of *n*-BuLi, then dimethyl formamide; (ii) $C_{11}H_{15}NOSSi$ from thien-2-ylC(=NOH)CH_2Br and $Me_3SiC\equiv CLi$, then K_2CO_3/MeOH; (iii) $C_{11}H_{12}N_2$ from MeCOC(=NNHPh)Me with $(EtO)_2POCH_2SEt$/*n*-BuLi.

8. Suggest a structure for the heterocyclic product, $C_7H_{13}NOSi$, formed by reaction of $Me_3SiC\equiv CC\equiv CSiMe_3$ and hydroxylamine.

References
[1] 'Progress in pyrazole chemistry', Kost, A. N. and Grandberg, I. I., *Adv. Heterocycl. Chem.*, **1966**, *6*, 347; 'The Azoles', Schofield, K., Grimmett, M. R. and Keene, B. R. T., Cambridge University Press, **1976**.

[2] (a) 'Isothiazoles', Hübenett, F., Flock, F. H., Hansel, W., Heinze, H. and Hofmann, H., *Angew. Chem., Int. Ed. Engl.*, **1963**, *2*, 714; (b) 'Isothiazoles', Slack, R. and Wooldridge, K. R. H., *Adv. Heterocycl. Chem.*, **1965**, *4*, 107; 'Recent advances in the chemistry of mononuclear isothiazoles', Wooldridge, K. R. H., *ibid.*, **1972**, *14*, 1.

[3] 'Recent developments in isoxazole chemistry', Kochetkov, N. K. and Sokolov, S. D., *Adv. Heterocycl. Chem.*, **1963**, *2*, 365; 'Isoxazole chemistry since 1963', Wakefield, B. J. and Wright, D. J., *ibid.*, **1979**, *25*, 147; 'Synthetic reactions using isoxazole compounds', Kashima, C., *Heterocycles*, **1979**, *12*, 1343.

[4] Trofimenko, S., Yap, G. P. A., Jove, F. A., Claramunt, R. M., García, M. A., Santa Maria, M. D., Alkorta, I. and Elguero, J., *Tetrahedron*, **2007**, *63*, 8104.

[5] Begtrup, M. and Vedso, P., *J. Chem. Soc., Perkin Trans. 1*, **1995**, 243.

[6] Reuther W., and Baus, V., *Liebig's Ann.*, **1995**, 1563.

[7] Chaplen, P., Slack, R. and Wooldridge, K. R. H., *J. Chem. Soc.*, **1965**, 4577.

[8] v. Auwers, K., Buschmann, W. and Heidenreich, R, *Justus Liebigs Ann. Chem.*, **1924**, *435*, 277.

[9] Ouk, S., Thiébaud, S., Borredon, E. and Chabaud, B., *Synth. Commun.*, **2005**, *35*, 3021.

[10] Pérez, E., Sotelo, E., Loupy, A., Mocelo, R., Suarez, M., Pérez, R. and Autié, M., *Heterocycles*, **1996**, *43*, 539.

[11] Katritzky, A. R., Rewcastle, G. W. and Fan, W.-Q., *J. Org. Chem.*, **1988**, *53*, 5685.

[12] Gérard, A.-L., Mahatsekake, C., Collot, V. and Rault, S., *Tetrahedron Lett.*, **2007**, *48*, 4123.

[13] Hüttel, R. and Kratzer, J., *Chem. Ber.*, **1959**, *92*, 2014; Williams, J. K., *J. Org. Chem.*, **1964**, *29*, 1377.

[14] Finar, I. L. and Lord, G. H., *J. Chem. Soc.*, **1957**, 3314.

[15] Kashima, C., *Heterocycles*, **2003**, *60*, 437.

[16] Bernatowicz, M. S., Wu, Y. and Matsueda, G. R., *J. Org. Chem.*, **1992**, *57*, 2497.

[17] Drake, B., Patek, M., and Lebl, M., *Synthesis*, **1994**, 579.

[18] Hüttel, R., Büchele, F. and Jochum, P., *Chem. Ber.*, **1955**, *88*, 1577.

[19] Caton, M. P. L., Jones, D. H., Slack, R. and Woolridge, K. R. H., *J. Chem. Soc.*, **1964**, 446.

[20] Reiter, L. A., *J. Org. Chem.*, **1987**, *52*, 2714.

[21] Suzuki, H. and Nonoyama, N., *J. Chem. Res. (S)*, **1996**, 244.

[22] Olah, G. A., Narang, S. C. and Fung, A. P., *J. Org. Chem.*, **1981**, *46*, 2706.

[23] Woodward, R. B., Olofson, R. and Mayer, H., *J. Am. Chem. Soc.*, **1961**, *83*, 1010.

[24] Pain, D. L. and Parnell, E. W., *J. Chem. Soc.*, **1965**, 7283.

[25] Knorr, L., *Justus Liebigs Ann. Chem.*, **1894**, *279*, 188.

[26] Hüttel, R., Schäfer, O. and Jochum, P., *Justus Liebigs Ann. Chem.*, **1955**, *593*, 200.

[27] Lipp., M., Dallacker, F. and Munnes, S., *Justus Liebigs Ann. Chem.*, **1958**, *618*, 110.

[28] Finley, J. H. and Volpp, G. P., *J. Heterocycl. Chem.*, **1969**, *6*, 841.

[29] Pino, P., Piacenti, F. and Fatti, G., *Gazz. Chim. Ital.*, **1960**, *90*, 356.

[30] Blount, J. F., Coffen, D. L. and Katonak, D. A., *J. Org. Chem.*, **1978**, *43*, 3821.

[31] Li, G., Kakarla, R. and Gerritz, S. W., *Tetrahedron Lett.*, **2007**, *48*, 4595.

[32] Juffermans, J. P. H. and Habraken, C. L., *J. Org. Chem.*, **1986**, *51*, 4656.

[33] Tojahn, C. A., *Chem. Ber.*, **1922**, *55*, 291.

[34] Benary, E., *Chem. Ber.*, **1926**, *59*, 2198; Holland, A., Slack, R., Warren, T. F. and Buttimore, *J. Chem. Soc.*, **1965**, 7277; Quilico, A. and Stagno d'Alcontres, G., *Gazz. Chim. Ital.*, **1949**, *79*, 654.

[35] Kashima, C., Takahashi, K. and Hosomi, A., *Heterocycles*, **1994**, *37*, 1075.

[36] Zhou, J., Yang, M., Akdag, A. and Schneller, S. W., *Tetrahedron*, **2006**, *62*, 7009.

[37] Dornow, A. and Teckenburg, H., *Chem. Ber.*, **1960**, *93*, 1103; Hatchard, W. R., *J. Org. Chem.*, **1964**, *29*, 660.

[38] Jones, R. G., Mann, M. J. and McLaughlin, K. C., *J. Org. Chem.*, **1954**, *19*, 1428.

[39] e.g. Hartshorne, C. M. and Steel, P. J., *Aust. J. Chem.*, **1995**, *48*, 1587.

[40] Elguero, J., Jaramillo, C. and Pardo, C., *Synthesis*, **1997**, 563.

[41] Almena, I., Díez-Barra, E., de la Hoz, A., Ruiz, J. and Sánchez-Migallón, A., *J. Heterocycl. Chem.*, **1998**, *35*, 1263.

[42] 'Synthesis and reactions of lithiated monocyclic azoles containing two or more hetero-atoms. Part I: Isoxazoles', Iddon, B., *Heterocycles*, **1994**, *37*, 1263; 'Part III: Pyrazoles', Grimmett, M. R. and Iddon, B., *ibid.*, **1994**, *37*, 2087; 'Part V: Isothiazoles and thiazoles', Iddon, B., *ibid.*, **1995**, *41*, 533.

[43] Heinisch, G., Holzer, W. and Pock, S., *J. Chem. Soc., Perkin Trans. 1*, **1990**, 1829.

[44] Gérard, A.-L., Bouillon, A., Mahatsekake, C., Collot, V. and Raul, S., *Tetrahedron Lett.*, **2006**, *47*, 4665.

[45] Katritzky, A. R., Lue, P. and Akutagawa, K., *Tetrahedron*, **1989**, *45*, 4253.

[46] Effenberger, F., Roos, M., Ahmad, R. and Krebs, A., *Chem. Ber.*, **1991**, *124*, 1639.

[47] Yagi, K., Ogura, T., Numata, A., Ishii, S. and Arai, K., *Heterocycles,*, **1997**, *45*, 1463.

[48] Vedso, P. and Begtrup, M., *J. Org. Chem.*, **1995**, *60*, 4995.

[49] Bunch, L., Krogsgaard-Larsen, P. and Madsen, U., *J. Org. Chem.*, **2002**, *67*, 2375.

[50] Paulson, A. S., Eskildsen, J., Vedso, P. and Begtrup, M., *J. Org. Chem.*, **2002**, *67*, 3904.

[51] Hessler, E. J., *J. Org. Chem.*, **1976**, *41*, 1828.

[52] Claisen, L. and Stock, R., *Chem. Ber.*, **1891**, *24*, 130.

[53] Di Nunno, L., Vitale, P., Scilimati, A., Simone, L. and Capitelli, F., *Tetrahedron*, **2007**, *63*, 12388.

[54] Alzeer, J., Nock, N., Wassner, G. and Masciadri, R., *Tetrahedron Lett.*, **1996**, *37*, 6857.

[55] Konoike, T., Kando, Y. and Araki, Y., *Tetrahedron Lett.*, **1996**, *37*, 3339.

[56] Layton, A. J. and Lunt, E., *J. Chem. Soc.*, **1968**, 611.

[57] Béringer, M., Prijs, B. and Erlenmeyer, H., *Helv. Chim. Acta*, **1966**, *49*, 2466.

[58] Kaae, B. H, Krogsgaard-Larsen, P. and Johansen, T. N., *J. Org. Chem.*, **2004**, *69*, 1401.

[59] Pavlik, J. W. and Kurzweil, E. M., *J. Heterocycl. Chem.*, **1992**, *29*, 1357.

[60] Hahn, M., Heinisch, G., Holzer, W. and Schwarz, H., *J. Heterocycl. Chem.*, **1991**, *28*, 1189.

[61] Guilloteau, F. and Miginiac, L., *Synth. Commun.*, **1995**, *25*, 1383.

[62] Caddick, S. and Joshi, S., *Synlett*, **1992**, 805.

[63] Thoms, H. and Schnupp, J., *Justus Liebigs Ann. Chem.*, **1923**, *434*, 296.

[64] Adams, A. and Slack, R., *J. Chem. Soc.*, **1959**, 3061.

[65] Baraldi, P. G., Barco, A., Benetti, S., Moroder, F., Pollini, G. and Simoni, D., *J. Org. Chem.*, **1983**, *48*, 1297.

[66] Shaw, G. and Sugowdz, G., *J. Chem. Soc.*, **1954**, 665.

[67] Sviridov, S. I., Vasil'ev, A. A. and Shorshnev, S. V., *Tetrahedron*, **2007**, *63*, 12195.

[68] Auricchio, S., Biri, A., Pastormerlo, E. and Truscello, A. M., *Tetrahedron*, **1997**, *53*, 10911.

[69] Barltrop, J. A., Day, A. C., Mack, A. G., Shahrisa, A. and Wakamatsu, S., *J. Chem. Soc., Chem. Commun.*, **1981**, 604.

[70] Tiefenthaler, H., Dörscheln, W., Göth, H. and Schmid, H., *Helv. Chim. Acta*, **1967**, *50*, 2244; Pavlik, J. W., Kebede, N., Bird, N. P., Day, A. C. and Barltrop, J. A., *J. Org. Chem.*, **1995**, *60*, 8138.

[71] Maeda, M., Kawahara, A., Kai, M. and Kojima, M., *Heterocycles*, **1975**, *3*, 389.

[72] Pavlik, J. W., Pandit, C. R., Samuel, C. J. and Day., A. C., *J. Org. Chem.*, **1993**, *58*, 3407.

[73] Singh, B. and Ullman, E. F., *J. Am. Chem. Soc.*, **1967**, *89*, 6911.

[74] 'Mononuclear heterocyclic rearrangements', Ruccia, M., Vivona, N. and Spinelli, D., *Adv. Heterocycl. Chem.*, **1981**, *29*, 141.

[75] Turchi, S., Giomi, D. and Nesi, R., *Tetrahedron*, **1995**, *51*, 7085.

[76] Dang, Q., Brown, B. S. and Erion, M. D., *J. Org. Chem.*, **1996**, *61*, 5204.

[77] Hofmann, H., *Justus Liebigs Ann. Chem.*, **1965**, *690*, 147.

[78] Kashima, C., Yamamoto, Y., Tsuda, Y. and Omote, Y., *Bull. Chem. Soc. Jpn.*, **1976**, *49*, 1047.

[79] Xia, X., Knerr, G. and Natale, N. R., *J. Heterocycl. Chem.*, **1992**, *29*, 1297.

[80] Parnell, E. W., *Tetrahedron Lett.*, **1970**, 3941; Begtrup, M., Larsen, P. and Vedso, P., *Acta Chem. Scand.*, **1992**, *46*, 972.

[81] Begtrup, M. and Vedso, P., *J. Chem. Soc., Perkin Trans. 1*, **1992**, 2555.

[82] Alberola, A., Calvo, L., Rodriguez, T. R. and Sañudo, C., *J. Heterocycl. Chem.*, **1995**, *32*, 537.

[83] 'The reaction of isoxazolium salts with nucleophiles', Woodward, R. B. and Olofson, R., *Tetrahedron, Suppl. No. 7*, **1966**, *7*, 415.

[84] 'Isothiazolium salts and their use as components for the synthesis of S,N-heterocycles', Wolf, J. and Schulze, B., *Adv. Hewterocycl. Chem.*, **2007**, *94*, 215.

[85] Cuadrado, P., González, A. Ma. and Pulido, Fco. J., *Synth. Commun.*, **1988**, *18*, 1847.

[86] Naito, T., Nakagawa, S., Okumura, J., Takahashi, K. and Kasai, K., *Bull. Chem. Soc. Jpn.*, **1968**, *41*, 959; Fagan, P. J., Neident, E. E., Nye, M. J., O'Hare, M. J. and Tang, W.-P., *Can. J. Chem.*, **1979**, *57*, 904.

[87] 'Pyrazol-3-ones. Part I: Synthesis and applications', Varvounis, G., Fiamegos, Y. and Pilidis, G., *Adv. Heterocycl. Chem.*, **2001**, *80*, 74; 'Part II: reactions of the ring atoms', *ibid.*, **2004**, *87*, 142.

[88] Westöö, G., *Acta Chem. Scand.*, **1952**, *6*, 1499.

[89] Mashraqui, S. H., Mudalian, C. D. and Hariharasubrahmanian, H., *Tetrahedron Lett.*, **1997**, *38*, 4865.

[90] Iseki, T., Sugiura, T., Yasunaga, S. and Nakasina, M., *Chem. Ber.*, **1941**, *74*, 1420.

[91] Ledrut, J., Winternitz, F. and Combes, G., *Bull. Soc. Chim. Fr.*, **1961**, 704.

[92] Knorr, L., *Chem. Ber.*, **1896**, *29*, 249.

[93] Oliva, A., Castro, I., Castillo, C. and León, G., *Synthesis*, **1991**, 481.

[94] 'Amino isoxazoles: preparations and utility in the synthesis of condensed systems', Kislyi, V. P., Danilova, E. B. and Semenov, V. V., *Adv. Heterocycl. Chem.*, **2007**, *94*, 173.

[95] Dorn, H. and Dilcher, H., *Justus Liebigs Ann. Chem.*, **1967**, *707*, 141.

[96] Morgan, G. T. and Reilly, J., *J. Chem. Soc.*, **1914**, 435.

[97] Patel, H. P. and Tedder, J. M., *J. Chem. Soc.*, **1963**, 4589.

[98] 'Synthesis of pyrazoles and condensed pyrazoles', Makino, K., Kim, H. S. and Kurasawa, Y., *J. Heterocycl. Chem.*, **1999**, *36*, 321.

[99] Hübenett, F., Flock, F. H. and Hofmann, H., *Angew. Chem., Int. Ed. Engl.*, **1962**, *1*, 508.

[100] Wille, F., Capeller, L. and Steiner, A., *Angew. Chem., Int. Ed. Engl.*, **1962**, *1*, 335.

[101] Wiley, R. H. and Hexner, P. E., *Org. Synth., Coll Vol. IV*, **1963**, 351.

[102] Martins, M. A. P., Freitag, R., Flores, A. F. C. and Zanatta, N., *Synthesis*, **1995**, 1491.

[103] Hoffmann, M. G., *Tetrahedron*, **1995**, *51*, 9511.

[104] Brederick, H., Sell, R. and Effenberger, F., *Chem. Ber.*, **1964**, *97*, 3407; Domínguez, E., Ibeas, E., Martínez de Marigorta, E., Palacios, J. K. and SanMartin, R., *J. Org. Chem.*, **1996**, *61*, 5435.

[105] McNab, H., *J. Chem. Soc., Perkin Trans. 1*, **1987**, 653.

[106] Knorr, L., *Chem. Ber.*, **1884**, *17*, 546.

[107] Rupe, H. and Grünholz, J., *Helv. Chim. Acta*, **1923**, *6*, 102.

[108] Longemann, W., Almirante, L. and Caprio, L., *Chem. Ber.*, **1954**, *87*, 1175; Tupper, D. E. and Bray, M. R., *Synthesis*, **1997**, 337.

[109] Dorn, H. and Zubek, A., *Org. Synth., Coll. Vol. V*, **1973**, 39.

[110] Dorn, H., Zubek, A. and Hilgetag, G., *Angew. Chem., Int. Ed. Engl.*, **1966**, *5*, 665.

[111] Nightingale, D. and Wadsworth, F., *J. Am. Chem. Soc.*, **1945**, *67*, 416; Bowden, K. and Jones, E. R. H., *J. Chem. Soc.*, **1946**, 953.

[112] Smith, C. D., Tchabanenko, K., Adlington, R. M. and Baldwin, J. E., *TetrahedronLett.*, **2006**, *47*, 3209.

[113] Cuadrado, P., González-Nogal, A. and Valeo, R., *Tetrahedron*, **2002**, *58*, 4975.

[114] Gosselin, F., O'Shea, P. D., Webster, R. A., Reamer, R. A., Tillyer, R. D. and Grabowski, E. J. J., *Synlett*, **2006**, 3267.

[115] Barnes, R. P. and Dodson, L. B., *J. Am. Chem. Soc.*, **1943**, *65*, 1585.

[116] Claisen, L., *Chem. Ber.*, **1926**, *59*, 144.

[117] Bunnelle, W. H., Singam, P. R., Narayanan, B. A., Bradshaw, C. W. and Lion, J. S., *Synthesis*, **1997**, 439.

[118] Alberola, A., González-Ortega, A., Sáadaba, L. L. and Sañudo, M. C., *J. Chem. Soc., Perkin Trans. 1*, **1998**, 4061; Valduga, C. J., Braibante, H. S. and Braibante, M. E. F., *J. Heterocycl. Chem.*, **1998**, *35*, 189.

[119] Meegalla, S. K., Doller, D., Liu, R., Sha, D., Soll, R. M. and Dhanoa, D. S., *Tetrahedron Lett.*, **2002**, *43*, 8639.

[120] 'The Nitrile Oxides', Grundmann, C. and Grünanger, P., Springer-Verlag, Berlin and New York, **1971**; 'Cycloaddition reactions of nitrile oxides with alkenes', Easton, C. J., Merrîcc, C., Hughes, M., Savage, G. P. and Simpson, G. W., *Adv. Heterocycl. Chem.*, **1994**, *60*, 261.

[121] For the use of (Boc)2O/DMAP for this see Basel, Y. and Hassner, A., *Synthesis*, **1997**, 309.

[122] McMurray, J. E., *Org. Synth., Coll. Vol. VI*, **1988**, 592.

[123] Hiraoka, T., Yoshimoto, M. and Kishida, Y., *Chem. Pharm. Bull.*, **1972**, *20*, 122; Barco, A., Benetti, S., Pollini, G. P. and Baraldi, P. G., *Synthesis*, **1977**, 837.

[124] For dehydrogenation of pyrazolines with 'clayfen' see Bougrin, K., Sonfiaoni, M. and El Yazid, M., *Tetrahedron Lett.*, **1995**, *36*, 4065.

[125] Iddon, B., Suschitzky, H., Thompson, A. W., Wakefield, B. J. and Wright, D. J., *J. Chem. Res.*, **1978**, (S) 174; (M) 2038.

[126] Sasaki, T. and Yoshioka, T., *Bull. Chem. Soc. Jpn.*, **1968**, *41*, 2212; Christl, M., Huisgen, R. and Sustmann, R., *Chem. Ber.*, **1973**, *106*, 3275.

[127] Hanson, R. N. and Mohamed, F. A., *J. Heterocycl. Chem.*, **1997**, *34*, 345.

[128] Oh, L. M., *Tetrahedron Lett.*, **2006**, *47*, 7943.

[129] Sakamoto, T., Shiga, F., Uchiyama, D., Kondo, Y. and Yamanaka, H., *Heterocycles*, **1992**, *33*, 813.

[130] Hanamoto, T., Egashira, M., Ishizuka, K., Furuno, H. and Inanaga, J., *Tetrahedon*, **2006**, *62*, 6332.

[131] Aggarwal, V. K., de Vicente, J. and Bonnert, R. V., *J. Org. Chem.*, **2003**, *68*, 5381.

[132] Barber, G. N. and Olofson, R. A., *J. Org. Chem.*, **1978**, *43*, 3015.

[133] He, Y. and Liu, N.-H., *Synthesis*, **1994**, 989.
[134] Nitz, T. J., Volkots, D. L., Aldous, D. J. and Oglesby, R. C., *J. Org. Chem.*, **1994**, *59*, 5828.
[135] Dang, T. T., Albrecht, U. and Lange, P., *Synthesis*, **2006**, 2515.
[136] Short, K. M. and Ziegler, C. B., *Tetrahedron Lett.*, **1993**, *34*, 75.
[137] Almirante, N., Cerri, A., Fedrizzi, G., Marazzi, G. and Santagostino, M., *Tetrahedron Lett.*, **1998**, *39*, 3287; Almirante, N., Benicchio, A., Cerri, A., Fedrizzi, G., Marazzi, G. and Santagostini, *Synlett*, **1999**, 299.

26

Benzanellated Azoles: Reactions and Synthesis

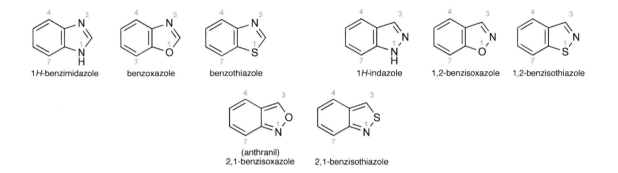

1*H*-benzimidazole   benzoxazole   benzothiazole   1*H*-indazole   1,2-benzisoxazole   1,2-benzisothiazole

(anthranil)
2,1-benzisoxazole   2,1-benzisothiazole

There is only one way in which a benzene ring can be fused to each of the three 1,3-azoles, generating 1*H*-benzimidazole,[1] benzoxazole and benzothiazole. Indazole[2] is the only possibility for the analogous fusion to a pyrazole; it exists as a 1*H* tautomer – the 2*H*-tautomer cannot be detected, though 2-substituted 2*H*-indazoles are known. Two distinct isomers each are possible for the other two benzo-fused 1,2-azoles: 1,2-benzoisothiazole and 2,1-benzoisothiazole,[3] 1,2-benzisoxazole and 2,1-benzisoxazole,[4] respectively.

26.1 Reactions with Electrophilic Reagents
26.1.1 Addition at Nitrogen
26.1.1.1 Protonation
Benzimidazole is somewhat weaker as a base and stronger as an N–H acid, than imidazole. These trends are echoed in the other benzo-azoles: the bicyclic systems are weaker bases than the corresponding monocyclic heterocycles and indazole is a slightly weaker acid than pyrazole.

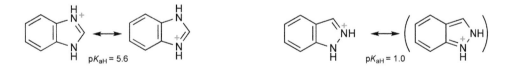

26.1.1.2 Alkylation at Nitrogen
The neutral heterocycles form quaternary salts by reaction at nitrogen with alkyl halides. Normally, indazoles react at the imine nitrogen, N-2.[5]

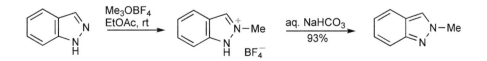

Alkylation of 6-methylindazole with dihydropyran and acid produces the *N*-1-tetrahydropyranyl deriva-tive;[6] as in N-1-acylation of indazoles, this 1-substituted product may be a thermodynamic product resulting from reversible alkylation.

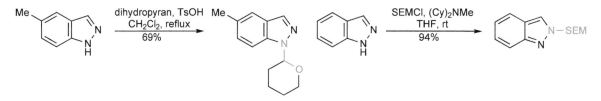

A very useful alternative protection of indazoles is achieved using trimethylsilyl chloride in the presence of a hindered base, producing 2-trimethylsilylindazole: it is the neutral indazole which is silylated by reac-tion at the imine nitrogen, the hindered base being present to deprotonate the initially formed salt.[7]

26.1.1.3 *Acylation at Nitrogen*

Acid anhydrides, with or without pyridine, bring about *N*-acylation of benzimidazoles, *N*-1-acylation of indazoles, and 1,3-diacylation of benzimidazol-2-one.[8] *N*-2-Acylation of indazole can be achieved in the presence of a hindered base.[7]

26.1.2 Substitution at Carbon

The only known *C*-substitutions in the *heterocyclic* rings of any benzo-azole are the 2-bromination of benzimidazole with *N*-bromosuccinimide,[9] the 2-substitution of benzothiazole with bromine at 450 °C[10] and the 3-nitration of indazole.[11] The general rule is that electrophilic nitrations and halogenations can be achieved only in the benzene ring at the 5-, or 6- or 7-positions, for example 5-bromobenzimidazole can be obtained in high yield from the unsubstituted heterocycle.[12] The efficient conversion of indazole into 3-iodoindazole[13] is achieved in the presence of base and probably involves iodination of the indazolyl anion (see 26.3).

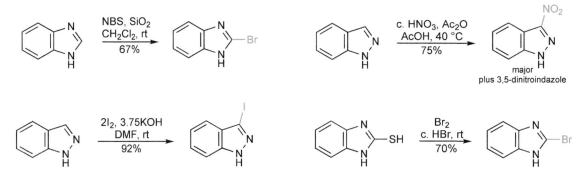

2-Bromobenzimidazole can be obtained from the reaction of benzimidazole-2-thiol with bromine,[14] but in general, hetero-ring halides are prepared from the corresponding oxygen-substituted heterocycle by treatment with thionyl chloride or phosphoryl chloride.[6]

The reluctance to undergo electrophilic *C*-substitution in the five-membered ring can be overcome using silylated derivatives, as illustrated.[15]

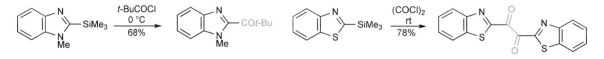

26.2 Reactions with Nucleophilic Reagents

The only position in these heterocycles where nucleophilic displacement of a leaving group is activated is that on the heterocyclic ring; conversion to thiol using thiourea as the nucleophile,[16] and amine displacements on 2-chloro-benzoxazoles[17] show the mildness of the conditions required.

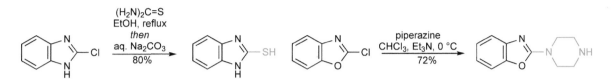

Both oxygen and sulfur substituents can be easily displaced from benzoxazoles[18] or benzothiazoles[19] by organolithiums or amines.

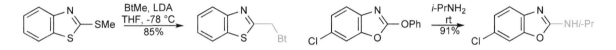

Susceptibility to nucleophilic displacement is central to a method for removing benzothiazol-2-ylsulfonyl protecting groups – reaction with a thiol at the heterocyclic ring 2-position releases the amine (either primary or secondary) as indicated, the sequence also being a reminder of the usual addition/elimination sequence for nucleophilic displacements at imine carbon.

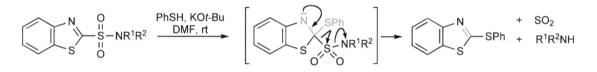

26.3 Reactions with Bases
26.3.1 Deprotonation of *N*-Hydrogen and Reactions of Benzimidazolyl and Indazolyl Anions

The pK_a for benzimidazole is 12.3 and for indazole, the value is 13.9. The benzimidazolyl and indazolyl anions react straightforwardly on nitrogen with electrophiles, though mixtures of N-1- and N-2-substituted products can result in the latter case. For example, amination with hydroxylamine *O*-sulfonic acid gives a 2:1 ratio of 1-amino-1*H*-indazole and 2-amino-2*H*-indazole[20] or to take another example, the ratio of N-1- to N-2-ethylated products from methyl indazol-3-ylcarboxylate can vary from 1:1 to 18:1 depending on the base and the solvent.[21] The *N-arylation* of benzimidazoles and indazoles can be achieved with palladium or copper catalysis (See 4.2.10).

26.4 Ring Metallation and Reactions of *C*-Metallated Derivatives

Benzothiazoles and benzimidazoles (blocked or protected on nitrogen) lithiate at the hetero-ring 2-position. This allows subsequent reaction with the usual range of electrophiles; the examples below show the introduction of iodine[22] and a silicon substituent;[23] 2-lithio-benzimidazoles also react efficiently with simple esters to give ketones.[24]

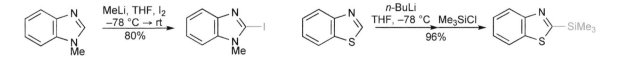

Lithiation of benzoxazole leads to a mixture of the 2-lithiated species and a ring-opened *ortho*-isocyano-phenolate. This difficulty can be avoided by magnesation and good yields of 2-substituted products obtained thereby.[25]

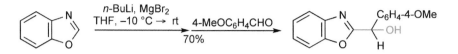

1,2-Benzisothiazole and 1,2-benzisoxazole have not been lithiated in the heterocyclic ring, no doubt because attempts so to do would lead to fragmentation of the heterocyclic ring in a way analogous to that observed on exposure of isoxazoles to lithiating conditions (25.5.1), and when indazole-3-acids are heated in quinoline.[26] However, 3-bromoindazole can be converted into an *N,C*-dithio species – this takes advantage of the fact that following deprotonation of the *N*-hydrogen, N-1 is no longer a leaving group.[27]

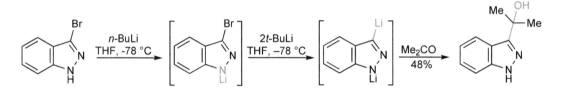

Lithiation of 2-alkyl-2*H*-indazoles proceeds straightforwardly and thus 3-substituted 2*H*-indazoles can be obtained[28] and comparable lithiations of 2-trimethylsilyl-indazoles, followed by removal of the silicon substituent, produce 3-substituted indazoles.

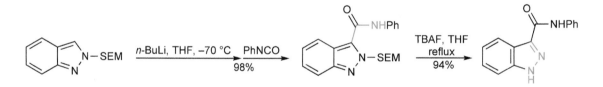

26.5   Reactions with Reducing Agents
The selective reduction of the hetero-ring of benzo-1,3-azoles or the benzo-1,2-azoles has not been reported, though benzazolium salts can be easily reduced to 1,2-dihydro derivatives with sodium borohydride or lithium aluminium hydride.[29]

26.6   Electrocyclic Reactions
2,1-Benzisothiazole and 2,1-benzisoxazole seem not to display the tendency to act as aza-dienes that might have been expected on the basis of comparison with the typical reactivity of isoindoles, benzo[*c*]thiophenes and isobenzofurans (cf. 22.2). In a different sense, electron-deficient 2-alkenyl-benzothiazoles react as 1-aza-1,3-dienes with electron-rich alkenes.[30]

26.7   Quaternary Salts
Benzo-1,3-azolium salts are susceptible to nucleophilic addition at C-2, for example, they are converted into the corresponding *ortho*-substituted benzene, with loss of C-2, by aqueous base, a process that must involve addition of hydroxide at C-2 as an initiating step.[31]

By a sequence involving 2-lithiation of benzothiazole, then reaction with an aldehyde, quaternisation, C-2-*addition* of an alkyllithium, and finally silver-promoted ring cleavage of the resulting dihydro-benzothiazole, the heterocycle can be made the means for the construction of α-hydroxy-ketones.[32] Lithium enolates also add smoothly to benzothiazolium salts.[33]

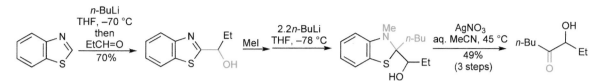

Reissert-type adducts (cf. 9.13) can be obtained from benzothiazole, benzoxazole and indazole, as exemplified.[34]

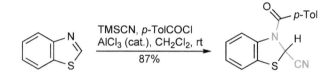

N-Heterocyclic carbenes[35] (cf. 24.10) can be obtained from 1,3-dialkyl-benzimidazole quaternary salts by 2-deprotonation,[36] or from 1,3-dialkyl benzimidazole-2-thione using sodium/potassium alloy.[37]

26.8 Oxy- and Amino-Benzo-1,3-Azoles

The benzo-1,3-azol-2-ones exist in the carbonyl tautomeric forms. Indazol-3-one, however, at least in dimethylsulfoxide solution, is largely in the hydroxy tautomeric form,[38] in contrast to 1,2-benzisothiazol-3-one, which is wholly in the carbonyl form, in the solid.[39]

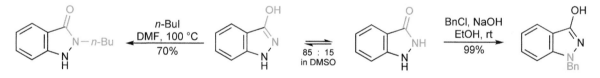

It is possible to be quite selective in the introduction of alkyl groups onto one of the two nitrogens of 3-hydroxyindazole. Reaction with a halide in neutral solution, no doubt involving the imine tautomer, is selective for N-2;[40] the anion reacts at N-1.[41]

Hot aqueous base brings about hydrolytic cleavage of the heterocyclic ring of benzo-1,3-azol-2-ones giving the corresponding *ortho*-disubstituted benzene.[42] Alkylation in basic medium generally leads to *N*- and *O*-substitution; thiones alkylate on the thione sulfur.[43]

26.9 Synthesis
26.9.1 Ring Synthesis of Benzo-1,3-Azoles
26.9.1.1 From ortho-*Heteroatom-Substituted Arenes*
The most important strategy for the synthesis of benzothiazoles, benzimidazoles and benzoxazoles is the insertion of C-2 into a precursor with *ortho*-heteroatoms on a benzene ring. The component that is required for this purpose usually has the future C-2 at the oxidation level of an acid, but many variants on this have been described.

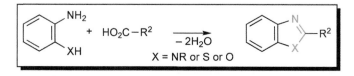

In the standard form of this route, a carboxylic acid is heated with the *ortho*-disubstituted benzene: *ortho*-amino-phenol, -thiophenol or -aniline. An iminoether will react at a much lower temperature[44] and iminochlorides, generated *in situ* using the acid, triphenylphosphine and hot carbon tetrachloride, can be used for benzimidazole or benzoxazole synthesis.[45] Orthoesters with a KSF clay is a highly recommended variant and can be used for the synthesis of all three unsubstituted benzo-1,3-azoles.[46]

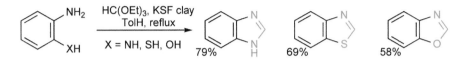

Important variants for the synthesis of benzimidazoles allow the use of aldehydes, rather than acids: ytterbium triflate[47] (best for aliphatic aldehydes) or scandium triflate[48] catalyse the condensation and the subsequent air oxidation of the dihydrobenzimidazole immediate product. Incorporating nitrobenzene in the reaction mixture as the oxidant can also be used.[49] *ortho*-Nitroanilines can be used by incorporating a reducing agent in the reaction mixture – e.g. hydrogen over palladium[50] or sodium dithionite[51] – for *in situ* generation of the *ortho*-phenylenediamine. 2-Aryl-benzothiazoles can be made from 2-aminobenzenethiol, aromatic aldehydes and air, in the presence of activated carbon.[52]

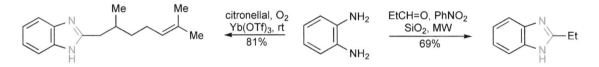

There are efficient ways in which to use starting materials that have the carboxylic acid component already installed on *both* heteroatoms: conversion to bis(silyloxy) derivatives,[53] or simply heating with *p*-toluenesulfonic acid.[54] An excellent route to mono-acylated precursors utilises mixed anhydrides.[31] A very mild method for the dehydrative ring closure of *ortho*-hydroxyarylamino-amides utilises typical Mitsunobu conditions – triphenylphosphine and diethyl azodicarboxylate.[55]

A device that has been frequently used to produce an *ortho*-acylamino-phenol synthon is to carry out a Beckmann rearrangement on an *ortho*-hydroxyaryl ketone, the Beckmann intermediate cyclising *in situ*, as illustrated below.[56]

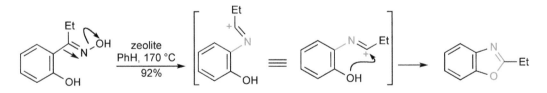

Microwave irradiation can be used to make oxazolo[4,5-*b*]pyridines[57] and allows reaction with amides in lieu of acids; when urea or thiourea are used, 2-ones (2-thiones) are obtained; carbon disulfide with potassium hydroxide also leads to 2-thiones. The use of triphosgene and 2,3-diamino-pyridines can be used to made the analogous imidazo[4,5-*b*]pyridin-2-ones.[58] Reaction with cyanogen bromide gives 2-amino-benzimidazoles.[16]

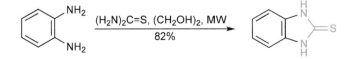

26.9.1.2 Other Methods

Chief amongst alternative methods, and of importance in that it does not require an *ortho*-diheteroatom starting material, is the oxidative ring closure of arylamine-thioamides giving benzothiazoles. Typically,[59] potassium ferricyanide or bromine is utilized; DDQ[60] or Dess–Martin periodinane,[61] both at room temperature, also function to bring about oxidative ring closure. An oxidative ring closure giving benzimidazoles results when *N*-aryl-amidines are reacted with iodobenzene diacetate.[62]

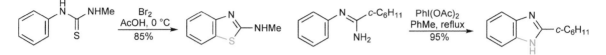

In substrates where a halogen is located *ortho* to nitrogen, ring closure to benzimidazoles can be achieved by palladium(0)-catalysed[63] or copper-catalysed[64] sequences; similar formation of benzothiazoles from *ortho*-haloarylamine-thioamides, does not require a metal catalyst.[65]

26.9.2 Ring Synthesis of Benzo-1,2-Azoles

26.9.2.1 Ring Synthesis of 1H-Indazoles, 1,2-Benzisothiazoles and 1,2-Benzisoxazoles

The earliest syntheses of 1,2-benzisoxazoles depended on the cyclisation of an *ortho*-haloaryl-ketone oxime, typical conditions are shown below and involve an intramolecular nucleophilic substitution of the halide.[66] Only one geometrical isomer of the oxime will ring close. Applying this approach to amidoximes is easier, because the two imine geometrical isomers in such compounds are easily interconvertible.[67] Comparable reaction with hydrazones produces indazoles.[68]

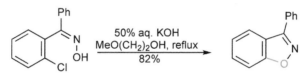

A number of routes involve formation of the bond between the two heteroatoms: typical is the conversion of di(2-cyanophenyl) disulfide into 3-chlorobenzoisothiazole with chlorine[69] and into 3-amino-benzoisothiazoles with magnesium amides, as shown, one 'half' of the starting material is converted directly into the heterocycle, the second 'half' requiring oxidation.[70]

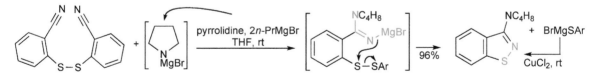

Generally speaking, one of the heteroatoms must carry a leaving group – the hydroxyl of oximes has served this purpose either *via* protonation[71] or acetylation.[72] The generation of an aryl chlorosulfide serves the purpose in the opposite sense.[73]

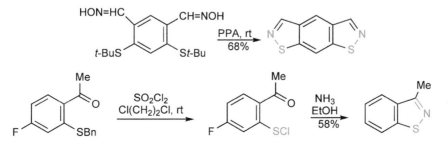

An excellent protocol for the cyclisation of salicylaldehyde oximes (or *ortho*-hydroxyaryl-ketoximes) is the application of Mitsunobu-type conditions;[74] the combination triphenylphosphine/DDQ is also very efficient.[75] Phenyliodine(III) bis(trifluoroacetate) (PIFA) can be used to effect ring closure of *ortho*-amino-arylcarboxamides giving indazolin-3-ones[76] or of *ortho*-thiol-arylcarboxamides, to give benzisothiazol-3-ones.[77]

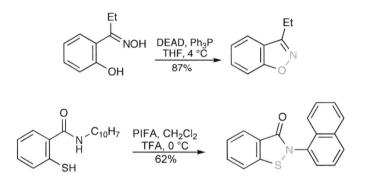

Salicylaldehydes react with hydrazine to give 1*H*-indazoles,[78] and indazoles can also be formed from the reaction of *ortho*-fluoro-benzaldehydes with hydrazine.[79] Rather similar ring closures of *N*-aryl-*N′*-(*ortho*-bromobenzyl)-hydrazines require palladium catalysis.[80]

Electrophilic cyclisation onto the aromatic ring achieves the synthesis of 3-amino-indazoles when aryl-hydrazines are reacted with *N*-(dichloromethylene)-*N,N*-dimethylammonium chloride, as shown.[81]

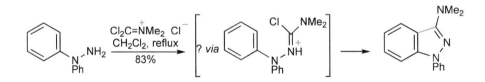

A classical,[82] but still used,[6,83] route to indazoles involves *N*-nitrosation of an acetanilide, followed by cyclisation onto an *ortho*-alkyl group – even unactivated methyl groups enter into reaction, though in the example shown below the *ortho*-substituent carries an activating ester. The sequence may involve a diazo-acetate as intermediate, as shown, with the acetate as leaving group. 3-Methylindazole can be obtained *via* diazotisation of 2-ethylaniline, then ring closure with a mild base.[84]

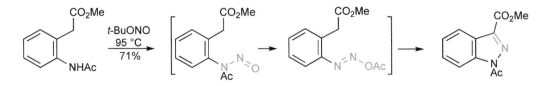

Diazotisation of anthranilic acid and immediate reduction of the diazo group is a very simple route to indazolin-3-one.[2] Hydrolysis of the amide unit of isatin, then diazotisation and reduction produces indazole-3-carboxylic acid;[85] adding the diazonium salt solution to the reducing mixture minimizes the formation of byproduct indazolin-3-ones.[86]

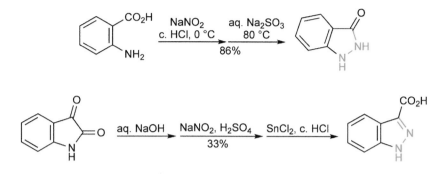

A route to 3-trimethylsilyl-indazoles involves the interaction of lithium trimethylsilyldiazomethane with benzynes, generated *in situ* from a halo-benzene.[87]

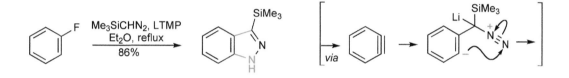

26.9.2.2 Ring Synthesis of 2,1-Benzisothiazoles and 2,1-Benzisoxazoles

Appropriate oxidations of *ortho*-aminoaryl ketones[88] or esters[89] produce 2,1-benzisoxazoles; these ring closures may involve the intermediate formation of a nitrene.

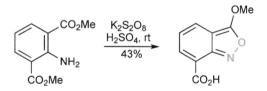

Involvement of a nitrene seems very likely in sequences where an *ortho*-azidoaryl ketone is decomposed thermally producing 3-substituted 2,1-benzisoxazoles.[90] In the opposite sense, reduction of *ortho*-nitro-araldehydes is a very efficient route to 3-unsubstituted 2,1-benzisoxazoles. Both the zinc and the 2-bromo-2-nitropropane are essential components of the reducing mixture. Note the survival of the aromatic halogen in the example shown.[91]

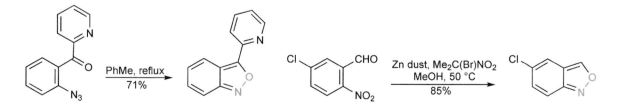

ortho-Amino-toluenes can be converted into 2,1-benzisothiazoles by reaction with thionyl chloride[92] or with *N*-sulfinylmethanesulfonamide.[93]

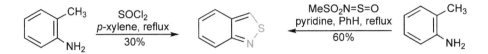

References

1. (a) 'The chemistry of benzimidazoles', Wright, J. B., *Chem. Rev.*, **1951**, *48*, 397; 'Synthesis, reactions and spectroscopic properties of benzimidazoles', Preston, P. N., *ibid.*, **1974**, *74*, 279; (b) '2-Aminobenzimidazoles in organic synthesis', Rastogi, R. and Sharma, S., *Synthesis*, **1983**, 861; (c) 'Imidazole and Benzimidazole Synthesis', Grimmett, M. R., Academic Press, London, **1997**.
2. 'Synthesis, properties and reactions of 1*H*-indazol-3-ols and 1,2-dihydro-3*H*-indazol-3-ones', Baiocchi, L., Corsi, G. and Palazzo, G., *Synthesis*, **1978**, 633.
3. 'Benzisothiazoles', Davis, M., *Adv. Heterocycl. Chem.*, **1972**, *14*, 43; 'Advances in the chemistry of benzisothiazoles and other polycyclic isothiazoles', idem, *ibid.*, **1985**, *38*, 105.
4. 'Benzisoxazoles (indoxazenes and anthranils)', Wünsch, K. H. and Boulton, A. J., *Adv. Heterocycl. Chem.*, **1967**, *8*, 277.
5. Cheung, M., Boloor, A. and Stafford, J. A., *J. Org. Chem.*, **2003**, *68*, 4093.
6. Sun, J.-H., Teleha, C. A., Yan, J.-S., Rogers, J. D. and Nugiel, D. A., *J. Org. Chem.*, **1997**, *62*, 5627.
7. Luo, G., Chen, L. and Dubowchik, G., *J. Org. Chem.*, **2006**, *71*, 5392.
8. e.g. Harrison, D., Ralph, J. T. and Smith, A. C. B., *J. Chem. Soc.*, **1963**, 2930
9. Mistry, A. G., Smith, K. and Bye, M. R., *Tetrahedron Lett.*, **1986**, *27*, 1051.
10. Jansen, H. E. and Wibaut, J. P., *Recl. Trav. Chim.*, **1937**, *56*, 699.
11. Cohen-Fernandes, P. and Habraken, C. L., *J. Org. Chem.*, **1971**, *36*, 3084.
12. Das, B., Venkateswarlu, K., Krishnaiah, M. and Holla, H., *Tetrahedron Lett.*, **2006**, *47*, 8693.
13. Crestey, F., Collot, V., Stiebing, S. and Rault, S., *Synthesis*, **2006**, 3506.
14. Ellingboe, J. W., Spinelli, W., Winkley, M. W., Nguyen, T. T., Parsons, R. W., Moubanak, I. F., Kitzen, J. M., Engen, D. V. and Bagli, J. F., *J. Med. Chem.*, **1992**, *35*, 705.
15. Jutzi, P. and Gilge, U., *J. Heterocycl. Chem.*, **1983**, *20*, 1011.
16. Harrison, D. and Ralph, J. T., *J. Chem. Soc.*, **1965**, 3132.
17. Yamada, M., Sato, Y., Kobayashi, K., Konno, F., Soneda, T. and Watanabe, T., *Chem. Pharm. Bull.*, **1998**, *46*, 445.
18. Kövér, J., Timár, T. and Tompa, J., *Synthesis*, **1994**, 1124.
19. Katritzky, A. R., Ghiviriga, I. and Cundy, D. J., *Heterocycles*, **1994**, *38*, 1041.
20. Adams, D. J. C., Bradbury, S., Horwell, D. C., Keating, M., Rees, C. W. and Storr, R. C., *Chem. Commun.*, **1971**, 828; Kleim, J. T., Davis, L., Olsen, G. E., Wong, G. S., Huger, F. P., Smith, C. P., Petko, W. W., Cornfeldt, M., Wilker, J. C., Blitzer, R. D., Landau, E., Haroutunian, V., Martin, L. L. and Effland, R. C., *J. Med. Chem.*, **1996**, *39*, 570.
21. Harada, H., Morie, T., Hirokawa, Y., Terauchii, H., Fugiwara, I., Yoshida, N. and Kato, S., *Chem. Pharm. Bull.*, **1995**, *43*, 1912.
22. Park, S. B. and Alper, H., *Org. Lett.*, **2003**, *5*, 3209.
23. Chikashita, H. and Itoh, K., *Heterocycles*, **1985**, *23*, 295.
24. Asakawa, K.-i., Dannenberg, J. J., Fitch, K. J., Hall, S. S., Kadowaki, C., Karady, S., Kii, S., Maeda, K., Marcune, B. F., Mase, T., Miller, R. A., Reamer, R. A. and Tschaen, D. M., *Tetrahedron Lett.*, **2005**, *46*, 5081.
25. Bayh, O., Awad, H., Mongin, F., Hoarau, C., Bischoff, L., Trécourt, F., Quéguiner, G., Marsais, F., Blanco, F., Abarca, B. and Ballesteros, R., *J. Org. Chem.*, **2005**, *70*, 5190.
26. Gale, D. J. and Wilshire, J. F. K., *Aust. J. Chem.*, **1973**, *26*, 2683.
27. Welch, W. M., Hanan, C. E. and Whalen, W. M., *Synthesis*, **1992**, 937.
28. Bunnell, A., O'Yang, C., Petrica, A. and Soth, M. J., *Synth. Commun.*, **2006**, *36*, 285.
29. Imidazolium, e.g. Hausner, S. H., Striley, C. A. F., Krause-Bauer, J. A. and Zimmer, H., *J. Org. Chem.*, **2005**, *70*, 5804; indazolium: Schmidt, A., Habeck, T., Snovydovych, B. and Eisfeld, W., *Org. Lett.* **2007**, *9*, 3515.
30. Sakamoto, M., Nagano, M., Suzuki, Y., Satoh, K. and Tamura, O., *Tetrahedron*, **1996**, *52*, 733.
31. Quast, H. and Schmitt, E., *Chem. Ber.*, **1969**, *102*, 568.
32. Chikashita, H., Ishibaba, M., Ori, K. and Itoh, K., *Bull. Chem. Soc. Jpn.*, **1988**, *61*, 3637.
33. Chikashita, H., Takegami, N., Yanase, Y. and Itoh, K., *Bull. Chem. Soc. Jpn.*, **1989**, *62*, 3389.
34. Uff, B. C., Ho, Y.-P., Brown, D. S., Fisher, I., Popp, F. D. and Kant, J., *J. Chem. Res., (S) 346, (M) 2652*.
35. 'Stable heteroaromaric carbenes of the benzimidazole and 1,2,4-triazole series', Korotkikh, N. I., Shvaika, O. P., Rayenko, G. F., Kiselyov, A. V., Knishevitsky, A. V., Cowley, A. H., Jones, J. N. and MacDonald, C. L. B., *Arkivoc*, **2005**, *iii*, 10.
36. Starikova, O. V., Dolgushin, G. V., Larina, L. I., Komarova, T. N. and Lopyrev, V. A., *Arkivoc*, **2003**, *xiii*, 119.
37. Hahn, F. E., Wittenbecher, L., Boese, R. and Bläser, D., *Chem. Eur. J.*, **1999**, 1931.
38. Ballesteros, P., Elguero, J., Claramunt, R. M., Faure, R., de la Concepción Foces-Foces, M., Hernández Caro, F. and Rousseau, A., *J. Chem. Soc., Perkin Trans. 2*, **1986**, 1677.
39. Cavalca, L., Gaetani, A., Mangia, A. and Pelizza, G., *Gazz. Chim. Ital.*, **1970**, *100*, 629.

[40] Arán, V. J., Diez-Barra, E., de la Hoz, A. and Sánchez-Verdú, P., *Heterocycles*, **1997**, *45*, 129.

[41] Gordon, D. W., *Synlett*, **1998**, 1065.

[42] Quast, H. and Schmitt, E., *Chem. Ber.*, **1969**, *102*, 568.

[43] Katritzky, A. R., Aurrecoechea, J. M. and Vazquez de Miguel, L. M., *Heterocycles*, **1987**, *26*, 427.

[44] Costanzo, M. J., Maryanoff, B. E., Hecker, L. R., Schott, M. R., Yabut, S. C., Zhang, H.-C., Andrade-Gordon, P., Kauffman, J. A., Lewis, J. M., Krishnan, R. and Tulinsky, A., *J. Med. Chem.*, **1996**, *39*, 3039.

[45] Ge, F., Wang, Z., Wan, W., Lu, W. and Hao, J., *Tetrahedron Lett.*, **2007**, *48*, 3251.

[46] Villemin, D., Hammadi, M. and Martin, B., *Synth. Commun.*, **1996**, *26*, 2895.

[47] Curini, M., Epifano, F., Montanari, F., Rosati, O. and Taccone, S., *Synlett*, **2004**, 1832.

[48] Itoh, T., Nagata, K., Ishikawa,H. and Ohsawa, A., *Heterocycles*, **2004**, *63*, 2769.

[49] Beu-Alloum, A., Bakkas, S. and Soufiaoui, M., *Tetrahedron Lett.*, **1998**, *39*, 4481.

[50] Hornberger, K. R., Adjabeng, G. M., Dickson, H. D. and Davis-Ward, R. G., Tetrahedron Lett., **2006**, *47*, 5359.

[51] Surpur, M. P., Singh, P. R., Patil, S. B. and Samant, S. D., *Synth. Commun.*, **2007**, *37*, 1375.

[52] Kawashita, Y., Ueba, C. and Hayashi, M., *Tetrahedron Lett.*, **2006**, *47*, 4231.

[53] Rigo, D., Valligny, D., Taisne, S. and Couturier, D., *Synth. Commun.*, **1988**, *18*, 167.

[54] De Luca, M. R. and Kerwin, S. M., *Tetrahedron*, **1997**, *53*, 457.

[55] Wang, T. and Hanske, J. R., *Tetrahedron Lett.*, **1997**, *38*, 6529.

[56] Bhawal, B. M., Mayabhate, S. P., Likhite, A. P. and Deshmukh, A. R. A. S., *Synth. Commun.*, **1995**, *25*, 3315.

[57] Myllymäki, M. and Koskinen, A. M. P., *Tetrahedron Lett.*, **2007**, *48*, 2295.

[58] Kuethe, J. T., Wong, A. and Davies, I. W., *J. Org. Chem.*, **2004**, *69*, 7752.

[59] Roe, A. and Tucker, W. P., *J. Heterocycl. Chem.*, **1965**, *2*, 148; Dreikorn, B. A. and Unger, P., *J. Heterocycl. Chem.*, **1989**, *26*, 1735; Ambati, N. B., Anand, V. and Hanumanthu, P., *Synth. Commun.*, **1997**, *27*, 1487; Jordan, A. D., Luo, C. and Reitz, A. B., *J. Org. Chem.*, **2003**, *68*, 8693.

[60] Bose, D. S., Idrees, M. and Srikanth, B., *Synthesis*, **2007**, 819.

[61] Bose, D. S. and Idrees, M., *J. Org. Chem.*, **2006**, *71*, 8261.

[62] Ramsden, C. A. and Rose, H. L., *J. Chem. Soc., Perkin Trans. 1*, **1995**, 615.

[63] Brain, C. T. and Steer, J. T., *J. Org. Chem.*, **2003**, *68*, 6814.

[64] Zheng, N. and Buchwald, S. L., *Org. Lett.*, **2007**, *9*, 4749.

[65] Bernardi, D., Ba, L. A. and Kirsch, G., *Synlett*, **2007**, 2121.

[66] King, J. F. and Durst, T., *Can. J. Chem.*, **1962**, *40*, 882.

[67] Fink, D. M. and Kurys, B. E., *Tetrahedron Lett.*, **1996**, *37*, 995.

[68] Walser, A., Flynn, T. and Mason, C., *J. Heterocycl. Chem.*, **1991**, *28*, 1121.

[69] Beck, J. R. and Yahner, J. A., *J. Org. Chem.*, **1978**, *43*, 1604.

[70] Nakamura, T., Nagata, H., Muto, M. and Saji, I., *Synthesis*, **1997**, 871.

[71] Meth-Cohn, O. and Tarnowski, B., *Synthesis*,. **1978**, 58.

[72] Saunders, J. C. and Williamson, W. R. N., *J. Med. Chem.*, **1979**, *22*, 1554; McKinnon, D. M. and Lee, K. R., *Can. J. Chem.*, **1988**, *66*, 1405.

[73] Fink, D. M. and Strupczewski, J. T., *Tetrahedron Lett.*, **1993**, *34*, 6525.

[74] Poissonnet, G., *Synth. Commun.*, **1997**, *27*, 3839.

[75] Iranpoor, N., Firouzabadi, H. and Nowrouzzi, N., *Tetrahedron Lett.*, **2006**, *47*, 8247.

[76] Correa, A., Tellitu, I., Domínguez, E. and SanMartin, R., *J. Org. Chem.*, **2006**, *71*, 3501.

[77] Correa, A., Tellitu, I., Domínguez, E. and San Martin, R., *Org. Lett.*, **2006**, *8*, 4811.

[78] Lokhande, P. D., Raheem, A., Sabale, S. T., Chabukswar, A. R. and Jagdale, S. C., *Tetrahedron Lett.*, **2007**, *48*, 6890.

[79] Lukin, K., Hsu, M. C., Fernando, D. and Leanna, M. R., *J. Org. Chem.*, **2006**, *71*, 8166.

[80] Song, J. J. and Yee, N. K., *Tetrahedron Lett.*, **2001**, *42*, 2937.

[81] Hervens, F. and Viehe, H. G., *Angew. Chem., Int. Ed. Engl.*, **1973**, *12*, 405.

[82] Huisgen, R. and Bast, K., *Org. Synth, Coll. Vol. V*, **1973**, 650; Rüchardt, C. and Hassman, V., *Justus Liebigs Ann. Chem.*, **1980**, 908.

[83] Yoshida, T., Matsuura, N., Yamamoto, K., Doi, M., Shimada, K., Morie, T. and Kato, S., *Heterocycles*, **1996**, *43*, 2701.

[84] Crestey, F., Collot, V., Stiebing, S. and Rault, S., *Tetrahedron*, **2006**, *62*, 7772.

[85] Snyder, H. R., Thompson, C. B. and Hinman, R. L., *J. Am. Chem. Soc.*, **1952**, *74*, 2009; Norman, M. H., Navas, F., Thompson, J. B. and Rigdon, G. C., *J. Med. Chem.*, **1996**, *39*, 4692.

[86] Johnson, B. L. and Rodgers, J. D., *Synth. Commun.*, **2005**, *35*, 2681.

[87] Shoji, Y., Hari, Y. and Aoyama, T., *Tetrahedon Lett.*, **2004**, *45*, 1769.

[88] Prakarh, O., Saini, R. K., Singh, S. P. and Varma, R. S., *Tetrahedron Lett.*, **1997**, *38*, 3147.

[89] Chauhan, M. S. and McKinnon, D. M., *Can. J. Chem.*, **1975**, *53*, 1336.

[90] Ning, R. Y., Chen, W. Y. and Sternbach, L. H., *J. Heterocycl. Chem.*, **1974**, *11*, 125; Smalley, R. K., Smith, R. H. and Suschitzky, H., *Tetrahedron Lett.*, **1978**, 2309.

[91] Kim, B. H., Jun, Y. M., Kim, T. K., Lee, Y. S., Baik, W. and Lee, B. M., *Heterocycles*, **1997**, *45*, 235.

[92] Davis, M. and White, A. W., *J. Org. Chem.*, **1969**, *34*, 2985.

[93] Singerman, G. M., *J. Heterocycl. Chem.*, **1975**, *12*, 877.

27

Purines: Reactions and Synthesis

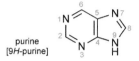

purine
[9H-purine]

Purines are of great interest for several reasons, but in particular, together with certain pyrimidine bases, they are constituents of DNA and RNA, and consequently of fundamental importance in life processes (see 32.4). Additionally, as nucleosides and nucleotides they act as hormones and neurotransmitters and are present in some co-enzymes. The interconversion of mono-, di- and triphosphate esters of nucleosides is at the heart of energy transfer in many metabolic systems and is also involved in intracellular signalling. This central biological importance, together with medicinal chemists' search for anti-tumour and anti-viral (particularly anti-AIDS) agents, have resulted in a rapid expansion of purine chemistry (33.6.3).

There are significant lessons to be learnt from the chemistry of purines since their reactions exemplify the interplay of its constituent imidazole and pyrimidine rings, just as the properties of indole show modified pyrrole and modified benzene chemistry. Thus purines can undergo both electrophilic and nucleophilic attack at carbon in the five-membered ring and mainly nucleophilic reactions at carbon in the six-membered ring.

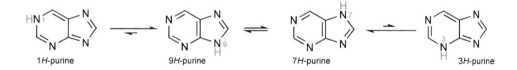

1H-purine   9H-purine   7H-purine   3H-purine

The numbering of the purine ring system is anomolous and reads as if purine were a pyrimidine derivative. There are, in principle, four possible tautomers of purine containing an N-hydrogen; in the crystalline state, purine exists as the 7H-tautomer, however in solution both 7H- and 9H-tautomers are present in approximately equal proportions; the 1H- and 3H-tautomers are not significant.[1]

The majority of reported purine chemistry pertains to the naturally occurring purines, which are amino and/or oxygenated substances and, as a consequence, reactions of the simpler purines, such as in other chapters are given as typical, have received limited attention. A nomenclature related to the natural purines has evolved and is in general usage. A 'nucleoside' is an N-sugar derivative of a purine (generally 9-(riboside) or 9-(2'-deoxyriboside)), for example adenosine is the 9-(riboside) of adenine, itself the generally used trivial name for 6-aminopurine. A nucleotide is a 5'-phosphate (or di- or tri-phosphate) of a nucleoside – adenosine 5'-triphosphate (ATP) is an example. Note that there are comparable pyrimidine nucleosides and nucleotides (32.4). (**NOTE**: In some diagrams, to save space, for β-D-2-deoxyribofuranosyl, β-D-ribofuranosyl and a generalized sugar, we use ⓓⓡ Ⓡ Ⓢ, respectively.)

Heterocyclic Chemistry 5th Edition   John Joule and Keith Mills
© 2010 Blackwell Publishing Ltd

27.1 Reactions with Electrophilic Reagents
27.1.1 Addition at Nitrogen
27.1.1.1 Protonation

Purine is a weak base, pK_{aH} 2.5. ^{13}C NMR studies suggest that all three protonated forms are present in solution, but the predominant cation is formed by N-1-protonation.[2] In strong acid solution a dication is formed by protonation at N-1 and on the five-membered ring.[3]

The presence of oxygen functionality does not seem to affect purine basicity to any great extent, thus hypoxanthine has a pK_{aH} of 2.0. Amino groups increase the basicity, as illustrated by the pK_{aH} of adenine, 4.2, and oxo groups reduce the basicity of amino-purines, thus guanine has a pK_{aH} of 3.3; the position of protonation of guanine in the solid state has been established by X-ray analysis – it is on the five-membered ring, which nicely illustrates the extremely subtle interplay of substituents and ring heteroatoms, for although the 2-amino group increases the basicity of the purine to which it is attached, this does not necessarily mean that it is the associated N-3 that is protonated.

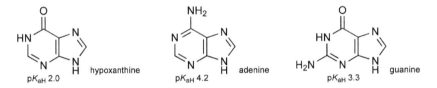

Purine itself slowly decomposes in aqueous acid, to the extent of about 10% in 1N sulfuric acid at 100 °C. The stability of oxy-purines to aqueous acid varies greatly, for example xanthine is stable to aqueous 1N sulfuric acid at 100 °C, whereas purin-2-one is converted into 5,6-diaminopyrimidin-2-one under the same conditions.

27.1.1.2 Alkylation at Nitrogen

As would be expected from systems containing four nitrogen atoms, *N*-alkylation of purines is complex and can take place on the neutral molecule or *via* an *N*-anion. Purine reacts with iodomethane to give a 7,9-dimethylpurinium salt.[4]

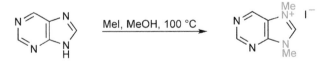

Adenine gives mainly 3-alkylated products under neutral conditions, but 7/9-substitution when there is a base present. Adenosine derivatives on the other hand usually give 1-alkylated products, presumably due to hindrance to N-3-attack by the *peri*-9-ribose substituent. That attack can still occur at C-3 is shown by the intramolecular quaternisation of N-3, which is an important side reaction when 5′-halides are subjected to displacement conditions.

An effective method for effectively alkylating the 6-amino group of adenosine is to bring about a Dimroth rearrangement of a 1-alkyl-adenosinium salt, as illustrated below (see also 31.2.2).[5,6]

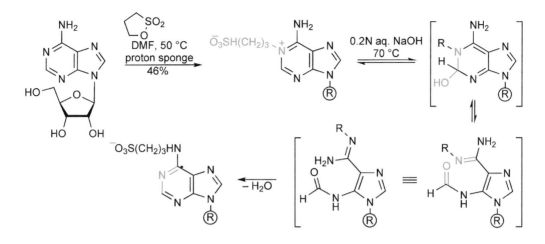

Alkylation of oxygenated purines in alkaline media, for example hypoxanthine, tends to occur both at amidic nitrogen and also at a five-membered ring nitrogen, making selectivity a problem. Under neutral conditions, xanthines give 7,9-dialkylated quaternary salts. The alkylation of 6-chloropurine illustrates the complexity: in basic solution both 7- and 9-substitution occurs,[7] whereas reaction with dihydrofuran under acidic conditions is selective for N-9, perhaps as a thermodynamic product.[8]

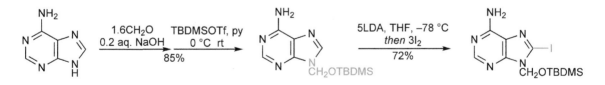

An *N-t*-butyldimethylsilyloxymethyl is a useful protecting group for adenines as it confers good solubility in organic solvents. It is introduced by stepwise conversion into the 9-hydroxymethyl derivative, by reaction with formaldehyde and base, followed by *O*-silylation.[9]

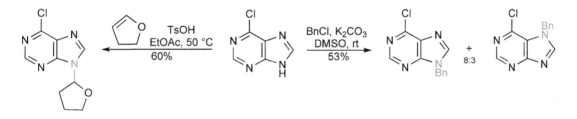

The ratio of N-9 to N-7 alkylation is also influenced by the size of a 6-substituent, larger groups at C-6 lead to increased percentages of 9- *versus* 7-alkylation.[10]

In suitable cases, where N-7/N-9 selectivity is poor, alkylation can be directed to N-9 by a bulky protecting group installed on a C-6 substituent.[11] Alternatively, a 6-amino can be converted into an azole leaving group, such as a 1,2,4-triazol-4-yl (29.1.1.2), which is a very effective hindering group for N-7. Imidazoles can also be used similarly[12] – unsubstituted imidazole or a 2-butylimidazole are very effective, whereas the apparently much bulkier 2,5-diphenylimidazole is much less so. This is explained by the simple imidazole (or triazole) being coplanar with the purine ring (and therefore very close to N-7), but a diphenylimidazole substituent is rotated away from the nitrogen.[13]

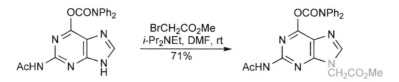

The N-9:N-7 ratio of products varies with time when alkylations employ a Michael acceptor like methyl acrylate, for here the alkylation is reversible and the concentration of thermodynamic product can build up.[14] Regiospecific 7-alkylations can be achieved *via* the quaternisation of a 9-riboside followed by hydrolytic removal of the sugar residue, as illustrated by a reaction with ethylene oxide.[15] Alkylation on N-7 in nucleic acids is the mechanism of mutagenesis/carcinogenesis by some natural toxins, such as aflatoxin.[16]

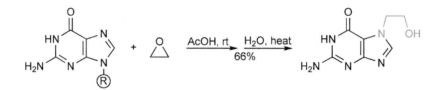

In the ribosylation of purines, in addition to the question of regioselectivity on the purine, there is the possibility of forming epimeric products at the linking C-1' of the ribose, and this is often the more difficult to control. A great deal of work has been done and many different conditions shown to be effective in specific cases, but conditions that are generally effective have not been defined.[17] These alkylations usually employ acylated or halo-ribosides in conjunction with a mercury,[18] silicon[18] or sodium[19] derivative of the purine, and stereoselective displacements of halide can sometimes be achieved.

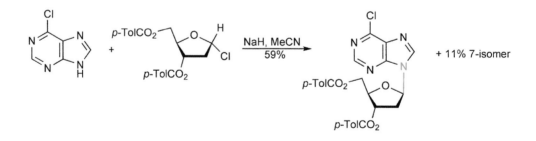

Other methods of controlling stereochemistry include the use of the size of an isopropylidene protecting group to shield one face of the sugar[20] or, as shown below, anchimeric assistance from a 2'-benzoate.[21]

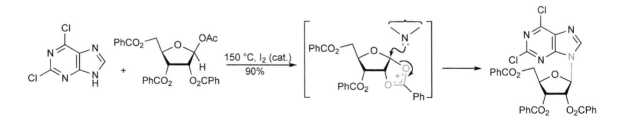

When base-sensitive alkylating agents are used, such as cycloalkyl derivatives for the preparation of nucleoside analogues, the use of equilibrium bases, such as potassium carbonate, often give poor results. Here, the outcome is improved by using pre-formed (neutral) salts, such as tetra-*n*-butylammonium,[22] DBU or phosphazenes.[23]

27.1.1.3 Acylation at Nitrogen

Purines react with acylating agents such as chloroformates or diethyl pyrocarbonate[24] to give non-isolable *N*⁺-acyl salts, which can suffer various fates following nucleophilic addition; products of cleavage of either ring have been observed, as have recyclisation products.[25]

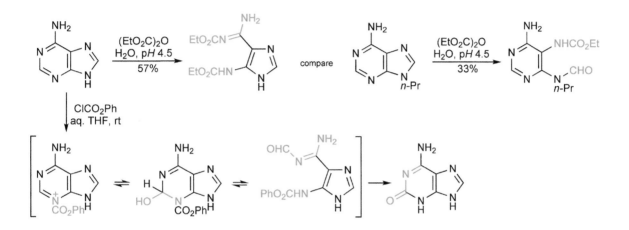

27.1.1.4 Oxidation at Nitrogen

Peracid *N*-oxidation of purines gives 1- and/or 3-oxides, depending on exact conditions.[26] Adenine and adenosine give 1-oxides, whereas guanine affords the 3-oxide.[27] The 3-oxide of purine itself has been obtained *via* oxidation of 6-cyanopurine (at N-3), then hydrolysis and decarboxylation,[27] the relatively easy loss of carbon dioxide echoing the analogous process discussed for pyridine α-acids (8.11). The *N*-7-oxide of adenine can be prepared by oxidation of 3-benzyl-3*H*-adenine, followed by deprotection.[28]

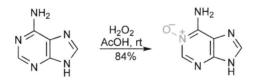

27.1.2 Substitution at Carbon

Typical electrophilic aromatic substitution reactions have not been reported for purine or simple alkyl derivatives, but oxy and amino derivatives generally react readily at C-8.

27.1.2.1 Halogenation

Purine itself simply forms an N⁺–halogen complex, but does not undergo *C*-substitution, however adenosine,[29] hypoxanthine and xanthine derivatives[30] undergo fluorination,[31] chlorination and bromination at C-8. However, there is the possibility that these substitution products arise *via* *N*-halo-purinium salts, *nucleophilic* addition of bromide anion to these at C-8, then elimination of hydrogen halide.

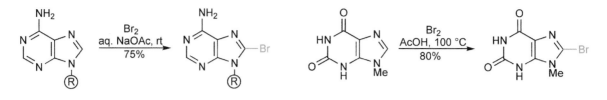

27.1.2.2 Nitration

Xanthines undergo 8-nitration, though under fairly vigorous conditions.[32]

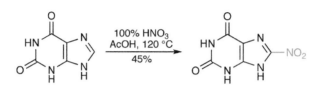

Nitration of 6-substituted purines at C-2, using a mixture of tetra-*n*-butylammonium nitrate and trifluoroacetic anhydride, is an exceptionally useful functionalisation of the purine ring system. The reaction works for both electron-rich (adenosine), 6-alkoxypurines and electron-poor (6-chloropurine) substrates, but full protection of all OH and NH groups is required. This is not a simple electrophilic substitution – the mechanism has been shown, using 6-chloro-9-Boc purine, to involve sequential nitration of N-7, addition of trifluoroacetoxy at C-8 and then migration of the nitro group to C-2. The final, re-aromatisation, step involves elimination of trifluoroacetic acid.[33] Displacement of a 2-nitro group, thus introduced, by fluoride as nucleophile (see 27.5 for nucleophilic substitutions) can be made the means to synthesise 2-fluoroadenosine.[34]

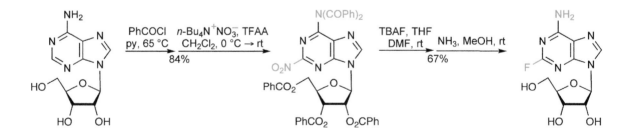

27.1.2.3 Coupling with Diazonium Salts

Amino- and oxy-purines couple at their 8-position; a weakly alkaline medium is necessary, so it seems likely that the reactive entity is an *N*-anion.[35]

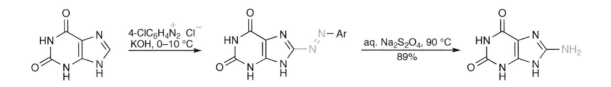

27.2 Reactions with Radicals

Purines react readily with hydroxyl, alkyl, aryl and acyl radicals, usually at C-6,[36] or at C-8 (or C-2) if the 6-position is blocked. Both reactivity and selectivity for C-8 are increased when the substitution is conducted at lower pH.[37] In nucleosides, a radical generated at C-5′ cyclises rapidly onto C-8, but can be trapped before cyclisation by using a large excess of acrylonitrile.[38]

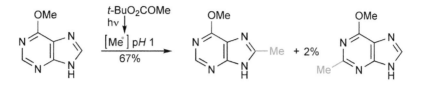

27.3 Reactions with Oxidising Agents

There are few significant oxidations of purines apart from *N*-oxidations (27.1.1.4), but dimethyldioxirane gives good yields of 8-oxo compounds, possibly *via* the intermediacy of a 9,8- or 7,8-oxaziridine.[39] C-8-Oxidation[40] is an important process *in vivo*, for example with the oxomolybdoenzyme xanthine oxidase, where oxygen is introduced at C-8 *via* a mechanism about which there is still debate.

27.4 Reactions with Reducing Agents

The reduction of substituted purines is very complex and ring-opened products are often obtained. 1,6-Dihydropurine is formed by catalytic or electrochemical[41] reduction of purine, but this is unstable. Stable reduced compounds can be obtained by reduction in the presence of *N*-acylating agents.[42] 7/9-Quaternary salts are easily reduced by borohydride, in the five-membered ring, producing 7,8-dihydro derivatives.[43]

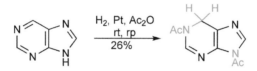

27.5 Reactions with Nucleophilic Reagents

Nucleophilic displacement of leaving groups from C-2, C-6 and C-8 is the most common means of preparation of substituted purines. Halides are the most popular leaving groups, particularly chlorides, but fluorides are generally the most reactive, although slightly less accessible. Bromides and iodides can be used similarly, but offer little advantage for simple nucleophilic substitutions. Sulfonates, sulfoxides, sulfones, quaternary ammonium, diphenylphosphonyloxy[44] and nitro[33] are also highly reactive leaving groups.

Although fluoro is intrinsically more reactive than chloro, this does not overcome the natural bias of the purine ring, so 6-chloro-2-fluoro-purines react selectively at C-6 with a number of nucleophiles.[45]

Relatively easy nucleophilic displacement, *via* the usual addition/elimination sequence, takes place at all three positions with a wide range of nucleophiles, such as alkoxides,[46] sulfides, amines, azide, cyanide and malonate and related carbanions.[47]

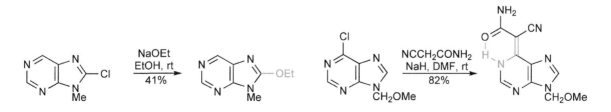

In 9-substituted purines, the relative reactivity of halides is 8 > 6 > 2, but strongly influenced by the presence of other substituents. In 9*H*-purines this is modified to 6 > 8 > 2, the demotion of the 8-position being associated with anion formation in the five-membered ring. Conversely, in acidic media the reactivity to nucleophilic displacement at C-8 is enhanced: protonation of the five-membered ring facilitates the nucleophilic addition step.[47] The relative reactivities of the 2- and 6-positions are nicely illustrated by the conditions required for the reaction of the respective chlorides with hydrazine, a relatively good nucleophile.[48] It is worth noting the parallelism between the relative positional reactivity here with that in halopyrimidines where it is 4 > 2.

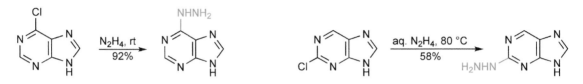

In 2,6-dichloropurine, reactivity at C-6 is enhanced relative to 6-chloropurine by the inductive effect of the second halogen, whereas the presence of electron-releasing substituents, such as amino, somewhat deactivates halogen to displacement, but, conversely, oxygenated purines, probably because of their carbonyl tautomeric structures, react easily.[49]

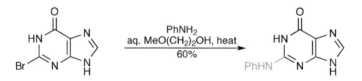

The generation of an *N*-anion by deprotonation in the five-membered ring is given as the reason why 8-chloropurine reacts with sodamide to give adenine: inhibition of attack at C-8 allows the alternative addition to C-6 to lead eventually to the observed major product.[50]

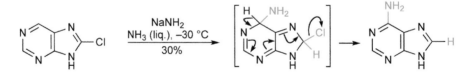

Direct conversion of inosines into 6-amino derivatives, without the intermediacy of a halo-purine, can be achieved by heating with a mixture of phosphorus pentoxide and the amine hydrochloride[51] or using the amine with *p*-toluenesulfonic acid and a silylating agent (HMDS),[52] or the amine with iodine and triphenylphosphine.[53]

Even where a nucleophilic displacement of halide is feasible, the use of transition-metal catalysis, such as with copper (for iodides) or palladium, generally offers much milder conditions (4.2.10).[54]

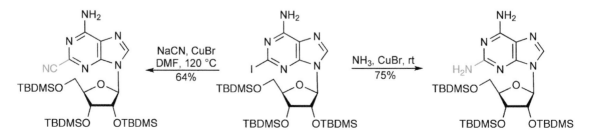

Other useful leaving groups in purine chemistry include sulfoxide,[55] triflate,[56] and aryl- or alkylthio.[57] Sulfones are highly reactive in some nucleophilic substitutions, and are also the reactive intermediates in sulfinate-catalysed displacements of halide.[58]

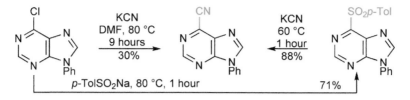

Displacement of halides can be catalysed by amines – trimethylamine, pyridine[59] and DABCO[60] have been used. Mechanistically, the catalysis involves formation of an intermediate quaternary ammonium salt that is more reactive towards nucleophiles than the starting halide. The intermediate quaternary salts can be isolated, if required. Trimethylamine gives the most reactive quaternary salt, but DABCO can be more convenient. The relative reactivities for nucleophilic displacement at C-6 are: trimethylamine:DABCO: chlorine = 100:10:1.[61] Cyano[62] and fluorine[63] are amongst the groups that have been introduced in this way.

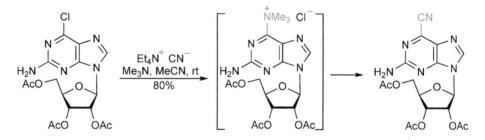

Arylamines can be particularly unreactive as nucleophiles and for these, the use of fluorine[64] or a sulfone[53] as a leaving group, or palladium-assisted displacement (4.2.10) of bromine[65] may be necessary. A 2-chlorine, deactivated by the presence of a 6-amino substituent, can be efficiently displaced by arylamines with trimethylsilyl chloride in butanol.[66] The displacements of fluorine, chlorine and butyl sulfone by anilines are greatly accelerated by carrying out the reactions in 2,2,2-trifluoroethanol, with the addition of excess trifluoroacetic acid.[67]

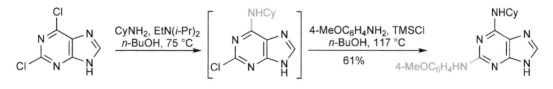

Amino groups can be converted into good leaving groups by incorporation into a 1,2,4-triazole.[12,68] Imidazoles can be used similarly.[53]

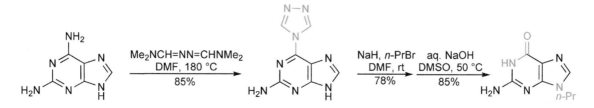

27.6 Reactions with Bases

27.6.1 Deprotonation of *N*-Hydrogen and Reactions of Purinyl Anions

Purine, with a pK_a of 8.9, is slightly more acidic than phenol and much more acidic than imidazole or benzimidazole (pK_as 14.2 and 12.3, respectively). This relatively high acidity is probably a consequence of extensive delocalisation of the negative charge over four nitrogens, however alkylation of the anion takes place in the five-membered ring, since attack at N-1 or N-3 would generate less aromatic products.

Oxy-purines are even more acidic, due to more extensive delocalisation involving the carbonyl groups: xanthine has a pK_a of 7.5 and uric acid, 5.75.

27.7 *C*-Metallation and Reactions of *C*-Metallated Purines

27.7.1 Direct Ring C–H Metallation

The rapid deuteration of purine at C-8[69] in neutral water at 100 °C probably involves 8-deprotonation of a concentration of purinium cation to give a transient ylide (cf. 1,3-azole 2-H-exchange, 24.1.2.1). 9-Alkylated purines undergo a quite rapid exchange in basic solution involving direct deprotonation of the free heterocycle.

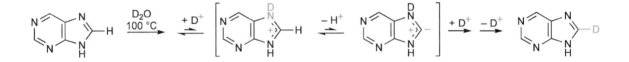

Preparative lithiation of purines requires the protection of the 7/9-position; lithiation then takes place at C-8.[70] 9-Blocked purines can be deprotonated at C-8 with strong bases, such as LDA, even in the presence of *N*-hydrogen in the other ring.[71] Good yields of 8-halo-purines can be obtained by reaction with a variety of halogen donors; 8-lithiation of *O*-silyl-protected 9-ribofuranosyl-purines can be achieved using about three mole equivalents of lithium diisopropylamide.[72]

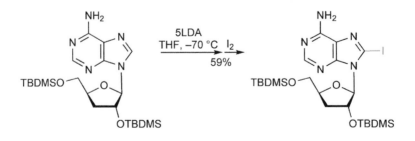

After selective lithiation at C-8 in a 6-chloro-purine riboside, quenching with a stannyl or silyl chloride leads to the isolation of the 2-substituted compound, *via* rearrangement of a 2-anion formed by a second lithiation of the initial 8-substituted product, as illustrated below.[73]

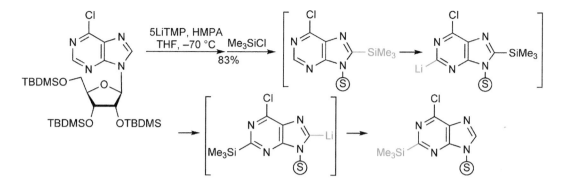

27.7.2 Metal–Halogen Exchange

Purines lithiated at C-2 or C-6 can be generated by way of halogen exchange with alkyllithiums, but it is important to maintain a very low temperature in order to avoid subsequent equilibration to the more stable 8-lithiated species.[74] 2-Chloro-6-iodo-9-alkyl- and 6-chloro-2-iodo-9-alkyl-purines each undergo exchange selectively at the iodine position with *iso*-propylmagnesium chloride at −80 °C.[75]

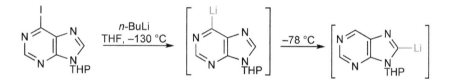

Purine aminocuprates, formed either *via* direct magnesiation at C-8 with TMPMgCl.LiCl, or *via* six-membered-ring lithio derivatives (formed from the iodides then transmetallated to the magnesio derivatives), on oxidation undergo (reductive) elimination of copper to give amino-purines; comparable processes also apply to pyrimidines.[76]

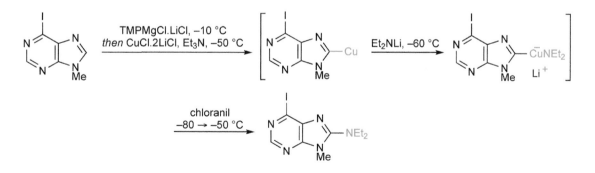

27.8 Oxy- and Amino-Purines

These are tautomeric compounds that exist predominantly as carbonyl and amino structures, thus falling in line with the analogous pyrimidines and imidazoles.

27.8.1 Oxy-Purines

27.8.1.1 Alkylation

The amide-like *N*-hydrogen in oxy-purines is relatively acidic; the acidity is readily understood in terms of the phenolate-like resonance contributor to the anion. Alkylation of such anions takes place at nitrogen, not oxygen.[77]

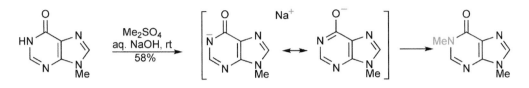

27.8.1.2 Acylation

In contrast to alkylation, acylation and sulfonylation frequently occur at oxygen; the resulting *O*-acylated products are relatively unstable, but can be utilised, for example, conducting the acylation in pyridine, as solvent, produces a pyridinium salt, resulting from displacement of acyloxy by pyridine. Both *O*-acylated-purines, and the corresponding pyridinium salts, can in turn be reacted with a range of nucleophiles[78] to allow the overall replacement of the amide-like oxygen; this is an important alternative to activation of the carbonyl by conversion into halogen (below).

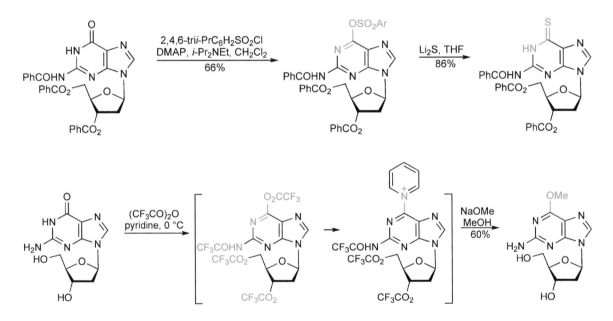

A closely related conversion utilises a silylating agent, in the presence of the desired nucleophile, and presumably involves *O*-silylation, then displacement of silyloxy.[79]

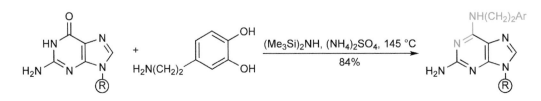

27.8.1.3 *Replacement by Chlorine*[80]

This is a very important reaction in purine chemistry and has been widely utilised to allow subsequent introduction of nucleophiles (27.5), including replacement with hydrogen by chemical (HI) or catalytic hydrogenolysis. Most commonly, phosphoryl chloride is used, neat, or in solution (especially when there is a ribose present); thionyl chloride is an alternative reagent. 2-Deoxy compounds are more sensitive to acid and with these, milder reagents (carbon tetrachloride with triphenylphosphine) must be used to convert oxo into chloro.[81]

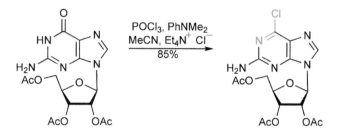

Syntheses of adenine and guanine from uric acid illustrate well the selective transformations to which the halo-purines, prepared from a precursor oxy-purine,[82] can be put.

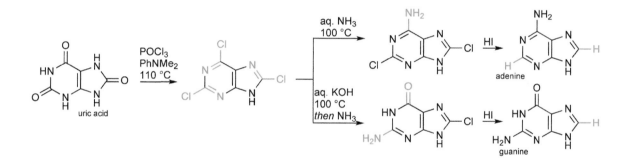

27.8.1.4 *Replacement by Sulfur*

Replacement by sulfur[83] can be achieved *via* a halo-purine, or directly using a phosphorus sulfide.

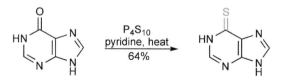

27.8.2 Amino-Purines

27.8.2.1 *Alkylation*

Alkylation under neutral conditions involves attack at a nuclear nitrogen; Dimroth rearrangement (27.1.1.2) of these salts affords side-chain-alkylated purines. However, direct introduction of substituents onto a side-chain nitrogen can be achieved by reductive *N*-alkylation.[84] A related method involves reduction of an isolated benzotriazolyl intermediate, which allows more control over the process.[85]

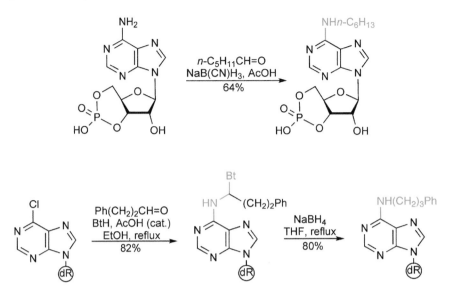

27.8.2.2 Acylation

Amino-purines behave just like anilines with anhydrides and acid chlorides, giving amides, but these are somewhat more easily hydrolysed. Both mono- and diacylation can be utilised as a protecting-group strategy.

27.8.2.3 Diazotisation

The reaction of 2- and 6-amino groups with nitrous acid is similar to that of 2-amino-pyridines, in that diazonium salts are produced but, relative to phenyldiazonium salts, these are unstable. Despite this, they can be utilised for the introduction of groups such as halide[86] or, oxygen by reaction with water, with loss of nitrogen. 8-Diazonium salts are considerably more stable.[87] In 2-aminoadenosine, selective diazotisation occurs at the 2-amino group.[52]

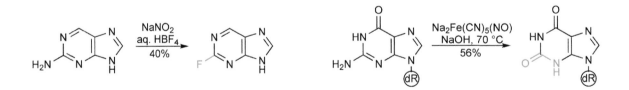

Diazotisation can also be carried out in basic solution and in this way acid-sensitive deoxyribosides can be tolerated.[88] A nucleophilic displacement of amino by hydroxy can be effected enzymatically using adenosine deaminase; this is a useful practical method because it is a very selective transformation under mild conditions.[89] Chemical hydrolysis requires more vigorous conditions.

The related reaction with alkyl nitrites generates purinyl radicals, which will efficiently abstract halogen from halogenated solvents, and this procedure is generally to be preferred for the transformation of an amino-purine into a halo-purine.[90] These reactions can be improved/enhanced by the addition of antimony trichloride.[91] Comparably, the use of dimethyl disulfide produces methylthio-purines,[92] and trimethylsilyl azide, azido-purines.[93]

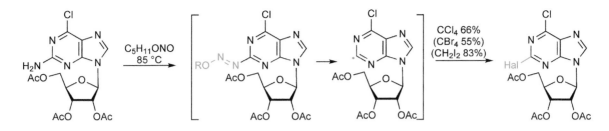

27.8.3 Thio-Purines

Thio-purines are prepared from halo-purines or oxy-purines or by ring synthesis. One method for displacement of bromine with sulfur utilises TMS(CH$_2$)$_2$SH as a nucleophile, TBAF removal of the sulfur substituent then reveals the thione.[94] In contrast with oxy-purines, in alkaline solution they readily alkylate on sulfur, rather than nitrogen.[95]

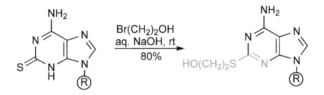

Thiols are also useful sources of the corresponding bromo compounds, by reaction with bromine and hydrobromic acid.[96] Alkylthio substituents can be displaced by the usual range of nucleophiles, but the corresponding sulfones are more reactive.[59,97]

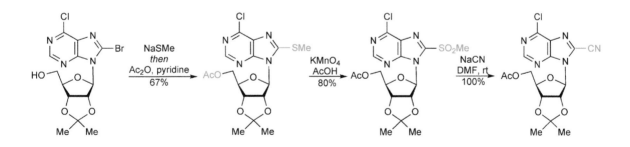

A useful conversion of a nucleoside 2,6-dithione into a 6-methylamino-adenosine *via* oxidation with dimethyldioxirane, illustrates several instructive points. The presumed intermediates are sulfinic acids: the 2-sulfinic acid loses sulfur dioxide to leave hydrogen at C-2, and nucleophilic displacement of the 6-sulfinic acid (or possibly the sulfonic acid after further oxidation) introduces the amino group. Similar reactions can be carried out on pyrimidine thiones.[98] The scheme shows intermediates derived from a disulfinic acid – it is not clear in what order oxidations/loss of sulfur dioxide/displacements take place.

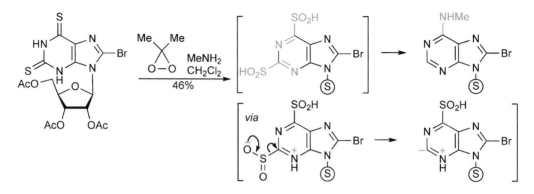

27.9 Alkyl-Purines

Comparatively little information is available concerning purine methyl groups, but it seems[99,100] that their reactivity is comparable to pyridine α- and γ-methyl substituents, undergoing base-catalysed condensations and selenium dioxide oxidation to aldehydes.

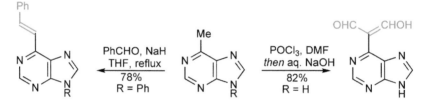

27.10 Purine Carboxylic Acids

Here again, comparatively little systematic information is available, but in parallel with pyridine α-acids, it can again be implied that purine acids undergo decarboxylation on heating – the 6-acid at 195 °C for example.[101]

27.11 Synthesis of Purines

Because of the ready availability of nucleosides from natural sources, a frequently used route to substituted purines is *via* the manipulation of one of these.

27.11.1 Ring Synthesis

There are two general approaches to the construction of the purine ring system. Additionally, a category which can be defined as 'one pot' methods, are adaptations of the type of process that probably took place in prebiotic times, when simple molecules, such as hydrogen cyanide and ammonia, are believed to have combined to give the first purines.

27.11.1.1   From 4,5-Diamino-pyrimidines
4,5-Diamino-pyrimidines react with carboxylic acids, or derivatives, to give purines, the 'carboxyl' carbon corresponding to C-8.

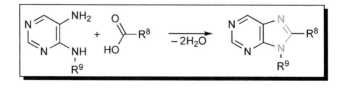

Traube Synthesis

8-Unsubstituted purines can be prepared simply by heating 4,5-diaminopyrimidines with formic acid,[102] but formamide[103] (or formamidine[104]) is better. The reaction proceeds *via* cyclising dehydration of an intermediate formamide; this usually takes place *in situ* using formamide, but generally requires a second, more forcing step when formic acid is employed initially. Purine itself can be prepared by this route.[105] Aldehydes react with 5,6-diaminouracils in the presence of bromodimethylsulfonium bromide to give the 8-substituted xanthines.[106]

8-Substituted purines are comparably prepared using acylating agents corresponding to higher acids; in most cases the amide is isolated and separately cyclised.[107] The diamino-pyrimidines required are usually prepared by the coupling of a 4-aminopyrimidine with an aryldiazonium ion (or by nitrosation[108]), then reduction, or by ring synthesis.[109]

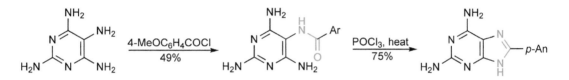

Precursors to 9-substituted purines, therefore requiring a substituent on the pyrimidine-4-amino group, are available from the reaction of a 4-chloropyrimidine with a primary amine.

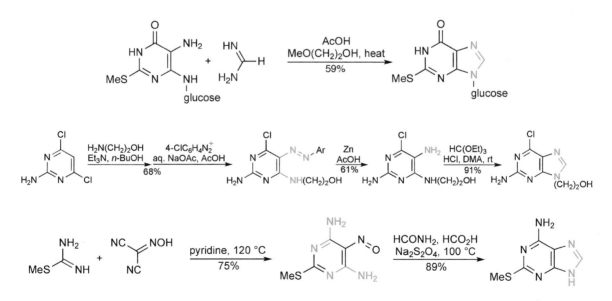

When milder conditions are required for the cyclisation, perhaps because of the presence of a sugar residue, an orthoester[110] (often activated[111] with acetic anhydride) or an acetal-ester[112] (illustrated below) can be used.

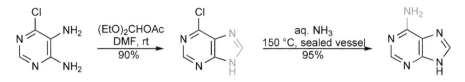

A related reaction is the oxidative cyclisation of anils, originally under vigorous conditions such as heating in nitrobenzene,[113] but now achievable at lower temperatures using diethyl azodicarboxylate.[114]

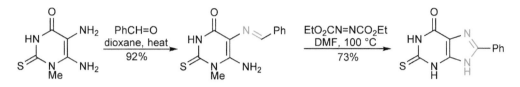

The formation of 8-oxo- or 8-thio-purines requires one-carbon components at a higher oxidation level: urea and thiourea are appropriate. The products of chloroformate-initiated five-membered ring cleavage of purines (27.1.1.3) can be recyclised to produce 8-oxo-purines.[115]

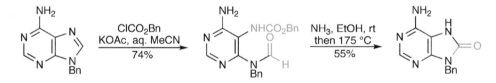

27.11.1.2 From 5-Aminoimidazole-4-Carboxamide or -Nitrile[116]
5-Aminoimidazole-4-carboxamides (or -nitriles) interact with components at the carboxylic acid oxidation level giving purines, the 'carboxyl' carbon becoming C-2.

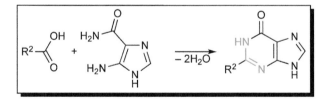

Biosynthetically, purines are built up *via* formation of the imidazole ring first, from glycine and formate, and thence to hypoxanthine and then the other natural purines. In the laboratory, most imidazole-based purine syntheses start with 5-aminoimidazole-4-carboxylic acid, particularly its amide (known by the acronym AICA), which as well as its riboside, is commercially available from biological sources. The use of 5-aminoimidazole-4-carbonitrile in this approach results in the formation of 6-amino-purines, as in a synthesis of adenine itself.[117]

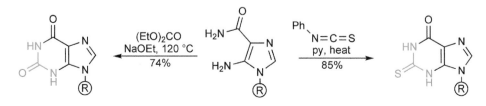

Conversion into 2-alkyl- or -aryl-purines requires the insertion of one carbon to create the six-membered ring, and this is usually effected by condensation with esters in the presence of base,[118] although amides[119] are occasionally utilised. The use of an isothiocyanate leads to a 2-thiopurine.[120]

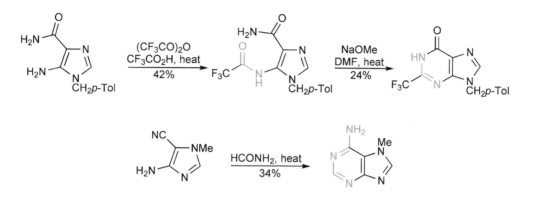

There are a few examples of purine ring syntheses which start from simpler imidazoles, for example a 5-aminoimidazole, generally prepared and utilised *in situ*.[121]

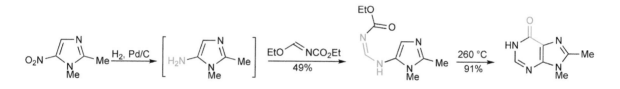

7-Substituted purines can be obtained from 4-aminoimidazole-5-carbaldehyde oximes after conversion into imino ethers and reaction with ammonia, as shown below.[122]

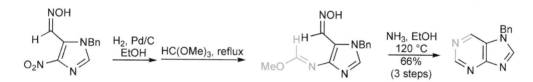

27.11.1.3 By Cycloadditions
Triethyl 1,3,5-triazine-2,4,6-tricarboxylate serves as an azadiene in reaction with 5-aminopyrazoles to produce purine isosteres, pyrazolo[3,4-*d*]pyrimidines.[123] In order to overcome the relative instability of 5-aminoimidazoles, required for analogous synthesis of purines, 5-aminoimidazole-4-carboxylic acids can be used, *in situ* decarboxylation producing the required dienophile.[124] Exactly comparable reaction with 2-amino-4-cyanopyrroles produces pyrrolo[2,3-*d*]pyrimidines.[125]

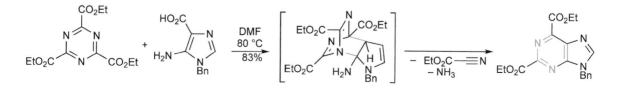

27.11.1.4 'One-step' Syntheses

It is amazing that relatively complex molecules, such as purines, can be formed by the sequential condensation of very simple molecules, such as ammonia and hydrogen cyanide. That the intrinsic reactivity embodied in these simple molecules leads 'naturally' to purines must surely be relevant to the evolution of a natural system that relies on these 'complex' molecules. In other words it seems highly likely that purines existed before the evolution of life and were incorporated into its mechanism because they were there and, of course, because they have appropriate chemical properties.

Adenine, $C_5H_5N_5$, is formally a pentamer of hydrogen cyanide and indeed can be produced in the laboratory by the reaction of ammonia and hydrogen cyanide, although not with great efficiency. A related and more practical method involves the dehydration of formamide.[126] Purine itself can also be obtained from formamide.[127]

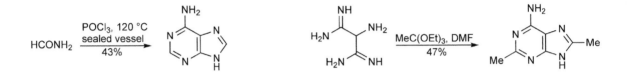

Methods derived from this fundamental process involve the condensation of one-, two- and three-carbon units such as amidines, amino-nitriles and carboxamides, which represent intermediate stages of the ammonia/hydrogen cyanide reaction. Pyrimidines or imidazoles are usually intermediates.[128]

27.11.2 Examples of Notable Syntheses Involving Purines

27.11.2.1 Aristeromycin

A synthesis of the antibiotic aristeromycin[129] makes use of the displacement of a pyrimidine 4-chloride to allow introduction of the amine and the generation of the 4,5-diamino-pyrimidine for subsequent closure of the five-membered ring.

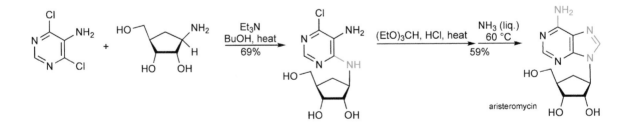

27.11.2.2 Adenosine

Adenosine[130] has also been synthesised using the pyrimidine → purine strategy. In this synthesis the sugar was also introduced at an early stage, but here *via* condensation with a 4-amino group.

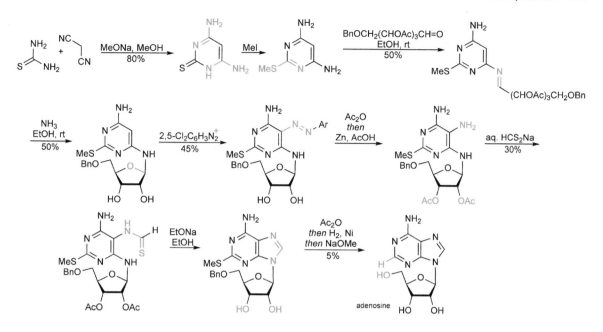

27.11.2.3 Sildenafil (Viagra™)

A synthesis of sildenafil, which contains a bicyclic system (a 1*H*-pyrazolo[4,3-*d*]pyrimidine) isomeric with that of a purine, starts with a routine synthesis of a pyrazole (cf. 25.12.1.1) followed by *N*-methylation and ring nitration. Functional group manipulation provides a pyrazole equivalent to AICA (27.11.1.2) from which the pyrimidone ring is formed *via* reaction with an aromatic acid chloride.

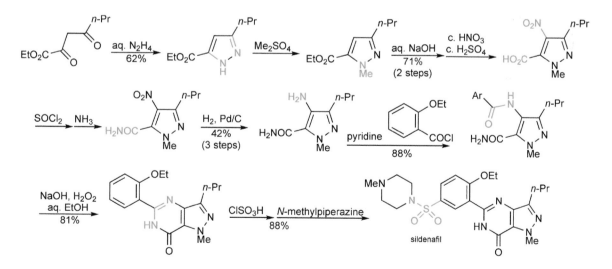

Exercises

Straightforward revision exercises (consult Chapter 27):

(a) What are the structures of the purine bases involved in DNA and RNA?

(b) How does the Dimroth rearrangement allow the synthesis of 6-alkylamino-purines from 6-amino-purines?

(c) What is the order of reactivity towards nucleophilic displacement of the 2-, 6- and 8-halo-purines? How does the inclusion of a tertiary amine in such nucleophilic displacements facilitate them?
(d) At what position does strong base deprotonation of 9-substituted purines take place?
(e) Name three types of compound which will react with 4,5-diamino-pyrimidines to produce purines.
(f) How could one synthesise a 2-thiopurine from 5-aminoimidazole-4-carboxamide?

More advanced exercises:
1. What are the structures of the intermediates and final product of the following sequence: guanosine 2′,3′,5′-triacetate reacted with $POCl_3 \rightarrow C_{16}H_{18}ClN_5O_7$, then this with t-BuONO/$CH_2I_2 \rightarrow C_{16}H_{16}ClIN_4O_7$, this product with NH_3/MeOH $\rightarrow C_{10}H_{12}IN_5O_4$ and finally this compound with $PhB(OH)_2$/$Pd(PPh_3)_4$/Na_2CO_3, giving $C_{16}H_{17}N_5O_4$. How could this same purine be prepared from AICA-riboside in four steps?
2. Suggest a sequence for the transformation of adenosine into 8-phenyladenosine.
3. Give structures and explain the following: adenosine with $Me_2SO_4 \rightarrow C_{11}H_{15}N_5O_4$, this with aq. HCl produces $C_6H_7N_5$ and finally aq. NH_3 on this last compound gives an isomer, $C_6H_7N_5$.
4. Write structures for the purines produced by the following reactions: (i) heating 4,5,6-triaminopyrimidine with formamide; (ii) treating 2-methyl-4,5-diaminopyrimidin-6-one with sodium dithioformate, then heating in quinoline.

References

[1] Dreyfus, M., Dodin, G., Bensaude, O. and Dubois, J. E., *J. Am. Chem. Soc.*, **1975**, *97*, 2369.
[2] Coburn, W. C., Thorpe, M. C., Montgomery, J. A. and Hewson, K., *J. Org. Chem.*, **1965**, *30*, 1110.
[3] Schumacher, M. and Günther, H., *Chem. Ber.*, **1983**, *116*, 2001.
[4] Taylor, E. C., Maki, Y. and McKillop, A., *J. Org. Chem.*, **1969**, *34*, 1170.
[5] Brookes, P. and Lawley, P. D., *J. Chem. Soc.*, **1960**, 539; Carrea, G., Ottolina, G., Riva, S., Danieli, B., Lesma, G. and Palmisano, G., *Helv. Chim. Acta*, **1988**, *71*, 762.
[6] 'The Dimroth rearrangement in the adenine series: a review updated', Fujii, T. and Itaya, T., *Heterocycles*, **1998**, *48*, 359.
[7] Montgomery, J. A. and Temple, C., *J. Am. Chem. Soc.*, **1961**, *83*, 630.
[8] Lewis, L. R., Schneider, F. H. and Robins, R. K., *J. Org. Chem.*, **1961**, *26*, 3837.
[9] Lang, P., Magnin, G., Mathis, G., Burger, A. and Biellmann, J.-F., *J. Org. Chem.*, **2000**, *65*, 7825.
[10] Geen, G. R., Grinter, T. J., Kincey, P. M. and Jarvest, R. L., *Tetrahedron*, **1990**, *46*, 6903.
[11] Breipohl, G., Will, D. W., Peyman, A. and Uhlmann, E., *Tetrahedron*, **1997**, *53*, 14671.
[12] Alarcon, K., Martelli, A., Demeunynck, M. and Lhomme, J., *Tetrahedron Lett.*, **2000**, *41*, 7211.
[13] Zhong, M. and Robins, M. J., *J. Org. Chem.*, **2006**, *71*, 8901.
[14] Geen, G. R., Kincey, P. M. and Choudary, B. M., *Tetrahedron Lett.*, **1992**, *33*, 4609.
[15] Sessler, J. L., Magda, D. and Furuta, H., *J. Org. Chem.*, **1992**, *57*, 818.
[16] Iyer, R. S., Voehler, M. W. and Harris, T. M., *J. Am. Chem. Soc.*, **1994**, *116*, 8863.
[17] 'Nucleoside syntheses, organosilicon methods', Lukevics, E. and Zablocka, A., Ellis Horwood, **1991** and references therein.
[18] Baker, B. R., Hewson, K., Thomas, H. J. and Johnson, J. A., *J. Org. Chem.*, **1957**, *22*, 954; Townsend, L. B., Robins, R. K., Loeppky, F. N. and Leonard, N. J., *J. Am. Chem. Soc.*, **1964**, *86*, 5320.
[19] Kazimierczuk, Z., Cottam, H. B., Revankar, G. R. and Robins, R. K., *J. Am. Chem. Soc.*, **1984**, *106*, 6379; Jhingan, A. K. and Meehan, T., *Synth. Commun.*, **1992**, *22*, 3129.
[20] Hanna, N. B., Ramasamy, K., Robins, R. K. and Revankar, G. R., *J. Heterocycl. Chem.*, **1988**, *25*, 1899.
[21] Imai, K., Nohara, A. and Honjo, M., *Chem. Pharm. Bull.*, **1966**, *14*, 1377.
[22] Bisacchi, G. S., Singh, J., Godgrey, J. D., Kissick, T. P., Mitt, T., Malley, M. F., Di Marco, J. D., Gougoutas, J. Z., Mueller, R. H. and Zahler, R., *J. Org. Chem.*, **1995**, *60*, 2902.
[23] Lukin, K. A., Yang, C., Bellettini, J. R. and Narayanan, B. A., *Nucleosides, Nucleotides Nucleic Acids*, **2000**, *19*, 815.
[24] Leonard, N. J., McDonald, J. J., Henderson, R. E. L. and Reichman, M. E., *Biochemistry*, **1971**, *10*, 3335.
[25] Pratt, R. F. and Kraus, K. K., *Tetrahedron Lett.*, **1981**, *22*, 2431.
[26] Giner-Sorolla, A., Gryte, C., Cox, M. L. and Parham, J. C., *J. Org. Chem.*, **1971**, *36*, 1228.
[27] Stevens, M. A., Magrath, D. I., Smith, H. W. and Brown, G. B., *J. Am. Chem. Soc.*, **1958**, *80*, 2755.
[28] 'The 7-*N*-oxides of purines related to nucleic acids: their chemistry, synthesis and biological evaluation', Fujii, T., Itaya, T. and Ogawa, K., *Heterocycles*, **1997**, *44*, 573.
[29] Ikehara, M. and Kaneko, M., *Tetrahedron*, **1970**, *26*, 4251.
[30] Fischer, E., *Justus Liebigs Ann. Chem.*, **1882**, *213*, 316; ibid., *221*, 336; Bruhns, G., *Chem. Ber.*, **1890**, *23*, 225; Beaman, A. G. and Robins, R. K., *J. Org. Chem.*, **1963**, *28*, 2310.
[31] Barrio, J. R., Namavari, M., Phelps, M. E. and Satyamurthy, N., *J. Am Chem. Soc.*, **1996**, *118*, 1408.
[32] Mosselhi, M. A. and Pfleiderer, W., *J. Heterocycl. Chem.*, **1993**, *30*, 1221.

[33] Deghati P. Y. F., Wanner, M. J. and Koomen, G.-J., *Tetrahedron Lett.*, **2000**, *41*, 1291; Rodenko, B., Koch, M., Van der Burg, A. M., Wanner, M. J. and Koomen, G.-J., *J. Am. Chem. Soc.*, **2005**, *127*, 5957.

[34] Braendvang, M. and Gundersen, L.-L., *Synthesis*, **2006**, 2993.

[35] Jones, J. W. and Robins, R. K., *J. Am. Chem. Soc.*, **1960**, *82*, 3773.

[36] Desaubry, L. and Bourguignon, J.-J., *Tetrahedron Lett.*, **1995**, *36*, 7875.

[37] Zady, M. F. and Wong, J. L., *J. Org. Chem.*, **1979**, *44*, 1450.

[38] Maria, E. J., Fourrey, J.-L., Machado, A. S. and Robert-Gero, M., *Synth. Commun.*, **1996**, *26*, 27.

[39] Saladino, R., Crestini, C., Bernini, R., Mincione, E. and Ciafrino, R., *Tetrahedron Lett.*, **1995**, *36*, 2665.

[40] Madyastha, K. M. and Sridhar, G. R., *J. Chem. Soc., Perkin Trans. 1*, **1999**, 677.

[41] Smith, D. L. and Elving, P. J., *J. Am. Chem. Soc.*, **1962**, *84*, 1412.

[42] Butula, I., *Justus Liebigs Ann. Chem.*, **1969**, *729*, 73.

[43] Hecht, S. M., Adams, B. L. and Kozarich, J. W., *J. Org. Chem.*, **1976**, *41*, 2303.

[44] Tanji, K.-i., Yokoi, T. and Sugimoto, O., *Heterocycles*, **2003**, *60*, 413.

[45] Ding, S., Gray, N. S., Ding, Q. and Schultz, P. G., *J. Org. Chem.*, **2001**, *66*, 8273.

[46] Barlin, G. B., *J. Chem. Soc. (B)*, **1967**, 954.

[47] Hamamichi, N. and Miyasaka, T., *J. Heterocycl. Chem.*, **1990**, *27*, 2011; *ibid.*, **1991**, *28*, 397.

[48] Montgomery, J. A. and Holum, L. B., *J. Am. Chem. Soc.*, **1957**, *79*, 2185.

[49] Focher, F., Hildebrand, C., Freese, S., Ciarrocchi, G., Noonan, T., Sangalli, S., Brown, N., Spadari, S. and Wright, G., *J. Med. Chem.*, **1988**, *31*, 1496.

[50] Kos, N. J., van der Plas, H. C. and van Veldhuizen, A., *Recl. Trav. Chim. Pays-Bas*, **1980**, *99*, 267.

[51] Motawia, M. S., Meldal, M., Sofan, M., Stein, P., Pedersen, E. B. and Nielsen, C., *Synthesis*, **1995**, 265.

[52] Krolikiewicz, K. and Vorbrüggen, H., *Nucleosides Nucleotides*, **1994**, *13*, 673.

[53] Lin, X. and Robins, M. J., *Org. Lett.*, **2000**, *2*, 3497.

[54] Nair, V. and Sells, T. B., *Tetrahedron Lett.*, **1990**, *31*, 807.

[55] Xu, Y.-Z., *Tetrahedron*, **1998**, *54*, 187.

[56] Edwards, C., Boche, G., Steinbrecher, T. and Scheer, S., *J. Chem. Soc., Perkin Trans. 1*, **1997**, 1887.

[57] Flaherty, D., Balse, P., Li, K., Moore, B. M. and Doughty, M. B., *Nucelosides Nucleotides*, **1995**, *14*, 65.

[58] Miyashita, A., Suzuki, Y., Ohta, K. and Higashino, T., *Heterocycles*, **1994**, *39*, 345.

[59] De Napoli, L., Montesarchio, D., Piccialli, G., Santacroce, C. and Varra, M., *J. Chem. Soc., Perkin Trans. 1*, **1995**, 15.

[60] Linn, J. A., McLean, E. W. and Kelley, J. L., *J. Chem. Soc., Chem. Commun.*, **1994**, 913.

[61] Lembicz, N. K., Grant, S., Clegg, W., Grifin, R. J., Heath, S. L. and Golding, B. T., *J. Chem. Soc., Perkin Trans. 1*, **1997**, 185.

[62] Herdewin, P., Van Aerschot, A. and Pfleiderer, W., *Synthesis*, **1989**, 961.

[63] Kiburis, J. and Lister, J. H., *J. Chem. Soc. (C)*, **1971**, 3942.

[64] Lee, H., Luna, E., Hinz, M., Stezowski, J. J., Kiselyov, A. S. and Harvey, R. G., *J. Org. Chem.*, **1995**, *60*, 5604.

[65] Lakshman, M. K., Keeler, J. C., Hilmer, J. H. and Martin, J. Q., *J. Am. Chem. Soc.*, **1999**, *121*, 6090.

[66] Ciszewski, L., Waykole, L., Prashad, M. and Repic, O., *Org. Proc. Res. Dev.*, **2006**, *10*, 799.

[67] Whitfield, H. J., Griffin, R. J., Hardcastle, I. R., Henderson, A., Meneyrol, J., Mesguiche, V., Sayle, K. L. and Golding, B. T., *Chem. Commun.*, **2003**, 2802.

[68] Miles, R. W., Samano, V. and Robins, M. J., *J. Am. Chem. Soc.*, **1995**, *117*, 5951; Clivio, P., Fourrey, J.-L. and Fauvre, A., *J. Chem. Sooc., Perkin Trans. 1*, **1993**, 2585.

[69] Wong, J. L. and Keck, J. H., *J. Chem. Soc., Chem. Commun.*, **1975**, 125.

[70] Hayakawa, H., Haraguchi, K., Tanaka, H. and Miyasaka, T., *Chem. Pharm. Bull.*, **1987**, *35*, 72.

[71] Hayakawa, H., Tanaka, H., Sasaki, K., Haraguchi, K., Saitoh, T., Takai, F. and Miyasaka, T., *J. Heterocycl. Chem.*, **1989**, *26*, 189.

[72] Nolsoe, J. M. J., Gundersen, L.-L. and Rise, F., *Synth. Commun.*, **1998**, *28*, 4303.

[73] Kato, K., Hayakawa, H., Tanaka, H., Kumamoto, H., Shindoh, S., Shuto, S. and Miyasaka, T., *J. Org. Chem.*, **1997**, *62*, 6833.

[74] Leonard, N. J. and Bryant, J. D., *J. Org. Chem.*, **1979**, *44*, 4612.

[75] Tobrman, T. and Dvorák, D., *Org. Lett.*, **2006**, *8*, 1291.

[76] Boudet, N., Dubbaka, S. R. and Knochel, P., *Org. Lett.*, **2008**, *10*, 1715.

[77] Elion, G. B., *J. Org. Chem.*, **1962**, *27*, 2478.

[78] Waters, T. R. and Connolly, B. A., *Nucleosides and Nucleotides*, **1992**, *11*, 1561; Fathi, R., Goswani, B., Kung, P.-P., Gaffney, B. L. and Jones, R. A., *Tetrahedron Lett.*, **1990**, *31*, 319.

[79] Vorbrüggen, H. and Krolikiewicz, K., *Justus Liebigs Ann. Chem.*, **1976**, 745.

[80] Gerster, J. F., Jones, J. W. and Robins, R. K., *J. Org. Chem.*, **1963**, *28*, 945; Robins, M. J. and Uznanski, B., *Can. J. Chem.*, **1981**, *59*, 2601.

[81] De Napoli, L., Messere, A., Montesarchio, D., Piccialli, G., Santacroce, C. and Varra, M., *J. Chem. Soc., Perkin Trans. 1*, **1994**, 923.

[82] Elion, G. B. and Hitchings, G. H., *J. Am. Chem. Soc.*, **1956**, *78*, 3508; Davoll, J. and Lowy, B. A., **1951**, *73*, 2936.

[83] Beaman, A. G. and Robins, R. K., *J. Am. Chem. Soc.*, **1961**, *83*, 4038.

[84] Kataoka, S., Isono, J., Yamaji, N., Kato, M., Kawada, T. and Imai, S., *Chem. Pharm. Bull.*, **1988**, *36*, 2212.

[85] El-Kafrawy, S. A., Zahran, M. A. and Pedersen, E. B., *Acta Chem. Scand.*, **1999**, *53*, 280.

[86] Montgomery, J. A. and Hewson, K., *J. Am. Chem. Soc.*, **1960**, *82*, 463.

[87] Jones, J. W. and Robins, R. K., *J. Am. Chem. Soc.*, **1960**, *82*, 3773.

[88] Moschel, R. C. and Keefer, L. K., *Tetrahedron Lett.*, **1989**, *30*, 1467.

[89] Orozco, M., Canela, E. I. and Franco, R., *J. Org. Chem.*, **1990**, *55*, 2630.

[90] Nair, V. and Richardson, S. G., *J. Org. Chem.*, **1980**, *45*, 3969; *idem, Synthesis*, **1982**, 670.

[91] Francom, P., Janeba, Z., Shibuya, S. and Robins, M. J., *J. Org. Chem.*, **2002**, *67*, 6788.

[92] Nair, V. and Hettrick, B. J., *Tetrahedron*, **1988**, *44*, 7001.

[93] Wada, T., Mochizuki, A., Higashiya, S., Tsuruoka, H., Kawhara, S., Ishikawa, M. and Sekine, M., *Tetrahedron Lett.*, **2001**, *42*, 9215.

[94] Chambert, S., Gautier-Luneau, I., Fontecave,M. and Décout, J.-L., *J. Org. Chem.* **2000**, *65*, 249.

[95] Kikugawa, K., Suehiro, H. and Aoki, A., *Chem. Pharm. Bull.*, **1977**, *25*, 2624.

[96] Beaman, A. G., Gerster, J. F. and Robins, R. K., *J. Org. Chem..*, **1962**, *27*, 986.

[97] Matsuda, A., Nomoto, Y. and Ueda, T., *Chem. Pharm. Bull.*, **1979**, *27*, 183; Wetzel, R. and Eckstein, F., *J. Org. Chem.*, **1975**, *40*, 658; Yamane, A., Matsuda, A. and Ueda, T., *Chem. Pharm. Bull.*, **1980**, *28*, 150.

[98] Saladino, R., Mincione, E., Crestini, C. and Mezzetti, M., *Tetrahedron*, **1996**, *52*, 6759.

[99] Brown, D. M. and Giner-Sorolla, A., *J. Chem. Soc. (C)*, **1971**, 128.

[100] Tanji, K., Satoh, R. and Higashino, T, *Chem. Pharm. Bull.*, **1992**, *40*, 227.

[101] Mackay, L. B. and Hitchings, G. H., *J. Am. Chem. Soc.*, **1956**, *78*, 3511.

[102] Traube, *Chem. Ber.*, **1900**, *33*, 1371; *ibid.*, 3035.

[103] Robins, R. K., Dille, K. J., Willits, C. H. and Christensen, B. E., *J. Am. Chem. Soc.*, **1953**, *75*, 263.

[104] Melguizo, M., Nogueras, M. and Sanchez, A., *Synthesis*, **1992**, 491.

[105] Albert, A. and Brown, D. J., *J. Chem. Soc.*, **1954**, 2060.

[106] Dong, M., Sitkovsky, M., Kallmerten, A. E. and Jones, G. B., *Tetrahedron Lett.*, **2008**, *49*, 4633.

[107] Albert, A. and Brown, D. J., *J. Chem. Soc.*, **1954**, 2066; Montgomery, J. A., *J. Am. Chem. Soc.*, **1956**, *78*, 1928; Young, R. C., Jones, M., Milliner, K. J., Rana, K. K. and Ward, J. G., *J. Med. Chem.*, **1990**, *33*, 2073; Elion, G. B., Burgi, E. and Hitchings, G. H., *J. Am. Chem. Soc.*, **1951**, *73*, 5235.

[108] Elzein, E., Kalla, R. V., Li, X., Perry, T., Gimbel, A., Zeng, D., Lustig, D., Leung, K. and Zablocki, J., *J. Med. Chem.*, **2008**, *51*, 2267.

[109] Taylor, E. C., Vogl, O. and Cheng, C. C., *J. Am. Chem. Soc.*, **1959**, *81*, 2442.

[110] Párkányi, C. and Yuan, H. L., *J. Heterocycl. Chem.*, **1990**, *27*, 1409.

[111] Goldman, L., Marsico, J. W. and Gazzola, A. L., *J. Org. Chem.*, **1956**, *21*, 599.

[112] Orji, C. C., Kelly, J., Ashburn, D. A. and Silks, R. A., *J. Chem. Soc., Perkin Trans. 1*, **1996**, 595.

[113] Jerchel, D., Kracht, M. and Krucker, K., *Justus Liebigs Ann. Chem.*, **1954**, *590*, 232.

[114] Nagamatsu, T., Yamasaki, H. and Yoneda, F., *Heterocycles*, **1992**, *33*, 775.

[115] Altman, J. and Ben-Ishai, D., *J. Heterocycl. Chem.*, **1968**, *5*, 679.

[116] 'Annelation of a pyrimidine ring to an existing ring', Albert, A., *Adv. Heterocycl. Chem.*, **1982**, *32*, 1.

[117] Ferris, J. P. and Orgel, L. E., *J. Am. Chem. Soc.*, **1966**, *88*, 1074 and 3829.

[118] Yamazaki, A., Kumashiro, I. and Takenishi, T., *J. Org. Chem.*, **1967**, *32*, 3258.

[119] Kelley, J. L., Linn, J. A. and Selway, J. W. T., *J. Med. Chem.*, **1989**, *32*, 218; Prasad, R. N. and Robins, R. K., *J. Am. Chem. Soc.*, **1957**, *79*, 6401.

[120] Imai, K., Marumoto, R., Kobayashi, K., Yoshioka, Y., Toda, J. and Honjo, M., *Chem. Pharm. Bull.*, **1971**, *19*, 576.

[121] Al-Shaar, A. H., Gilmour, D. W., Lythgoe, D. J., McClenaghan, I. and Ramsden, C. A., *J. Chem. Soc., Chem. Commun.*, **1989**, 551.

[122] Ostrowski, S., *Synlett*, **1995**, 253.

[123] Dang, Q., Brown, B. S. and Erion, M. D., *J. Org. Chem.*, **1996**, *61*, 5204.

[124] Dang, Q., Liu, Y. and Erion, M. D., *J. Am. Chem. Soc.*, **1999**, *121*, 5833.

[125] Dang, Q. and Gomez-Galeno, J. E., *J. Org. Chem.*, **2002**, *67*, 8703.

[126] Ochiai, M., Marumoto, R., Kobayashi, S., Shimazu, H. and Morita, K., *Tetrahedron*, **1968**, *24*, 5731.

[127] Yamada, H. and Okamoto, T., *Chem. Pharm. Bull.*, **1972**, *20*, 623.

[128] Richter, E., Loeffler, J. E. and Taylor, E. C., *J. Am. Chem. Soc.*, **1960**, *82*, 3144.

[129] Arita, M., Adachi, K., Ito, Y., Sawai, H. and Ohno, M., *J. Am. Chem. Soc.*, **1983**, *105*, 4049.

[130] Kenner, G. W, Taylor, C. W. and Todd, A. R., *J. Chem. Soc.*, **1949**, 1620.

28

Heterocycles Containing a Ring-Junction Nitrogen (Bridgehead Compounds)

In addition to the biologically important purines and pteridines, and the major benzo-fused heterocycles, such as indole, many other aromatic, fused heterocyclic ring systems are known, and of these, the most important are those containing a ring-junction nitrogen – that is, *where a nitrogen is common to two rings*.[1] The vast majority of these systems do not occur naturally, but they have been the subject of many studies from the theoretical viewpoint, for the preparation of potentially biologically active analogues, and for other industrial uses. For reasons of space, only combinations of five- and six-membered rings are considered here, and only some of these possibilities, though other combinations are possible and are known.

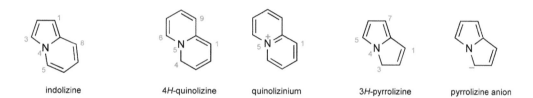

indolizine   4*H*-quinolizine   quinolizinium   3*H*-pyrrolizine   pyrrolizine anion

Of the parent systems that have the ring-junction nitrogen as the only heteroatom, only indolizine (often 'pyrrocoline' in the older literature) has a neutral, fully conjugated 10-electron π-system, comprising four pairs of electrons from the four double bonds and a pair from nitrogen, much as in indole. 4*H*-Quinolizine is not aromatic – there is a saturated atom interrupting the conjugation – but the cation, quinolizinium, formed formally by loss of hydride from quinolizine, does have an aromatic 10-π-electron system: it is completely isoelectronic with naphthalene, the positive charge resulting from the higher nuclear charge of nitrogen *versus* carbon. Similarly, pyrrolizine, which is already aromatic in being a pyrrole (with an α-vinyl substituent), on conversion into its conjugate anion, attains a 10-electron π-system.

28.1 Indolizines[2]

The aromatic character of indolizine is expressed by three main mesomeric contributors, two of which incorporate a pyridinium moiety; other structures (not shown) incorporating neither a complete pyrrole nor a pyridinium are less important.

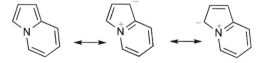

Heterocyclic Chemistry 5th Edition   John Joule and Keith Mills
© 2010 Blackwell Publishing Ltd

28.1.1 Reactions of Indolizines

Indolizine is an electron-rich system and its reactions are mainly electrophilic substitutions, which occur about as readily as for indole, and go preferentially at C-3, but may also take place at C-1. Consistent with their similarity to pyrroles, rather than pyridines, indolizines are not attacked by nucleophiles, nor are there examples of nucleophilic displacement of halide.

Indolizine, pK_{aH} 3.9,[3] is much more basic than indole (pK_{aH} −3.5) and the implied relative stability of the cation makes it less reactive, and thus indolizines are resistant to acid-catalysed polymerisation (cf. 20.1.1.9). Indolizine protonates at C-3, but 3-methylindolizine protonates mainly (79%) at C-1; the delicacy of the balance is further illustrated by 1,2,3-trimethyl- and 3,5-dimethylindolizines, each of which protonate exclusively at C-3. Electrophilic substitutions such as acylation,[4] Vilsmeier formylation[5] and diazo-coupling[6] all take place at C-3.

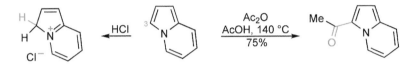

Nitration of 2-methylindolizine under mild conditions results in substitution at C-3,[7] but under strongly acidic conditions it takes place at C-1,[8] presumably *via* attack on the indolizinium cation.

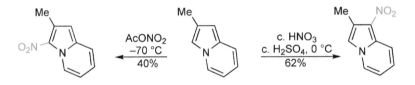

Indolizine and its simple alkyl derivatives are sensitive to light and to aerial oxidation, which lead to destruction of the ring system. Catalytic reduction in acidic solution (reduction of the indolizinium cation) selectively saturates the pyrrole ring, giving a pyridinium salt;[9] complete saturation, affording indolizidines, results from reductions over platinum.[10]

Despite its 10-electron aromatic π-system, indolizine apparently participates as an 8-electron system in its reaction with diethyl acetylenedicarboxylate, though the process may be stepwise and not concerted. By carrying out the reaction in the presence of a noble metal as catalyst, the initial adduct is converted into an aromatic cyclazine (28.5).[11]

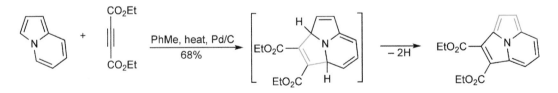

Indolizines normally lithiate at C-5,[4] but 5-methylindolizine undergoes lithiation at the side-chain methyl.[12]

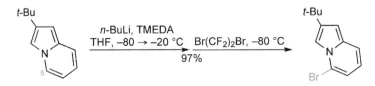

Of its functional derivatives, worth noting is the easy cleavage of carboxyl and acyl groups on heating with aqueous acid, and the instability of amino derivatives, which cannot be diazotised, but which can be converted into stable acetamides.

28.1.2 Ring Synthesis of Indolizines[13]

The traditional route to indolizines is the Chichibabin synthesis,[14] which involves quaternisation of a 2-alkyl-pyridine with an α-halo-ketone, followed by base-catalysed cyclisation *via* deprotonation of the pyridinium α-methyl,[15] rendered easy by the quaternisation, then intramolecular attack on the carbonyl group.[16]

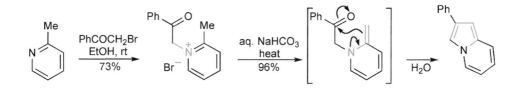

A variation of the classical procedure utilizes 2-methoxy-6-methylpyridine as a starting material and leads eventually to 5-aminoindolizines; the intermediate oxazolo[3,2-*a*]pyridinium salts are isolated as perchlorate salts (**CAUTION**).[17]

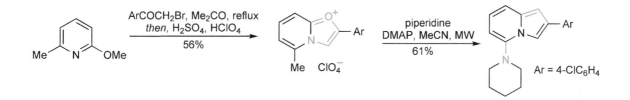

Another useful method involves the intermediacy of a pyridinium ylide as a 1,3-dipole in a cycloaddition.[18]

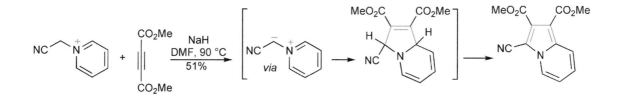

An important feature of this type of reaction (which can also be used in an analogous fashion to prepare aza-indolizines) is that although a dihydroindolizine is the logical product from a mechanistic viewpoint, the fully aromatic compound is usually obtained. The mechanism of aromatisation is not clear: it could be by air oxidation during work-up, or *via* hydride transfer to some other component in the reaction mixture. When dihydro compounds are isolated they can be easily aromatised using the usual reagents – palladium/charcoal or quinones. An extension of this analysis is the production of aromatic indolizines from reaction of an ylide with an alkene (which would be expected to give the tetrahydro product) *in the presence* of a suitable oxidant such as the cobalt chromate complex[19] shown below.[20]

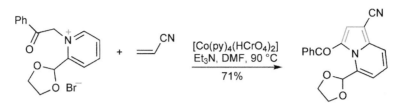

Aromatisation can alternatively occur by loss of HX (HCN in the example shown) when a leaving group is present in one of the reactants.[21]

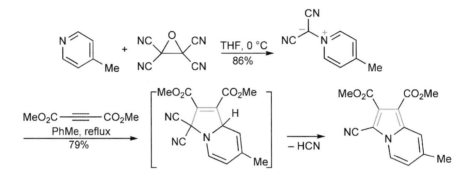

In several variants, pyridines carrying either a 2-alkynylmethyl or a 2-alkenylmethyl substituent can be ring closed to indolizines. Iodine[22] and silver tetrafluoroborate[23] have been used to close precursors available conveniently using the 2-lithiated pyridine. As in several situations in this chapter, comparably substituted imine units in other starting materials, can be ring closed by this route, for example 2-alkynylmethylquinoxalines give pyrrolo[1,2-a]quinoxalines and 2-alkynylmethylthiazoles give pyrrolo[2,1-b]thiazoles.[23]

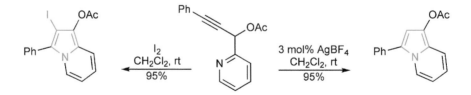

In a one-pot three-component sequence, a 2-alkynylmethylpyridine intermediate is generated *in situ*, pyridine-2-carbaldehyde being the starting material and 1-aminoindolizines the products.[24]

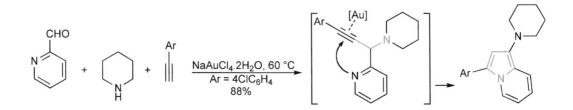

28.2 Aza-Indolizines

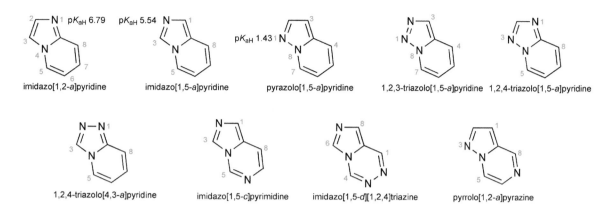

imidazo[1,2-*a*]pyridine   imidazo[1,5-*a*]pyridine   pyrazolo[1,5-*a*]pyridine   1,2,3-triazolo[1,5-*a*]pyridine   1,2,4-triazolo[1,5-*a*]pyridine

1,2,4-triazolo[4,3-*a*]pyridine   imidazo[1,5-*c*]pyrimidine   imidazo[1,5-*d*][1,2,4]triazine   pyrrolo[1,2-*a*]pyrazine

Note: as can be seen from the examples above, numbering sequences vary with the number and disposition of the nitrogen atoms.

Seven monoaza- and many more polyaza-indolizines (some are shown above) are possible, indeed compounds with up to six nitrogen atoms have been reported. Despite the rarity of such systems in nature, there is much interest in aza-indolizines stemming from their structural similarity to both indoles and purines.

Apart from pyrrolo[1,2-*b*]pyridazine, all the monoaza-indolizines protonate on the second (non-ring-junction) nitrogen, rather than on carbon.[3,25] Alkylation similarly goes on nitrogen, however other electrophilic reagents attack with regioselectivity similar to indolizine itself – they effect substitution of the five-membered ring at positions 1 and 3 (where these are carbon).

28.2.1 Imidazo[1,2-*a*]pyridines

Electrophilic substitutions such as halogenation, nitration etc. go at C-3, or at C-5 if position 3 is blocked.[26] Acylation does not require a catalyst.[27]

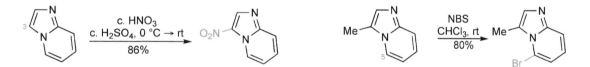

Of all the positional chloro isomers, nucleophilic displacement reactions are known only for the 5-isomer; 7-chloroimidazolo[1,2-*a*]pyridine, where one might have anticipated similar activation, is not reactive in this sense,[28] but, of course, transition-metal-catalysed replacement of halogen (4.2) can be carried out at any position.[29]

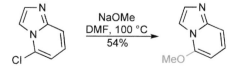

Base-catalysed deuterium exchange proceeds at C-3 and C-5;[30] preparative lithiation occurs at C-3, or if C-3 is blocked, at C-5 or C-8 depending on other substituents,[31] but the 2,6-dichloro compound reacts selectively at C-5, even in the presence of hydrogen at C-3.[32]

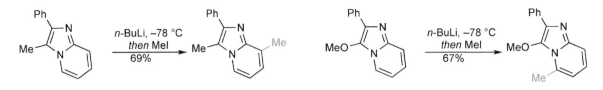

Amino-imidazo[1,2-*a*]pyridines exist as amino tautomers, but are even more unstable than amino-indolizines. 2- and 5-oxo compounds are in the carbonyl tautomeric form and react with phosphoryl chloride yielding chloro compounds.[20]

The traditional ring synthesis of imidazo[1,2-*a*]pyridines is based on the Chichibabin route to indolizines (28.1.2), but using 2-aminopyridines instead of 2-alkyl-pyridines. The initial reaction with the halo-ketone is regioselective for the ring nitrogen, so isomerically pure products are obtained.[33,34] 2-Oxo-imidazo[1,2-*a*] pyridines are the products when an α-bromo-ester is used instead of a ketone.[35]

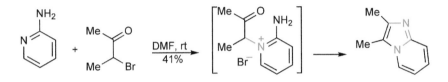

The 1,2-bis-electrophiles generated from glyoxal, benzotriazole and an amine can be used to produce imidazo[1,2-*a*]pyridines as illustrated below, but also imidazo[1,2-*a*]pyrimidines, and imidazo[1,2-*c*] pyrimidines.[36]

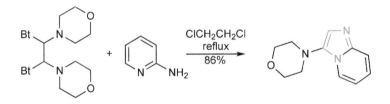

2-Aminopyridines can be used in other ways: copper-catalysed interaction with diazo-ketones[37] and three-component reactions[38,39,40] involving an isonitrile and an aldehyde also give imidazo[1,2-*a*]pyridines; such condensations can be promoted using an ionic liquid[41] or scandium triflate with microwave heating.[42] In the former route, 2-substituted products result and from the latter, 3-aminoimidazo[1,2-*a*]pyridines are formed.

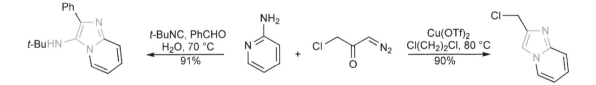

Here again other starting materials having an amidine sub-unit, as in 2-aminopyridine, react comparably: 2-aminothiazoles give imidazo[2,1-*b*]thiazoles,[40] 2-aminopyrazines generate imidazo[1,2-*a*]pyrazines,[38] 4-aminopyrimidines lead to imidazo[1,2-*c*]pyrimidines[43] and 2-aminopyrimidines afford imidazo[1,2-*a*] pyrimidines (illustrated below).[44]

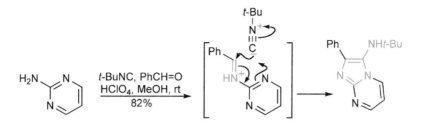

28.2.2 Imidazo[1,5-*a*]pyridines

Electrophilic substitution in this system again occurs in the five-membered ring, at C-1, or at C-3 if the former position is occupied.[45,46] Reaction with bromine gives a 1,3-dibromo product.[47]

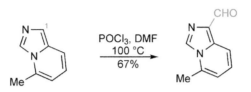

Benzoylation provides an instructive example: under normal conditions *C*-substitution occurs at C-1, however in the presence of triethylamine, 3-benzoylimidazo[1,5-*a*]pyridine is the product.[48] This can be explained by assuming the intermediacy of an ylide formed by deprotonation of an initial *N*⁺-benzoyl salt (cf. 24.1.2.5).

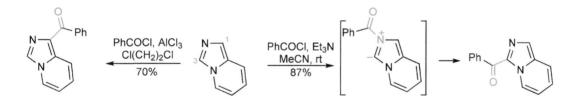

Five-membered ring cleavage occurs relatively easily: hot aqueous acid converts these heterocycles into 2-aminomethyl-pyridines.

Lithiation, by direct analogy with imidazole, involves the 3-proton,[5] but 5-lithiation occurs on comparable treatment of its 3-ethylthio derivative, the substituent both blocking attack at C-3 *and* assisting lithiation at the *peri*-position; the ethylthio group can, of course, be subsequently easily removed.[49]

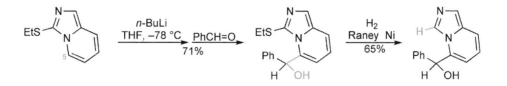

Imidazo[1,5-*a*]pyridines are synthesised by the dehydrative cyclisation of *N*-acyl-2-aminomethyl-pyridines[46] or of thioamides,[50] which can be prepared and ring closed under very mild conditions.[51] 3-Amino[52] -oxy[53] and -thio[54] derivatives are available *via* related cyclisations.

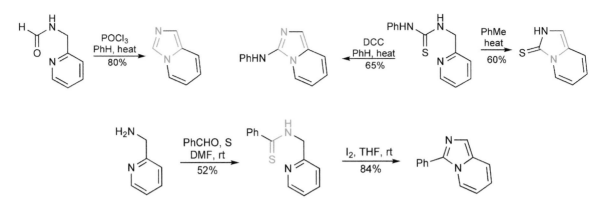

28.2.3   Pyrazolo[1,5-*a*]pyridines

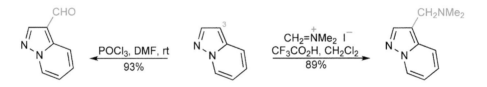

In this system, electrophilic substitution occurs at C-3[55] and lithiation takes place at C-7.[56]

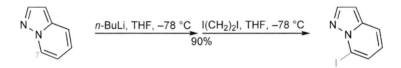

Pyrazolo[1,5-*a*]pyridines can be prepared by cycloaddition of pyridinium *N*-imides (produced by *N*-deprotonation of *N*-amino-pyridinium salts with base[57]) with alkynes[58] or *N*-amination of 2-alkynyl-pyridines.[59] The cycloaddition of 3-benzyloxypyridinium *N*-imide involves preferentially the more hindered C-2.[60]

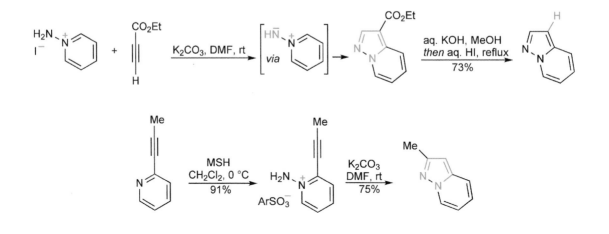

28.2.4 Triazolo-[61] and Tetrazolo-Pyridines

1,2,3-Triazolo[1,5-*a*]pyridine could theoretically be in equilibrium with its ring-opened diazo tautomer[62] and, although it actually exists in the closed form, its reactions tend to reflect this potential equilibrium: reaction with electrophiles can take two courses. Acylation and nitration occur normally, at C-1, but reagents such as bromine lead to a very easy ring cleavage.[63] Aqueous acid similarly brings about ring cleavage and the formation of 2-hydroxymethylpyridine.

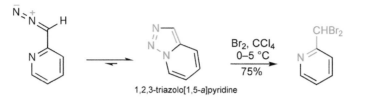

2-Azido-azines are in equilibrium with fused tetrazoles, the position of the equilibrium being very sensitive to substituent influence, for example for tetrazolo[1,5-*a*]pyridine, the equilibrium lies predominantly towards the closed form, whereas the analogous 5-chloro compound is predominantly open.[64]

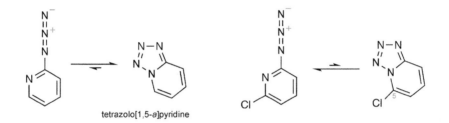

Direct lithiation of 1,2,3-triazolo[1,5-*a*]pyridines occurs with ease, at C-7, subsequent reaction with electrophiles being unexceptional, for example conversion into the 7-bromo derivative then allows nucleophiles to be introduced *via* displacement of bromide, thus providing, overall, a route to 2,6-disubstituted pyridines.[65]

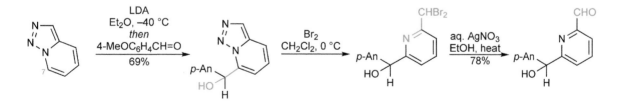

1,2,4-Triazolo[1,5-*a*]pyridine seems to be resistant to electrophilic attack, but can be lithiated at C-5; in contrast, 1,2,4-triazolo[4,3-*a*]pyridine readily undergoes electrophilic substitution at C-3.[66]

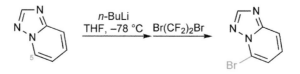

1,2,3-Triazolo[1,5-*a*]pyridines can be synthesised by oxidation of pyridine 2-carboxaldehyde hydrazones, presumably by way of the diazo species,[67] or *via* diazo-transfer reactions (*tosyl azide presents a significant explosion hazard – safer alternatives are available (5.4))*.[68]

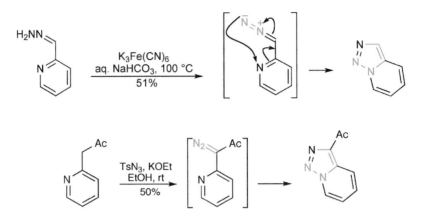

The 1,2,4-triazolo[4,3-*a*]pyridine nucleus can be accessed by cyclocondensation of 2-hydrazino-pyridines; the synthesis of the antidepressant trazodone, shown below, is an example.[69] Condensation of 2-hydrazinopyridines with carboxylic acids, promoted by polymer-supported triphenylphosphine with trichloroacetonitrile and heated with microwave irradiation, is also efficient.[70]

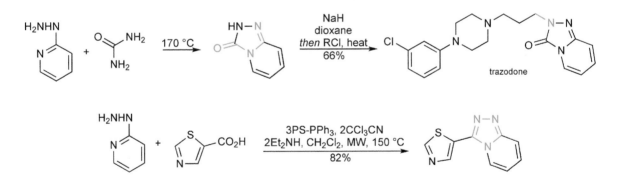

Oxidative closure of pyridin-2-yl hydrazones also produces 1,2,4-triazolo[4,3-*a*]pyridines.[71]

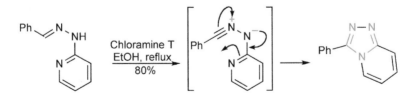

1,2,4-Triazolo[1,5-*a*]pyridines can be prepared *via* oxidative cyclisation of amidines[72] or acid-catalysed cyclisation of amidoximes,[73] including formamidoximes, which give rise to 2–unsubstituted compounds[74] – each of these is illustrated below.

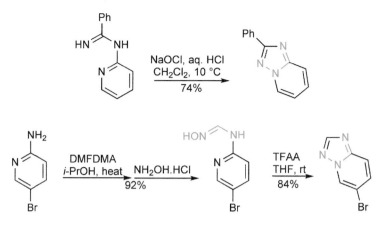

28.2.5 Compounds with an Additional Nitrogen in the Six-Membered Ring

In addition to the propensity for electrophilic substitution at C-1/C-3 (see above), the main feature of this class of heterocycle is that they undergo relatively easy nucleophilic attack in the six-membered ring,[75] which is now distinctly electron-deficient through the incorporation of imine units – the analogy with the ease of nucleophilic addition to diazines *versus* pyridines, is obvious. Some are so susceptible to nucleophilic addition that they form 'hydrates' even on exposure to moist air.[62] Preparative lithiations can, however, be carried out using less nucleophilic bases.

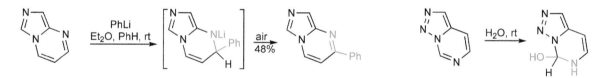

Ring synthesis of such molecules can proceed from diazines using methods analogous to those described for the synthesis of azolo-pyridines from pyridines, for example the dipolar cycloaddition of an *N*-imide to an alkyne,[76] or the bromine-promoted ring closure of hydrazones of 2-hydrazino-pyridines, -pyridazines (shown) or -pyrimidines.[77]

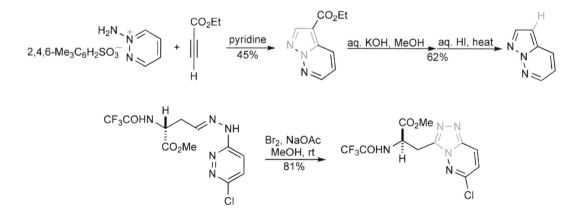

Various routes from the five-membered ring component are also possible – some representative chemistry involving pyrroles[78,79] is shown below.

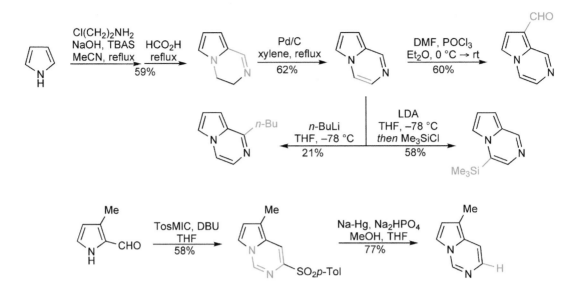

Using pyrazoles or imidazoles naturally leads to bicyclic systems with more than one nitrogen in *each* ring: 5-amino-pyrazoles[80] (in condensations with 1,3-dicarbonyl synthons; cf. 14.13.2.1), 1-amino-imidazoles,[81] and 4-amino-imidazoles[82] have been used as shown below. 3-Amino-pyrazoles[83] and 3-amino-1,2,4-triazoles[84] have been elaborated into purine analogues with four or five nitrogen atoms in the systems.

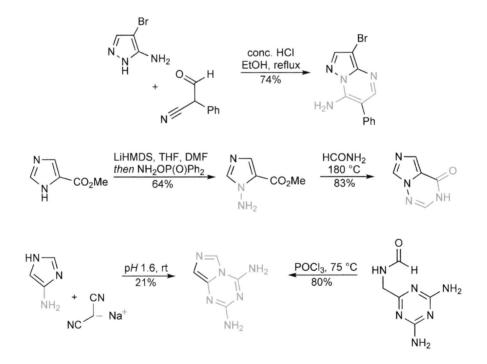

Condensation of 3,6-dichloropyridazine with hydrazides, best with microwave heating, produces [1,2,4]triazolo[4,3-*b*]pyridazines, in which the remaining chlorine can be displaced with amines.[85]

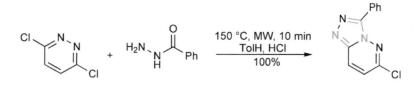

28.3 Quinolizinium[86] and Related Systems

Practically all the reactions of quinolizinium ions are similar to those of pyridinium salts, thus they are resistant to electrophilic attack, but readily undergo nucleophilic addition, the initial adducts undergoing spontaneous electrocyclic ring opening to afford, finally, 2-substituted pyridines;[87] however the susceptibility of the cations to nucleophiles is not extreme – like simpler pyridinium salts they are stable to boiling water.

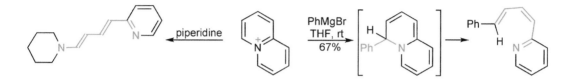

Quinolizine derivatives are usually prepared by cyclisations onto the nitrogen in a precursor pyridine.[88] Quinazolones *can* be made to undergo electrophilic substitution, at C-1/C-3,[89] there being a clear analogy with the reactivity of pyridones.

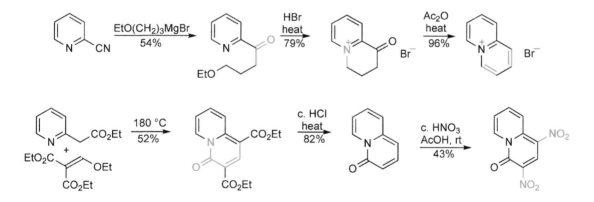

28.4 Pyrrolizine and Related Systems

The relatively high pK_a of 29 for deprotonation of 3*H*-pyrrolizine (cf. indene pK_a 18.5) indicates that formation of the 10π-electron pyrrolizine anion adds only minor stabilisation relative to the simple pyrrole originally present. Its reactions are those of a highly reactive carbanion, for example benzophenone condenses to generate a fulvene-like product.[90]

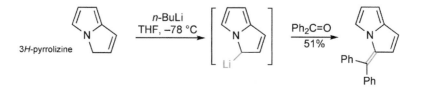

Isoelectronic replacement of a carbanionic carbon by a heteroatom gives much more stable compounds, and such 5,5-bicyclic aromatic systems have received considerable attention. In these compounds, sulfur, nitrogen and oxygen can also be incorporated into fully conjugated systems, unlike the 5,6-compounds, where only nitrogen can be used. Because of the variety of such systems, it is difficult to generalise about reactivity, but electrophilic substitution, which can take place in either ring, has been most widely reported with occasional examples of nucleophilic displacements and lithiations. Some representative reactions and self-explanatory syntheses are shown below.[91,92]

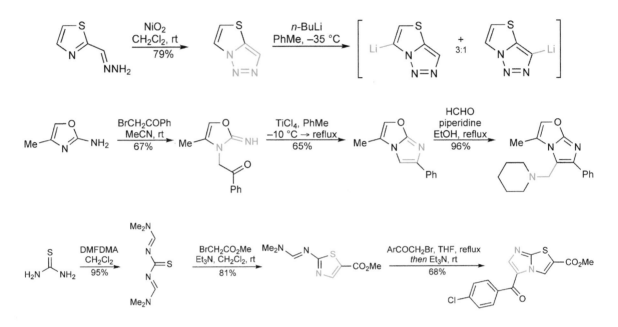

28.5 Cyclazines

The cyclazines (a trivial name) are tricyclic fused molecules containing a central nitrogen and a peripheral π-system. The definition of aromaticity in these compounds is not as straightforward as for the simple bicyclic molecules discussed above, and a more detailed analysis of the molecular orbitals may be required.

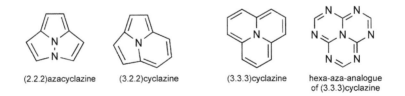

| (2.2.2)azacyclazine | (3.2.2)cyclazine | (3.3.3)cyclazine | hexa-aza-analogue of (3.3.3)cyclazine |

(3.2.2)Cyclazine is a stable aromatic system with a ring current, has a 10-electron annular π-system (excluding nitrogen), and is stable to light and air but, unlike its close analogue indolizine, is non-basic indicating the much weaker interaction between the nitrogen lone pair and the peripheral π-system. It does, however, react as an electron-rich aromatic, undergoing electrophilic substitution readily.

In contrast, (3.3.3)cyclazine has no aromatic resonance stabilisation and is unstable and highly reactive, displaying some diradical character. However, its hexa-aza analogue is extremely stable, this stabilisation being attributed to perturbation of the molecular orbitals by the electronegative atoms leading to a much larger separation of the HOMO and LUMO.[93] The double-bridgehead nitrogen system, (2.2.2)azacyclazine is isoelectronic with (3.2.2)cyclazine and is similarly a stable system.

Cyclazines can be prepared by cyclisation of bicyclic precursors, for example (3.2.2)cyclazine is prepared *via* a cycloaddition reaction on indolizine (28.1), or by cyclocondensation.[94]

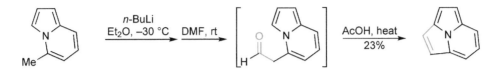

Exercises
Straightforward revision exercises (consult Chapter 28):
(a) At what position(s) does indolizine undergo electrophilic substitution? Why that position(s)?
(b) At what position does indolizine undergo strong base deprotonation?
(c) How could 2-methylpyridine be converted into 3-methylindolizine?
(d) What would be the product of reacting 2-aminopyridine with methyl bromoacetate?
(e) Draw resonance contributors to the quinolizinium cation to rationalise the position at which nucleophiles add to it.
(f) Is pyrrolizine aromatic? If so, how many electrons are there in the aromatic π-system?

More advanced exercises:
1. Suggest a structure for the final, monocyclic product of the following sequence: quinolizinium bromide with LiAlH$_4$ and then H$_2$/Pd giving C$_9$H$_{13}$N.
2. Write down the structures of the intermediates in the following synthesis of the quinolizinium cation: 2-methylpyridine was reacted with LDA, then EtO(CH$_2$)$_2$CH=O to give C$_{11}$H$_{17}$NO$_2$, which was heated with HI (→ C$_9$H$_{12}$NO$^+$ I$^-$); this salt was then heated with Ac$_2$O (→ C$_9$H$_{10}$N$^+$ I$^-$) and this finally heated with Pd–C to afford quinolizinium iodide.
3. Which indolizines would be formed from the following combinations: (i) 2-picoline with: (a) BrCH$_2$COMe/NaHCO$_3$, (b) MeCHBrCHO/NaHCO$_3$? (ii) What would be the products if the 2-picoline was replaced by 2-aminopyridine?
4. Deduce the structures of intermediates and final product in the following sequence: 5-methoxy-2-methylpyridine reacted with KNH$_2$/*i*-AmONO → C$_7$H$_8$N$_2$O$_2$, then this with Zn/AcOH → C$_7$H$_{10}$N$_2$O and finally this with HCO$_2$Me/PPE (polyphosphate ester) → C$_8$H$_8$N$_2$O.
5. Imidazo[1,5-*a*]pyridine, on reaction with aqueous HNO$_2$ gave 3-(pyridin-2-yl)-1,2,4-oxadiazole. Suggest a mechanism. What product would be obtained by reaction of indolizine with nitrous acid?
6. Give the structures of the bicyclic compounds formed by the following reactions: (i) 2-hydrazinothiazole with nitrous acid → C$_3$H$_2$N$_4$S; (ii) 2-aminothiazole with BrCH$_2$COPh → C$_{11}$H$_8$N$_2$S.

References
[1] 'Heterocyclic systems with bridgehead nitrogen atoms' in 'The Chemistry of Heterocyclic Compounds', Vol. 15, Ed. Mosby, W. L. Series Ed. Weissberger, A., Wiley-Interscience, **1961**; 'Special topics in heterocyclic chemistry' in 'The Chemistry of Heterocyclic Compounds', Vol. 30, Series Eds. Weissberger, A. and Taylor, E. C., Wiley-Interscience, **1977**.

[2] 'The chemistry of the pyrrocolines and the octahydropyrrocolines', Borrows, E. T. and Holland, D. O., *Chem. Rev.*, **1948**, *42*, 611; 'Advances in indolizine chemistry', Swinbourne, F. J., Hunt, J. H., and Klinkert, G., *Adv. Heterocycl. Chem.*, **1978**, *23*, 103.

[3] Armarego, W. L. F., *J. Chem. Soc.*, **1964**, 4226.

[4] Scholtz, M., *Chem. Ber.*, **1912**, *65*, 1718; Kuznetsov, A. G., Bush, A. A. and Babaev, E. V., *Tetrahedron*, **2008**, *64*, 749.

[5] Rossiter, E. D. and Saxton, J. E., *J. Chem. Soc.*, **1953**, 3654; Fuentes, O. and Paudler, W. W., *J. Heterocycl. Chem.*, **1975**, *12*, 379.

[6] Holland, D. O. and Nayler, J. H. C., *J. Chem. Soc.*, **1955**, 1504.

[7] Hickman, J. A. and Wibberly, D. G., *J. Chem. Soc., Perkin Trans. 1*, **1972**, 2954.

[8] Borrows, E. T., Holland, D. O. and Kenyon, J., *J. Chem. Soc.*, **1946**, 1077; Greci, L. and Ridd, J. H., *J. Chem. Soc., Perkin Trans. 2*, **1979**, 312.

[9] Lowe, O. G. and King, L. C., *J. Org. Chem.*, **1959**, *24*, 1200.

[10] Walter, L. A. and Margolis, P., *J. Med. Chem.*, **1967**, *10*, 498.

[11] Galbraith, A., Small, T., Barnes, R. A. and Boekelheide, V., *J. Am. Chem. Soc.*, **1961**, *83*, 453.

[12] Windgassen, R. J., Saunders, W. H. and Boekelheide, V., *J. Am. Chem. Soc.*, **1959**, *81*, 1459.

[13] 'Methods for the construction of the indolizine nucleus', Uchida, T. and Matsumoto, K., *Synthesis*, **1976**, 209.

[14] Chichibabin, A. E., *Chem. Ber.*, **1927**, *60*, 1607.

[15] Borrows, E. T., Holland, D. O. and Kenyon, J., *J. Chem. Soc.*, **1946**, 1069.

[16] Bragg, D. R. and Wibberly, D. G., *J. Chem. Soc.*, **1963**, 3277.

[17] Tielmann P. and Hoenke, C., *Tetrahedron Lett.*, **2006**, *47*, 261.

[18] Henrick, C. A., Ritchie, E. and Taylor, W. C., *Aust. J. Chem.*, **1967**, *20*, 2467.

[19] Wei, J., Hu, Y., Li, T. and Hu, H., *J. Chem. Soc., Perkin Trans. 1*, **1993**, 2487.

[20] Hou, J., Hu, Y. and Hu, H., *Synth. Commun.*, **1998**, *28*, 3397.

[21] Matsumoto, K., Ogasawara, A., Kimura, S., Hayashi, N. and Machiguchi, T., *Heterocycles*, **1998**, *48*, 861.

[22] Kim, I., Choi, J., Wo, H. K. and Lee, G. H., *Tetrahedron Lett.*, **2007**, *48*, 6863.

[23] Seregin, I. V., Schammel, A. W. and Gevorgyan, V., *Org. Lett.*, **2007**, *9*, 3433.

[24] Yan, B. and Liu, Y., *Org. Lett.*, **2007**, *9*, 4323.

[25] Fraser, M. *J. Org. Chem.*, **1971**, *36*, 3087; *ibid.*, **1972**, *37*, 3027.

[26] Paolini, J. P. and Robins, R. K., *J. Org. Chem.*, **1965**, *30*, 4085.

[27] Chayer, S., Schmitt, M., Collot, V. and Bourgignon, J.-J., *Tetrahedron Lett.*, **1998**, *39*, 9685.

[28] Paolini, J. P. and Robins, R. K., *J. Heterocycl. Chem.*, **1965**, *2*, 53.

[29] e.g. Cai, L., Cuevas, J., Peng, Y.-Y. and Pike, V. W., *Tetrahedron Lett.*, **2006**, *47*, 4449.

[30] Paudler, W. W. and Helmick, L. H., *J. Org. Chem.*, **1968**, *33*, 1087.

[31] Paudler, W. W. and Shin, H. G., *J. Org. Chem.*, **1968**, *33*, 1638; Guildford, A. J., Tometzki, M. A. and Turner, R. W., *Synthesis*, **1983**, 987.

[32] Gudmundsson, K. S., Drach, J. C. and Townsend, L. B., *J. Org. Chem.*, **1997**, *62*, 3453.

[33] Chichibabin, A. E., *Chem. Ber.*, **1924**, *57*, 1168; Kröhnke, F., Kickhöfen, B. and Thoma, C., *ibid.*, **1955**, *88*, 1117.

[34] Gueiffier, A., *et al.*, *J. Med. Chem.*, **1998**, *41*, 5108.

[35] Chichibabin, A. E., *Chem. Ber.*, **1924**, *57*, 2092.

[36] Katritzky, A. R., Xu, Y.-J. and Tu, H., *J. Org. Chem.*, **2003**, *68*, 4935.

[37] Yadav, J. S., Reddy, B. V. S., Rao, Y. G., Srinivas, M. and Narsaiah, A. V., *Tetrahedron Lett.*, **2007**, *48*, 7717.

[38] Blackburn, C., Guan, B., Fleming, P., Shiosaki, K. and Tsai, S., *Tetrahedron Lett.*, **1998**, *39*, 3635

[39] Rousseau, A. L., Matlaba, P. and Parkinson, C. J., *Tetrahedron Lett.*, **2007**, *48*, 4079;

[40] Adib, M., Mahdavi, M., Noghani, M. A. and Mirzaei, P., *Tetrahedron Lett.*, **2007**, *48*, 7263.

[41] Shaabani, A., Ebrahim Soleimani, E and Maleki, A., *Tetrahedron Lett.*, **2006**, *47*, 3031.

[42] Ireland, S. M., Tye, H. and Whittaker, M., *Tetrahedron Lett.*, **2003**, *44*, 4369.

[43] Umkerer, M., Ross, G., Jäger, N., Burdack, C., Kolb, J., Hu, H., Alvim-Gaston, M. and Hulme, C., *Tetrahedron Lett.*, **2007**, *48*, 2213.

[44] Bienayme, H. and Bouzid, K., *Angew. Chem., Int. Ed. Engl.*, **1998**, *37*, 2234.

[45] Fuentes, O. and Paudler, W. W., *J. Heterocycl. Chem.*, **1975**, *12*, 379.

[46] Bower, J. D. and Ramage, G. R., *J. Chem. Soc.*, **1955**, 2834.

[47] Paudler, W. W. and Kuder, J. E., *J. Org. Chem.*, **1967**, *32*, 2430.

[48] Hlasta, D. J. and Silbernagel, M. J., *Heterocycles*, **1980**, *48*, 101.

[49] Blatcher, P. and Middlemiss, D., *Tetrahedron Lett.*, **1980**, *21*, 2195.

[50] Moulin, A., Garcia, S., Martinez, J. and Fehrentz, J.-A., *Synthesis*, **2007**, 2667.

[51] Shibahara, F., Kitagawa, A., Yamaguchi, E. and Murai, T., *Org. Lett.*, **2006**, *8*, 5621.

[52] Bourdais, J. and Omar, A.-M. M. E., *J. Heterocycl. Chem.*, **1980**, *17*, 555.

[53] Iwao, M. and Kuraishi, T., *J. Heterocycl. Chem.*, **1977**, *14*, 993.

[54] Bourdais, J., Rajniakova, O. and Povazanec, F., *J. Heterocycl. Chem.*, **1980**, *17*, 1351.

[55] Tanji, K., Sasahara, T., Suzuki, J. and Higashino, T., *Heterocycles*, **1993**, *35*, 915; Miki, Y., Yagi, S., Hachiken, H. and Ikeda, M., *ibid.*, **1994**, *38*, 1881.

[56] Aboul-Fadl, T., Löber, S. and Gmeiner, P., *Synthesis*, **2000**, 1727.

[57] Krischke, R., Grashey, R. and Huisgen, R., *Justus Liebigs Ann. Chem.*, **1977**, 498.

[58] Boekelheide, V. and Fedoruk, N. A., *J. Org. Chem.*, **1968**, *33*, 2062.

[59] Tsuchiya, T., Sashida, H. and Konoshita, A., *Chem. Pharm. Bull.*, **1983**, *31*, 4568.

[60] Miki, Y., Tasaka, J., Uemura, K., Miyazeki, K. and Yamada, J., *Heterocycles,*, **1996**, *43*, 2249.

[61] 'The chemmistry of the triazolopyridines', Jones, G. and Sliskovic, D. R., *Adv. Heterocycl. Chem.*, **1983**, *34*, 79.

[62] Maury, G., Paugam, J.-P. and Paugam, R., *J. Heterocycl. Chem.*, **1978**, *15*, 1041.

[63] Jones, G. and Sliskovic, D. R., *J. Chem. Soc., Perkin Trans. 1*, **1982**, 967; Jones, G., Mouat, D. J. and Tonkinson, D. J., *ibid.*, **1985**, 2719.

[64] 'Some aspects of azido-tetrazolo isomerisation', Tishler, M., *Synthesis*, **1973**, 123.

[65] Abarca, B., Ballesteros, R., Jones, G. and Mojarrad, F., *Tetrahedron Lett.*, **1986**, *27*, 3543.

[66] Finkelstein, B. L., *J. Org. Chem.*, **1992**, *57*, 5538.

[67] Bower, J. D. and Ramage, G. R., *J. Chem. Soc.*, **1957**, 4506.

[68] Regitz, M., *Chem. Ber.*, **1966**, *99*, 2918.

[69] Kauffman, T., Vogt, K., Barck, S. and Schulz, J., *Chem. Ber.*, **1966**, *99*, 2593.

[70] Wand, Y., Sarris, K., Sauer, D. R. and Djuric, S. W., *Tetrahedron Lett.*, **2007**, *48*, 2237.

[71] Bourgeois, P., Cantegril, R., Chene, A., Gelin, J., Mortier, J. and Moyroud, J., *Synth. Commun.*, **1993**, *23*, 3195.

[72] Grenda, V. J., Jones, R. E., Gal, G. and Sletzinger, M., *J. Org. Chem.*, **1965**, *30*, 259.

[73] Polanc, S., Vercek, B., Sek, B., Stanovnik, B. and Tisler, M., *J. Org. Chem.*, **1974**, *39*, 2143.

[74] Huntsman, E. and Balsells, J., *Eur. J. Org. Chem.*, **2005**, 3761.

[75] Paudler, W. W., Chao, C. I. P. and Helmick, L. S., *J. Heterocycl. Chem.*, **1972**, *9*, 1157.

[76] Rees, C. W., Stephenson, R. W. and Storr, R. C., *J. Chem. Soc., Chem. Commun.*, **1974**, 941; Beswick, P., Bingham, S., Bountra, C., Brown, T., Browning, K., Campbell, I., Chessell, I., Clayton, N., Collins, S., Corfield, J., Guntrip, S., Haslam, P. C., Lambeth, P., Lucas, F., Mathews, N., Murkit, G., Naylor, A., Pegg, N., Pickup, E. and Player, H., *Bioorg. Med. Chem. Lett.*, **2004**, *14*, 5445; Whitehead, A. J., Ward, R. A. and Jones, M. F., *Tetrahedron Lett.*, **2007**, *48*, 911.

[77] Svete, J., Stanovnik, B. and Tisler, M., *J. Heterocycl. Chem.*, **1994**, *31*, 1259; Turk, C., Golobic, A., Golic, L. and Stanovnik, B., *ibid.*, **1998**, *35*, 513.

[78] Minguez, J. M., Castellote, M. I., Vaquero, J. J., Garcia-Navio, J. L., Alvarez-Builla, J. and Castano, O., *J. Org. Chem.*, **1996**, *61*, 4655.

[79] Minguez, J. M., Vaquero, J. J., Garcia-Navio, J. L. and Alvarez-Builla, J., *Tetrahedron Lett.*, **1996**, *37*, 4263.

[80] Frey, R. R., Curtin, M. L., Albert, D. H., Glaser, K. B., Pease, L. J., Soni, N. B., Bouska, J. J., Reuter, D., Stewart, K. D., Marcotte, P., Bukofzer, G., Li, J., Davidsen, D. K. and Michaelides, M. R., *J. Med. Chem.*, **2008**, *51*, 3777.

[81] Heim-Riether, A. and Healy, J., *J. Org. Chem.*, **2005**, *70*, 7331.

[82] Wang, Z., Huynh, H. K., Han, B., Krishnamurthy, R. and Eschenmoser, A., *Org. Lett.*, **2003**, *5*, 2067.

[83] Dolzhenko, A. V., Dolzhenko, A. V. and Chui, W.-K., *Heterocycles*, **2008**, *75*, 1575.

[84] '1,2,4-Triazolo[1,5-*a*][1,3,5]triazines (5-azapurines): synthesis and biological activity', Dolzhenko, A. V., Dolzhenko, A. V. and Chui, W.-K., *Heterocycles*, **2006**, *68*, 1723.

[85] Aldrich, L. N., Lebois, E. P., Lewis, L. M., Naywajko, N. T., Niswender, C. M., Weaver, C. D., Conn, P. J. and Lindsley, C. W., *Tetrahedron Lett.*, **2009**, *50*, 212.

[86] 'Aromatic quinolizines', Thyagarajan, B. S., *Adv. Heterocycl. Chem.*, **1965**, *5*, 291; 'Aromatic quinolizines', Jones, G., *ibid.*, **1982**, *31*, 1.

[87] Miyadera, T., Ohki, E. and Iwai, I., *Chem. Pharm. Bull.*, **1964**, *12*, 1344; Kröhnke, F. and Mörler, D., *Tetrahedron Lett.*, **1969**, 3441.

[88] Boekelheide, V. and Lodge, J. P., *J. Am. Chem. Soc.*, **1951**, *73*, 3681; Glover, E. E. and Jones, G., *J. Chem. Soc.*, **1958**, 3021.

[89] Thyagarajan, B. S. and Gopalakrishnan, P. V., *Tetrahedron*, **1964**, *20*, 1051 and **1965**, *21*, 945; Forti, L., Gelmi, M. L., Pocar, D. and Varallo, M., *Heterocycles*, **1986**, *24*, 1401.

[90] Okamura, W. H. and Katz, T. J., *Tetrahedron*, **1967**, *23*, 2941.

[91] Jones, G., Ollivierre, H., Fuller, L. S. and Young, J. H., *Tetrahedron*, **1991**, *47*, 2851 and 2861.

[92] Mekonnen, B., Crank, G. and Craig, D., *J. Heterocycl. Chem.*, **1997**, *34*, 589; Mekonnen, B. and Crank, G., *Tetrahedron*, **1997**, *53*, 6959.

[93] Farquhar, D., Gough, T. T. and Leaver, D., *J. Chem. Soc., Perkin Trans. 1*, **1976**, 341.

[94] Windgassen, R. J., Saunders, W. H. and Boekelheide, V., *J. Am. Chem. Soc.*, **1959**, *81*, 1459.

29

Heterocycles Containing More Than Two Heteroatoms

In systems that contain more than two heteroatoms in the same ring, the trends in properties that this book describes are taken to further extremes. In particular, the additional heteroatoms, in both six- and five-membered systems, lead to a suppression of electrophilic substitution and a slowing of electrophilic addition to nitrogen. On the other hand, further increases in tendencies for nucleophilic substitution and addition, and in the five-membered compounds further increases in acidities of *N*-hydrogen, are found.

Multi-heteroatom heterocycles are comparatively rare in nature, however in medicinal chemistry they are of very considerable significance, for example analogues of the pyrimidine and purine nucleosides have been extensively studied (see examples in Chapter 33).

29.1 Five-Membered Rings
29.1.1 Azoles
The triazoles are numbered to indicate the relative positions of the nitrogen atoms; tetrazole and pentazole are unambiguous names. In considering the chemistry of systems containing three or more nitrogens, stability and therefore safety must always be a consideration, particularly for tetrazoles. 1,2,3-Triazoles are, however, surprisingly stable, when one considers that they contain three directly linked nitrogen atoms, though on flash vacuum pyrolysis at 500 °C, they do lose nitrogen to give 2*H*-azirines, probably *via* the 1*H*-isomer.[1] Benzotriazole is similarly relatively stable and has been distilled *in vacuo* at 200 °C, though explosions have been reported during this process. Although many tetrazoles are relatively stable, some are explosive, often designed as such. The pentazole ring system is only known in a few aryl derivatives that generally decompose (possibly explosively) at or below room temperature,[2] the most stable being 4-dimethylaminophenylpentazole. The pentazolate anion has been detected by mass spectrometry.[3]

The additional heteroatoms make these systems less basic, but more acidic than comparable 1,2- and 1,3-azoles. Each is subject to the same kind of tautomerism as discussed for the 1,2- and 1,3-azoles (Chapter 25 Introduction and 24.1.1.1), in which the tautomers are equivalent. Also, in these systems, tautomerism generates different arrangements of imine nitrogen and *N*-hydrogen components.

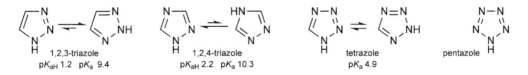

There is a far from complete set of comparisons of relative reactivities[4] in nucleophilic displacement of halide. From a study of reactions of bromo-*N*-methyl-tetrazoles, -triazoles and -imidazoles with piperidine, it seems that the presence of two doubly bonded nitrogens is necessary to overcome electron release from the singly bonded nitrogen to approach the reactivity of 2-bromopyridine.[5]

Heterocyclic Chemistry 5th Edition   John Joule and Keith Mills
© 2010 Blackwell Publishing Ltd

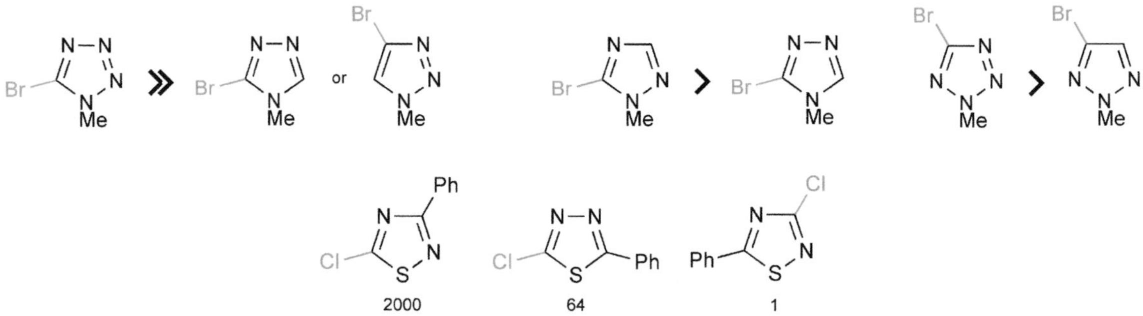

Relative rates of reaction with piperidine in ethanol

29.1.1.1 1,2,3-Triazole[6]

1,2,3-Triazole is fairly resistant to *N*-alkylation under neutral conditions, however both acylations and alkylations involving *N*-anions occur readily, but mixtures of 1- and 2-substituted products are often obtained.[7,8] Trimethylsilylation produces 2-trimethylsilyl-1,2,3-triazole, and this can be alkylated unambiguously at N-1, with concomitant loss of silicon.[9] An equilibrium mixture of *N*-acetyl-1,2,3-triazoles contains predominantly the 2-acetyl isomer. The 2*H*-tautomer is also dominant in 1,2,3-triazole itself,[10] which may reflect unfavourable adjacent imine lone-pair/lone-pair interactions in the 1*H*-isomer and is in agreement with calculations that suggest that the 2*H*-isomer is more aromatic.[11] Heating in sulfolane at 150 °C converts the *N*-acyl compounds into oxazoles in a synthetically useful transformation.[12]

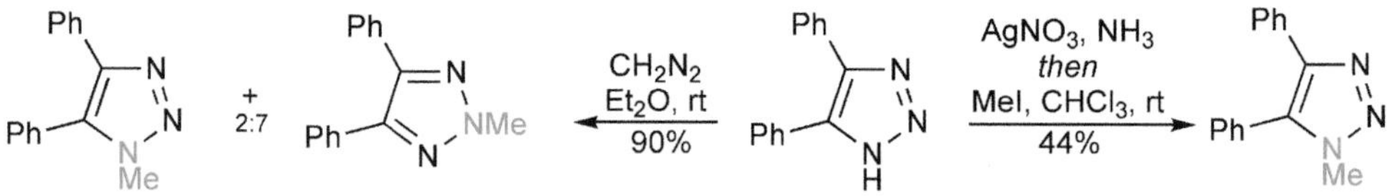

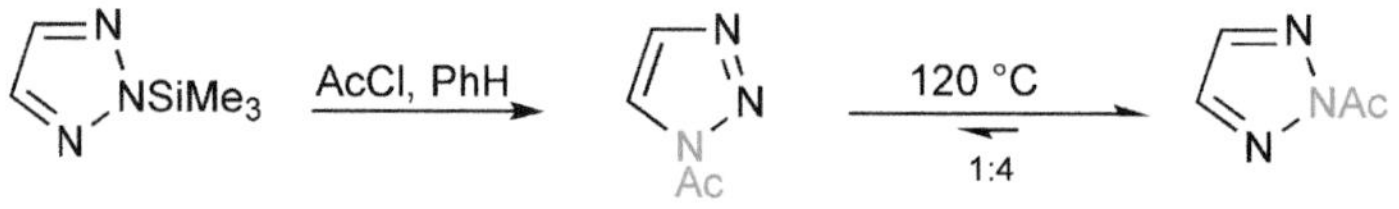

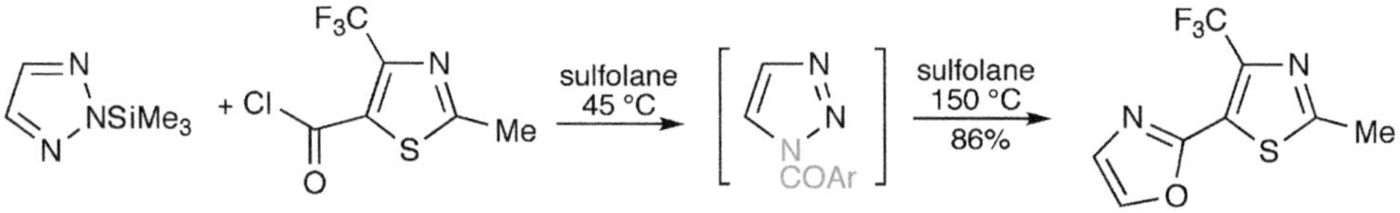

1,2,3-Triazole itself forms a 4,5-dibromo derivative in high yield with bromine at 50 °C.[13] 1-Methyl-1,2,3-triazole can be mono-brominated at C-4, but the 2-methyl isomer is less reactive, requiring the use of an iron catalyst;[14] the lower reactivity of the latter is probably related to the presence of two imine units. Nitration of 2-phenyl-1,2,3-triazole proceeds first on the benzene ring, but then does bring about hetero-ring substitution.[15]

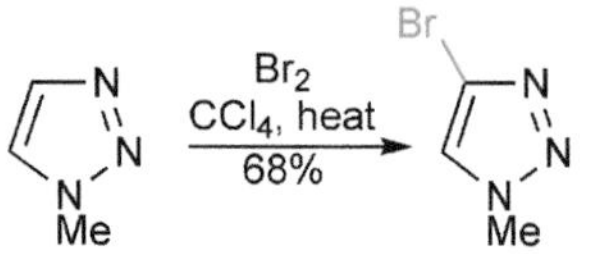

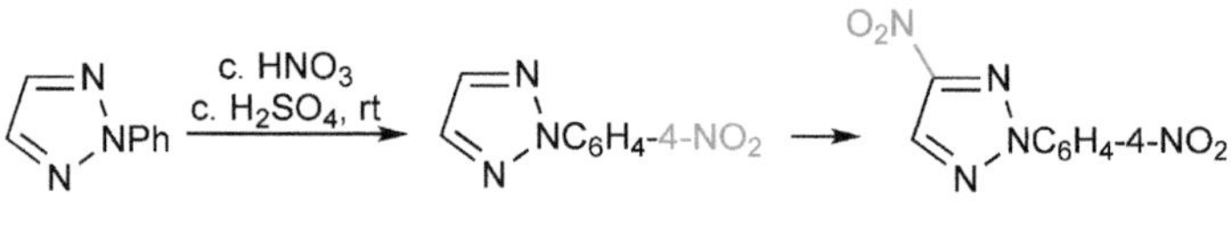

The ring system is relatively resistant to both oxidation and reduction, as exemplified below.[16]

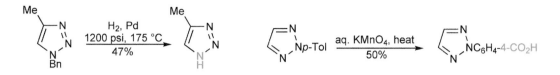

N-Substituted 1,2,3-triazoles can be lithiated directly at carbon, but low temperatures must be maintained to avoid ring cleavage by cycloreversion.[17,18]

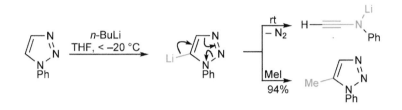

4,5-Dibromo-1-methoxymethyl-1,2,3-triazole forms the 5-lithio compound with *n*-butyllithium at −80 °C and 4,5-dibromo-2-methoxymethyl-1,2,3-triazole undergoes a comparable exchange.[15] 1-Phenyl-1,2,3-triazoles participate in Diels–Alder cycloadditions with dimethyl acetylenedicarboxylate using aluminium chloride catalysis, producing 1-phenylpyrazoles.[19]

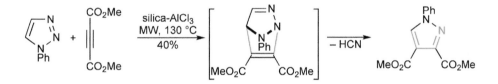

29.1.1.2 *1,2,4-Triazole*[20]

N-Alkylations and -acylations generally occur at N-1, reflecting the higher nucleophilicity of N–N systems (cf. 14.1.1.2), however 4-alkyl derivatives can be prepared *via* quaternisation of 1-acetyl-1,2,4-triazole[21] or the acrylonitrile or crotononitrile adducts[22] (note that N-1 and N-2 are equivalent until substitution occurs). In alkylations of 1,2,4-triazole, generally a ratio of 9 : 1, N-1:N-3 isomers is present in the crude product, but the minor product is very easily removed; even simple distillation completely separates the isomers, illustrating the very large difference in physical properties between isomers.[23] Direct hydroxymethylation of 3,4-disubstituted 1,2,4-triazoles occurs at C-5 by heating with formalin or paraformaldehyde.[24]

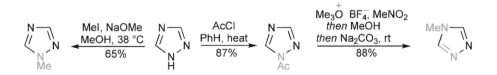

Bromination occurs readily in alkaline solution giving 3,5-dibromo-1,2,4-triazole;[25] the 3-monochloro derivative can be obtained by thermal rearrangement of the *N*-chloro isomer;[26] an analogous N→C 1,5-sigmatropic shift followed by tautomerisation converts the 1- into the 3-nitro compound.[27]

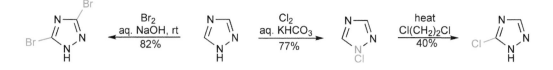

Secondary radical Minisci alkylations of 1-alkyl-1,2,4-triazoles take place at C-5; primary radical substitutions are known in an intramolecular sense.[28]

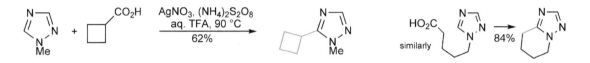

C-Lithiations can be easily effected on *N*-1-protected 1,2,4-triazoles, the resulting 5-lithio derivatives being much more stable than *C*-lithiated 1,2,3-triazoles.[29] 5-Silylation can even be achieved using triethylamine with trimethylsilyl bromide – deprotonation of an N^+-trimethylsilyl triazolium cation is presumably involved.[30] Exactly comparable silylations can be achieved with 2-aryl-1,3,4-oxa- and -thia-diazoles; application of this regime to 1-phenyltetrazole produced phenyl trimethylsilyl carbodiimide in 90% yield!

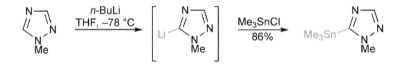

3-Amino-1,2,4-triazole can be diazotised normally: the resulting diazonium salt has been used for the production of azo dyes, and also loses nitrogen with easy replacement by nucleophiles. The bromo- and nitro-triazoles, which can be thus prepared are themselves substrates for nucleophilic displacement reactions.[21,31] 5-Bromo and 5-nitro[32] groups are good leaving groups in 1-alkyl-1,2,4-triazoles for hetero nucleophiles; for carbon nucleophiles, methylsulfone is a better leaving group.[33]

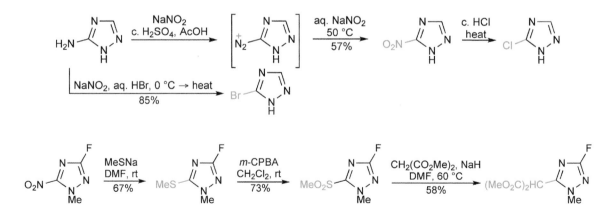

The oxidative desulfurisation of 1,2,4-triazole thiones is a type of reaction common to other electron-deficient nitrogen heterocycles; nitric acid is the oxidant in the example below. The process involves loss of sulfur dioxide from an intermediate sulfinic acid.[34]

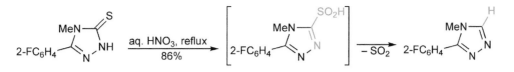

4-Phenyl-1,2,4-triazoline-3,5-dione (often PTAD or 'Cookson's reagent') is a highly reactive dieno-phile;[35] it has often been used to trap unstable dienes or characterize dienes as adducts. Two examples of its reactivity are shown below.[36,37]

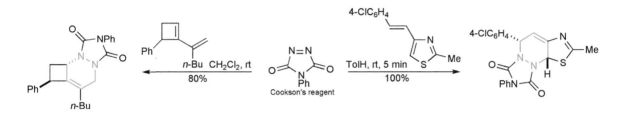

29.1.1.3 Tetrazoles[38]

Noting the similarity of (NH) tetrazole pK_as to those of carboxylic acids, tetrazoles have often been used as bioequivalent replacements for CO_2H, and as general variants, in pharmacologically active compounds (Chapter 33). The acid replacement extends to the tetrazole analogue of proline (p. 567), which is more soluble than proline itself, but retains its catalytic properties for condensation reactions.

Stability

Tetrazoles are generally surprisingly stable, although tetrazole itself (mp 158 °C, decomposes > 180 °C) is classified as an explosive, at least for shipping purposes. Chloro- and alkylthio-tetrazoles decompose at somewhat lower temperatures. The decomposition and explosive properties of certain tetrazole derivatives are very important from an applications and safety viewpoint, although, as noted above, tetrazoles are also important (and stable) components of drugs.

Tetrazole and 5-aminotetrazole are components of mixtures (with oxidants) used as high-speed inflators in car air bags, *via* a controlled explosive liberation of nitrogen and other gases. Other tetrazole derivatives are definitely explosives – 'Tetrazene' (a commercial name, not to be confused with the tetrazine ring system) has been used as a primer for percussion caps. Salts of 5-nitrotetrazole are promising safer (less sensitive) and less toxic replacements for lead azide primers – they are described as 'green explosives' because they generate non-toxic products![39]

Certain intermediates, such as diazotized 5-aminotetrazole (see below) present a serious risk of explosion. Introduction of large substituents generally reduces, but may not eliminate, explosive hazards, therefore caution must always be used. Unexpected hazardous by-products may also turn up, for example 5-azidotet-razole was inadvertently formed in an attempt to repeat a literature preparation of 1,5-diaminotetrazole *via* nitrosation of diaminoguanidine, resulting, on work-up, in an explosion that injured two chemists.[40] 5-Azidotetrazole is highly sensitive to friction, shock and electrostatic discharge; the 1-phenyl derivative is significantly less sensitive, but explodes violently on rapid heating.[41]

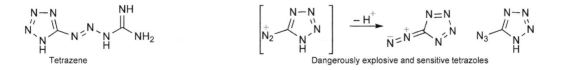

Reactions

Tetrazoles alkylate and acylate on N-1(4) and/or N-2(3), the regioselectivity depending, in part, on the substituent at C-5. The formation of mixtures is the usual outcome for alkylation and this is a significant problem in medicinal chemistry. Although a range of specifically N-1-substituted tetrazoles are available by ring synthesis, the direct synthesis of N-2-substituted compounds is more problematic.

In some cases, highly regioselective N-1-alkylation can be carried out by first introducing a bulky labile group at N-2, then quaternisation followed by cleavage of the 2-substituent. Both tri-*n*-butylstanyl[42] and *t*-butyl[43] have been used for this purpose.

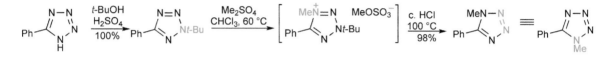

In addition to steric effects, the presence of an electron-withdrawing group at C-5 favours N-2-alkylation. The best directing substituents are 4-nitrophenyl (ratio 5.8:1) and cyano (5.4:1).[44] Out of a number of 5-thio-substituted tetrazoles, the 4-nitrophenylsulfone is the only one to give highly selective 2-alkylation.[45]

The use of a benzyl dithiocarbonate alkylating agent gives only the 2-benzyl isomer, in very high yield,[46] while the methyl analogue gives a mixture, although still with very good N-2 selectivity (7:1).[47]

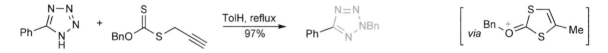

The difference in physical and chemical properties between 1- and 2-substituted tetrazoles can be large. The preparation of pure 2-methyltetrazole *via* alkylation (below) and the lithiation properties described later in this section, nicely demonstrate this differential reactivity, which can be put to good preparative use.

The relatively selective methylation (85:15, N-2:N-1) of 5-cyanotetrazole can be used to prepare pure 2-methyltetrazole by utilising the very large difference in stabilities of 1- and 2-methyl-tetrazole-5-carboxylic acids. The mixture of methylated nitriles can be hydrolysed to the carboxylate salts and, upon acidification, the 1-methyl acid decarboxylates spontaneously. The resulting 1-methyltetrazole and 2-methyltetrazole-5-carboxylic acid can be easily separated, then the latter decarboxylated at 200 °C to give pure 2-methyl tetrazole.[48]

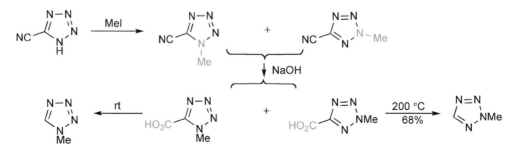

Remarkably, some *C*-electrophilic substitutions such as bromination, mercuration and even Mannich reactions, but not nitration, can be achieved, though the mechanisms for these substitutions may not be of the conventional type.

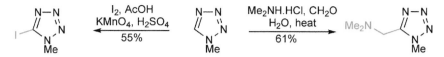

5-Halide and sulfonyl groups in 1-substituted tetrazoles are readily displaced by nucleophiles, methyl-sulfonyl being more reactive than chlorine.[49] 5-Bromo-1-methyltetrazole is considerably more reactive than the 2-methyl isomer due to more effective delocalization of the negative charge in the intermediate adduct.[50]

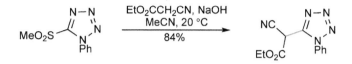

Metallation

C-Lithiation of 1-substituted tetrazoles occurs readily and the resulting lithio derivatives can be trapped with electrophiles, but there is a strong tendency for decomposition of the lithio compounds, which depends on the 1-substituent. 5-Lithiated 2-substituted tetrazoles appear to be significantly more stable than the 1-isomers, as shown below, by a comparison of the two *N*-benzyloxymethyl (BOM)-protected isomers.[51] *p*-Methoxybenzyl can be removed finally by hydrogenation or oxidation with cerium(IV) ammonium nitrate.

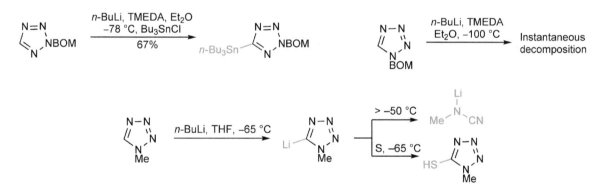

5-Alkyl groups in 1-substituted tetrazoles can be lithiated, but in a 5-alkyl-2-methyl-tetrazole it is the *N*-methyl that is metallated; 5-methyl-2-trityltetrazole lithiates normally at the *C*-methyl. The methylene group in 5-(benzotriazolylmethyl)tetrazole is sufficiently activated that *N*-protection is unnecessary – the benzotriazole can then act as a leaving group for displacement by Grignard reagents (29.3), as shown below.[52]

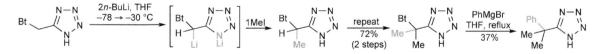

A tetrazole can also act as an *ortho*-directing group, as in the lithiation of 5-phenyltetrazole.[53,54] A tetrazol-5-yl ether similarly directs *ortho*-metallation, but here the tetrazole migrates from the oxygen to the lithiated carbon.[55]

Flash vacuum pyrolysis of 5-aryl-tetrazoles generates aryl-carbenes,[56] but heating the pyrimidinyl-phenyltetrazole shown below in refluxing decalin (180 °C) results in the formation of an intermediate nitrene, which then cyclises onto the pyrimidine nitrogen, forming a [1,2,4]triazolo[1,5-*a*]-pyrimidine.[57]

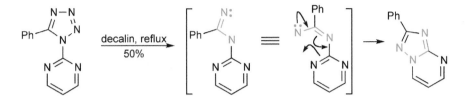

Hypobromite oxidation of 5-benzylaminotetrazole provides a useful preparation of benzyl-isonitrile.[58]

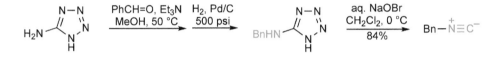

1,3,4-Oxadiazoles are formed on heating tetrazoles with acylating agents, *via* rearrangement of first-formed 2-acyl derivatives.[59]

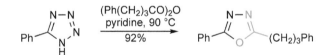

Diazotization of 5-Aminotetrazole

This is a useful, but very hazardous preparative procedure for a number of 5-substituted tetrazoles[60] and also, upon treatment with base, diazotetrazole,[61] a source of atomic carbon. *It cannot be overemphasized how dangerous this diazotisation reaction is!* The low limit for possible spontaneous detonation in *water* is between 1 and 5%! Even under very well-controlled conditions, during the synthesis of 5-nitrotetrazole by diazotization in the presence of excess nitrite, microdetonations occur in splashes on the walls of the vessel and often cause significant damage. However, the use of micro flow reactors (cf. 5.3) allow the reaction to be carried out safely.[62]

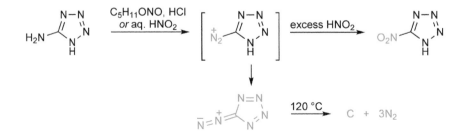

29.1.1.4 Pentazoles

The only known pentazoles are *N*-aryl derivatives, the most stable being those bearing an electron-releasing group on the aryl ring, although the pentazolate anion has been detected by mass spectrometry and NMR (see below). 1-(4-Dimethylaminophenyl)pentazole is stable at room temperature for several hours and its structure was determined by X-ray crystallography at 128 K.[63] *N*-Aryl-pentazoles are prepared by the reaction of aryl diazoniums with azide anion.

An experimental and theoretical study was based on the preparation and degradation of the tetra-*n*-butylammonium phenolate salt of 1-(4-hydroxyphenyl)pentazole, using ^{15}N-labelled compounds. Attempted removal of the aryl 'protecting group' by ozonolysis only resulted in degradation of the pentazole ring.[64] 1-(4-chlorophenyl)pentazole was observed (by ^{15}N NMR) as an unstable intermediate in the reaction of the aryl diazonium salt with azide to give the aryl azide.[65] Pentazole (as the zinc derivative) is observed by NMR following ceric ion oxidation of 1-anisylpentazole at –40 °C.[66]

29.1.1.5 Ring Synthesis of Azoles
1,2,3-Triazoles

1,2,3-Triazoles are generally prepared by the cycloaddition of an alkyne with an azide, which can be promoted thermally or by metal catalysis.[67] This cycloaddition reaction of an azide and an alkyne is a typical 'Click' reaction,[68] that is, a highly efficient and reliable general reaction that can be used for linking moieties for various purposes, such as in combinatorial chemistry. Opinion is divided on whether 'Click' is a useful or necessary concept[69] but it seems to have become associated, almost synonymously, but incorrectly, solely with this triazole synthesis.

The hazardous nature (see 'Tetrazoles' below) of some alkyl azides limits the method in these cases, however *N*-alkyl- (benzyl) -1,2,3-triazoles can be obtained by preparing the azide *in situ*, for example using a mixture of sodium azide, the alkyne and a benzyl halide;[70] *in situ* preparation of aryl azides is also feasible.[71] An alternative method in which the alkyl azide is also generated *in situ*, is to use a diazo transfer to a primary amine; the diazo transfer requires Cu(II) and the cycloaddition needs Cu(I) so a reducing agent is added together with the alkyne following the first phase.[72]

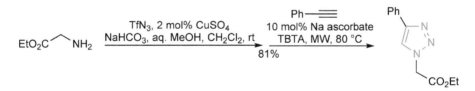

A convenient synthesis, which leads to *N*-hydrogen-1,2,3-triazoles, utilises the stable (and relatively safe) trimethylsilyl azide.[73] Alternatively, by conducting a cycloaddition in the presence of formaldehyde, *N*-hydroxymethyl-triazoles are formed (mainly the 2-isomer from isomerisation of initially formed 1-hydroxymethyl-triazole) from which *N*-unsubstituted heterocycles can be easily obtained using base; azidomethanol is formed *in situ* and is the entity that adds to the alkyne.[74]

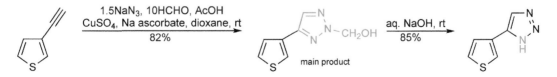

For *C*-unsubstituted 1,2,3-triazoles, rather than using gaseous ethyne, it is much more convenient to use vinyl acetate as starting material; or in general, an enamine, an enol ether[75] or a vinyl sulfone[76] as alkyne equivalents.

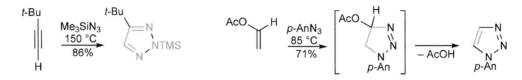

It is now usual to promote these cycloadditions by catalysts:[77] for example, reaction with *N*-tosyl-ynamides, using ruthenium[78] or copper[79] catalysts, giving 1-substituted 5- and 4-amino triazoles, respectively; the formation of the 1,4-substitution pattern with copper catalysis[77] and 1,5-pattern with ruthenium catalysis[80] seems to be general. The latter metal will also promote addition to internal alkynes.[81]

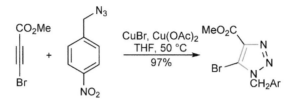

Another variation in which the interaction is between an acetylenic Grignard reagent and the azide, and is probably a stepwise sequence, is shown below, the immediate organometallic product being available for subsequent reaction with an electrophile.[82]

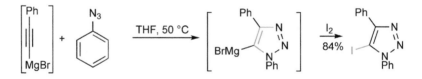

Other useful syntheses of 1,2,3-triazoles include diazo-transfer to enamino-ketones from either sulfonyl azides[83] or 3-diazo-oxindole,[84] and reaction of dichloroacetaldehyde tosylhydrazone with amines, and each of these is illustrated below.[85]

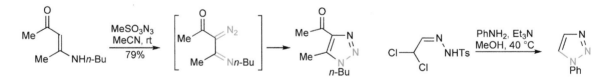

1,2,4-Triazoles

1,2,4-Triazoles are available *via* cyclodehydration reactions of *N,N′*-diacyl-hydrazines with amines, although the conditions are often quite vigorous.[86] An interesting variant utilises *sym*-triazine (1,3,5-triazine) as an equivalent of HN(CHO)$_2$.[87] Condensations of aminoguanidine with esters, or an ortho-formate,[88] give 3-amino-1,2,4-triazoles.[89]

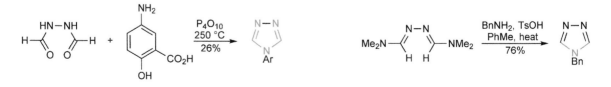

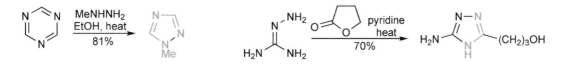

1,2,3-Triazolin-3-ones can be formed from primary amines in reaction with methyl (*E*)-*N*-(ethoxymethylene) hydrazinecarboxylate,[90] and the synthesis of the much-used 4-phenyl-1,2,4-triazoline-3,5-dione also starts from methyl carbazate, a final oxidation of 4-phenylurazole giving the carmine red crystalline product.[91]

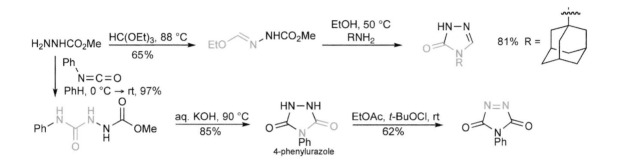

Acyl-hydrazides react with *S*-methyl-isothioureas producing 3-amino-1,2,4-triazoles,[92] or with imidoyl chlorides[93] forming, 1,2,4-triazoles.

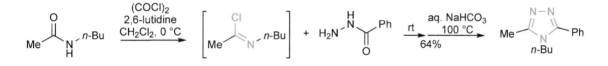

Tetrazoles

CAUTION: *Sodium azide is highly toxic and environmentally damaging.* There is also a possibility of forming sensitive explosive azides when certain metals – copper, mercury and, possibly, zinc – are used to catalyse some of the following reactions, and also if they are flushed into drains where they can react with metal pipes.

A procedure for destroying residual sodium azide has been described, as part of a detailed preparation (80 mmol scale) of (*S*)-5-pyrrolidin-2-yltetrazole, a useful enantio-catalyst.[94]

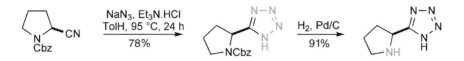

Tetrazoles are usually prepared by the reaction of an azide with a nitrile, or an activated amide; tri-*n*-butyltin azide and trimethylsilyl azide are more convenient and safer reagents than azide anion in some cases; copper(I) oxide catalysis in the trimethylsilyl azide protocol is very efficient for the production of *N*-unsubstituted tetrazoles,[95] and arylsulfonyl cyanides react with organic azides very efficiently giving rise to 1-substituted 5-arylsulfonyl-tetrazoles.[96] Zinc bromide can be used to catalyse the reaction between sodium azide and nitriles in hot water.[97] Intramolecular examples involving cyanamides proceed in hot DMF.[98] In additions to nitriles, one can include triethylammonium chloride (instead of ammonium chloride)

to avoid the possible sublimation of potentially explosive azides;[99] micelles can also be employed as reaction media.[100]

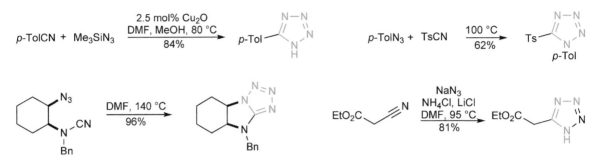

For reaction with azides, amides can be activated with trifluoromethanesulfonic anhydride[101] or benzotriazole,[102] or *via* formation of the thioamide,[103] or by the use of triphenylphosphine with diethyl azodicarboxylate. Alternatively, imidochlorides can be used under phase-transfer conditions.[104]

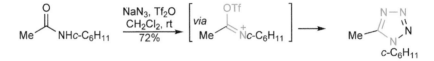

Related methods can be used to prepare 5-hetero-substituted compounds: isonitriles with *N*-halosuccinimides and azide give halo derivatives,[105] aryl isothiocyanates with azide give the 5-thiols,[106] which can be converted into 5-unsubstituted tetrazoles by oxidation with hydrogen peroxide[107] or chromium trioxide.[108]

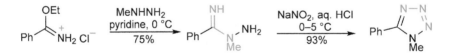

N-Nitrosation of amidrazones is a method that avoids the use of azide and also offers a regiocontrolled synthesis of 1-substituted[109] or 2,5-diaryl-tetrazoles.[110]

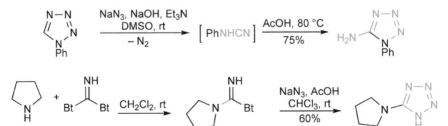

The synthesis of 5-amino-tetrazoles can be achieved *via* the reaction of azide anion with benzotriazolylimines.[111] Another route utilizes *N*-aryl-cyanamides generated *in situ*, from 1-aryl-tetrazoles, with sodium azide.[112]

29.1.2 Oxadiazoles and Thiadiazoles

Only one divalent heteroatom can be incorporated into a simple five-membered, aromatic heterocycle. These systems are named with the non-nitrogen atom numbered as 1, and the positions of the nitrogen atoms shown with reference to the divalent atom.

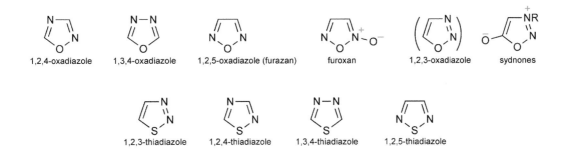

1,2,4-oxadiazole   1,3,4-oxadiazole   1,2,5-oxadiazole (furazan)   furoxan   1,2,3-oxadiazole   sydnones

1,2,3-thiadiazole   1,2,4-thiadiazole   1,3,4-thiadiazole   1,2,5-thiadiazole

1,2,4-Oxadiazoles,[113] 1,3,4-oxadiazoles[114] and 1,2,5-oxadiazoles are well known, but the 1,2,3-oxadiazole system, which calculations indicate to be unstable relative to its ring-open diazo-ketone tautomer,[115] is known only as a benzo-fused derivative (in solution) and in mesoionic substances, known as 'sydnones',[116] which have been well investigated. 'Furoxans',[117] which are formed by the dimerisation of nitrile oxides, have also been extensively studied. 1,2,3-Thiadiazoles, 1,2,4-thiadiazoles,[118] 1,3,4-thiadiazoles[119] and 1,2,5-thiadiazoles[120] are all represented by well-characterised compounds. Estimates of aromaticity, based on bonds lengths and NMR data produce the following relative order.[121]

As with the azoles, oxa- and thiadiazoles are very weak bases due to the inductive effects of the extra heteroatoms, although *N*-quaternisation reactions can be carried out. For similar reasons, electrophilic substitutions on carbon are practically unknown, apart from a few halogenations and mercurations[122] – it is an intriguing paradox that mercurations, with what is generally thought of as a weak electrophile, are often successful in electron-poor heterocycles. Another important difference from other azoles is of course the absence of *N*-hydrogen, so that *N*-anion-mediated reactions are not available.

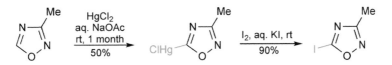

All these systems are susceptible to nucleophilic attack, particularly the oxadiazoles, which often undergo ring cleavage with aqueous acid or base unless both carbon positions are substituted. Similarly, leaving groups are generally displaced easily; there is substantial differential positional reactivity: in both 1,2,4-oxa- and -thiadiazoles a 5-chlorine is displaced much more easily than a 3-chlorine, no doubt due to the more effective stabilisation of the intermediate anionic adduct in the former situation.

Base-catalysed proton exchange occurs readily, but decomposition *via* cycloreversion or β-elimination in the anion often competes,[123] for example a 4-aryl-1,2,3-thiadiazole is deprotonated at C-5 by carbonate as base, leading firstly, *via* loss of nitrogen, to an arylalkynyl-thiolate,[124] this being a preferred route for their synthesis.[125] Direct lithiations at carbon are generally easy,[126] but the resulting lithio derivatives vary greatly in stability.[127]

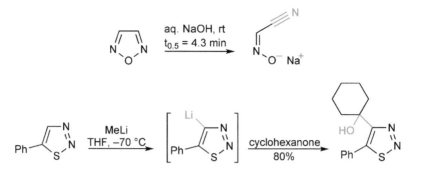

Hydrogens on side-chain alkyl groups are acidified by delocalisation of the charge in the deprotonated species onto ring nitrogens. There is an interesting difference between 1,2,5-oxa- and 1,2,5-thiadiazoles in this context: in the former, smooth metallation of a 3-methyl occurs with *n*-butyllithium, but for the latter, lithium diisopropylamide must be used to avoid competing nucleophilic addition to the sulfur, leading then to ring decomposition.[128]

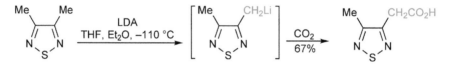

4-Substituted 5-chloro-1,2,3-thiadiazoles react with simple hetero nucleophiles by displacement of the chlorine, but reaction with aryl- and alkyllithiums gives alkynyl-thioethers *via* attack at sulfur and then ring cleavage with loss of nitrogen. A similar ring cleavage occurs, but by a different mechanism, when the 5-unsubstituted analogue is treated with base.[129]

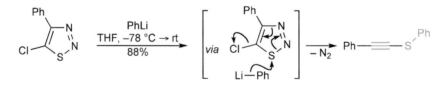

Generally, amines can be diazotised and converted, for example, into halides, but in some cases the intermediate, *N*-nitroso compound is stable, and only then subsequently converted into a diazonium salt by treatment with strong acid – this may reflect the lower stability of a positively charged group attached to an electron-deficient ring.[130]

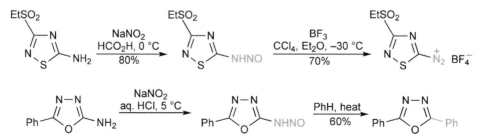

An interesting and fairly general type of reaction in ring systems such as these is ring interconversion *via* intramolecular attack on nitrogen (cf. 25.8).[131]

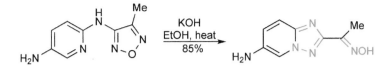

1,2,4-Oxadiazoles are useful as masked amidines, where such strongly basic groups would be incompatible with reaction conditions – the amidine is easily liberated by hydrogenation, as illustrated below.[132]

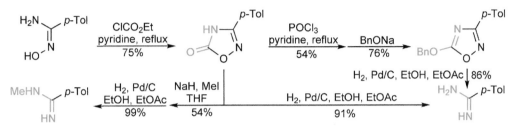

29.1.2.1 Ring Synthesis of Oxadiazoles and Thiadiazoles
1,2,4-Oxadiazoles

1,2,4-Oxadiazoles can be prepared by reaction of amidoximes with activated acids,[135,136] acid chlorides or anhydrides,[137] esters[138] or β-keto-esters.[139]

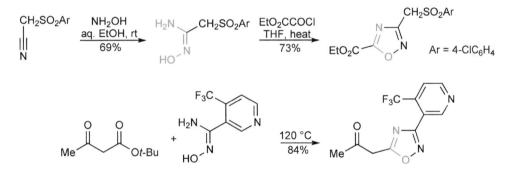

1,2,4-Oxadiazoles can also be prepared from amides *via* acylamidines,[140] or *via* the cycloaddition of nitrile oxides to nitriles, as illustrated.[141] 3-Alkylamino-1,2,4-oxadiazoles result from the reaction of *N*-acyl-1-benzotriazolyl-1-carboximidamides with hydroxylamine.[142]

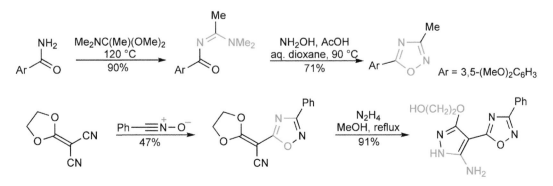

1,3,4-Oxadiazoles

1,3,4-Oxadiazoles are available by cyclodehydration of *N,N'*-diacyl-hydrazines or their equivalents,[143,144] for example closure of *N*-acyl-thiosemicarbazides generates 2-amino-1,3,4-oxadiazoles.[145] It is possible to conduct the preparation of the *N,N'*-diacyl-hydrazine and the ring closure in one pot, thus reaction of a hydrazide with a trialkyl orthoalkanoate produces 1,3,4-oxadiazoles (2-unsubstituted 1,3,4-oxadiazoles using triethyl orthoformate; 1,3,4-thiadiazoles using P_4S_{10}/Al_2O_3 to effect ring closure)[146] or by treating a hydrazide with an acid with microwave heating and a dehydrating agent,[147] or by a peptide coupling followed by a cyclising dehydration with triphenylphosphine with carbon tetrabromide.[148] Starting from a lower oxidation level, aldehyde *N*-acyl-hydrazones give the aromatic heterocycles on treatment with cerium(IV) ammonium nitrate.[149]

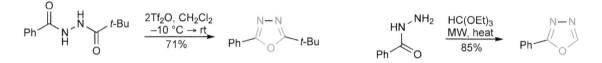

These heterocycles are also available from tetrazoles (29.1.1.3) or by oxidative cyclisation of *N*-acyl-hydrazones.[150]

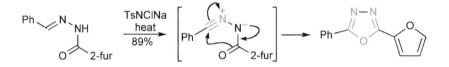

1,2,5-Oxadiazoles

1,2,5-Oxadiazoles result from the dehydration of 1,2-bisoximes.

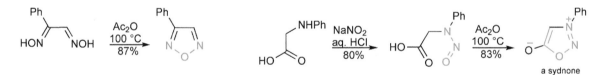

a sydnone

Sydnones

Sydnones are normally prepared by the dehydration of *N*-nitroso α-amino acids.[151]

1,2,3-Thiadiazoles

1,2,3-Thiadiazoles are prepared by reaction of a hydrazone, containing an acidic methylene group, with thionyl chloride.[124,152] The 5-thiol can be prepared by reaction of chloral tosylhydrazone with polysulfide, as indicated below.[153]

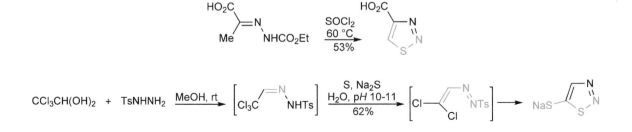

Benzotriazolyl is a useful leaving group for nucleophilic substitutions in 1,2,3-thiadiazoles.[154]

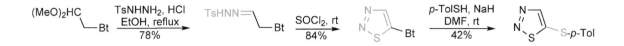

1,2,4-Thiadiazoles

1,2,4-Thiadiazoles carrying identical groups at the 3- and 5-positions are obtained by the oxidation of thioamides;[155] 5-chloro-1,2,4-thiadiazoles result from the reaction of amidines with perchloromethyl mercaptan.[156] 5-Amino-1,2,4-thiadiazoles can be produced by heterocyclisation of thioacylguanidines, using diisopropyl azodicarboxylate, and generated in the same pot from amidines and an isothiocyanate.[157]

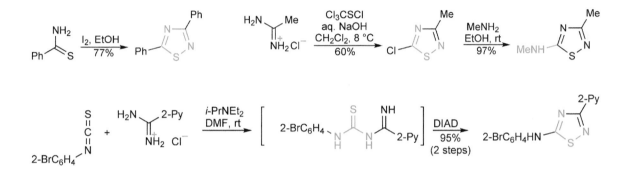

1,3,4-Thiadiazoles

1,3,4-Thiadiazoles are available by a number of convenient general routes including cyclisation of *N,N'*-diacyl-hydrazines, or 1,3,4-oxadiazoles, with phosphorus sulfides.[158] 2-Amino-1,3,4-thiadiazoles are prepared *via* acylation of thiosemicarbazides[159] and the parent compound is easily obtained from hydrogen sulfide and dimethylformamide azine.[160]

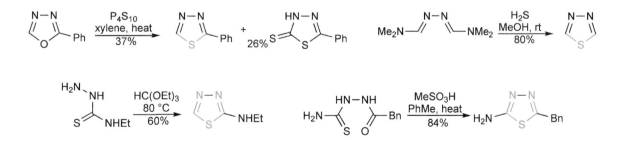

1,2,5-Thiadiazoles

1,2,5-Thiadiazoles can be prepared by the oxidative cyclisation of 1,2-diamines or aminocarboxamides.[161] Condensation of sulfamide ($SO_2(NH_2)_2$) with 1,2-diketones gives 1,2,5-thiadiazole 1,1-dioxides.[162] A good general method is the reaction of trithiazyl trichloride with activated alkenes and alkynes; this method is also useful for the fusion of a 1,2,5-thiadiazole onto other heterocycles, such as pyrroles. The reaction possibly proceeds *via* cycloaddition to an N–S–N unit in the trithiazine ring.[163]

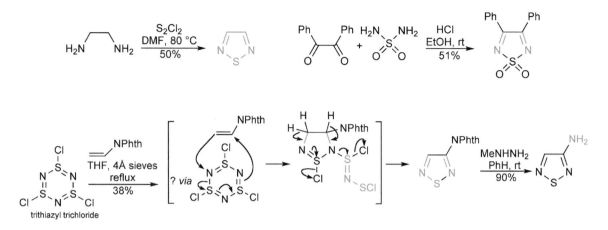

29.1.3 Other Systems

Of the higher aza-compounds, only derivatives of 1,2,3,4-thiatriazole[133] are well defined, but even here alkyl derivatives decompose at or below 0 °C, though 5-aryl and amino derivatives are generally fairly stable. Many other derivatives are, however, dangerously explosive, for example the 5-chloro and 5-thiolate derivatives. The controlled decomposition of 5-alkoxy-1,2,3,4-thiatriazoles (for example the 5-ethoxy derivative in ether at 20 °C) has been recommended as the best preparation of pure alkyl cyanates; thermal decomposition of 5-aryl compounds gives the corresponding nitrile.[134]

A one-pot conversion of amines into 1,2,3,4-thiatriazoles, *via* reaction with thiocarbonyldiimidazole and azide, works well for both primary and secondary amines. With simple amines, the intermediate thio-imidazolide requires quaternisation with methyl iodide for the final azide reaction, but with amino-acid derivatives, this additional activation is not always necessary.[164]

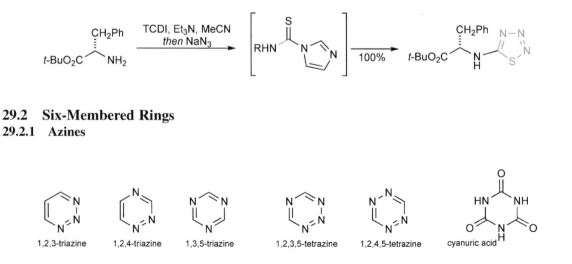

29.2 Six-Membered Rings
29.2.1 Azines

Simple uncharged six-membered aromatic heterocycles cannot contain a divalent heteroatom. The azines are numbered to indicate the relative positions of the nitrogen atoms. 1,2,3,4-Tetrazine, pentazine and hexazine are unknown, however, a number of fused 1,2,3,4-tetrazines, primarily *N*-oxides and *N*-aryl quaternary salts, are known,[165] but of monocyclic compounds, only a few di-*N*-oxides have been prepared.[166] Of the other systems, 1,2,3,5-tetrazine is unknown, although theoretically it could be moderately stable, but fused derivatives include the drug temozolomide (see 33.7). Derivatives of 1,3,5-triazine are very well known and available in large quantities, indeed they are amongst the oldest known heterocycles: the trioxy-compound ('cyanuric acid') was first prepared in 1776 by Scheele by the pyrolysis of uric acid.

The thermal stabilities of the parent systems vary from 1,2,3-triazine, which decomposes at about 200 °C, to 1,3,5-triazine, which is stable to over 600 °C – at this temperature it decomposes to give hydrogen cyanide, of which it is formally a trimer. Melamine (2,4,6-triamino-1,3,5-triazine) is an important industrial intermediate (34.6.2). Its high nitrogen content has led to its illegal misuse in foodstuffs to achieve higher nitrogen analyses.

In comparison with the diazines, the inductive effects of the 'extra' nitrogen(s) leads to an even greater susceptibility to nucleophilic attack and, as a result, all the parent systems and many derivatives react with water, in acidic or basic solution. Similarly, simple electrophilic substitutions do not occur; some apparent electrophilic substitutions, such as the bromination of 1,3,5-triazine, probably take place *via* bromide nucleophilic addition to an N^+–Br triazinium salt.[167] Attempted direct *N*-oxidation of simple tetrazines with the usual reagents generally results in ring cleavage, however it can be achieved satisfactorily with methyl(trifluoromethyl)dioxirane.[168]

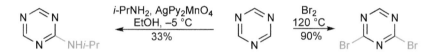

The examples shown are illustrative of the many easy nucleophilic additions to the polyaza-azines: both 3-phenyl-1,2,4,5-tetrazine[169] and 1,3,5-triazine itself[170] add ammonia and simple amines (contrast the requirement for hot sodamide (Chichibabin reaction) for pyridine (8.3.1.2)) and thus amino and alkyamino derivatives can be obtained *via* oxidative trapping with permanganate.

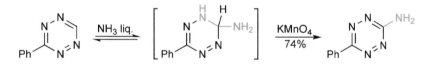

The easy addition at C-5 of 1,2,4-triazines[171] is shown by the VNS (3.3.3) reaction of the 3-methylthio derivatives in the absence of activating groups; a closely related addition of nitro-alkanes represents a very useful nucleophilic acylation.[172] The ready displacement of methylthio from the same compound is also significant.[173]

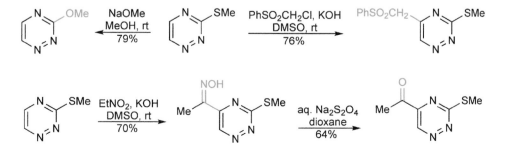

Nucleophilic displacement of methylthio in 1,2,4-triazines and 1,2,4,5-tetrazines by alkoxide and amines is very easy. Mono-displacement can be carried out on 3,6-bis(methylthio)-1,2,4,5-tetrazine, but the reaction using methoxide requires careful control of reaction conditions to avoid formation of the dimethoxy derivative.[174,175]

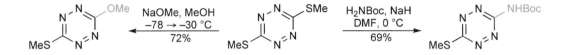

In triazine chemistry, sulfone is a better leaving group than halide for displacement with carbanions.[176]

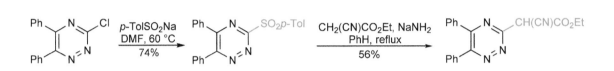

The susceptibility of 1,3,5-triazine to nucleophilic attack with ring opening makes it a synthetically useful equivalent of formate, or formamide, particularly for the synthesis of other heterocycles, such as imidazoles and triazoles[177] (see above). Despite the high susceptibility of 1,2,4-triazines to nucleophilic addition, 3-substituted-6-methoxy-1,2,4-triazines can be successfully lithiated.[178]

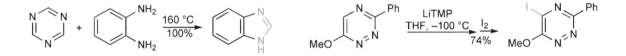

Reaction of 1,2,3-triazine with nucleophiles usually leads to ring opening *via* attack at C-4. However, silyl enol ethers react with chloroformate/1,2,3-triazine salts to give 5-substituted 2,5-dihydro-1,2,3-triazines, which can be re-aromatised using cerium(IV) ammonium nitrate. In this case, initial addition of the electrophile takes place at N-2, leading to the specific activation of C-5.[179]

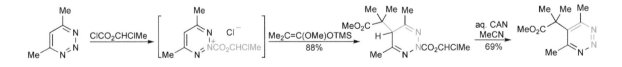

An interesting variant of the Minisci reaction has been reported for 1,2,3-triazine, which is unstable to the usual acidic conditions: here, activation of the heterocycle to attack by the nucleophilic radical is brought about by the agency of a dicyanomethine ylide.[180]

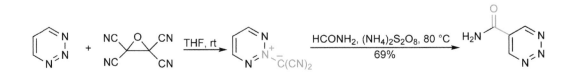

The reductive removal of hydrazine substituents under oxidising conditions from azines and some azoles is conceptually related to oxidative removal of thiols (e.g. 29.1.1.2). In this case, the intermediacy of a diimide seems likely, as illustrated below.[181]

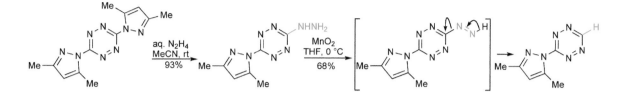

Probably the most useful and general reaction of all these systems[182] is the inverse-electron-demand Diels–Alder reaction with acetylenes (or equivalents) to produce either pyridines or diazines *via* elimination of hydrogen cyanide, a nitrile or nitrogen.[183]

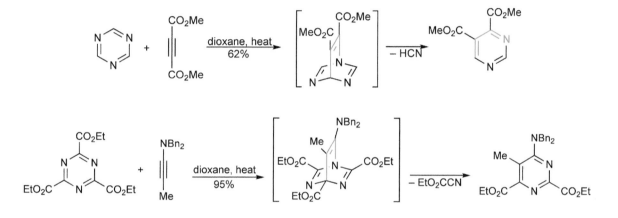

Enamines have most often been used as acetylene equivalents, the amine fragment being lost at a late stage to generate the aromatic heterocycle. Prior synthesis of the enamine is unnecessary: these processes can also be achieved by heating together a ketone, a secondary amine and the heterocycle.[184]

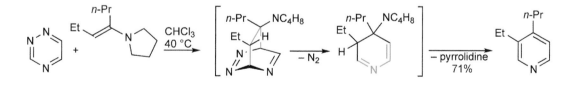

The best equivalent of ethyne itself for these reactions is norbornadiene, cyclopentadiene being lost in the final stage.[185]

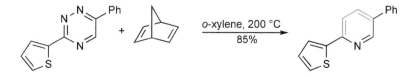

In reactions of 1,2,4,5-tetrazines, sulfoxide substituents convey a greater reactivity to these inverse-electron-demand process than the corresponding sulfides;[186] similarly, ester groupings increase the reactivity.[187]

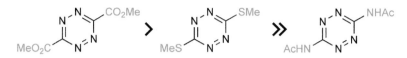

The high regioselectivity of addition to sulfoxide-bearing 1,2,4,5-tetrazines is not, however, what one might anticipate, on the basis of the polarity of the components.

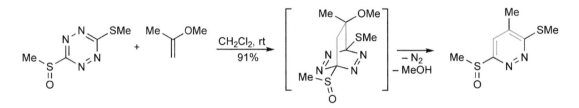

29.2.1.1 Ring Syntheses
1,2,3-Triazines
Using various reagents, 1,2,3-triazines, including 1,2,3-triazine itself, can be prepared by the oxidation of 1-aminopyrazoles with NiO_2[188] or periodate.[189]

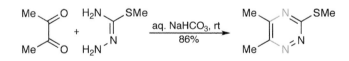

1,2,4-Triazines
1,2,4-Triazines can be prepared by the condensation of amidrazones with diketones[190,191,192] or of hydrazides with 2-halo-ketones.[185] It is convenient to oxidize 2-hydroxy-ketones to the dicarbonyl component with manganese(IV) oxide, in the presence of the amidrazone, to produce the heterocycles.[193] Hydrazides can also be utilized with 1,2-diketones, by including an ammonia source in the reagent mix.[194,195]

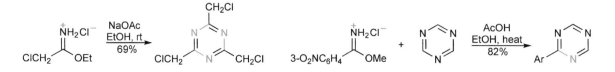

1,3,5-Triazines
1,3,5-Triazines are usually most easily obtained by substitution reactions on 2,4,6-trichloro-1,3,5-triazine,[196] but the ring system can also be synthesised by cyclocondensation reactions. Trimerisation of nitriles[197] (a common industrial method) or imidates[198] gives symmetrically substituted compounds; mono-substituted-1,3,5-triazines can be obtained *via* reaction of imidates with 1,3,5-triazine itself.[199]

A route that allows the synthesis of 1,3,5-triazines with different substituents at each carbon is exemplified below – an *N'*-acyl-*N,N*-dimethylamidine reacts with an amidine (shown) or guanidine to form a 1,3,5-triazine.[200]

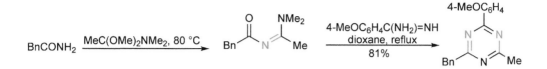

1,2,4,5-Tetrazines

1,2,4,5-Tetrazines can be produced by condensation of hydrazine with carbonyl compounds at acid oxidation level, followed by oxidation of the dihydro products: this generally produces 3,6-identically substituted derivatives, crossed condensation reactions being inefficient.[160,201]

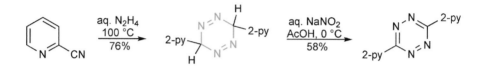

The much-used dimethyl 1,2,4,5-tetrazine-3,6-dicarboxylate is generated by the dimerisation of ethyl diazoacetate, then oxidation.[202]

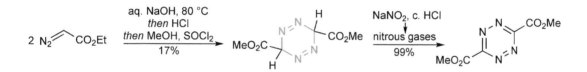

29.3   Benzotriazoles

The chemistry of benzotriazole has been developed to the point where it now finds extensive use for heterocyclic[203] and general synthesis.[204,205,206,207] A useful set of properties give it this role: (i) α-carbanions are stabilised to the same extent as at the benzylic position of a benzene compound; (ii) α-carbocations are also stabilised; (iii) the benzotriazolyl anion is a good leaving group with a combination of good reactivity and stability/ease of handling of the benzotriazole starting materials. Sequential combinations of these reactivities have been applied to the synthesis of a wide variety of molecules – some illustrative examples are shown below.

The starting benzotriazole derivatives are usually prepared from benzotriazole itself by *N*-alkylation with a halide, *N*-acylation with an acid or acid chloride, or *via* reaction with aldehydes or acetals; in each case, these can lead to mixtures of 1- and 2-substituted benzotriazoles; however, the reactivities of the two isomers are similar. For clarity, only reactions of 1-substituted compounds are shown here.

N-Acyl-benzotriazoles, which are stable and easily handled (cf. acid chlorides or anhydrides) and can be prepared from a large variety of acids, including *N*-protected α-amino acids, have been widely used as *N*-, *O*-, *S*- and *C*-acylating agents, for example to make acyl azides.[208] The benzotriazole derivatives of *N*-protected-α-amino acids have been extensively developed as agents for peptide coupling.[209]

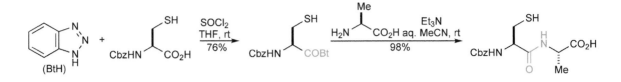

Alkylation of lithiated 1-(1-ethoxyprop-2-enyl)benzotriazole leads to enones after hydrolytic removal of the heterocycle; addition of the lithiated species to cyclohexenone then hydrolytic cleavage of the heterocycle produces an unsaturated 1,4-diketone.[210] Addition of the same anion to methyl but-2-enoate generates an anion in which the benzotriazole is displaced intramolecularly and a cyclopropane results.[211]

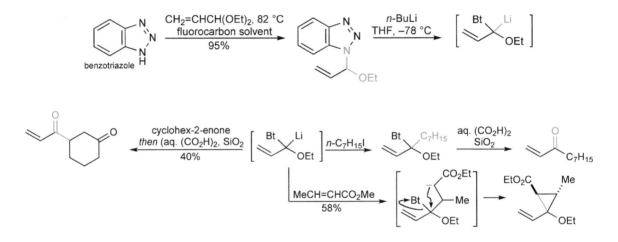

In the next example, the benzotriazole unit facilitates benzylic lithiation and in the final step acts as a leaving group.[212]

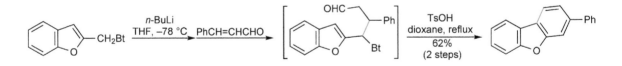

The ability of benzotriazole to stabilise a cation allows 1-hydroxymethylbenzotriazole to alkylate indole; the product can then be lithiated to allow substitution by an electrophile and then finally the benzotriazole unit can be displaced by a Grignard nucleophile.[213]

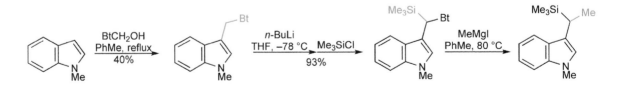

Tris(benzotriazol-1-yl)methane anion will add to nitrobenzene in a VNS process in which the benzotriazole unit is both the anion-stabilising unit and also the leaving group;[214] another use for this compound is illustrated by alkylation of its anion, then hydrolysis, forming an acid.[215]

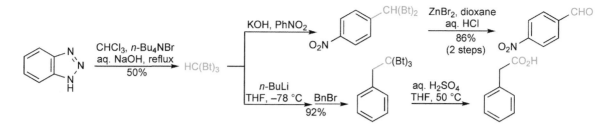

Benzotriazoles also have interesting reactivity in their own right and can be used to form other heterocyclic compounds *via* various ring cleavage reactions, leading to reactive intermediates, such as benzynes,[216] aryllithiums,[217] diradicals[218] and arylhydrazones,[219] some illustrated below.

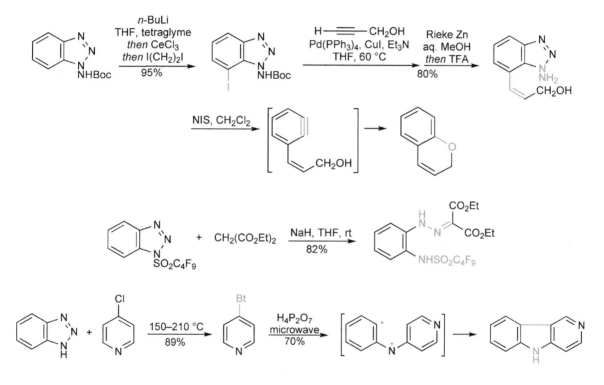

Exercises

1. What are the products of the following (Diels–Alder) reactions: (i) 1-pyrrolidinylcyclopentene with: (a) 1,3,5-triazine, (b) 1,2,4-triazine; (ii) 3-phenyl-1,2,4,5-tetrazine with 1,1-diethoxyethene?
2. Thiophosgene (S=CCl$_2$) reacts at low temperature with sodium azide to give a product which contains no azide group; on subsequent reaction with methylamine this compound is converted into C$_2$H$_4$N$_4$S – suggest structures.
3. What are the products of the reaction of PhCONH$_2$ with DMFDMA then: (a) N$_2$H$_4$ and (b) H$_2$NOH?
4. 1,3,5-Triazine reacts with: (i) aminoguanidine to give 2-amino-1,3,4-triazole and with (ii) diethyl malonate to give ethyl 4-hydroxypyrimidine 5-carboxylate. Write mechanisms for these transformations.

References

[1] Gilchrist, T. L., Gymer, G. E. and Rees, C. W., *J. Chem. Soc., Perkin Trans. 1*, **1973**, 555; *ibid.*, **1975**, 1.
[2] 'Pentazoles', Ugi, I., *Adv. Heterocycl. Chem.*, **1964**, *3*, 373.

[3] Östmark, H., Wallin, S., Brinck, T., Carlqvist, P., Claridge, R., Hedlund, E. and Yudina, L, *Chem. Phys. Lett.*, **2003**, *379*, 539.

[4] Alemagna, A., Bacchetti, T. and Beltrame, P., *Tetrahedron*, **1968**, *24*, 3209.

[5] Barlin, G. B., *J. Chem. Soc. (B), Phys. Org.*, **1967**, 641.

[6] '1,2,3-Triazoles', Gilchrist, T. L. and Gymer, G. E., *Adv. Heterocycl. Chem.*, **1974**, *16*, 33.

[7] 'Synthesis and reactions of lithiated monocyclic azoles containing two or more hetero-atoms. Part VI: Triazoles, tetrazoles, oxadiazoles and thiadazoles', Grimmett, M. R. and Iddon, B., *Heterocycles*. **1995**, *41*, 1525.

[8] Pederdsen, C., *Acta Chem. Scand.*, **1959**, *13*, 888.

[9] Ohta, S., Kawasaki, I., Uemura, T., Yamashita, M., Yoshioka, T. and Yamaguchi, S., *Chem. Pharm. Bull*,. **1997**, *45*, 1140.

[10] Birkofer, L. and Wegner, P., *Chem. Ber.*, **1967**, *100*, 3485.

[11] Bird, C. W., *Tetrahedron*, **1985**, *41*, 1409.

[12] Williams, E. L., *Tetrahedron Lett.*, **1992**, *33*, 1033.

[13] Iddon, B. and Nicholas, M., *J. Chem. Soc., Perkin Trans. 1*, **1996**, 1341.

[14] Hüttel, R. and Welzel, G., *Justus Liebigs Ann. Chem.*, **1955**, *593*, 207.

[15] Lynch, B. M. and Chan, T.-L., *Can. J. Chem.*, **1963**, *41*, 274.

[16] Wiley, R. H., Hussung, K. F. and Moffat, J., *J. Org. Chem.*, **1956**, *21*, 190; El-Khadem, H., El-Shafei, Z. M. and Meshreki, M. H., *J. Chem. Soc.*, **1961**, 2957.

[17] Raap, R., *Can. J. Chem.*, **1971**, *49*, 1792.

[18] Ghose, S. and Gilchrist, T. L., *J. Chem. Soc., Perkin Trans. 1*, **1991**, 775.

[19] Díaz-Ortiz, Á., de Cózar, A., Prieto, P., de la Hoz, A. and Moreno, A., *Tetrahedron Lett.*, **2006**, *47*, 8761.

[20] 'Approaches to the synthesis of 1-substituted 1,2,4-triazoles', Balasubramanian, M., Keay, J. G. and Scriven, E. F., *Heterocycles*, **1994**, *37*, 1951.

[21] Olofson, R. A. and Kendall, R. V., *J. Org. Chem.*, **1970**, *35*, 2246.

[22] Horvath, A., *Synthesis*, **1995**, 1183.

[23] Bulger, P. G., Cottrell, I. F., Cowden, C. J., Davies, A. J. and Dolling, U.-H., *Tetrahedron Lett.*, **2000**, *41*, 1297.

[24] Ivanova, N. V., Sviridov, S. I, Shorshnev, S. V. and Stepanov, A. E., *Synthesis*, **2006**, 156.

[25] Kröger, C.-F. and Miethchen, R., *Chem. Ber.*, **1967**, *100*, 2250.

[26] Grinsteins, V. and Strazdina, A., *Khim. Geterotsikl. Soedin.*, **1969**, *5*, 1114 (*Chem. Abstr.*, **1970**, *72*, 121456).

[27] Habraken, C. L. and Cohen-Fernandes, P., *J. Chem. Soc., Chem. Commun.*, **1972**, 37.

[28] Hansen, K. B., Springfield, S. A., Desmond, R., Devine, P. N., Grabowski, E. J. J. and Reider, P. J., *Tetrahedron Lett.*, **2001**, *42*, 7353.

[29] Anderson, D. K., Sikorski, J. A., Reitz, D. B. and Pilla, L. T., *J. Heterocycl. Chem.*, **1986**, *23*, 1257; Jutzi, P. and Gilge, K., *J. Organomet. Chem.*, **1983**, *246*, 163.

[30] Zarudnitskii, E. V., Pervak, I. I., Merkulov, A. S., Yurchenko, A. A.,Tolmachev, A. A. and Pinchuk, A. M., *Synthesis*, **2006**, 1279.

[31] Browne, E. J., *Aust. J. Chem.*, **1969**, *22*, 2251; Bagal, L. I., Pevzner, M. S., Frolov, A. N. and Sheludyakova, N. I., *Khim. Geterotsikl. Soedin.*, **1970**, 259 (*Chem. Abstr.*, **1970**, *72*, 11383); Bagal, L. I., Pevzner, M. S., Samarenko, V. Ya. and Egorov, A. P., *ibid.*, 1701 (*Chem. Abstr.*, **1971**, *74*, 99948).

[32] Sano, S., Tanba, M. and Nagao, Y., *Heterocycles*, **1994**, *38*, 481.

[33] Zumbrunn, A., *Synthesis*, **1998**, 1357.

[34] Kane, J. M., Dalton, C. R., Staeger, M. A. and Huber, E. W., *J. Heterocycl. Chem.*, **1995**, *32*, 183; 'Facile desulfurization of cyclic thioureas by hydrogen peroxide', Grivas, S. and Ronne, E., *Acta Chem. Scand.*, **1995**, *49*, 225.

[35] 'Azodicarbonyl compounds in heterocyclic synthesis', Moody, C. J., *Adv. Heterocycl. Chem*, **1982**, *30*, 1; '1,2,4-Triazoline-3,5-diones', Rádl, S., *ibid.*, **1987**, *67*, 119.

[36] Alajarín, M., Cabrera, J., Pastor, A., Sábnchez-Andrada, P. and Bautista, D., *J. Org. Chem.*, **2008**, *73*, 963.

[37] Debledes, O. and Campagne, J.-M., *J. Am. Chem. Soc.*, **2008**, *130*, 1562.

[38] 'Recent advances in tetrazole chemistry', Butler, R. N., *Adv. Heterocycl. Chem.*, **1977**, *21*, 323.

[39] Huynh, M. H. V., Coburn, M. D., Meyer, T. J. and Wetzler, M., *PNAS*, **2006**, *103*, 10322.

[40] Hiskey, M. A., *Chem. Eng. News*, **2005**, *83*, 5.

[41] Hammerl, A., Klapötke, T. M., Mayer, P., Weigand, J. J. and Holl, G., *Propellants, Explosives, Pyrotechnics*, **2005**, *30*, 17.

[42] Isida, T., Akiyama, T., Nabika, K., Sisido, K. and Kozima, S., *Bull. Chem. Soc. Jpn.*, **1973**, *46*, 2176; Takach, N. E., Holt, E. M., Alcock, N. W., Henry, R. A. and Nelson, J. H., *J. Am. Chem. Soc.*, **1980**, *102*, 2968.

[43] Koren, A. O., Gaponik, P. N., Ivashkevich, O. A. and Kovalyova, T. B., *Mendeleev Commun.*, **1995**, 10.

[44] Spear, R. J., *Aust. J. Chem.*, **1984**, *37*, 2453.

[45] Alam, L. M. and Koldobskii, G. I., *Russ. J. Org. Chem.*, **1997**, *33*, 1149.

[46] Fauré-Tromeur, M. and Zard, S. Z., *Tetrahedron Lett.*, **1998**, *39*, 7301.

[47] Boivin, J., Henriet, E. and Zard, S. Z., *J. Am. Chem. Soc.*, **1994**, *116*, 9739.

[48] Moderhack, D. and Lembcke, A., *Chem. Ztg.*, **1984**, *108*, 188.

[49] Gol'tsberg, M. A. and Koldobskii, G. I., *Chem. Het. Comps.*, **1996**, *32*, 1300.

[50] Barlin, G. B., *J. Chem. Soc. (B)*, **1967**, 641.

[51] Booker, B. C., *Tetrahedron Lett.*, **2000**, *41*, 2805.

[52] Katritzky, A. R., Aslan, D., Shcherbakova, I. A., Chen, J. and Belyakov, S. A., *J. Heterocycl. Chem.*, **1996**, *33*, 1107.

[53] Raap, R., *Can. J. Chem.*, **1971**, *49*, 2139.

[54] Flippin, L. A., *Tetrahedron Lett.*, **1991**, *32*, 6857.

[55] Dankwardt, J. W., *J. Org. Chem.*, **1998**, *63*, 3753.

[56] Golden, A. H. and Jones, M., *J. Org. Chem.*, **1996**, *61*, 4460; Kumar, A., Narayanan, R. and Shechter, H., *ibid.*, 4462.

[57] Kamala, K., Rao, P. J. and Reddy, K. K., *Bull. Chem. Soc. Jpn.*, **1988**, *61*, 3791.

[58] Hoffle, G. and Lange, B., *Org. Synth., Coll. Vol. VII*, **1992**, 27.

[59] Jursic, B. S. and Zdravkovski, Z., *Synth. Commun.*, **1994**, *24*, 1575.

[60] Renz, R. N., Williams, M. D. and Fronabarger, J. W., US patent 7253288 (publ. 08/07/2007).

[61] Kammula, A. and Shevlin, P. B., *J. Am. Chem. Soc.*, **1974**, *76*, 7830.

[62] Kralj, J. G., Murphy, E. R., Jensen, K. F., Williams, M. D. and Renz, R. 41st AIAA/ASME/SAE/ASEE Joint Propulsion Conference and Exhibit, 10–13 July **2005**, Tucson, Arizona. Paper AIAA 2005-3516.

[63] Wallis, J. D. and Dunitz, J. D., *J. Chem. Soc., Chem. Commun.*, **1983**, 910.

[64] Benin, V., Kaszynski, P. and Radziszewski, J. G., *J. Org. Chem.*, **2002**, *67*, 1354.

[65] Butler, R. N., Fox, A., Collier, S. and Burke, L. A., *J. Chem. Soc., Perkin Trans. 2*, **1998**, 2243.

[66] Butler, R. N., Stephens, J. C. and Burke, L. A., *Chem. Commun.*, **2003**, 1016.

[67] 'Click chemistry – what's in a name? Triazole synthesis and beyond', Gil, M. V., Arévalo, M. J. and López, O., *Synthesis*, **2007**, 1589.

[68] Kolb, H. C., Finn, M. G. and Sharpless, K. B., *Angew. Chem. Int. Ed.*, **2001**, 2004.

[69] Ball, P., *Chem. World*, **2007**, *4*, *issue* 4, 46.

[70] Wang, Z.-X. and Zhao, Z.-G., *J. Heterocycl. Chem.*, **2007**, *44*, 89.

[71] Andersen, J., Bolvig, S. and Liang, X., *Synlett*, **2005**, 2941.

[72] Beckmann, H. S. G. and Wittman, V., *Org. Lett.*, **2007**, *9*, 1.

[73] Birkofer, L. and Wegner, P., *Chem. Ber.*, **1966**, *99*, 2512.

[74] Kalisiak, J., Sharpless, K. B. and Fokin, V. V., *Org. Lett.*, **2008**, *10*, 3171.

[75] Roque, D. R., Neill, J. L., Antoon, J. W. and Stevens, E. P., *Synthesis*, **2005**, 2497.

[76] Gao, Y. and Lam, Y., *Org. Lett.*, **2006**, *8*, 3283.

[77] 'Catalytic azide-alkyne cycloaddition: reactivity and applications', Wu, P. and Fokin, V. V., *Aldrichimica Acta*, **2007**, *40*, 7.

[78] Oppilliart, S., Mousseau, G., Zhang, L., Jia, G., Thuéry, P., Rousseau, B. and Cintrat, J.-C., *Tetrahedron*, **2007**, *63*, 8094.

[79] IJsselstijn, M. and Cintrat, J.-C., *Tetrahedron*, **2006**, *62*, 3837.

[80] Zhang, L., Chen, X., Xue, P., Sun, H. H. Y., Williams, I. D., Sharpless, K. B., Fokin, V. V. and Jia, G., *J. Am. Chem. Soc.*, **2005**, *127*, 15988.

[81] Kuijpers, B. H. M., Dijkmans, G. C. T., Groothuys, S., Quaedflieg, P. J. L. M., Blaauw, R. H., van Delft, F. L. and Rutjes, F. P. J. T., *Synlett*, **2005**, 3059; Majireck, M. M. and Weinreb, S. M., *J. Org. Chem.*, **2006**, *71*, 8680.

[82] Krasinski, A., Fokin, V. V. and Sharpless, K. B., *Org. Lett.*, **2004**, *6*, 1237.

[83] Romeiro, G. A., Pereira, L. O. R., deSouza, M. C. B. V., Ferreira, V. F. and Cunha, A. C., *Tetrahedron Lett.*, **1997**, *38*, 5103.

[84] Augusti, R. and Kascheres, C., *J. Org. Chem.*, **1993**, *58*, 7079.

[85] Harada, K., Oda, M., Matsushita, A. and Shirai, M., *Heterocycles*, **1998**, *48*, 695.

[86] Bartlett, R. K. and Humphrey, I. R., *J. Chem. Soc., C*, **1967**, 1664.

[87] Grundmann, C. and Rätz, R., *J. Org. Chem.*, **1956**, *21*, 1037.

[88] Zhang, Q., Peng, Y., Wang, X. I., Keenan, S. M., Arora, S. and Welsh, W. J., *J. Med. Chem.*, **2007**, *50*, 749.

[89] Ried, W. and Valentin, J., *Chem. Ber.*, **1968**, *101*, 2117.

[90] Shao, N., Wang, C., Huang, X., Xiao, D., Palani, A., Aslanian, R. and Shih, N.-Y., *Tetrahedron Lett.*, **2006**, *47*, 6743.

[91] Cookson, R. C., Gupte, S. S., Stevens, I. D. R. and Watts, C. T., *Org. Synth.*, **1971**, *51*, 121.

[92] Batchelor, D. V., Beal, D. M., Brown, T. B., Ellis, D., Gordon, D. W., Johnson, P. S., Mason, H. J., Ralph, M. J., Underwood, T. J. and Wheeler, S., *Synlett*, **2008**, 2421.

[93] Lindström, J. and Johansson, M. H., *Synth. Commun.*, **2006**, *36*, 2217.

[94] Aureggi, V., Franckevicius, V., Kitching, M. O., Ley, S. V., Longbottom, D. A., Oelke, A. J. and Sedelmeier, G., *Org. Synth.*, **2008**, *85*, 72.

[95] Jin, T., Kitahara, F., Kamijo, S. and Yamamoto, Y., *Tetrahedron Lett.*, **2008**, *49*, 2824.

[96] Demko, Z. P. and Sharpless, K. B., *Angew. Chem. Int. Ed.*, **2002**, *41*, 2110.

[97] Demko, Z. P. and Sharpless, K. B., *J. Org. Chem.*, **2001**, *66*, 7945.

[98] Demko, Z. P. and Sharpless, K. B., *Org. Lett.*, **2001**, *3*, 4091.

[99] Koguro, K., Oga, T., Mitsui, S. and Orita, R., *Synthesis*, **1998**, 910.

[100] Jursic, B. S. and LeBlanc, B. W., *J. Heterocycl. Chem.*, **1998**, *35*, 405.

[101] Thomas, E. W., *Synthesis*, **1993**, 767.

[102] Katritzky, A. R., Cai, C. and Meher, N. K., *Synthesis*, **2007**, 1204

[103] Lehnhoff, S. and Ugi, I., *Heterocycles*, **1995**, *40*, 801.

[104] Artamonova, T. V., Zhivich, A. B., Dubinskii, M. Yu. and Koldobskii, G. I., *Synthesis*, **1996**, 1428.

[105] Collibee, W. L., Nakajima, M. and Anselme, J.-P., *J. Org. Chem.*, **1995**, *60*, 468.

[106] Lieber, E. and Ramachandran, J., *Can. J. Chem.*, **1959**, *37*, 101.

[107] Markgraf, J. H. and Sadighi, J. P., *Heterocycles*, **1995**, *40*, 583.

[108] Kauer, J. C. and Sheppard, W. A., *J. Org. Chem.*, **1967**, *32*, 3580.

[109] Boivin, J., Husinec, S. and Zard, S. Z., *Tetrahedron*, **1995**, *51*, 11737.

[110] Chattaway, F. D. and Parkes, G. D., *J. Chem. Soc.*, **1926**, 113.

[111] Katritzky, A. R., Rogovoy, B. V. and Kovalenko, K. V., *J. Org. Chem.*, **2003**, 4941.

[112] Vorobiov, A. N., Gaponik, P. N., Petrov, P. T. and Ivashkevich, O. A., *Synthesis*, **2006**, 1307.

[113] '1,2,4-Oxadiazoles', Clapp, L. B., *Adv. Heterocycl. Chem.*, **1976**, *20*, 65.

[114] 'Recent advances in 1,3,4-oxadiazole chemistry', Hetzhein, A. and Möckel, K., *Adv. Heterocycl. Chem.*, **1966**, *7*, 183.

[115] Nguyen, M. T., Hegarty, A. F. and Elguero, J., *Angew. Chem., Int. Ed. Engl.*, **1985**, *24*, 713.

[116] 'The chemistry of sydnones', Stewart, F. H. C., *Chem. Rev.*, **1964**, *64*, 129; Huisgen, R., Grashey, R., Gotthardt, H. and Schmidt, R., *Angew. Chem., Int. Ed. Engl.*, **1962**, *1*, 48.

[117] 'Furoxans and benofuroxans', Gasco, A. and Boulton, A. J., *Adv. Heterocycl. Chem.*, **1981**, *29*, 251; 'The chemistry of furoxans', Sliwa, W. and Thomas, A., *Heterocycles*, **1985**, *23*, 399.

[118] '1,2,4-Thiadiazoles', Kurzer, F., *Adv. Heterocycl. Chem.*, **1982**, *32*, 285.

[119] 'Recent advances in the chemistry of 1,3,4-thiadiazoles', Sandström, J., *Adv. Heterocycl. Chem.*, **1968**, *9*, 165.

[120] 'The 1,2,5-thiadiazoles', Weinstock, L. M. and Pollak, P. I., *Adv. Heterocycl. Chem.*, **1968**, *9*, 107.

[121] Bak, B., Nygaard, L., Pedersen, E. J. and Rastrup-Anderson, J., *J. Mol. Spectrosc.*, **1966**, *19*, 283.

[122] Moussebois, G. and Eloy, F., *Helv. Chim., Acta*, **1964**, *47*, 838.

[123] Olofson, R. A. and Michelman, J. S., *J. Org. Chem.*, **1965**, *30*, 1854.

[124] Androsov. D. A. and Neckers, D. C., *J. Org. Chem.*, **2007**, *72*, 5368.

[125] Sandrinelli, F., Boudou, C., Caupène, C., Averbuch-Pouchot, M.-T., Perrio, S. and Metzner, P., *Synlett*, **2006**, 3289.

[126] Thomas, E. W. and Zimmerman, D. C., *Synthesis*, **1985**, 945.

[127] 'Ring-opening of five-membered heteroaromatic anions', Gilchrist, T. I., *Adv. Heterocycl. Chem.*, **1987**, *41*, 41.

[128] Boger, D. L. and Brotherton, C. E., *J. Heterocycl. Chem.*, **1981**, *18*, 1247.

[129] Voets, M., Smet, M. and Dehaen, W., *J. Chem. Soc., Perkin Trans. 1*, **1999**, 1473.

[130] Goerdeler, J. and Deselaers, K., *Chem. Ber.*, **1958**, *91*, 1025; Goerdeler, J. and Rachwalsky, H., *ibid.*, **1960**, *93*, 2190; Butler, R. N., Lambe, T. M., Tobin, J. C. and Scott, F. L., *J. Chem. Soc., Perkin Trans. 1*, **1973**, 1357.

[131] 'Ring transformations of five-membered heterocycles', Vivona, N., Buscemi, S., Frenna, V. and Cusmano, G., *Adv. Heterocycl. Chem.*, **1993**, *56*, 49; Cusmano, G., Macaluso, G. and Gruttadauria, M., *Heterocycles*, **1993**, *36*, 1577.

[132] Bolton, R. E., Coote, S. J., Finch, H., Lowden, A., Pegg, N. and Vinader, M. V., *Tetrahedron Lett.*, **1995**, *36*, 4471.

[133] '1,2,3,4-Thiatriazoles', Holm, A., *Adv. Heterocycl. Chem.*, **1976**, *20*, 145.

[134] Jensen, K. A. and Holm, A., *Acta Chem. Scand.*, **1964**, *18*, 826.

[135] Liang, G.-B. and Feng, D. D., *Tetrahedron Lett.*, **1996**, *37*, 6627.

[136] Wang, Y., Miller, R. L., Sauer, D. R. and Djuric, S. W., *Org. Lett.*, **2005**, *7*, 925.

[137] Tamura, M., Ise, Y., Okajima, Y., Nishiwaki, N. and Ariga, M., *Synthesis*, **2006**, 3453.

[138] Amarasinghe, K. K. D., Maier, M. B., Srivastava, A. and Gray, J. L., *Tetrahedon Lett.*, **2006**, *47*, 3629.

[139] Du, W., Truong, Q., Qi, H., Guo, Y., Chobanian, H. R., Hagmann, W. K. and Hale, J. J., *Tetrahedron Lett.*, **2007**, *48*, 2231.

[140] Lin, Y., Lang, S. A., Lovell, M. F. and Perkinson, N. A., *J. Org. Chem.*, **1979**, *44*, 4160.

[141] Neidlein, R. and Li, S., *J. Heterocycl. Chem.*, **1996**, *33*, 1943; *idem, Synth. Commun.*, **1995**, *25*, 2379.

[142] Makara, G. M., Schell, P., Hanson, K. and Moccia, D., *Tetrahedron Lett.*, **2002**, *43*, 5043.

[143] Ainsworth, C., *J. Am. Chem. Soc.*, **1955**, *77*, 1148.

[144] Dolman, S. J., Gosselin, F., O'Shea, P. D. and Davies, I. W., *J. Org. Chem.*, **2006**, *71*, 9548.

[145] Chekler, E. L. P., Elokdah, H. M. and Butera, J., *Tetrahedron Lett.*, **2008**, *49*, 6709.

[146] Polshettiwar, V. and Varma, R. S., *Tetrahedron Lett.*, **2008**, *49*, 879.

[147] Wang, Y., Sauer, D. R. and Djuric, S. W., *Tetrahedron Lett.*, **2006**, *47*, 105.

[148] Rajapakse, H. A., Zhu, H., Young, M. B. and Mott, B. T., *Tetrahedron Lett.*, **2006**, *47*, 4827.

[149] Dabiri, M., Salehi, P., Baghbanzadeh, M. and Bahramnejad, M., *Tetrahedron Lett.*, **2006**, *47*, 6983.

[150] Jedlovska, E. and Lesko, J., *Synth. Commun.*, **1994**, *24*, 1879.

[151] Thoman, C. J. and Voaden D. J., *Org. Synth., Coll. Vol. V*, **1973**, 962.

[152] Hurd, C. D. and Mori, R. I., *J. Am. Chem. Soc.*, **1955**, *77*, 5359.

[153] Harada, K., Inoue, T. and Yoshida, H., *Heterocycles*, **1997**, *44*, 459.

[154] Katritzky, A. R., Tymoshenko, D. O., and Nikonov, G. N., *J. Org. Chem.*, **2001**, *66*, 4045.

[155] Cronyn, M. W. and Nakagawa, T. W., *J. Am. Chem. Soc.*, **1952**, *74*, 3693.

[156] Goerdeler, J., Groschopp, H. and Sommerlad, U., *Chem. Ber.*, **1957**, *90*, 182.

[157] Wu, Y.-J. and Zhang, Y., *Tetrahedron Lett.*, **2008**, *49*, 2869.

[158] Ainsworth, C., *J. Am. Chem. Soc.*, **1958**, *80*, 5201.

[159] Whitehead, C. W. and Traverso, J. J., *J. Am. Chem. Soc.*, **1955**, *77*, 5872; Kress, T. J. and Costantino, S. M., *J. Heterocycl. Chem.*, **1980**, *17*, 607.

[160] Föhlisch, B., Braun, R. and Schultze, K. W., *Angew. Chem., Int. Ed. Engl.*, **1967**, *6*, 361.

[161] Weinstock, L. M., Davis, P., Handelsman, B. and Tull, R., *J. Org. Chem.*, **1967**, *32*, 2823.

[162] Wright, J. B., *J. Org. Chem.*, **1964**, *29*, 1905.

[163] Duan, X.-G., Duan, X.-L. and Rees, C. W., *J. Chem. Soc., Perkin Trans. 1*, **1997**, 2831; Duan, X.-G., Duan, X.-L., Rees, C. W. and Yue, T.-Y., *ibid.*, 2597; Duan, X.-G. and Rees, C. W., *ibid.*, 2695.

[164] Ponzo, M. G., Evindar, G. and Batey, R. A., *Tetrahedron Lett.*, **2002**, *43*, 7601.

[165] 'Progress in 1,2,3,4-tetrazine chemistry', Churakov, A. M. and Tartakovsky, V. A., *Chem. Rev.*, **2004**, *104*, 2601.

[166] Tyurin, A. Y., Churakov, A. M., Strelenko, Y. A., Ratnikov, M. O. and Tartakovsky, V. A., *Russ. Chem. Bull., Int. Ed.* **2006**, *55*, 1648.

[167] Grundmann, C. and Kreutzberger, A., *J. Am. Chem. Soc.*, **1955**, *77*, 44.

[168] Adam, W., van Barneveld, C. and Golsch, D., *Tetrahedron*, **1996**, *52*, 2377.

[169] Counotte-Potman, A. and van der Plas, H. C., *J. Heterocycl. Chem.*, **1981**, *18*, 123.

[170] Gulevskaya, A. V., Maes, B. U. W. and Meyers, C., *Synlett*, **2007**, 71.

[171] 'Behaviour of monocyclic 1,2,4-triazines in reactions with C-, N-, O- and S-nucleophiles', Charushin, V. N., Alexeev, S. G., Chupahkin, O. N. and van der Plas, H. C., *Adv. Heterocycl. Chem.*, **1989**, *46*, 76.

[172] Rykowski, A. and Lipinska, T., *Synth. Commun.*, **1996**, *26*, 4409; Rykowski, A., Branowska, D., Makosza, M. and van Ly, P., *J. Heterocycl. Chem.*, **1996**, *33*, 1567.

[173] Makosza, M., Golinski, J. and Rykowski, A., *Tetrahedron Lett.*, **1983**, *24*, 3277; Paudler, W. W. and Chen, T.-K., *J. Heterocycl. Chem.*, **1970**, *7*, 767.

[174] Sakya, S. M., Groskopf, K. K. and Boger, D. L., *Tetrahedron Lett.*, **1997**, *38*, 3805.

[175] Boger, D. L., Schaum, R. P. and Garbaccio, R. M., *J. Org. Chem.*, **1998**, *63*, 6329.

[176] Konno, S., Yokoyama, M., Kaite, A., Yamasuta, I., Ogawa, S., Mizugaki, M. and Yamanaka, H., *Chem. Pharm. Biull.*, **1982**, *30*, 152.

[177] 'Synthesis with s-triazine', Grundmann, C., *Angew. Chem., Int. Ed. Engl.*, **1963**, *2*, 309.

[178] Plé, N., Turck, A., Quéguiner, G., Glassi, B. and Neunhoeffer, H., *Liebigs Ann. Chem.*, **1993**, 583.

[179] Itoh, T., Matsuya, Y., Hasegawa, H., Nagata, K., Okada, M. and Ohsawa, A., *J. Chem. Soc., Perkin Trans. 1*, **1996**, 2511.

[180] Nagata, K., Itoh, T., Okada, M., Takahashi, H. and Ohsawa, A., *Heterocycles*, **1991**, *32*, 855.

[181] Chavez, D. E. and Hiskey, M. A., *J. Heterocycl. Chem.*, **1998**, *35*, 1329.

[182] 'Recent advances and applications in 1,2,4,5-tetrazine chemistry', Saracoglu, N., *Tetrahedron*, **2007**, *63*, 4199.

[183] 'Diels-Alder reactions of heterocyclic azadienes: scope and applications', Boger, D. L., *Chem. Rev.*, **1986**, *86*, 781; 'Hetero Diels-Alder methodology in organic synthesis', Ch. 10, Boger, D. L. and Weinreb, S. M., Academic Press, **1987**; '1-Azadienes in cycloaddition and multicomponent reactions towards N-heterocycles', Groenendaal, B., Ruijter, E. and Orru, V. A., *Chem. Commun.*, **2008**, 5474.

[184] Sainz, Y. F., Raw, S. A. and Taylor, R. J. K., *J. Org. Chem.*, **2005**, *70*, 10086.

[185] Kozhevnikov, V. N., Ustinova, M. M., Slepukhin, P. A., Santoro, A., Bruce, D. W. and Kozhevnikov, D. N., *Tetrahedron Lett.*, **2008**, *49*, 4096.

[186] Hamasaki, A., Ducray, R. and Boger, D. L., *J. Org. Chem.*, **2006**, *71*, 185.

[187] Boger, D. L. and Sakya, S. M., *J. Org. Chem.*, **1988**, *53*, 1415.

[188] Osawa, A., Kaiho, T., Ito, T., Okada, M., Kawabata, C., Yamaguchi, K. and Igeta, H., *Chem. Pharm. Bull.*, **1988**, *36*, 3838; Itoh, T., Nagata, K., Okada, M. and Ohsawa, A., *ibid.*, **1990**, *38*, 1524.

[189] Mättner, M. and Neunhoeffer, H., *Synthesis*, **2003**, 413.

[190] Paudler, W. W. and Chen, T.-K., *J. Heterocycl. Chem.*, **1970**, *7*, 767.

[191] Paudler, W. W. and Barton, J. M., *J. Org. Chem.*, **1966**, *31*, 1720.

[192] Gehre, A., Stanforth, S. P and Tarbit, B., *Tetrahedron Lett.*, **2007**, *48*, 6974.

[193] Laphookhieo, S., Jones, S., Raw, S. A., Sainz, F. Y. and Taylor, R. J. K., *Tetrahedron Lett.*, **2006**, *47*, 3865.

[194] Zhao, A., Leister, W. H., Strauss, K. A., Wisnoski, D. D. and Lindsley, C. W., *Tetrahedron Lett.*, **2003**, *44*, 1123.

[195] Potewar, T. M., Lahoti, R. J., Daniel, T. and Srinivasan, K. V., *Synth. Commun.*, **2007**, *37*, 261.

[196] 'Recent applications of 2,4,6-trichloro-1,3,5-triazine and its derivatives in organic synthesis', Blotny, G., *Tetrahedron*, **2006**, *62*, 9507.

[197] Kostas, I. D. Andreadaki, F. J., Medlycott, E. A., Hanan, G. S. and Monflier, E., *Tetrahedron Lett.*, **2009**, *50*, 1851.

[198] Schaefer, F. W. and Peters, G. A., *J. Org. Chem.*, **1961**, *26*, 2778.

[199] Schaefer, F. W. and Peters, G. A., *J. Org. Chem.*, **1961**, *26*, 2784.

[200] Chen, C., Dagnino, R. and McCarthy, J. R., *J. Org. Chem.*, **1995**, *60*, 8428.

[201] Geldard, J. F. and Lions, F., *J. Org. Chem.*, **1965**, *30*, 318.

[202] Boger, D. L., Panek, J. S. and Patel, M., *Org. Synth.*, **1992**, *70*, 79.

[203] 'Heterocyclic synthesis with benzotriazole', Katritzky, A. R., Henderson, S. A. and Yang, B., *J. Heterocycl. Chem.*, **1998**, *35*, 1123.

[204] 'Properties and synthetic utility of *N*-substituted benzotriazoles', Katritzky, A. R., Lan, X., Yang, J. Z. and Denisko, O. V., *Chem. Rev.*, **1998**, *98*, 409.

[205] 'The continuing magic of benzotriazole: an overview of some recent advances in synthetic methodology', Katritzky, A. R., *J. Heterocycl. Chem.*, **1999**, *36*, 1501.

[206] 'Benzotriazole: an ideal synthetic auxiliary', Katritzky, A. R. and Rogovoy, B. V., *Chem. Eur. J.*, **2003**, *9*, 4586.

[207] 'Benzotriazole mediated amino-, amido-, alkoxy- and alkthio-alkylation', Katritzky, A. R., Manju, K., Singh, S. K. and Meher, N. K., *Tetrahedron*, **2005**, *61*, 2555.

[208] Katritzky, A. R.,Widyan, K. and Kirichenko, K., *J. Org. Chem.*, **2007**, *72*, 5802.

[209] Katritzky, A. R., Angrish, P., Hür, D. and Suzuki, K., *Synthesis*, **2005**, 397.

[210] Katritzky, A. R., Zhang, G. and Jiang, J., *J. Org. Chem.*, **1995**, *60*, 7589.

[211] Katritzky, A. R. and Jiang, J., *J. Org. Chem.*, **1995**, *60*, 7597.

[212] Katritzky, A. R., Fali, C. N. and Li, J., *J. Org. Chem.*, **1997**, *62*, 8205.

[213] Katritzky, A. R., Xie, L. and Cundy, D., *Synth. Commun.*, **1995**, *25*, 539.

[214] Katritzky, A. R. and Xie, L., *Tetrahedron Lett.*, **1996**, *37*, 347.

[215] Katritzky, A. R., Yang, Z. and Lam, J. N., *Synthesis*, **1990**, 666.

[216] Knight, D. W. and Little, P. B., *Tetrahedron Lett.*, **1998**, *39*, 5105.

[217] Katritzky, A. R., Zhang, G., Jiang, J. and Steel, P. J., *J. Org. Chem.*, **1995**, *60*, 7625.

[218] Molina, A., Vaquero, J. J., Garcia-Navio, J. L., Alvarez-Builla, J., de Pascual-Teresa, B., Gago, F., Rodrigo, M. M. and Ballesteros, M., *J. Org. Chem.*, **1996**, *61*, 5587.

[219] Anwar, M. U., Tragl, S., Ziegler, T. and Subramanian, L. R., *Synlett*, **2006**, 627.

30

Saturated and Partially Unsaturated Heterocyclic Compounds: Reactions and Synthesis

This book is principally concerned with the chemistry of aromatic heterocycles, however mention must be made of the large body of remaining heterocycles, including those with small rings[1] (three- and four-membered). Most of the reactions of saturated and partially unsaturated heterocyclic compounds are so closely similar to those of acyclic or non-heterocyclic analogues that a full discussion is not appropriate in this book, however in this chapter we discuss briefly those aspects in which they do differ – perhaps the most obvious aspect in which they differ from aromatic heterocycles is in having sp[3] hybridised atoms, and consequently the possible exhibition of stereoisomerism.[2] They exhibit differences from 'normal' saturated compounds because they are cyclic, which opens up various possibilities for reactivity, particularly where ring opening is involved, and this is seen mainly in the smaller rings, where strain (energy) is more likely to induce such reactivity.

Tetrahydrofuran (THF) and dioxane are well-known solvents for organic reactions. N-Methylpyrrolidone (NMP) and sulfolane are useful dipolar aprotic solvents, with characteristics like those of dimethylformamide (DMF) and dimethyl sulfoxide (DMSO). Saturated and partially unsaturated heterocycles occur widely as components of natural products (Chapter 32).

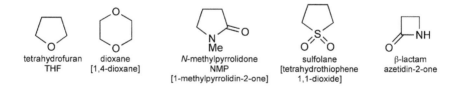

| tetrahydrofuran THF | dioxane [1,4-dioxane] | N-methylpyrrolidone NMP [1-methylpyrrolidin-2-one] | sulfolane [tetrahydrothiophene 1,1-dioxide] | β-lactam azetidin-2-one |

The four-membered β-lactam ring is the essential biologically active component of the penicillin and cephalosporin antibiotics (see 33.6.2). Epoxides (three-membered saturated oxygen-containing rings) are components of epoxy resins and occur in some natural products. Epoxides, because of their alkylating properties, can be carcinogenic – the biologically active metabolites of carcinogenic hydrocarbons are examples. Aziridines (three-membered saturated nitrogen-containing rings) are found, for example, in the mitomycins, anti-tumour agents from *Streptomyces lavendulae*. Thiiranes (three-membered saturated sulfur-containing rings) also occur naturally, as plant products, such as thiirane-2-carboxylic acid, isolated from asparagus.

Heterocyclic Chemistry 5th Edition   John Joule and Keith Mills
© 2010 Blackwell Publishing Ltd

30.1 Five- and Six-Membered Rings
30.1.1 Pyrrolidines and Piperidines

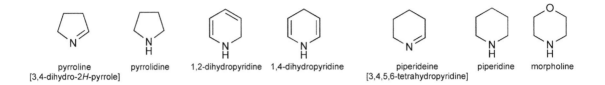

pyrroline
[3,4-dihydro-2*H*-pyrrole]

pyrrolidine

1,2-dihydropyridine

1,4-dihydropyridine

piperideine
[3,4,5,6-tetrahydropyridine]

piperidine

morpholine

The main chemical aspect in which compounds with a nitrogen in a five- or six-membered ring differ from their acyclic counterparts is that they can be dehydrogenated to the corresponding aromatic system. Dihydro-aromatic systems naturally show the greatest tendency to aromatise, indeed one of the important reducing coenzymes, NADPH (a 1,4-dihydropyridine), makes use of this tendency – it is a 'hydride donor' (32.2.1).

Dihydro compounds are often useful synthetic intermediates showing different reactivity patterns to the parent, aromatic heterocycle. For example, indolines (2,3-dihydroindoles) can be used to prepare indoles with substituents in the carbocyclic ring, *via* electrophilic substitution then aromatisation (20.16.1.17), and similarly, electrophilic substitutions of dihydropyridines, very difficult in simple pyridines, followed by aromatisation, can give substituted pyridines. Dehydrogenation of tetra- and hexahydro-derivatives requires more vigorous conditions.

Pyrrolidine and piperidine are better nucleophiles than diethylamine, principally because the lone pair is less hindered – in the heterocycles the two alkyl 'substituents', i.e. the ring carbons, are constrained back and away from the nitrogen lone pair, and approach by an electrophile is thus rendered easier than in diethylamine, where rotations of the C–N and C–C bonds interfere. The pK_{aH} values of pyrrolidine (11.27) and piperidine (11.29) are typical of amine bases; they are slightly stronger bases than diethylamine (10.98). Morpholine (8.3) is a somewhat weaker base than piperidine.

Piperidines, like cyclohexanes, adopt a preferred chair conformation, where both *N*-hydrogen and *N*-alkyls take up equatorial conformations, though in the former case the equatorial isomer is favoured by only a small margin.[3] However, alkylations at nitrogen do not necessarily reflect these ground-state conformational populations.

In the days before spectroscopy, structure determination of natural products involved degradative methods. Many alkaloids incorporate saturated nitrogen rings, so degradations were used that gave information about the environment of the basic nitrogen atom. The classical method for doing this was the 'Hofmann exhaustive methylation' procedure. This is illustrated below as it would be applied to piperidine. What the method does is to cleave N–C bonds and eventually remove the nitrogen. If one cycle removes the nitrogen then it can be concluded that it was originally *not* part of a ring; if two cycles are required, as in the piperidine example, then the nitrogen must originally have been part of a ring. A third cycle would be necessary if the nitrogen had been originally a component of two rings. At the end of the process a nitrogen-free fragment is left, for study to determine the carbon skeleton.

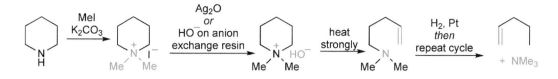

30.1.2 Piperideines and Pyrrolines

Generally speaking, piperideines and pyrrolines exist predominantly in the imine form and not in the tautomeric enamine form. These cyclic imines are resistant to hydrolytic fission of the C=N bond, in strong contrast with acyclic imines, but nonetheless they are susceptible to nucleophilic addition at the azomethine carbon. An example of this is that both piperideine and pyrroline exist as trimers formed by the nucleophilic addition of nitrogen of one molecule to the azomethine carbon of a second molecule, etc. The trimerisation of pyrroline can be prevented by forming a complex, $(C_4H_7N)_2ZnI_2$, from which the imine can be regenerated using ammonia.[4]

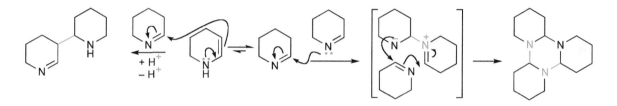

The presence of some enamine, at equilibrium, is demonstrated by the conversion of piperideine into a dimer, indeed, the ability of these two systems to serve as both imines and enamines in such condensations is at the basis of their roles in alkaloid biosynthesis. Formed in nature by the oxidative deamination and decarboxylation of ornithine and lysine, they become incorporated into alkaloid structures by condensation with other precursor units.[5] Hygrine is a simple example in which the pyrroline has condensed with acetoacetate, or its equivalent.

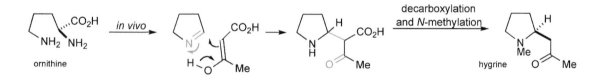

Controlled oxidation of *N*-acyl-piperidines and -pyrrolidines can be used to prepare 2-alkoxy derivatives or the equivalent enamides, which are useful general synthetic intermediates.[6] The former are susceptible to nucleophilic substitution under Lewis-acid catalysis, *via* Mannich-type intermediates, and the latter can undergo electrophilic substitution at C-3 or addition to the double bond.

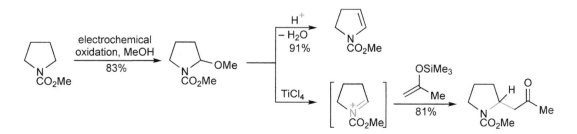

Enol phosphates derived from *N*-acyl-piperidin-2-one (or 2-oxoazepane) can be utilized in cross-coupling processes.[7]

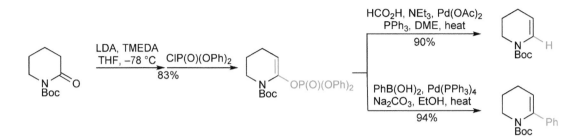

Various oxidants have been used to convert pyrrolidine and piperidine into nitrones (imine oxides), of value for 1,3-dipolar cycloadditions; in one method hydrogen peroxide is the oxidant in the presence of a catalyst.[8]

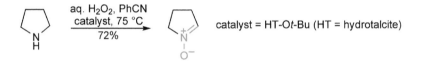

30.1.3 Pyrans and Reduced Furans

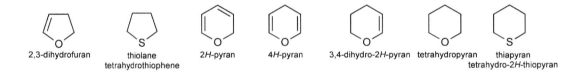

3,4-Dihydro-2*H*-pyran and 2,3-dihydrofuran behave as enol ethers, the former being widely used to protect alcohols,[9] with which it reacts readily under acidic catalysis, producing acetals that are stable to even strongly basic conditions, but easily hydrolysed back to the alcohol under mildly acidic aqueous conditions. Each can also serve as equivalents of 5-(4-)-hydroxy-aldehydes (see below).

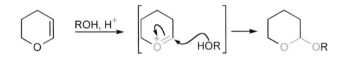

A great deal is known about hydroxylated tetrahydrofurans and tetrahydropyrans because such ring systems occur in sugars and sugar-containing compounds – sucrose and RNA (32.4) are examples.[10]

Tetrahydropyran, like piperidine, adopts a chair conformation. One of the interesting aspects to emerge from studies of alkoxy-substituted tetrahydropyrans is that when located at C-2, alkoxyl groups prefer an axial orientation (the 'anomeric effect'[11]). The reason for this is that in an equatorial orientation there are unfavourable dipole–dipole interactions between lone pairs on the two oxygen atoms, and the energy gain when these are relieved in a conformation with the C-2-substituent axial, more than offsets the unfavourable 1,3-diaxial interactions that are introduced at the same time.

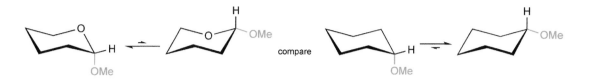

Glucose is an equilibrium mixture of cyclic forms (hemiacetals containing a tetrahydropyran), and a small concentration of acyclic polyhydroxyaldehyde, which is responsible for many of the observed chemical reactions. This illustrates the inherent stability of chair conformers of saturated six-membered systems. The propensity for cyclisation is a general one: 5-hydroxy-aldehydes, -ketones and -acids all easily form six-membered oxygen-containing rings – lactols and lactones respectively.

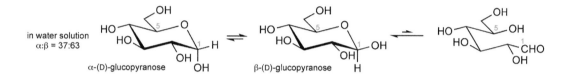

Five-membered rings, too, are relatively easy to form: depending on conditions, glucose derivatives can easily be formed in the furanose form, i.e. based on tetrahydrofuran. The polyfunctional and stereo-defined sugars can be used to prepare sterochemically related nitrogen heterocycles.[12]

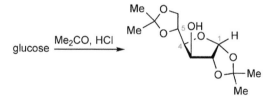

Saturated five- and six-membered cyclic ethers are, like acyclic ethers, rather inert, requiring strong conditions for C–O bond cleavage;[13] this contrasts strongly with oxetanes and oxiranes (30.2 and 30.3).

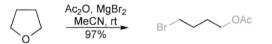

S-Alkylthiolanium compounds, even *S*-methylthiolanium salts, react with nucleophiles mainly at C-2, leading to ring-opened products.[14]

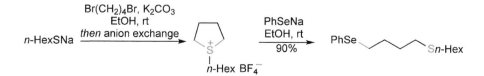

30.2 Three-Membered Rings
30.2.1 Three-Membered Rings with One Heteroatom

Δ-2-Unsaturated three-membered systems are unknown as stable molecules because they would have a four-electron π-system, and thus be antiaromatic.[15] 1*H*-Azirines occur as reactive intermediates and there is evidence for the existence of 2-thiirene in a low-temperature matrix.[16] Azirines,[17] by contrast, are well-known stable compounds. Thiirene *S,S*-dioxides are also stable molecules, probably best likened to cyclopropenones.[18] The chemistry of saturated three-membered heterocycles is, however, very extensive – epoxides (oxiranes), and to a lesser extent, aziridines are important intermediates in general synthesis.

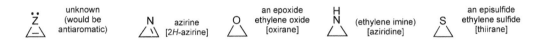

A major advance was the development of an efficient synthesis of epoxides of high optical purity from allylic alcohols and related systems (the *Sharpless epoxidation*) (below); such epoxides have been used extensively for the synthesis of complex natural products in homochiral form.

The pK_{aH} of aziridine (7.98) shows it to be an appreciably weaker base than azetidine (11.29), the four-membered analogue, which has a value 'normal' for acyclic amines and for five- and six-membered saturated amines. The low basicity is mirrored in the oxygen series, as measured by the ability of oxiranes to form hydrogen bonds. The explanation is probably associated with the strain in the three-membered compounds, meaning that the lone pair is in an orbital with less p-character than a 'normal' sp^3 nitrogen or oxygen orbital, and is therefore held more tightly. The rate of pyramidal inversion of the saturated nitrogen in azirines is very slow compared with simpler amines. This is because there is a further increase in angle strain when the nitrogen rehybridises ($\rightarrow$ sp^2) in the transition state for inversion.

The chemical reactions of three-membered heterocycles are a direct consequence of the strain inherent in such small rings, which, combined with the ability of the heteroatom to act as a leaving group, means that most of the chemical properties involve ring-opening reactions. Most epoxide ring-openings occur by S$_N$2 nucleophilic displacements at carbon, and a very wide range of carbanion and heteroatom nucleophiles have been shown to react in this way, including amines,[19] alcohols, thiols, hydride (LiAlH$_4$), malonate anions,[20] etc. Assistance by protic solvents or *O*-coordinating metal cations (Lewis acids) that help to further weaken the C–O bond can dramatically increase the rate of reaction. Additives such as alumina,[21] titanium alkoxides[22] and lithium perchlorate,[23] and reagents such as tributyltin azide,[24] which is itself a Lewis acid (coordination to 'Bu$_3$Sn$^+$'), but also contains a nucleophilic function (N$_3^-$), are useful in this respect.

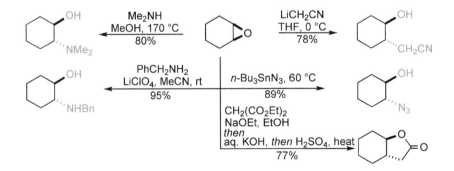

'Harder' organometallic nucleophiles such as alkyllithiums often give rise to side reactions, but their use with boron trifluoride at −78 °C gives very clean and efficient ring-opening reactions.[25]

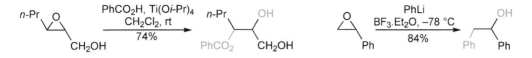

The regiochemistry of ring opening is determined mainly by steric, and, to a lesser extent, by inductive and electronic effects. Where strong Lewis acids are used or where a highly stabilised (incipient) carbonium ion can be formed, such as when an α-aryl substituent is present, reaction can occur mainly at the most substituted position, an extreme case being the solvolysis of 2-furyloxirane in neutral methanol;[26] however, selective substitution at the most highly substituted position of even simple, alkyl epoxides can be achieved with an allyl-titanium reagent.[27]

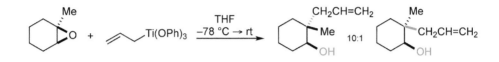

Regioselective opening of Sharpless epoxides, that is those (enantio-enriched) derived from allylic alcohols, has major synthetic significance. The usual outcome is opening at C-3, but the reaction with nucleophiles such as azide, cyanide and thiophenoxide, in the presence of trialkyl borates, can be highly selective for attack at C-2.[28]

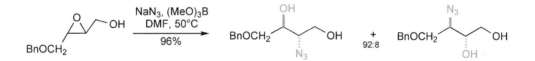

The Payne rearrangement of epoxy-alcohols is a special case of an intramolecular nucleophilic opening of epoxides and is of particular synthetic utility when it is applied to epoxides from the Sharpless procedure.[29]

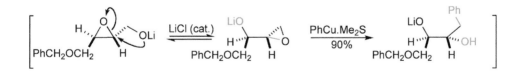

Ring-opening of epoxides by β-elimination, on reaction with strong bases, such as lithium amides, or combinations of trimethylsilyl triflate with diazabicycloundecane,[30] is a useful synthetic method to prepare allylic alcohols, particularly as it can be carried out enantioselectively.[31]

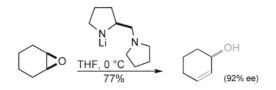

The relative stereochemistry of epoxides can be inverted by equilibration with cyanate anion.[32]

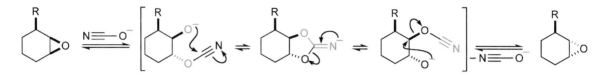

Acid-catalysed opening of aziridines is usually quite rapid, but simple nucleophilic reactions, without acid catalysis, are very slow, due to the much poorer leaving ability of negatively charged nitrogen, however *N*-acyl- or *N*-sulfonyl-aziridines have reactivity similar to epoxides.[33]

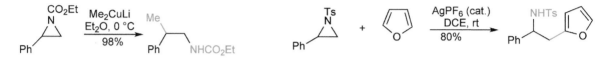

A number of catalysts have been used to promote ring opening of *N*-tosyl-aziridines, such as phospho-molybdic acid and silica gel, for azide, cyanide and alcohols[34] and tri-*n*-butylphosphine for thiols and amines.[35] Opening with 'iodide' occurs at room temperature with iodine and thiophenol in the presence of air.[36] In the nucleophilic ring opening of *N*-tosyl-aziridines, silver ion catalysis facilitates reactions with electron-rich arenes or hetarenes.[37]

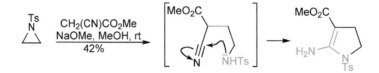

N-Tosyl-aziridines (and *N*-tosyl-azetidines) react with cyclic enol ethers and enamides, under Lewis acid catalysis to give synthetically useful bicycles.[38]

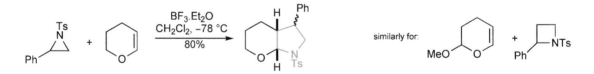

Methylene aziridines have some use as multi-functional synthons. They can be metallated on the ring carbon or ring opened with nucleophiles, giving enaminates that can be further transformed.[39]

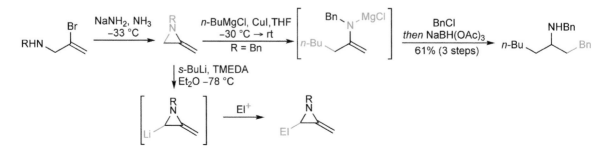

Thiiranes similarly undergo ring-opening reactions with nucleophiles, such as amines,[40] but attack at sulfur can also occur with lithium reagents.[41]

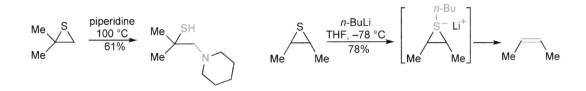

The heteroatom in a three-membered heterocycle can be eliminated *via* various cycloreversion reactions, for example by nitrosation of aziridines,[42] or by the reaction of thiiranes with trivalent phosphorus compounds.

A related elimination of sulfur dioxide occurs during the Ramberg–Bäcklund synthesis[43] of alkenes, which generates an episulfone as a transient intermediate, although episulfones are isolable under controlled conditions.[44]

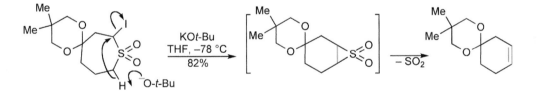

Substituted derivatives of all three systems are able to undergo a highly stereospecific concerted thermal ring opening, generating ylides that can be utilised (trapped) in [3 + 2] cycloaddition reactions, for example using aziridines provides a route to pyrrolidines.[45]

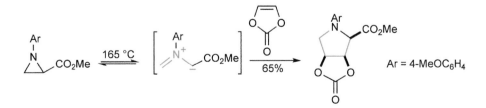

Azirines with an ester group on the imine carbon, will take part in cycloadditions, with the imine unit as the dienophile, as illustrated below.[46]

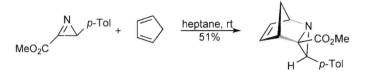

30.2.2 Three-Membered Rings with Two Heteroatoms
Diaziridines, diazirines and dioxiranes are all relatively stable, isolable systems, although some dioxiranes are explosive.

Three-membered rings with two heteroatoms are usually encountered only as reagents. Diazirines are useful carbene precursors[47] – they are generally more stable than the equivalent isomeric diazo compounds, though they are sometimes explosive in the pure state. They can be prepared by oxidation of diaziridines that, in turn, are available *via* the condensation of a ketone or aldehyde with ammonia and chloramine.[48] Chloro-diazirines, from the reaction of amidines with hypochlorite, will undergo S_N2 or S_N2' displacement reactions.[49]

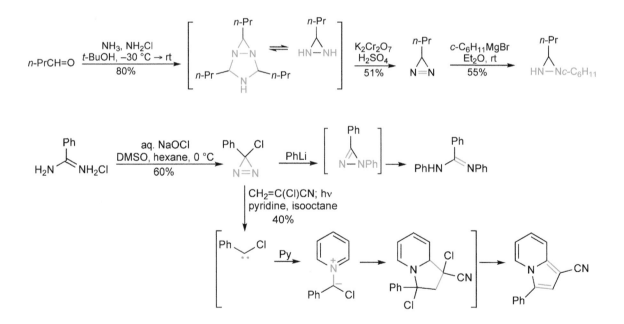

Dimethyldioxirane is a relatively strong oxidant, but can show good selectivity: its reactivity is similar to that of a peracid, but it has the advantage of producing a neutral byproduct (acetone). Methyl(trifluoromethyl)dioxirane is a more powerful oxidant that can insert oxygen into C–H bonds with retention of configuration, as shown below.[50] Dioxiranes are obtained by reaction of ketones with *OXONE*® ($2KHSO_5.KHSO_4.K_2SO_4$).[51] (**NOTE:** *Dioxiranes are explosive and are usually handled in dilute solution.*)

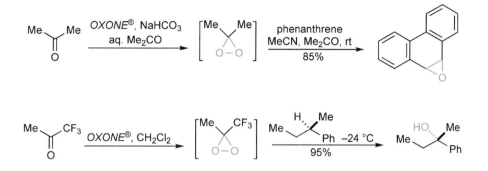

Oxaziridines, prepared by the oxidation of imines,[52] are selective oxygen-transfer reagents.[53] In particular, the camphor-derived reagent is widely used for enantioselective oxygenation of enolates[54] and other nucleophiles.

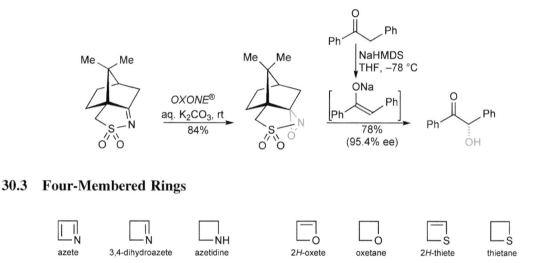

30.3 Four-Membered Rings

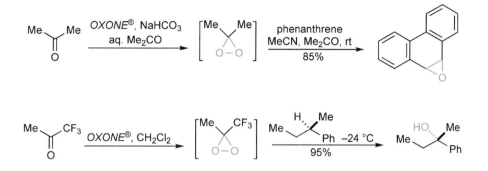

azete	3,4-dihydroazete	azetidine	2H-oxete	oxetane	2H-thiete	thietane

Derivatives of azete are only known as unstable reaction intermediates. Oxetane and azetidine are considerably less reactive than their three-membered counterparts (oxetane reacts with hydroxide anion 10^3 times more slowly than does oxirane), but nonetheless do undergo similar ring-opening reactions, for example oxetane reacts with organolithium reagents[25] in the presence of boron trifluoride, or with cuprates,[55] and azetidine is opened on heating with concentrated hydrochloric acid. Azetidinium ions react much more easily with nucleophiles.[56]

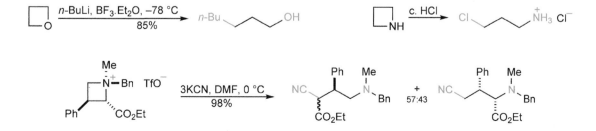

The most important four-membered system is undoubtedly the β-lactam ring[57] that is present in, and essential for the biological activity of, the penicillin and cephalosporin antibiotics. β-Lactams are very susceptible to ring-opening *via* attack at the carbonyl carbon – in stark contrast to the five-membered analogues (pyrrolidones) or acyclic amides, which are relatively resistant to nucleophilic attack at carbonyl carbon. In addition, β-lactams are hydrolysed by a specific enzyme, β-lactamase, the production of which enzyme is a mechanism by which bacteria become resistant to such antibiotics. Although the β-lactam ring is easily cleaved by nucleophiles, both *N*- and *C*-alkylation (α to carbonyl) can be achieved using bases to deprotonate; it is even possible to carry out Wittig reactions at the 'amide' carbonyl without ring opening.[58] Substitution of the acetoxy group in a 4-acetoxy-azetidinone by nucleophiles is an important synthetic method; the reaction proceeds *via* an imine or an iminium intermediate, rather than by direct displacement.[59]

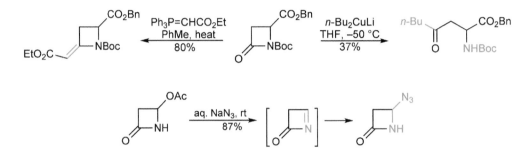

1-Tosyloxyazetidin-2-ones undergo an interesting nucleophilic substitution at C-3, the mechanism being thought to involve an allylic-type 1,3-displacement *via* an enolised intermediate.[60]

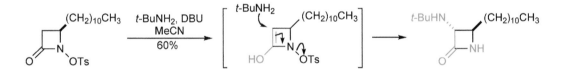

β-Lactones (propiolactones)[61] too are readily attacked at the carbonyl carbon, for example they are particularly easily hydrolysed, but a second mode of nucleophilic attack – S$_N$2 displacement of carboxylate *via* attack at C-4 – occurs with many nucleophiles.[62] The example shows the use of a homochiral β-lactone, available from serine.

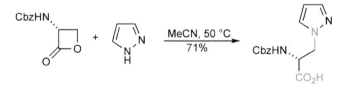

30.4 Metallation

3,4-Dihydro-2*H*-pyran and 2,3-dihydrofuran can be metallated in the same way as their acyclic analogues, i.e. at the enol-ether carbon adjacent to the oxygen. Tetrahydrofuran can also be lithiated adjacent to the oxygen, by warming with *n*-butyllithium, but the lithio derivative produced then undergoes a cycloreversion, generating ethene and the lithium enolate of ethanal.[63] This process represents a convenient preparation of this enolate, but can also be a significant, unwanted side-reaction during lithiation reactions using tetra-hydrofuran as solvent.

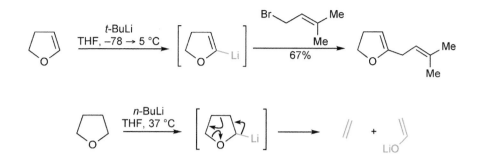

Three-membered rings have not been metallated directly in the absence of anion-stabilising substituents but simple lithio derivatives of aziridines can be prepared by exchange from the corresponding stannane.[64] *N*-*t*-Butylsulfonyl-aziridines can be substituted, *via* non-stabilised lithio intermediates, by reaction with LiTMP, usually in the presence of the electrophile.[65]

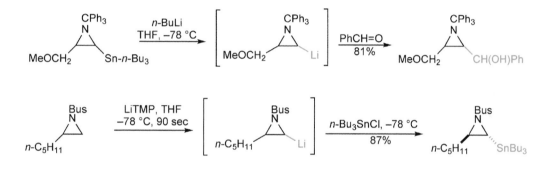

2-Trifluoromethyloxirane can be lithiated at C-2 using *n*-butyllithium at −102 °C, the product after reaction with electrophiles retaining configuration.[66]

30.5 Ring Synthesis

Five- and six-membered saturated rings can be prepared by reduction of the corresponding aromatic compounds, but the most general method for making all ring sizes is by cyclisation of an ω-substituted amine, alcohol or thiol *via* an intramolecular nucleophilic displacement. As an illustration, the rate of cyclisation of ω-halo-amines goes through a minimum at the four-membered ring size; the five and six-membered rings are by far the easiest to make. The relative rates for formation of 3-, 4-, 5- and 6-membered rings respectively are 72:1:6000:1000.[67] A factor which influences the rate of 3-*exo-tet*-cyclisations is the degree of substitution at the carbon carrying the heteroatom: increasing substitution increases the rate of cyclisation, because in the small ring product there is some relief of steric crowding for the substituents compared with the acyclic starting material.[68]

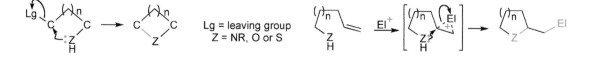

Related cyclisations involving heteroatom attachment to an alkene *via* π-complexes with cations, such as Br⁺, I⁺, Hg⁺ and Pd⁺, are useful methods because they give products with functionalised side-chains for further transformations.

30.5.1 Aziridines and Azirines

The main routes to aziridine include alkali-catalysed cyclisation of 2-halo-amines or of a 2-hydroxyamine sulfonate ester, as illustrated,[69] or by additions to alkenes or imines.

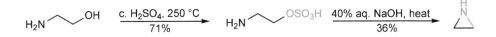

Various homochiral aziridines can be easily obtained from serine;[70] such substances can be transformed into a range of polyfunctional homochiral intermediates and products.

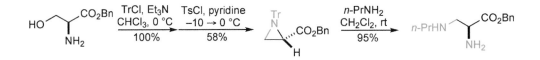

Reversing the sense of cyclisation, *N*-chloro-amines, with a suitable C–H-acidifying substitutent, can be ring closed efficiently producing *N*-alkyl-aziridines.[71]

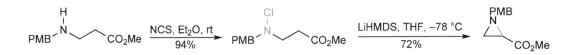

Aziridines can be obtained from alkenes using iodine isocyanate[72] or iodine azide.[73] The product from the latter reaction can be converted into the aziridine *via* reduction, or into an azirine *via* elimination of hydrogen iodide and pyrolysis.[74]

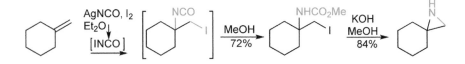

N-Tosyl-aziridines can be obtained directly from alkenes by reaction with Chloramine T (TsN(Cl)Na),[75] or directly from sulfonamides using *t*-butyl hypochlorite with sodium iodide[76] or with iodosylbenzene and CuI.[77]

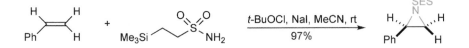

Aziridines can also be prepared by addition of nitrenes to alkenes,[78] or by the use of nitrogen-transfer agents analogous to epoxidising agents.[79]

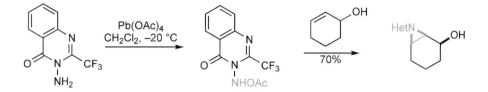

Addition to imines is a further obvious way in which to construct an aziridine, as illustrated below.[80]

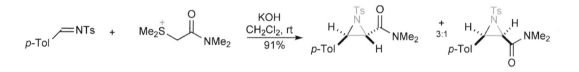

Azirines can be synthesised, enantioselectively if required, using a natural alkaloid as base, from the *O*-tosyl derivatives of the oximes of 1,3-keto-esters; in this synthesis the carbon is the nucleophilic centre and it is the nitrogen that is attacked, with departure of tosylate.[81]

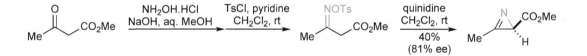

A related reaction using α-benzotriazolyl-oximes is particularly useful because the benzotriazolyl unit in the product can be displaced by nucleophiles.[82]

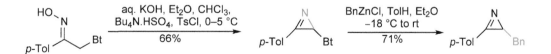

Intramolecular addition, catalysed by copper, produces bicyclic aziridines effectively from sulfonamides.[83]

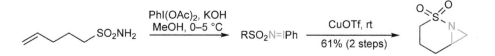

30.5.2 Azetidines and β-Lactams

Azetidines can be obtained by cyclisations of 3-halo-amines, but yields are generally not as good as those for the formation of aziridines. The generation of the bifunctional precursors for cyclisation to azetidines has been achieved in a number of ways.[84]

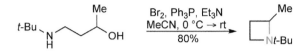

Synthesis of 1-azabicyclo[1.1.0]butane, which contains both a four-membered and two three-membered nitrogen-containing rings (!), follows the general route described above.[85] As one would anticipate, ring-opening reactions, one of which is illustrated, lead to products with an azetidine unit, rather than an aziridine unit.

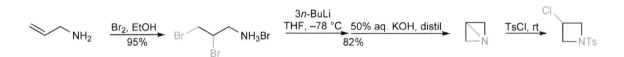

Many methods have been developed for β-lactam synthesis,[57,86] including cyclisation of the corresponding amino acids. The most widely used methods are two-component couplings,[47,87] which occur *via* concerted cycloaddition or two-step mechanisms. Another simple route to 3-functionalised azetidinones is the reaction of aziridine-2-carboxylic acid sodium salt with oxalyl chloride or thionyl chloride.[88]

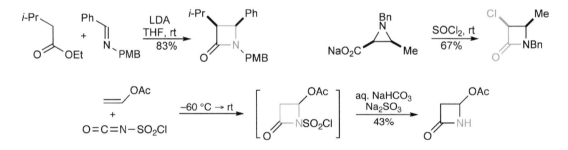

30.5.3 Pyrrolidines

A very neat method for the synthesis of pyrrolidines does not require a difunctionalised starting material, but relies on the Hofmann–Löffler–Freytag reaction[89] – which is a radical process – to introduce the second functional group. The six-membered size of the cyclic transition state leads selectively to a 1,4-halo-amine, and thence to pyrrolidines.

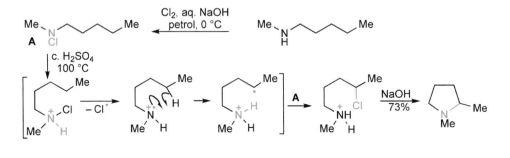

The cycloaddition of azomethine ylides to alkenes is another elegant entry to pyrrolidines. The required 1,3-dipoles can be produced in a number of ways; the example below is one of the most simple, wherein a trimethylsilylmethylamine, an aldehyde and the alkene are simply heated in tetrahydrofuran.[90]

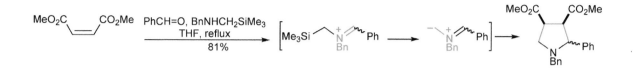

Normally, 5-*endo-trig*-cyclisations are disfavoured geometrically (Baldwin's rules), however formation of pyrrolidines does take place with tosylamide anion as nucleophile in a situation where 5-*exo-tet*-attack at a trihalomethyl group is inhibited electrostatically.[91]

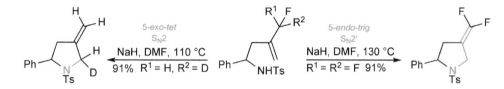

In another example, 5-*endo-trig*-cyclisation of a 2-(2-tosylaminoethyl)acrylate occurs with a phenyl ester, but 5-*exo-trig* with an ethyl ester: the relative leaving abilities of groups from the 5-*exo-trig*-intermediate allows the 5-*endo-trig*-pathway to dominate (TsN$^-$ > EtO$^-$ but PhO$^-$ > TsN$^-$) rationalizes this dichotomy.

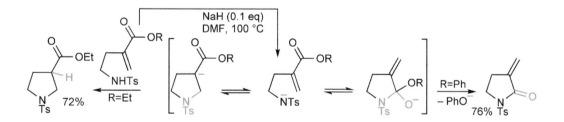

30.5.4 Piperidines[92]

Closure of 5-halo-amines by nucleophilic displacements of halogen, is complemented by the conversion of *N*-chloro-pent-4-enylamines into 3-chloropiperidines by exposure to tetra-*n*-butylammonium iodide.[93] The process is believed to involve conversion to an *N*-iodo-amine, which serves as an electrophilic source of iodine, cyclisation to a 2-(iodomethyl)-pyrrolidine, ring closure to a bicyclic aziridinium ion and ring opening *via* attack by chloride to produce the six-membered heterocycle.

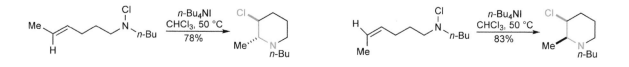

A construction of functionalised glutarimides relies on the doubly nucleophilic dianion that can be generated by C- and N-deprotonation, from a 2-tosyl-acetamide.[94] Regioselective reduction to a piperidone can be achieved *via* formation of the stabilized enolate.

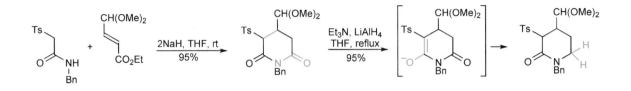

A particularly useful general method for the synthesis of saturated nitrogen- and oxygen-containing partially unsaturated heterocycles, from 5-membered to medium ring-sized,[95] is the Grubbs olefin metathesis[96] applied, for example, to acyclic dialkenyl-amines, as illustrated by syntheses of a dihydropyrrole[97] and a tetrahydropyridine.[98]

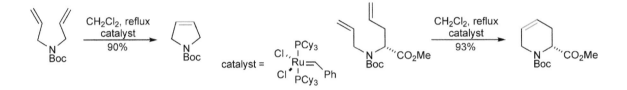

The superacid HF-SbF$_5$ can also bring about cyclisation of *N*-protected bis(allyl)-amines, producing 4-fluoropiperidines.[99]

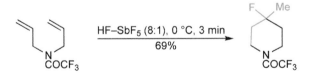

30.5.5 Saturated Oxygen Heterocycles

The most widely used method for the preparation of epoxides involves oxidation of an alkene by a peracid,[100] *via* a direct one-step transfer of an oxygen atom. More highly (alkyl) substituted alkenes react fastest showing that electronic effects are more important than steric effects in this reaction. Steric effects do, however, control the facial selectivity of epoxidation; conversely hydrogen-bonding groups, such as OH and NH, can direct the reaction to the *syn* face.

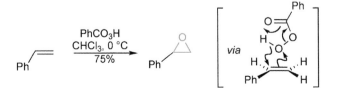

Several other direct oxygen-transfer reagents have been developed of which by far the most important is Sharpless' reagent – a mixture of a hydroperoxide with titanium isopropoxide and a dialkyl tartrate.[101] The structure of the reagent is complex, but it reacts readily with alkenes containing polar groups, for example allylic alcohols, which can coordinate the metal. The most important feature of this process is that when homochiral tartrate esters are used, a highly ordered asymmetric reactive site results, leading in turn to high optical induction in the product.

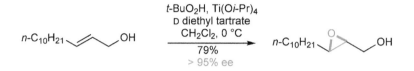

Epoxides and oxetanes can also be prepared by cyclisation of 1,2-halohydrins and 1,3-halo-alcohols.[102]

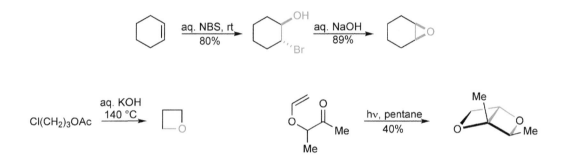

Oxetanes have often been prepared by the *Paternò–Büchi reaction*,[103] in which a compound containing a carbon–carbon double bond cycloadds to an aldehyde or ketone under the influence of light.[104]

30.5.6 Saturated Sulfur Heterocycles

Thiiranes can be prepared by cyclisation of 2-halo-thiols, but the most common method is *via* reaction of an epoxide with thiocyanate,[105] thiourea,[106] a phosphine sulfide or with dimethylthioformamide.[107]

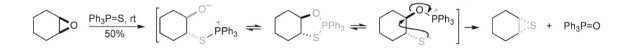

Thietanes, tetrahydrothiophenes and tetrahydrothiapyrans can all be prepared by the reaction of the appropriate 1,ω-dihalide with sulfide anion.

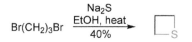

References

1. 'Ethylenimine and other aziridines', Derner, O. C. and Ham, G. E., Academic Press, **1969**; 'Thiiranes', Sander, M., *Chem. Rev.*, **1966**, *66*, 297.
2. 'Stereochemistry of Heterocyclic Compounds', Parts 1 and 2, Armarego, W. L. F., Wiley-Interscience, New York, **1977**; 'The Conformational Analysis of Heterocyclic Compounds', Riddell, F. G., Academic Press, **1980**.
3. 'Conformation of piperidine and of derivatives with additional ring heteroatoms', Blackburne, I. D., Katritzky, A. R. and Takeuchi, Y., *Acc. Chem. Res.*, **1975**, *8*, 300; '*N*-Methyl inversion barriers in six-membered rings', Katritzky, A. R., Patel, R. C. and Riddell, F. G., *Angew. Chem., Int. Ed. Engl.*, **1981**, *20*, 521.
4. Baxter, G., Melville, J. C. and Robins, D. J., *Synlett*, **1991**, 359.
5. 'Introduction to the Alkaloids. A Biogenetic Approach', Cordell, G. A., Wiley-Interscience, **1981**.
6. Matsumura, Y., Terauchi, J., Yamamoto, T., Konno, T. and Shono, T., *Tetrahedron*, **1993**, *49*, 8503.
7. Lepifre, F., Clavier, S., Bouyssou, P. and Coudert, G., *Tetrahedron*, **2001**, *57*, 6969.
8. Choudary, B. M., Reddy, Ch. V., Prakash, B. V., Bharathi, B., Kantam, M. L., *J. Mol. Catal. A. Chemical*, **2004**, *217*, 81.
9. 'Protective groups in organic synthesis', Greene, T. W. and Wuts, P. G. M., Wiley, **1999**.
10. 'Stereochemistry of Carbohydrates', Stoddart, J. F., Wiley-Interscience, **1971**.
11. 'Stereoelectronic Effects in Organic Chemistry', Deslongchamps, P., Pergamon Press, **1983**.
12. 'Synthesis of Naturally Occurring Nitrogen Heterocycles from Carbohydrates', El Ashry, E. S. H. and El Nemr, A., Blackwell, **2005**.
13. Goldsmith, D. J., Kennedy, E. and Campbell, R. G., *J. Org. Chem.*, **1975**, *40*, 3571.
14. Krief, A., Dumont, W. and Robert, M., *Synlett* **2006**, 2601; Eliel, E. L., Hutchins, R. O., Mebane, R. and Willer, R. L., *J. Org. Chem.*, **1976**, *41*, 1052.
15. Dewar, M. J. S. and Ramsden, C. A., *J. Chem. Soc., Chem. Commun.*, **1973**, 688.
16. Torres, M., Clement, A., Bertie, J. E., Gunnig, H. E. and Strausz, O. P., *J. Org. Chem.*, **1978**, *43*, 2490.
17. 'Synthesis of heterocycles *via* cycloadditions to 1-azirines' anderson, D. J. and Hassner, A., Synthesis, **1975**, 483; '1-Azirine ring chemistry', Nair, V. and Hyup Kim, K., *Heterocycles*, **1977**, *7*, 353.
18. Carpino, L., McAdams, L. V., Rynbrandt, R. H. and Spiewak, J. W., *J. Am. Chem. Soc.*, **1971**, *93*, 476.
19. Taguchi, T., *J. Pharm. Soc. Jpn.*, **1952**, *72*, 921.
20. Newman, M. S. and VanderWerf, C. A., *J. Am. Chem. Soc.*, **1945**, *67*, 233.
21. Posner, G. H. and Rogers, D. Z., *J. Am. Chem. Soc.*, **1977**, *99*, 8208 and 8214.
22. Chong, J. M. and Sharpless, K. B., *J. Org. Chem.*, **1985**, *50*, 1557.
23. Chini, M., Crotti, P. and Macchia, F., *Tetrahedron Lett.*, **1990**, *31*, 4661.
24. Saito, S., Yamashita, S., Nishikawa, T., Yokoyama, Y., Inaba, M. and Moriwake, T., *Tetrahedron Lett.*, **1989**, *30*, 4153.
25. Eis, M. J., Wrobel, J. E. and Ganem, B., *J. Am. Chem. Soc.*, **1984**, *106*, 3693.
26. Alcaide, B., Biurran, C. and Plumet, J., *Tetrahedron*, **1992**, *48*, 9719.
27. Tanaka, T., Inoue, T., Kamei, K., Murakami, K. and Iwata, C., *J. Chem. Soc., Chem. Commun.*, **1990**, 906.
28. Sasaki, M., Tanino, K., Hirai, A. and Miyashita, M., *Org. Lett.*, **2003**, *5*, 1789.
29. Behrens, C. H., Ko, S. Y., Sharpless, K. B. and Walker, F. J., *J. Org. Chem.*, **1985**, *50*, 5687; Bulman Page, P. C., Rayner, C. M. and Sutherland, I. O., *J. Chem. Soc., Chem. Commun.*, **1988**, 356.
30. Murata, S., Suzuki, M. and Noyori, R., *J. Am. Chem. Soc.*, **1979**, *101*, 2738.
31. Asami, M., *Chem. Lett.*, **1984**, 829.
32. Jankowski, K. and Daigle, J.-Y., *Can. J. Chem.*, **1971**, *49*, 2594.
33. Kozikowski, A. P., Ishida, H. and Isobe, K., *J. Org. Chem.*, **1979**, *44*, 2788; Stamm, H. and Weiss, R., *Synthesis*, **1986**, 392 and 395; Lehmann, J. and Wamhoff, H., *Synthesis*, **1973**, 546.
34. Kishore Kumar, G. D. and Baskaran, S., *Synlett*, **2004**, 1719.
35. Hou, X.-L., Fan, R.-H. and Dai, L.-X., *J. Org. Chem.*, **2002**, 5295.
36. Wu, J., Sun, X., Sun, W. and Ye, S., *Synlett*, **2006**, 2489.
37. Bera, M. and Roy, S., *Tetrahedron Lett.*, **2007**, *48*, 7144.
38. Ungureanu, I., Klotz, P., Schoenfelder, A. and Mann, A., *Tetrahedron Lett.*, **2001**, *42*, 6087.
39. 'Methyleneaziridines: unusual vehicles for organic synthesis', Shipman, M., *Synlett*, **2006**, 3205.
40. Snyder, H. R., Stewart, J. M. and Ziegler, J. B., *J. Am. Chem. Soc.*, **1947**, *69*, 2672.
41. Trost, B. M. and Ziman, S. D., *J. Org. Chem.*, **1973**, *38*, 932.
42. Clark, R. D. and Helmkamp, G. K., *J. Org. Chem.*, **1964**, *29*, 1316; Lee, K. and Kim, Y. H., *Synth. Commun.*, **1999**, *29*, 1241.
43. 'The Ramberg-Bäcklund reaction', Taylor, R. J. K. and Casy, G., *Org. React.*, **2003**, *62*, 357.
44. Ewin, R. A., Loughlin, W. A., Pyke, S. M., Morales, J. C. and Taylor, R. J. K., *Synlett*, **1993**, 660.
45. De Shong, P., Kell, D. A. and Sidler, D. R., *J. Org. Chem.*, **1985**, *50*, 2309; De Shong, P., Sidler, D. R., Kell, D. A. and Aronson, N. N., *Tetrahedron Lett.*, **1985**, *26*, 3747.
46. Bhuller, P., Gilchrist, T. L. and Maddocks, P., *Synthesis*, **1997**, 271.
47. Bonneau, R., Liu, M. T. H. and Lapouyade, R., *J. Chem. Soc., Perkin Trans. 1*, **1989**, 1547.
48. Schmitz, E. and Ohme, R., *Chem. Ber.*, **1962**, *95*, 795.
49. Padwa, A. and Eastman, D., *J. Org. Chem.*, **1969**, *34*, 2728.
50. Adam, W., Asensio, G., Curci, R., González-Núñez, M. E. and Mello, R., *J. Org. Chem.*, **1992**, *57*, 953.
51. Murray, R. W. and Jeyaraman, R., *J. Org. Chem.*, **1985**, *50*, 2847.
52. Towson, J. C., Weismiller, M. C., Lal, G. S., Sheppard, A. C. and Davis, F. A., *Org. Synth.*, **1990**, *69*, 158.
53. 'Applications of oxaziridines in organic synthesis', Davis, F. A. and Sheppard, A. C., *Tetrahedron*, **1989**, *45*, 5703; 'Asymmetric hydroxylation of enolates with *N*-sulfonyloxaziridines', Davis, F. A. and Chen, B.-C., *Chem. Rev.*, **1992**, *9*, 919.
54. Davis, F. A., Sheppard, A. C., Chen, B.-C. and Haque, M. S., *J. Am. Chem. Soc.*, **1990**, *112*, 6679.

[55] Huynh, C., Derguini-Boumechal, F. and Linstrumelle, G., *Tetrahedron Lett.*, **1979**, *20*, 1503.

[56] Couty, F., David, O. and Drouillat, B., *Tetrahedron Lett.*, **2007**, *48*, 9180.

[57] 'The organic chemistry of β-lactams', Georg, G. I., Ed., VCH, New York, **1993**.

[58] Baldwin, J. E., Edwards, A. J., Farthing, C. N. and Russell, A. T., *Synlett*, **1993**, 49; Baldwin, J. E., Adlington, R. M., Godfrey, C. R. A., Gollins, D. W., Smith, M. L. and Russel, A. T., *ibid.*, 51.

[59] Clauss, K., Grimm, D. and Prossel, G., *Justus Liebigs Ann. Chem.*, **1974**, 539.

[60] Durham, T. B. and Miller, M. J., *J. Org. Chem.*, **2003**, *68*, 27.

[61] 'Recent advances in β-lactone chemistry', Pommur, A. and Pons, J.-M., *Synthesis*, **1993**, 441.

[62] Arnold, L. D., Kalantar, T. H. and Vederas, J. C., *J. Am. Chem. Soc.*, **1985**, *107*, 7105.

[63] Jung, M. E. and Blum, R. B., *Tetrahedron Lett.*, **1977**, 3791.

[64] Vedejs, E. and Moss, W. O., *J. Am. Chem. Soc.*, **1993**, *115*, 1607.

[65] Hodgson, D. M., Humphreys, P. G. and Ward, J. G., *Org. Lett.*, **2005**, *7*, 1153.

[66] Yamauchi, Y., Katagiri, T. and Uneyama, K., *Org. Lett.*, **2002**, *4*, 173.

[67] Galli, C., Illuminati, G., Mandolini, L. and Tamborra, P., *J. Am. Chem. Soc.*, **1977**, *99*, 2591.

[68] 'Effective molarities for intramolecular reactions', Kirby, A. J., *Adv. Phys. Org. Chem.*, **1980**, *17*, 183.

[69] Allen, C. F. H., Spangler, F. W. and Webster, E. R., *Org. Synth., Coll. Vol. IV*, **1963**, 433.

[70] 'Serine derivatives in organic synthesis', Kulkarni, Y. S., *Aldrichimica Acta*, **1999**, *32*, 18.

[71] Bew, S. P., Hughes, D. L., Palmer, N. J., Savic, V., Soapi, K. M. and Wilson, M. A., *Chem. Commun.*, **2006**, 4338.

[72] Hassner, A., Lorber, M. E. and Heathcock, C., *J. Org. Chem.*, **1967**, *32*, 540.

[73] Fowler, F. W., Hassner, A. and Levy, L. A., *J. Am. Chem. Soc.*, **1967**, *89*, 2077; Hassner, A. and Fowler, F. W., *J. Org. Chem.*, **1968**, *33*, 2686.

[74] Smolinsky, G., *J. Org. Chem.*, **1962**, *27*, 3557.

[75] Jeong, J. U., Tao, B., Sagasser, I., Henninges, H. and Sharpless, K. B., *J. Am. Chem. Soc.* **1998**, *120*, 6844; Minakata, S., Kano, D., Fukuoka, R., Oderaotoshi, Y. and Komatsu, M., *Heterocycles*, **2003**, *60*, 289.

[76] Minakata, S., Morino, Y., Oderaotoshi, Y. and Komatsu, M., *Chem. Commun.*, **2006**, 3337.

[77] Chang, J. W. W., Ton, T. M. U., Zhang, Z., Xu, Y. and Chan, P. W. H., *Tetrahedron Lett.*, **2009**, *50*, 161.

[78] 'Nitrenes', Ed. Lwowski, W., Interscience, **1970**.

[79] Atkinson, R. S., Coogan, M. P. and Cornell, C. L., *J. Chem. Soc., Chem. Commun.*, **1993**, 1215.

[80] Li, A.-H., Dai, L.-X. and Hou, X.-L., *J. Chem. Soc., Perkin Trans. 1*, **1996**, 2725; Zhou, Y.-G., Li, A.-H., Hou, X.-L. and Dai, L.-X., *Tetrahedron Lett.*, **1997**, *38*, 7225.

[81] Verstappen, M. M. H., Ariaans, G. J. A. and Zwannenburg, B., *J. Am. Chem. Soc.*, **1996**, *118*, 8491.

[82] Katritzky, A. R., Wang, M., Wilkerson, C. R. and Yang, H., *J. Org. Chem.*, **2003**, *68*, 9105.

[83] Dauban, P. and Dodd, R. H., *Org. Lett.*, **2000**, *2*, 2327.

[84] Wadsworth, D. H., *Org. Synth., Coll. Vol. VI*, **1988**, 75; Freeman, J. P. and Mondron, P. J., *Synthesis*, **1974**, 894; Szmuszkovicz, J., Kane, M. P., Laurian, L. G., Chidester, C. G. and Scahill, T. A., *J. Org. Chem.*, **1981**, *46*, 3562.

[85] Dave, P. R., *J. Org. Chem.*, **1996**, *61*, 5453; Hayashi, K., Sato, C., Kumagai, T., Tamai, S., Abe, T. and Nagao, Y., *Tetrahedron Lett.*, **1999**, *40*, 3761.

[86] 'Synthesis of B-lactams', Mukerjee, A. K. and Srivastava, R. C., *Synthesis*, **1973**, 327.

[87] Gluchowski, C., Cooper, L., Bergbreiter, D. E. and Newcomb, M., *J. Org. Chem.*, **1980**, *45*, 3413.

[88] Sharma, S. D., Kanwar, S. and Rajpoot, S., *J. Heterocycl. Chem.*, **2006**, *43*, 11.

[89] 'Cyclisation of *N*-halogenated amines. (The Hofmann–Loffler reaction)', Wolff, M. E., *Chem. Rev.*, **1963**, *63*, 55.

[90] Torii, S., Okumoto, H. and Genba, A., *Chem. Lett.*, **1996**, 747.

[91] Ichikawa, J., Lapointe, G. and Iwai, Y., *Chem. Commun.*, **2007**, 2698.

[92] 'Stereoselective synthesis of piperidines', Laschat, S. and Dickner, T., *Synthesis*, **2000**, 1781.

[93] Noack, M. and Göttlich, R., *Eur. J. Org. Chem.*, **2002**, 3171.

[94] Tsai, M.-R., Hung, R.-C., Chen, B.-F., Cheng, C.-C. and Chang, N.-C., *Tetrahedron*, **2004**, *60*, 10637.

[95] 'Formation of medium-ring heterocycles by diene and enyne metathesis', Chattopadhyaya, S. K., Karmakar, S., Biswas, T., Majumdar, K. C., Rahaman, H. and Roy, B., *Tetrahedron*, **2007**, *63*, 3919.

[96] 'Evolution and applications of second-generation ruthenium olefin metathesis cartalysts', Schrodi, Y. and Pederson, R. L., *Aldrichimica Acta*, **2007**, *40*, 45.

[97] Ferguson, M. L., O'Leary, D. J. and Grubbs, R. H., *Org.Synth.*, **2003**, *80*, 85.

[98] Rutjes, F. P. J. T. and Schoemaker, H. E., *Tetrahedron Lett.*, **1997**, *38*, 677.

[99] Vardelle, E., Gamba-Sanchez, D., Martin-Mingot, A., Joannetaud, M.-P., Thibaudeau, S. and Marrot, J., *Chem. Commun.*, **2008**, 1473.

[100] 'Epoxidation and hydroxylation of ethylenic compounds with organic peracids', Swern, D., *Org. Reactions*, **1953**, *7*, 378; Rebek, J., Marshall, L., McManis, J. and Wolak, R., *J. Org. Chem.*, **1986**, *51*, 1649.

[101] Katsuki, T. and Sharpless, K. B., *J. Am. Chem. Soc.*, **1980**, *102*, 5974; 'Asymmetric epoxidation of allylic alcohols: the Sharpless reaction', Pfenninger, A., *Synthesis*, **1986**, 89; 'Mechanism of asymmetric epoxidation; 1. Kinetics', Woodward, S. S., Finn, M. G. and Sharpless, K. B., *J. Am. Chem. Soc.*, **1991**, *113*, 106; '2. Catalyst structure', Finn, M. G. and Sharpless, K. B., *ibid.*, 113.

[102] Searles, S. and Gortatowski, M. J., *J. Am. Chem. Soc.*, **1953**, *75*, 3030.

[103] Paternò, E. and Chieffi, G., *Gazz. Chim. Ital.*, **1909**, *39*, 341; Büchi, G., Inman, C. G. and Lipinski, E. S., *J. Am. Chem. Soc.*, **1954**, *76*, 4327; 'Stereoselective intermolecular [2 + 2]-photocycloaddition reactions and their application in synthesis', Bach, T., *Synthesis*, **1998**, 683.

[104] Dalton, J. C. and Tremont, S. J., *Tetrahedron Lett.*, **1973**, 4025.

[105] Bouda, H., Borredon, M. E., Delmas, M. and Gaset, A., *Synth. Commun.*, **1987**, *17*, 943; Iranpoor, N. and Kazemi, F., *Synthesis*, **1996**, 821.

[106] Bouda, H., Borredon, M. E., Delmas, M. and Gaset, A., *Synth. Commun.*, **1989**, *19*, 491.

[107] Ettlinger, M. G., *J. Am. Chem. Soc.*, **1950**, *72*, 4792; Chan, T. H. and Finkenbine, J. R., *J. Am. Chem. Soc.*, **1972**, *94*, 2880; Takido, T., Kobayashi, Y. and Itabashi, K., *Synthesis*, **1986**, 779.

31

Special Topics

31.1 Synthesis of Ring-Fluorinated Heterocycles

This section deals only with compounds where fluorine is attached to the heterocyclic ring. Compounds with fluorine in a benzo-fused ring or in side-chains, for example as trifluoromethyl, are usually prepared by methods similar to those used for standard analogues and thus do not require special treatment.[1] Several general reviews of organofluorine chemistry are available.[2]

The syntheses of ring-fluorinated heterocycles are collected together in this section as a 'special topic' as they are significantly different to those of other halo-heterocycles. Fluorine substituents are also much more important than the other halogens as stable modifying groups in drugs and are also of interest in other areas such as PET (31.2.3) and as very good leaving groups for nucleophilic substitutions in azoles and azines (3.3.2). Such fluorine-containing compounds are only occasionally prepared by the standard methods for the other halides, i.e. direct halogenation or replacement of oxy groups, in fact the latter is almost unknown for fluorine.

(Direct fluorinations with elemental fluorine and other powerful fluorinating agents can be carried out, but they are, to say the least, inconvenient. They are also potentially very hazardous for the inexperienced chemist and are best left to a fluorine specialist. Such reactions are, of course, carried out industrially, but using dedicated special equipment with rigorous safety control.)

31.1.1 Electrophilic Fluorination

CAUTION: *Both fluorine gas and some other electrophilic fluorinating agents are highly toxic and may react violently with organic materials. Hydrogen fluoride, a common by-product of electrophilic fluorinations, is highly corrosive.*

The reaction of organic materials with neat fluorine gas is very vigorous and can result in explosions. Fluorine diluted with an inert gas (nitrogen or helium) is much safer, although even this can, for example, ignite paper, even at very low concentrations.

Other powerful fluorinating agents, such as trifluoromethyl hypofluorite, perchloryl fluoride and acetyl hypofluorite (CF_3OF, $FClO_3$, $AcOF$) have been used successfully, but are not easy to obtain, are toxic and the last two are intrinsically explosive. Some newer, milder, but more limited, fluorinating agents are discussed later in this section.

Direct reaction with fluorine, can be carried out successfully and selectively on robust electron-rich ring systems and has been used commercially for the synthesis of 5-fluorouracil, an important anti-cancer drug. This reaction proceeds *via* 5,6-addition of F and the solvent, elimination of the 6-substituent occurring on heating the reaction mixture.[3]

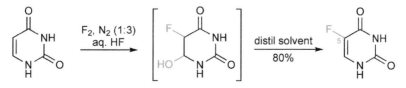

Heterocyclic Chemistry 5th Edition   John Joule and Keith Mills
© 2010 Blackwell Publishing Ltd

Similarly, direct fluorination is a useful method for oxy- and amino-purines,[4] but yields are often only modest.

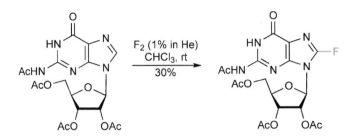

With other electron-rich systems, such as pyrrole and thiophene, the reaction may be difficult to control and polyfluorinated and/or non-aromatic products are often formed.

Pyridines react with fluorine by formation of *N*-fluoropyridinium fluorides, which are unstable above 0 °C. These can be converted, by exchange with sodium salts, into more stable salts, such as tetrafluoroborates, some of which are commercially available as mild electrophilic fluorinating agents for highly reactive substrates, such as enolates.[5] Addition of *N*-fluoropyridinium tetrafluoroborates to excess triethylamine results in the formation of the 2-fluoropyridine(s), *via* a carbene intermediate.[6]

Reaction of pyridine with iodine fluoride (a mixture of F_2 and I_2) gives 2-fluoropyridine directly, but this involves an addition-elimination mechanism on *N*-iodopyridinium fluoride, rather than electrophilic fluorination.[7]

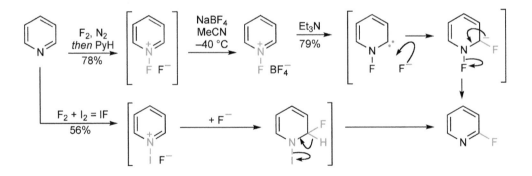

Several milder electrophilic fluorinating agents are commercially available, the most notable being various *N*-fluoro-amides or *N*-fluoro quaternary salts, examples being NFSI and Selectfluor™.

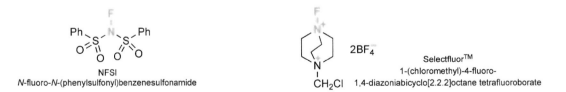

These can be used for direct C-fluorination[8] or reaction with lithio-heterocycles or other organometallic derivatives. In a process for the synthesis of a 5-fluorothiazole using a lithio derivative, although the yield was moderate, the reaction could be carried out on a large scale.[9]

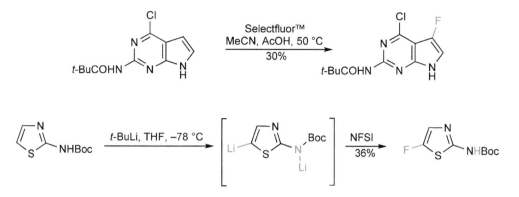

Cesium fluoroxysulfate (CsSO₄F) is moderately stable, although potentially explosive, and is particularly useful for *ipso*-fluorination of stannanes and boronic acids.[10]

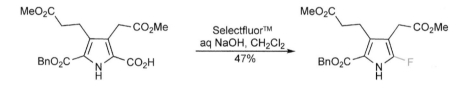

Ipso-replacement of a carboxylic acid group can also be carried out.[11]

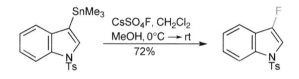

Electrochemical (anodic) fluorinations can be carried out, but may be difficult to control and over-fluorination and/or fluorination of substituents often results. The mechanism involves conversion of the substrates into radical cations, which are then trapped by fluoride, rather than electrophilic fluorination.[12] Again, the method is more suited to robust systems such as pyrimidinones and purines.

31.1.2 The Balz–Schiemann Reaction

The Balz–Schiemann reaction, the classic synthesis of fluoro-aromatic compounds, involves diazotisation of an aromatic amine, isolation of the diazonium fluoroborate or hexafluorophosphate, then thermal decomposition of the dry salt, usually diluted with sand for safety. It is a useful method, with application to a number of heterocyclic systems, for example methyl 3-aminothiophene-2-carboxylate.[13]

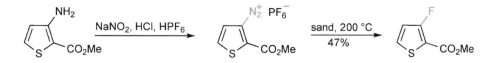

An industrial method involves diazotisation in anhydrous HF. This hazardous method has advantages for large-scale production, but is not suitable for normal laboratory-scale work!

A variation on the standard reaction avoids isolation of the diazonium salt by its generation, under anhydrous conditions, at its decomposition temperature. The only heterocyclic example so far described is the preparation of 3-fluoroquinoline,[14] but the method could no doubt be applied to other systems.

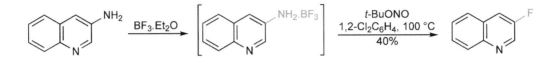

A method using polyvinylpyridinium hydrofluoride (PVPHF), as fluoride source, at low temperature was suitable for the relatively sensitive 2-deoxy-nucleosides.[15] This low temperature reaction was probably successful due to the instability, i.e. high reactivity, of diazonium salts at the α-position of azines, and may not be general for other heterocycles.

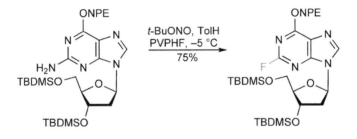

31.1.3 Halogen Exchange (Halex) Reactions

These reactions, involving nucleophilic displacement of other halogens by fluoride anion are important industrially and of some use in the laboratory for electron-deficient rings. The exchange/equilibrium of ^{19}F with ^{18}F is of use in PET (see 31.2.3). Nitro is also a good leaving group for this process in a number of systems: purines (27.5), pyridines (8.3.2) and 1,2,4-triazoles (29.1.1.2).

31.1.4 Ring Synthesis Incorporating Fluorinated Starting Materials

The most promising approach for the efficient preparation of fluoro-heterocycles is by ring synthesis from fluorine-containing starting materials and intermediates.

CAUTION: *Monofluoroacetates are intermediates for some literature syntheses of fluorinated heterocycles. Fluoroacetic acid and its derivatives, such as fluoroacetamide and ethyl fluoroacetate, and compounds which could be metabolically converted into fluoroacetate, such as 2-fluoroethanol, are extremely toxic and have no antidote. A lethal dose of the acid in humans may be as low as 100 mg. They should be avoided if at all possible, but if their use is essential, they must only be used under rigorous control.*

There are, perhaps surprisingly, many complex polyfluorinated starting materials available commercially and some apparently quite exotic reagents and intermediates can be easily (and often cheaply) prepared from these. These intermediates may simply be the fluorinated analogues of standard intermediates, for example, 1,3-dicarbonyl compounds or their equivalents, such as an iminium salt synthon for fluoromalonaldehyde.[16]

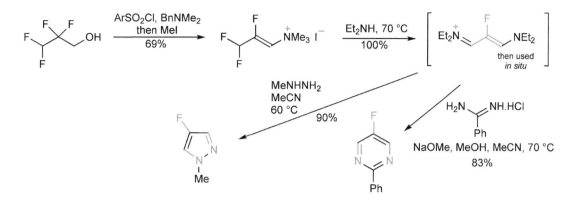

A related method giving oxy compounds involves displacement of the 'activated' fluorine in 2,3-difluoroacrylates.[17]

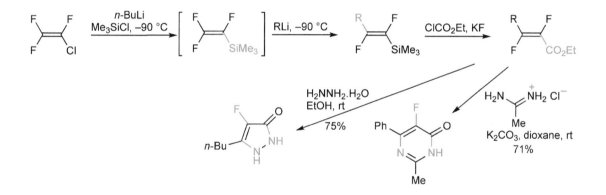

The other approach, with great potential, involves cyclisation reactions of polyfluoro compounds, with *in situ* aromatisation *via* loss of one or more fluorines. The reaction mode involves nucleophilic addition to *gem*-difluoro-methylidene intermediates. This moiety is highly susceptible to nucleophilic addition to the fluorine-bearing carbon, even in the absence of other activation. A transient intermediate anion is thought to be stabilised by hyperconjugation and inductive effects from the two β-fluorines, this being a much more favourable situation than an anion α to the fluorines.[18]

Reactions involving quinone methide intermediates do not need this special activation, but the aromatisation follows a comparable route. The synthesis of 4-fluoroquinolines *via* reaction of 2-trifluoromethylaniline with enolates illustrates, in part, the *gem*-difluoromethylidene concept. Here elimination of hydrogen fluoride by the enolate acting as a base, initiates the sequence, generating a highly reactive quinone methide.[19]

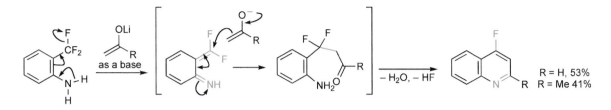

Intramolecular addition of carbon nucleophiles can also be used. In this case a benzylic anion brings about the first fluoride elimination.[20]

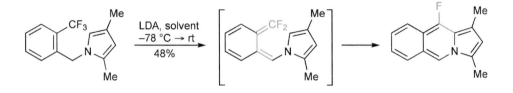

5-*Endo-trig*-cyclisations, which are normally difficult, are successful with assistance from 'difluoro activation' as demonstrated by the synthesis of the 2-fluoroindole shown below. The corresponding 3-unsubstituted 2-fluoroindole, benzofurans and benzothiophenes can also be prepared in similar yields.[21] The versatile, intermediate aniline, can also be converted into the corresponding isonitrile, which, in turn, can be used to generate a lithio-quinoline, which reacts well with a variety of electrophiles.[22] The isonitrile can also be reacted with Grignard reagents to give 2-alkyl-quinolines and a similar addition to the corresponding nitrile leads to isoquinolines.[23]

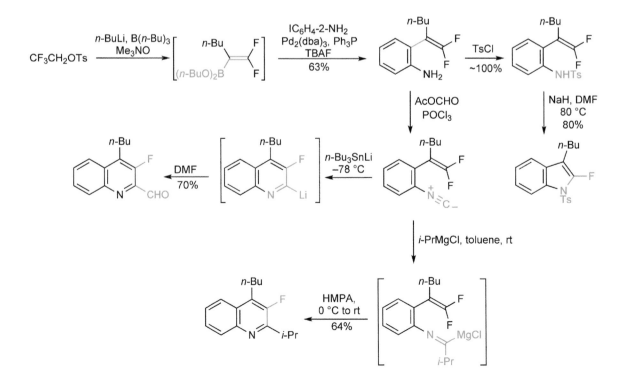

It is worth noting that these (5-*endo-trig*) cyclisations also occur readily in simple aliphatic systems, giving dihydro-heterocycles.[21]

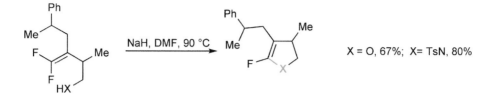

Another cyclisation of this type gives a dihydroisoquinoline, which aromatises *in situ*, by elimination of sulfinate, if potassium hydride is used as base. The intermediate dihydroisoquinoline can be isolated if NaH is used as base.[24]

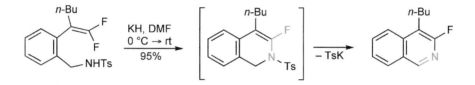

Difluoromethylene moieties can be incorporated into stable reagents and intermediates where, eventually, loss of one of the fluorines results from a key aromatisation step.[25]

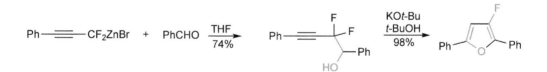

gem-Difluoro-dihydro-heterocycles can be isolated in some cases, such as in the iodosilyl-ketone reaction shown below.[26]

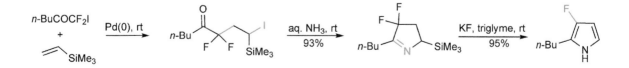

Dipolar additions onto perfluoro-alkenes is followed by aromatisation *via* two fluoride/halide eliminations.[27]

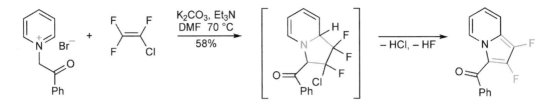

31.2   Isotopically Labelled Heterocycles[28]

The preparation of heterocycles containing isotopically labelled atoms, either in the ring or as substituents, may be required for a number of purposes, for example mechanistic studies of reactions or biosynthesis, spectroscopic studies or for medical research and diagnosis.

[**NOTE**: The correct term for an individual isotope of an element, for example ^{13}C, is 'nuclide', but the word 'isotope' is in loose common usage. The use of 'isotope', without further qualification, is equivalent to saying that '1-butene is an isomer', without saying of what.]

The nuclides that are commonly used in heterocyclic systems are:

1. Stable: ^{2}H (D), ^{13}C, ^{15}N
2. Relatively long-lived, beta-emitting, radionuclides: ^{3}H (T) ($t_{1/2}$ 12.3 years), ^{14}C ($t_{1/2}$ 5,700 years)
3. Short-lived, positron-emitting, radionuclides: ^{18}F ($t_{1/2}$ 110 min), ^{11}C ($t_{1/2}$ 20.4 min).

31.2.1   Hazards Due to Radionuclides

^{3}H and ^{14}C present a relatively low external risk, as the beta radiation is very easy to shield, so normal laboratory techniques can be used, with appropriate containment. However, ingestion or inhalation of any radioactive material, even the weak beta emitters, is very hazardous. The short-lived positron-emitting nuclides generate intense gamma radiation, which requires heavy shielding with automated remote handling.

31.2.2   Synthesis

The syntheses used for labelled compounds are often not those that would be considered the best from a purely chemical point of view, but those that use the expensive, and possibly hazardous, nuclide most efficiently and safely.

[**NOTE**: Quoting yields in this area is more complex than usual. Normal chemical yields can be given, but for radionuclides, the radiochemical yield, which shows the efficiency of incorporation of the nuclide into the final product, is more meaningful. The situation is more complex for the very short-lived (PET) nuclides, where the yields have to be corrected for the very substantial decay that occurs throughout the whole process.]

Deuterium and tritium can be introduced at certain positions of heterocyclic rings by acid- or base-catalysed exchange of hydrogen, or protonolysis of organometallics, and this has long been used in mechanistic studies.

Incorporation of labelled atoms into and onto heterocyclic rings can also be carried out *via* standard heterocyclic ring syntheses using commercially available labelled organic building blocks, but the range is limited and prices generally high. More commonly, a synthesis is devised that uses simple, and relatively cheap, labelled inorganic components. Common ^{13}C or ^{14}C sources are cyanide and carbon dioxide (as, for example, from barium carbonate) and ^{15}N from ammonium salts and nitrite.

A common approach is to take a final product or key intermediate and degrade it to a compound that can be built up again by insertion of the labelled atom. The chemical yield of the degradation is not important, the most important factor being the efficient use of the replacement nuclide. A nice example of this is a synthesis of ^{14}C-2 sumatriptan.[29]

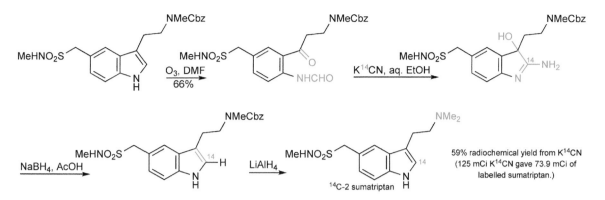

The synthesis of ^{15}N-3 uridine triacetate was carried out by direct exchange of the nitrogen isotopes, *via* nitration of N-3, then reaction with ^{15}NH$_3$, *via* a double nucleophilic replacement of nitramide (NH$_2$NO$_2$) (14.9.2.2). Note the modified conditions for efficient use of the labelled reagent.[30]

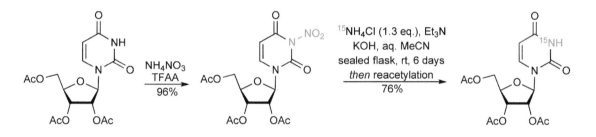

A modified Dimroth rearrangement (27.1.1.2) allowed an efficient labelling of N-1 of adenosine, *via* initial exocyclic introduction of the labelled nitrogen by nucleophilic substitution at C-6.[31]

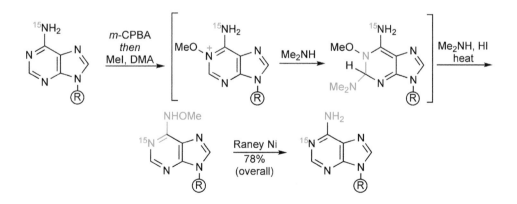

31.2.3 PET (Positron Emission Tomography)[32]

One of the most significant medical applications of radiolabelled compounds is for PET scans. These are important diagnostic and research tools that can be used to study the distribution of drugs or biomolecules in the body, and for imaging of organs.

Positrons are the antimatter counterparts of electrons and are emitted from a number of low-atomic-number radionuclides. The procedure begins with administration of a compound containing a very short-lived, positron-emitting, nuclide to the patient. The emitted positron undergoes matter–antimatter annihilation with an electron, which happens within a very short distance, generating two gamma photons at 180 ° to one another, which can be detected synchronously outside the body. An array of gamma detectors gives a 3D image of the distribution of the labelled compound. The dose of radiation to the patient is very small – about the same as an X-ray. The chemist is more at risk than the patient!

The nuclides ^{18}F and ^{11}C are by far the most commonly used positron emitters for complex organic molecules. Unfortunately, from the heterocyclic point of view, ^{13}N has a half-life that is just too short (10 min) for sensible organic chemistry.

31.2.3.1 Synthesis of Compounds for PET

This is a challenging area for heterocyclic synthesis, as the very short half-lives of these nuclides give rise to two problems – the need for very rapid synthesis and purification, and the high gamma emission, which requires the use of remote handling. The amounts of radioactive materials produced are minute (sub-microgram quantities); it should be borne in mind that one nanogram of ^{11}C is about as radioactive as one gram of radium.

Another unique requirement is that the radionuclides themselves must be prepared by nuclear reactions, brought about by bombardment of a target of ^{14}N or ^{18}O, usually in the presence of a co-reactant, with high energy (18 MeV) protons in a cyclotron. Some very simple reagents can then be prepared from the initial products.

Because of the very short half-lives, all the work, including the nuclear synthesis, must be done close enough to the patient – often in the same building – that sufficient activity to carry out the procedure remains in the sample when it is used. This is a more acute requirement for ^{11}C than for [18]F. In the former case, 87% of the material is lost in one hour, and 98% in two hours.

The development of an efficient synthesis of labelled phosgene was the main determinant in a synthesis of [11]C-2 thymine.[33]

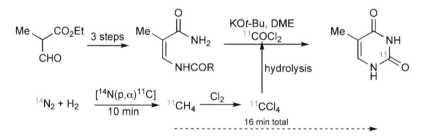

Simple reactions are best, but the complex conversion carried out using sequential enzymic transformations for the synthesis of labelled 5-hydroxytryptophan, for use in neuroendocrine tumour imaging, is notable.[34]

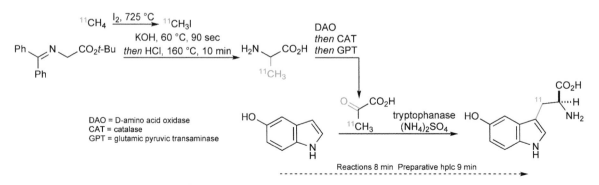

The nuclide ^{18}F can be introduced using a number of standard methods using fluorine, HF or fluoride anion. In contrast to normal chemistry, [18]F$_2$ (as a dilute mixture in argon or neon) is a convenient reagent in the PET context, being prepared directly by proton bombardment of [18]O$_2$ in argon, or deuteron bombardment of neon. An example of its use is the preparation of labelled 5-fluorouracil and its ribosides by direct electrophilic fluorination.[35]

[18]F is often introduced by nucleophilic displacement using potassium fluoride with the cryptand Kryptofix 2.2.2. The substrates can be halides, but nitro and quaternary ammonium are the preferred leaving groups in azines and purines.[36] Such reactions are standard, but apparently un-activated substrates, such as the nitro-oxindole shown below, have also been used successfully.[37]

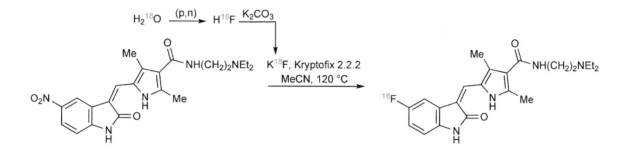

31.3 Bioprocesses in Heterocyclic Chemistry[38]

The use of biological methods has a small, but significant niche in synthetic heterocyclic chemistry, being used both on a research scale and for fine-chemicals production. The processes may use isolated enzymes or whole microorganisms, the main reactions being oxidations of a heterocyclic nucleus or of side-chains, but other reaction types are also used, for example enzymatic catalysis has been used to ribosylate purines and related bases by reaction with a 7-alkylated nucleoside.[39]

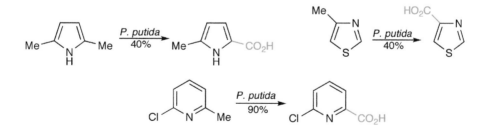

A particular advantage of biological methods is their potential regio-, stereo- and enantioselectivity, which may not be attainable using chemical reagents. On the other hand, non-selective reactions have their uses: for example, the subjection of natural products to non-selective oxidations will generate a series of starting materials for the preparation of a wider range of analogues for biological evaluation.

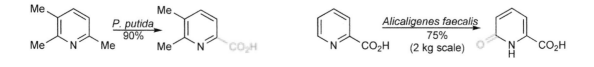

The oxidation of pyridines to pyridones[40] and the selective oxidation of a side-chain in alkyl-pyridines and other azines, have been well studied.[41]

The enantioselective *cis*-dihydroxylation[42] of benzothiophenes and benzofurans in the heterocyclic ring, by *Pseudomonas putida*, is analogous to well-known conversions of simple benzenoid compounds,[43] but in the heterocyclic context, hydroxyl groups introduced at an α-carbon easily epimerise. Indole gives indoxyl probably *via* dehydration of an intermediate 2,3-diol. In contrast, *cis*-dihydroxylation of quinolines, or of 2-phenylpyridines, takes place selectively in the benzene ring.[44]

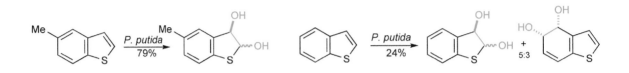

The introduction of an amino acid side-chain onto 4-, 5-, 6- and 7-azaindoles by an enzyme-catalysed alkylation with serine is an impressive demonstration of the power of biological methods.[45]

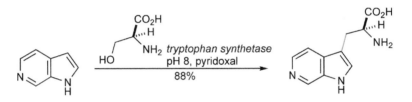

31.4   Green Chemistry

'Green chemistry' is currently a fashionable topic, and a variation – 'sustainable chemistry' – is also in vogue. However, although its aims are laudable, the term has been overused and frequently abused, often as a tag to aid publication and grant applications. Green chemistry can be simply and accurately, but broadly, defined as chemistry that can be carried out in a *relatively* environmentally friendly or benign manner; alternatively, carrying out a chemical process in the most environmentally friendly way that is reasonably achievable, as no chemical reaction or process can be totally 'green'. 'Official' checklist definitions have been stated and may include aspects of reaction safety, which is actually separate and more important,[46] but common sense is more useful than a clipboard in this area.

Atom economy[47] – the incorporation of as much of the starting materials as possible into the product – is a useful concept, but it should not overshadow the whole.

No reaction can be green in its own right until *all* the aspects of the process in which it is involved have been analysed, which includes complete analysis of the 'green-ness' of the origin of starting materials and reagents.

Many aspects of green chemistry have been in use for years, because efficient chemical processes with minimum waste are also economically preferable, and in industry are also confined by legal controls on polluting practices, at least in some countries.

31.5   Ionic Liquids[48]

The development of ionic liquids (ILs) was a significant addition to the range of chemical tools and is of particular interest here because many of them are heterocycles.[49] The name is self-explanatory and can include a vast range of types, but is usually taken to mean salts that are liquid at or slightly above room temperature and are thus sometimes referred to as 'room temperature ionic liquids' (RTILs). They are either protic or, more usually quaternary, salts of organic bases, most prominently 1,3-dialkyl-imidazolium salts, although pyridinium and higher azolium salts are also used. More diverse types are used industrially. Somewhat surprisingly, some types can be distilled *in vacuo*.[50]

Their normal laboratory applications are as specific solvents, catalysts and phase tags (5.1.5.2).

e.g. R^1 = *n*-Bu, R^3 = Me, X = Cl
[BMIM][Cl] or [bmim][Cl]

X = Cl, Br, PF$_6$, BF$_4$, Tf$_2$N, HSO$_4$

Typical heterocyclic ionic liquids

Their abbreviations, for use in reaction schemes, are not universally defined, but amongst the common styles are, for example, 1-*n*-**b**utyl-3-**m**ethyl**im**idazolium chloride = [BMIM][Cl] or [bmim][Cl].

The particular advantages of ILs are that they are good, but selective, solvents with very low vapour pressures, low flammability and high temperature stability, although the latter two features depend on the anion. Significantly, their properties can be fine-tuned by variations of the core, alkyl groups and anions. The anion is chemically significant – halides, which are good complex-forming anions with Lewis acids, and the triflimide (Tf$_2$N$^-$), which is very poorly coordinating, are extremes and can have a major effect on reactivity.[51]

ILs have, unfortunately, become strongly associated with green chemistry to such an extent that merely using an IL as solvent is often claimed to make a reaction 'green', even when the work-up involves dilution with water and extraction with large amounts of normal solvents! The main claim to green-ness for ILs is their low volatility, in contrast to common solvents. However, although this gives scope for them to be used in green processes, they can have some very 'un-green' features. The fact that they are more complex structures than the usual solvents means that significant chemistry is required for their preparation, and therefore they start with a green deficit, which has to be counterbalanced by significant gains in other areas, particularly efficient recycling/re-use. When certain counter ions, such as nitrate, are present, they can present a significant fire hazard[52] and many ionic liquids have significant water solubility and significant toxicity to aquatic organisms.[53]

While it can be questioned how green ILs are, they are certainly useful chemically. They have numerous uses as Lewis or protic acid catalysts, for example aluminium chloride in [emim][Cl] is a superior catalyst for the Friedel–Crafts acylation of indoles bearing electron-withdrawing groups and of azaindoles.[54]

Ionic liquid bisulfate salts are good catalysts for the esterification of acids with neopentanol, where the starting materials are soluble in the ionic liquid, but the ester insoluble, allowing an easy isolation of the product.[55]

Both pyridinium and imidazolium ionic liquids have been used as solvents for palladium-catalysed reactions, such as the Suzuki coupling, but only the use of 1-(3-cyanopropyl)pyridinium triflimide salt (mp −64.5 °C) allows very efficient retention and recycling of the palladium catalyst.[56]

31.6 Applications and Occurrences of Heterocycles

Apart from their academic interest and occurrences in natural substances (Chapter 32), heterocycles find extensive applications in industrial and fine chemicals, medicines (Chapter 33), analysis, agriculture, and in many day-to-day activities and in many technological fields.[57]

31.6.1 Toxicity

This not really an 'application' of heterocycles, but is essential knowledge for the practical chemist. The occurrence of explosive hazards is also of major importance (see 5.4 and 29.1.1.5).

The hazards of all common solvents and reagents are well known and those of new materials can often be inferred from structural properties and similarities. For example, toxic effects are to be expected from certain classes of heterocycle, particularly those that are subject to easy nucleophilic replacement of halogen and therefore are alkylating agents in their effect on biomolecules. For example,the fungicide davicil (2,3,5,6-tetrachloro-4-methylsulfonylpyridine) and the useful intermediate 5-bromo-2,4-dichloropyrimidine, are allergenic substances, the latter very strongly so.

However, structures do not always warn of toxicity, so vigilance must always be exercised – as in drugs, minor structural variations may lead to large changes in biological activity. These unexpected toxic effects are usually discovered by their actions being observed on the chemists who have first made them! An example is the apparently innocuous-looking heterocycle (related to the anti-psychotic drug loxapine) 'CR gas', a powerful lachrymator, and a severe and very unpleasant skin irritant. It has been developed, controversially, as a riot control agent. (It is not, of course, a 'gas'.)

The group of fused triazoles shown below were implicated, although the precise agent was not established, as chloracnegens, with similar effects to the notorious (heterocyclic) pollutant tetrachlorodioxin.[58] This type of effect does not relate to any chemical reactivity, as they exert their effect by physical binding to DNA, and they may produce severe effects when present as only minor impurities.

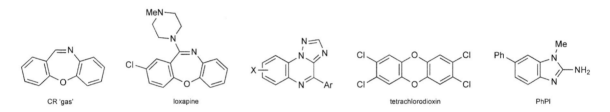

CR 'gas'   loxapine   tetrachlorodioxin   PhPI

'PhPI'[59] is one of a number of related (potential) carcinogens, known vaguely as 'heterocyclic amines' (HCAs), which are formed during cooking, particularly frying, of meat.[60] Interestingly, pre-microwave treatment of beef burgers before frying dramatically reduces the formation of these HCAs by removing their precursors – 'creatine, creatinine, amino acids (*sic*), and glucose', together with water and fat![61]

31.6.2 Plastics and Polymers

Melamine condenses with formaldehyde to give a widely used plastic with good heat resistance, most familiar in laminates for kitchen worktops and as housewares, and known as Formica.

Polybenzimidazole fibre forms one of the most fire-resistant textiles (mp 760 °C; usable to 540 °C) and, although very expensive, is used for high-tech applications, such as protective clothing for fire fighters, astronauts and motor-racing drivers.

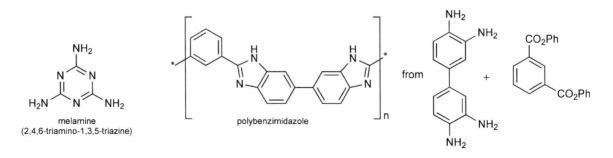

melamine
(2,4,6-triamino-1,3,5-triazine)   polybenzimidazole   from

31.6.3 Fungicides and Herbicides

Heterocycles are contained in a large number of 'agrochemicals'. Triazoles, such as cyproconazole, are effective plant fungicides and have mechanisms of action – inhibition of steroid synthesis – similar to those used medicinally. Other triazoles have limited antifungal activity, but are useful plant-growth regulators, for example paclobutrazol. Another widely used fungicide is davicil.

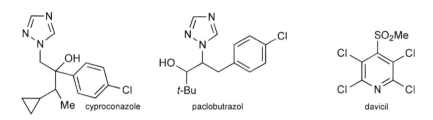

Useful selective herbicides include paraquat, pyridate, maleic hydrazide and atrazine, the last being somewhat controversial as it is widely used in the US, but banned in the EU.

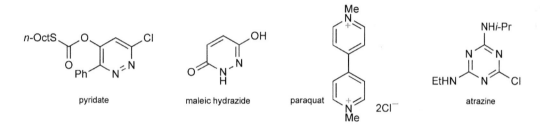

31.6.4 Dyes and Pigments

(A pigment can be any colouring material but is often, particularly when synthetic, an insoluble substance, whereas a dye is a soluble colorant applied to a substrate for which it has an affinity. Many dyes are also used as specific stains for biological tissues for use in microscopy, for example methylene blue. Pigments and dyes absorb certain wavelengths of white light, the non-absorbed (reflected or transmitted) wavelengths being the colour. This is a contrast with fluorescent substances, which emit specific wavelengths (see below).

Heterocycles are very important plant pigments, for example porphyrins and anthocyanins (32.3 and 32.5.6), but are also the basis of numerous synthetic pigments and dyes. Naturally derived dyes include substances such as indigo (20.13.2), and synthetic pigments include the very important copper phthalocyanins (22.3), which are used for many purposes, including in 'cyan' inks.

The process of dying can involve carrying out heterocyclic chemistry, when using 'reactive dyes', which form a strong chemical bond to substrates that contain nucleophilic groups, for example cellulose-based materials. Reactive dyes are composed of a standard dye attached to a reactive linker containing leaving groups. The most common linkers are halo-azines, a typical example being Reactive Blue 4.

Other potential organic pigments are those that show metallic 'colours', such as the dithienyl pyrroles above, which form high melting (>300 °C) golden lustrous crystals.[62]

31.6.5 Fluorescence-Based Applications

Fluorescence – the absorption of a photon, followed very rapidly by emission of a photon at longer wavelength – has many practical applications. The intrinsic fluorescent properties of a molecule can be utilised, or fluorescent tags can be attached.

Fluorescein is probably the best known fluorescent compound and is used for many purposes, such as tracking water leaks. A more specific fluorescent reagent (for use in fluorescence microscopy) is DAPI, which binds selectively to DNA, giving characteristic changes in its emission spectra. (Note that this strong binding to DNA results, not surprisingly, in its being mutagenic.)

31.6.5.1 Sensors

Fluorescent sensors have been widely investigated for the detection of many types of compounds and a particularly fruitful field is the selective detection of metal ions: for example, a series (Zinpyr) of zinc sensors, based on fluorescein-containing attached chelating ligands, which can be used for quantitative determination of Zn or imaging of zinc-containing biological structures.[63]

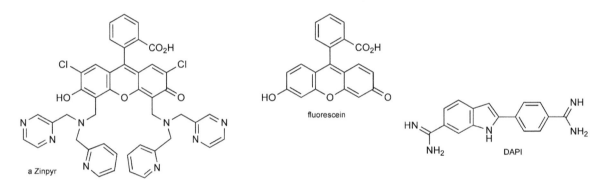

31.6.5.2 Scintillators

Scintillators for detection of ionizing radiation (usually beta) can be incorporated into plastic blocks, or used in solution for liquid scintillation counting for the determination of ^{3}H and ^{14}C. In the latter case, they are often used as a 'scintillation cocktail' – a mixture of an organic solvent, such as toluene, a primary scintillator (fluor), such as PBD or PPO, and a wavelength shifter, such as POPOP. The majority of the radiation is absorbed by the solvent, which transfers the energy to the primary scintillator (fluor), which then emits UV light. The function of the wavelength shifter is to absorb this UV light and re-emit at a more efficient wavelength (blue) for a photomultiplier to convert into an electronic signal.

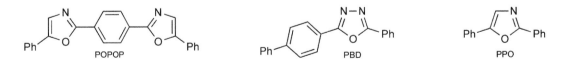

31.6.5.3 Optical Brighteners

These are added to paper, plastics and detergents to increase the 'whiteness' by absorbing UV light and emitting blue light. Many optical brighteners are stilbene derivatives, but a number of heterocyclic ring types are also important, for example the oxazole and triazole derivatives shown below, 1,3,5-triazines and coumarins.[64]

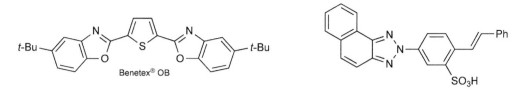

Benetex® OB

31.6.6 Electronic Applications[65]

An important area where technology and heterocyclic chemistry combine is that of electroactive organic materials. The potential applications of these materials are based mainly on their properties as conductors and semiconductors. Here they have potential advantages over conventional metal conductors and inorganic semiconductors, being available in essentially unlimited quantities and free from problems (political, geographic and environmental) associated with the supply of (often rare) inorganic materials. They also have very desirable fabrication properties, such as flexibility and the ability to build electronic circuits by printing. Although there has been a large amount of research and specialist applications, large-scale commercialisation has proved to be relatively slow, but still has great promise and is developing.

The mechanism of electronic conduction involves charge transfer through and between long or medium chains of conjugated molecules and/or π-stacked structures. A detailed discussion of the theory of conduction is beyond the scope of this book; this section is restricted to demonstrating the range of applications of heterocycles in this area.

The types of compound used in organic conductors cover a wide range of unsaturated molecules, such as poly(acetylene) and poly(aniline), but of particular significance from the heterocyclic viewpoint are poly(pyrrole), poly- and oligo(thiophene) and π-stacked structures derived from tetrathiafulvalenes. An advantage of using heterocycles is that a wide range of electron-rich, electron-poor and mixed systems can be easily prepared, allowing for tuning of the electronic and electrical properties of such materials.

31.6.6.1 Poly(pyrrole) and Poly(thiophene)[66]

Pyrrole, thiophene and their derivatives can be oxidatively polymerised, either electrochemically or chemically, for example using iron(III) chloride, to give mainly 2,5-coupled polymers. The initial neutral polymers are non-conducting, but on further oxidation are converted partially into cation radicals or dications, with incorporation of counter ions from the reaction medium – a process known as 'doping' – giving conducting materials. Reductive doping is also possible in other systems.

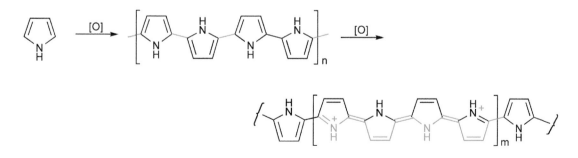

The conducting and physical properties can be modified by the use of 3-/4-substituents, or *N*-substituents in the case of pyrrole. The counter-ions can be incorporated into a side-chain (self-doping), as in the polymer of 3-(thien-3-yl)propanesulfonic acid. Oligo(thiophenes) are also useful in these applications and have been specifically synthesised up to 27 units long by palladium(0)-catalysed couplings or *via* the diacetylene synthesis (17.12.1.1).[67]

A number of polythiophenes, such as the Clevios series and Plexcore are commercially available in quantity. Monomeric ethylenedioxythiophene can be polymerized *in situ* by Fe(III) salts.

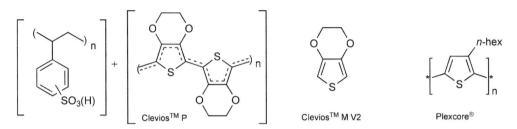

In addition to the straightforward application of conductors and semi-conductors, there are some interesting examples of other uses. Polymers derived from 3,4-ethylenedioxythiophene are produced commercially as anti-static agents and substituted compounds have found particular application in electrochromic devices – substances that change colour on application of an electric current.

31.6.6.2   Tetrathiafulvalenes[68]

These unusual heterocycles and their analogues, such as with selenium and tellurium replacing sulfur, have been intensively studied since the discovery that single crystals of TTF.TCNQ (tetrathiafulvalene.tetracyanoquinodimethane) show electrical conductivity. The bis(ethylenedithio) analogue of TTF, usually known as BEDT-TTF, is particularly useful. Incidentally, the electron-donating ability of TTF allows its use as a radical initiator for diazonium salts.[69]

Tetrathiafulvalenes can be prepared in a number of ways, for example *via* 1,3-dithiole-2-thione-4,5-dithiolate[70] – the simple salts of this dianion are not very stable, but can be stored as a zinc complex, or as the dithiobenzoate[71] shown below.

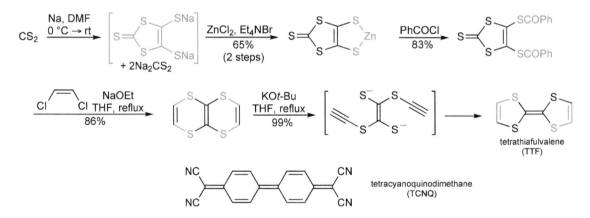

Most chemical transformations of TTF are based on lithiations – even a tetralithio derivative is easily formed. Palladium(0)-catalysed couplings utilising trialkyltin derivatives can also be carried out without difficulty.[72]

References

[1] 'Recent Advances in Fluoroheterocyclic Chemistry', Silvester, M. J., *Adv. Heterocycl. Chem.*, **1994**, *59*, 1.

[2] 'Chemistry of Organic Fluorine Compounds', 2nd Edn., Hudlicky, M., Ellis Horwood/PTR Prentice Hall, **1992**; 'Fluorine-containing synthons', Soloshonok, V. A., **2005**, ACS Symposium Series No. 911; 'Ring constructions by the use of fluorine substituent as activator and controller', Ichikawa, J., *Pure Appl. Chem.* **2000**, *72*, 1685; Uneyama, K., *Organofluorine Chemistry*, Blackwell, **2006**; 'Recent trends in the chemistry of fluorinated five- and six-membered heterocycles', Erian, A. W., *J. Heterocycl. Chem.*, **2001**, *38*, 793.

3   'Kirk-Othmer Encyclopedia of Chemical Technology', 4th Edn., Vol. 11, p. 606.

4   Barrio, J. R., Namavari, M., Phelps, M. E. and Satyamurthy, N., *J. Org. Chem.*, **1996**, *61*, 6084.

5   Umemoto, T., Tomita, K. and Kawada, K., *Org. Synth.*, **1993**, *Coll. Vol. 8*, 286.

6   Umemoto, T. And Tomizawa, G., *J. Org. Chem.*, **1989**, *54*, 1726.

7   Chambers, R. D., Hutchinson, J. and Sandford, G., *J. Fluorine Chem.*, **1999**, *100*, 63.

8   Seela, F., Xu, K. and Chittepu, P., *Synthesis*, **2006**, 2005; also give ref for pyrrole

9   Briner, P. H., Fyfe, M.C.T., Martin, P., Murray, P. J., Naud, F. and, Procter, M. J., *Org. Process Res. Dev.*, **2006**, *10*, 346.

10 Hodson, H. F., Madge, D. J. and Widdowson, D. A., *Synlett*, **1992**, 831.

11 Wang J. and Scott, A. I., *J. Chem. Soc., Chem Commun.*, **1995**, 2399.

12 Riyadh, S. M. and Fuchigami, T., *J. Org. Chem.* **2002**, *67*, 9379; Tajima, T., Nakajima, A. and Fuchigami, T., *J. Org. Chem.*, **2006**, *71*, 1436; Dawood, K. M. and Fuchigami, T., *Synlett*, **2003**, 1631.

13 Kiryanov, A. A., Seed, A. J. and Sampson, P., *Tetrahedron Lett.*, **2001**, *42*, 8797.

14 Garel L. and L. Saint-Jalmes, J., *Tetrahedron Lett.*, **2006**, *47*, 5705.

15 Adib, A., Potier, P. F., Doronina, S., Huc, I. and Beh, J.-P., *Tetrahedron Lett.*, **1997**, *38*, 2989.

16 Shi, X., Ishihara, T., Hamanaka, H. and Gupton, J. T., *Tetrahedron Lett.*, **1995**, *36*, 1527.

17 Q. Zhang and L. Lu, *Tetrahedron Lett.*, **2000**, *41*, 8545.

18 Leriche, C., He, X., Chang, C.-w. T. and Liu, H.-w., *J. Am. Chem. Soc.*, **2003**, *125*, 6348.

19 Strekowski, L., Lin, S.-Y., Lee, H. and Mason, J. C., *Tetrahedron Lett.*, **1996**, *37*, 4655; Strekowski, L., Kiselyov, A. S. and Hojjat, M., *J. Org. Chem.*, **1994**, *59*, 5886.

20 Kiselyov, A. S., *Tetrahedron*, **2006**, *62*, 543.

21 Ichikawa, J., Nadano, R., Mori, T. and Wadda, Y., *Org. Synth.*, **2006**, *83*, 111.

22 Ichikawa, J., Mori, T., Miyazaki, H. and Wada, Y., *Synlett*, **2004**, 1219.

23 Ichikawa, J., Wada, Y., Miyazaki, H., Mori, T. and Kuroki, H., *Org. Lett.*, **2003**, *5*, 1455.

24 Ichikawa, J., Sakoda, K., Moriyama, H. and Wada, Y., *Synthesis*, **2006**, 1590.

25 Hing L. S. and Betebenner,D. A., *J. Chem. Soc., Chem. Commun.*, **1991**, 1134.

26 Zai-Ming Q. and Burton, D. J., *Tetrahedron Lett.*, **1995**, *36*, 5119.

27 Wu, K. and Chen, Q.-Y., *Synthesis*, **2003**, 35.

28 'Radiochemistry and Nuclear Chemistry', 3rd Edn., **2002**, Butterworth-Heinemann, Choppin, G. R., Liljenzin, J. and Rydberg, J.; 'Handbook of Radiopharmaceuticals: Radiochemistry and Applications', Ed. Welch, M. J. and Redvanly, C. S., Eds., John Wiley & Sons Ltd, **2002**.

29 Waterhouse, I., Cable, K. M., Fellows, I., Wipperman, M. D. and Sutherland, D. R., *J. Label. Compd. Radiopharm.*, **1996**, *38*, 1021.

30 Ariza, X. and Vilarrasa, J., *J. Org. Chem.* **2000**, *65*, 2827.

31 Pagano, A. R., Zhao, H., Shallop, A. and Jones, R. A., *J. Org. Chem.*, **1998**, *63*, 3213.

32 'PET tracers and radiochemistry', Schlyer, D. J., Ann. Acad. Med. Singapore, **2004**, *33*, 146; 'Chemistry of β^+-emitting compounds based on fluorine-18', Lasne, M., Perrio, C., Rouden, J., Barre, L., Roeda, D., Dolle, F. and Crouzel, C., *Top. Curr. Chem.*, **2002**, *222*, 201.

33 Ohkura, K., Nishijima, K., Sanoki, K., Kuge, Y., Tamaki, N. and Seki, K., *Tetrahedron Lett.*, **2006**, *47*, 5321.

34 Neels, O. C., Jager, P. L., Koopmans, K. P., Eriks, E., de Vris, E. G. E., Kema, I. P. and Elsinga, P. H., *J. Label. Compd. Radiopharm.* **2006**, *49*, 889.

35 Ishiwata, K., Ido, T., Kawashima, K., Murakami, M. and Takahashi, T., *Eur. J. Nucl. Med.*, **1984**, *9*, 185.

36 Karramkam, M., Hinnen, F., Vaufrey, F. and Dollé, F., *J. Label. Compd. Radiopharm.*, **2003**, *46*, 979; Irie, T., Fukushi, K. and Ido, T., *Int. J. Appl. Rad. Isot.*, **1982**, *33*, 445.

37 Wang, J., Miller, K. D., Sledge, G. W. and Zheng, Q., *Bioorg. Med. Chem. Lett.*, **2005**, *15*, 4380.

38 'Biocatalysis', Petersen, M. and Kiener, A., *Green Chem.*, **1999**, *1*, 99; 'Biotransformations for fine chemical production', Meyer, H. P., Kiener, A., Imwinkelried, R. and Shaw, N., *Chimia*, **1997**, *51*, 287; 'Biotechnological processes in the fine chemicals industry', Birch, O. M., Brass, J. M., Kiener, A., Robins, K., Schmidhalter, D., Shaw, N. M. and Zimmermann, T., *Chim. Oggi*, **1995**, *13*, 9; 'Biosynthesis of functionalised aromatic N-heterocycles', Kiener, A., *Chemtech*, **1995**, *25*, 31.

39 Hennen, W. J. and Wong, C.-H., *J. Org. Chem.*, **1989**, *54*, 4692.

40 Kiener, A., Glocker, R. and Heinzmann, K., *J. Chem. Soc., Perkin Trans. 1*, **1993**, 1201.

41 Kiener, A., *Angew. Chem., Int. Ed. Engl.*, **1992**, *31*, 774.

42 Boyd, D. R., Sharma, N. D., Boyle, R., McMurray, B. T., Evans, T. A., Malone, J. F., Dalton, H., Chima, J. and Sheldrake, G. N., *J. Chem. Soc., Chem. Commun.*, **1993**, 49; Boyd, D. R., Sharma, N. D., Brannigan, I. N., Haughey, S. A., Malone, J. F., Clarke, D. A. and Dalton, H., *Chem. Commun.*, **1996**, 2361.

43 'Enzymatic dihydroxylation of aromatics in enantioselective synthesis: expanding asymmetric methodology', Hudlicky, T., Gonzalez, D. and Gibson, D. T., *Aldrichimica Acta*, **1999**, *32*, 35.

44 Boyd, D. R., Sharma, N. D., Sbircea, L., Murphy, D., Belhocine, T., Malone, J. F., James, S. L., Allen, C. C. R. and Hamilton, J. T.. G., *Chem. Commun.*, **2008**, 5535.

45 Sloan, M. J. and Phillips, R. S., *Bioorg. Med. Chem. Lett.*, **1992**, *2*, 1053.

46 http://www.epa.gov/greenchemistry/pubs/principles.html

47 Trost, B. M., *Science*, **1991**, *254*, **1471**; Trost, B. M., *Angew. Chem. Int. Ed. Engl.*, **1995**, *34*, 259.

48 ACS Symposium Series (2007), 950 (Ionic Liquids in Organic Synthesis); *Acc. Chem. Res.*, **2007**, special issue, *40 (11)*, 1077–1277.

49 'Ionic liquids in synthesis', 2nd. Edn., Wasserscheid, P. and Welton, T., Wiley VCH, **2007**; 'Ionic liquids in heterocyclic synthesis', Martins, M. A. P., Frizzo, C. P., Moreira, D. N., Zanatta, N. and Bonacorso, H. G., *Chem. Rev.*, **2008**, *108*, 2015 (This review contains a substantial general discussion of ionic liquids, including their environmental qualities.)

50 Earle, M. J., Esperança, J. M. S. S., Gilea, M. A., Canongia Lopez, J. N., Rebelo, L. P. N., Magee, J. W., Seddon, K. R. and Widegren, J. A., *Nature*, **2006**, *439*, 831.

51 Pinto, A. C., Moreira Lapis, A. A., Vasconcellos da Silva, B., Bastos, R. S., Dupont, J. and Neto, B. A. D., *Tetrahedron Lett.*, **2008**, *49*, 5639.

[52] Smiglak, M., Reichert, W. M., Holbrey, J. D., Wilkes, J. S., Sun, L., Thrasher, J. S., Kirichenko, K., Singh, S. and Katritzky, A. R., *Chem. Commun.*, **2006**, 2554.

[53] 'Toxicity of ionic liquids', Zhao, D., Liao, Y. and Zhang, Z., *Clean*, **2007**, *35*, 42.

[54] Yueng, K.-S., Farkas, M. E., Qui, Z. and Yang, Z., *Tetrahedron Lett.*, **2002**, *43*, 5793; Yueng, K.-S., Qui, Z., Farkas, M. E., Xue, Q., Regueiro-Ren, A., Yang, Z., Bender, J. A., Good, A. C. and Kadow, J. F. *Tetrahedron Lett.*, **2008**, *49*, 6250.

[55] Arfan, A. and Bazureau, J. P., *Org. Proc. Res. Dev.*, **2005**, *9*, 743.

[56] Zhao, D., Fei, Z., Geldbach, T. J., Scopelliti, R. and Dyson, P. J., *J. Am. Chem. Soc.*, **2004**, *126*, 15876.

[57] 'Heterocycles in Life and Society. An Introduction to Heterocyclic Chemistry and Biochemistry and the Role of Heterocycles in Science, Technology, Medicine and Agriculture', Pozharskii, A. F., Soldatenkov, A. T. and Katritzky, A. R., John Wiley & Sons, Ltd, **1997**.

[58] MacKenzie, A. R. and Brooks, S., *Chem. Ind.*, **1998**, 902.

[59] Bavetta, F. S., Caronna, T., Pregnolato, M. and Terreni, M., *Tetrahedron Lett.*, **1997**, *38*, 7793.

[60] 11th Report on Carcinogens, National Toxicology Program, US Department of Health and Social Services.

[61] Felton, J. S., Fultz, E., Dolbeare, F. A. and Knize, M. G., *Food Chemical Toxicology*, **1994**, *32*, 897.

[62] Ogura, K., Zhao, R., Jiang, M., Akazome, M., Matsumoto, S. and Yamaguchi, K., *Tetrahedron Lett.*, **2003**, *44*, 3595.

[63] Zhang, X.-a, Hayes, D., Smith, S. J., Friedle, S. and Lippard, S. J., *J. Am. Chem. Soc.*, **2008**, *130*, 15788; 'Small-molecule fluorescent sensors for investigating zinc metalloneurochemistry', Nolan, E. M. and Lippard, S. J., *Acc. Chem. Res.*, **2009**, *42*, 193.

[64] 'Fluorescent Whitening Agents', Kirk-Othmer Encyclopedia of Chemical Technology, 4th Edn., Vol. 11, p. 227.

[65] 'Handbook of Conducting Polymers' 2nd Edtn., Ed. Skotheim, T. A. and Reynolds, J. R., Taylor & Francis, **2007**.

[66] 'Handbook of Oligo- and Polythiophenes', Ed. Fichou, D., John Wiley & Sons, Ltd, **1998**.

[67] 'Methods for the synthesis of oligothiophenes', Lukevics, E., Arsenyan, P. and Pudova, O., *Heterocycles*, **2003**, *60*, 663.

[68] 'Tetrathiafulvalenes, oligoacenenes and their buckminsterfullerene derivatives: the bricks and mortar of organic electronics', Bendikov, M., Wudl, F. and Perepichka, D. F., *Chem. Rev.*, **2004**, *104*, 4891; 'The organic chemistry of 1.3-dithiole-2-thione-4,5-dithiolate (DMIT)', Svenstrup, N. and Becher, J., *Synthesis*, **1995**, 215.

[69] Fletcher, R. J., Lampard, C., Murphy, J. A. and Lewis, N., *J. Chem. Soc., Perkin Trans. 1*, **1995**, 623.

[70] Meline, E. L. and Elsenbaumer, R. L., *J. Chem. Soc., Perkin Trans. 1*, **1998**, 2467.

[71] Hansen, T. K., Becher, J., Jorgensen J. T., Varma, K. S., Khedekar, R. and Cava, M. P., *Org. Synth. Col. Vol. 9*, 203.

[72] Iyoda, M., Fukuda, M., Yoshida, M. and Sasaki, S., *Chem. Lett.*, **1994**, 2369; Lovell, J. M. and Joule, J. A., *J. Chem. Soc., Perkin Trans. 1*, **1996**, 2391.

32

Heterocycles in Biochemistry; Heterocyclic Natural Products

32.1 Heterocyclic Amino Acids and Related Substances

There are four amino acids, amongst the 20 that make up proteins, that have an aromatic side-chain, and of these, two have a heteroaromatic side-chain – histidine, with an imidazole, and tryptophan, with an indole. Both of these are amongst the 'essential amino acids', i.e. they need to be part of the diet since they cannot be biosynthesised by human beings. Decarboxylation of histidine produces the hormone histamine. Proline is the only heterocyclic DNA-coded α-amino acid – it is based on pyrrolidine; hydroxyproline is an essential component of collagen, the fibrous structural protein that supports tissues and is the main component of cartilage.

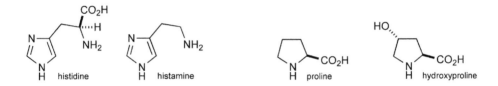

Decarboxylated tryptophan is called tryptamine. The phenol, 5-hydroxytryptamine (5-HT or serotonin) and histamine are important neurotransmitters (33.2). N-Acetyl-5-methoxytryptamine, known as melatonin, is produced by the pineal gland, a pea-sized gland at the base of the brain. It is involved in controlling the natural daily cycle of hormone release in the body – the circadian rhythm. The secretion of melatonin is triggered by the dark and is suppressed by natural daylight, therefore controlling periods of sleepiness and wakefulness. Indol-3-ylacetic acid (IAA) is an auxin – a plant growth stimulant. Brassinin, isolated from turnips, is a phytoalexin – one of a group of compounds produced by plants as a defense mechanism against attack by microorganisms.

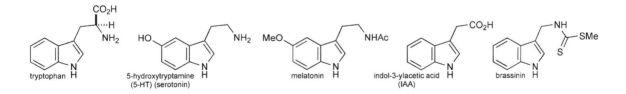

The ability of an imidazole (24.1.1.1 and 24.4.1) to act as both an acid (N-hydrogen) and a base (the imine nitrogen) is put to good use in the active sites of several enzymes in which the imidazole rings of appropriately placed histidines effectively 'shuffle' protons from one place to another. One example is the

Heterocyclic Chemistry 5th Edition   John Joule and Keith Mills
© 2010 Blackwell Publishing Ltd

digestive enzyme chymotrypsin, which brings about the hydrolysis of protein amide groups in the small intestine: the enzyme provides a 'proton' at one site, while it accepts a 'proton' at another, making use of the ambiphilic character of the imidazole ring to achieve this. Effectively, the imidazole activates a serine alcoholic hydroxyl by removing the proton as the oxygen attacks the amide bond. Subsequently, that same proton is delivered to the cleaving amide nitrogen as the tetrahedral intermediate breaks down.

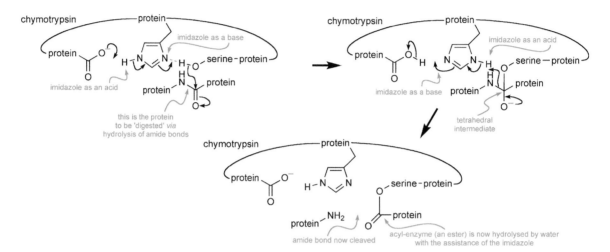

The role of histidine in chymotrypsin hydrolysis of peptides

32.2 Enzyme Co-Factors; Heterocyclic Vitamins; Co-Enzymes[1]

Many enzymes require a component, other than the protein portion, for their catalytic activity – a co-factor. If the co-factor is removed, the remaining protein (apoenzyme) has no catalytic activity. A co-factor that is firmly bound to the apoenzyme is termed a prosthetic group and most contain a metal centre. A co-factor that is bound loosely to the apoenzyme and can be readily separated from it is called a co-enzyme. Co-enzymes play a critical role in the catalysis.

Vitamins are substances essential for a healthy life; humans must ingest vitamins *via* their diet because there is no mechanism for their biosynthesis in the body. There are 14 vitamins – the name was coined when the first vitamin chemically identified (vitamin B_1 in 1910) turned out to be an amine – a *vital amine*. A typical vitamin is folic acid, a complex molecule in which the functionally important unit is the bicyclic pyrazino[2,3-*d*]pyrimidine (pteridine) ring system, and its arylaminomethyl substituent. Folic acid is converted in the body into tetrahydrofolic acid (FH_4) which is crucial in carrying one-carbon units, at various oxidation levels, for example in the biosynthesis of purines, and is mandatory for healthy development of the foetus during pregnancy. Other essential co-factors that contain pteridine units must and can be biosynthesised in humans – without them we cannot survive – and are incorporated into oxygen-transfer enzymes based on molybdenum, in which the metal is liganded by a complex ene-dithiolate.

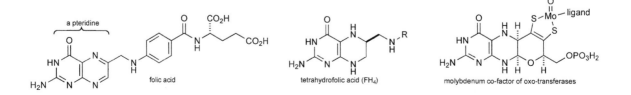

folic acid   tetrahydrofolic acid (FH_4)   molybdenum co-factor of oxo-transferases

Several highly significant vitamins are water-soluble and heterocyclic in nature and further, their utility in the co-enzymes into which they are incorporated, can only be understood on the basis of their intrinsic heterocyclic reactivity. We consider in detail firstly the two important pyridine-containing vitamins – vitamin B_3 (niacin or nicotinamide) and vitamin B_6 (pyridoxine) and then the thiazole-containing thiamin (vitamin B_1).

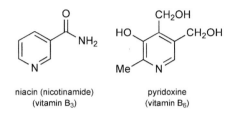

niacin (nicotinamide)
(vitamin B_3)

pyridoxine
(vitamin B_6)

32.2.1   Niacin (Vitamin B_3) and Nicotinamide Adenine Dinucleotide Phosphate (NADP⁺)

Nicotinamide adenine dinucleotide phosphate (NADP⁺) is a large complicated co-enzyme, but the significant part for its role in oxidation/reduction processes is the pyridinium ring – to understand the mechanism one can think of it as simply an *N*-alkyl pyridinium salt of nicotinamide. The positively charged nitrogen acts as an electron sink and allows this co-enzyme to accept two electrons and a proton, i.e. effectively, hydride. In line with typical pyridinium reactivity (8.12) the hydride adds at a γ-position, thus producing a 1,4-dihydropyridine (NADPH), the process being feasible because NADPH is a stabilised 1,4-dihydropyridine in which the ring nitrogen is conjugated to the carbonyl of the 3-substitutent (cf. Hantzsch synthesis products, 8.14.1.2). In the reverse sense, NADPH is a vital reducing agent in biosynthesis – it is nature's sodium borohydride. The rationale for the reverse process is the regain of aromaticity in the co-enzyme product – a pyridinium ion.

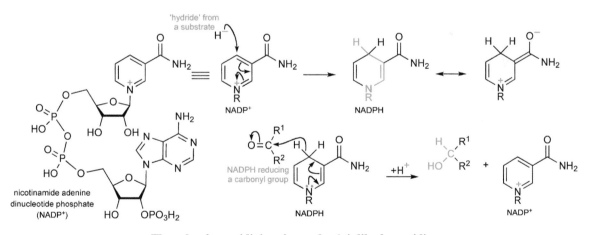

**The role of a pyridinium ion and a 1,4-dihydropyridine
in enzyme-catalysed oxidation and reduction processes**

32.2.2   Pyridoxine (Vitamin B_6) and Pyridoxal Phosphate (PLP)

The vitamin is transformed into pyridoxal phosphate (PLP), which, as a pyridine, is basic at the ring nitrogen, and in the active form, is *N*-protonated. Enzymes containing PLP have various functions, all connected with amino acids. Amongst other activities, PLP-containing enzymes can: (i) effect transfer of an amino group from an α-amino acid to an α-keto acid; (ii) bring about decarboxylation of an α-amino acid or (iii) bring about de-amination of an α-amino acid. In each case, the chemistry of the process depends critically on the intrinsic reactivity of a pyridinium salt. The scheme shows decarboxylation of an α-amino acid: condensation of the pyridine-4-aldehyde with the amino group generates an imine, stabilised by hydrogen bonding with the adjacent phenolic hydroxyl group. The decarboxylation is promoted by the flow of electrons from the breaking C–C bond through to the positively charged pyridine nitrogen, generating an extensively conjugated enamine/imine system, which regains the aromaticity of the pyridine ring by *C*-protonation. Finally, a standard hydrolysis of the new imine link produces the amine corresponding to the original amino acid, together with the regenerated co-enzyme.

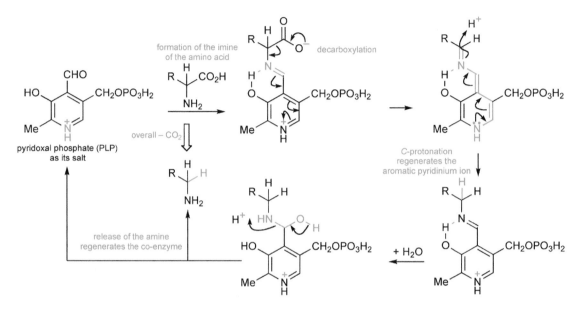

The role of a pyridinium-4-aldehyde in enzyme-catalysed amino acid decarboxylation

32.2.3 Riboflavin (Vitamin B₂)

Riboflavin is incorporated into another complex co-enzyme, flavin adenine dinucleotide (FAD). This is involved in enzyme-catalysed reductions of carbon–carbon double bonds, and the reverse. By accepting two hydrogens, the co-enzyme is converted into a dihydro derivative (FADH₂), the driving force for this being the relief of the unfavoured interaction between the polarised, opposed C=N bonds.

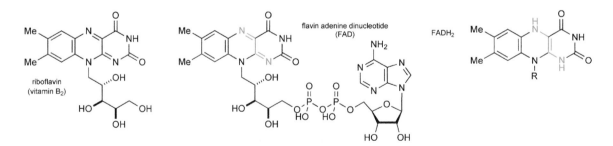

32.2.4 Thiamin (Vitamin B₁) and Thiamine Pyrophosphate

Thiamin pyrophosphate acts as a coenzyme in several biochemical processes and, in each case, its mode of action depends on the intermediacy of a C-2-deprotonated species – an ylide (24.1.2.1 and 24.10). For example, in the later stages of alcoholic fermentation, which converts glucose into ethanol and carbon dioxide, the enzyme pyruvate decarboxylase catalyses the conversion of pyruvate into ethanal (acetaldehyde) and carbon dioxide, the former then being converted into ethanol by the enzyme alcohol dehydrogenase. Thiamin pyrophosphate, in the form of its ylide, adds to the ketonic carbonyl group of pyruvate; this is followed by loss of carbon dioxide then the release of ethanal by expulsion of the original ylide as a leaving group, to continue the cycle.

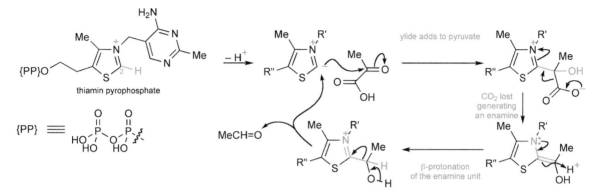

The role of a thiazolium ylide in the enzyme-catalysed decarboxylation of pyruvate

Aromatic thiophenes play no part in animal metabolism, however aromatic thiophenes do occur in some plants, in association with polyacetylenes with which they are biogenetically linked. Biotin (vitamin H), is a tetrahydrothiophene.

After the identification of the 13th vitamin, vitamin B_{12}, in 1948, there was a gap of 55 years before the 14th, a pyrroloquinoline quinone (PQQ), previously known and named methoxatin as a redox enzyme co-factor in bacteria, was shown, in 2003, to be a human dietary requirement.

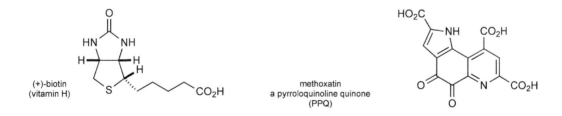

32.3   Porphobilinogen and the 'Pigments of Life'

Ultimately, all life on Earth depends on the incorporation of atmospheric carbon dioxide into carbohydrates. The energy for this highly endergonic process is sunlight, and the whole is called photosynthesis. The very first step in the complex sequence is the absorption of a photon by pigments, of which the most important in multicellular plants is the green pigment chlorophyll *a*. This photonic energy is then used chemically to achieve a crucial carbon–carbon bonding reaction to carbon dioxide, in which ultimately oxygen is liberated. Thus, formation of the by-product of this process, molecular oxygen, allowed the evolution of aerobic organisms, of which man is one.

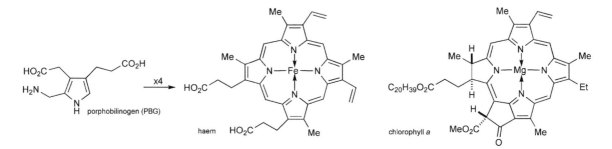

Haemoglobin is the agent that carries oxygen from lung to tissue in the arterial bloodstream in mammals; it is made up of the protein globin associated with a prosthetic group, the red pigment haem (heme). The very close structural similarity of haem with chlorophyll is striking, strongly suggesting a common evolutionary origin. In oxygenated haemoglobin, the iron is six-coordinate iron(II) with a ring nitrogen of a protein histidine residue as ligand on one side of the plane of the macrocycle, and on the other, molecular oxygen. Haem without the ferrous iron is called protoporphyrin IX and the unsubstituted macrocycle is called porphyrin. Haem is also the active site of the cytochromes, which are enzymes concerned with electron transfer. Another porphobilinogen-derived system is vitamin B_{12}, which is involved in the metabolism of all cells in the body.

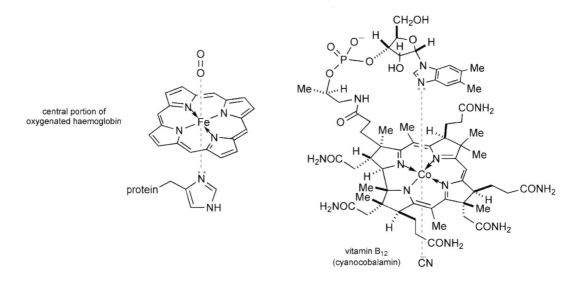

central portion of oxygenated haemoglobin

vitamin B_{12}
(cyanocobalamin)

All the tetrapyrrole pigments are biosynthesised by the combination of four molecules of porphobilinogen (PBG), this in turn is made from 5-aminolevulinic acid ($H_2NCH_2CO(CH_2)_2CO_2H$), and this from glycine and succinic acid. There are complexities late in the tetramerisation sequence that we do not go into here, but the essence of the process can be easily understood on the basis of simple pyrrole chemistry: the first step is typical. Protonation of the amino group of PBG converts it into a leaving group, generating an electrophilic azafulvene, (cf. 16.11) for attachment *via* a nucleophilic atom, X, on the enzyme. This first enzyme-bound pyrrole now undergoes electrophilic substitution, as is typical for pyrroles, at an α-position (Chapter 15), the electrophile being a second azafulvenium ion.

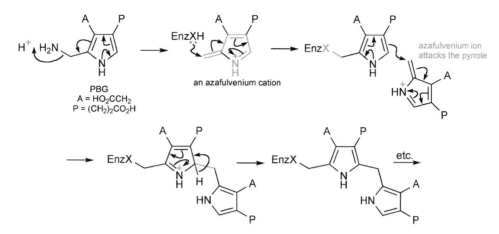

The first steps in the biosynthesis of porphyrins from porphobilinogen (PBG)

Catabolism of haemoglobin leads to the yellow bilirubin, which is excreted as a diglucuronide (to solubilise it). The catabolic sequence is complex and malfunction can occur at several stages. For example, liver disease can be diagnosed by measurements of bilirubin levels in blood, which rise with defective operation. Bilirubin is responsible for the yellow colour of bruises and the yellow discolouration in jaundice.

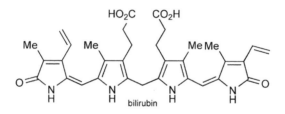

32.4 Ribonucleic Acid (RNA) and Deoxyribonucleic Acid (DNA); Genetic Information; Purines and Pyrimidines

Nucleic acids[2] are high molecular weight, mixed polymers of mononucleotides, in which chains are formed by monophosphate links between the 5′-position of one nucleoside and the 3′-position of the next.[3] They occur in every living cell. The polymer is known as **ribo**nucleic **a**cid (RNA) when the sugar is ribose, and **d**eoxyribonucleic **a**cid (DNA) when the sugar is 2-deoxyribose. Deoxyribonucleic acid (DNA), the carrier of all genetic information, consists of two intertwining helices. Each intertwining strand has a backbone, on the outside, consisting of alternating ribose and dialkyl phosphate units. From each ribose, protruding inwards, there is one of four heterocyclic bases: two purines, adenine (A) and guanine (G), and two pyrimidines, thymine (T) and cytosine (C), linked from C-1 of the ribose to N-1 of the pyrimidine bases or N-9 of the purine bases. The close association of the two strands is based on very specific hydrogen bonding between an A residue of one strand and a T residue in the precisely opposite section of the other strand, and between a C residue on one strand and a G residue on the other. This pairing is absolutely specific: adenine cannot form multiple hydrogen bonds with guanine or cytosine, and cytosine cannot form multiple hydrogen bonds with thymine or adenine. It is the *sequence of the bases* along the chain that carries the information – particular sets of three bases code for a particular amino acid – the genetic information content comes down simply to two sets of hydrogen bonds! The hydrogen bonding serves, not only to hold the two strands together, *but also* to transfer information, since, when the two strands separate, each strand acts as a template to form a new strand through the specific AT and CG base pairings, of DNA *or* of RNA (ribonucleic acid). The information in messenger RNA (mRNA) thus produced is translated into protein.

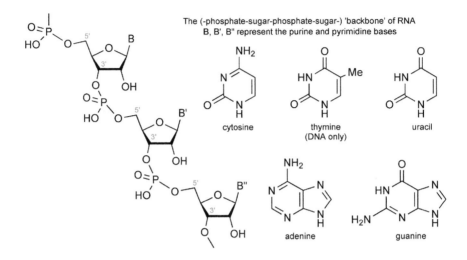

In order to understand the specific hydrogen bonding that is the genetic basis of all life, we need to recall the tautomeric forms that six-membered nitrogen heterocyles with amino and hydroxy substituents adopt:[4] the amino-heterocycles *exist as amino tautomers*; the heterocycles with a potential hydroxyl α or γ to the nitrogen *exist in the carbonyl tautomeric form*. These two preferences are illustrated in the structures for the four DNA bases shown above. Only when this propensity is recalled can the hydrogen bonding interactions be understood. Ring imine nitrogens and carbonyl oxygens are hydrogen-bond acceptors and N-hydrogens are donors.

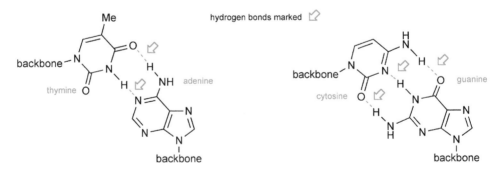

The only effective hydrogen bonding pairings in DNA: adenine/thymine (AT) and guanine/cytosine (GC)

There is another purine derivative of crucial biochemical importance – adenosine triphosphate (ATP). This substance is a carrier of energy, for when a phosphate link is broken, a large amount of energy is released. Note some trivial nomenclature: the moieties produced by linking one of the heterocyclic bases to a ribose or 2′-deoxyribose sugar, are known as 'nucleosides' (e.g. adenosine, guanosine, cytidine, thymidine). A 'nucleotide' is a 5′-phosphate (or di- or tri-phosphate) of a nucleoside – ATP is a nucleotide.

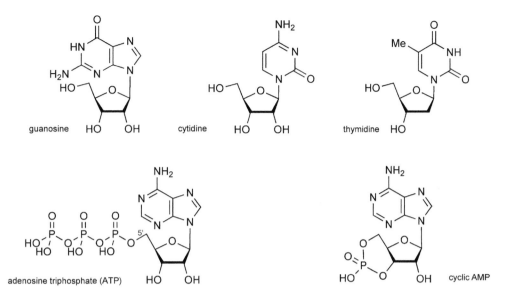

Caffeine (1,3,7-trimethylxanthine) is the well-known stimulant present in tea and coffee. Uric acid, the end product of nucleic acid catabolism in humans, birds and reptiles (uric acid was one of the first heterocyclic compounds to be isolated as a pure substance, by the Swedish chemist Carl Scheele in 1776) is formed by the action of the enzyme xanthine oxidase. In cases of excess uric acid, deposition of crystals of uric acid can occur, leading to the joint pain known as gout, usually initially in the big toe and usually in males.

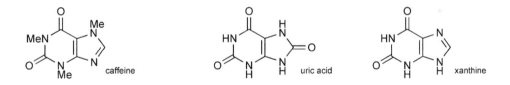

32.5 Heterocyclic Natural Products
In addition to substances involved in primary metabolism, there are a whole host of compounds that are termed secondary metabolites, the isolation and structural determination of which formed the major part of early science of organic chemistry. Many such substances are heterocyclic and we give here just a small selection to illustrate the enormous range and variety of structures produced by living organisms, often with no obvious purpose or benefit to the organism! Many such substances do have very considerable biological effects on mammals, notably on man, and plant extracts containing them were used in ancient medicine.

32.5.1 Alkaloids[5]
Plants of many genera produce compounds called alkaloids (alkali-like), and indeed all the thousands of known alkaloids contain nitrogen, by definition, and most are basic, and many are also toxic. Nicotine, a structurally simple example of an alkaloid, is a highly toxic substance, and is the major active component in tobacco (*Nicotiana* sp.), and amongst the most addictive drugs – an extraordinary contrast to the vital role in life played by nicotinic acid amide (32.2.1). Coniine, the active ingredient of hemlock (*Conium maculatum*), is another structurally simple example.

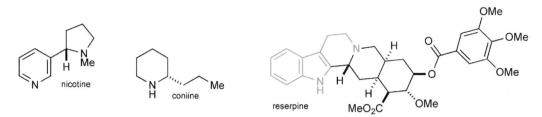

Much more complex are the thousands of alkaloids that include an indole (or 2,3-dihydroindole) sub-unit[6] and in each of these one can discern the tryptamine unit of the biosynthetic precursor tryptophan: strychnine (33.3.2) (*Strychnos nux vomica*), vincristine (33.7) (*Catharanthus roseus*), and the antipsychotic and anti-hypertensive reserpine (*Rauwolfia serpentina*) are examples. A group of amides of lysergic acid, for example, ergotamine, occur in ergot fungi (e.g. *Claviceps purpurea*), the remainder of the molecule comprising proline, phenylalanine and alanine units. Lysergic acid diethylamide is the notorious LSD.

Aromatic indolizines (28.1) are very rare in nature, but the fully reduced (indolizidine) nucleus is widespread in alkaloids, of which swainsonine is a typical example. There are many saturated or partially saturated pyrrolizidine alkaloids; senecionine is an example.

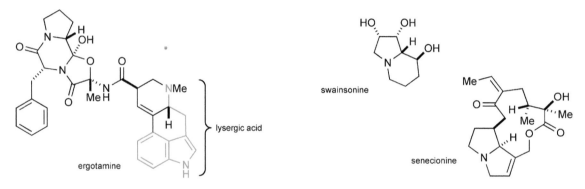

The quinolizinium ion (28.3) occurs naturally only rarely, for example as a fused ylide in the alkaloid sempervirine, however there are hundreds of indole alkaloids that contain the same tetracyclic system, but with the quinolizine at an octahydro level, as exemplified above (reserpine). In addition, many simpler quinolizidine alkaloids, such as lupinine, are known.

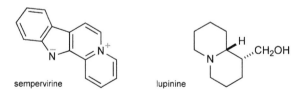

There are also many alkaloids based on isoquinoline, but fewer on quinoline nuclei: papaverine and morphine from the opium poppy (*Papaver somniferum*) and quinine (*Cinchona officinalis*) are typical.

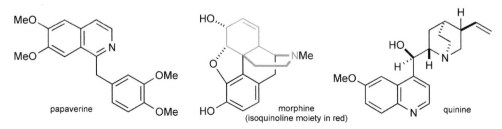

32.5.2   Marine Heterocycles

Investigation of animals from the marine environment has thrown up an intriguing variety of poly-heterocyclic products (sometimes called sea alkaloids): five representatives are variolin B[7] (sponge, *Kirkpatrickia varialosa*), lamellarin B[8] (mollusc, *Lamellaria* sp.), ascididemine[9] (tunicate, *Didemnum* sp.), wakayin[10] (ascidian, *Clavelina* sp.) and dendrodoine[11] (tunicate, *Dendroda grossular*), a rare natural example of a multi-heteroatom compound.

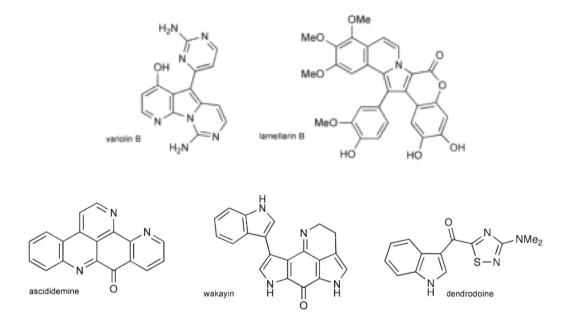

The development of synthetic routes to such systems has prompted new thinking in the construction of poly-heterocyclic systems, as a preamble to the preparation of synthetic analogues of these 'lead compounds', for biological evaluation.

32.5.3   Halogenated Heterocycles

Halogenated natural products occur in land organisms, but are especially prevalent in the marine environment, yielding a multitude of structural types and many of these are heterocyclic. Given the typical susceptibility of pyrroles and indoles to electrophilic substitution, it is not surprising that biohalogenation in the sea produces substances ranging from tetrabromopyrrole, hexabromo-2,2′-bipyrrole, to *N*-methyldibromoisophakellin and, amongst the halogenated indoles, from 4,7-dibromo-2,3-dichloroindole to hinckdentine A and chartelline A (tryptamine residue in red).[12]

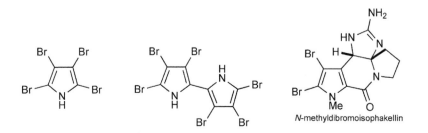

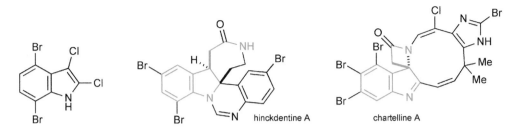

32.5.4 Macrocycles Containing Oxazoles and Thiazoles

An extraordinary group of natural products, often again from the marine environment, having oxazole and/ or thiazole rings, often macrocyclic, are derived from biosynthetic cyclisation of precursor α-amino acid units in a small peptide chain, and include substances such as ulapualide A[13] (nudibranch, *Hexabranchus sanguineus*), telomestatin[14] (*Streptomyces anulatus*), and diazonamide A[15] (ascidian, *Diazona chinensis*).

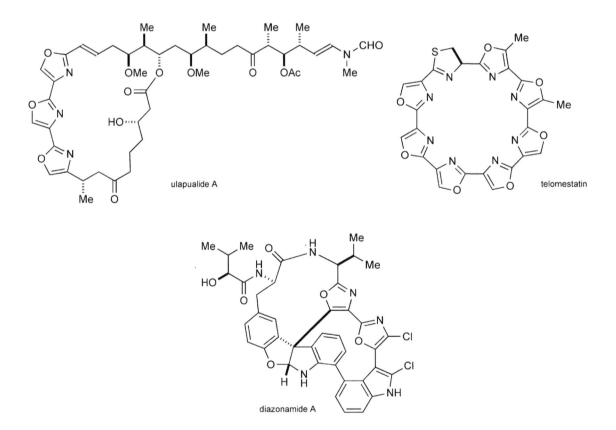

32.5.5 Other Nitrogen-Containing Natural Products

The pyrazine ring system is found in the fungal metabolite aspergillic acid and in dihydro form in the luciferins of several beetles, including the firefly, *Cypridina hilgendorfii*, and is responsible for the chemi-luminescence[16] of this ostracod. Quite simple methoxy-pyrazines are very important components of the aromas of many fruits and vegetables, such as peas and capsicum peppers, and also of wines.[17] Although present in very small amounts, they are extremely odorous and can be detected at concentrations as low as

0.00001 ppm. Related pyrazines, probably formed by the pyrolysis of amino acids during the process of cooking, are also important in the aroma of roasted meats. Several polyalkyl pyrazines are insect pheremones, for example 2-ethyl-3,6-dimethylpyrazine is the major component of the trail pheremone of the South American leaf-cutting ant.

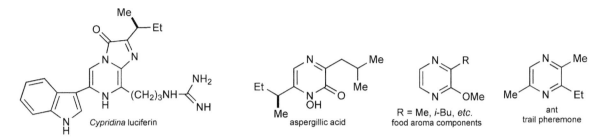

Cypridina luciferin

aspergillic acid

R = Me, *i*-Bu, *etc.*
food aroma components

ant
trail pheremone

Some fungal metabolites from *Streptomyces* species include the quaternary salt pyridazinomycin. Divicine, present as a glucoside in broad beans and related plants, is the toxic agent responsible for 'favism', a serious hemolytic reaction in genetically susceptible people (of Mediterranean origin).

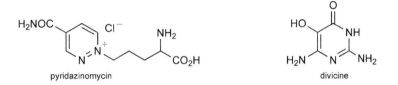

pyridazinomycin

divicine

32.5.6 Anthocyanins and Flavones

There certainly is a biological role for the anthocyanidins and natural flavones, which, grouped together, are known as the flavonoids.[18] These occur in flower pigments – as attractants for pollinating insects. The anthocyanidins are polyhydroxyflavylium salts and occur in a large proportion of the red to blue flower pigments and in fruit skins, for example grapes and therefore in red wines made therefrom.[19] Anthocyanidins are generally bound to sugars, and these glycosides are known as anthocyanins.[20] As an example, cyanin (isolated as its chloride) is an anthocyanin that occurs in the petals of the red rose (*Rosa gallica*), the poppy (*Papaver rhoeas*) and very many other flowers. Another example is malvin chloride, which has been isolated from many species, including *Primula viscosa*, a mauvy-red alpine primula.

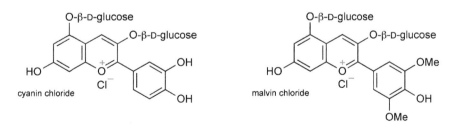

cyanin chloride

malvin chloride

In the living cell these compounds exist in more complex bound forms, interacting with other molecules, for example flavones,[21,22] and the actual observed colour depends on these interactions.[23] However, it is interesting that even *in vitro*, simple pH changes bring about extreme changes in the electronic absorption of these molecules. For example cyanidin is red in acidic solution, violet at intermediate pH and blue in weakly alkaline solution, the deep colours being the result of extensive resonance delocalisation in each of the structures.

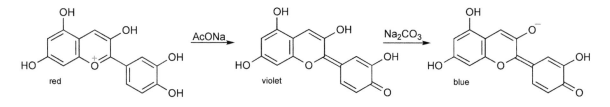

The naturally occurring flavones are yellow and are very widely distributed in plants. They accumulate in almost any part of a plant, from the roots to the flower petals. Unlike the anthocyanins, which are too reactive and short-lived, the much more stable flavones have, from time immemorial, been used as dyes, for they impart various shades of yellow to wool. As an example, the inner bark of one of the North American oaks, *Quercus velutina*, was a commercial material known as quercitron bark and much used in dyeing: it contains quercitrin. The corresponding aglycone, quercetin, is one of the most widely occurring flavones, found, for example, in *Chrysanthemum* and *Rhododendron* species, horse chestnuts, lemons, onions and hops.

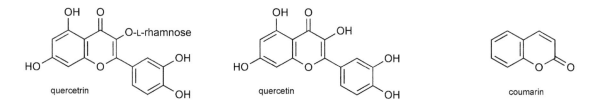

quercetrin   quercetin   coumarin

Coumarin[24] has the sweet scent of newly mown hay, and is used in perfumes. The official birthdate of the synthetic fragrance industry is held to be[25] the discovery of a synthetic route to coumarin by W. H. Perkin in 1868[26] using the reaction we now call the Perkin condensation (12.3.2.2).

Benzofuran occurs in a range of plant- and microbial-derived natural products, ranging in complexity from 5-methoxybenzofuran, through the orange 'aurones', a group of plant pigments isomeric with co-occurring flavones (12.2), to griseofulvin, from *Penicillium griseofulvum*, used in medicine as an antifungal agent.

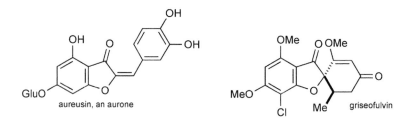

aureusin, an aurone   griseofulvin

References

[1] 'Introduction to Enzyme and Coenzyme Chemistry', Bugg, T., Blackwell, **2004**.

[2] 'Nucleic acids in chemistry and biology', Ed. Blackburn, G. M. and Gait, M. J., Oxford University Press, **1996**.

[3] 'Oligo- and poly-nucleotides: 50 years of chemical synthesis', Reese, C. B., *Org. Biomol. Chem.*, **2005**, *3*, 3851.

[4] 'Tautomerism and electronic structure of biological pyrimidines', Kwiatkowski, J. S. and Pullman, B., *Adv. Heterocycl. Chem.*, **1975**, *18*, 200.

[5] 'The Alkaloids; Chemistry and Biology', Volume 69, Ed. Cordell, G. A., **2010**, Elsevier, and previous volumes in the series.

[6] 'Monoterpenoid indole alkaloids' in 'Indoles' in 'The Chemistry of Heterocyclic Compounds', Series Ed. Taylor, E. C., Vol. 25, Part 4, Ed. Saxton, J. E., Wiley-Interscience, **1983** and supplement, **1994**.

[7] Perry, N. B., Ettouati, L., Litaudon, M., Blunt, J. W., Munro, M. H. G., Parkin, S. and Hope, H., *Tetrahedron*, **1994**, *50*, 3987; Trimurtulu, G., Faulkner, D. J., Perry, N. B., Ettouati, L., Litaudon, M., Blunt, J. W., Munro, M. H. G. and Jameson, G. B., *Tetrahedron*, **1994**, *50*, 3993.

[8] Anderson, R. J., Faulkner, D. J., Cun-heng, H., Van Duyne, G. D. and Clardy, J., *J. Am. Chem. Soc.*, **1985**, *107*, 5492.

[9] Kobayashi, J., Cheng, J., Nakamura, H., Ohizumi, Y., Hirata, Y., Sasaki, T., Ohta, T. and Nozoe, S. *Tetrahedron Lett.*, **1988**, *29*, 1177.

[10] Copp, B. R., Ireland, C. M. and Barrows, L. R., *J. Org. Chem.*, **1991**, *56*, 4596.

[11] Heitz, S., Durgeat, M., Guyot, M., Brassy,C. and Bachet, B., *Tetrahedron Lett.*, **1980**, *21*, 1457.

[12] 'Naturally occurring halogenated pyrroles and indoles', Gribble, G. W., *Prog. Heterocycl. Chem.*, **2003**, *15*, 58.

[13] Roesener, J. A. and Scheuer, P.J., *J. Am. Chem. Soc.*, **1986**, *108*, 846; Pattenden, G., Ashweek, N. J., Baker-Glenn, C. A. G., Kempson, J., Walker, G. M. and Yee, J. G. K., *Org. Biomol. Chem.*, **2008**, *6*, 1478.

[14] Shin-ya, K., Wierzba, K., Matsuo, K., Ohtani, T.,Yamada, Y., Furihata, K., Hayakawa, Y. and Seto, H., *J. Am. Chem. Soc.*, **2001**, *123*, 1262.

[15] Li, J., Burgett, A. W. G., Esser, L., Amezcua, C. and Harran, P. G., *Angew. Chem. Int. Ed.*, **2001**, *40*, 4770; Lachia, M. and Moody, C. J., *Nat. Prod. Rep.*, **2008**, *25*, 227.

[16] 'Introduction to beetle luciferases and their applications', Wood, K. V., Lam, Y. A., McElvoy, W. D., *J. Biolumin. Chemilumin.*, **1989**, *4*, 289.

[17] 'Odor threshold of some pyrazines', Shibamoto, T., *J. Food Sci.*, **1986**, *51*, 1098.

[18] 'The flavonoids', Harborne, J. B., Mabry, T. J. and Mabry, H., Eds., Chapman and Hall, London, **1975**; 'Flavonoids: chemistry and biochemistry', Morita, N. and Arisawa, M., *Heterocycles*, **1976**, *4*, 373; 'The Flavonoids: Advances in Research Since 1980', Ed. Harborne, J. B. and Mabry, T. J., Chapman and Hall, London and NY, **1988**; 'Flavonoids: Chemistry, Biochemistry and Applications', Andersen, O. M. and Jordheim, M., CRC Press, **2006**.

[19] 'A curious brew', Allen, M., *Chem. Brit.*, **1996** (May), 35.

[20] 'The Chemistry of Plant Pigments', Ed. Chichester, C. O., Academic Press, NY, **1972**; 'The chemistry of anthocyanins, anthocyanidins and related flavylium salts', Iacobucci, G. A. and Sweeny, J. E., *Tetrahedron*, **1983**, *39*, 3005.

[21] 'Naturlich vorkommende chromone', Schmid, H., *Fortschr. Chem. Org. Naturst.*, **1954**, *11*, 124.

[22] 'Structure, stability and color variation of natural anthocyanins', Goto, T., *Fortschr. Chem. Org. Naturst.*, **1987**, *52*, 113; Goto, T. and Kondo, T., 'Structure and molecular stacking of anthocyanins – flower colour variation', *Angew. Chem., Int. Ed. Engl.*, **1991**, *30*, 17; 'The chemistry of rose pigments', Eugster, C. H. and Märki-Fischer, E., *ibid.*, 654; 'Nature's palette', Haslam, E., *Chem. Brit.*, **1993** (Oct.), 875.

[23] Kondo, T., Toyama-Kato, Y. and Yoshida, K., *Tetrahedron Lett.*, **2005**, *46*, 6645.

[24] 'Synthesis of coumarins with 3,4-fused ring systems and their physiological activity', Darbarwar, M. and Sundaramurthy, V., *Synthesis*, **1982**, 337; 'Naturally occurring coumarins', Dean, F. M., *Fortschr. Chem. Org. Naturst.*, **1952**, *9*, 225; 'Naturally occurring plant coumarins', Murray, R. D. H., *Fortschr. Chem. Org. Naturst.*, **1978**, *35*, 199; 'The natural coumarins: occurrence, chemistry and biochemistry', Murray, R. D. H., Méndez, J. and Brown, S. A., Chichester, Wiley, **1982**.

[25] 'Perfumes: The Guide', Turin, L. and Sanchez, T., Profile Books, **2008**.

[26] Perkin, W. H., *J. Chem. Soc.*, **1868**, 53.

33

Heterocycles in Medicine

The drugs used in human medicine[1] cover the whole range of chemical structure types, but a majority are heterocyclic small molecules or have heterocyclic structural components. Heterocyclic alkaloids were the active ingredients in many natural remedies before the development of modern chemistry and some are still used today, for example morphine derivatives. Rather than using complex systematic names, drugs are given 'trivial' generic names and drugs acting on the same pharmacological basis often have related names, particularly by the word ending, for example ondansetron and granisetron (see 33.4.3), losartan and candesartan (33.4.5). The manufacturer will also give a proprietary (trade) name to a drug, which is treated as a proper noun and capitalised. In this chapter, we give first the generic name with, where appropriate, the proprietary name in parenthesis. Note that proprietary names may be different in different countries.

In this book we are concerned with heterocycles and so the coverage in what follows does not necessarily reflect the relative medical importance of different areas. Complete coverage of all areas of significance is obviously impossible, but the aim is to give an appreciation of the wide-ranging importance of heterocycles in medicine. Although heterocycles occur in drugs for all areas, they are particularly prominent, for obvious reasons, for systems involving the heterocyclic neurotransmitters.

Of the top 10 best-selling (prescription) drugs by value in the year June 2006 – June 2007, seven were (small molecule) heterocycles:[2] atorvastatin (Lipitor; $13.5bn; a statin for cholesterol reduction); esomeprazole (Nexium; $6.9bn; a proton-pump inhibitor for reduction of gastric acid); clopidogrel (Plavix; $5.8bn; an anti-platelet agent to prevent blood clots); olenzapine (Zyprexa; $4.9bn; an anti-schizophrenic); risperidone (Risperdal; $4.8bn; an anti-schizophrenic); amlodipine (Norvasc; $4.5bn; an anti-hypertensive agent); quetiapine (Seroquel; $4.2bn; for treatment of schizophrenia and bipolar disorder). Of the three remaining compounds, two were biomolecules – modified polypeptides: carbepoetin (Arancep; $5.1bn) and etanercept (Enbrel; $4.9bn) and the remaining one – seretide (Advair; $6.7bn; an anti-asthmatic) – is the only non-heterocyclic, small-molecule-based drug of the 10, a mixture of a steroid with a beta stimulant. Note, however, that these are not the best-selling drugs by volume – those include over-the-counter (OTC) drugs such as aspirin and paracetamol, which are available in supermarkets for very low prices (of which a good proportion is for the packaging!).

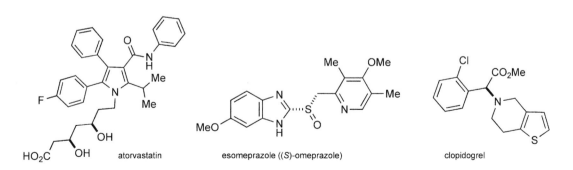

atorvastatin   esomeprazole ((S)-omeprazole)   clopidogrel

Heterocyclic Chemistry 5th Edition   John Joule and Keith Mills
© 2010 Blackwell Publishing Ltd

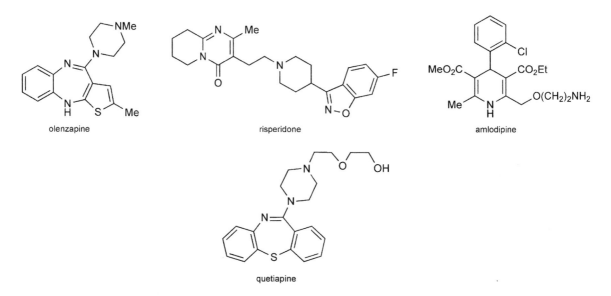

33.1 Mechanisms of Drug Actions[3]

Drugs exert their effects by a number of mechanisms, the most important being the following.

33.1.1 Mimicking or Opposing the Effects of Physiological Hormones or Neurotransmitters

Compounds that mimic the effect of the natural agent are known as *agonists*; those that oppose are known as *antagonists*. These drugs act by binding at a *receptor* – the site where the natural physiological agent binds. This receptor is usually a specific domain of a protein molecule comprising part of a cell. Agonists activate the receptor and antagonists block it. Sometimes the word 'blocker' is used in reference to antagonists, as in 'beta-blocker'. There are variants on these simple definitions, for example *partial agonists* that cannot achieve the same maximum effect as full agonists, but can still compete with them for receptor binding, therefore acting partly as antagonists.[2]

33.1.2 Interaction with Enzymes

Interaction with enzymes usually involves inhibition, but enzyme induction also has its uses, or can occur unintentionally. Interactions generally occur by binding at the *active site* of the enzyme (where the substrate binds), in a very similar manner to receptor binding – the comments below, concerning receptors, can equally apply to enzyme interactions with drugs.

33.1.3 Physical Binding with, or Chemically Modifying, Natural Macromolecules

Physical binding with, or chemically modifying, natural macromolecules such as DNA and RNA, can involve either direct interaction or the incorporation of synthetic analogues of structural components (e.g. one of the heterocyclic bases) into the polymer.

Binding at the receptor may be *competitive* (i.e. the drug is in equilibrium/exchange with the natural agent) or it may be *non-competitive* – firmly bound with no exchange.

Receptors usually exist as a number of sub-types, often with further sub-divisions, which mediate different physiological effects by interaction with the single natural agent. The key to a successful drug very often depends on devising a molecule that has a selective interaction with just one sub-type.

Sometimes, binding at a site other than the normal receptor (*allosteric binding*) is the mechanism of action – the remote binding alters the shape of the protein as a whole and hence influences the receptor. Binding/action at the receptor is often only the start of a complex cascade of actions within the cell.

The mechanisms of binding to the receptor can be essentially physical (H-bonding or attractions *via* ionic or Van der Waals forces, etc.), with several points of interaction, or it may be covalent (e.g. *N*-alkylation or *O*-acylation), which is often irreversible.

33.2 The Neurotransmitters

The nervous system is made up of two parts: the *central nervous system* (CNS) and the *peripheral nervous system*. The former operates in the brain and spinal cord and the latter in the rest of the body.

The CNS and the peripheral nervous systems are separated by the blood–brain barrier, which stands between the bloodstream and the brain tissue – it is essentially the walls of the brain's blood vessels, which are different from those in the periphery. A complex combination of physical factors, such as lipophilicity/polarity, molecular weight and active transport systems determines the ability of molecules to cross the blood–brain barrier and thus gain access to the CNS. The consideration as to whether a drug can or cannot reach the brain is always an important design consideration – drugs acting on the CNS obviously need to reach it, but for peripherally acting drugs, the opposite is required – keeping them out may be essential to avoid side-effects.

The conduction of signals along nerves is dependent on the release of neurotransmitters at synapses.

Some of the major neurotransmitters are heterocyclic (histamine, 5-hydroxytryptamine), others are not heterocyclic (catecholamines: adrenaline, noradrenaline, dopamine; acetylcholine). These neurotransmitters are present in both the peripheral and central systems, although dopamine acts mainly in the latter. The CNS also has other neurotransmitters, including the amino acids glycine, glutamic acid and gamma-aminobutyric acid (GABA).

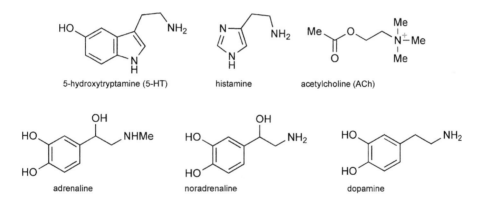

Some receptors are named after the natural agent, e.g. histamine and 5HT, but others after natural products, the interactions with which were the only means of characterisation before the structures of the true transmitters were known, e.g. opioid after opium (morphine) – the natural ligands here are peptides (enkephalins and endorphins); cannabinoid after cannabis – the natural ligands are amides of arachadonic acid.

33.3 Drug Discovery and Development

33.3.1 Stages in the Life of a Drug

1. Discovery: finding a series of potential candidates.
2. Optimisation: finding the best biological candidate(s) on the basis of safety and effectiveness.
3. Chemical development: finding a suitable medium-scale synthesis (maybe 1 kg) for initial clinical trials and later a large-scale synthesis for production.
4. Clinical trials (several stages): initially, proof of the biological concept and finally demonstration of clinical effectiveness and safety in a substantial number of patients.

5. Approval by the regulatory authorities, such as the US Food and Drug Administration (FDA) and the European Medicines Agency (EMEA), as having clinical advantages over current treatments and a suitable safety profile.
6. Release for marketing, with appropriate post-marketing surveillance to detect low-level and unexpected side-effects.
7. 'Me-too' competition: competitor companies try to develop related compounds, not covered by the original patents, that may have a sufficiently improved profile to allow approval by the regulatory authorities.
8. Expiry of patent: generic free-for-all: anyone free to make and market the compound, although processes may be protected by the original inventor company.
9. Over-the-counter (OTC): Certain drugs, for example sumatriptan, ranitidine, omeprazole and various anti-histamines, after extensive usage under prescription, showing a low incidence of side-effects, may be approved for 'over-the-counter' sales without a doctor's prescription. These are generic drugs, but the original proprietary brand names are still owned by the inventing company and may still be preferred by some customers.

33.3.1.1 Me-too drugs
This stage of drug 'development' starts as soon as the identity of the lead drug has been disclosed. It involves competitor companies investigating related molecules that are not covered by the inventor's patent, in the hope that they will have advantages over the original. Although cynics regard this as simply a money-making exercise, which is true to some degree, it has played an important role in optimising particular classes of drug and is actually an important stage of development. It often happens that apparently similar compounds have different side-effect profiles. Thus the existence of competitor drugs allows the eventual emergence of the best, and of alternatives when the original drug shows unacceptable levels of side-effects a significant length of time after release, as has often occurred.

In the absence of 'me-too' drugs, there would be many fewer therapeutic options available.

33.3.1.2 Orphan Drugs
Research and development is very expensive, particularly as the majority of drugs in development fail to reach the market place, and requires sufficient income from sales of successful drugs to fund it. However, some diseases have such a low incidence that income from potential sales of drugs developed under the standard scheme cannot cover these enormous costs. Here, the regulatory authorities may grant orphan drug status, which allows a less onerous version of the approval process. Safety is still checked rigorously, but larger-scale trials are waived, not least because there may not be enough patients available. The manufacturers are also given financial incentives.

33.3.1.3 Prodrugs
A number of drugs, for example proguanil (33.6.1) and omeprazole (33.4.1), are not active in their own right, but undergo activation *in vivo*. Sometimes, however, an active drug is deliberately converted into a protected, possibly inactive, form – a prodrug (esters are quite common), so as to give better absorption or protection from breakdown, for example in the stomach. It does, of course, require a mechanism for re-activation, possibly an enzymatic conversion, at the site of action.

33.3.1.4 Pharmacogenetics
It has always been known that in any group of patients there will be responders and non-responders to many drugs, and that there are differences in susceptibilities to side-effects. If a drug is only effective in a sub-population of patients, the benefits for this minority may be lost during statistical analysis if the (non-targeted) trials have involved the general population. Genetic variations can be very important in this respect and genetic targeting is a current 'hot' topic.

33.3.1.5   Risk–Benefit

No medicine is perfect; all have side-effects. Some side-effects, because they are relatively uncommon, will not become statistically significant until a large number of patients have been treated, and are therefore not apparent until after general release. There is always a risk to be balanced with a benefit, higher risks (side-effects up to possible death) being acceptable, the more serious and less treatable the target disease is. This is inevitable and must be accepted by the informed patient as part of the 'deal'.

33.3.2   Drug Discovery

A traditional logical approach to drug design involves synthesis of compounds that are structurally similar to the natural agent, with alterations of substituents or variations in the electronic nature or precise shape of any aromatic rings (a key asset of heterocycles!). Modifications using heterocycles can also be used to give favourable physical properties, for example replacing a benzene ring by a pyridine ring may improve aqueous solubility, or other variations may be used to facilitate or prevent access to the CNS.

A surprising number of drugs bear only a tenuous (or no!) structural similarity to the natural agent/ transmitter. An excellent illustration is the well-known indole alkaloid strychnine, which is a competitive antagonist of the CNS neurotransmitter glycine.

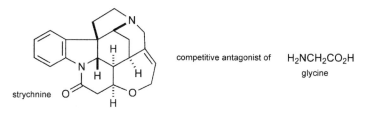

competitive antagonist of $H_2NCH_2CO_2H$
glycine

Other routes to drug discovery include screening of compounds ('natural products') from natural sources that may have been reported (but not necessarily – random screening of these natural compounds is more frequently used) in folklore to have medicinal properties, fortuitous discovery of unexpected activity in a compound being investigated in other areas (Viagra is a prime example) and the crude but reasonably effective method of screening large numbers of random compounds synthesised by combinatorial chemistry. This last method is very popular at present, as it is amenable to automation, as is much biological testing.

When a *lead compound* has been obtained by any of these methods, more rational optimisation is carried out.

33.3.3   Chemical Development

(**NOTE**: The active compound may be referred to as the API (active pharmaceutical ingredient) or DS (drug substance). This is formulated with various additives into the final medicine.)

One key to commercialising a drug is a good chemical synthesis, which is highly reproducible, both in chemistry and product purity. It must also be produced to a target price that varies greatly depending on a number of factors, including potency i.e. it relates to the amount of the chemical substance used in the final formulation (tablets etc). Most of the total cost is in discovery, development and testing rather than manufacture. The cost of production of the API is a relatively minor part of the final cost, but still substantial, so production costs are controlled, requiring a synthetic route that must be as efficient and cheap as possible.

This is the most demanding area for the chemist. The requirements for scale-up and production are very different from those in medicinal chemistry/discovery and need a very different approach to chemistry. Yields for steps in a process are very frequently increased dramatically by intensive process research – improving literature reactions with reported yields of 20–30% up to a usable 80–90% is not uncommon.

The number of steps in the synthesis is a major determinant of cost. Standard preliminary targets are for 90% hplc yields in each stage, with purification by (usually) crystallisation; chromatographic purification on a production scale, which may be in ton quantities, is very rare.

A crucial aspect is the quality of the final product, with very rigorous criteria for reproducible levels of purity and of levels of potentially toxic trace impurities, particularly metals. All impurities down to ca. 0.1% must be identified and possibly reference samples of the impurities synthesised, which often provides another interesting challenge for the chemist. Some metal derivatives, such as organotin compounds and mercury must be at essentially undetectable levels; others, such as palladium, may be acceptable at very low levels, depending on dose. These requirements can greatly restrict the synthetic chemistry available for the production route – many elegant and ingenious reactions found, for example, in this book, though suitable for general synthesis and preliminary phases of drug discovery, may be of no use for production due to residues of reagents in the final product. A very clean synthesis from a chemical point of view could have 0.1% of a heavy-metal impurity, but this corresponds to 1000 ppm for something that may only be allowed at 10 ppm or less in a drug.

Minimisation of environmental impact must also be considered – there are legal constraints on effluents of various types, as well as costs of disposal. These factors must therefore be taken into account when considering the production synthesis of a drug.

33.3.4   Good Manufacturing Practice (GMP)
(The term 'cGMP' ('current good manufacturing practice'), found in some publications, can be even more confusing for chemists for whom GMP can mean guanosine monophosphate and cGMP, cyclic guanosine monophosphate!)

GMP is the enforcement of rigorous control of purity and consistency of the production of any substance that is to be administered to a human. It is designed to avoid human error that could lead to contamination of the final product. It does not apply to process *research*, but does when any process on any scale is used for the preparation of any batch of drug for clinical use. It is the responsibility of the manufacturer, under potential scrutiny from the regulatory authorities, and adds dramatically to the cost of the process, particularly at the small-scale early stages.

33.4   Heterocyclic Drugs[4,5]
For convenience, the following sections are grouped, where possible but with flexibility, according to the natural neurotransmitter or enzyme to which the drugs relate and may include central and peripheral actions. A separate section on CNS-specific drugs with more complex mechanisms is included later. Later major sections, for example infection and cancer, are, of necessity, related to disease areas.

33.4.1   Histamine
Four histamine receptors have been identified, but two – H_3 and H_4 – have no clinical relevance at present.

Histamine, acting at the H_1 receptor, is mainly known for producing allergic responses, for example hay fever and skin reactions, such as urticaria ('nettle rash'). Antagonists of this receptor, which suppress these actions of histamine, are well known to the layman, under the loose description 'anti-histamines', as hay fever remedies.

Typical H_1 antagonists are not particularly structurally reminiscent of histamine! Early compounds often had side-effects caused by actions at other receptors such as that for acetylcholine, but their main drawback is antagonism of H_1 receptors in the CNS, leading to drowsiness. Promethazine (Phenergan) has such strong effects in the CNS that it has found use as a sedative and is also of use for treatment of motion sickness. Others, such as chlorpheniramine (Piriton), show only moderate sedative effects. More recently, compounds such as loratidine (Clarityn) have been devised that do not enter the CNS and are therefore free from sedative effects.

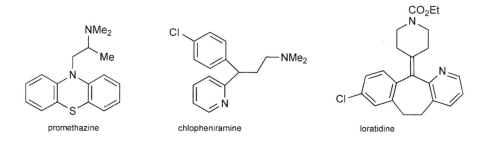

The H_2 receptor is part of a complex system within parietal cells, involved in mediating release of acid into the stomach. Although this is an essential part of digestion, excessive amounts of acid can lead to simple indigestion, but can also cause or aggravate the more serious medical problem of peptic ulcers. The development, in the 1970s, of selective H_2 antagonists that inhibit the release of gastric acid was a major advance in medicine, as they are very effective in treating ulcers, are very safe, and have almost eliminated the need for surgery for this illness. They were an enormous financial success because of this. The first compound developed was cimetidine (Tagamet), which is closely related, structurally, to histamine and as such, a more 'logically' designed compound than the H_1 antagonists. Other H_2 antagonists, particularly ranitidine (Zantac) followed, which had an even more favourable side-effect profile. The dimethylamino-furan unit in ranitidine can be visualised as replacing the imidazole ring at the receptor. A closer imidazole substitute is the thiazole ring found in famotidine (Pepcid).

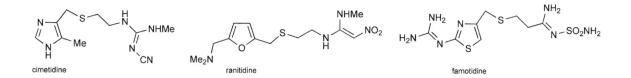

Another type of antacid is the *proton-pump inhibitor*. As mentioned above, H_2 receptors initiate a complex cascade within a cell resulting in the production of gastric acid, but the actual final release of the acid from the cell is *via* an enzyme – H^+,K^+-ATPase, the 'proton pump' – which exchanges protons from inside the cell with potassium ions outside. Omeprazole (Losec) was designed to inactivate this enzyme and is even more effective than the H_2 antagonists, almost completely stopping the production of acid. The mechanism of action of omeprazole is very different from the competitive H_2 antagonism. The drug is converted *in vivo* into a highly selective sulfenylating agent for the SH of a specific cysteine residue in the enzyme. This disulfide formation irreversibly deactivates the enzyme so effectively that recovery of acid production requires the biosynthesis of new enzyme. This activation of omeprazole is acid-catalysed and may involve initial protonation of the benzimidazole nitrogen and/or the sulfoxide oxygen. Because of this susceptibility to acid, omeprazole is formulated as gastro-resistant granules, so as to avoid activation and decomposition in the stomach before it can be absorbed from the more alkaline part of the gastro-intestinal (GI) tract.

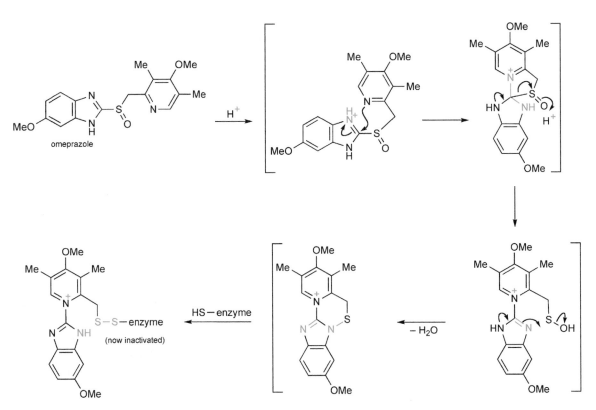

Mechanism of action of omeprazole

33.4.2 Acetylcholine (ACh)

Acetylcholine mediates two different types of activity *via* corresponding receptors – *muscarinic* and *nicotinic* – which relate to the pharmacological activities of two natural products, muscarine and nicotine. The former type of action occurs in nerve synapses and the latter at neuromuscular junctions and peripheral ganglia. The term *cholinergic* is used for the general effects of acetylcholine.

Clinically useful muscarinic agents tend to be simple choline derivatives, although the alkaloid pilocarpine is a muscarinic agonist used in the treatment of glaucoma. The muscarinic antagonist pirenzepine (Gastrozepin) is an alternative to the H_2 antagonists for the treatment of peptic ulcers and some cholinergic side-effects caused by radiotherapy. The best-known nicotinic agonist is nicotine! Nicotinic antagonists include the natural bis-quaternary isoquinoline curare alkaloid tubocurarine, the paralytic agent from curare (a South American arrow poison), which is used in surgery as a muscle-relaxant.

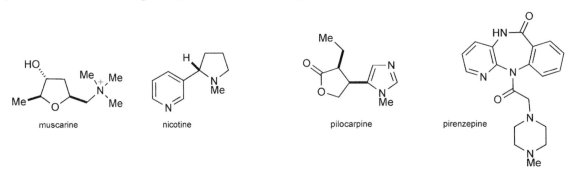

33.4.2.1 Anticholinesterase Agents

The physiological activity of acetylcholine relies on local release, stimulation of the receptor, then rapid hydrolysis (deacetylation) by acetylcholinesterase, which results in deactivation. The indole alkaloid physostigmine, from the West African calabar bean, and the relatively simple synthetic compound pyridostigmine, which has a more obvious relationship to choline, are reversible inhibitors of acetylcholinesterase. Controlled inhibition of the enzyme by such drugs, which results in a build-up of ACh, is useful in conditions such as myasthenia gravis, a muscle weakness, which is caused by insufficient production of ACh.

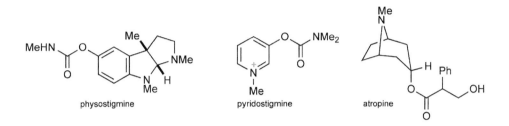

Irreversible inhibition of acetylcholinesterase is the mechanism of action of poisoning by nerve gases, such as sarin and, to a lesser degree, by other organophosphates, such as those used as insecticides, leading to persistent, widespread and quite possibly fatal, excessive cholinergic effects. The inhibition is due to phosphorylation of a serine OH at the active site of the enzyme – this OH is the nucleophile that attacks the acetyl group during physiological functioning of the enzyme. It is possible to reactivate the enzyme, provided treatment is given promptly, by use of pralidoxime (2-PAM). Here, the *N*-methylpyridinium binds *via* electrostatic forces to the same site as the choline trimethylammonium, bringing the very nucleophilic oxime oxygen close enough to attack the phosphoryl group, releasing the serine OH. The (saturated heterocyclic) alkaloid atropine, a muscarinic antagonist, is also administered.

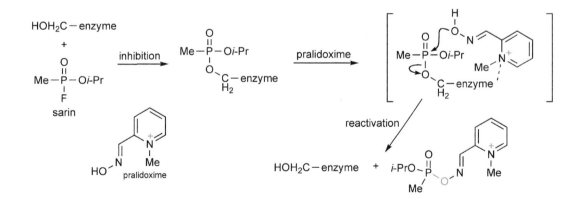

33.4.3 5-Hydroxytryptamine (5-HT)

5-Hydroxytryptamine has at least 14 receptors and sub-types. Compounds acting on these receptors are drugs for the treatment of disorders of the cardiovascular, gastrointestinal and central nervous systems. Sumatriptan (Imigran), a 5-HT$_{1D}$ agonist, in part, causes vasoconstriction selectively in intracranial blood vessels, opposing the vasodilation that is the pathological basis of migraine. The discovery of the triptan class of drugs was a major advance in the treatment of migraine.

The 5-HT$_3$ antagonists ondansetron (Zofran) and granisetron (Kytril) are effective in relieving the nausea and vomiting that are side-effects of radiotherapy and treatment with cytotoxic drugs. They probably function by a combination of central and peripheral actions.

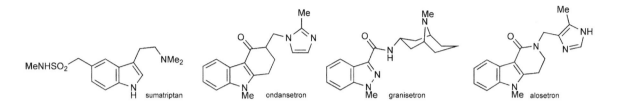

Alosetron (Lotronex) is effective in certain variants of irritable bowel syndrome (IBS), but was withdrawn from use because it may cause serious side-effects. However, it was later re-instated for restricted use following petitions from patients.

33.4.4 Adrenaline and Noradrenaline

Adrenaline and noradrenaline are prominent throughout the cardiovascular system, acting through α- and β-adrenergic receptors, which exist as a number of sub-types. The interaction of effects from these receptors is quite complex. Many common adrenergic drugs are simple carbocyclic analogues of adrenaline, but a few of the important compounds are heterocyclic, for example prazosin (Hypovase) and indoramin (Doralese) are α-antagonists, used for the treatment of hypertension and benign prostatic hypertrophy; timolol (Blocaderen) is an anti-hypertensive, anti-angina and anti-arrythmic β-antagonist.

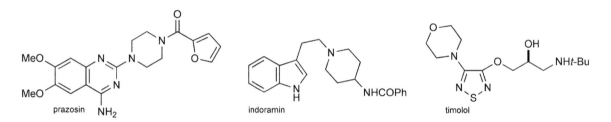

33.4.5 Other Significant Cardiovascular Drugs

Amlodipine (Norvasc) is one of a number of 1,4-dihydropyridines, the mechanism of action of which involves blocking calcium channels (they are said to be 'calcium antagonists'), resulting in a relaxation of vascular smooth muscle. They are useful for hypertension and angina. Diazoxide (Proglycem) is a vasodilator used for intravenous administration in hypertensive emergency. Hydralazine (Apresoline) and minoxidil (Rogaine) are vasodilators used for chronic hypertension, the latter being particularly useful for cases resistant to other drugs. Minoxidil is also one of the few drugs of proven value for the treatment of baldness (alopecia).

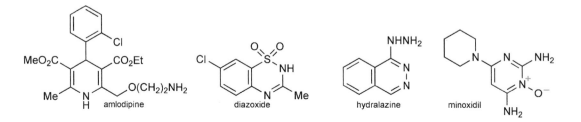

The angiotensin system also affects blood pressure. Angiotensin-I (inactive) is converted by angiotensin converting enzyme (ACE) into angiotensin-II, which causes vasoconstriction and therefore increased blood pressure. These effects of angiotensin-II can be controlled, either by use of an inhibitor of ACE or an angiotensin-II antagonist. ACE inhibitors are generally derived from amino acids, but the angiotensin antagonists are heterocyclic compounds, which notably contain tetrazoles as replacements for carboxyl groups, resulting in increased potency. Losartan (Cozaar) was a widely used angiotensin-II antagonist, but more potent compounds such as candesartan cilexetil (Amias) (a complex-ester prodrug) are now more popular.

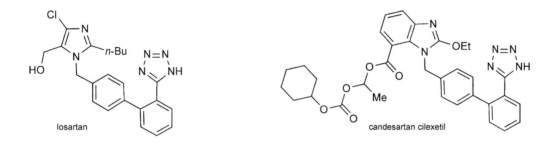

losartan

candesartan cilexetil

33.4.6 Drugs Affecting Blood Clotting

Blood clotting is an essential function to avoid excessive bleeding, but its malfunction is also of major importance as a cause of disease. The mechanism of blood clotting is a complex sequence of events. It is initiated by platelet activation, leading to their aggregation and formation of plug at the site of damage. This aggregation also causes the release of substances that trigger the rest of the cascade, involving a series of clotting factors.

Controlled inhibition of blood clotting is a very important therapeutic strategy for prevention and treatment of heart attacks and occlusive stoke. It is usually combined with drugs to reduce blood pressure and statins to control cholesterol levels.

Low-dose aspirin is the drug most commonly used to inhibit platelet aggregation. It acts by inhibiting the enzyme that initiates synthesis of pro-aggregatory prostaglandins. Proton-pump inhibitors or H_2-blockers are often co-administered with aspirin to prevent gastric bleeding. Other drugs inhibit platelet action by different mechanisms. Ticlopidine (Ticlid) was the first of its class, acting as an ADP antagonist, but clopidogrel (a close structural analogue; see page 645) has a better side-effect profile and, as was mentioned above, is extremely successful commercially. It is also used in combination with aspirin following angioplasty.

Cilostazol (Pletal), a selective inhibitor of PDE III, is used for the treatment of intermittent claudication, an occlusive disease of blood vessels in the legs, which causes pain on walking. It acts as a vasodilator and inhibitor of platelet aggregation. Warfarin, initially developed as a rat poison, and a number of similar compounds, are effective anti-clotting agents by their action as vitamin K antagonists.

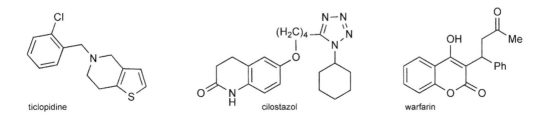

ticlopidine

cilostazol

warfarin

33.4.7 Other Enzyme Inhibitors

COX (cyclooxygenase) enzymes are involved in the first stage of prostoglandin synthesis and their inhibition is the basis of action of the *Non-Steroidal Anti-Inflammatory Drugs* (*NSAIDS*). There are two types of enzyme, COX-1 and COX-2 – inhibition of the latter results in anti-inflammatory and analgesic effects while inhibition of the former is responsible for side-effects, particularly gastric bleeding. Most older NSAIDS have both COX-1 and COX-2 activity (indomethacin (Indocin) is an example), although etodolac (Eccoxolac) shows very good selectivity for COX-2. Highly selective COX-2 inhibitors such as celecoxib (Celebrex) are now at the forefront, although another COX-2 inhibitor (Vioxx) was withdrawn due to cardiovascular problems. Ketorolac (Acular) has a different balance of analgesic and anti-inflammatory properties and is particularly useful for topical administration to the eye.

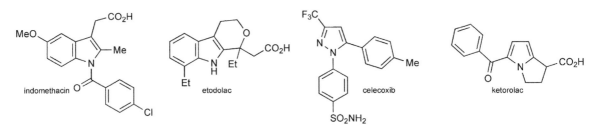

Statins reduce the amount of cholesterol in the blood by inhibition of one of the enzymes (HMG CoA reductase) involved in its biosynthesis and are therefore very useful for prevention of heart disease. The original statins were complex aliphatic fungal metabolites, such as simvastatin (Zocor), but a number of synthetic heterocyclic analogues carry only a small portion (comprising the ring-opened lactone) of the original structures. Atorvastatin (Lipitor) (page 645 for structure) is the most significant drug in this class by value, and is particularly useful due to its lower incidence of the serious side-effects sometimes found for other drugs of this class. Cerivastatin (Baycol), another heterocyclic analogue (with improved solubility) was widely used, but later withdrawn from sale due to an unacceptable level of these serious side-effects.

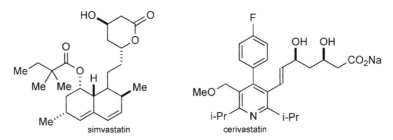

Other useful enzyme inhibitors include acetazolamide (Diamox) (carbonic anhydrase inhibitor) for the treatment of glaucoma, congestive heart failure, epilepsy and motion sickness (a very useful drug!). Sildenafil (Viagra), famous for treatment of impotence, inhibits a phosphodiesterase enzyme (PDE5) that is involved in the breakdown of ATP.

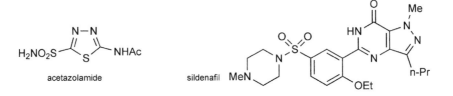

Leflunomide (Arava) is an anti-proliferative drug that inhibits dihydroorotate dehydrogenase, an enzyme essential for the synthesis of pyrimidines such as uracil, *via* aromatisation of dihydroorotic acid. It also has some immunomodulatory effects and is useful for intractable cases of rheumatoid arthritis, despite being liable to cause a number of serious side-effects. It is actually a prodrug, being converted into the active compound, a nitrile formed by cleavage of the isoxazole ring (cf. 25.5.1).

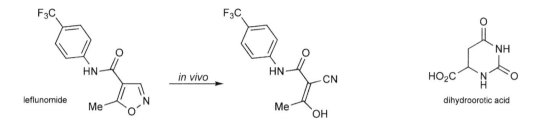

33.4.7.1 Cytochrome P450 Enzymes
The enzymes in this very large group have active sites containing haem, where oxidation–reduction reactions are governed by interconversion of Fe(II) and Fe(III). They are very widely distributed in all forms of life from animals down to the simplest organisms. They are important medically as they are responsible for much metabolic degradation of drugs and are also involved in the biosynthesis of biomolecules, such as steroids. Sub-types, which may be genetically determined, may lead to different rates of metabolism of drugs in different patient groups, causing problems in establishing the correct dose.

Inhibitors can be used therapeutically, but also occur in unwanted situations. They cover a range of structural types, but often are heterocycles containing imine groups (azines and azoles), in which the nitrogen coordinates to the metal. This binding can be quite sensitive to steric hindrance, allowing control in the design of drugs where inhibition is either wanted or not.

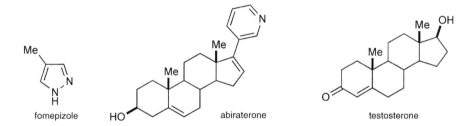

Fomepizole is used for the treatment of poisoning by ethylene glycol or methanol. It acts by inhibiting alcohol dehydrogenase, the enzyme that converts these two compounds into toxic metabolites. Abiraterone, which inhibits enzymes involved in the biosynthesis of testosterone, shows great promise in the treatment of prostate cancer. The logic of the design of this drug is clear – add the 3-pyridyl inhibitor group to a structure closely related to the target biomolecule. The anti-fungal azoles (see 33.6.1) have a related mechanism as they selectively inhibit enzymes involved in the biosynthesis of steroids necessary for fungal growth.

On the downside, grapefruit juice contains inhibitors such as bergamotin and naringin, which can drastically reduce the rate of metabolism of many drugs, leading to toxic effects from overdose! However, there may be some potential here for improving drugs that are subject to excessively rapid metabolism.

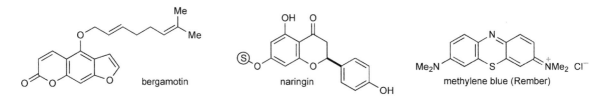

bergamotin naringin methylene blue (Rember)

33.4.8 Enzyme Induction

Although most enzyme-related drugs are inhibitors, some are enzyme inducers – they switch on the gene responsible for synthesis of the enzyme, increasing its production. An interesting example is methylene blue, originally used as a biological stain, but also having a complex combination of biological actions. Its best-established medical role is for the treatment of methemoglobinemia*, a condition in which a portion of hemoglobin exists in the Fe(III) state, which is unable to transport oxygen. Methylene blue induces production of the enzyme diaphorase II, also known as NADPH methemoglobin reductase, which is self-explanatory – it reduces Fe(III) to Fe(II), the oxygen-carrying form.

Other actions of methylene blue include monoamine oxidase *inhibition*, which can cause dangerous interactions with certain other drugs. However, it also has potential to treat malaria and Alzheimer's disease, the latter as a formulation named 'Rember'. Its activity in Alzheimer's disease is thought to depend on converting an abnormal 'tangled' form of a protein in the brain back to the normal form, although the mechanism for this is not clear.

33.5 Drugs Acting on the CNS

The CNS contains a wide variety of neurotransmitters and high concentrations of receptors. Mechanisms of action of many drugs are often complex combinations of receptor-based actions. Some of the most widely used (and abused) drugs are hypnotics/sedatives, for treatment of insomnia. Barbiturates such as amylo-barbitone have been used for many years, but suffer from side-effects and are addictive. Thiopentone sodium salt (sodium pentothal), however, is very useful as a short-acting intravenous anaesthetic. The benzodiazepines, such as diazepam (Valium) and alprazolam (Xanax), are safer drugs for insomnia and also can be used for treatment of anxiety and muscle spasms. Zolpidem (Ambien) is a newer and more selective hypnotic.

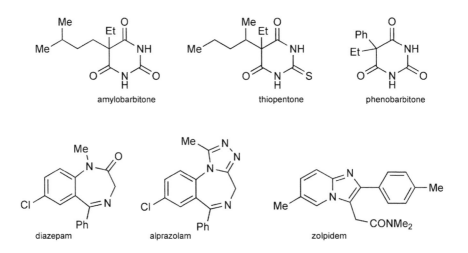

amylobarbitone thiopentone phenobarbitone

diazepam alprazolam zolpidem

*Pronounced 'met-hemoglobinemia', not 'meth-', which could imply methylation.

Another barbiturate (phenobarbitone (Luminal)) and a newer drug lamotrigine (Lamictal) are used to treat epilepsy. Trazodone (Molipaxin) is a useful anti-depressant with fewer side-effects than some of the older drugs. The action of varenicline (Champix), used to treat smoking withdrawal, has a straightforward mechanism, being a nicotinic partial agonist.

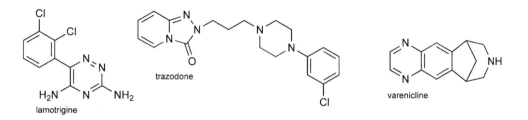

trazodone

lamotrigine

varenicline

Anti-psychotic agents, acting primarily as dopamine and 5-HT antagonists, are used mainly to treat schizophrenia. Newer agents such as olanzapine (Zyprexa) and risperidone (Risperidal) have fewer side-effects than the traditional phenothiazines, such as chlorpromazine (Thorazine).

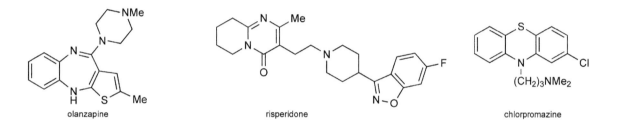

olanzapine

risperidone

chlorpromazine

The neurodegenerative states present a challenge for medicinal chemistry. Parkinson's disease is caused primarily by a deficiency of dopamine and so dopamine agonists, such as ropinirole (Requip) and pramipexole (Mirapexin), are effective in alleviating some of the symptoms. Both of these are also used to treat 'restless legs' syndrome. Riluzole (Rilutek) is the only drug so far developed to have a significant effect in treatment of motor neurone disease.

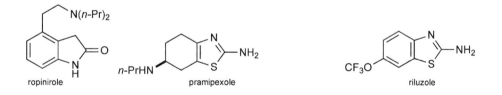

ropinirole

pramipexole

riluzole

33.6   Anti-Infective Agents
This section covers parasitic (protozoa and helminths ('worms')), bacterial and viral diseases. The mechanisms of action of many of these agents are complex and diverse and, in some cases, not fully understood, but enzyme inhibition is common to many.

33.6.1   Anti-Parasitic Drugs
The most important protozoan infection is malaria and this was an early target for medicinal chemical research. The traditional treatment was an extract of the bark of the cinchona tree, which contained the

alkaloid quinine. Current emphasis is on prevention by continuous dosing during exposure, using synthetic anti-malarial drugs, which must be under constant review as resistance eventually develops. Many of these drugs are quinolines with obvious similarities to quinine, for example mefloquine (Lariam). Proguanil (Paludrine) is of particular interest because the active species, formed *in vivo*, is a triazine (cycloguanil).

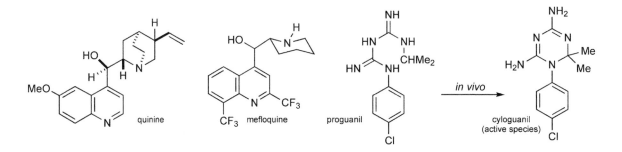

Metronidazole (Flagyl) is used for both bacterial (commonly for dental infections) and some protozoan infections, such as amoebic dysentery. Benzimidazoles, such as mebendazole (Pripsen), are an important group of anthelmintics. The most common anti-fungal agents are triazoles such as fluconazole (Diflucan).

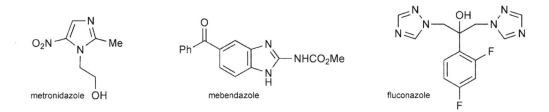

33.6.2 Anti-Bacterial Drugs

Among anti-bacterial agents, the simplest must be isoniazide (Rimifon), an important anti-tuberculosis drug. Heterocyclic contributions to mainstream antibiotics include a group of quinolones e.g. the broad-spectrum anti-bacterial ciprofloxacin (Cipro) perhaps best known in connection with terrorist use of anthrax and the diamino-pyrimidine trimethoprim (Triprim). Many of the sulfonamides, the first synthetic antibiotics, contain heterocyclic residues.

Co-trimoxazole (Septrin) is a well-known combination of a sulfonamide (sulfamethoxazole) with trimethoprim. This combination inhibits enzymes at two points of folic acid (32.2) utilisation – the sulfonamide inhibits incorporation of *p*-aminobenzoic acid during bacterial folic acid synthesis, and trimethoprim inhibits its conversion into tetrahydrofolate. The overall result is *synergistic*, i.e. there is a greater activity than the sum of the two components.

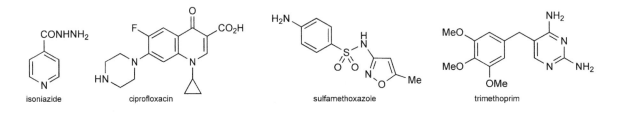

Penicillins and cephalosporins are saturated heterocycles; many aromatic heterocycles are included in a multitude of their side-chain variants.

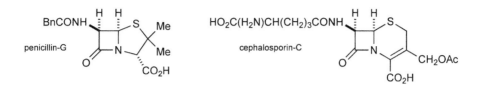

33.6.3 Anti-Viral Drugs

Many antiviral[6] agents are simply modified nucleosides, often only in the ribose residue, such as the anti-herpes drug acyclovir (Zovirax), the anti-HIV protease inhibitor zidovudine (Retrovir) (AZT) and lamivudine (Epivir) (3-TC), which is active against HIV and hepatitis. Heterocycle-modified compounds include the anti-herpes idoxuridine and a general anti-viral compound ribavirin (Rebetol). These generally work by disrupting the synthesis of viral DNA following incorporation, in competition with natural nucleosides, at an early stage. Non-nucleoside compounds, such as delavirdine (Rescriptor), a reverse transcriptase inhibitor, and saquinavir (Invirase), a protease inhibitor, (both anti-HIV drugs) act by binding close to the active site of an enzyme, altering its conformation and thereby deactivating it.

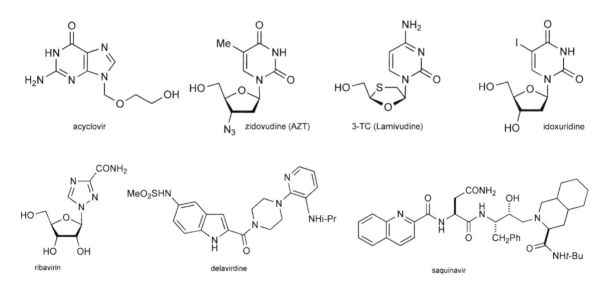

33.7 Anti-Cancer Drugs

Anti-cancer drugs generally act by disrupting the growth of cells and therefore oppose the excessive and abnormal growth that is the basis of cancer pathology. They are often referred to as cytotoxic or anti-metabolites. One of several mechanisms can be operative: disrupting the synthesis of DNA either by *N*-alkylating one of the bases (adenine, guanine, cytosine or thymine) or by incorporation of a modified base or nucleoside by competition with the natural bases for the DNA-synthesising enzymes. Alkylating agents are usually simple chemical 'reagents' or their precursors, an example of the latter being temozolomide (Temodal), which breaks down *in vivo* to give the powerful alkylating agent diazomethane. Possibly the resemblance of temozolomide to a nucleic acid base allows it to come into close contact with its target. It is used for glioma and melanoma. 6-Mercaptopurine and 5-azacytidine (Vidaza) are anti-leukemics; gemcitabine (Gemzar) is used against breast and lung cancers.

Some cytotoxic agents, such as azathioprine (Imuran), are used as immunosuppressive drugs rather than for treating cancer, acting by similar mechanisms, but with different balances of effects. These drugs would typically be used to prevent organ rejection following transplantation or to treat severe inflammatory diseases due to autoimmune reactions. Other mechanisms include interference with the function of folic acid, an important growth factor. Methotrexate (MTX) is a well-established and widely used anti-cancer and immunosuppressive agent that acts in this manner.

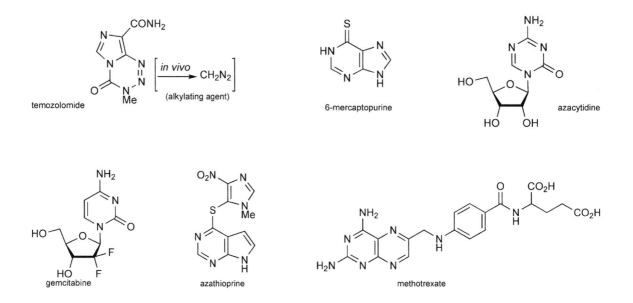

5-Fluorouracil (5-FU) is a cytotoxic agent used to treat solid tumours, such as those of the breast and colon. However, its prodrug capecitabine (Xeloda) is much better absorbed orally. 5-FU, applied as a cream, is also very useful for the treatment of certain skin cancers. It is significant that 8% of the population lack an enzyme that can metabolise 5-FU. In these people, administration of normal doses of the drug can lead to dangerously high toxic levels in the blood.

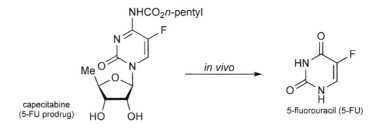

A number of inhibitors of various tyrosine kinase enzymes are important new cancer drugs: sunitinib (Sutent) is used for the treatment of certain kidney and GI cancers; imatinib (Glivec) for chronic myeloid leukemia and GI tumours, and lapatinib (Tykerb/Tykerv) for breast and lung cancers. Bortezomib (Velcade), a proteasome inhibitor is used to treat multiple myeloma and is notable for being a boronic acid.

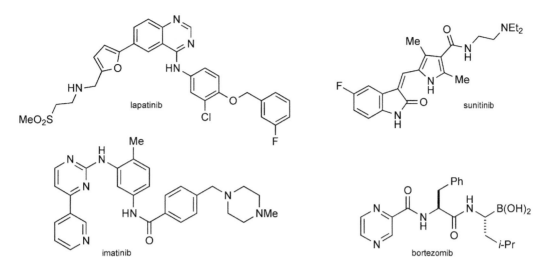

Irinotecan (Camptosar), a more selective synthetic analogue of camptothecin (a natural, but toxic, anticancer alkaloid), acts by inhibiting topoisomerase I, an enzyme involved in ordering the strands of DNA. Some compounds act by physical binding with vital natural polymers – the complex indole alkaloid vincristine is a classic example – it binds to tubulin, a protein essential to cell division.

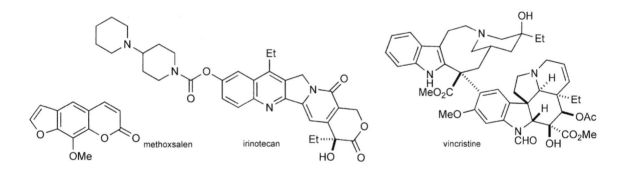

33.8   Photochemotherapy

Treatment of certain diseases, particularly of the skin, by ultraviolet irradiation (including sunshine) is a very old technique (phototherapy). A more modern approach involves administering photosensitising agents followed by irradiation with ultraviolet or visible light (photochemotherapy). The ideal situation is that selectivity is achieved by the sensitising agent concentrating in the target organ, but accurately targeted irradiation using a laser is also possible. Note, the laser is not a burning laser, just a high intensity source for activation of the sensitiser.

33.8.1   Psoralen plus UVA (PUVA) Treatment

Oral administration of a psoralen such as methoxsalen (8-methoxypsoralen), followed by exposure to ultraviolet light (UV-A) can be used to treat psoriasis, eczema, vitiligo and a number of other skin diseases. There is a small risk of inducing relatively easily treatable, non-melanoma, skin cancers and possibly, generally only after extensive treatment, a very small risk of malignant melanoma. Similar psoralens are the components, somewhat controversially, of some tanning aids.

33.8.2 Photodynamic Therapy (PDT)

Photodynamic therapy is a slightly different type of photochemotherapy and is used to treat cancers and pre-malignant lesions, both of the skin and internal organs, the latter using an endoscopic laser. It usually involves intravenous administration of certain porphins, followed by laser irradiation at visible wavelengths. A common agent used is porfimer (Photofrin is porfimer sodium), a linked porphin oligomer (MW ca. 10 000) prepared from natural materials, but 5-aminolevulinic acid (ALA) can be applied topically to some accessible (skin) lesions, where it becomes a precursor for biosynthesis of protoporphyrin IX. This therapy can only be used for tumours that lie relatively close to the surface of the tissue, that is, to the depth that the light can penetrate.

Verteporfin (Visudyne) is used similarly for the treatment of the 'wet' form of macular degeneration (the variant characterised by excessive vascularisation of the retina) and other eye diseases.

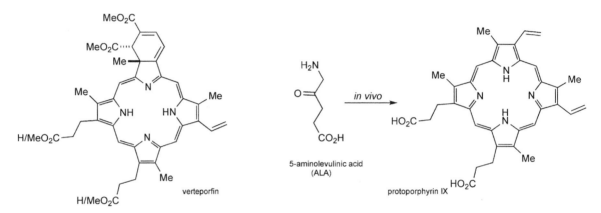

5-aminolevulinic acid
(ALA)

verteporfin

protoporphyrin IX

The mechanism of PUVA and PDT both involve the formation of singlet oxygen, which is cytotoxic, and the former also probably also includes some direct action on DNA.

References

[1] 'Comprehensive Medicinal Chemistry II', Ed. Triggle, D. and Taylor, J., Elsevier, **2006**.
[2] Chem. World, Jan **2008**, p.15.
[3] 'Goodman & Gilman's The Pharmacological Basis of Therapeutics', Ed. Brunton, L. L., Lazo, J. S., and Parker, K. L., 11th Edtn., McGraw-Hill, **2005**. This is the standard textbook, which is subject to frequent revision.
[4] 'Molecules and Medicine', Corey, E. J., Czakó, B., and Kürti, L., John Wiley & Sons, Ltd, **2007**. This is a useful general discussion from a chemical/biochemical viewpoint of major drugs of all structural types.
[5] We do not give references for the individual drugs. General information is contained in references 1,3 and 4, but an internet search will give up-to-date information. Wikipedia is often a good start as it usually gives chemical structures (unlike many medical papers), however further verification should be sought.
[6] 'AIDS-Driven nucleoside chemistry', Huryn, P. M. and Okabe, M., *Chem. Rev.*, **1992**, *92*, 1745.

Index

Heterocyclic Chemistry 5th Edition   John Joule and Keith Mills
© 2010 Blackwell Publishing Ltd

Printed and bound by CPI Group (UK) Ltd, Croydon, CR0 4YY